P. Slovic

Environmental Impact Assessment,
Technology Assessment, and Risk Analysis

NATO ASI Series

Advanced Science Institutes Series

A series presenting the results of activities sponsored by the NATO Science Committee, which aims at the dissemination of advanced scientific and technological knowledge, with a view to strengthening links between scientific communities.

The Series is published by an international board of publishers in conjunction with the NATO Scientific Affairs Division

A Life Sciences	Plenum Publishing Corporation
B Physics	London and New York
C Mathematical and Physical Sciences	D. Reidel Publishing Company Dordrecht, Boston and Lancaster
D Behavioural and Social Sciences	Martinus Nijhoff Publishers Boston, The Hague, Dordrecht and Lancaster
E Applied Sciences	
F Computer and Systems Sciences	Springer-Verlag Berlin Heidelberg New York Tokyo
G Ecological Sciences	

Series G: Ecological Sciences No. 4

Environmental Impact Assessment, Technology Assessment, and Risk Analysis

Contributions from the Psychological and Decision Sciences

Edited by

Vincent T. Covello

U.S. National Science Foundation, Washington, D.C., USA

Jeryl L. Mumpower

State University of New York at Albany, Albany, New York, USA

Pieter J.M. Stallen

Centre for Technology and Policy Studies, Apeldoorn, The Netherlands

V.R.R. Uppuluri

Oak Ridge National Laboratory, Oak Ridge, Tennessee, USA

Springer-Verlag Berlin Heidelberg New York Tokyo

Published in cooperation with NATO Scientific Affairs Division

Proceedings of the NATO Advanced Study Institute on Environmental Impact Assessment, Technology Assessment, and Risk Analysis held at Les Arcs Conference Center and Resort Complex (Bourgh-St. Maurice, France) August 21–31, 1983

ISBN 3-540-15684-4 Springer-Verlag Berlin Heidelberg New York Tokyo
ISBN 0-387-15684-4 Springer-Verlag New York Heidelberg Berlin Tokyo

Library of Congress Cataloging in Publication Data. NATO Advanced Study Institute on Environmental Impact Assessment, Technology Assessment, and Risk Analysis (1983 : Les Arcs, Bourg-Saint-Maurice, France) Environmental impact assessment, technology assessment, and risk analysis. (NATO ASI series. Series G, Ecological sciences ; no. 4) Includes index. 1. Environmental impact analysis—Congresses. 2. Environmental impact analysis—Social aspects—Congresses. 3. Risk management—Congresses. 4. Risk management—Social aspects—Congresses. I. Covello, Vincent T. II. Title. III. Series. TD194.6.N373 1983 333.7'1 85-26089
ISBN 0-387-15684-4 (U.S.)

Printed in Germany

Printing: Beltz Offsetdruck, Hemsbach; Bookbinding: J. Schäffer OHG, Grünstadt
2131/3140-543210

PREFACE

This volume is the outcome of a recent NATO Advanced Study Institute (ASI) on "Technology Assessment, Environmental Impact Assessment, and Risk Analysis: Contributions from the Psychological and Decision Sciences." The Institute was held in Les Arcs, France and functioned as a high level teaching activity during which scientific research results were presented in detail by eminent lecturers. Support for the Institute was provided by grants from the NATO Division of Scientific Affairs, the U.S. Office of Naval Research, and the Russell Sage Foundation.

The Institute covered several areas of research, including quantitative studies on decision and judgmental processes, studies on human intellectual limitations, studies on risk attitudes and perceptions, studies on factors contributing to conflicts and disputes about hazardous technologies and activities, studies on factors influencing forecasts and judgments by experts, studies on public preferences for decisionmaking processes, studies on public responses to technological hazards, and case studies applying principles and methods from the psychological and decision sciences in specific settings.

The volume is divided into five sections. The papers in Section I provide an overview of the field of inquiry. Section II focuses on theoretical and methodological contributions from psychology, while Section III focuses on contributions from the decision and system sciences. Several case studies are included in each section. The papers in Section IV provide cultural and philosophical perspectives on the three central topics covered by the Institute. The last section includes several overview papers written from the perspective of policy analysis, administrative science, and the law.

Finally, we would like to express our appreciation and thanks to the following persons for their support and assistance: Craig Sinclair, Marshall Robinson, Steven Del Sesto, Randy Simpson, Joshua Menkes, David Seaver, Ellen Pierce, and Patty Taylor.

ASI Directors: Vincent T. Covello
Jeryl L. Mumpower
Peter J.M. Stallen
V.R.R. Uppuluri

CONTENTS

SECTION I. OVERVIEW

SECTION II. Psychological Perspectives

Part 1: Theoretical and Methodological Contributions

Social and Behavioral Research on Risk: Uses in Risk Management Decisionmaking*

Vincent T. Covello

National Science Foundation, Washington, D.C. 20550

INTRODUCTION

In response to rising concern about health and environmental risks, a concerted effort is being made to improve risk analysis methods and risk management approaches. As part of this effort, social and behavioral scientists have produced a substantial body of knowledge on public risk perceptions, attitudes, and behavior (see Covello, 1983). The purpose of this paper is to identify and describe five critical policy uses for such information (see Table 1). A basic assumption of that paper is that policy analysts and decisionmakers can benefit from a better understanding of how the public thinks and makes

Table 1

FIVE USES OF SOCIAL AND BEHAVIORAL RESEARCH IN HEALTH AND ENVIRONMENTAL DECISIONMAKING

Uses in:

1. Identifying the nature and extent of public concerns.

2. Structuring public debate and resolving conflict.

3. Anticipating public responses to new technologies.

4. Educating and informing the public.

5. Designing and implementing health and environmental policies and systems.

---

* The views expressed in this paper do not necessarily represent the views of the National Science Foundation, but are exclusively those of the author.

NATO ASI Series, Vol. G4
Environmental Impact Assessment, Technology Assessment, and Risk Analysis. Edited by V.T. Covello et al.

decisions about technological risks. Without such understanding, well-intended policies may be ineffective or even counterproductive.

USES IN IDENTIFYING THE NATURE AND EXTENT OF PUBLIC CONCERNS

For effective decisionmaking, a risk manager needs accurate and reliable information about the nature and extent of public concerns. This person also needs information that will validate or invalidate unsubstantiated assertions by those who might intentionally or unintentionally misrepresent such concerns. Probably the most effective tool for acquiring such information is the social science survey. In recent years, however, critics have attacked the usefulness of such surveys, arguing that they do nothing more than report the obvious at an exorbitant cost. Upon reading the results of a survey, a common response is: "Of course that's the way things are. Do we really need to spend taxpayer money in such a wasteful manner?"

Understanding the origin and prevalence of this point of view is important to the present discussion. Consider the following typical survey finding: People living near a nuclear power facility are less supportive of nuclear power than those living farther away.

Since this finding is obvious--people living near such a facility are clearly subjected to the greatest objective risk and therefore logically should be the most concerned--why spend taxpayer money to establish a trivial result? Would it not be wiser to take the proposition for granted and use the resultant savings for other purposes? This might be the case except for one interesting point. The finding is the direct opposite of what has actually been found by researchers.

People living near a nuclear power plant facility tend to be more supportive of nuclear power than those living farther away (Office of Technology Assessment, 1984, p. 222). Once an explanation is offered, the finding itself becomes obvious. First, utilities are usually extremely careful in selecting sites for new nuclear power plants. Sites in economically depressed areas, in rural areas, and in areas where the population is already receptive tend to be favored. Second, utilities often embark on massive public relations campaigns once a site is selected. Third, the population living closest to the plant often receives direct economic benefits, such as jobs and tax revenues. Fourth, those living closest to the plant are likely to experience cognitive dissonance if they truly believe that the plant is dangerous. One way to resolve the dissonance is to change one's beliefs about the safety of nuclear power. Finally, the survey population is likely to be biased. At least some of the people against nuclear power may have moved out of the area and are therefore not represented in a survey of the local population.

If the actual survey result had been mentioned first, it is likely that at least some readers would have labelled this result as "obvious." As social and behavioral scientists have noted on numerous occasions, what this suggests is perhaps there is something wrong with the entire argument of "obviousness." Since every kind of behavior or attitude is conceivable, and since it is hard to find a human behavior or attitude that has not already been observed somewhere, it is of great importance to know which behaviors and attitudes actually occur most frequently and under what conditions. Only then will we advance beyond anecdotes and "war stories" to more advanced understanding and more enlightened policy.

STRUCTURING PUBLIC DEBATE AND RESOLVING CONFLICT

Public hearings and meetings dealing with controversial health, safety, and environmental issues are often characterized by emotional and bitter disagreements and conflicts. Many of these meetings are unproductive and more often than not lead to recriminations and to a hardening of positions. Given our lack of knowledge in this area, social scientists (e.g., Mazur, 1981; Nelkin, 1979) have examined several recent conflicts in order to better understand the sources of such conflict and to

Table 2

TYPOLOGY OF SOURCES OF CONFLICT AND DEBATE

1. Disagreements about data and statistics.

2. Disagreements about risk estimates and probabilities.

3. Disagreements about assumptions and definitions

4. Disagreements about risk/cost/benefit tradeoffs.

5. Disagreements about the distribution of risks, costs, and benefits

6. Disagreements about basic values and ideologies.

suggest improved mechanisms for conflict resolution. One set of studies (von Winterfeldt et al., 1980; von Winterfeldt and Edwards, 1982; Otway and von Winterfeldt, 1982), for example, found that disagreements between participants in conflicts about technological hazards can be broken down into six different types of disagreements (Table 2). Using the debate about nuclear power as an example, a disagree- ment about data and statistics might involve disagreement about the actual number of nuclear power plant accidents that have occurred or about the amount of radiation that leaked from the nuclear reactor at Three Mile Island. A disagreement about estimates and probabilities might involve disagreement about the probability of a nuclear reactor meltdown. A disagreement about assumptions and definitions might involve disagreement about whether risk should be defined in quantitative terms (e.g., as probability times consequences) or in qualitative terms (e.g., as voluntary and controllable or involuntary and uncontrollable). A disagreement about risk-cost-benefit tradeoffs might involve disagreement about tradeoffs between the value of lives saved and the cost of additional safety measures.

Such disagreements may become particularly intense when one party claims that the value of a human life is infinite and therefore no price can be assigned to it. A disagreement about the distribution of risks and benefits might involve disagreement about inequities between the benefits of cheap and reliable electricity for the present generation and the risks from nuclear waste for future generations. Finally, a disagreement about basic values and ideologies might involve a disagreement about centralized versus decentralized power (e.g., nuclear power versus dollar power), or about the value of rapid technological progress and economic growth.

Based on the experience of researchers who have applied the typology in actual public debates and conflicts (e.g., Edwards, 1983), there is some evidence that such an approach and others related to it may be useful in structuring public debate and resolving conflicts. It has been found, for example, that the potential for conflict resolution increases if the participants in the conflict can be made to focus on one specific aspect at a time. All too often, opponents in public debates talk past one another, each focusing on a different aspect of the problem. Typologies of the type described above can help avoid this problem by providing a framework for disaggregating disagreements and for focusing the debate.

USES IN ANTICIPATING PUBLIC RESPONSES TO NEW TECHNOLOGIES

One of the most important findings to emerge from the social and behavioral literature on risk is that the public takes a complex array of qualitative and quantitative factors into consideration in evaluating the seriousness of a risk (Slovic, et. al, 1980; Vlek and Stallen, 1981; Litai et. al., 1983; Renn, 1981). Research suggests that perceptions of risk are influenced not only by mortality and morbidity rates but a host of other factors. As summarized in Table 3, these include:

(1) Catastrophic potential: The public is more concerned about fatalities and injuries that are grouped in time and space (e.g., airplane crashes) than about fatalities and injuries that are scattered or random in time and space (e.g., automobile accidents).

(2) Familiarity: The public is more concerned about risks that are unfamiliar (e.g., ozone depletion due to emissions of fluorocarbons) than about risks that are familiar (e.g., household accidents).

(3) Understanding: The public is more concerned about activities characterized by poorly understood exposure mechanisms or processes (e.g., exposure to radiation) than about activities characterized by apparently well-understood exposure mechanisms or processes (e.g., pedestrian accidents or slipping on ice).

(4) Uncertainty: The public is more concerned about risks that are scientifically unknown or uncertain (e.g., recombinant DNA experimentation) than about risks that are relatively known to science (e.g., actuarial data on automobile accidents).

(5) Controllability: People are more concerned about risks that they perceive to be not under their personal control (e.g., travelling as a passenger in an airplane or in an automobile) than about risks that they perceive to be under their personal control (e.g., driving an automobile or riding a bicycle).

(6) Voluntariness of exposure: People are more concerned about risks that they perceive to be involuntary (e.g., exposure to unlabelled food additives or to air or water pollutants) than about risks that they perceive to be voluntary (e.g., smoking, sunbathing, or mountain climbing).

(7) Effects on children: People are more concerned about activities that put children specifically at risk (e.g., school bus accidents or exposure to toxic chemicals by pregnant women)

than about activities that do not put children specifically at risk (e.g., adult smoking).

(8) Effects on future generations: People are more concerned about activities that pose risks to future generations (e.g., genetic effects due to exposure to radiation) than to risks that pose no special risks to future generations (e.g., skiing accidents).

(9) Victim identity: People are more concerned about risks to identifiable victims (e.g., a yachtsman lost at sea or a trapped coal miner) than about risks to statistical victims (e.g., statistical profiles of automobile accident victims).

(10) Dread: People are more concerned about risks that are dreaded and evoke a response of fear, terror, or anxiety (e.g., exposure to potential carcinogens from toxic waste dumps or to nuclear radiation) than to risks that are not especially dreaded and do not evoke a special response of fear, terror, or anxiety (e.g., common colds or household accidents).

(11) Trust in institutions: People are more concerned about situations characterized by a lack of trust in the responsible risk management institutions (e.g., criticisms of the Nuclear Regulatory Agency for its perceived close ties to industry) than they are about situations characterized by trust in the responsible risk management institutions (e.g., trust in the management of recombinant DNA risks by universities and by the National Institutes of Health).

(12) Media attention: People are more concerned about risks that receive much media attention (e.g., airline crashes) than about risks that receive little media attention (e.g., on-the-job accidents).

(13) Accident history: People are more concerned about activities that have a history of major and sometimes minor accidents (e.g., nuclear power plant accidents such as the accident at Three Mile Island) than about activities that have a history of no major or minor accidents (e.g., recombinant DNA experimentation).

(14) Equity: People are more concerned about activities that are characterized by a perceived inequitable distribution of risks and benefits (e.g., off-shore oil exploration) than about activities characterized by a perceived equitable distribution of risks and benefits (e.g., vaccination).

(15) Benefits: People are more concerned about risky activities that are perceived to have unclear benefits (e.g., nuclear

power generation) than about risky activities that are perceived to have clear benefits (e.g., automobile driving).

(16) Reversibility: People are more concerned about activities characterized by potentially irreversible adverse effects (e.g., acid rain) than about activities characterized by reversible adverse effects (e.g., injuries from sports or household accidents).

(17) Personal involvement: People are more concerned about activities that they believe place them (or their families) personally and directly at risk (e.g., drinking contaminated water due to local dumping of hazardous waste) than about activities that do not place them (or their families) personally and directly at risk (e.g., dumping of hazardous waste at sea or in other remote sites).

Three qualifications need to be made concerning these factors. First, most of the reported findings are based on surveys of small, highly specialized, and unrepresentative groups. As a result, the findings are more suggestive than definitive. Second, in many cases only correlations and associations exist, not causal relationships. The relationship between media attention and public concern, for example, is highly complex (in some cases media coverage reflects existing public concerns; in other cases media attention is a major factor in putting a health, safety, or environmental issue on the political agenda). Third, although the items in Table 3 have been listed separately, the various factors clearly interact, either to negate or to reinforce one another. In recent years, attempts have been made to describe these interactions (see Slovic et. al., 1980). However, there appears to be a growing consensus among researchers that this is a vast and difficult undertaking requiring a substantial amount of additional data and study.

Researchers in this area of study have proposed at least two possible uses for the type of information contained in Table 3 (see Hohenemser, et. al., 1983; Fischhoff et al., 1983). First, the information can be used to anticipate public responses to new hazards (such as recombinant DNA research or toxic shock syndrome). Second, it can be used to develop a risk profile or index for each hazard that captures the complexity of the hazard better that a simple mortality index. The information can then be used by decisionmakers to compare hazards and establish safety priorities, risk management strategies, and budgets. While some recent attempts have been made in this direction (Fischhoff, et al., 1983), probably the best known example is the proposal still under debate by the Nuclear Regulatory Commission's Advisory Committee on Reactor

TABLE 3. FACTORS INVOLVED IN PUBLIC RISK PERCEPTION

FACTOR	CONDITIONS ASSOCIATED WITH INCREASED PUBLIC CONCERN	CONDITIONS ASSOCIATED WITH DECREASED PUBLIC CONCERN
CATASTROPHIC POTENTIAL	FATALITIES AND INJURIES GROUPED IN TIME AND SPACE	FATALITITES AND INJURIES SCATTERED OR RANDOM IN TIME AND SPACE
FAMILIARITY	UNFAMILIAR	FAMILIAR
UNDERSTANDING	MECHANISMS OR PROCESS NOT UNDERSTOOD	MECHANSIMS OR PROCESS UNDERSTOOD
UNCERTAINTY	RISKS SCIENTIFICALLY UNKNOWN OR UNCERTAIN	RISKS KNOWN TO SCIENCE
CONTROLLABILITY (PERSONAL)	UNCONTROLLABLE	CONTROLLABLE
VOLUNTARINESS OF EXPOSURE	INVOLUNTARY	VOLUNTARY
EFFECTS ON CHILDREN	CHILDREN SPECIFICALLY AT RISK	CHILDREN NOT SPECIFICALLY AT RISK
EFFECTS ON FUTURE GENERATIONS	RISK TO FUTURE GENERATIONS	NO RISK TO FUTURE GENERATIONS
VICTIM IDENTITY	IDENTIFIABLE VICTIMS	STATISTICAL VICTIMS
DREAD	EFFECTS DREADED	EFFECTS NOT DREADED
TRUST IN INSTITUTIONS	LACK OF TRUST IN RESPONSIBLE INSTITUTIONS	TRUST IN RESPONSIBLE INSTITUTIONS
MEDIA ATTENTION	MUCH MEDIA ATTENTION	LITTLE MEDIA ATTENTION
ACCIDENT HISTORY	MAJOR AND SOMETIMES MINOR ACCIDENT	NO MAJOR OR MINOR ACCIDENT
EQUITY	INEQUITABLE DISTRIBUTION OF RISKS AND BENEFITS	EQUITABLE DISTRIBUTION OR RISKS AND BENEFITS
BENEFITS	UNCLEAR BENEFITS	CLEAR BENEFITS
REVERSIBILITY	EFFECTS IRREVERSIBLE	EFFECTS REVERSIBLE
PERSONAL INVOLVEMENT	INDIVIDUAL PERSONALLY AT RISK	INDIVIDUAL NOT PERSONALLY AT RISK

Safeguards to assign a greater weight to potential catastrophic deaths than to ordinary deaths. According to one formula, 1,000 catastrophic deaths would be considered equivalent in seriousness to 4,000 ordinary deaths (Advisory Committee on Reactor Safeguards, 1980).

USES IN EDUCATING AND INFORMING THE PUBLIC

One of the most difficult questions faced by risk managers is "How safe is safe enough?" Even more difficult is the question "Who is to decide?" -- especially in cases (1) where the data are uncertain; (2) where a substantial segment of the population believes that the benefits outweigh the risks; and (3) where a substantial segment of the population believes that regulation poses a significant threat to personal freedom. Although a large number of hazards meet one or more of the conditions, the recent debates about saccharin, motorcycle helmets, and automobile passive constraints serve as good examples.

It is often argued that in such cases the most effective risk management strategy is to let the consumer decide. Although attractive in theory, the weakness of this approach is that we know very little about how consumers make decision under conditions of risk and uncertainty (see Morris, Mazis, and Bardofsky, 1980). As the failure of the "buckle up for safety" campaign indicates (Robertson, 1976), we know even less about how to convince people to take protective action in their own best interests (10-15 percent of the U.S. population currently wear seat belts).

Probably the most serious problem faced by any informational or educational effort is that people often do not cope well when confronted with risk problems and decision (Kahneman, et. al., 1983; Tversky and Kahneman, 1974; Tversky and Kahneman, 1981). Intellectual limitations and the need to reduce anxiety often lead to denial that risk exists and to unrealistic oversimplifications of essentially complex problems. To simplify risk problems, people use a number of inferential or judgmental rules that often cause risk information campaigns and educational efforts to work at cross purposes (Tversky and Kahneman, 1981; Slovic, et. al., 1980).

Information may, for example, heighten the imaginability and consequently the perceived probability of an event, even when the information is designed to assure individuals that the event is unlikely. A package insert listing all the risks of taking a drug, or a published report describing safety precautions at a hazardous waste dump, may serve only to increase concern about the substance or activity. By identifying previously unknown ways in which things can go wrong, the information provider takes the chance that people will incorrectly assess the information (i.e., that they will consider the event more likely as a result of increased awareness and sensitivity).

In order to confront these problems, researchers in several countries are currently studying how best to inform people about risks. Several studies have found, for example, that people are highly influenced by the way the risk is expressed. In one set of studies (Slovic, et. al., 1978; Schwalm and Slovic, 1982), researchers reported that people are more likely to wear seat belts when presented data about the cumulative lifetime risks of driving (40,000 trips). The study found little effect when people were presented data about the probability of death from a single trip.

Another study found that the phrasing of a risk analysis report can have a substantial impact on a decisionmaker's choice among various risk management strategies. More specifically, researchers have reported that systematic reversals of preferences can be obtained when probabilistic data is expressed in terms of lives saved as opposed to lives lost. Arrow (1982), for example, noted that one study compared two therapies for the treatment of lung cancer--surgery, and radiation therapy. Different groups of physicians were presented with the probabilities of survival during treatment for each of the two therapies. With these data, 84 percent of the physicians preferred surgery and 15 percent preferred radiation therapy. Then another group was presented the same data expressed differently: the probabilities of dying were give instead of the probabilities of survival. The probability of dying is, of course, merely one minus the probability of survival, so that the two formulations are not merely logically equivalent, but can be transformed into each other by a trivial calculation. Yet the proportion of physicians choosing surgery over radiation therapy dropped from 84 percent to 50 percent.

USES IN DESIGNING AND IMPLEMENTING RISK MANAGEMENT POLICIES AND SYSTEMS

The effectiveness and efficiency of any risk management policy or system will be substantially influenced by a decisionmaker's ability to forecast public responses. For example, it is clear that any automobile safety program that relies on seat belts needs to take into account statistics on actual seat belt usage by the public.

Probably the clearest example of a failure to consult the social and behavioral literature on risk when designing a risk management system was the U.S. flood insurance program. Under this program, flood insurance was made available at a highly subsidized rate to persons living in 100 year flood plains. Unfortunately, few people bought the insurance. When the program was evaluated, several important conclusions emerged (Kunreuther, et. al., 1978; Slovic et. al., 1977). First, unless a person had been made graphically aware of flood risks, typically through past experience, he was unlikely to buy the insurance. Second, most people relied more on information from friends and neighbors than on official communications in deciding whether to buy the insurance. Third, people had difficulty understanding and interpreting probabilistic information that was provided to them.

Residents of a 100 year flood plain typically believed that a recent severe flood precluded the possibility of another severe flood in the near future. As a result, they believed that insurance, even at low cost, was unnecessary. These findings and cases similar to it strongly suggest that risk managers can avoid serious mistakes and inefficiencies by consulting the relevant social and behavioral research during the early stages of policy or project development.

CONCLUSION

Although the uses described in this paper may seem apparent after they have been enumerated, it is surprising how little actual use is made of social and behavioral research in risk management practice Some of the fault lies with the research community and the low priority assigned by many researchers to problems of communication, dissemination, and policy utilization. Some of the fault also lies with risk managers, many of whom assume that common sense and anecdotal evidence can substitute for systematic scientific research. If progress is to be made toward the common goal of better understanding, better policy, and better protection of public health and safety, all parties need to make a serious effort to incorporate social and behavioral research on risk in policy formulation and risk management decisionmaking.

REFERENCES

Advisory Committee on Reactor Safeguards (1980) "An Approach to Quantitative Goals for Nuclear Power Plants," NUREG-0739, Nuclear Regulatory Commission, Washington, D.C.

Arrow, K. (1982) "Risk Perception in Psychology and Economics," Economic Inquiry, 20, 2-9.

Covello, V. (1983) "The Perception of Technological Risks: A Literature Review," Technological Forecasting and Social Change, 23, 285-297.

Edwards, W. (1983) "An Application of Multiattribute Utility Analysis to an Energy Dispute," Paper presented at the NATO Advanced Study Institute on Technology Assessment, Risk Analysis, and Environmental Impact Assessment, Les Arcs, France, Aug. 21-31, 1983.

Fischhoff, B., Watson, S., and Hope, C. (1984) "Defining Risk," Technical Report, Decision Research, Eugene, Oregon.

Hohenemser, C., Kates, R., and Slovic P. (April 22, 1983), "The Nature of Technological Hazard," Science, 220 378-384.

Kahneman, D., Slovic, P., and Tversky, A. eds., (1982) "Judgment Under Uncertainty: Heuristics and Biases," Cambridge University Press, New York.

Kunreuther, H., Ginsberg, R., Miller, L., Sagi, P., Slovic, P., Borkan, B., and Katz, N. (1978) "Disaster Insurance Protection: Public Policy Lessons," Wiley, New York.

Litai, D., Lanning, D., and Rasmussen, N. (1983) "The Public Perception of Risk," in Covello, V., Flamm, W., Rodericks, J., and Tardiff, R., eds., "The Analysis of Actual Versus Perceived Risks," Plenum Press, New York, 213-224.

Mazur, A. (1981) "The Dynamics of Technological Controversy," Communications Press: Washington, D.C.

Morris, L., Mazis, M., and Barofsky, I., eds. (1980) "Product Labeling and Health Risks," Banbury Report 6, Cold Spring Harbor Laboratory, Cold Spring Harbor, N.Y.

Nelkin, D. (eds) (1979) "Controversy: The Politics of Technical Decisions," Beverly Hills: Sage.

Office of Technology Assessment (1984) "Nuclear Power in An Age of Uncertainty," U.S. Government Printing Office, Washington, D.C.

Otway, H. J. and von Winterfeldt, D. (1982) "Beyond Acceptable Risk: On the Social Acceptability of Technologies," Policy Sciences, 14, 247-256.

Renn, O. (June 1981) "Man, Technology, and Risk: A Study of Intuitive Risk Assessment and Attitudes towards Nuclear Power," Report Jul-Spez, Julich, Federal Republic of Germany, Nuclear Research Center.

Robertson, L. S. (1976) "The Great Seat Belt Campaign flop," Journal of Communication, 26, 41-45.

Schwalm, N. D., and Slovic, P. (January 1982) "Development and Test of a Motivational Approach and Materials for Increasing Use of Motor-Vehicle Occupant Restraints," Perceptronics Technical Report PD FTR 1100-82-1, Eugene, Oregon.

Slovic, P., Fischhoff B., and Lichtenstein, S. (1980) "Facts and Fears: Understanding Perceived Risk" R. Schwing and W. Albers, Jr., eds., "Societal Risk Assessment: How Safe is Safe Enough?", Plenum Press, New York, 181-216.

Slovic, P., Fischhoff, B., Lichtenstein, S., Corrigan, B., and Combs, B. (1977) "Preference for Insuring Against Probable Small Losses: Implications for the Theory and Practice of Insurance," Journal of Risk and Insurance, 44, 237-258.

Slovic, P., Fischhoff, B., and Lichtenstein, S. (1978) "Accident Probabilities and Seat Belt Usage: A Psychological Perspective," Accident Analysis and Prevention, 10, 281-285.

Tversky, A., and Kahneman, D. (Sept. 27, 1974) "Judgment Under Uncertainty: Heuristics and Biases," Science, 185, 1124-1131.

Tversky, A., and Kahneman, D. (1981) "The Framing of Decisions and the Psychology of Choice," Science, 211, 1453-1458.

Vlek, C. A. J., and Stallen, P. J. (1980) "Rational and Personal Aspects of Risk," Acta Psychologica, 45, 273-300.

Vlek, C., and Stallen, P. J. (Oct 1981) "Judging Risks and Benefits in the Small and in the Large," Organizational Behavior and Human Performance, 38.

von Winterfeldt, D., Edwards, W., Anson, J., Stillwell, W., and Slovic, P. (May 1980) "Development of a Methodology to Evaluate Risks from Nuclear Electric Power Plants: Phase I--Identifying Social Groups and Structuring Their Values and Concerns," Final Report to Sandia National Laboratories, Albuquerque, N.M.

von Winterfeldt, D. and Edwards, W. (1982) "Patterns of Conflict about Risky Technologies," Technical report, Social Sciences Research Institute, University of Southern California, Los Angeles, California.

Risk Theory and the Environmental Assessment Process

James E. Dooley*

INTRODUCTION

Two kinds of assessment have evolved in recent years: risk assessment (RA) and environmental impact assessment (EIA). These have developed along parallel paths, at times drawing procedures from each other, and sometimes being used in the same inquiry. Both are concerned with the well-being of the environment and with the human condition, so it is natural to make comparisons between them looking for similarities and differences. At the present time EIA has become more institutionalized than RA so it is useful to consider ways that evolving risk theory, assessment, and management can enhance the EIA process and where RA needs a different perspective. This is particularly important in cases where the EIA process is applied to a project in which the dominant concerns are the risks to humans, the natural environment, or valued material objects.

A logical difficulty arises insofar as RA and EIA cannot be contrasted until a distinction is made between them in their respective definitions and yet the constrasts are an essential part of such definitions. To get around this difficulty, an arbitrary choice was made to treat RA as applying to impacts

---

*The financial assistance for the preparation and presentation of this paper provided by the National Research Council of Canada (Grant A9300) and the National Academy of Sciences is acknowledged with thanks. Material on the Beaufort Sea case and the regulatory aspects of EIA in Canada was generously provided by J. Herity of the Federal Environmental Assessment Review Office. Finally, thanks to Mrs. Rosalita Jones who typed the notes in record time and with a remarkable ability for improving on the accuracy of the original.

NATO ASI Series, Vol. G4
Environmental Impact Assessment, Technology Assessment, and Risk Analysis. Edited by V.T. Covello et al.

having uncertainty in some respect and EIA as applying to impacts that, for practical purposes, can be treated as predictable. Furthermore, environmental impacts are assumed to arise primarily from projects whereas risks arise from a broader range of sources. This choice is not entirely satisfactory but generally is consistent with practice and does allow the discussion to proceed without the introduction of new terminology. The term management will include both the assessment process and the actions implemented to control the impacts. Management of uncertain impacts will be referred to as risk management (RM) and management of predictable impacts will be called environmental management (EM).

The paper begins by a brief review of the present legislative and regulatory basis for EIA in Canada both federally and provincially. A brief review is also included on risk management in Canada but the responsibility is dispersed and complex so the review is indicative rather than comprehensive. A number of case studies are then examined looking for comparisons in the treatment of risk and predictable impacts. The following are the cases that were selected to provide a variety of risk and inquiry types:

- Beaufort Sea Hydrocarbon Development Proposal (federal EIA)
- Royal Commission on Electric Power Planning in Ontario (provincial inquiry)
- Cluff Lake Board of Inquiry into the Expansion of Uranium Mining in Saskatchewan (provincial EIA)
- Commission of Inquiry into the Safety of Aluminum Wire (Ontario provincial inquiry)
- Royal Commission into the Non-medical Use of Drugs (federal inquiry)
- Inquiry into the Mississauga Railway Accident (federal inquiry)

The Beaufort Sea Hydrocarbon Production Proposal (Beaufort Sea) had an EIA prepared under federal guidelines. The interesting feature is that the EIA was generic simply because the specific drill locations were not known at the time of the EIA for this large area. Also, a lot of design remained to be done so that project details could not be assessed. The project contains both important prevention and response features. The Royal Commission on Electric Power Planning (Porter) was charged with investigating broad issues in electric power planning. The inquiry did not deal with a specific EIA or RA nor was it required to do so. This case was included because it illustrates how risk emerged and how it was treated in a case where no regulations or guidelines were imposed. The inquiry was focused on the event of bringing out the issues and on educating the public and other participants. The Cluff Lake inquiry is an example of a project-specific EIA that was conducted prior to a formal act in the Province of Saskatchewan. The terms of reference were the only guidelines provided and those were quite general. The inquiry was conducted using a quasi-judicial procedure with the need to arrive at a recommendation on whether or not to proceed.

The next inquiry, the Aluminum Wiring Inquiry, was selected because it was ordered following the emergence and recognition of a fire hazard from aluminum wire use after it had been approved and used for some time. This inquiry was mainly technical and economic with little other value considerations. The interesting feature is the way that this inquiry arose and was treated. In contrast, the Royal Commission on Non-Medical Use of Drugs (Le Dain) was highly charged with emotions and the viewpoints depended largely on the value system of the participants. The latter two inquiries represent societal risks but are not created by projects (there are many other risks of this kind). Their inclusion enables simple comparisons to be made between risks that arise from projects and those that arise from other sources. The final inquiry, Mississauga, arose from an accident and illustrates some features of responsive actions.

Many of these inquiries have been examined at length in a study of public inquiry processes by Salter and Slaco (1981). In their examination they observe that risk is a central issue: "...most of the inquiries chosen for study took as their scientific question the determination of risk..." (Salter and Slaco, 1981; 150).

The examination of the cases leads to the identification of a number of analogous concepts in EIA and RA but they have important differences. An exploration of these concepts is the primary focus of the paper. The concepts from EIA that are discussed and contrasted are:

- Impact/hazard emergence
- Baseline data
- Mitigation
- Monitoring
- Compensation
- Compliance
- Assessment
- Management framework

In the final section a number of suggestions are made on how risk treatment might be enhanced within or outside of the EIA framework. These suggestions follow in a natural way from the contrasts made in the preceding section and from observations from the cases.

It is observed that traditional EIA could be taken to include risk as a special class of impacts. Likewise RA could be taken to include predictable impacts as a special class (those having occurrence probability of one).

Assessment of risk has been dealt with in a number of workshops and conferences with published proceedings. See, for example, those edited by Conrad, 1980; Dierkes, 1980; Schwing, 1980; Goodman and Rowe, 1979; Burton et al, 1982; Rogers and Bates 1983; and Kunreuther and Ley, 1982.

ENVIRONMENTAL IMPACT ASSESSMENT IN CANADA

Environmental impact assessment in Canada began to be treated more formally and comprehensively when the federal Environmental Assessment and Review Process was established by Cabinet decision in 1973. This was followed by the Environmental Assessment Act in Ontario, passed in 1975, and subsequently by various acts in the remaining provinces. Prior to these acts EIA had been carried out on a much more informal basis as directed by a number of agencies. Since the acts were passed a great deal of experience has been gained and now the process is well established in Canada although not necessarily conducted in an ideal way (Doern, 1977, 1981; Beanlands & Duinker, 1983).

The history of EIA in Canada, some of the characteristics of the Canadian nation, and a survey of the process is given in Couch et al, 1981. Briefly, Canada is made up of four Atlantic provinces (Newfoundland, Prince Edward Island, Nova Scotia, and New Brunswick). Their economy and outlook is dominated by the sea and by resource industries, principally forest-based. The next province, moving westward, is Quebec which is predominantly French-speaking with civil law based on the Napoleonic code. The civil law in the other provinces is based on English civil law based on the Napoleonic code. The civil law in the other provinces is based on English civil law. Ontario is the next province and in contrast, is predominantly English speaking. Both Quebec and Ontario have economies based on a mixture of industry (resource and manufacturing), intensive agriculture, and service industries. The three prairie provinces (Manitoba, Saskatchewan, and Alberta) are more sparsely populated deriving wealth from large-scale agricultural operations and resource industries. There has been a recent emphasis on secondary manufacturing in these provinces to broaden their economic base. The final province moving westward to the Pacific is British Columbia which is densely populated in a few small areas but is otherwise sparsely populated and mountainous. The wealth of British Columbia derives from forest and other resource industries and from fishing in the Pacific Ocean. In addition there are two territories, with an area of approximately 4.9 million square km. (40 percent of Canada's total) and a population of only 75,000 (1 percent of the 24 million total in Canada). The territories, administered by the federal government, have a large native population whose special rights require different treatment in environmental assessments as explored in the Berger Inquiry (Berger, 1977).

There is a fairly well-defined division of powers between the federal and provincial governments. These were spelled out in the British North America Act of 1867 under which Canada was formed as a separate nation. One power granted to the provinces that has an important bearing on EIA's was the power over resources within their respective boundaries. The specific details of the EIA process in the federal government and in the provinces has been summarized recently by the Federal Environmental Assessment Review Office, the agency responsible for the federal EIA process (Couch, 1983). A summary of these procedures from Couch, 1983, is included as Appendix A of this paper.

The federal process has been subjected to an intensive investigation to see how the science of ecology could contribute to the design and conduct of assessment studies (Beanlands and Duinker, 1983). The investigation used a number of workshops with broad participation to review what has been done and what should be done in performing an EIA.

Out of their investigations, Beanlands and Duinker developed a set of requirements for organizing and conducting ecological impacts studies. The requirements are sufficiently general that they could be extended to include other aspects of EIA and RA. The requirements suggested by Beanlands and Duinker are summarized here with some modifications to incorporate risk. Their implementation is aimed at planning and conducting an EIA or RA that is both comprehensive and acceptable:

- o The valued ecosystem and societal components should be identified early to provide a focus for subsequent activities. This scoping exercise should provide for input by interested parties.

- A context should be defined within which changes in significant ecosystem and societal components can be determined. The context would include such descriptions as the uncertainty, the spatial and temporal contexts, reversability, rarity of species and so on as a basis for judging the significiance of changes. The context also provides boundaries for the study.

- A study strategy should be developed and implemented for investigating the interactions between the project and the valued components. The study strategy should show how the individual studies are related.

- The EIA or RA should state explicitly the predicted impacts and the basis on which they are made.

- The assessment should detail and make a commitment to a well-defined program for monitoring the impacts (or hazardous situation).

It is interesting to note that none of the acts governing federal or provincial EIA processes ask specifically for a treatment of risk to the author's knowledge. Nevertheless, the general intent of the acts could be interpreted as requiring an examination of the risk, and risk has often been considered either by request in project-specific EIA guidelines or by having been insisted upon in public hearings.

RISK MANAGEMENT IN CANADA

Risk management is done by a large number of agencies in Canada and in most other countries. Certainly government agencies have responsibility, but responsibility is also assumed by professional associations, corporations, unions, and public interest groups. The significant point is that responsibility is dispersed so widely that it is difficult to visualize a comprehensive framework for risk management. A discussion of some of the government departments involved will illustrate this point.

Risks from aircraft use are administered by the federal Department of Transport. Motor vehicle risks are the responsibility of various provincial ministries. Occupational health and safety is generally the responsibility of provincial departments of labor. Some risks have also been dealt with by EIAs and in a variety of public inquiries.

It would simplify matters if the risk from a comprehensive set of situations could be listed together with the agencies responsible for their management. Unfortunately, in Canada (and probably in other countries) there is no such comprehensive assignment. In a study of hazardous materials it was found that there is considerable jurisdictional confusion (Doern, 1977; 152). Production of a material is generally under the jurisdiction of the Department of Consumer and Corporate Affairs but if the material is hazardous to health then the Department of Health and Welfare becomes involved. The provincial departments of labor who generally deal with occupational health and safety are also involved. If the material is radioactive then the Atomic Energy Control Board has responsibility for setting exposure limits. When the material is transported then the Department of Transportation and Communications has responsibility. As a second example, health is the responsibility of the federal Department of Health and Welfare but in each province there is a health ministry of some sort. The complicating factor is that more and more diseases are being found to arise from product use or from air and water pollution (not too many years ago typhoid was thought to come from the smell of sewage, not from drinking it). These brief examples indicate some of the difficulties in making an assignment of responsibility for managing risk. To begin with, a classification system for risks by cause and consequence is needed. Then there needs to be a proper risk accounting system so that risks can be placed in perspective. A beginning toward this was made by Miller and Ridgeway (Burton et al., 1982, Ch. 3) who developed a risk profile for Canada. A more complete risk accounting system would aid considerably in establishing a comprehensive risk management framework for Canada. In the meantime, a very large number of government departments, labor unions, universities, corporations, professions, and public interest groups are contributing to risk management.

PUBLIC INQUIRIES

Beaufort Sea

Oil was believed to exist in the Beaufort Sea--MacKenzie Delta region since before exploratory drilling began in 1965. Since that time about 100 onshore wells and 38 offshore wells have been drilled from 19 artificial islands and 15 drillships. Seven onshore and six offshore discoveries have been made (Beaufort Sea, 1982, Volume 6). The result of these discoveries is the estimate that there are very large reserves of gas and oil in the area.

The area is, at the same time, harsh and environmentally sensitive. The temperature ranges from -50°C in winter to 30°C in summer. In winter there are extended periods of darkness and in summer extended periods of light. The water is ice-covered from November to June. In addition there is land fast ice all year round, seasonal pack ice, ice island fragments, and polar pack ice. One of the main hazards for offshore drilling is the presence of icebergs.

The area is sensitive in that the native people (Eskimos) have a very tentative existence living off the land. The permafrost (permanently frozen muskeg swamp) on which land-based operations will be located can begin to thaw even when disturbed by driving vehicles over the land.

A primary hazard to the environment is oil spills. These can occur due to subsea or platform blowout, storage facility spill, tanker collision, tanker grounding, and tanker fire and explosion. The degree of hazard, the potential consequences, and the response measures will depend on the location of these events. Several accident scenarios with the consequences and countermeasure strategies are given in Beaufort Sea, 1982, Volume 6.

The Beaufort Sea Hydrocarbon Production Proposal was received by the Minister of Indian Affairs and Northern development who referred it to the Minister of the Environment (Federal). In accordance with the normal procedure a set of guidelines for the preparation of the environmental impact statement was issued by a panel which was created to deal with the proposal (Beaufort Sea Environmental Assessment Panel, 1982).

The guidelines (40 pages in total) called for the environmental and socio-economic issues to be dealt with. They called for a treatment of risks which included major oil and chemical spills, their probability, clean-up capability, and impact potential. An environmental baseline description was asked for but no "risk baseline." In a more detailed list of requirements the guidelines asked for information on blow-out prevention, design against "environmental threats" such as moving ice, ice scour, and storms. An interesting point is that the environment poses a threat to the project which in turn would pose a threat to other parts of the environment. Many other components were called for in a list of 25 items and many could be regarded as risk items but only specifically mentioned risk:

"Information of risk analysis for all major production and transportation components, singly and within overall development scenarios with the inclusion of possible future fail safe devices." (Beaufort Sea Environment Assessment Panel, 1982; 10).

In addition the guidelines asked for information on environmental hazard protection systems which were to include surveillance and prediction systems, (for example iceberg tracking). Further on the guidelines included a subsection on risk analysis of major oil spills or other pollutants and their potential impact using "worst case" scenarios and the limitations of planned or possible countermeasures. The requirement to deal with risk was specified directly in the EIA guidelines. The comparative attention given to RA and conventional EIA certainly favors the latter but the RA requirements are sufficiently general to insist on a comprehensive treatment.

The proponents' response to the guidelines is contained in Beaufort Sea, 1982, Volumes 1 to 7 with Volume 6 containing the response on spills. A review of the risk and risk analysis treatment in the report (Volumes 1, 2, 4, 6, and 7) was undertaken for the panel by Lemberg, 1983. The review is limited to environmental risk analysis and does not examine social or economic risk.

Lemberg raises a number of issues on the proponents' report for the panel to consider. These are specific for the report in question but some are sufficiently general to warrant discussion here. The first issue is concerned with subsequent design and operations decisions on countermeasures which the proponent will make using some form of cost-benefit analysis. It is not clear at the assessment stage which detailed choices the proponent will find feasible. The decisions will take place after the EIA process. This problem is typical of projects. It is dealt with in the nuclear industry by having inspections at the design and operations phases with authority to withold licenses for non-compliance.

A number of spill scenarios have been developed by the proponent but Lemberg suggests that those with the largest impacts have not been considered. The general question arising from this issue is the extent to which scenarios should be used and by what process will these be deemed to be sufficient for the authorities to make decisions on the project. Another issue raised is whether or not the proponent should be required to submit "estimates or calculations of accident probabilities, severities, and spill expsoures, and the degree to which countermeasures may reduce all or some of these." (Lemberg, 1982; 2-11).

In the Beaufort Sea case, regardless of precautions, accidents will happen and countermeasures become an important issue. Lemberg raises several questions on this issue dealing with:

- Manpower and equipment availability
- Government determination of response adequacy
- Whether the operator or government will ask for industry assistance
- Necessity for and adequacy of government resources
- Who pays for clean-up
- Assurance of adequate industry funds

Each of these points could entail considerable investigation at the assessment stage but each also requires monitoring. Consequently, the countermeasure preparedness and execution must be monitored to assure that the basis for project approval is not violated.

The final issues raised by Lemberg are on the transportation mode to markets. The options are by icebreaker tanker around Alaska or overland by pipeline through the MacKenzie Valley. Each has very different risks and impacts. The MacKenzie Valley pipeline proposal was the subject of a major inquiry (Berger, 1977). The difficulty is to decide when risk has received adequate treatment so that a decision can be made. The remainder of Lemberg's report examines the degree of compliance with the guidelines.

Porter Commission

The Royal Commission on Electric Power Planning, under the chairmanship of Dr. Arthur Porter, (the Porter Commission) was commissioned under an Order-in-Council of the Ontario Government on July 17, 1975 to examine long-range electric power planning in Ontario from 1983 to 1993. The inquiry originated primarily because of a concern that the capital requirements for Ontario Hydro's planned expansion program would strain the borrowing power of the province. The annual growth rate was forecast at that time to be about 7 percent and expansion would come primarily from nuclear power. The Porter Commission was not charged specifically with conducting an environmental impact assessment or a risk assessment of specific projects.

Nevertheless, the terms of reference were broad enough to allow the examination of a broad range of issues. The Porter Commission undertook a program of inquiry consisting of preliminary meetings, public information hearings, and debate stage hearings. Emphasis was placed on exploring the issues and on multi-way education between the public, those involved in the industry, and the government. An examination of the Porter Commission's treatment of risk was undertaken by the Institute for Environmental Studies, University of Toronto (Dooley and Robinson, 1982) and forms the basis for the comments on the Porter Commission in this section.

It is interesting that, whereas the term "risk" does not appear in the terms of reference, risk emerged as a major issue and became the subject of the first of a series of issue papers published by the Commission following the preliminary hearings. As attention became more focused on the nuclear debate a further Order-in-council was issued requesting the Commission to complete its examination of nuclear power. The Commission issued an interim report on this phase of its investigation (Royal Commission on Electric Power Planning, 1978). The interim report dealt with health, environmental, and safety considerations (the technical dimension); demand, economic, and social impacts (the social dimension); and social, ethical, and political issues (the political dimension). A final report was issued in February 1982 dealing with the full inquiry but containing further material on nuclear risk (Royal Commission on Electric Power Planning, 1980).

The early submissions by Ontario Hydro voluntarily raised the question of risk but treated risk primarily as an engineering matter concerned with reliability and compliance with the Atomic Energy Control Board (AECB) regulations for nuclear plant design, construction, and operation. This view is limited but it should be stressed that Ontario Hydro has an excellent safety record and has achieved much lower radioactive emission rates than are prescribed under the regulations. They felt, with this record, that they were justified in taking a limited technological view (Ontario Hydro, 1976).

The early submissions of the AECB stressed their continuing review of the regulations and care in licensing (AECB is the federal nuclear regulatory agency). They also emphasized their concern with all aspects of the fuel cycle from mining through to waste handling. The board described their cooperation with the International Commission for Radiation Protection (ICRP) and noted that AECB regulations call for lower exposures than ICRP standards.

Atomic Energy of Canada Ltd. (AECL) took the view that nothing is without risk and that there has been no significant increase in risk to the public as a result of nuclear plants compared to other risks (Atomic Energy of Canada Ltd., 1976). They also noted that the design limits on emissions (and those so far achieved) are far lower than those allowed by AECB (Snell, 1979). In general, AECL subscribed to the type of analysis used by Rasmussen.

The opponents took the view that nuclear weapons proliferation is a natural extension of a nuclear industry. There was concern that nuclear energy presents unique types of risk, particularly with respect to the magnitude of possible consequences of accidents. They argued that there should be a moratorium on new plants until the waste management problem is solved. They tended to disagree with the reliability estimates provided by the industry. The opponents also raised the question about sabotage and its possible consequences. The opponents challenged the argument that nuclear power is economic on the grounds that the economics are unknown until the waste management problem is solved.

In summary, it can be said that both the proponents and opponents raised the issue of risk in nuclear power generation. The proponents tended to take an engineering/scientific view and argued that the risks were "acceptably" low. The opponents enlarged the scope of risk treatment to include a broader spectrum of social and ethical considerations and argued for conservation and in favor of alternative ways of producing energy.

The Porter Commission had the opportunity to examine both traditional environmental and social impacts as well as issues of risk. Risk was treated as a separate issue as discussed above and as part of environmental and health implications (Royal Commission on Electric Power Planning, 1980, Volume 6). It must be stressed that the Porter Commission was not guided by environmental assessment legislation, regulations, or guidelines. In that sense, they could react to the issues as raised by the various participants in the debate. They did not have to make any hard decisions, such as approval or not of a nuclear project, so they could concentrate on presenting the diversity of views. For the purpose of this paper that was useful because it provides a case in which the relationship between traditional EIA and RA has been explored outside of official procedures.

Cluff Lake

The Cluff Lake Board of Inquiry was appointed by Order-in-Council of The Government of Saskatchewan (1st Feb. 1977) with terms of reference to:

> "...review all available information on the probable environmental, health, safety, social, and economic effects of the proposed uranium mine and mill at Cluff Lake..." (The Cluff Lake Board of Inquiry Final Report, 1978; 308).

The terms of reference also ask for provision of information to the public, implications of expanding the uranium industry, assessment of the protection measures proposed by the proponent, Amok Limited, and a recommendation on whether the project should proceed, not proceed, or proceed subject to conditions. The inquiry was ordered by the Lieutenant Governor in Council on the recommendation of the Saskatchewan Minister of the Environment. This inquiry is a case of an EIA that was conducted at the request of the government but before the formal provincial EIA act of 1980. The procedures adopted were dictated by:

> "...the terms of reference, the provisions of the Public Inquiries Act, the nature of the relevant questions, the type of inquiry, common sense and the rules of common justice..." (The Cluff Lake Board of Inquiry Final Report, 1978; 8).

In the event the participants divided into two groups -- those who supported and those who did not support nuclear energy (this is a common occurrence).

The report provides a comprehensive assessment of the probable environmental, health, safety, social, and economic effects and so treats risk as a major focus of the EIA. The Cluff Lake inquiry was conducted as a semi-judicial process. Critics have regarded the inquiry as thorough and in some sense ideal as a process for reaching a decision. Nevertheless, Salter and Slaco, 1981, point out that inquiry fatigue can result and that the participants can become worn down. The proponent in this case had already invested $30 million in the project so had a vested interest in continuing until the opposition gave up.

Risk from radioactive material was treated in the Cluff Lake inquiry by referring to the Canadian and Saskatchewan standards for radiation exposure. It dealt with the adequacy of the standards and made comparisons with other industries based on expected annual fatalities (per million workers per year). Comparisons of annual fatality rates were also made with several recreational activities. The inquiry went on to examine the genetic effects. The inquiry concluded that the standards are adequate. The examination of precautions proposed by Amok resulted in several recommendations on details. Safety, in the sense of conventional mine safety, was also considered. A comparison was made with occupational safety in Ontario mines. The nuclear proliferation question also received detailed attention reaching the conclusion that safeguards are not adequate in certain third-world countries. This debate proceeded to the moral issues of nuclear technology.

In summary, the Cluff Lake inquiry was a formal EIA that explored quite thoroughly the issues in uranium mining and milling. The scope of the debate extended well beyond local issues. Risk was treated as a major component of the inquiry. Risks were estimated and compared with both regulations and with societal risk profiles.

Aluminum Wiring

Aluminum wiring was introduced as an alternative to copper for branch circuits in residential construction in about 1950. A number of problems began to appear during the 1960s which suggested that aluminum wiring could lead to fires. The risk in this case was brought to public attention by the fire investigations of the insurance underwriters in Canada and the United States and by reports of wiring failures. Homeowner groups were formed and applied pressure for an inquiry. Following various proposals by government departments, standards associations, and Ontario Hydro, inquiry was commissioned by the Ontario government in April 1977. The terms of reference called for an investigation of safety and reliability, public hearings, and recommendations.

In this case there was no prior EIA or RA but the inquiry arose as a result of evidence and public pressure. It is difficult to say whether the risks from aluminum wire use would have been uncovered if an EIA or RA had been carried out before it had been approved for use.

Le Dain Inquiry

Throughout the 1960s drug use increased and became more visible as a result of its more extensive use by the middle class and media attention. The general belief became widespread that drugs presented a risk to individuals and to society. The traditional ways of dealing with drugs (including the cr status for distribution and use) came into question. Th mary legislation covering drugs in Canada is the Narcotics Control Act (opiate narcotics) and the Food and Drug Act (hallucinogens, barbituates, amphetamines) administered by the Federal Department of Health and Welfare.

The emergence of risk as an issue was a result of publicity about the negative effects of its use for the individual and society. A second reason was the inconsistency of the legal implications such as the use of marijuana being a criminal offense in a society that was coming to accept its use as normal. Risks also appeared to the society arising from a fear of widespread use and the means of distribution by the criminal element.

The government's response was to establish the Royal Commission of Inquiry into the Non-Medical Use of Drugs under the chairmanship of G. Le Dain in 1969. The terms of reference called for a scientific examination into the possible dangers to individuals and to society. Tobacco and alcohol were omitted from the terms of reference. This caused subsequent difficulties for the commision.

The commission contracted a number of studies to gain scientific evidence on the effects of drug use. There was a paradoxical aspect to gathering scientific evidence from users in the form of testimony because a public inquiry is required to make the evidence public and yet to do so the users would be making a public admission of criminal acts. Therefore special provisions for hearings, in camera, had to be made.

In spite of the attempt at a scientific treatment of RA, in this case the scientific evidence was not used effectively. (Salter and Slaco, 1981, Chapter VII). Nevertheless, the final report of some 1,100 pages devoted considerable space to an assembly of scientific information on drug use (Commission of Inquiry into the Non-medical Use of Drugs, 1973).

Mississauga Inquiry

On Saturday, November 10, 1979, a freight train derailed in the City of Mississauga just west of Toronto. In all, 24 cars came off the rails, propane tank cars ruptured, and several bleves (boiling liquid expanding vapor explosions) occurred. At the heart of the jumble of cars was a car containing 90 tons of chlorine. The chlorine car ruptured and early arrivals at the site could detect the smell of chlorine. The blazing propane was very spectacular and extremely hot thus denying access to the chlorine car so the extent of the rupture could not be determined. Eventually, 225,000 people were evacuated for up to seven days. Remarkably no deaths have been attributed directly to the accident.

Risk emergence was dramatic in the case. The fire presented one hazard but was far enough from buildings that its spread was not a serious problem. The more subtle, uncertain, and dangerous factor was the presence of chlorine gas.

The immediate treatment of risk was the countermeasures to contain the cause and to avoid its consequences by evacuation. A longer term component of risk treatment was the Grange inquiry which was established on December 4, 1979 by Order-in-Council of the federal government to ascertain the cause, the steps needed to reduce risk of recurrence, the adequacy of existing regulations and operating practices, and enforcement (Report of the Mississauga Railway Accident Inquiry, 1980).

Engineering and observer evidence was taken which finally led to the conclusion that the derailment has been caused by a journal failure brought on by a "hot box" (overheated bearing). A number of recommendations were made without evident resort to cost/benefit analysis or technical considerations. One was the recommendation to use roller bearings for all cars on trains carrying hazardous material. A second was the recommendation to use hot box detectors. The interesting problem here is that journal bearings fail slowly and create a hot box during failure wich can be detected but journal bearings normally fail quickly and defy detection. Another recommendation calls for Transport Canada to:

> "...Make available at the scene of, and within hours of, an accident, a person capable of directing the clean-up of that accident and of protecting the populace. He will lend all assistance to the local or provincial authorities and will take charge of the scene if no such authorities are evident..." (Ibid, 199).

The intent of the recommendation is not misdirected but the practical implementation is hard to imagine in a large country, often sparsely populated, and with thousands of dangerous goods requiring different treatments being transported.

The Grange inquiry made an attempt at risk management in the transport of dangerous goods by rail but is only a modest beginning and ignored much of the technical material available. No formal risk assessment was made.

The insight provided by these and other cases, was used to make comparisons between EM and RM. This is contained in the next section.

RISK AND ENVIRONMENTAL MANAGEMENT

For discussion purposes, EIA was taken to apply to the impacts that are fairly predictable and generally arising from projects. In contrast, RA was taken to apply to uncertain impacts from projects and other sources. Risk and environmental management are distinguished similarly. This distinction facilitates the discussion of concepts common to RM and EM but needing different treatments. This section contains a discussion of the concepts, their similarities and differences, and some of the implications of RM. None of the comparisons are treated exhaustively nor is the extensive literature on each concept cited. That is beyond the scope of this paper. The intent has been to make a preliminary examination of some of the contrasting ideas and practice in environmental and risk management.

It may appear that EIA has been oversimplified in comparison to RA. This is partly because of the a priori restriction of EIA for discussion purposes to encompass the more deterministic impacts of projects and programs. The apparent simplicity of EIA is also due to the length of time that the EIA process has been in place and the nature of the legislation and its implementation which has assured a somewhat restricted view of EIA and a more routine approach to assessment. Nevertheless, the treatment of risk, with its uncertainty and pervasiveness in society, makes it inherently more difficult to deal with.

An EIA framework does exist in Canada and provides a useful one for dealing with some of the risks. Throughout the comparisons that follow there are several places where the differences between treating deterministic and uncertain impacts from projects indicate where concepts from risk need special treatment within the EIA framework.

Impact/Hazard Emergence

Emergence refers to the way that the environmental impact or risk gets onto the agenda of the assessment or inquiry process. In the early days of environmental concern there were many unsuspected impacts (e.g., DDT) but now, in its more mature phase, the mechanisms are better understood. Through a combination of expert advice, public interest groups (e.g., Pollution Probe in Ontario), and such aids as checklists, few impacts should be overlooked. The tendency for EIAs to be applied to projects often isolates the locality and many impacts then become implicit in the project description (e.g., amount of forest cleared for the project).

It was found in the cases that the specific project and generic studies identify and treat risks insofar as these can be predicted at the time of the inquiry. Although there may be disagreement between the proponents and the opponents on how risks should be treated, nevertheless most risks emerge and get managed at some level of competence. In the Beaufort Sea case, the issues were raised initially by the panel, and later by Lemberg in response to the proponents' EIA report. The Cluff Lake inquiry explored carefully the proponents' risk management plans and made recommendations where the inquiry felt provisions were inadequate. In the Porter case, the public forced a broader consideration of risk than was originally envisaged by the nuclear industry. There will be those who say the risks are too great to let these projects proceed and those who disagree but the collection of risks have certainly been identified. The formality of EIA processes and the public inquiry, where used, provide ample opportunity for the known or suspected risks to get on the agenda. The adequacy of their management is another matter not dealt with here.

What about other risks? In the aluminum wire case, unsuspected risks arose after approval of the product and got on the agenda as a result of public pressure. Still less clear is the emergence of risk in the Le Dain case on the non-medical use of drugs. Drugs of certain types have been known to be harmful for a long time. In this case it was a growing concern for individuals and society brought on by greater use that led to a formal inquiry. The hesitancy by which the inquiry proceeded is evidence of the ill-defined nature of the risks.

The conclusion is that where formal EIA frameworks exist, such as in the EIA procedures of the federal and provincial governments, or, for example, in the case of the Food and Drug Act administered by the Federal Department of Health and Welfare, the risks are identified reasonably well. The emergence of risk is less well managed where there is no formal routine investigation of the siutation involving risk and the jurisdiction is not clear. A more comprehensive treatment of risk will require some form of risk monitoring of the societal risk profile to detect risks formerly unsuspected.

The checklist has been an effective device for identifying impacts. The same device can be used to identify risks but not as effectively, due to the more pervasive nature of risk. It would be useful to have a comprehensive world-wide checklist of impacts and risks, dynamically updated as new ones are identified. A useful arrangement would be to have a system accessible by computer with a retrieval system so that appropriate information for a particular case could be obtained.

Baseline Data

The notion of baseline data is well established in EIAs. The state of the environment without the project is described. Changes in the state of the environment with the project can then be compared with the baseline to provide an assessment of the relative impact of the project. To be realistic, the baseline is predicted over time with and without the project to account for changes from other causes. (Munn, 1979, 9, Oppenheimer, 1976).

Is there a corresponding notion in risk assessment? It is suggested that there is. For RM the appropriate baseline is the risk profile for the segment at risk and for the society at large. The impact of a particular risk can then be evaluated by examining the net impact on the risk profile. A comparison of project risks with other risks was made by the proponents in the Porter case and was dealt with by the inquiry in the Cluff Lake case. Two major technical problems have to be solved to achieve an adequate risk profile. The first problem is to develop appropriate measurements of risk so that comparisons can be made. The second is to develop an adequate risk classification system so that proper accounting of risks can be accomplished.

Mitigation

In traditional EM, mitigation deals with the measures taken to restore, control, or counterbalance the impact of the project on the environment. One simple example is the installation of fish ladders around a dam to enable fish to travel upstream for spawning.

The corresponding notion in risk management is countermeasures. The uncertainties that countermeasures have to deal with may be an unsuspected consequence (e.g., thalidamide, aluminum wiring), or uncertainty about the location, time and magnitude of the consequence (e.g., transport of hazardous material) and so on. One form of countermeasure attempts to reduce the hazard probability. This can be done by design and operations choices in the case of projects. In drug use the countermeasures might deal with drug distribution or publicity on drugs' negative aspects. This is now happening in the case of smoking. A second form of countermeasure is the prepardness for, and response to, unpredictable events if, when, and where they take place. For regardless of the effort to reduce risk probabilities, hazardous events will occur. In the Beaufort Sea case there were several hazardous events postulated in scenarios with the corresponding countermeasures. The appraisal of Lemberg, 1983, felt that these were inadequate and raised questions about who should evaluate and pay for preparedness. Responsive action was also discussed briefly in the Mississauga derailment case. The problem of preparedness and response decisionmaking in crisis situations is discussed in Dooley, 1982.

Monitoring

Monitoring can refer to either the follow-up on a project to see that the proposed mitigation and operations have complied with regulations (see compliance below) or can refer to environmental surveillance to detect changes and problems. The current work on monitoring the ozone layer and CO_2 in the stratosphere are examples of surveillance. Monitoring is quite similar in environmental and risk management in that certain variables are chosen and monitored. In both cases there is a need to look for unsuspected changes. The major difference is that risk covers a broader spectrum of causes and consequences and so demands a more extensive monitoring mechanism.

The baseline data and risk profile provide the norm against which to monitor for changes but predictive models and procedures are also needed.

Compensation

Compensation refers here to the payment to those who are disadvantaged by the project, program, or activity. The compensation works to offset their disadvantage and to gain their acceptance. A common form of compensation is to pay for land expropriated to make room for a project. There is usually a fair market value for the land expropriated. The social impact of a project is less easily evaluated. One way to approach this is by social impact analysis (Whyte, 1977; Lang and Arman, 1981; Bowles, 1981). Bidding is a way to arrive at the amount of compensation. In one example in Ontario, a fixed sum is paid into the municipal treasury for each ton of waste dumped into a local waste disposal site.

The corresponding notion to compensation in risk has two forms. The first is the compensation for the anxiety created as a result of the hazard potential. The second is the compensation to those recipients of the consequences when a hazardous event occurs. Insurance is one way that those at risk are assured that specified compensation will be available in case a hazardous event occurs.

There is a need to explore compensation further to discover the ramifications, and to develop procedures for determining fair levels of compensation. There are also questions about who should be held liable and their capability of providing compensation particularly for realized consequences.

Compliance

If the impacts are deterministic and a combination of mitigation measures and guidelines (e.g., emission rates or water quality) have been specified and agreed on in the EIA process, then assessment of compliance should be fairly straightforward. It may not be straightforward when for instance, the impact is a pollutant and there are a number of contributors. An example is acid rain in which there are many contributors but there are arguments about the dispersion of each contribution and its impact.

In contrast, risk management has to be judged by whether the event probabilities are greater than those agreed upon or not. To illustrate the point, the emission of radioactive materials from a nuclear plant can be measured and complaince with guidelines can be assessed. It is much more difficult to determine (both initially and subsequently) that the chance of a particular accident continues to be less than say one in a million. The discussion of such probabilities has been caried on extensively in the nuclear area by such people as Rasmussen, Lewis, and Flowers, and still no firm agreement has been reached.

The management problems remain those of obtaining agreement on hazardous event probabilities and controlling for compliance. One approach would be to monitor the design and operation of a project to see whether or not the design and operating conditions that were assumed as a basis for estimating probabilities are followed. The requirements of Beanlands and Duinker discussed in Section 2 are relevant. They argue for a monitoring program as part of the EIA so that subsequent complaince can be assessed. Their point is equally valid for RA.

Assessment

A very difficult part of risk management that has emerged is the difference between the scientific and social perception of risks of various kinds as exhibited by behavior (Dinman, 1980; Pochin, 1975; Otway, 1980). The expressed perception and acceptance has been studied extensively by Slovic, Fischhoff, and Lichtenstein in a number of articles and books (see for example Fischhoff et al, 1981; Slovic et al, 1982; Lowrance, 1976). The difference between the objectively estimated risk and the perceived risk (or its acceptance) is the core of many of the disputes that arise in public hearings as, for example, in the Porter Commission. One view is that the resolution of these differences in the decision process is an extension of the democratic principle (Fagnani in Greer-Wotten, 1980).

Management Framework

The EIA process is well-established by legislation and use although it is by no means problem-free. Nevertheless, there is an existing framework for managing the environment. The temptation is to suggest that risk should be managed within the existing environmental legislation, regulations, and procedures. This view is reinforced by the evidence that risk has been treated within this framework on many occasions. The Beaufort Sea and the Cluff Lake cases both dealt with risk within the EIA process. Nevertheless, the nature of risk is such that a broader framework seems to be necessary. In fact, there are those who argue that environmental assessment should be subsumed under a risk management framework (O'Riordan, 1979). The question of whether EM should subsume RM or the converse is unresolved. Regardless of which becomes the dominant framework, there are differences in the way that predictable and unpredictable impacts need to be treated for some of the reasons discussed in this paper.

One major problem is to allocate resources for risk reduction. There are those who argue that resources should be allocated first to the activity that will produce the most cost-effective risk reduction (however that is measured). In the nuclear industry, for example, it is argued that further safety measures to reduce the risk (expected value) are more expensive per life saved on average than on other risk reduction activities (Siddall, 1980). This view is contrary to the expressed wishes of society who have an abhorrence for nuclear accidents but accept traffic deaths as the norm. Of course, one difference between a nuclear accident and traffic deaths is that the standard deviation is so much greater in the nuclear case. It seems that a much better understanding of risk and the cost of risk reduction is needed before the several people involved can allocate resources more appropriately.

CONCLUSION

This paper set out to draw comparisons between environmental and risk assessment and management. It was assumed, and largely confirmed, that predictability versus uncertainty is the essential feature of impacts that necessitates a contrasting treatment. To facilitate discussion, assessment of essentially deterministic, project-oriented impacts were labelled EIA and uncertain impacts were labelled RA. This distinction proved useful but it was noted that the EIA process has often included an evaluation of risk in practice.

Nevertheless, it was found in the cases that there is more assurance about how deterministic impacts should be treated than uncertain ones. The paper has sought to explore the similarities and differences between EIA (and EM) and RA (and RM). These were discussed in the last section under a number of topics. Part of the discussion was to suggest some of the areas needing further work to facilitate better risk management. Some of the topic areas are listed below without elaboration or review of the work already done in these areas.

(a) A risk classificiation is needed for developing baseline data (risk profile) for making comparisons between project alternatives, and for assigning responsibility for risk management.

(b) Risk and environmental baseline information system are needed. Ideally these should be world-wide, dynamically updated, and should have computerized retrieval facilities.

(c) Procedures need to be developed further to resolve differences between technological and behavioral assessment of risk. It seems to be forgotten sometimes that all parties are part of the same society and all have an interest in risk reduction.

(d) Theoretically sound and behaviorally-acceptable procedures are needed to compare risks, to evaluate the net effect of specific changes that affect risk, and to allocate resources to risk management activities.

(e) The concepts of prior compensation and insurance need further theoretical development and procedures need to be worked out for determining the appropriate form and level of compensation and its implementation in specific instances.

(f) Compensation for the consequences of a hazardous event that has occurred needs to be investigated from several perspectives. First is the amount of compensation. This is typically handled in the courts after the event. Then there is a need to determine who should be held responsible which needs to be determined ahead of time so that adequate financial provisions can be made through such means as insurance. A related question is the ability of those at fault to provide the compensation.

(g) Countermeasures are an important part of risk management. There are emergency plans in place to deal with many of the risks that we experience. There is a need to examine the preparedness of responsive actions of various kinds and in various locations. There is also a need to monitor their adequacy following the approval stage.

REFERENCES

1. Atomic Energy of Canada Ltd. (1976), Nuclear Power in Canada: The Canadian Issues, Ottawa.

2. Beanlands, Gordon E., and Duinker, Peter N. (1983), An Ecological Framework for Environmental Impact Assessment in Canada, Institute for Resource and Environmental Studies, Dalhousie University, Halifax, Nova Scotia, and Federal Environmental Assessment Review Office.

3. Beaufort Sea (1982), Environmental Impact Statement for Hydrocarbon Development in the Beaufort Sea - MacKenzie Delta Region, Vol. 1 to 7. Prepared by Dome Petroleum Ltd., Esso Resources Canada Ltd., and Gulf Canada Resources Ltd.

4. Beaufort Sea Environmental Assessment Panel (1982), Guidelines for the Preparation of an Environmental Impact Statement for the Beaufort Sea Hydrocarbon Proposal, Canada, Environmental Assessment Review.

5. Berger, Mr. Justice Thomas R. (1977), Northern Frontier, Northern Homeland. The report of the MacKenzie Valley Pipeline Inquiry, Volume 1, Ottawa, Department of Supply and Services.

6. Bowles, Roy T. (1981). Social Impact Assessment in Small Communities, Scarborough, Ontario, Butterworths.

7. Burton, I.; Fowle, C.D.; and McCullough, R.S. (1982), Living with Risk: Environmental Risk Management in Canada, Toronto, Institute for Environmental Studies, University of Toronto, Environmental Monograph No. 3.

8. Commission of Inquiry into the Non-Medical Use of Drugs (1973), Final Repot, Chairman G. LeDain, Ottawa.

9. Conrad, J. (ed.) (1980), Society, Technology, and Risk Assessment, London, Academic Press.

10. Couch, William J. (ed.) (1983), Environmental Assessment in Canada: 1983 Summary of Current Practice, Ottawa, Federal Environmental Assessment Review Office.

11. Couch, W.J.; Herity, J.F.; and Munn, R.E. (1983), Environmental Impact Assessment in Canada. Contained in PADC, 1983.

12. Crouch, Edmund A.C., and Wilson, Richard (1982), Risk/Benefit Analysis, Cambridge, Massachusetts, Ballinger Pub. Co.

13. Dierkes, Meinolf; Edwards, Sam; and Coppock, Rob, (ed.) (1980), Technological Risk, Cambridge, Massachusetts, Oelgeschlager, Gunn & Hain Pub.

14. Dinman, B.D. (1980), The Reality and Acceptance of Risk, J. American Medical Association, 244 (11):1226-1228.

15. Doern, G. Bruce (1977), Regulatory Processes and Jurisdiction Issues in the Regulation of Hazardous Products in Canada, Ottawa, Science Council of Canada, Background Study No. 41.

16. Doern, G. Bruce (1981), The Peripheral Nature of Scientific and Technological Controversy in Federal Policy Formation, Ottawa, Science Council of Canada, Background Study No. 46.

17. Dooley, James E. (1982), Decision Making in Crisis Situations, in Kunreuther and Ley (1982).

18. Dooley, James E., and Robinson, John B. (1982), Nuclear Risk Management in Canada: An Examination of the Royal Commission on Electric Power Planning, Report to the Beijer Institute of The Royal Swedish Academy of Sciences and the Swedish Energy Research and Development Commission.

19. Dooley, James E., and Burton, Ian (1983), "Risk Assessment for the Transport of Hazardous Materials." Published in the Proceedings of a Symposium on the Assessment and Perception of Risk to Human Health in Canada. Sponsored by The Royal Society of Canada and The Science Council of Canada.

20. Douglas, Mary, and Wildavsky (1982), Risk and Culture, Berkeley, University of California Press.

21. Fairfax, S. (1978), A Disaster in the Environmental Movement, Science 199, 743-748.

22. Fischhoff, Baruch; Lichtenstein, Sarah; Slovic, Paul; Derby, Stephen L.; and Keeney, Ralph (1981), Acceptable Risk, Cambridge, Cambridge University Press.

23. Goodman, Gordon T., and Rowe, Wm. D. (ed.) (1979), Energy Risk Management, London, Academic Press.

24. Greer-Wootten, Bryn (1980), Context, Concept and Consequence in Risk Assessment Research: A Comparative Overview of North American and European Approaches in the Social Sciences; published in Conrad, 1980, pp. 67-101.

25. Inhaber, Herbert (1982), Energy Risk Assessment, New York, Gordon and Breach.

26. Kahneman, D., Slovic, P., and Tversky, A. (ed.) (1982), Judgment Under Uncertainty: Heuristics and Biases, Cambridge, Cambridge University Press.

27. Kunreuther, Howard C., and Ley, Eryl V. (ed.) (1982), The Risk Analysis Controversy: An Institutional Perspective, Berlin, Springer-Verlag. Proceedings of a Summer Study on Decision Processes and Institutional Aspects of Risk held at IIASA, Laxenburg, Austria, 22-26 June 1981.

28. Lang, Reg, and Armour, Audrey (1981), The Assessment and Review of Social Impacts, Ottawa, Federal Environmental Assessment Review Office.

29. Lemberg, Ray (1983), A Critical Risk Analysis Evaluation of the Environmental Impact Assessment: Beaufort Sea Hydrocarbon Production Proposal. Report preapred for the Beaufort Sea Environmental Assessment Panel, Lemberg Consultants (Canada) Inc.

30. Lowrance, William W. (1976), Of Acceptable Risk, Los Atlos, California, William Kaufmann Inc.

31. Oppenheimer, C.H.; Oppenheimer, D.; and Brogden, William B. (1976), Environmental Data Management, Proceedings of the Conference on Environmental Data Management held in the facilities of the Houston Museum of Natural Science, Houston, Texas, sponsored by the NATO Special Program Panel on Ecology.

32. O'Riordan, Timothy (1979), Environmental Impact Analysis and Risk Assessment in a Management Perspective, Goodman & Rowe, 1979, pp. 21-36.

33. Otway, Harry (1980), The Perception of Technological Risks: A Psychological Perspective, Ch. 5 of Dierkes et al., 1980.

34. Otway, Harry (1980), The Perception of Technological Risks: A Psychological Perspective, Ch. 5 of Dierkes et al., 1980.

35. PADC Environmental Impact Assessment and Planning Unit (1983), Environmental Impact Assessment, The Hague, Martinus Nijhoff Pub.

36. Pochin, E.E. (1975), The Acceptance of Risk, British Medical Journal, 31(3):184-190.

37. Proceedings of NATO Advanced Study, Institute on Environmental Impact Assessment, Toulouse, France, 30 August - 12 September 1981.

38. Report of the Mississauga Railway Accident Inquiry (1980), The Honorable Mr. Justice Samuel G.M. Grange, Commissioner, Canada, Ministry of Supply and Services.

39. Rogers, J.T. and Bates, D.V. (ed.) (1983), A Symposium on the Assessment and Perception of Risk to Human Health in Canada, Ottawa, Royal Society of Canada.

40. Royal Commission on Electric Power Planning (1978), A Race Against Time, Interim Report on Nuclear Power in Ontario, Arthur Porter, Chairman, Ontario.

41. Royal Commission on Electric Power Planning (1980), Final Report, Volume 1 to 9, Arthur Porter, Chairman, Ontario.

42. Salter, L. and Slaco, D. (1981), Public Inquiries in Canada, Ottawa, Science Council of Canada.

43. Schwing, Richard C., and Albers, Walter A. (ed.) (1980), Societal Risk Assessment: How Safe is Safe Enough, New York, Plenum Press. Proceedings of the General Motors Symposium on Societal Risk Assessment, held in Warren, Michigan, October 7-9, 1979.

44. Siddall, E. (1980), Control of Spending on Nuclear Safety, Nuclear Safety, 21(4):451-460.

45. Slovic, Paul; Fischhoff, B.; and Lichtenstein, Sarah, Facts versus Fears: Understanding Perceived Risk. Published in Kahneman et al. (ed.), 1982.

46. Snell, V.G. (1979), "Safety of CANDU Nuclear Power Stations," Atomic Energy of Canada Ltd. Report AECL-6329.

47. The Cluff Lake Board of Inquiry Final Report (1978), Mr. Justice E.D. Bayda, Chairman, Saskatchewan.

48. Tribe, Lawrence H.; Schelling, Corinne S.; and Voss, John (ed.) (1976), When Values Conflict: Essays on Environmental Analysis, Discourse, and Decision, Cambridge, Massachussetts, Ballinger Pub. Co. Published for the American Academy of Arts and Sciences.

49. Whyte, Anne V.T. (1977), Guidelines for Field Studies in Environmental Perception, Paris, United Nations Educational, Scientific and Cultural Organization.

APPENDIX A

Environmental Assessment in Canada Today*

Environmental assessment methodology in Canada has evolved very quickly during the last decade. Most jurisdictions have systems in place and are largely concerned with improving what is now working. In spite of Canada's geographic, economic, social, and institutional diversity, the structures and areas of concern which have emrged across the nation bear much similarity to each other. The statements below reflect these similarities and suggest emerging trends.

Legal Authorities for Environmental Assessment: Departments and agencies derive their authority to carry out environmental assessments from a variety of legal bases. Ontario was the first province to pass an environmental assessment act in 1975. In 1978, Quebec passed explicit environmental assessment legislation in an amendment to the Environment Quality Act. Saskatchewan and Newfoundland passed environmental assessment acts in 1980. In 1981, British Columbia supplemented environmental assessment procedures within existing statutes through the passage of the Environment Management Act. The authority for environmental assessment in Alberta and Nova Scotia is derived from provincial statutes. The Federal Government, New Brunswick, and Manitoba each have a process based on government policy directives.

Administration: In most jurisdictions, the developer has initial responsibility for the environmental assessment of activities. There is usually a lead agency appointed to co-ordinate the review of the environmental assessment on behalf of the government. These administrative agencies are usually located in the department responsible for the environment. The Federal Environmental Assessment Review Office (FEARO) operates at arm's length from the federal Department of the Environment and reports directly to the federal Minister of the Environment. In British Columbia, aspects of environmental assessment have been transferred to the Minister of Environment from an interministerial steering committee.

---

*Source: Couch, W.D. (Ed.), Environmental Assessment in Canada: 1983 Summary of Current Practice, Ottowa, Federal Environmental Assessment Review Office, 1983 (Reprinted with permission).

Screening: In all jurisdictions, decisionmaking occurs in discrete steps. As the number of projects potentially subject to environmental assessment is usually large, a screening mechanism of a fairly decentralized nature exists, whereby small routine projects with no substantial or significant impacts are screened out and allowed to proceed with a minimum loss of time and expense.

Those projects which may have a significant adverse environmental impact or may be of a sensitive nature are usually submitted to the administrative agency for a more searching, and possibly more visible and structured review. Control by the administrative agency usually becomes more firm at this point. It may issue guidelines and give general or specific direction to the proponent. It usually coordinates needed reciprocal reviews. Major projects could be subject to a public review by an independent board panel.

The Environmental Impact Statement: The environmental impact statement (EIS) is an administrative tool used in most jurisdictions during the second more formal phase of assessment. The EIS's contents and format are similar in each jurisdiction: need and alternatives, a description of the project, a description of the environment without the project, an analysis and prediction of how the project will impact on the environment, a presentation of proposed mitigating measures, and a statement of residual adverse effects. It is usually prepared by the project proponent and may be reviewed by the adminstrative agency, specialized technical agencies, or a public review board or panel.

Public Participation: Public participation is an integral part of environmental assessment in Canada. The two have matured together. Systems have been refined, and the quality and significance of public participation has grown in all jurisdictions. People want to participate. In response, governments encourage participation. All processes based on environmental assessment acts include explicit provision for public participation within each act. Public participation is part of those processes which are derived from legislation. And all processes which are based on administrative policy require public participation.

Public Hearings: The degree of formality during public reviews varies among jurisdictions, from a strict adherence to judicial practice to informal community meetings. The ultimate decisionmaker as to whether or not there will be a formal public review is usually a Cabinet Minister or Cabinet as a whole. The review board or panel is usually empowered to submit only recommendations to the minister responsible for the environment. Cabinet retains the prerogative to accept, in whole or in part, or reject the recommendations. In jurisdictions where review bodies do not have decisionmaking authority, this lack of authority of the reviewing bodies has become more apparent than real. The openness of public hearings, the wide publicity hearings may receive, and the fact that recommendations are usually made public, have combined to give these bodies more "de facto" authority than they possess in theory. Thus in view of pressures created by public expectations, that the recommendations of these public review bodies be followed by decision-makers, there has not been a widespread demand to give boards, panels, and similar agencies additional power in law.

Funding Public Participation: There is growing pressure on governments at all levels to provide financial assistance for public participation. Advocates of funding contend that public participation improves government decisions and that since there is a real cost to society in the effort expended, these costs should be borne by government. It is still unclear whether this position will be accepted in all jurisdictions. Some federal and provincial regulatory agencies have clauses in their legislation or regulations permitting payment of part of participants' expenses. Saskatchewan has included a provision for funding in its Environmental Assessment Act. The federal government is also providing intervenor funding in one major review.

Social Impact Assessment: The trend to increased public participation has led to increased consideration of the social impacts of proposals in the context of environmental assessment. Where there is an environmental assessment act, social and community considerations are included in the definition of the environment. In jurisdications where there is not an explicit environmental assessment act, this broad interpretation is also generally accepted.

Considerable effort is currently being made across Canada, including under the auspices of the CCREM, to define parameters for social impact assessment. It is clear from these studies that social impact assessment is a field which must continue to evolve before widely-accepted and tested analytical frameworks are possible.

With a recent trend to very large-scale projects, new relations are being formed among governments, industry, and the public, resulting in complex service requirements. This is also leading, notably in British Columbia, to the development of systems for the coordinated management of impacts.

Land Use Planning: In the beginning, it was broadly assumed that project-oriented impact studies would, in sum, protect the environment. It is being learned, through experience with interactive effects, that this is not necessarily the case. Also, there is a growing awareness of the importance of framing project decisions in a broad planning and policy context. Comprehensive land use planning is now an overall priority in British Columbia. Federally there is increased emphasis being placed on ensuring the existence of a policy framework in which to examine individual proposals, particularly in the North. In some other jurisdictions there is strong interaction between land use planning and environmental assessment. The concept is that project-oriented environmental assessments should not be the leading force in planning. It is also recognized that both are necessary. Once proposals are placed in an overall policy context, following land use and resource planning, environmental assessment provides a linkage for integration of regional concepts and a verification of compatibility with the planning framework.

Cost Allocation: Cost allocation for the various elements in each process are similar across the country. The environmental assessment administrative agency or designated unit within the government pays for the assessment process and the approval of regulatory permits and licenses, including public reviews. The proponent pays for the preparation of the EIS. Monitoring costs are normally cost-shared between proponent and government.

CONCLUSION

This overview has attempted to catch some of the main features of environmental assessment in Canada. The remainder of the document looks at environmental assessment in each jurisdiction. But, even as the various sections were being written, changes--or plans for change--were occurring. Change, growth, and adaptation still remain central features of this dynamic phenomenon in government decisionmaking.

Assessing Long-Range Cumulative Impacts[1]

Evan Vlachos

Professor of Sociology and Director, Environmental Resources Center, Colorado State University, Fort Collins, CO 80523

INTRODUCTION

This particular paper is written as a combination of two different topics, i.e., "cumulative impacts" and "sociological dimensions of EIA, TA, and RA." While the emphasis is on the first, a number of pertinent remarks are made with regard to EIA, TA, and RA in order to exemplify an underlying thrust as to "assessment" efforts. At the same time, heavy emphasis is placed on assessment associated with natural resources management, public projects, and environmental hazards.

The material that follows is organized around five categories of concern:

1) The background of "assessment" efforts, commonalities of epistemological perspectives and the search for unifying theory of social change;

2) What the cumulative impact assessment process entails;

3) Definitions, types, and fundamental issues concerning cumulative impacts;

4) Broad approaches and methodological strategies and tactics in accounting for cumulative impacts; and,

5) Risk and uncertainty in the context of cumulative impacts.

---

1 I would like to thank Gene Stakhiv and Gerry Horak for their commentary, support, and insights.

NATO ASI Series, Vol. G4
Environmental Impact Assessment, Technology Assessment, and Risk Analysis. Edited by V. T. Covello et al.

THE BACKGROUND OF "ASSESSMENT" EFFORTS

One may begin the topic by noticing the significant presence of the letter "A" in such prevalent acronyms as EIA, TA, and RA. While some texts may refer to it as "analysis," most writers and researchers (as well as legal practice) view the last element as "assessment." The term "assessment," according to the dictionary, implies to "evaluate," to "estimate or determine the significance, importance or value of." It is interesting that the original Latin word assidere means to sit beside, to assist (in office). Thus, one can see in the evolution of the term assessment (of sitting beside) the important semantic connotation to be combined with the evaluation of any impacts of given actions. Presumably, when the time comes the decisionmaker will have somebody who "sits beside him" and, therefore, is able to assist him in passing an evaluation about a given action, technology, or project and about the decision to proceed with a given course of action.

In its broadest meaning, assessment becomes a procedure for anticipating effects of purposive action and, thereby, forestalling or upsetting adverse consequences to which a particular project may give rise. In this sense, assessment may be viewed as a hedge against uncertainty in the planning process.

More specifically, impact assessment can be broadly understood as an appraisal of the effects of actions on the environment. Operationally, impact assessment can be described as compliance with the law; conceptually, however, it becomes a beginning point for an integrated, long-range planning process.

As has been observed by many writers, no federal legislation of recent years bears greater potential significance for the management of the environment in the United States than the National Environmental Policy Act of 1969. The purpose of NEPA is to "encourage productive and enjoyable harmony between man and his environment and stimulate the health and welfare of man." NEPA recognizes throughout the profound impact of man's activity on the interrelations of all components of the natural environment and prescribes the use of all practicable means and measures to create and maintain conditions of productive harmony with the environment and fulfillment of social, economic, and other requirements of present and future generations.

Yet this concern with the assessment of the effects of public projects' activity or program implementation did not start in 1969. It can be traced to a variety of interests, concerns, and legislative actions that tried to find out the effects of proposed interventions. These efforts resulted in the succinct imperatives of NEPA and of similar documents which made more urgent than ever the need to account for environmental changes. At the same time, distinctively social impacts have tended to be implicit, indirect, and qualitative. It is only recent administrative regulations, legal interpretations, and public pressures that have broadened and deepened the social content implicit in many of the acts and the concern with higher order, unanticipated consequences of technological actions.

Thus, one of the most important developments in recent years in the U.S. is the important imperatives and requirements that have come about as a result of legislation that attempted to either arrest or improve the quality of the surrounding physical environment. The National Environmental Policy Act of 1969, as well as a host of similar "environmental" legislation and the 1972 Technology Assessment Act, have provided the basis for not only a new planning ethos, but also (as in the case of NEPA) a definite legal imperative to plan with new considerations of long-range consequences and appropriate mechanisms for correcting (in time) adverse project manifestations. In language and spirit, NEPA and similar documents, through such generic terms as "impact assessment," have further promulgated the need for a process and methodological expertise associated with the larger context of technology assessment and social forecasting. In essence, current legal mandates and the theory and practice of what one may call collectively "prospective studies" have produced important conceptual and methodological advances related to a promising process of "assessment," evaluation, and decisionmaking.

Given the magnitude, intensity, and varying degrees of presence of public projects it has always been difficult to describe, categorize, or otherwise classify, and in a meaningful way evaluate, the perturbations from such large scale facilities. The principal concern in the past was with what one may describe as first order consequences, or immediate (primarily economic) benefits to be gained by developing and widely applying innovative technologies and large-scale facilities. As a matter of fact in pre-technological societies there was little, if any, concern with higher order and unanticipated consequences since most of the changes in the environment were incremental and gradual with few products introduced at a time. Today, however, in a highly technological society all conditions of the past have been demolished because of the unprecedented rate of change, visible negative effects, and formerly unrecognized harmful consequences of a seemingly good technology.

Taking into account, therefore, a systematic thinking which recognizes the intricate interaction of population, environment, technology, and social organization, technological options are increasingly considered as significant social matters and their future implications need to be plotted to the best of our ability. Again, the National Environmental Policy Act may be used as an example and as an important document exemplifying the shift and concern over an appropriate and careful consideration of all environmental aspects of proposed technological actions. Such an Act was made possible by the fact that the negative effects of natural perturbations became more visible, the concern over the high momentum and the fast pace of technological changes has become widespread, and a pervasive feeling that each member of the public has been only a secondary part to every decision concerning the use of technology. At the same time, the present market forces did not seem to be satisfactory in allocating secondary costs, and the legal system seems to have been relatively ineffective in coping with technological damages.

The 1970s, then, became an important decade for elaborating the need for "assessment." Legal doctrines, administrative regulations, and emerging policies were all part of a necessary reconsideration of the prevailing notion of quite a few engineering designs which were based upon the assumption of closed systems fully described by the technological inner constraints and marked by clear boundaries decoupling the system from other parts of the environment.

It is important at this point to provide a few introductory remarks about EIA, TA, and RA that can serve as a backdrop for our main theme, namely the search for cumulative impact assessment.

To start with, Environmental Impact Assessment came about as a result of the crucial impetus of the legal requirement in NEPA that a "detailed statement of environmental impact" be prepared by the responsible agency for every "major action significantly affecting the environment." As a result three positions or interpretations of this requirement followed. The initial position was that EIA was a new procedure separate from planning and decisionmaking, directed solely toward the preparation of a new document (the EIS), whose function was to provide public information and to meet judicial documentation requirements. A second interpretation concentrated on the understanding that EIA can add new information to the planning process about a range of possible consequences of alternative actions. Finally, a third view (not espoused by many) considers EIA as a new methodology, a new thought process that permits more informed choice through a combination of private citizens, experts, and public officials. In shorter form, EIA was viewed as a full disclosure Act motivated by a new planning ethos and by increased policy capability.

EIA and subsequent developments in TA and RA brought about a certain euphoria as to the potential role of a systematic, comprehensive, forward-looking, critical and open process of assessment. But the euphoria and promises of the early 1970s, were succeeded by sober reflections of the 1980s. No doubt there has been considerable consciousness-raising, a flurry of legal challenges, increased form of public participation, as well as major conceptual and methodological advances. However, the ultimate social and political effectiveness of assessment efforts has been questioned and the contradictions between theoretical constructs and practice became quite noticeable. But the fact remains that NEPA and the host of environmental legislation provided the basis for important legal and administrative references and exemplified in the most dramatic way subsequent assessment efforts.

The above remarks relate to the unfolding efforts of progressively moving from an understanding of past trends, through an explication of present conditions, to finally, a projection of probable, possible, or even plausible futures. Our present analysis can, thus, relate the current imperative of impact analysis with the broader quest and policy analysis of technology assessment, and with the capability of social forecasting, all of them as means of helping the planning process and efforts of "reasonable" decisionmaking. Ideally, measurement of effects and technology assessment should culminate into an expanding planning process which includes also analysis of future developments through social forecasting and evaluation of as yet not perceived, uncertain future environments, as well as questions of decisionmaking in terms of risk assessment.

The three key approaches of impact analysis, TA, and social forecasting can be used here as a springboard of a holistic, integrative approach outlined in Fig. 1. Furthermore, all three of them, either as part of specific laws, or imperatives can be seen as overlapping circles aimed at better social control of technology and at strategic management of alternative options. At the same time, they also provide a locus for risk analysis and assessment (Fig. 2).

In practice, however, both the expertise brought in the fields of "assessment" as well as substantive approaches have been replete with confusion as to where impact ends, assessment begins, evaluation follows, or forecasting takes place. In any case, these overlapping contentions add richness to the whole

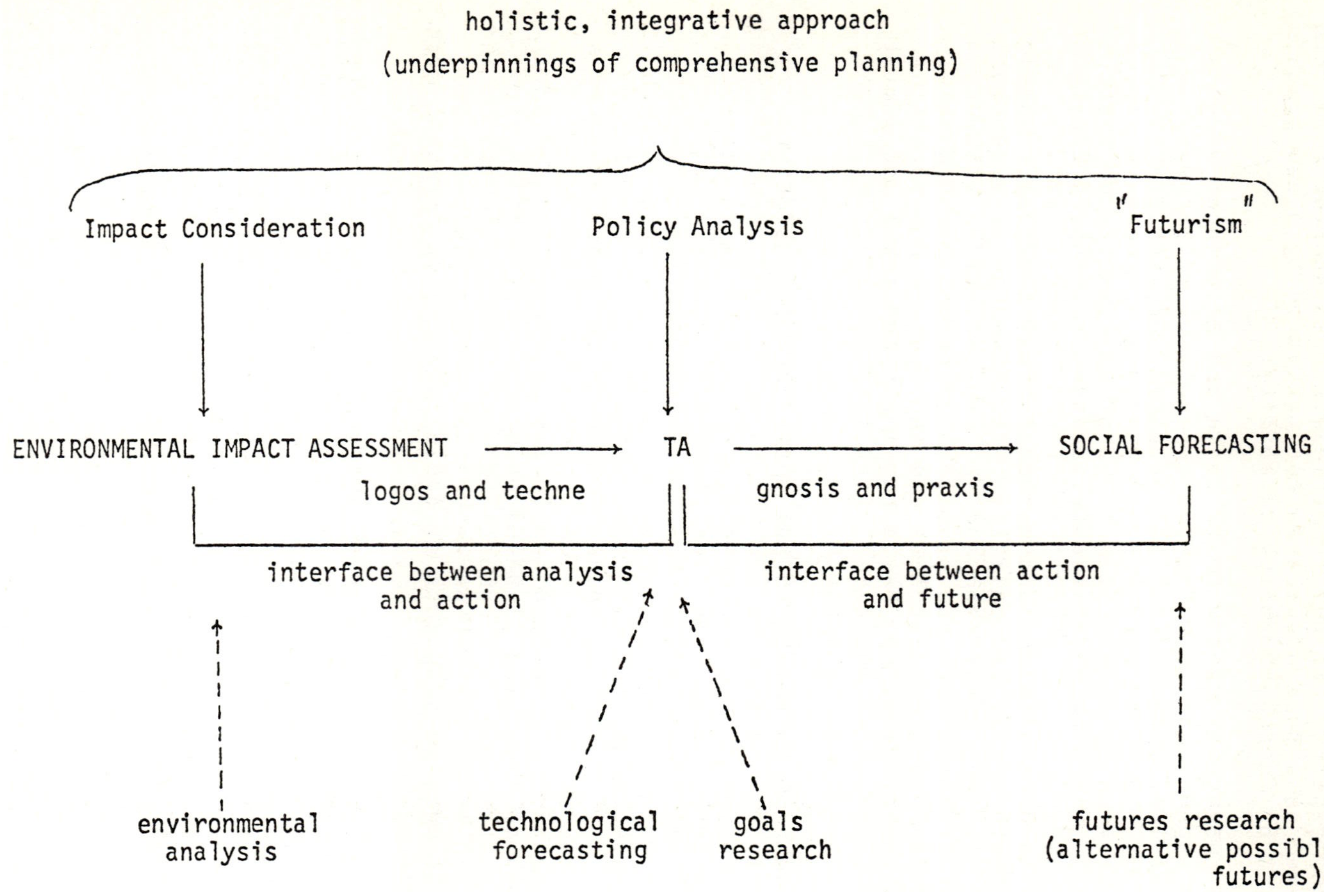

Fig. 1: EIA, TA and Social Forecasting.

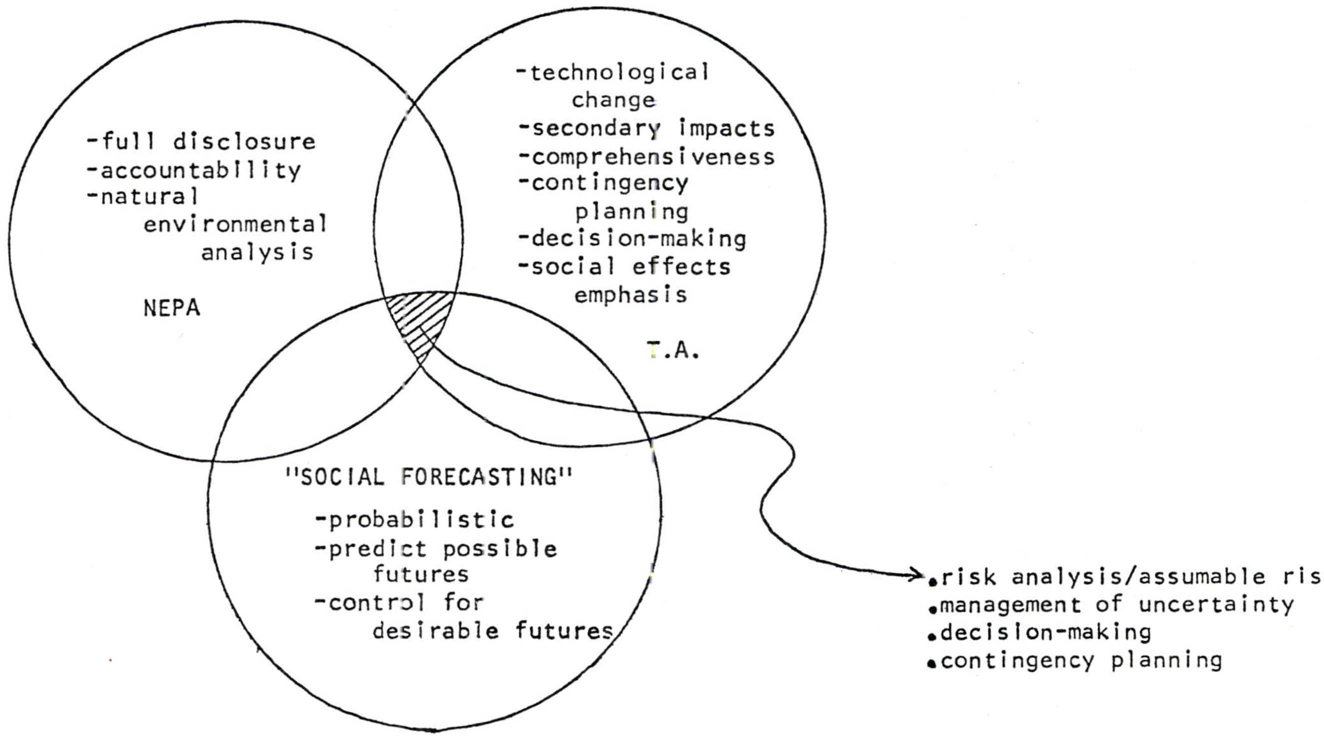

Fig. 2: Interrelationship of Impact Analysis, Technology Assessment and Social Forecasting.

quest of managing technology either through legal means (EIS); or through the prediction of possible futures (social forecasting); or through an advanced policy analysis of benefits and risks involved in the use of new or expanded technological capabilities.

It would be impossible, anyway, to separate all such elements or to provide a sharp distinction between environmental analysis and technology assessment, other than to note in particular that each of them has appealed by and large to different professional and political constituencies. Whatever the argument here, technology and environmental assessment have been mutually supportive, although the first seems to be much more form-free. It is macro-forecasting, especially as exemplified in "futures research," that is for the time being in a legal limbo. Although a worldwide movement and practiced as part of technology assessment, it is sometimes looked upon as an exercise in disciplined prophecy and conjecture, rather than as part of the more serious efforts of technology assessment (and yet the borrowing of methods, concepts, and of resources from each other is quite remarkable...).

As suggested in all definitions, especially those of the Congressional Committee and by NSF itself, TA is not intended to be the new political panacea or a new scientific discipline. Part of the commitment of technology assessment is to develop an applied, comprehensive, analytic strategy, and if you wish, a contingency planning function in order to achieve a more responsive and responsible management of technological change.

It is difficult by now to trace the exact evolution of the concept and imperatives of technology assessment since the movement now has caught fire and has quite a few claimers of fame. A number of articles have already tried to point out the evolution of this particular office and the variety of legislative acts, committees, advisory boards, studies, and assorted efforts that have brought about the need for technology assessment.

Thus, the technology assessment "movement" is part of a continuous evolution of an increasing legislative mandate within both the government and outside it and a result of circumstances that have continuously accentuated the concern over adverse effects, almost to such an extent that a number of advisors have even warned that technology assessment may even end up as "technology arrestment."

Given on the one hand the major transformations, the apprehensions as to far-reaching effects of technology; and, on the other, the dangers involved in controlling the autonomy of technology, one may ask why is something like technology assessment needed. The literature seems to agree that what is new and what is needed today, in some form of "technology assessment," stems from the fact that problems have become more numerous, more severe, and more complex. If nothing else, the popular demand to cope and the general public awareness of these problems have also become much more acute, and the insistence that something must be done has become not only more vocal but a political mandate for change in our era.

Similarly, in recent years public attention has focused on an array of technological hazards that can pose significant threats to human health, safety, and the environment. The need for societal risk assessment has become more urgent as new and developing technologies accentuate far-reaching problems of social choice. The determination of risks, the specification of levels of risk and their acceptability have become major issues. The central question here is whether a product or technology is acceptably safe. More important, compared to EIA and TA, risk assessment raises broader policy questions and more intricate aspects of individual and collective good life.

By now it becomes apparent that "assessment" is not only a major area of concern with regard to potential spillover effects of interventions, but also an important planning imperative for all federal and state agencies. NEPA was used as a beginning point to exemplify the shifting concern over an appropriate and careful consideration of all environmental aspects of proposed technological actions. Such a legislative act and a subsequent number of federal, state, and other documents and guidelines were made possible because negative effects of a variety of public projects have become more visible while, at the same time, the concern over the high momentum of technological change has also become widespread.

In sum, four developments concerning resources, the environment, and society characterize "assessment" efforts in recent years:

1) Impact assessment

2) Technology assessment

3) Forecasting and futurism

4) Risk analysis, management of uncertainty, and decisionmaking

All of them, as part of policy analysis also underscore the need for conceptual and methodological improvements in social cost/benefit estimation, consideration of indirect, long-range consequences, and the comparison between tangible and intangible effects of public projects. In doing so, one can attempt to relate in some "social calculus" what is environmentally sound, technically safe, economically viable, legally pertinent, socially acceptable, and politically feasible.

All in all, in such approaches the underlying preoccupation can be summarized in three key interrelated questions:

1) How do we balance, equitably the costs and benefits involved in the alterations of the surrounding environment?

2) How do we make appropriate changes and transitions to new states without unacceptable disruption to all systems?

3) How do we measure in a valid, reliable, and comparable manner, impacts and consequences and, therefore, provide guidance for alternatives and reasonable decisionmaking?

Such questions serve also as a backdrop for a more generalized concern with the development of new approaches to assessment, planning, and decisionmaking that would:

- Place emphasis on stronger interdisciplinary approaches integrating a variety of "environments" and disciplines in any assessment effort

- Underscore the search for a "social calculus" of project effects and the establishment of a wider range of alternatives

- Expand the time horizon from a narrow enumeration of immediate impacts, to a long-range forecasting of higher order consequences

- Combine the historical experience of the past, the constraints of the present, and the visions of both possible and desirable futures.

Central in a proposed combination of new approaches and methodologies is the emphasis on far-reaching, indirect, secondary, cumulative effects of projects or technological interventions. The basic problem to be tackled here has to do with the question of how can we account for the incremental impacts of a given action, when added to other past, present, and reasonably foreseeable future actions. This sounds like a straightforward question. But it becomes quite complicated as a problem when one also considers five dimensions that are usually associated with changes in the surrounding environment:

1) Societal complexity and interdependence (and hence increasing vulnerability)

2) The magnitude of change, i.e., increasing intensity, severity, or duration of impacts

3) The rate of change (timing and "future shock")

4) The distribution of effects (equity and fair access)

5) The demand and ability to cope with change (impact absorption, adaptation, accomodation)

Such dimensions bring us, then, to the point of searching for commonalities of all "assessment" efforts (ETA, TA, RA). Before we do so, however, let us recapitulate key points and definitions provided above:

- EIA = emphasis on environmental impacts; changes on surrounding environments; full disclosure Act ("what happens when...")

- TA = management of technological change; emphasis on unintended consequences; extended time horizon (if...then...")

- RA = hazard (especially technological) assessment and management; emphasis on decision making and social choices

Whatever the definition and practice, there are certain commonalities that run through the "assessment" emphasis of these processes. Such commonalities include the commitment to:

- Account for full range of impacts (comprehensiveness)

- Delineate consequences in advance (anticipatory vs. reactive thinking)

- Control events (especially technology) and the environment for the common good

- Interact with our destiny in such a way as to be part of it.

Finally, at the heart of all such "assessment" efforts is the search for a unifying theory of social change (critical theory) that calls for:

- Prediction of trends and developments
- Understanding of timing and rate of change
- Explication of the full range of impacts and consequences
- Forecasting of future environments
- Ways of coping with or accommodating change

WHAT DOES THE CUMULATIVE IMPACT ASSESSMENT ENTAIL?

As related previously, the practice of impact analysis, assessment, and evaluation has significantly improved during the 1970s, following the pivotal provisions of the U.S. National Environmental Policy Act (NEPA). Impact analysis and assessment have become a central part of the process of evaluating the effects of large-scale public projects, especially by contrasting short-term gains against long-term enhancement. At the same time, as a result of an expanded time horizon, a new multi-dimensional, multi-objective long-range model of impact assessment has emerged incorporating additional considerations concerning the far-reaching consequences of technological inducements.

Yet, despite all the writing, the myriad of environmental impact statements (EISs), and technology assessments (TAs), a key problem remains, namely the frustrating task of specifying some generally agreed upon methodologies for carrying out a sweeping assessment mandate and for accounting for the far-reaching, aggregate effects of projects and technologies. This task becomes particularly difficult in view of the lack of existing appropriate literature and of the paucity of concrete methods for carrying out cumulative impact assessment in a manageable and relevant manner.

The cumulative issue has been recognized for years. Many governmental units in the 1960s and 1970s began to address this issue. Probably, the single most influential piece of legislation was NEPA which was signed into law on January 1, 1970. This law required the preparation of environmental reports which were supposed to examine the effects of projects in an integrated and far-reaching manner.

More explicit guidelines for conducting NEPA-required environmental analyses were provided by the Council on Environmental Quality (CEQ) in 1978. Here, Federal agencies required to comply were instructed to address the direct, indirect, and cumulative effects of Federal and non-Federal actions of natural resources. The majority of cumulative impact assessments, then, are a result of these regulations. However, the vast majority of impact assessments and statements have addressed direct impacts, one at a time, on individual components of the surrounding environment.

But the fact remains that legal interpretations, administrative guidelines, and the practice of "assessment" all started paying increased attention to an underlying intent of NEPA, namely to clearly address all possible impacts, especially secondary, unanticipated consequences. At the same time a parallel continuous preoccupation has been accentuated, i.e., that modern society, because of its size, complexity, diversity, structural differentiation, environmental alterations, or whatever, contains a comparatively high amount of potential disorganizational strain that is accelerating or subsiding. What is important for the present argument is the observation by a variety of authors that the core source of intrasocietal contradictions and strains are such large social morphological revolutions as the population explosion, spatial imbalances, diversification, the rapid pace of technological change, and environmental spillovers.

Cumulative impact analysis can be viewed as a system of concentric webs of causative factors, disturbances, responses, and impacts, each potentially capable of initiating a new stream or web of effects. One of the aspects of any type of impact analysis is the need to exercise a great deal of professional judgement in selecting the significant causative factors and tracing significant effects, as well as knowing which branches to pursue in a reasonable analysis of cumulative impacts.

The central concept of a cumulative impact assessment involves the development of an approach which recognizes that although individual actions may have insignificant effects, e.g., effects of one boat dock on a river or bay, in their aggregate (especially through interaction), they do become significant. Would it not be enough to simply "add them up" and ascertain the total impact? Unfortunately, such an approach depends on whether the actions are repetitive, aggregative, continuous, or "delayed." This means that different causes may exist depending on whether actions are "repetitive" and, therefore, produce incremental effects; or whether various other activities are taking place concurrently or in the future, and thus collectively affect a particular attribute; or, finally, whether effects may be discernible after some time delay, i.e., after they accumulate over many years to a measureable level (assisted perhaps by some "trigger" factor or event--the proverbial "last straw").

The obvious problem here is that, so far, there are no accepted rules for solving the problem of aggregation, although a number of existing methodological and conceptual advances can be used as a springboard for considering cumulative impacts in the surrounding environment. Again, the ultimate significance and meaning of cumulative impacts rests against some "ideal" environment or state which is being disturbed by the collective effects of a given project or action. The judgment of whether impacts are significant requires a look beyond both the spatial limits of a given area and the narrow goals of a piecemeal or individual action.

Apart from the normal considerations of significance, magnitude, and incidence of impacts, cumulative impacts possess the key characteristic of "synergism," or interactive reinforcement of effects acting over some future time horizon. This notion must be particularly emphasized because it is central in the differentiation of cumulative impacts from both primary and secondary impacts (which depend heavily on additive approaches). The general principle of synergism underscores the observation that the whole is more than the sum of its parts in magnitude, severity, intensity, or complexity. Synergism is not simple aggregation but interaction, combination, and new patterns.

In essence, the previous remarks have pointed out that cumulative impact assessment should be based on a holistic perspective; should differentiate from traditional treatments of impacts; and should be based on a perspective that emphasizes dynamic changes. Also, cumulative impacts must not be viewed as only adverse; they can also be beneficial.

In such a context, the process of cumulative impact assessment becomes part of a broader framework of analyzing environmental and socioeconomic transformations. Four critical elements must be recognized:

1) The nature of inducing actions
2) The scale or extent of on-going transformations
3) The rate and timing of change

4) The arena of operation or the physical setting where the various actions take place.

All such elements combine to produce forms of change and transformation that seriously test the resiliency, stability, tolerance, and recovery of surrounding environments.

In the simplest form possible, cumulative impacts may be defined as the integration of the interaction of effects of all current and reasonably foreseeable actions over time and space. This definition is supported by the explicit wording of the CEQ regulations of 20 November 1978 (effective date 30 July 1979) (40CFR Parts 1508.7 and 1508.8) where it is stated,

> "Cumulative impact" is the impact on the environment which results from the incremental impact of the action when added to other past, present, and reasonably foreseeable future actions regardless of what agency (Federal or non-Federal) or person undertakes such actions. Cumulative impacts can result from individually minor but collectively significant actions taking place over a period of time.

A specific definition of cumulative impacts should contain three distinct elements:

1) The sum incremental effects on the surrounding environments resulting from multiple human activities (including total losses by attrition)

2) Effects that are synergistic in nature

3) Impacts which could remain relatively nonsignificant until some undetermined threshold values are reached.

Cumulative impacts can be, then, defined as the totality of interactive impacts over time, i.e., the sum incremental, synergistic effects caused by all current and reasonable future actions over time and space. A key word in this definition is "future." As a matter of fact, the Council of Environmental Quality regulations have used the word "foreseeable" in place of future; whereas, other opinions use "certain."

Figure 3 represents a "map" of cumulative impacts. At the center is a series of identical actions, whose direct effects have been traditionally measured as "primary impacts." Such actions may cause ripple effects and extend (over time and

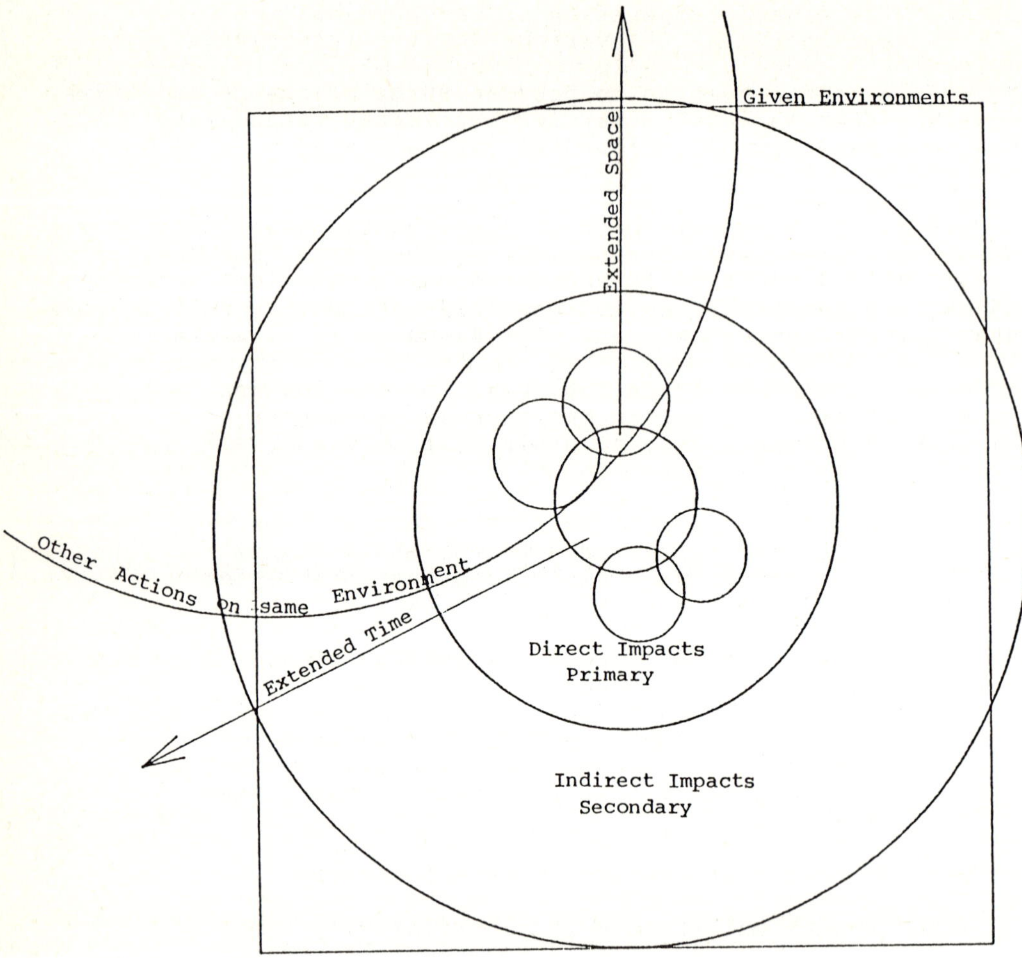

Fig. 3. A "map" of cumulative impacts

space) the influence of those actions through indirect, "secondary impacts." The totality of effects (illustrated by the encompassing solid circle) represent the concept of "cumulative impacts." Such "cumulative impacts" may not be confined within a given environment, but spill over to larger geographical areas. Finally, cumulative impacts include other actions on the same environment, in essence, the totality of interactive effects over time.

A requisite flexibility of the cumulative impact assessment process can be enhanced if the assessment incorporates a combination of probability statements and imagination. Often it is assumed that "the worst" will not happen; however, one real use of impact assessment, and particularly cumulative assessment, is to prevent "worst case" scenarios from occurring. Realistic but imaginative ("what if?") future projections (short of the doomsday approach) are a necessary component of the assessment process.

Cumulative impact assessment as an ongoing process should be able to provide adaptations to changes as they arise. It is recognized that uncertainty always exists and that monitoring offers the opportunity to refine estimates as more information and data are obtained. Processes within a system may repeat themselves, or cause chain reactions--all of which require impact management flexibility. Therefore, a cumulative assessment requires "midstream correction," adaptive management, and techniques that are dynamic and flexible. This capability allows new explanatory and predictive systems to be advanced when new data are collected or existing data are viewed in a different light.

THE REALM OF CUMULATIVE IMPACTS

It has been emphasized that the complexity of laws and regulations does not necessarily allow for a very clear delineation of the substantive process of cumulative impact assessment. However, laws and regulations provide a general framework by continuously expressing a concern about impacts that arise out of the combined effects of small, individually insignificant activities in a given environment.

Inherent in such an approach are a number of concepts that should be associated with any systematic assessment of cumulative impacts, such as carrying capacity, thresholds, causal paths, system alterations, loading of the system, etc. Also basic to the legal definition of CIs are two concepts, namely, extended time horizon and interactions among critical component parts of the surrounding environment.

The quest for conceptual clarity and the search for relevant typologies must address four realistic questions encountered in practice:

1) What criteria should be used to determine thresholds, sensitivity, or resilience of the surrounding environment?

2) How does a practitioner determine what measurements to use for relevant critical variables?

3) How is the result of an action to be determined? (What are the different impacts from varying sequences?)

4) How can the effects of such actions (present or foreseeable) be summed, and how do actions interact?

Figure 4 further exemplifies the dimensions that characterize cumulative impacts. In this figure cumulative impacts are to be found in the lower right quadrant, denoting the central emphasis of interactive totality of impacts over time. The other categories of this particular figure outline other dimensions associated with the impacts of disturbances on the surrounding environment ranging from the direct primary to the induced secondary effects. At the same time this figure comprises two major axes that can help in the search for methodologies directed toward an accounting of cumulative impacts: the "causal chain" or traceability of effects (to the extent that they are "caused" or "induced" by specific actions); and the "aggregative axis" that separates piecemeal from cumulative or aggregative impacts.

In the definition of cumulative impacts provided above, three interrelated conceptual dimensions can be distinguished: the aggregative dimension emphasizes addition or summing of impacts of identical or related actions, the interactive dimension underscores the cross-effects of related actions, and finally, a diachronic dimension underlines the influence and importance of an extended time horizon. All three dimensions relate not only to what has been broadly labelled "synergism" of actions and impacts, but also to significant "transformations" of the surrounding environments. These transformations can be understood in terms of important environmental changes, altered spatial relations, structural differentiations and/or socio-

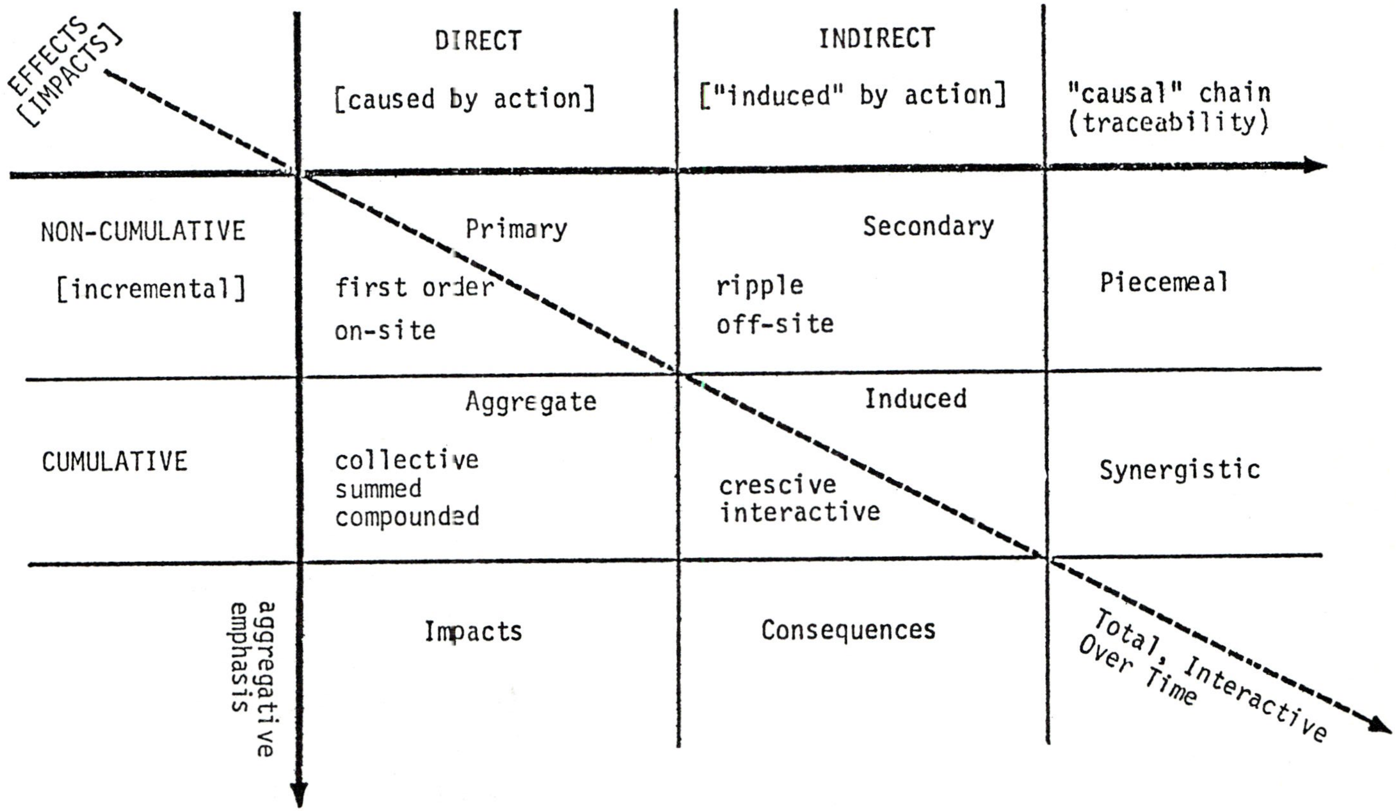

Fig. 4. Dimensions of cumulative impacts.

cultural changes. Taking into account these differentiations we may, then, provide an initial summary of these dimensions, their basic thrust, and some overall pertinent methods (Fig. 5).

All in all, cumulative impact assessment represents a higher order of analysis and evaluation of cause-effect linkages and their interactions than the more commonly accepted "environmental impact assessment." In essence, the far-reaching consequences of a series of specific actions or programs which may preordain a whole new environment are traced. Such observations point to a central argument that runs through cumulative impact assessment, namely to a shift of emphasis in "thinking paradigms." Such a shift in emphasis can be summarized as follows:

Established Procedures (Traditional)	New Emphasis (Alternative)
o Species-oriented	o Community/ecosystem oriented
o Linear/extrapolative	o Non-linear/nonmonotonic
o Causal	o Interactive/mutual causation
o Individualistic/segmented	o Holistic/integrative
o "Snapshot"	o Evolving/dynamic
o Hierarchical/classificational	o Contextural/relevance-selecti
o Structural	o Functional

Thus, new conceptual frameworks and alternative interpretative schemes can provide additional ways of understanding and proceeding with cumulative impact assessment. Such a plurality of frameworks can unite categorical and synthetic thinking. Rather than providing exhaustive lists of impacts and definitive solutions to impact assessment questions, the frames of reference associated with cumulative impacts underline significantly the context of relative value judgments and of desired futures (without ignoring, at the same time, such very real questions as carrying capacity, ecological thresholds, and levels of tolerance.)

Finally, a major hazard in this process is the expectation that traditional, deterministic procedures can be transferred from current practices to fulfill the requirements of cumulative impact assessment. The demand for cumulative impact assessment requires a complete restructuring of the problem itself; an articulation of the assumptions driving the assessment; new techniques and tools for aggregating diverse impacts; and a search for standards or criteria of significance in order to judge overall, long-range impacts.

	Dimensions of CIs	Thrust	Method
Structural Emphasis	Additive (A)	- Add up all types of effects (primary + secondary) · - Consider induced/secondary impacts - Extrapolate future impacts	- 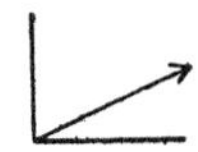 linear projection - Addition = Primary + Induced { Sensitivity { Constraint mapping

------------------------------------"SIGNIFICANCE THRESHOLD"------------------------------------

[legal significance, technical significance, social significance]

	Dimensions of CIs	Thrust	Method
Functional Emphasis	Interactive (I)	- Through appropriate interaction matrices consider joint effects - Reinforcing influences - Consensus as to what are mutually affected - Satisfy multiple objective	- Structural modeling - Matrices  - Delphi/Consensus { Multi-objective technique { Optimization
	Diachronic (D)	- Consider host of carrying capacity concepts (threshold, tolerance, resilience) - Establish standards + criteria - Consider range of potential futures and alternative scenarios	- 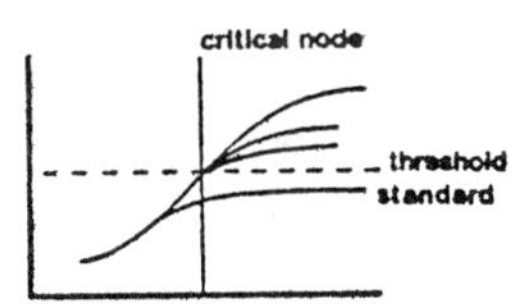 - Alternative futures/scenarios - Normative planning

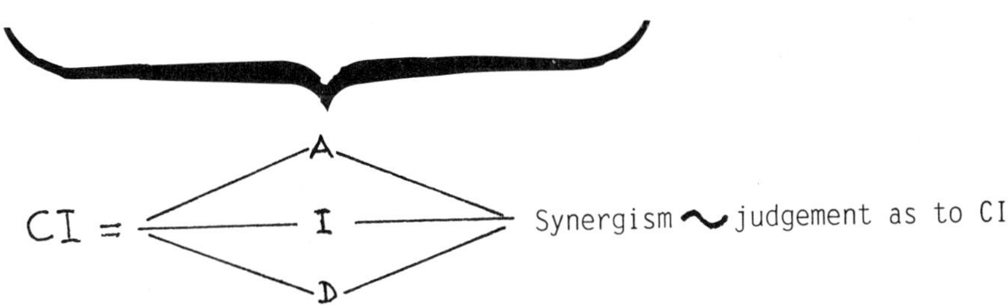

Fig. 5. Dimensions of cumulative impacts and their particular emphasis

METHODOLOGIES FOR CUMULATIVE IMPACTS

In the existing literature and ongoing efforts, there seem to be two prevailing methodological practices with regard to cumulative impacts. The first may be labelled the "sensitivity approach." Here, in advance, constraint maps can be drawn that reflect a variety of values and limits up to which the surrounding environment can be "stretched" (such as habitats of endangered species, agricultural lands, stressed environments, threshold estimates, developmental "hotspots," community development goals and plans, etc.). Thus, beyond a certain tolerance line (again, specified in advance) impacts become cumulative, i.e., they exceed the parameters established and produce an unacceptable disruption to the surrounding environment. At that point the cumulative impacts have become significant.

The second methodological approach does not necessarily begin with such advance considerations or well-articulated constraint maps. In this approach the emphasis is on an additive procedure. Somehow, impacts through aggregation (and presumably through interacton over time) reach an "unbearable" point. That point can be judged against some articulated "threshold" denoting the arrival at a condition which threatens tolerance or carrying capacity. Such a rational model that is causal, sequential, and additive conforms to quantitative reasoning that can be related to familiar concepts such as carrying capacity of ecosystems, multipliers, inducers, etc. Indeed, it is this type of approach that can also be related to the search for secondary impacts. There, "inducers" or multipliers are derived through projections of such critical variables as land and demography. Future impacts, then, are seen as linear extrapolations of a well-articulated baseline condition.

Such approaches are not exclusive and quite often they overlap in both assumptions and procedures. Their common concepts are carrying capacity and tolerance of the surrounding environment either conceived in advance (sensitivity maps) or arrived at through an additive manner (threshold emphasis). What we need to incorporate, however, in the form of a synthesis is a "judgment model" that recognizes qualitative, preferential, non-sequential considerations. Such an additional element to the above approaches recognizes contextual and normative factors, policy considerations, political goals, and preferable futures (as contrasted to probable or plausible futures).

The strategy and tactics for cumulative impact assessment come from a variety of efforts and do not necessarily emanate from a single authoritative study. These efforts tend to converge, in part, from at least seven major "tributaries" which help produce insights from parallel efforts that can be of help to an analysis concentrating on growth, impact assessment, and social and ecological changes. These seven contributing areas include:

1) Broad methodological considerations
2) Technology assessment
3) Forecasting
4) Ecosystems analysis
5) Land-use and growth
6) Impact studies
7) Risk assessment studies

So far, studies and reports related to cumulative impact assessment tend to be extensions of more or less traditional EIAs and TAs, i.e., based on lists of critical variables, modeling of important environmental dimensions, and linear projects (on enlarged space and extended time). However, whatever form such current efforts may take, it is important to emphasize again the needed combination of both rational and analytical approaches, inductive and deductive strategies, and historical and normative emphases. We can further classify relevant methodological tactics as falling within three categories:

1) Retrospective assessment--emphasizing historical information and development of present data basis (including constraint mapping)

2) Prospective assessment--based on prediction, extrapolation, and forecasting (especially in terms of cross-impacts and scenarios)

3) Policy analysis--concentrating on questions of risk and uncertainty, decisionmaking, tradeoffs, and alternative considerations.

Whatever the basic model (or epistemological thrust) three broad approaches can be distinguished that characterize the theory and practice of cumulative impact assessment with an emphasis on prospective methodologies (as summarized in Fig. 6). The "linear" approach forecasts cumulative impacts as historical (extrapolative) projections of past trends and events. The "multipliers" approach concentrates mostly in relating functional aspects, or through relevant "weights," examines the implications of present conditions. Finally, a "future environments" approach emphasizes normative forecasting, i.e., examines potential cumulative effects against probable and/or desirable futures.

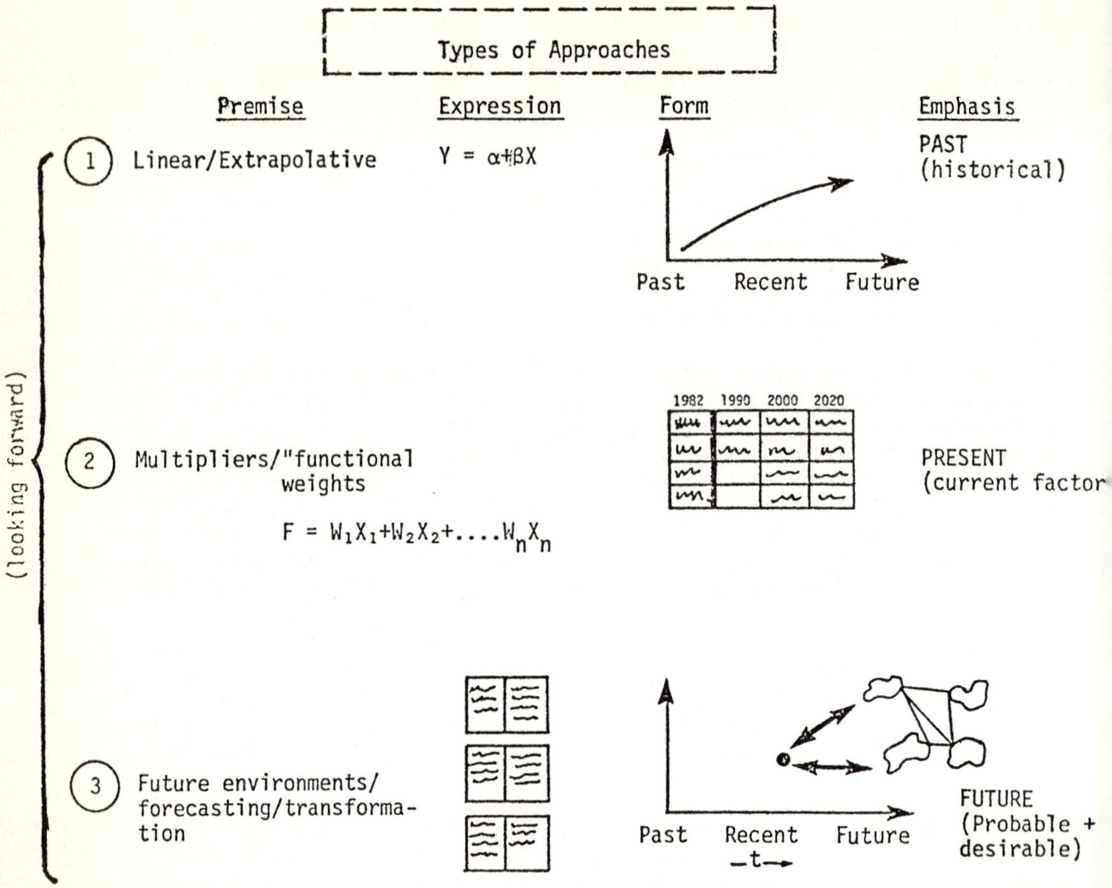

Fig. 6. Basic modes and prevailing approaches in cumulative impact assessment (prospective emphasis)

Table 1 also summarizes a number of forecasting techniques that seem to be particularly applicable to the process of cumulative impacts assessment.

One last remark should be made in summarizing prevailing methods and practices related to cumulative impacts. A basic distinction that must be drawn is the extent to which a particular approach tends to be more holistic or contextual as contrasted to logical or sequential. The first tends to view cumulative impacts in a more visionary, futuristic manner. The second tends to be more "objective" or "scientific" and is based on a more categorical thinking model. In other words, the first requires a futures perspective that is divergent, recognizes far-reaching consequences, and implies creative thinking and synergistic consequences against alternative scenarios. The second emphasizes thinking that tends to be convergent, additive, and based on the classical example of scientific thinking through deduction and understanding of component parts.

We can now summarize the previous discussion by relating the two prevailing practices of cumulative impact assessment in the context of two underlying emphases ("hard"----causal versus "soft"----imaginative) in the following figure (Fig. 7).

Although the assessment of cumulative impacts poses very difficult and complex conceptual and methodological obstacles and thereby has the potential for occupying an inordinate amount of time and resources for its resolution, it has been shown that there already exist in current practices many fail-safe, overlapping mechanisms for indirectly encompassing the concerns related to this process, even in the absence of a structured methodology. Thus, a preliminary strategy for dealing with cumulative impact assessment is to identify and strengthen existing overlapping policy considerations, along with methodologies and procedures which are currently successfully employed to reveal significant primary effects of actions. This approach is virtually dictated by the increasing dichotomy between the legislated requirements for comprehensive, multi-objective assessments of impacts and the actual practice and application of rudimentary analytical techniques.

CUMULATIVE IMPACTS AND RISK ASSESSMENT

Let us stop, then, for a minute and rethink some of the previous questions concerning cumulative impacts. What we have in our hands are such terms as "incremental," "additive," "aggregative," "total effects," and in some cases, "collective con-

Methodological Emphasis	Approach/Current Practice: Ex Ante (sensitivity, constraints)	Approach/Current Practice: Post Hoc (additive, ad hoc)
"soft" (non-Newtonian, Heisenbergian) Flexible Qualitative Holistic	Goal oriented Utopian Imaginative forecasting Adaptive management	Structural modeling Descriptive graphs Analogs Management of uncertainty
"hard" (Newtonian) Causal Linear Quantitative "Objective"	Mapping Parameter setting	Modeling Matrices Lists Risk assessment

Fig. 7. Methodological emphasis and practices in CI assessment

Table I. Types of forecasting applicable to cumulative impacts

1.	Intuitive/Holistic Methods	Conjecture Brainstorming Heuristic programming Consensus (Delphi)
2.	Metaphors and Scenarios	Growth Metaphors Historical Analogies Scenarios Alternative Futures
3.	Extrapolative/Time series	Historical surveys Social Trends Analysis Monitoring Correlation and regression Probabilistic Analysis Time series Economic Forecasting Models
4.	Models/Matrices	Structural Modeling Dynamic Modeling Cross-impacts Simulation Iterative System Projection Systems Analysis Graphic Models Input/Output Analysis
5.	Decision trees	Morphological Analysis Contextual Mapping Objective Trees Relevance Trees Judgment Theory

sequences." A way of restating the above would simply be to say that projects or actions lead to perturbations in the environment which may trigger a complex sequence of effects. These effects are made up of a number of component parts that proceed from one to another, that often interact, and that have a time span or a "maturation sequence." Often this web or profusion of effects arises from a single action or from several related actions which may multiply a single effect. At this point a beginning assessment process considers individual disturbances which in the easiest form (assuming that such disturbances are of the same kind) could perhaps be directly added. However, cumulative impacts are not necessarily results of simple addition. They may come not only as part of some incremental effects but also as an expression of loss by attrition. This problem is widely recognized and due attention is paid in all writings, namely that somehow there are some cumulative effects either of a single action over time or of the compounding of consequences as a result of separate actions that converge in a particular area; or, part of a complex network of interacting factors with the ability to pass along the effects of changes from one component to another and to perhaps amplify in a synergistic manner these effects throughout a particular system.

The problem remains, however, that so far there are no accepted rules for solving the problem of aggregation, although a number of existing methodological and conceptual advances can be used as a springboard for considering cumulative impacts in the surrounding environment. In a nutshell, the ultimate definition of cumulative impacts is really a social-political matter and not a simple technical one. This sounds like an evasive and perhaps difficult statement. What it entails, however, is that the ultimate significance and meaning of cumulative impacts rests against some "ideal" environment or state which is being disturbed by the collective effects of independent actions. The judgement, therefore, of whether impacts are significant requires a look beyond both the spatial limits of a given area and the narrow goals of a piecemeal or individual action. In other words, we have an expanded spatial and goal horizon which provides us with different criteria and more complex methodological requirements.

As repeatedly stated, the derived "cumulative" approach involves a different way of thinking about environmental impact problems. Cumulative impact assessment encompasses a combination of broad geographical scope, many inducers or actions, extended time horizon, and continuous monitoring. This type of assessment is more difficult to accomplish than the more traditional "snap-shot" approach, where existing conditions are described and impact dates are superimposed upon them. In addition, potential impacts (e.g., secondary effects) are calculated in a straight linear fashion or through some extrapolative functional relationships. A broad region that is continually evolving is difficult to measure and to monitor, especially when interaction and synergism are taken into account.

Perhaps one way of summarizing problems associated with a desired procedure of cumulative impact assessment is to explicate why cumulative impact is not usually done. Reasons include the lack of time, absence of systematic database, limited theoretical and methodological knowledge of the topic, and last but not least, lukewarm commitment to cumulative impact utilization and limited support for the far-reaching question that the search of cumulative impacts entail. No agency is given the responsibility to view questions at such a far-reaching scale and scope. There is no collective request among agencies for cumulative impact assessments.

As a result of many of the observations made above, a public policy process can become the vehicle for meeting three major problems or needs: conflicting and competing demands of individuals and collectivities as well as differential perceptions of assumable risks; strategic uncertainty inherent in future environments; and technological opportunities emanating from changing circumstances of the surrounding world.

Given these generic problems and needs of the socio-political arena, three corresponding issues of public policy also emerge: the resolution of conflict through a wider range of options; the reduction of uncertainty through risk assessment and improved decisionmaking; and the juxtaposition of opportunity costs vs. real costs, especially through benefit-cost analysis.

The concept of risk is central in assessment efforts and policy analysis. This is true not only in the differentiation of its types (technological/non-technological, voluntary/involuntary), but also in its perception by different segments of the population. The literature has identified three critical differentiations in risk perception:

1) Individual readiness or inducement to accept risk
2) Conditions influencing the acceptability of risk
3) Differences in the willingness to accept risk

For policy analysis, the pragmatic necessity to distribute risks in some socially meaningful manner is a driving force in policy formulation and execution. Indeed, the fundamental problem facing the policymaker is the reduction of those uncertainties to probabilities. Two processes, then, converge:

1) Conflict resolution, which has to do with unequal but legitimate distribution/redistribution of resources

2) Risk assessment, or the diminishing of uncertainty

The fact, however, remains that the future course of every public policy issue necessarily involves uncertainty. This uncertainty is a fundamental characteristic of our life, since the future is not fully predictable; the models of social change are inadequate; the social system is counter-intuitive; and many of the consequences of interventions cannot be understood until action has really occurred.

One of the key tasks of "assessment" is also the determination of the range of plausible consequences resulting from different (or combinations thereof) of policy choices. The articulation of policy options allows for the development of a "reasonable approach" and the design of an effective research strategy. The payoff of such an effort would be:

1) The identification of coherent policy scenarios

2) The provision of a consistent framework for data collection

3) The coupling or research strategy with a policy-relevant framework

In such an approach, the central concern has to do with the identification of those problematic areas that affect and are affected by policy issues before the assessment takes place. The unique task of such approaches as TA is on anticipatory analytical capability and on <u>a priori</u> bounding of the assessment. Policy analysis and intervention, then, can be seen as part of the strategic intelligence and design required to bridge the gap between technological shock and the inability of institutions to "cope" (adapt to or accommodate changes). In essence, a "reasonable approach" to decisionmaking, an emphasis on <u>phronesis</u>, (reason and imagination) and the anticipatory or contingency character of policy analysis can help in the provision of strategic intelligence that is relevant for decisions under different degrees of freedom, and for identifying policy scenarios that can better protect the decisionmaker.

Within such a context, the ultimate purpose of assessment is to provide an arena of negotiable discourse and contending values, a form of focused disagreement. Social groups can then form coalitions and utilize relatively ambiguous goals as opportunities for articulating or promoting their own interests.

But such broad considerations and concerns offer only general guidelines. Even after one has examined the armory of available EIA and TA techniques and has delineated some promising new approaches, there remain a host of pragmatic questions that are unique to cumulative impact assessment and part of the general ambiguity in this far-reaching, long-range, diachronic, interactive process of accounting for consequences of projects and actions.

Such concerns point out again the difficulties encountered by current cumulative impact assessment efforts. While most people have a general feeling that a lot of small actions may create significant problems in the long run (or provide us with new opportunities), there is widespread disagreement as to when significant thresholds are reached; what the tolerance/resiliency level of particular environments is; or, how positive or negative changes brought about by a transforming environment can be evaluated. But whatever the difficulties, the search for cumulative impacts is central to integrated, futures-oriented planning and to the quest for accounting for a wide gamut of indirect, secondary, aggregative effects of a multitude of actions characterizing a complex, interdependent society.

SELECTED BIBLIOGRAPHY

The following titles have been particularly useful in the formulation of the previous discussion.

Anderson, B.F., Cascaded Tradeoffs: A Multiple-Objective, Multiple-Publics Method for Alternatives Evaluation in Water Resources Planning. Washington, D.C. 1981.

Holling, C.S. (ed), Adaptive Environmental Assessment and Management. New York. Wiley-Interscience, 1978.

Kates, R., Risk Assessment of Environmental Hazards. New York. John Wiley & Sons, 1978.

Moreau, D.H. et al., Elicitation of Environmental Values in Multiple-Objective Decisionmaking. Chapel Hill, NC. University of North Carolina, Department of City and Regional Planning, 1981.

Porter, A.L. et al., A Guidebook for Technology Assessment and Impact Analysis. New York. North Holland, 1980.

Rowe, W.D., An Anatomy of Risk. New York. John Wiley & Sons, 1977.

Schwing, R. and W.A. Albers (eds), Societal Risk Assessment: How Safe is Safe Enough? New York. Plenum Press, 1980.

Vlachos, E. and D.W. Hendricks, Secondary Impacts and Consequences of Highway Projects. Washington, D.C. Department of Transportation, 1976.

The Impact of EIA on Decisionmaking

Timothy O'Riordan

School of Environmental Sciences, University of East Anglia, Norwich, UK

Figure 1 illustrates how most nations order goals according to what are judged to be key priorities. Military expenditures escalate when a country believes itself to be under threat or strategically vulnerable. Public health budgets grow when there is a major epidemic--especially when politically influential groups are endangered. Economic growth is the driving engine but its resources move upwards and leftwards under threat. Once a country is reasonably wealthy, redistribution becomes a major objective: first redistribution across space (regional development) followed by sharing of wealth (through welfare programs and progressive taxation) and, finally, power (via more active participation and possibly devolution of government). The environment generally takes third place, and even then monitoring of damage tends to take precedence over alleviation.

This was the picture throughout the 1970s and still holds true for some conservative political administrations and many developing countries. The diagram suggests that environmental considerations were very much "add ons" to the two upper goals, ready to be jettisoned in times of economic or military emergency. There was (and still is) little recognition of how environmental themes should infiltrate the other two goals--as suggested by the World Conservation Strategy (1) and by the somewhat idealistic dreams of environmentalists as depicted in Fig. 2. For a more detailed discussion of these themes, see O'Riordan (2).

Figure 3 illustrates what are now commonly accepted modes of ideological thought toward environment and development (2). These divide into two major orientations. First is "technocentrism," based on a man-centered view of the earth and a managerialist approach where the status quo of power and socioeconomic structure remains essentially unchanged. Second is "ecocentrism," based on a more humble man-nature view illustrated by the theme Gaia (3) and the Greek Goddess of the Earth

NATO ASI Series, Vol. G4
Environmental Impact Assessment, Technology Assessment, and Risk Analysis. Edited by V.T. Covello et al.

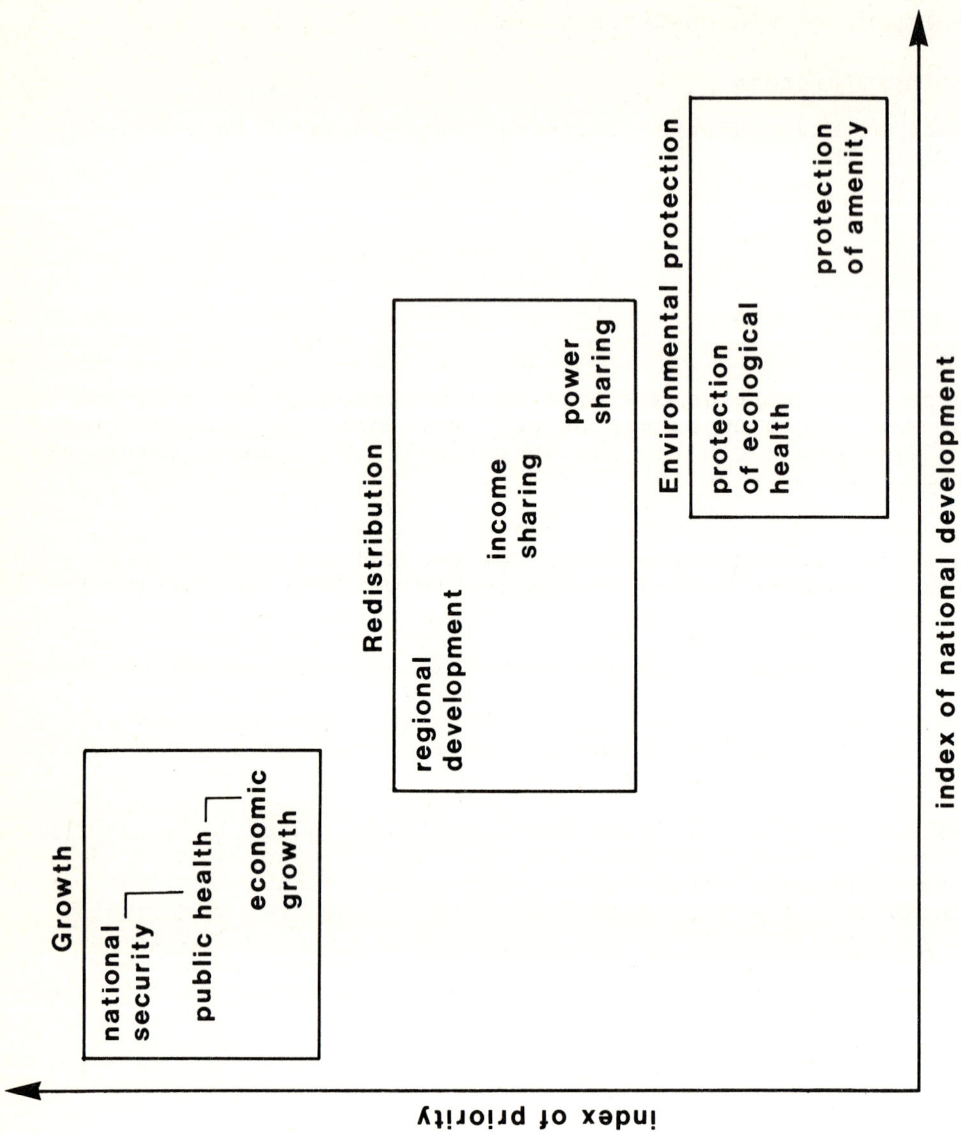
index of priority
index of national development
Growth
national security
public health
economic growth
Redistribution
regional development
income sharing
power sharing
Environmental protection
protection of ecological health
protection of amenity

Figure 1

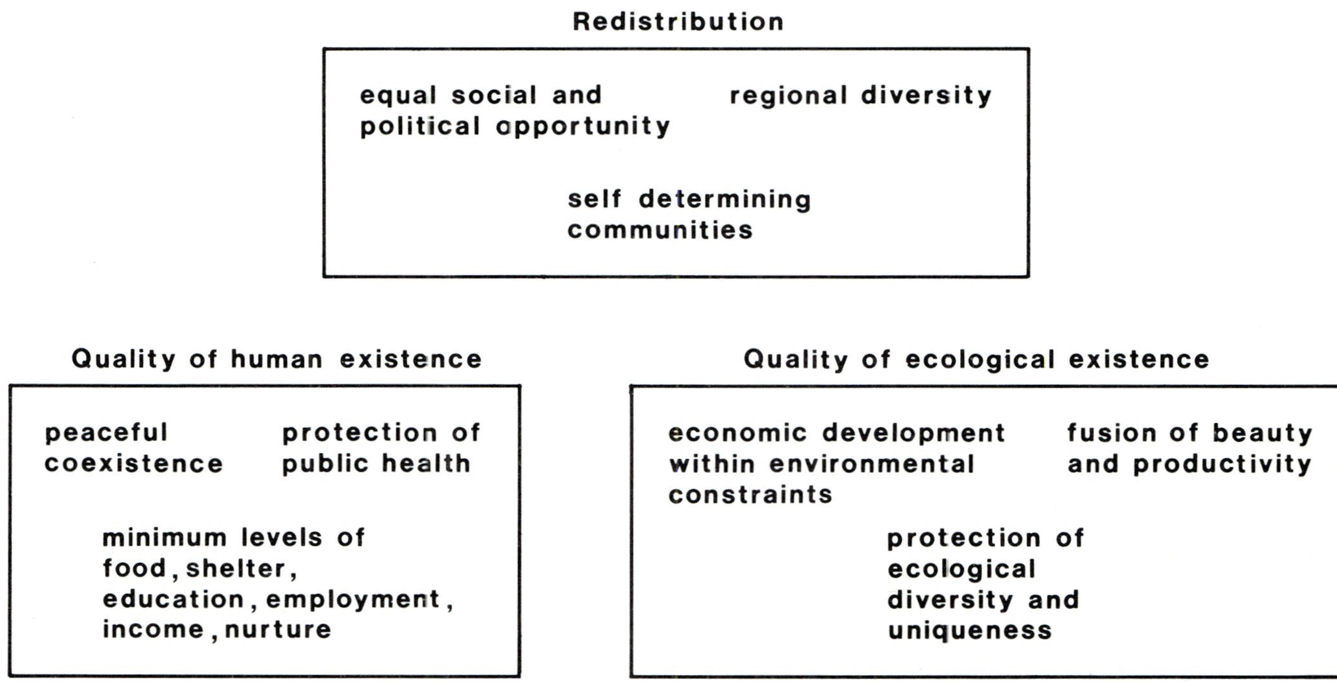

Redistribution
equal social and political opportunity
regional diversity
self determining communities
Quality of human existence
peaceful coexistence
protection of public health
minimum levels of food, shelter, education, employment, income, nurture
Quality of ecological existence
economic development within environmental constraints
fusion of beauty and productivity
protection of ecological diversity and uniqueness

Figure 2

Environmentalism

← Egocentrism — Technocentrism →

Deep environmentalists and **Self-reliance, soft technologists** (Egocentrism):

1. Lack of faith in modern large-scale technology and its associated demands on elitist expertise, central state authority, and inherently antidemocratic institutions
2. Implication that materialism for its own sake is wrong, and that economic growth can be geared to providing for the basic needs for those below subsistence levels

Deep environmentalists

- Intrinsic importance of nature for the humanity of man
- Ecological (and other natural) Laws dictate human morality
- Biorights-the right of endangered species or unique landscapes to remain unmolested

Self-reliance, soft technologists

3. Emphasis on smallness of scale and hence community identity in settlement, work, and leisure
4. Integration of concepts of work and leisure through a process of personal and communal improvement
5. Importance of participation in community affairs, and of guarantees of the rights of minority interests. Participation seen both as a continuing education and political function

Accommodaters

1. Belief that economic growth and resource exploitation can continue assuming
   (a) suitable economic adjustments to taxes, fees, etc.;
   (b) improvements in the legal rights to a minimum level of environmental quality;
   (c) compensation arrangements satisfactory to those who experience adverse environmental and/or social effects
2. Acceptance of new project-appraisal techniques and decision review arrangements to allow for wider discussion or genuine search for consensus among representative groups of interested parties
3. Provision of effective environmental management agencies at national and local levels

Cornucopians

1. Belief that man can always find a way out of any difficulties, either politically, scientifically or technologically
2. Acceptance that progrowth goals define the rationality of a project appraisal and of policy formulation
3. Optimistic about the ability of man to improve the lot of the world's people
4. Faith that scientific and technological expertise provides the basic foundation for advice on matters pertaining to economic growth, public health, and safety
5. Suspicious of attempts to widen the basis for participation and lengthy discussion in project appraisal and policy review
6. Belief that any impediments can be overcome given a will, ingenuity, and sufficient resources arising out of wealth

Figure 3. The pattern of environmentalist ideologies.

and a symbol of some superhuman order in this planet, which holds that only societal transformation with radical shifts in ethos, lifestyles and political power will achieve long lasting survival.

These two major themes can be subdivided into four subsidiary nodes. First is optimism, (or cornucopia) which holds that through managerial skills and the realism of economics growth and survival can be guaranteed. Second is accommodation--adjustment by those in positions of power and influence to the environmentalist challenge. This accommodation takes the form of accepting regulation, undertaking the three major kinds of assessment which form the central theme of this symposium, yet maintaining control over the forces and apparatus of power. This thesis postulates that EIA, TA, and RA are essentially tools for continuing a variation of business as usual, a very convenient lineage for people in positions of importance.

The two subsidiary orders on the left-hand side of the diagram involve radical change; hence, there is a major divide down the middle of the diagram. The communalism/soft technology mode argues for smaller scale patterns of living with decentralised power and suitable technology to fit human needs and local materials within recognised energy and resource limitations (4). The deep ecology (5) or post-materialist (6) theme is shared by those who preach for a more fundamental ethic toward the rights of nature, natural objects, and minority cultures, and who are prepared to sacrifice development to ensure that these rights are guaranteed.

Figure 4 indicates that bridging each of these four subsidiary themes are three sets of environmental activism. First, there are the private, self-serving groups blocking unpleasant developments in their neighborhood. Second, there are the public interest groups who challenge principles and espouse causes and who bridge the great ideological divide portrayed in Fig. 3. Third, there are the ecology, green, and alternative groups who preach a reformulated environmental ethic.

At present neither EIA, TA, nor RA breaches the barrier that runs through the middle of the diagram. But circumstances are changing--especially political disenchantment with certain technologies, with disruption of environmental cycles, and with apparently insensitive government and corporate power. As a result, EIA, TA, and RA are on the threshold of breaking through into the commercialism-soft technology mode, most particularly risk assessment. Other papers in this symposium explain why (7).

Figure 4 EIA TA AND RA WITHIN IDEOLOGICAL MODES OF WORLD VIEW

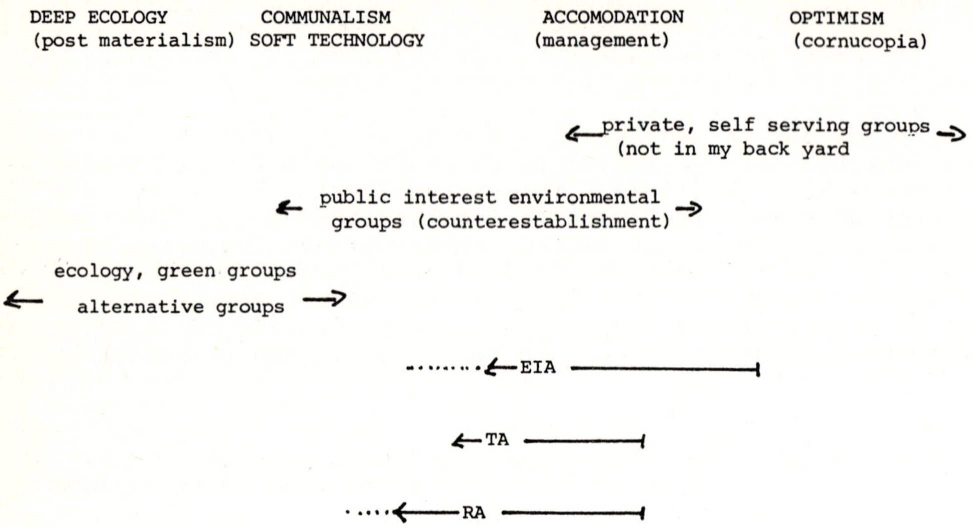

Figure 5 POSSIBLE EVOLUTIONS OF ENVIRONMENTAL IMPACT ASSESSMENT (EIA) TECHNOLOGY ASSESSMENT (TA) AND RISK ASSESSMENT (RA)

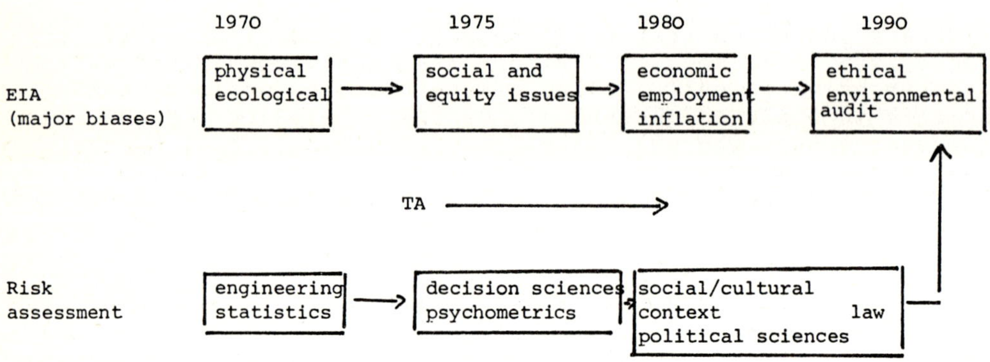

EIA itself is a tool almost uniquely confined to the US. It arose during a period of great national doubt and in a nation with notoriously ill-coordinated controls over development control and land use planning (8). It is more helpful to make a distinction between EIA and EA (environmental assessment). The latter is a concept much preferred in continental Europe for it implies (a) an existing land use planning apparatus, (b) flexibility over what assessments are actually undertaken and by whom they are guided, and (c) restrictions as to how much the public is informed. While EIA is appropriate for open government with fixed decision rules and adversarial regulation, EIA is more acceptable to closed governments with flexible decision rules, discretionary powers in regulation, and limited access for participation (9).

Figure 4 describes the evolution of both EIA and RA over the past 10 years. While the US Congress always meant the famous S 102(2)(c) of the 1969 National Environmental Policy Act to be a full disclosure, action-forcing, and comprehensive appraisal, in practice agencies tended to concentrate in the early years on physical and ecological considerations (see the Alaska pipeline EIS). Subsequently the practice of social impact assessment (SIA) with its own set of methodologies developed, coupled with its interest in "fragile" social values and the interests of minority groups (10). In the late 1970s with recession and unemployment rampant, agencies increasingly emphasized the inflationary and employment aspects of major developments--themes which should normally be incorporated in a cost-benefit analysis.

Risk assessment began from a different ancestry, the engineering and decision sciences (11). Nowadays risk assessment is being emphasized more in its political and cultural context (12) with stress being placed on the linkage between assessment and evaluation as clothed in social and political meaning. It is this last aspect that lends some hope to the view that RA may prove to be the technique that breaks through the ideological barrier depicted in Fig. 3.

Both EIA and EA, while warily disliked by developers and regulators initially, are now being incorporated by those interests to suit their own ends. Thus, major corporations (both public and private) now admit to the advantages of EIA/EA as a managerial tool to improve design efficiency and to avoid costly and/or embarrassing mistakes. Good use of EIA/EA can also smooth decisionmaking pathways and allow a degree of control over both regulators and decisionmakers (9). Also EIA/EA are now considered politically essential for both sensitive projects (such as nuclear power stations and major roadways) and for environmentally-sensitive areas (wetlands, designated areas).

Nevertheless, EIA/EA does expose issues that would otherwise be dormant, stirs up environmental groups, makes final judgments more explicitly political, and does result in often time-consuming and expensive delays--though savings in both time and cost are very possible given well done assessments (9).

This means that conventional decisionmaking models have to be reformulated. Figure 5 outlines the classic model together with a variation based on a more interventionist strategy. The latter, which is also developed elsewhere (2), suggests that decision sequences may follow a pattern of routineness (with limitations borne of policy paradoxes and agency myopia) followed by conflict leading to non-routine (a strategic) decision approaches in which various new styles of participation, consultation, and compensation can emerge. These approaches are of interest to this symposium because they suggest new avenues for incorporating assessments within the decision-making process. Of course, governments can (and do) ignore all this and push for a decision based on political expediency. However, the existence of EIA/TA/RA with their broader societal as well as analytical dimensions make this less likely, or at least more obvious. Democracy may be served as a consequence, though not necessarily best served, for "real world" decision tracks can be messy and frustrating. Routine decisionmaking has a pleasant habit of becoming more innovative and enlightened. EIA/RA and TA have certainly helped to push these processes along, but it is also a function of a more participatory society, of greater social divisions (when key decisions must be taken), and of a neo-enlightenment scepticism of expertise, officialdom, and the legitimacy of governments.

Figure 5 Models of Decision making

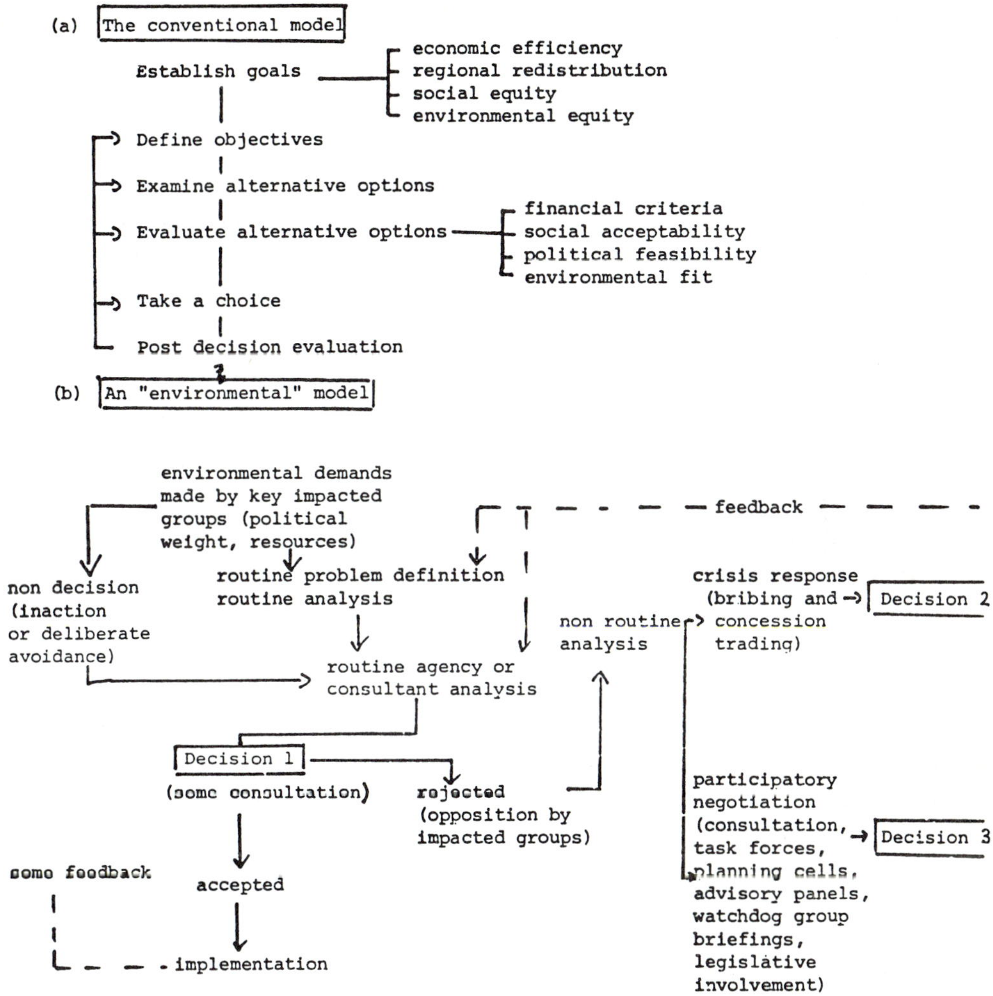

REFERENCES

1. International Union for the Conservation of Nature, World Conservation Strategy (IUCN, Geneva, 1980).

2. T. O'Riordan Environmentalism 2nd ed. (Pion and Methuen, London, 1981). Especially pp.1-19, 375-380.

3. J.E. Lovelock Gaia: A New Look at Life on Earth (Oxford University Press, Oxford, 1979); K. Redlar, The Quest for Gaia (Granada, London, 1979); P. Russell The Awakening Earth: Our Next Evolutionary Leap (Routledge Kegan Paul, London, 1982); J.D. Hughes "Gaia: an ancient view of our planet" The Ecologist 13 (2/3) (1983), pp.54-60.

4. A. Schneiberg "Saving the environment: from whom, for whom and by whom?" in Environmental Policies and Politics P. Knoefel and N. Watts eds. (Harper and Row, London) forthcoming.

5. W.B. Devall "The deep ecology movement" Natural Resources Journal 20(2) (1980) pp.299-322.

6. R. Inglehart "Changing values and the rise of industrialism in western societies" (International Institute for Environment and Society, Berlin Preprint 82-114, 1982).

7. See for example the papers by Renn, Jungermann, Clark and Wynn.

8. T. O'Riordan 'US experience in EIA: lessons for Britain' ECOS, 3(2), (1982), pp.6-11.

9. T. O'Riordan 'Environmental impact assessment in the context of economic recession' The Geographical Journal 148(3), pp.346-354.

10. E.J. Soderstrom Social Impact Assessment: Experimental Methods and Approaches (Praeger, New York, 1981).

11. T. O'Riordan "Risk perception studies and policy priorities" Risk Analysis 2(2) (1982), pp.95-100.

12. M. Douglas and A. Wildawsky Risk and Culture (University of California Press, Berkeley, 1982).

Opportunities for the Social Sciences in Risk Analysis--An Engineer's Viewpoint

Chris Whipple

Energy Study Center, Electric Power Research Institute

INTRODUCTION

The approach taken by engineers to manage technological risks has been changing in recent years. In the past, engineers dealt with risk as an engineering problem. Engineering safety factors were and still are a basic feature of the design of any structure, and these safety factors were institutionalized into codes and standards. Those risks which remained inherent in engineering designs were considered to be tolerable residuals, and were not explicitly analyzed. The question, "How safe is safe enough?," was dealt with implicitly through the use of safety factors, and over time, trial-and-error experience provided feedback for the adjustment of safety factors and for the recognition and correction of failure-prone designs.

For many technologies, this approach is no longer considered suitable. For many of the technologies which are objects of public concern, the qualitative character of the risk is unsuited for trial-and-error management. There are risks with slow feedback, e.g., exposures to carcinogens, as well as technologies with potentially catastrophic accidents, in which substantial public harm would occur if a trial-and-error approach was applied to the detection and correction of high-risk design features.

As a consequence of a heightened public concern with technological risk (perhaps due to the changing nature of the risk, perhaps due to other reasons), the engineer now faces a social expectation of trial without error. Further, the question, "How safe is safe enough?," is now viewed as a societal question rather than an engineering concern. The present conception among engineers accepts this current viewpoint: it is the engineer's task to design technologies based upon risk criteria that come from the society at large. An excellent illustration

NATO ASI Series, Vol. G4
Environmental Impact Assessment, Technology Assessment, and Risk Analysis. Edited by V.T. Covello et al.

of this viewpoint can be found in the work by John Wiggins (1) on earthquake safety standards for the city of Long Beach, California. Although he was asked to develop such standards, Wiggins provided the city council with an analysis of the estimated safety levels and costs associated with a variety of standards, and argued that it was not appropriately his role to make such a decision.

A final aspect of change in the social management of health and safety risks is the recent rise in concern by industry for liability claims arising from past actions, as lawsuits over asbestos health effects now illustrate. Until recently, regulation was a dominant force in safety engineering, and regulation poses known or predictable costs on an industry. By contrast, liability is a more difficult risk management issue to deal with. The costs are not known nor are they predictable. No simple rule of behavior can ensure avoidance of substantial liability costs, whereas compliance with regulation is usually straightforward. To some degree the rising importance of liability as a motivation for careful risk management increases the incentive to use societal risk criteria, because such criteria (in contrast to engineering safety factors or other criteria internal to a firm) can justify risk management decisions reviewed from hindsight.

RISK ASSESSMENT VERSUS SAFETY ENGINEERING

The changes in motion in safety engineering are leading engineers to try to develop increasingly explicit estimates of risk. This step is necessary for public (i.e., nontechnical) involvement in safety decisions, especially where technology choices are at issue. Attempts to quantify risk also reveal where the greatest uncertainties lie, and often are thought to provide valuable insights for risk management.

This step is not without its difficulties. Usually it is far easier to calculate that a system has been engineered to tolerate some multiple of the stresses it is likely to experience in actual service than it is to calculate the residual failure probability. When the technical system is complex and has a large number of failure modes, as with an aircraft or a nuclear power plant, the job of risk estimation is substantial. For reference, in the United States, a nuclear power plant risk assessment today usually costs in excess of $4 million.

SOCIAL SCIENCE RESEARCH AND RISK ASSESSMENT

The trend toward the greater use of explicit risk estimation in engineering has led to identification of a number of risk assessment problems in the domain of the social sciences. For example, risk assessments of complex engineered systems typically involve consideration of human behavior in a wide variety of circumstances. Whether one is analyzing aircraft or nuclear power plants, the risk is not usually an inherent feature of the engineering design, but is strongly affected by the operator of the system. Operators commit errors which figure in the overall chances of an accident, and operators can intervene creatively to halt accident sequences and limit the resultant damage. In fact, in the two most serious commercial nuclear power plant accidents that have occurred in the U.S., TMI and Browns Ferry, we find evidence of both kinds of behavior. There are other, more subtle types of human error which influence technological risk; for example, the same maintenance error performed on independent safety systems can produce a hidden common mode failure.

In some cases it is possible to avoid detailed behavioral assumptions because sufficient data are available. For example, one need not try to assess all possible errors in driving a car to have a reliable estimate of accident probabil ities. But in many cases the available data are insufficient. In such cases (e.g. nuclear power plants) operator error rates must be estimated from available information, and the effect of such errors must be projected from physical principles. This is a complicated task for several reasons. First, it is virtually impossible to consider all the possible errors that might be committed. Second, once a dangerous or off-design situation occurs, it is unlikely that operators will behave as they do under normal conditions. Despite some insights into behavior under normal and accident conditions (e.g., it is now widely accepted that prior simulation of accidents improves behavior during actual accidents) this remains a large source of uncertainty in many types of risk analysis.

Behavioral projections for risk analysis are not limited to pilots and nuclear plant operators. An additional area of social science research is the prediction of behavior by the public in response to warnings, calls for evacuation, and other danger signals. Such understanding is needed for accurate risk estimation, and more importantly for the design of warning systems. A small project in this area has been started at the Electric Power Research Institute (EPRI) to examine nuclear power plant warning systems. More generally, the term risk homeostasis has been applied to describe a behavioral adjustment to changing risk situations (2).

An additional area where social science research can aid in risk assessment is when the probabilities are not obvious from conclusive data, and when risk estimates have a significant judgmental component. In these cases, a working knowledge of the systematic biases and errors that occur when subjective estimates are given (e.g., overconfidence, anchoring, etc.) can lead to elicited probabilities that are more representative of an expert's views.

RISK ANALYSIS AND PUBLIC POLICY

Technology choices are strongly affected by risk considerations, although the social processes through which this occurs are not well understood. An element of the acceptance of or resistance to a technology by the public is the perception of its risk. This area, widely explored by psychologists, has been of increasing interest to engineers (particularly nuclear engineers) for several reasons. The initial exploration of this literature by nuclear risk analysts was to seek an understanding of strong public reaction generated in response to nuclear power risks, which, to the analysts, seem to be smaller than other well accepted risks.

Studies such as that conducted at Decision Research on the determinants of perceived risk (3) (i.e., the League of Women Voters Survey) have been quite helpful to the engineering risk analysis community in developing an understanding of some of the elements that influence public reaction to technologies and their risks. From such work, many engineers have learned of the significance of qualitative risk factors to the public, and in response these analysts have tried to develop risk measures and descriptions which permit understanding of risk characteristics from the public perspective. For example, those who conduct probabilistic risk assessments of nuclear plants are generally sensitive to provide more than expressions of expected losses (i.e., probability times consequence) given an apparent public sensitivity to the catastrophic potential of accidents. The method that is used to describe nuclear plant risk estimates is the complementary cumulative probability distribution; this indicates the probability of occurrence of an accident of greater than a specified magnitude.

RISK EXPRESSIONS

As psychological research has shown, the response to risk is sensitive to the expression used to describe the risk (4). In rough form, this result has been known intuitively by partisans in debates on technological risk; for example, nuclear power opponents frequently cite the estimated consequences of severe accidents, while proponents refer to probabilities. Risk comparisons have been used to provide a frame of reference for risk expressions, and at times such comparisons have implicitly argued for the acceptability or unacceptability of a technology, based on the assumption that a safer technology is preferable to a more hazardous one, all else being equal.

The sensitivity of response to risk information poses a problem for the risk analyst who wishes to describe risks to the public. Regardless of the perspective taken, the analyst can be charged with biasing the public by people who prefer an alternative expression. For example, Herman Kahn once described the reaction to his estimate that atmospheric nuclear weapon testing would produce 1,000 fatalities annually, and that this corresponded to an individual risk of one in three million per year (based upon a world population of three billion at the time). Half of his mail said "Why do you mention one chance in three million? Are you trying to corrupt the thousand?" And the other half said "Why do you mention a thousand? Are you trying to scare people? There's only one chance in three million." (5)

When the interpretation of estimates of energy production risks is considered, a number of issues of perspective arise (6). For example, it is common in the literature to compare the risks of various sources on the basis of the estimated risk from a plant of a given size, e.g., 1,000 MWe nuclear vs. coal vs. oil (7). Such an approach is logical from the framework of an electric utility which is choosing a plant (although clearly factors other than risk go into such a decision), but such estimates do not indicate how public risk would be affected by the addition of a new plant to the existing system, nor would these estimates appeal to those who argue that alternatives are available which make a new plant unnecessary. Perhaps most seriously, risk expressed on an average, per plant basis implies a linearity between risk and supply that may not exist. When one considers the issues of nuclear weapons proliferation as it may relate to a commercial nuclear system, the issues of CO_2-induced climate change and acid rain due to coal combustion, or shocks to the economy and heightened global tensions that can accompany dependence on oil, it seems clear that the expression of "risk per plant" fails to capture the essence of the issue.

As Kahneman and Tversky (1980) and others have noted (8), the framing of alternatives often can influence choice, so the question of how one neutrally (if such a thing as a neutral description exists expresses a risk relates directly to the public acceptance or rejection of a technology in consideration of its risk.

CRITERIA FOR TECHNOLOGICAL RISK

As engineers move from the use of engineering safety factors to more explicit estimates of and criteria for risk, it becomes clear that a method for selecting risk criteria is needed. Such criteria should arise through a process which provides for public input, but at the same time the criteria must be specific enough to provide guidance for engineering choices. The type of guidance typically provided by the U.S. Congress to the various health and safety regulatory agencies is usually too vague to provide such guidance: e.g., no undue risk, no unreasonable risk, to protect public health with an adequate margin of safety, etc.

Traditionally, engineering standards have been set by voluntary committees from the engineering community. This approach has worked well on the whole, with only rare abuses. But because of the closed nature of this decision process, and because many of the engineers who participate in standard setting activities are from an industry affected by such standards, this approach is vulnerable to criticism on the grounds that the process is controlled by interested parties.

While it appears plausible or even straightforward to develop engineering safety criteria which are consistent with public values and attitudes toward risk, there are limits to the degree to which one can theoretically define what is and is not an acceptable risk. When this problem is described as the development of risk criteria which will lead to the public acceptance of a technology (at least to the extent that risk plays a part in such acceptance), the problem seems more difficult. It is perhaps too much of an abstraction to think that the diverse concerns people hold towards the risks from technologies can be satisfied by a sufficiently stringent risk criterion. For example, the degree of public concern associated with a nuclear power plant seems unlikely to vary with the thickness of the containment building or other technical choices which affect risk.

One important consideration when risk criteria are proposed with public values in mind is the relative importance of the qualitative character of the risk versus the quantitative magnitude of risk. We know that qualitative factors such as voluntariness and catastrophic potential are important, and public responses may be dominated by qualitative factors rather than by quantitative estimates of risk. It is frequently observed that we accept risk levels in voluntary or self- controlled circumstances (e.g., smoking, automobiles) that are never approached for involuntary risk exposures.

Another important consideration is that the decision process has to be accepted by the public, especially when a decision has identifiable winners and losers. Not only must the numerical criteria define a risk level that is consistent with public values, but the process by which such numbers are selected must be accepted as being fair to all viewpoints. In the U.S., public referenda are widely accepted as being fair. Referenda have been held repeatedly in various states of the U.S. on risk-related issues such as nuclear power, water fluoridation, smoking in public places, and gun control. But such measures are used to decide "yes" or "no" questions, and seem unsuited for setting numerical engineering criteria. Similarly, Congressional action in the U.S. establishes the legitimacy of decisions. But Congress rarely (and many argue, wisely) gets involved in the level of detail needed to set engineering criteria. Congress has gotten involved in specific issues on occasion--for example, it voted to permit the continued sale of soft drinks and other products containing sodium saccharin, making an exception to the food safety laws.

PUBLIC INTERPRETATION OF RISK

Public perception of the risks and benefits of a technology may be the main determinants of its acceptance, yet such perceptions are not necessarily sensitive to numerical design criteria. It would be incorrect to assume, for those technologies where risk is apparently the major public issue, that a sufficiently stringent criteria can remove public anxiety. Public acceptance will be strongly affected by other factors, including confidence in the integrity and competency of those responsible for safety, the accident record, the availability of alternatives, and the perceived value of the benefits.

People do not develop favorable or unfavorable attitudes toward technologies on the basis of quantitative estimates made by risk analysts. As the above discussion notes, acceptance is affected by a number of other factors such as the qualitative character of the risk, the perception of the risk, and the process by which risk decisions are made. While all these factors are influential, there is a limit to how well one can explain the public reaction to a technology based on the factors above. It seems apparent that many people develop attitudes toward technologies and their risk by relying on the opinion of others: authority figures, friends, relatives, etc. There is a dynamic social process through which specific issues become raised to become public controversies.

Social science research appears to be addressing the problem of describing the factors which affect public acceptance, rejection, and anxiety with regard to technological risk. Undoubtedly the insights from this research will be useful to the engineering community in its approach to risk management. But such insights are likely to be general, conceptual, and qualitative. These insights are unlikely, in the near term, to provide the engineer with a method for designing criteria in accordance with "public values."

DESCRIPTIVE AND PRESCRIPTIVE MODELS

In economics and the social sciences, a distinction is frequently made between descriptive and prescriptive models and analyses. This distinction is important to social scientists (but not so important to engineers for reasons noted below) because there is a large difference between describing how people behave and telling people how they should behave if they are to be consistent with the axioms of rational choice and their expressed values.

By its very nature, the selection of risk criteria is prescriptive. Therefore, from a social science perspective, the selection process should meet a number of logical tests, and the values or utilities that are used in the decision should ideally reflect those of the public. But there is at present no accepted way to satisfy all these objectives. In fact, it is not clear that the diverse ways in which the public considers an issue can be represented by a single problem representation, and certainly not by a single set of values. Any approach to develop risk criteria by the inference of public values (e.g., by revealed or expressed preferences) will have deficiencies. No analytical approach is foreseeable which can satisfy the diverse requirements that it be logically sound, reflective of public values, responsive to public perceptions, and accepted as politically legitimate by all stakeholders in a decision.

In reference to proposed analytical approaches (such as those by Rowe (9) and Litai (10) which attempt to infer weights for risks of differing characteristics, A. V. Cohen (11) expresses a view held by many economists and social scientists regarding attempts to define acceptable risk analytically:

> Such a calculus would need to be used with caution. There is a danger otherwise that one man's set of utilities will be buried in submissions to decisionmakers and lost sight of. Typical of such dangers are attempts to put a value on life (the decisionmaker is not always reminded whether this includes or excludes subjective judgments of intrinsic values) or 'objective' assessments of risk which include 'unified indices of woe' of the kind I referred to earlier. Hidden values can present real problems in other areas too. What weight should be given at the strategic level to economic criteria (do we shut a firm down or tolerate a marginally greater risk for a bit longer? To what extent do differential perceptions of risk act as an effective tax on technological innovation? These are essentially political judgments, which must not and cannot be replaced by a calculus.

Other earlier attempts have been made to propose criteria for acceptable risk, such as that by Farmer (12) which took the form of an acceptability boundary on a basis of accident probability and consequences, and of Starr, (13) in which acceptability was based on individual risk and benefit, taking the voluntariness of the risk-taking into account. These approaches have been criticized (as have virtually all analytical attempts to define an acceptable risk level) for many of the reasons given in the above quote, and for other reasons as well. It has been noted, for example, that under the Farmer and Starr approaches, there is not a requirement that marginal risks equal marginal control costs at the target level.

One can easily understand the social scientists' dissatisfaction with these approaches if they are viewed as prescriptions for the level of risk a society should accept. From this perspective, these approaches resemble naive attempts at economic, psychological, and sociological analysis by engineers and scientists frustrated with the complexities and realities of political decisionmaking. But this is not in fact the engineer's perspective, and those who have proposed these approaches are well aware of the political nature of risk decisions.

ENGINEERING RISK CRITERIA

The approaches from the engineering community to analytically define an acceptable risk level are ment to fill a void. Currently the political decisions regarding the acceptability of risk are made after a technology has been developed to the point that some commercial development has begun. Thus, the design criteria for risk are not known except long after the plans and investments have been made. No political mechanism to provide legitimate public input to technology designers has emerged to replace the judgments that were traditionally made by engineers.

For a well established technology, the approach has generally been to design to safety levels which are at least as conservative as those in current practice. In practice, risks from most technologies decrease with technological progress. Often regulatory or industry standards have been developed which provide detailed guidance for designers in mature industries. Consequently, an automobile manufacturer generally sets design targets for performance (for braking, for example) which meet industry and government standards and which are at least as stringent as the performance targets of previous years.

For a new technology or for a radical departure from traditional designs in existing technologies (i.e., new chemicals), the design criteria are not generally so easily established. The designers can consider a number of rationales for selection of a risk design target, e.g., cost-effectiveness, comparison with alternative risks, etc., but in essence the approaches typically revert to an attempt to develop a calculus to predict (and later, to defend) a risk level that would eventually emerge from the political process. It is not that the engineers wish to substitute a simple calculus for political risk-benefit decisions, it is that risk targets are needed as design parameters. To the engineering community, the problem of setting risk criteria for new technologies seems a recurrent one that deserves a general approach.

From this perspective, the engineering approaches to define "acceptable" risk are meant to be generalizations of the apparent rules in the social decision process, useful for their descriptive validity. It seems reasonable under this view to project the societal acceptance or rejection of a technology or risk grounds based on a review of societal risk taking generally; this is philosophically just an extension or generalization of the principle that the design for the brakes on next year's automobile should be consistent with the risk accepted in brake designs previously.

SAFETY TRENDS

These approaches to the selection of design targets for risks have been criticized for enshrining the risks of the status quo, and for providing no mechanism for the evolution of safety. Further, it is questioned whether current risk levels are appropriate reference points for future risks when current levels may themselves be quite controversial, and, consequently, not clear indicators of acceptability (14). The implications in both these criticisms is that society actually prefers or will prefer in the future risk levels which are below those at present, and that to base risk criteria on revealed or historical preferences will lead to the selection of risk criteria which are insufficiently stringent.

Perhaps a stronger restatement of this view is that risks from virtually all sources have been declining dramatically since World War II, and expectations have been created both publicly and among risk analysts that the trend will continue. Certainly the trend of declining risk and increasing attention to risk has multiple causes, all of which are not understood. Improvements in technology, as for example in transportation systems, has played a role. Regulation has undoubtedly contributed as well; in fact, many risk regulatory agencies in the U.S. did not exist 20 years ago (e.g., Consumer Product Safety Commission, Occupational Safety and Health Administration, Environmental Protection Agency, etc.). Liability considerations have also increased safety incentives (15). In total, the public response to risk has changed substantially; health and safety are now higher priorities than in the past. Whether this is because the trend toward greater longevity has lowered the social "noise" level below which risks are ignored, or whether greater wealth has permitted greater expenditures for safety is not well understood. But the trend does exist.

Does the existence of a trend toward decreasing risk (at least as statistics reveal risks) invalidate the use of historical risk levels as a guide to future designs? The rates of change in risk rates appear small when compared against the precision of predictions of risk based on a priori engineering analyses. Risk rates may change a few percent a year; risk levels may be predicted to within an order of magnitude or less under favorable conditions. Given such uncertainties and the inability to avoid the use of conservative assumptions in most risk analyses, it appears unlikely that the use of historical levels to provide design guidance would lead to a stagnation of risk levels. More significantly, one can argue that the trend of risk reduction is itself a socially revealed preference relevant to the selection of risk criteria.

From the engineering perspective, the task remains to find better ways of developing risk criteria for developing technologies. For the reasons given here, there is no better source for practical guidance than current social risk management policies. While such an approach cannot guarantee "acceptability" of a technology, it does reveal the engineering emphasis on criteria that appear to be consistent with public values. Assistance in the development of such criteria by researchers in the social sciences would be much appreciated.

CONCLUSIONS

Social science risk research can improve risk analysis and management in a number of ways: 1) by providing insights into behavior that lead to better estimation of risk, 2) by identifying bias-prone analytical approaches, and error-prone risk management approaches, 3) by examining the efficiency of alternative means for reducing risk, descriptions which are appropriate for the decisions at issue; and 4) by providing conceptual and operational guidance for coping with uncertainty.

Despite these contributions from the social sciences, the engineering community is still faced with the difficult problem of selecting explicit criteria for the design and operation of potentially risky technologies. Such criteria have emerged over a period of time for mature technologies, but for relatively new technologies, the political process provides little operative guidance. In the absence of such guidance, engineers have tended to base design targets on estimated risks from accepted technologies and on calculations of marginal cost-effectiveness for risk reductions. Often such methods have been codified into ad hoc rules for risk judgments such as de minimis levels and acceptable levels.

These simple approaches have been criticized by social scientists, often for valid reasons, but also because of misconceptions regarding desriptive rules (e.g., revealed preferences) applied in a prescriptive manner. In spite of these criticisms, operational alternatives have not emerged. Whether alternative methods for selecting risk design targets are analytical or primarily defined by their process, they would be very welcome by the engineers who now must make these judgments. A solid descriptive theory of the social response to risky technologies would be a constructive step toward this end.

REFERENCES

1. J.H. Wiggins, Jr., "Earthquake Safety in the City of Long Beach Based on the Concept of Balanced Risk," in Perspectives on Benefit-Risk Decision Making, National Academy of Engineering, Washington, D.C., 1972.

2. G. J. S. Wilde, Risk Analysis, 2, 4, 1982.

3. B. Fischhoff, P. Slovic, S. Lichtenstein, S. Read, B. Combs, Policy Science, 9, 127, 1978.

4. P. Slovic, B. Fischhoff, S. Lichtenstein, Accident Analysis and Prevention, 10, 28, 1978.

5. H. Kahn, in the proceedings of the Symposium/Workshop on Nuclear and Nonnuclear Energy Systems: Risk Assessment and Governmental Decision Making, February 5-7, 1979, Washington, D.C., the MITRE Corporation, Metrek Division, MTR-79W00335, September 1979.

6. C. Whipple, Risk Analysis, 1, 1, 1981.

7. A. V. Cohen, D. K. Pritchard, Comparative Risks of Electricity Production Systems: A Critical Survey of the Literature, Research Paper 11, U.K. Health and Safety Executive, 1980.

8. A. Tversky, D. Kahneman, Science, 1981.

9. W. D. Rowe, An Anatomy of Risk, Wiley, New York, 1977.

10. D. Litai, A Risk Comparison Methodology for the Assessment of Acceptable Risks, Ph.D. thesis, MIT, January 1980.

11. A. V. Cohen, "The Nature of Decisions in Risk Management," in Dealing with Risk, R. F. Griffiths, Ed., Manchester University Press, 1981.

12. F. R. Farmer, Nuclear Safety, 8, 539, 1967.

13. C. Starr, Science, 165, 1232, 1969.

14. B. Fischhoff, P. Slovic, S. Lichtenstein, R. Keeney, S. Derby, Acceptable Risk, Cambridge University Press, 1981.

15. W. Hammer, Product Safety Management and Engineering, Prentice-Hall, 1980.

SECTION II: Psychological Perspectives

Part I: Theoretical and Methodological Contributions

The Framing of Decisions and the Psychology of Choice*

Amos Tversky** and Daniel Kahneman***

Explanations and predictions of people's choices, in everyday life as well as in the social sciences, are often founded on the assumption of human rationality. The definition of rationality has been much debated, but there is general agreement that rational choices should satisfy some elementary requirements of consistency and coherence. In this chapter, we describe decision problems in which people systematically violate the requirements of consistency and coherence, and we trace these violations to the psychological principles that govern the perception of decision problems and the evaluation of options.

A decision problem is defined by the acts of options among which one must choose, the possible outcomes or consequences of these acts, and the contingencies of conditional probabilities that relate outcomes to acts. We use the term <u>decision frame</u> to refer to the decisionmaker's conception of the acts, outcomes, and contingencies associated with a particular choice. The frame that a decisionmaker adopts is controlled partly by the formulation of the problem and partly by the norms, habits, and personal characteristics of the decisionmaker.

It is often possible to frame a given decision problem in more than one way. Alternative frames for a decision problem may be compared to alternative perspectives on a visual scene. Veridical perception requires that the perceived relative height of

---

* This work was supported by the Office of Naval Research under contract N00014-79-C-0077 to Stanford University. This material is reprinted with permission from Science, 1981, 211, 453-458.
** Stanford University, Stanford, CA
*** University of British Columbia, Vancouver, B.C.

NATO ASI Series, Vol. G4
Environmental Impact Assessment, Technology Assessment, and Risk Analysis. Edited by V.T. Covello et al.

two neighboring mountains, say, should not reverse with changes of the vantage point. Similarly, rational choice requires that the preference between options should not reverse with changes of frame. Because of imperfections of human perception and decision, however, changes of perspective often reverse the relative apparent size of objects and the relative desirability of options.

We have obtained systematic reversals of preference by variations in the framing of acts, contingencies, or outcomes. These effects have been observed in a variety of problems and in the choices of different groups of respondents. Here we present selected illustrations of preference reversals, with data obtained from students at Stanford University and at the University of British Columbia who answered brief questionnaires in a classroom setting. The total number of respondents for each problem is denoted by N, and the percentage who chose each option is indicated in brackets.

The effect of variations in framing is illustrated in problems 1 and 2.

> Problem 1 [N=152]. Imagine that the U.S. is preparing for the outbreak of an unusual Asian disease, which is expected to kill 600 people. Two alternative programs to combat the disease have been proposed. Assume that the exact scientific estimate of the consequences of the programs are as follows:
>
> If program A is adopted, 200 people will be saved. [72 percent]
>
> If program B is adopted, there is a one-third probability that 600 people will be saved, and two-thirds probability that no people will be saved. [28 percent]
>
> Which of the two programs would you favor?

The majority choice in this problem is risk-averse: The prospect of certainly saving 200 lives is more atrractive than a risky prospect of equal expected value, that is, a one-in-three chance of saving 600 lives.

A second group of respondents was given the cover story of problem 1 with a different formulation of the alternative programs, as follows:

> Problem 2 [N=155]. If program C is adopted, 400 people will die. [22 percent]

If program D is adopted, there is a one-third probability that nobody will die, and two-thirds probability that 600 people will die. [78 percent]

Which of the two programs would you favor?

The majority choice in problem 2 is risk taking: The certain death of 400 people is less acceptable than the two-in-three chance that 600 people will die. The preferences in problems 1 and 2 illustrate a common pattern: Choices involving gains are often risk-averse, and choices involving losses are often risk taking. However, it is easy to see that the two problems are effectively identical. The only difference between them is that the outcomes are described in problem 1 by the number of lives saved and in problem 2 by the number of lives lost. The change is accompanied by a pronounced shift from risk aversion to risk taking. We have observed this reversal in several groups of respondents, including university faculty and physicians. Inconsistent responses to problems 1 and 2 arise from the conjunction of a framing effect with contradictory attitudes toward risks involving gains and losses. We turn now to an analysis of these attitudes.

THE EVALUATION OF PROSPECTS

The major theory of decisionmaking under risk is the expected-utility model. This model is based on a set of axioms--for example, transitivity of preferences--that provide criteria for the rationality of choice. The choices of an individual who conforms to the axioms can be described in terms of the utilities of various outcomes for that individual. The utility of a risky prospect is equal to the expected utility of its outcomes, obtained by weighting the utility of each possible outcome by its probability. When faced with a choice, a rational decisionmaker will prefer the prospect that offers the highest expected utility (Fishburn, 1970; Raiffa, 1968; Savage, 1954; von Neumann and Morgenstern, 1947).

As will be illustrated below, people exhibit patterns of preference which appear incompatible with expected-utility theory. We have presented elsewhere (Kahneman and Tversky, 1979) a descriptive model, called <u>prospect theory</u>, which modifies expected-utility theory so as to accommodate these observations. We distinguish two phases in the choice process: an initial phase in which acts, outcomes, and contingencies are framed, and a subsequent phase of evaluation. (The framing phase includes various editing operations that are applied to simplify prospects; for example, by combining events or outcomes or by discarding negligible components.) For simplicity, we restrict the formal treatment of the theory to choices involving stated numerical probabilities and quantitative outcomes, such as money, time, or number of lives.

Consider a prospect that yields outcome x with probability p, outcome y with probability q, and the status quo with probability $1-p-q$. According to prospect theory, there are values $v(.)$ associated with outcomes, and decision weights $\pi(.)$ associated with probabilities, such that the overall value of the prospect equals $\pi(p)\ v(x) + \pi(q)\ v(y)$. A slightly different equation should be applied if all outcomes of a prospect are on the same side of the zero point. (Specifically, if $p + q = 1$ and either $x \rightarrow y \rightarrow 0$ or $x \leftarrow y \leftarrow 0$, the equation is replaced by $v(y) + \pi(p)[v(x) - v(y)]$, so that decision weights are not applied to sure outcomes.)

In prospect theory, outcomes are expressed as positive or negative deviations (gains or losses) from a neutral reference outcome, which is assigned a value of zero. Although subjective values differ among individuals and attributes, we propose that the value function is commonly S-shaped, concave about the reference point and convex below it, as illustrated in Fig. 1. For example, the difference in subjective value between gains of $10 and $20 is greater than the subjective difference between gains of $110 and $120. The same relation between value differences holds for the corresponding losses. Another property of the value function is that the response to losses is more extreme than the response to gains. The displeasure associated with losing a sum of money is generally greater than the pleasure associated with winning the same amount, as is reflected in people's reluctance to accept fair bets on a toss of a coin. Several studies of decision (Eraker and Sox, in press; Fishburn and Kochenberger, 1979; Kahneman and Tversky, 1979; Laughhunn, Payne, and Crum, 1980; Payne, Laughhunn, and Crum, 1980) and judgment (Galanter and Pliner, 1974) have confirmed these properties of the value function. The extension of the proposed value function to multiattribute options with or without risk, deserves careful analysis. In particular, indifference curves between dimensions of loss may be concave upward, even when the value functions for the separate losses are both convex, because of market subadditivity between dimensions.

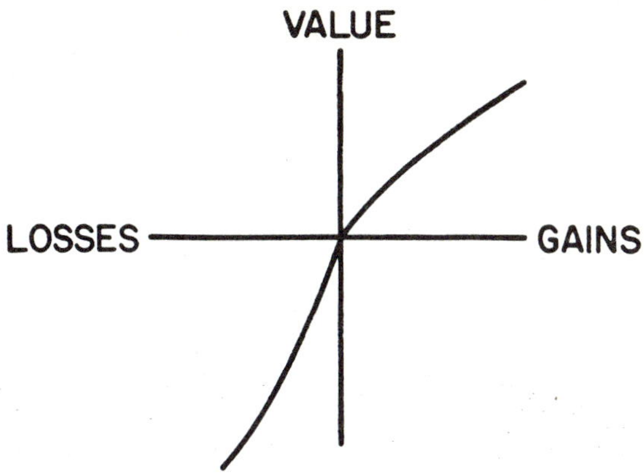

Figure 1. A Hypothetical Value Function

The second major departure of prospect theory from the expected-utility model involves the treatment of probabilities. In expected-utility theory, the utility of an uncertain outcome is weighted by its probability; in prospect theory, the value of an uncertain outcome is multiplied by a decision weight $\pi(p)$, which is a monotonic function of p but is not a probability. The weighting function π has the following properties. First, impossible events are discarded, that is, $\pi(0) = 0$, and the scale is normalized so that $\pi(1) = 1$, but the function is not well behaved near the endpoints. Second, for low probabilities $\pi(p) > p$, but $\pi(p) + \pi(1 - p) < 1$. Thus, low probabilities are overweighted, moderate and high probabilities are underweighted, and the latter effect is more pronounced than the former. Third, $\pi(pq)/\pi(p) < \pi(pqr)/\pi(pr)$ for all $0 < p, q, r \leq 1$. That is, for any fixed probability ratio q, the ratio of decision weights is closer to unity when the probabilities are low than when they are high, for example, $\pi(.1)/\pi(.2) > \pi(.4)/\pi(.8)$. A hypothetical weighting function that satisfies these properties is shown in Figure 2. The major qualitative properties of decision weights can be extended to cases in which the probabilities of outcomes are subjectively assessed rather than explicitly given. In these situations, however, decision weights may also be affected by other characteristics of an event, such as ambiguity or vagueness (Ellsberg, 1961; Fellner, 1965).

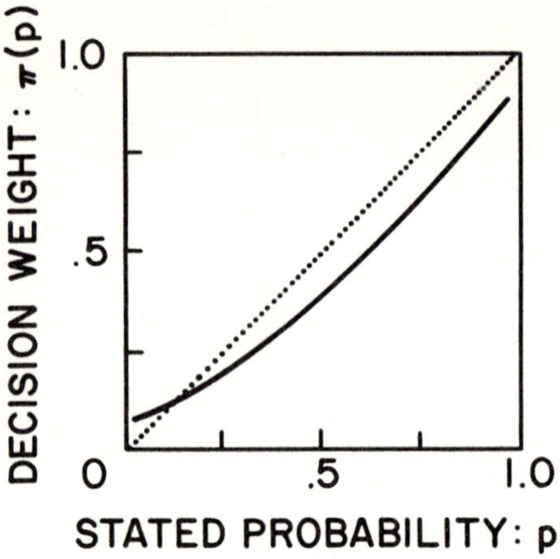

Figure 2. A Hypothetical Weighting Function

Prospect theory, and the scales illustrated in Figures 1 and 2, should be viewed as an approximate, incomplete, and simplified description of the evaluation of risky prospects. Although the properties of v and π summarize a common pattern of choice, they are not universal; the preferences of some individuals are not well described by an S-shaped value function and a consistent set of decision weights. The simultaneous measurement of values and decision weights involves serious experimental and statistical difficulties. The scaling of v and π by pair comparisons requires a large number of observations. The procedure of pricing gambles is more convenient for scaling purposes, but is subject to a severe anchoring bias: The ordering of gambles by their cash equivalents diverges systematically from the preference order observed in direct comparisons (Lichtenstein and Slovic, 1971).

If π and v were linear throughout, the preference order between options would be independent of the framing of acts, outcomes, or contingencies. Because of the characteristic nonlinearities of π and v, however, different frames can lead to different choices. The next three sections describe reversals of preference caused by variations in the framing of acts, contingencies, and outcomes.

> Problem 3 [N=150]. Imagine that you face the following pair of concurrent decisions. First examine both decisions, then indicate the options you prefer.
>
> Decision i. Choose between (A) a sure gain of $240 [84 percent] and (B) a 25 percent chance to gain $1,000, with a 75 percent chance to gain nothing [16 percent]
>
> Decision ii. Choose between (C) a sure loss of $750 [13 percent] and (D) a 75 percent chance to lose $1,000, with a 25 percent chance to lose nothing [87 percent]

The majority choice in decision i is risk-averse: A riskless prospect is preferred to a risky prospect of equal or greater expected value. In contrast, the majority choice in decision ii is risk taking: A risky prospect is preferred to a riskless prospect of equal expected value. This pattern of risk aversion in choices involving gains and risk seeking in choices involving losses is attributable to the properties of v and π. Because the value function is S-shaped, the value associated with a gain of $240 is greater than 24 percent of the value associated with a gain of $1,000, and the (negative) value associated with a loss of $750 is smaller than 75 percent of the value associated with a loss of $1,000. Thus, the shape of the value function contributes to risk aversion in decision i and to risk seeking in decision ii. Moreover, the underweighting of moderate and high probabilities contributes to the relative attractiveness of the sure gain in decision i and to the relative aversiveness of the sure loss in decision ii. The same analysis applies to problems 1 and 2.

Because decisions i and ii were presented together, the respondents had in effect to choose one prospect from the set: A and C, B and C, A and D, B and D. The most common pattern (A and D) was chosen by 73 percent of respondents, while the least popular pattern (B and C) was chosen by only 3 percent of respondents. However, the combination of B and C is definitely superior to the combination A and D, as is readily seen in problem 4.

> Problem 4 [N=86]. Choose between: (A and D) a 25 percent chance to win $240 and a 75 percent chance to lose $760 [0 percent] and (B and C) a 25 percent chance to win $250 and a 75 percent chance to lose $750 [100 percent]

When the prospects were combined and the dominance of the second option became obvious, all respondents chose the superior option. The popularity of the inferior option in problem 3 implies that this problem was framed as a pair of separate choices. The respondents apparently failed to entertain the possibility that the conjunction of two seemingly reasonable choices could lead to an untenable result.

The violations of dominance observed in problem 3 do not disappear in the presence of monetary incentives. A different group of respondents who answered a modified version of problem 3, with real payoffs, produced a similar pattern of choices. A new group of respondents (N=126) was presented with a modified version of problem 3, in which the outcomes were reduced by a factor of 50. The participants were informed that the gambles would actually be played by tossing a pair of fair coins and that one participant in 10 would be selected at random to play the gambles of his or her choice. To ensure a positive return for the entire set, a third decision, yielding only positive outcomes, was added. These payoff conditions did not alter the pattern of preferences observed in the hypothetical problem: 67 percent of respondents chose prospect A, and 86 percent chose prospect D. The dominated combination of A and D was chosen by 60 percent of respondents, and only 6 percent favored the dominant combination of B and C. Other authors have also reported that violations of the rules of rational choice, originally observed in hypothetical questions, were not eliminated by payoffs (Grether, 1980; Grether and Plott, 1979; Lichtenstein and Slovic, 1973; Lieblich and Lieblich, 1969).

We suspect that many concurrent decisions in the real world are framed independently, and that the preference order would often be reversed if the decisions were combined. The respondents in problem 3 failed to combine options, although the integration was relatively simple and was encouraged by instructions. (For other demonstrations of a reluctance to integrate concurrent options, see Payne and Braunstein, 1971; Slovic and Lichtenstein, 1968). The complexity of practical problems of concurrent decisions, such as portfolio selection, would prevent people from integrating options without computational aids, even if they were inclined to do so.

THE FRAMING OF CONTINGENCIES

The next trio of problems illustrates the framing of contingencies. Each problem was presented to a different group of respondents. Each group was told that one participant in 10, preselected at random, would actually be playing for money. Chance events were realized in the respondents' presence, by drawing a single ball from a bag containing a known proportion of balls of the winning color, and the winners were paid immediately.

> Problem 5 [N=77]. Which of the following options do you prefer: (A) a sure win of $30 (78 percent) or (B) an 80 percent chance to win $45? [22 percent]

> Problem 6 [N=85]. Consider the following two-stage game. In the first stage, there is a 75 percent chance to end the game without winning anything, and a 25 percent chance to move into the second stage. If you reach the second stage, you have a choice between (C) a sure win of $30 [74 percent] and (D) an 80 percent chance to win $45 [26 percent]

> Your choice must be made before the game starts; that is, before the outcome of the first stage is known. Please indicate the option you prefer.

> Problem 7 [N=81]. Which of the following options do you prefer: (E) a 25 percent chance to win $30 [42 percent] or (F) a 20 percent chance to win $45? [58 percent]

Let us examine the structure of these problems. First, note that problems 6 and 7 are identical in terms of probabilities and outcomes, because prospect C offers a .25 chance to win $30 and prospect D offers a probability of .25 x .80=.20 to win $45. Consistency therefore requires that the same choice be made in problems 6 and 7. Second, note that problem 6 differs from problem 5 only by the introduction of a preliminary stage. If the second stage of the game is reached, then problem 6 reduces to problem 5; if the game ends at the first stage, the decision does not affect the outcome. Hence, there seems to be no reason to make a different choice in problems 5 and 6. By this logical analysis, problem 6 is equivalent to problem 7 on the one hand and problem 5 on the other. The participants, however, responded similarly to problems 5 and 6 but differently to problem 7. This pattern of responses exhibits two phenomena of choice: the certainty effect and the pseudocertainty effect.

The contrast between problems 5 and 7 illustrates a phenomenon discovered by Allais (1953), which we have labeled the certainty effect: A reduction of the probability of an outcome by a constant factor has more impact when the outcome was initially certain than when it was merely probable (Allais, 1953; MacCrimmon and Larsson, 1979). Prospect theory attributes this effect to the properties of π. It is easy to verify by applying the equation of prospect theory to problems 5 and 7, that people for whom the value ratio $\underline{v}(30)/\underline{v}(45)$ lies between the weight ratios $\pi(.20)/\pi(.25)$, and $\pi(.80)/\pi(1.0)$ will prefer A to B and F to E, contrary to expected-utility theory. Prospect theory does not predict a reversal of preference for every individual in problems 5 and 7. It only requires that an individual who has no preference between A and B prefer F to E. For group data, the theory predicts the observed directional shift of preference between the two problems.

The first stage of problem 6 yields the same outcome (no gain) for both acts. Consequently, we propose, people evaluate the options conditionally, as if the second stage had been reached. In this framing, of course, problem 6 reduces to problem 5. Generally, we suggest that a decision problem is evaluated conditionally when there is a state in which all acts yield the same outcome, such as failing to reach the second stage of the game in problem 6, and the stated probabilities of other outcomes are conditional on the nonoccurrence of this state.

The striking discrepancy between the responses to problems 6 and 7, which are identical in outcomes and probabilities, could be described as a pseudocertainty effect. The prospect yielding $30 is relatively more attractive in problem 6 than in problem 7, as if it had the advantage of certainty. The sense of certainty associated with option C is illusory, however, since the gain is in fact contingent on reaching the second stage of the game.

Another group of respondents (N=205) was presented with all three problems, in different orders, without monetary payoffs. The joint frequency distribution of choices in problems 5, 6, and 7 was as follows: ACE, 22; ACF, 65; ADE, 4; ADF, 20; BCE, 7; BCF, 18; BDE, 17; BDF, 52. These data confirm in a within-subject design the analysis of conditional evaluation proposed in the text. More than 75 percent of respondents made compatible choices (AC or BD) in problems 5 and 6, and less than half made compatible choices in problems 6 and 7 (CE or DF) or 5 and 7 (AE or BF). The elimination of payoffs in these questions reduced risk aversion but did not substantially alter the effects of certainty and pseudocertainty.

We have observed the certainty effect in several sets of problems, with outcomes ranging from vacation trips to the loss of human lives. In the negative domain, certainty exaggerates the aversiveness of losses that are certain relative to losses that are merely probable. In a question dealing with the response to an epidemic, for example, most respondents found a sure loss of 75 lives more aversive than an 80 percent chance to lose 100 lives but preferred a 10 percent chance to lose 75 lives to an 8 percent chance to lose 100 lives, contrary to expected-utility theory.

We also obtained the pseudocertainty effect in several studies where the description of the decision problems favored conditional evaluation. Pseudocertainty can be induced either by a sequential formulation, as in problem 6, or by the introduction of causal contingencies. In another version of the epidemic problem, for instance, respondents were told that risk to life existed only in the event (probability .10) that the disease was carried by a particular virus. Two alternative programs were said to yield a sure loss of 75 lives or an 80 percent chance to lose 100 lives if the critical virus was involved and no loss of life in the event (probability .90) that the disease was carried by another virus. In effect, the respondents were asked to choose between a 10 percent chance of losing 75 lives and an 8 percent chance of losing 100 lives, but their preferences were the same as when the choice was between a sure loss of 75 lives and an 80 percent chance of losing 100 lives. A conditional framing was evidently adopted, in which the contingency of the noncritical virus was eliminated, giving rise to a pseudocertainty effect. The certainty effect reveals attitudes toward risk that are inconsistent with the axioms of rational choice, whereas the pseudocertainty effect violates the more fundamental requirement that preferences should be independent of problem description.

Many significant decisions concern actions that reduce or eliminate the probability of a hazard, at some cost. The shape of π in the range of low probabilities suggests that a protective action which reduces the probability of a harm from 1 percent to zero, say, will be valued more highly than an action that reduces the probability of the same harm from 2 percent to 1 percent. Indeed, probabilistic insurance, which reduces the probability of loss by half, is judged to be worth less than half the price of regular insurance that eliminates the risk altogether (Kahneman and Tversky, 1979).

It is often possible to frame protective action in either conditional or unconditional form. For example, an insurance policy that covers fire but not flood could be evaluated either as full protection against the specific risk of fire or as a reduction in the overall probability of property loss. The preceding analysis suggests that insurance should appear more attractive when it is presented as the elimination of risk than when it is described as a reduction of risk. As Slovic, Fischhoff, and Lichtenstein report, a hypothetical vaccine that reduces the probability of contracting a disease from .20 to .10 is less atarractive if it is described as effective in half the cases than if it is presented as fully effective against one of two (exclusive and equiprobable) virus strains that produce identical symptoms. In accord with the present analysis of pseudocertainty, the respondents valued full protection against an identified virus more than probabilistic protection against the disease.

The preceding discussion highlights the sharp contrast between lay responses to the reduction and the elimination of risk. Because no form of protective action can cover all risks to human welfare, all insurance is essentially probabilistic: It reduces but does not eliminate risk. The probabilistic nature of insurance is commonly masked by formulations that emphasize the completeness of protection against identified harms, but the sense of security that such formulations provide is an illusion of conditional framing. It appears that insurance is bought as protection against worry, not only against risk, and that worry can be manipulated by the labeling of outcomes and by the framing of contingencies. It is not easy to determine whether people value the elimination of risk too much or the reduction of risk too little. The contrasting attitudes to the two forms of protective action, however, are difficult to justify on normative grounds. (For further discussion of rationality in protective action, see Kunreuther and others, 1978).

THE FRAMING OF OUTCOMES

Outcomes are commonly perceived as positive or negative in relation to a reference outcome that is judged neutral. Variations of the reference point can therefore determine whether a given outcome is evaluated as a gain or as a loss. Because the value function is generally concave for gains, convex for losses, and steeper for losses than for gains, shifts of reference can change the value difference between outcomes and thereby reverse the preference order between options (Eraker and Sox, in press; Fishburn and Kochenberger, 1979; Laughhunn, Payne, and Crum, 1980; Payne, Laughhunn, and Crum, 1980). Problems 1 and 2 illustrate a preference reversal induced by a shift of reference that transformed gains into losses.

For another example, consider a person who has spent an afternoon at the racetrack, has already lost $140, and is considering a $10 bet on a 15:1 long shot in the last race. This decision can be framed in two ways, which corrrespond to two natural reference points. If the status quo is the reference point, the outcomes of the bet are framed as a gain of $140 and a loss of $10. However, it may be more natural to view the present state as a loss of $140 for the betting day, and accordingly frame the last bet as a chance to return to the reference point or to increase the loss to $150. Prospect theory implies that the latter frame will produce more risk seeking than the former. Hence, people who do not adjust their reference point as they lose are expected to take bets that they would normally find unacceptable. This analysis is supported by the observation that bets on long shots are most popular on the last race of the day (McGlothlin, 1956).

Because the value function is steeper for losses than for gains, a difference between options will loom larger when it is framed as a disadvantage of one option rather than as an advantage of the other option. An interesting example of such an effect in a riskless context has been noted by Thaler (1980). In a debate on a proposal to pass to the consumer some of the costs associated with the processing of credit card purchases, representatives of the credit card industry requested that the price difference be labeled a cash discount rather than a credit card surcharge. The two labels induce different reference points by implicitly designating as normal reference the higher or the lower of the two prices. Because losses loom larger than gains, consumers are less willing to accept a surcharge than to forego a discount. A similar effect has been observed in experimental studies of insurance: The proportion of respondents who preferred a sure loss to a larger probable loss was significantly greater when the former was called an insurance premium (Fischhoff, Slovic, and Lichtenstein, 1980; Hershey and Schoemaker, 1980).

These observations highlight the liability of reference outcomes, as well as their role in decisionmaking. In the examples discussed so far, the neutral reference point was identified by the labeling of outcomes. A diversity of factors determines the reference outcome in everyday life. The reference outcome is usually a state to which one has adapted; it is sometimes set by social norms and expectations; it sometimes corresponds to a level of aspiration, which may or may not be realistic.

We have dealt so far with elementary outcomes, such as gains or losses in a single attribute. In many situations, however, an action gives rise to a compound outcome, which joins a series of changes in a single attribute, such as a sequence of monetary gains and losses, or a set of concurrent changes in several attributes. To describe the framing and evaluation of compound outcomes, we use the notion of a psychological account, defined as an outcome frame which specifies both the set of elementary outcomes that are evaluated jointly and the manner in which they are combined, and a reference outcome that is considered neutral or normal. In the account that is set up for the purchase of a car, for example, the cost of the purchase is not treated as a loss, nor is the car viewed as a gift. Rather, the transaction as a whole is evaluated as positive, negative, or neutral, depending on such factors as the performance of the car and the price of similar cars in the market. A closely related treatment has been offered by Thaler (1980).

We propose that people generally evaluate acts in terms of a minimal account, which includes only the direct consequences of the act. The minimal account associated with the decision to accept a gamble, for example, includes the money won or lost in that gamble and excludes other assets or the outcome of previous gambles. People commonly adopt minimal accounts because this mode of framing simplifies evaluation and reduces cognitive strain, reflects the intuition that consequences should be causally linked to acts, and matches the properties of hedonic experience, which is more sensitive to desirable and undesirable changes than to steady states.

There are situations, however, in which the outcomes of an act affect the balance in an account that was previously set up by a related act. In these cases, the decision at hand may be evaluated in terms of a more inclusive account, as in the case of the bettor who views the last race in the context of earlier losses. More generally, a sunk-cost effect arises when a decision is referred to an existing account in which the current balance is negative. Because of the nonlinearities of the evaluation process, the minimal account and a more inclusive one often lead to different choices.

Problems 8 and 9 illustrate another class of situations in which an existing account affects a decision:

Problem 8 [N=183]. Imagine that you have decided to see a play where admission is $10 per ticket. As you enter the theater, you discover that you have lost a $10 bill.

Would you still pay $10 for a ticket for the play?
Yes [88 percent] No [12 percent]

Problem 9 [N=200]. Imagine that you have decided to see a play and paid the admission price of $10 per ticket. As you enter the theater, you discover that you have lost the ticket. The seat was not marked, and the ticket cannot be recovered.

Would you pay $10 for another ticket?
Yes [46 percent] No [54 percent]

The marked difference between the responses to problems 8 and 9 is an effect of psychological accounting. We propose that the purchase of a new ticket in problem 9 is entered in the account that was set up by the purchase of the original ticket. In terms of this account, the expense required to see the show is $20, a cost which many of our respondents apparently found excessive. In problem 8, however, the loss of $10 is not linked specifically to the ticket purchase, and its effect on the decision is accordingly slight.

The following problem, based on examples by Savage (1954, p. 103) and Thaler (1980), further illustrates the effect of embedding an option in different accounts. Two versions of this problem were presented to different groups of subjects. One group (N=93) was given the values that appear in parentheses, and the other group (N=88) was given the values shown in brackets.

Problem 10. Imagine that you are about to purchase a jacket for ($125) [$15], and a calculator for ($15) [$125]. The calculator salesman informs you that the calculator you wish to buy is on sale for ($10) [$120] at the other branch of the store, located twenty minutes' drive away. Would you make the trip to the other store?

The responses to the two versions of problem 10 were markedly different: 68 percent of the respondents were willing to make an extra trip to save $5 on a $15 calculator; only 29 percent were willing to exert the same effort when the price of the calculator was $125. Evidently, the respondents do not frame problem 10 in the minimal account, which involves only a benefit of $5 and a cost of some inconvenience. Instead, they evaluate the potential savings in a more inclusive account, which includes the purchase of the calculator but not the purchase of the jacket. By the curvature of v, a discount of $5 has a greater impact when the price of the calculator is low than when it is high.

A closely related observation has been reported by Pratt, Wise, and Zeckhauser (1979), who found that the variability of the prices at which a given product is sold by different stores is roughly proportional to the mean price of that product. The same pattern was observed for both frequently and infrequently purchased items. Overall, a ratio of 2:1 in the mean price of two products is associated with a ratio of 1.86:1 in the standard deviation of the respective quoted prices. If the effort that consumers exert to save each dollar on a purchase, for instance by a phone call, were independent of price, the dispersion of quoted prices should be about the same for all products. In contrast, the data of Pratt, Wise, and Zeckhauser (1979) are consistent with the hypothesis that consumers hardly exert more effort to save $15 on a $150 purchase than to save $5 on a $50 dollar purchase (Thaler, 1980). Many readers will recognize the temporary devaluation of money which facilitates extra spending and reduces the significance of small discounts in the context of a large expenditure, such as buying a house or a car. This paradoxical variation in the value of money is incompatible with the standard analysis of consumer behavior.

DISCUSSION

In this chapter, we have presented a series of demonstrations in which seemingly inconsequential changes in the formulation of choice problems caused significant shifts of preference. The inconsistencies were traced to the interaction of two sets of factors: variations in the framing of acts, contingencies, and outcomes, and the characteristic nonlinearities of values and decision weights. The demonstrated effects are large and systematic, although by no means universal. They occur when the outcomes concern the loss of human lives as well as in choices about money; they are not restricted to hypothetical questions and are not eliminated by monetary incentives.

Earlier, we compared the dependence of preferences on frames to the dependence of perceptual appearance on perspective. If while traveling in a mountain range you notice that the apparent relative height of mountain peaks varies with your vantage point, you will conclude that some impressions of relative height must be erroneous, even when you have no access to the correct answer. Similarly, one may discover that the relative attractiveness of options varies when the same decision problem is framed in different ways. Such a discovery will normally lead the decisionmaker to reconsider the original preferences, even when there is no simple way to resolve the inconsistency. The susceptibility to perspective effects is of special concern in the domain of decisionmaking because of the absence of objective standards, such as the true height of mountains.

The metaphor of changing perspective can be applied to other phenomena of choice, in addition to the framing effects with which we have been concerned here (Fischhoff, Slovic, and Lichtenstein, 1980). The problem of self-control is naturally construed in these terms. The story of Ulysses' request to be bound to the mast of the ship in anticipation of the irresistible temptation of the Sirens' call is often used as a paradigm case (Aimslie, 1975; Elster, 1979; Strotz, 1955-1956; Thaler and Shifrin, 1981). In this example of precommitment, an action taken in the present renders inoperative an anticipated future preference. An unusual feature of the problem of intertemporal conflict is that the agent who views a problem from a particular temporal perspective is also aware of the conflicting views that future perspectives will offer. In most other situations, decisionmakers are not normally aware of the potential effects of different decision frames on their preferences.

The perspective metaphor highlights four aspects of the psychology of choice. Individuals who face a decision problem and have a definite preference, first, might have a different preference in a different framing of the same problem; second, are normally unaware of alternative frames and of their potential effects on the relative attractiveness of options; third, would wish their preferences to be independent of frame; but, fourth, are often uncertain how to resolve detected inconsistencies (Slovic and Tversky, 1974). In some cases (such as problems 3 and 4 and perhaps problems 8 and 9) the advantage of one frame becomes evident once the competing frames are compared, but in other cases (problems 1 and 2 and problems 6 and 7) it is not obvious which preferences should be abandoned.

These observations do not imply that preference reversals or other errors of choice or judgment are necessarily irrational (Einhorn and Hogarth, 1981; Nisbett and Ross, 1980; Slovic, Fischhoff, and Lichtenstein, 1977; Tversky and Kahneman,

1974). Like other intellectual limitations, discussed by Simon (1955, 1956) under the heading of "bounded rationality," the practice of acting on the most readily available frame can sometimes be justified by reference to the mental effort required to explore alternative frames and avoid potential inconsistencies. However, we propose that the details of the phenomena described in this chapter are better explained by prospect theory and by an analysis of framing than by ad hoc appeals to the notion of cost of thinking.

The work described here has been concerned primarily with the descriptive question of how decisions are made, but the psychology of choice is also relevant to the normative question of how decisions ought to be made. In order to avoid the difficult problem of justifying values, the modern theory of rational choice has adopted the coherence of specific preferences as the sole criterion of rationality. This approach enjoins the decisionmaker to resolve inconsistencies but offers no guidance on how to do so. It implicitly assumes that the decisionmaker who carefully answers the question, "What do I really want?" will eventually achieve coherent preferences. However, the susceptibility of preferences to variations of framing raises doubt about the feasibility and adequacy of the coherence criterion.

Consistency is only one aspect of the lay notion of rational behavior. As noted by March (1978), the common conception of rationality also requires that preferences or utilities for particular outcomes should be predictive of the experiences of satisfaction or displeasure associated with their occurrence. Thus, a man could be judged irrational either because his preferences are contradictory or because his desires and aversions do not reflect his pleasures and pains. The predictive criterion of rationality can be applied to resolve inconsistent preferences and to improve the quality of decisions. A predictive orientation encourages the decisionmaker to focus on future experience and to ask "What will I feel then?" rather than "What do I want now?" The former question, when answered with care, can be the more useful guide in difficult decisions. In particular, predictive considerations may be applied to select the decision frame that best represents the hedonic experience of outcomes.

Further complexities arise in the normative analysis because the framing of an action sometimes affects the actual experience of its outcomes. For example, framing outcomes in terms of overall wealth or welfare rather than in terms of specific gains and losses may attentuate one's emotional response to an occasional loss. Similarly, the experience of a change for the worse may vary if the change is framed as an uncompensated loss or as a cost incurred to achieve some benefit. The framing of acts and outcomes can also reflect the acceptance or rejection of responsibility for particular consequences, and the deliberate manipulation of framing is commonly used as an instrument of self-control (Aimslie, 1975; Elster, 1979; Strotz, 1955-1956; Thaler and Shifrin, 1981). When framing influences the experience of consequences, the adoption of a decision frame is an ethically significant act.

REFERENCES

Ainslie, G. "Specious Reward: A Behavioral Theory of Impulsiveness and Impulse Control." Psychological Bulletin, 1975, 82, 463,496.

Allais, M. "Le Comportement de l'Homme Devant le Risque: Critique des Postulats et Axiomes de l'Ecole Americaine." Econometrica, 1953, 21, 503-546.

Einhorn, H.J. and Hogarth, R.M. "Behavioral Decision Theory: Processes of Judgment and Choice." Annual Review of Psychology, 1981, 32, 53-88.

Ellsberg, D. "Risk, Ambiguity, and the Savage Axioms." Quarterly Journal of Economics, 1961, 75, 643-669.

Elster, J. Ulysses and the Sirens: Studies in Rationality and Irrationality. London: Cambridge University Press, 1979.

Eraker, S.A. and Sox, H.C. "Assessment of Patients' Preferences for Therapeutic Outcomes." Medical Decision Making, in press.

Fellner, W. Probability and Profit: A Study of Economic Behavior Along Bayesian Lines. Homewood, Ill.: Irwin, 1965.

Fishburn, P.C. Utility Theory for Decision Making. New York: Wiley, 1970.

Fishburn, P.C. and Kochenberger, G.A. "Two-Piece von Neumann-Morgenstern Utility Functions." Decision Sciences, 1979, 10, 503-518.

Fischhoff, B., Slovic, P., and Lichtenstein, S. "Knowing What You Want: Measuring Labile Values." In T. Wallsten (Ed.), Cognitive Processes in Choice and Decision Behavior. Hillsdale, N.J.: Erlbaum, 1980.

Galanter, E. and Pinter, P. "Cross-Modality Matching of Money Against Other Continua." In H.R. Moskowitz and others (Eds.), Sensation and Measurement. Dordrecht: Reidel, 1974.

Grether, D.M. "Bayes Rule as a Descriptive Model: The Representativeness Heuristic." Quarterly Journal of Economics, 1980, 95, 537-557.

Grether, D.M. and Plott, C.R. "Economic Theory of Choice and the Preference Reversal Phenomenon." American Economic Review, 1979, 69, 623-638.

Hershey, J.C. and Schoemaker, P.J.H. "Risk Taking and Problem Context in the Domain of Losses: An Expected-Utility Analysis." Journal of Risk and Insurance, 1980, 47, 111-132.

Kahneman, D. and Tversky, A. "Prospect Theory: An Analysis of Decision Under Risk." Econometrica, 1979, 47, 263-291.

Kunreuther, H. and others. Disaster Insurance Protection: Public Policy Lessons. New York: Wiley, 1978.

Laughhunn, D.J., Payne, J.W., and Crum, R. "Managerial Risk Preferences for Below-Target Returns." Management Science, 1980, 26, 1238-1249.

Lichtenstein, S. and Slovic, P. "Reversals of Preference Between Bids and Choices in Gambling Decisions." Journal of Experimental Psychology, 1971, 98, 46-55.

Lichtenstein, S. and Slovic, P. "Response-Induced Reversals of Preference in Gambling: An Extended Replication in Las Vegas." Journal of Experimental Psychology, 1973, 101, 16-20.

Lieblich, I. and Lieblich, A. "Effects of Different Payoff Matrices on Arithmetical Estimation Tasks: An Attempt to Produce Rationality." Perceptual and Motor Skills, 1969, 29, 467-473.

MacCrimmon, K.R. and Larsson, S. "Utility Theory: Axioms Versus Paradoxes." In M. Allais and O. Hagen (Eds.), Expected-Utility Hypotheses and the Allais Paradox, Vol. 21. Doedrecht: Reidel, 1979.

March, J.G. "Bounded Rationality, Ambiguity, and the Engineering of Choice." Bell Journal of Economics, 1978, 9, 587-608.

McGlothlin, W.H. "Stability of Choices Among Uncertain Alternatives." American Journal of Psychology, 1956, 69, 604-615.

Nisbett, R.E. and Ross, L. Human Inference: Strategies and Shortcomings of Social Judgment. Englewood Cliffs, N.J.: Prentice-Hall, 1980.

Payne, J.W. and Braunstein, M.L. "Preferences Among Gambles with Equal Underlying Distributions." Journal of Experimental Psychology, 1971, 87, 13-18.

Payne, J.W., Laughhunn, D.J., and Crum, R. "Translation of Gambles and Aspiration-Level Effects in Risky Choice Behavior." Management Science, 1980, 26, 1039-1060.

Pratt, J.W., Wise, D., and Zeckhauser, R. "Price Differences in Almost Competitive Markets." Quarterly Journal of Economics, 1979, 93, 189-211.

Raiffa, H. Decision Analysis: Introductory Lectures on Choices Under Uncertainty. Reading, Mass.: Addison-Wesley, 1968.

Savage, L.J. The Foundations of Statistics. New York: Wiley, 1954.

Simon, H.A. "A Behavioral Model of Rational Choice." Quarterly Journal of Economics, 1955, 69, 99-118.

Simon, H.A. "Rational Choice and the Structure of the Environment." Psychological Review, 1956, 63, 129-138.

Slovic, P., Fischhoff, B., and Lichtenstein, S. "Behavioral Decision Theory." Annual Review of Psychology, 1977, 28, 1-39.

Slovic, P. and Lichtenstein, S. "Importance of Variance Preferences in Gambling and Decisions." Journal of Experimental Psychology, 1968, 78, 646-654.

Slovic, P. and Tversky, A. "Who Accepts Savage's Axiom?" Behavioral Science, 1974, 14, 368-373.

Strotz, R. "Myopia and Inconsistency in Dynamic Utility Maximization." Review of Economic Studies, 1955-1956, 23, 165-180.

Thaler, R. "Toward a Positive Theory of Consumer Choice." Journal of Economic Behavior and Organization, 1980, 1, 39-60.

Thaler, R. and Shifrin, H.M. "An Economic Theory of Self-Control." Journal of Political Economy, 1981, 89, 392-406.

Tversky, A. and Kahneman, D. "Judgment Under Uncertainty: Heuristics and Biases." Science, 1974, 185, 1124-1131.

von Neumann, J. and Morganstern, O. Theory of Games and Economic Behavior. Princeton, N.J.: Princeton University Press, 1947.

Rating The Risks: The Structure Of Expert And Lay Perceptions*

Paul Slovic, Baruch Fischhoff, and Sarah Lichtenstein

Decision Research, A Branch of Perceptronics, 1201 Oak Street, Eugene, OR 97401

INTRODUCTION

People respond to the hazards they perceive. If their perceptions are faulty, efforts at public and environmental protection are likely to be misdirected. In order to improve hazard management, a risk assessment industry has developed over the last decade which combines the efforts of physical, biological, and social scientists in an attempt to identify hazards and measure the frequency and magnitude of their consequences (1).

For some hazards extensive statistical data are readily available; for example, the frequency and severity of motor vehicle accidents are well documented. For other familiar activities, such as the use of alcohol and tobacco, the hazardous effects are less readily discernible and their assessment requires complex epidemiological and experimental studies. But in either case, the hard facts go only so far and then human judgment is needed to interpret the findings and determine their relevance for the future.

Other hazards, such as those associated with recombinant DNA research or nuclear power, are so new that risk assessment must be based on theoretical analyses such as fault trees (see Fig. 10.1), rather than on direct experience. While sophisticated, these analyses, too, include a large component of human judgment. Someone, relying on educated intuition, must determine the structure of the problem, the consequences to be considered, and the importance of the various branches of the fault tree.

Once performed, the analyses must be communicated to the various people, including industrialists, environmentalists, regulators, legislators, and voters, who are actually responsible

---

* Reprinted by permission from Environment, vol. 21 (April 1979).

NATO ASI Series, Vol. G4
Environmental Impact Assessment, Technology Assessment, and Risk Analysis. Edited by V. T. Covello et al.

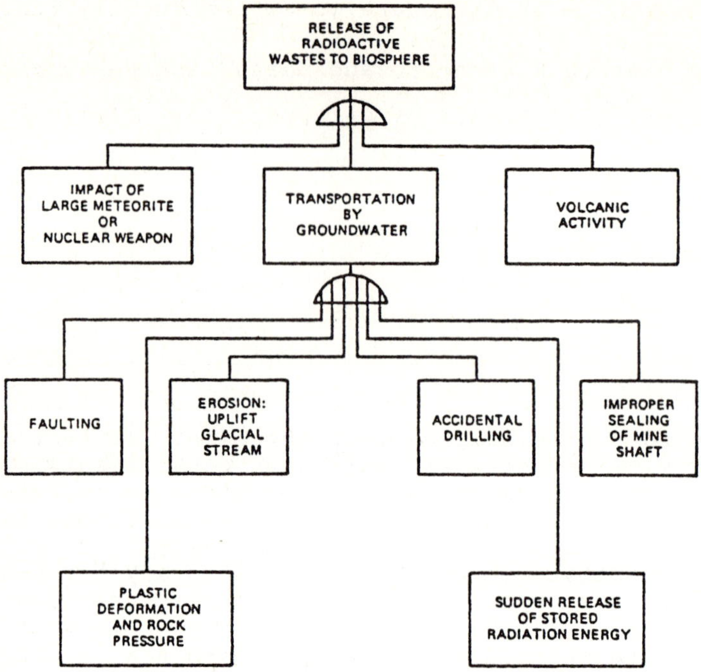

Figure 10.1. Illustration of a fault tree. Fault trees are used most often to characterize hazards for which direct experience is not available. The tree shown here indicates the various ways in which radioactive material might accidentally be released from nuclear wastes buried within a salt deposit. To read this tree, start with the bottom row of possible initiating events, each of which can lead to the transportation of radioactivity by groundwater. This transport can in turn release radioactivity to the biosphere. As indicated by the second level of boxes, release of radioactivity can also be produced directly (without the help of groundwater) through the impact of a large meteorite, a nuclear weapon, or a volcanic eruption. Fault trees may be used to map all relevant possibilities and to determine the probability of the final outcome. To accomplish this latter goal, the probabilities of all component stages, as well as their logical connections, must be completely specified. (Source: P. E. McGrath, "Radioactive Waste Management," Report EURFNR 1204, Karlsruhe, Germany, 1974.) Reprinted by permission.

for dealing with the hazards. If these people do not see, understand, or believe these risk statistics, then distrust, conflict, and ineffective hazard management may result.

JUDGMENTAL BIASES

When lay people are asked to evaluate risks, they seldom have statistical evidence on hand. In most cases they must rely on inferences based on what they remember hearing or observing about the risk in question. Recent psychological research has identified a number of general inferential rules that people seem to use in such situations (2). These judgmental rules, known technically as heuristics, are employed to reduce difficult mental tasks to simpler ones. Although valid in some circumstances, in others they can lead to large and persistent biases with serious implications for risk assessment.

Availability

One heuristic that has special relevance for risk perception is known as "availability" (3). People who use this heuristic judge an event as likely or frequent if instances of it are easy to imagine or recall. Frequently occurring events are generally easier to imagine and recall than rare events. Thus, availability is often an appropriate cue. Availability, however, is also affected by numerous factors unrelated to frequency of occurrence. For example, a recent disaster or a vivid film such as The China Syndrome may seriously distort risk judgments.

Availability-induced errors are illustrated by several recent studies in which we asked college students and members of the League of Women Voters to judge the frequency of various causes of death, such as smallpox, tornadoes, and heart disease (4). In one study, these people were told the annual death toll (50,000) for motor vehicle accidents in the United States; they were then asked to estimate the frequency of 40 other causes of death. In another study, participants were given two causes of death and asked to judge which of the two is more frequent. Both studies showed people's judgments to be moderately accurate in a global sense; that is, people usually knew which were the most and least frequent lethal events. Within this global picture, however, there was evidence that people made serious misjudgments, many of which seemed to reflect availability bias.

Figure 10.2 compares the judged number of deaths per year with the actual number according to public health statistics. If the frequency judgments were accurate, they would equal the actual death rates, and all data points would fall on the straight line making a 45-degree angle with the axes of the graph. In fact, the points are scattered about a curved line that sometimes lies above and sometimes below the line of accurate judgment. In general, rare causes of death were overestimated and common causes of death were underestimated. As a result, while the actual death toll varied over a range of one million, average frequency judgments varied over a range of only a thousand.

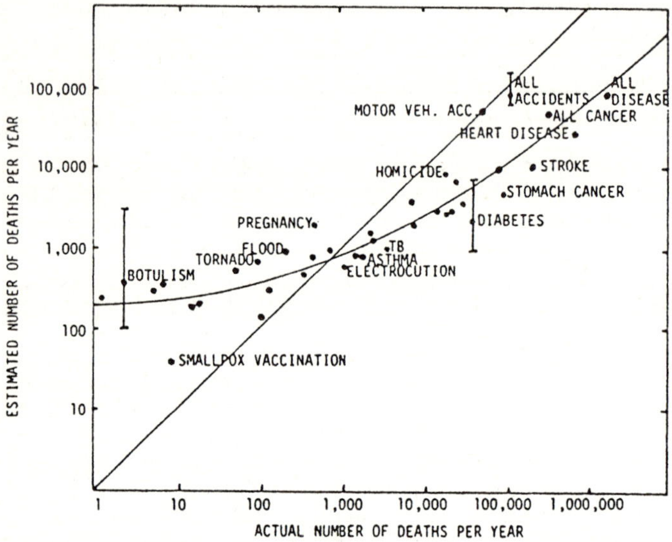

Figure 10.2. Relationship between judged frequency and the actual number of deaths per year for 41 causes of death. If judged and actual frequencies were equal, the data would fall on the straight line. The points, and the curved line fitted to them, represent the averaged responses of a large number of lay people. While people were approximately accurate, their judgments were systematically distorted. As described in the text, both the compression of the scale and the scatter of the results indicate this. To give an idea of the degree of agreement among subjects, vertical bars are drawn to depict the 25th and 75th percentile of individual judgment for botulism, diabetes, and all accidents. Fifty percent of all judgments fall between these limits. The range of responses for the other 37 causes of death was similar. From Lichtenstein et al.[4] Copyright 1978 by the American Psychological Association. Reprinted by permission.

In addition to this general bias, many important specific biases were evident. For example, accidents were judged to cause as many deaths as diseases, whereas diseases actually take about 15 times as many lives. Homicides were incorrectly judged to be more frequent than diabetes and stomach cancer.

Homicides were also judged to be about as frequent as stroke, although the latter actually claims about 11 times as many lives. Frequencies of death from botulism, tornadoes, and pregnancy (including childbirth and abortion) were also greatly overestimated.

Table 10.1 lists the lethal events whose frequencies were most poorly judged in our studies. In keeping with availability considerations, overestimated items were dramatic and sensational whereas underestimated items tended to be unspectacular events which claim one victim at a time and are common in a non-fatal form.

In the public arena the availability heuristic may have many effects. For example, the biasing effects of memorability and imaginability may pose a barrier to open, objective discussions of risk. Consider an engineer demonstrating the safety of subterranean nuclear waste disposal by pointing out the improbability of each branch of the fault tree in Fig. 10.1. Rather than reassuring the audience, the presentation might lead individuals to feel that "I didn't realize there were so many things that could go wrong." The very discussion of any low-probability hazard may increase the judged probability of that hazard regardless of what the evidence indicates.

Table 10.1 Bias in Judged Frequency of Death

Most Overestimated	Most Underestimated
All accidents	Smallpox vaccination
Motor vehicle accidents	Diabetes
Pregnancy, childbirth, and abortion	Stomach cancer
	Lightning
Tornadoes	Stroke
Flood	Tuberculosis
Botulism	Asthma
All cancer	Emphysema
Fire and flames	
Venomous bite or sting	
Homicide	

In some situations, failure to appreciate the limits of "available" data may lull people into complacency. For example, we asked people to evaluate the completeness of a fault tree showing the problems that could cause a car not to start when the ignition key was turned (5). Respondents' judgments of completeness were about the same when looking at the full tree as when looking at a tree in which half of the causes of starting failure were deleted. In keeping with the availability heuristic, what was out of sight was also out of mind.

Overconfidence

A particularly pernicious aspect of heuristics is that people are typically very confident about judgments based on them. For example, in a follow-up to the study on causes of death, participants were asked to indicate the odds that they were correct in their judgment about which of two lethal events was more frequent (6). Odds of 100:1 or greater were given often (25 percent of the time). However, about one out of every eight answers associated with such extreme confidence was wrong (fewer than one in 100 would have been wrong if the odds had been appropriate). About 30 percent of the judges gave odds greater than 50:1 to the incorrect assertion that homicides are more frequent than suicides. The psychological basis for this unwarranted certainty seems to be people's insensitivity to the tenuousness of the assumptions upon which their judgments are based (in this case, the validity of the availability heuristic). Such overconfidence is dangerous. It indicates that we often do not realize how little we know and how much additional information we need about the various problems and risks we face.

Overconfidence manifests itself in other ways as well. A typical task in estimating failure rates or other uncertain quantities is to set upper and lower bounds so that there is a 98 percent chance that the true value lies between them. Experiments with diverse groups of people making many different kinds of judgments have shown that, rather than two percent of true values falling outside the 98 percent confidence bounds, 20 percent to 50 percent do so (7). People think that they can estimate such values with much greater precision than is actually the case.

Unfortunately, experts seem as prone to overconfidence as lay people. When the fault tree study described above was repeated with a group of professional automobile mechanics, they, too, were insensitive to how much had been deleted from the tree. Hynes and Vanmarcke (8) asked seven "internationally known" geotechnical engineers to predict the height of an embankment that would cause a clay foundation to fail and to specify confidence bounds around this estimate that were wide enough to have a 50 percent chance of enclosing the true failure height. None of the bounds specified by these experts actually did enclose the true failure height. The multi-million dollar Reactor Safety Study (the "Rasmussen Report") (9), in assessing the probability of a core meltdown in a nuclear reactor, used a procedure for setting confidence bounds that has been found in experiments to produce a high degree of overconfidence. Related problems led a review committee, chaired by H. W. Lewis of the University of California, Santa Barbara, to conclude that the Reactor Safety Study greatly overestimated the precision with which the probability of a core meltdown could be assessed (10).

Another case in point is the 1976 collapse of the Teton Dam. The Committee on Government Operations has attributed this disaster to the unwarranted confidence of engineers who were absolutely certain they had solved the many serious problems that arose during construction (11). Indeed, in routine practice, failure probabilities are not even calculated for new dams even though about one in 300 fails when the reservoir is first filled. Further anecdotal evidence of overconfidence may be found in many other technical risk assessments. Some common ways in which experts may overlook or misjudge pathways to disaster include the following:

- Failure to consider the ways in which human errors can affect technological systems--Due to inadequate training and control room design, operators at Three Mile Island repeatedly misdiagnosed the problems of the reactor and took inappropriate actions.

- Overconfidence in current scientific knowledge--Use of DDT came into widespread and uncontrolled use before scientists had even considered the possibility of the side effects that today make it look like a mixed and irreversible blessing.

- Insensitivity to how a technological system functions as a whole--Though the respiratory risk of fossil-fueled power plants has been recognized for some time, the related effects of acid rains on ecosystems were largely missed until very recently.

- Failure to anticipate human response to safety measures--The partial protection offered by dams and levees gives people a false sense of security and promotes development of the flood plain. When a rare flood does exceed the capacity of the dam, the damage may be considerably greater than if the flood plain had been unprotected. Similarly, "better" highways, while decreasing the death toll per vehicle mile, may increase the total number of deaths because they increase the number of miles driven.

Desire for Certainty

Every technology is a gamble of sorts and, like other gambles, its attractiveness depends on the probability and size of its possible gains and losses. Both scientific experiments and casual observation show that people have difficulty thinking about and resolving the risk/benefit conflicts even in simple gambles. One way to reduce the anxiety generated by confronting uncertainty is to deny that uncertainty. The denial resulting from this anxiety-reducing search for certainty thus represents an additional source of over-confidence. This type of denial is illustrated by the case of people faced with natural hazards, who often view their world as either perfectly safe or as predictable enough to preclude worry. Thus, some flood victims interviewed by Kates (12) flatly denied that floods could ever recur in their areas. Some thought (incorrectly) that new dams and reservoirs in the area would contain all potential floods, while others attributed previous floods to freak combinations of circumstances unlikely to recur. Denial, of course, has its limits. Many people feel that they cannot ignore the risks of nuclear power. For these people, the search for certainty is best satisfied by outlawing the risk.

Scientists and policymakers who point out the gambles involved in societal decisions are often resented for the anxiety they provoke. Borch (13) noted how annoyed corporate managers get with consultants who give them the probabilities of possible events instead of telling them exactly what will happen. Just before a blue-ribbon panel of scientists reported that they were 95 percent certain that cyclamates do not cause cancer, Food and Drug Administration Commissioner, Alexander Schmidt, said, "I'm looking for a clean bill of health, not a wishy-washy, iffy answer on cyclamates" (14). Senator Edmund Muskie has called for "one-armed" scientists who do not respond "on the one hand, the evidence is so, but on the other hand..." when asked about the health effects of pollutants (15). As Gori warns in Chapter 11, such demands may tempt scientists to issue "certain" answers which, however convenient for regulators, are unsupportable by science.

The search for certainty is legitimate if it is done consciously, if the remaining uncertainties are acknowledged rather than ignored, and if people realize the costs. If a very high level of certainty is sought, those costs are likely to be high. Eliminating the uncertainty may mean eliminating the technology and foregoing its benefits. Often some risk is inevitable. Efforts to eliminate it may only alter its form. We must choose, for example, between the vicissitudes of nature on an unprotected flood plain and the less probable, but potentially more catastrophic, hazards associated with dams and levees (Coates, Chapter 3; Okrent, Chapter 13).

ANALYZING JUDGMENTS OF RISK

In order to be of assistance in the hazard management process, a theory of perceived risk must explain people's extreme aversion to some hazards, their indifference to others, and the discrepancies between these reactions and experts' recommendations. Why, for example, do some communities react vigorously against locating a liquid natural gas terminal in their vicinity despite the assurances of experts that it is safe? Why do other communities situated on flood plains and earthquake faults or below great dams show little concern for the experts' warnings? Such behavior is doubtless related to how people assess the quantitative characteristics of the hazards they face. The preceding discussion of judgmental processes was designed to illuminate this aspect of perceived risk. The studies reported below broaden the discussion to include more qualitative components of perceived risk. They ask, when people judge the risk inherent in a technology, are they referring only to the (possibly misjudged) number of people it could kill or also to other, more qualitative features of the risk it entails?

Quantifying Perceived Risk

In our first studies, we asked four different groups of people to rate 30 different activities and technologies according to the present risk of death from each (16). Three of these groups were from Eugene, Oregon; they included 30 college students, 40 members of the League of Women Voters (LOWV), and 25 business and professional members of the "Active Club." The fourth group was composed of 15 persons selected nation-wide for their professional involvement in risk assessment. This "expert" group included a geographer, an environmental policy analyst, an economist, a lawyer, a biologist, a biochemist, and a government regulator of hazardous materials.

All these people were asked, for each of the 30 items, "to consider the risk of dying (across all U.S. society as a whole) as a consequence of this activity or technology." In order to make the evaluation task easier, each activity appeared on a 3" x 5" card. Respondents were told first to study the items individually, thinking of all the possible ways someone might die from each (e.g., fatalities from non-nuclear electricity were to include deaths resulting from the mining of coal and other energy production activities as well as electrocution; motor vehicle fatalities were to include collisions with bicycles and pedestrians). Next, they were to order the items from least to most risky and then assign numerical risk values by giving a rating of 10 to the least risky item and making the other ratings accordingly. They were also given additional suggestions, clarifications, and encouragement to do as accurate a job as possible. For example, they were told "A rating of 12 indicates that that item is 1.2 times as risky as the least risky item (i.e., 20 percent more risky). A rating of 200 means that the item is 20 times as risky as the least risky item, to which you assigned a 10 . . ." They were urged to cross-check and adjust their numbers until they believed they were right.

Table 10.2 shows how the various groups ranked the relative riskiness of these 30 activities and technologies. There were many similarities between the three groups of lay persons. For example, each group believed that motorcycles, other motor vehicles, and handguns were highly risky, and that vaccinations, home appliances, power mowers, and football were relatively safe. However, there were strong differences as well. Active Club members viewed pesticides and spray cans as relatively much safer than did the other groups. Nuclear power was rated as highest in risk by the LOWV and student groups, but only eighth by the Active Club. The students viewed contraceptives and food preservatives as riskier and swimming and mountain climbing as safer than did the other lay groups. Experts' judgments of risk differed markedly from the judgments of lay persons. The experts viewed electric power, surgery, swimming, and X-rays as more risky than the other groups, and they judged nuclear power, police work, and mountain climbing to be much less risky.

What Determines Risk Perception?

What do people mean when they say that a particular technology is quite risky? A series of additional studies was conducted to answer this question.

<u>Perceived Risk Compared To Frequency Of Death</u>: When people judge risk, as in the previous study, are they simply estimating frequency of death? To answer this question, we collected the best available technical estimates of the annual number of deaths from each of the 30 activities included in our study. For some cases, such as commercial aviation and handguns, there is good statistical evidence based on counts of known victims. For other cases, such as the lethal potential of nuclear or fossil-fuel power plants, available estimates are based on uncertain inferences about incompletely understood processes. For still others, such as food coloring, we could find no estimates of annual fatalities.

Table 10.2. Ordering of Perceived Risk for 30 Activities and Technologies[a]

	Group 1 LOWV	Group 2 Coll. Stud.	Group 3 Active Club	Group 4 Experts
Nuclear power	1	1	8	20
Motor vehicles	2	5	3	1
Handguns	3	2	1	4
Smoking	4	3	4	2
Motorcycles	5	6	2	6
Alcoholic beverages	6	7	5	3
General (private) aviation	7	15	11	12
Police work	8	8	7	17
Pesticides	9	4	15	8
Surgery	10	11	9	5
Fire fighting	11	10	6	18
Large construction	12	14	13	13
Hunting	13	18	10	23
Spray cans	14	13	23	26
Mountain climbing	15	22	12	29
Bicycles	16	24	14	15
Commercial aviation	17	16	18	16
Electric power	18	19	19	9
Swimming	19	30	17	10
Contraceptives	20	9	22	11
Skiing	21	25	16	30
X-rays	22	17	24	7
High school & college football	23	26	21	27
Railroads	24	23	20	19
Food preservatives	25	12	28	14
Food coloring	26	20	30	21
Power mowers	27	28	25	28
Prescription antibiotics	28	21	26	24
Home appliances	29	27	27	22
Vaccinations	30	29	29	25

[a]The ordering is based on the geometric mean risk ratings within each group. Rank 1 represents the most risky activity or technology.

For the 25 cases for which we found technical estimates for annual frequency of death, we compared these estimates with perceived risk. Results for experts and the LOWV sample are shown in Fig. 10.3 (the results for the other lay groups were quite similar to those from the LOWV sample). The experts' mean judgments were so closely related to the statistical or calculated frequencies that it seems reasonable to conclude that they viewed the risk of an activity or technology as synonymous with its annual fatalities. The risk judgments of lay people, however, showed only a moderate relationship to the annual frequencies of death (17), raising the possibility that, for them, risk may not be synonymous with fatalities. In particular, the perceived risk from nuclear power was disproportionately high compared to its estimated number of fatalities.

Lay Fatality Estimates: Perhaps lay people based their risk judgments on annual fatalities but estimated their numbers inaccurately. To test this hypothesis, we asked additional groups of students and LOWV members "to estimate how many people are likely to die in the U.S. in the next year (if the next year is an average year) as a consequence of these 30 activities and technologies." We asked our student and LOWV samples to consider all sources of death associated with these activities.

The mean fatality estimates of LOWV members and students are shown in columns two and three of Table 10.3. If lay people really equate risk with annual fatalities, one would expect that their own estimates of annual fatalities, no matter how inaccurate, would be very similar to their judgments of risk. But this was not so. There was a moderate agreement between their annual fatality estimates and their risk judgments, but there were important exceptions. Most notably, nuclear power had the lowest fatality estimate and the highest perceived risk for both LOWV members and students. Overall, lay people's perceptions were no more closely related to their own fatality estimates than they were to the technical estimates (Fig. 10.3).

These results lead us to reject the idea that lay people wanted to equate risk with annual fatality estimates but were inaccurate in doing so. Instead, we are led to believe that lay people incorporate other considerations besides annual fatalities into their concept of risk.

Some other aspects of lay people's fatality estimates are of interest. One is that they were moderately accurate. The relationship between the LOWV members' fatality estimates and the best technical estimates is plotted in Fig. 10.4. The lay estimates showed the same overestimation of those items that cause few fatalities and underestimation of those resulting in

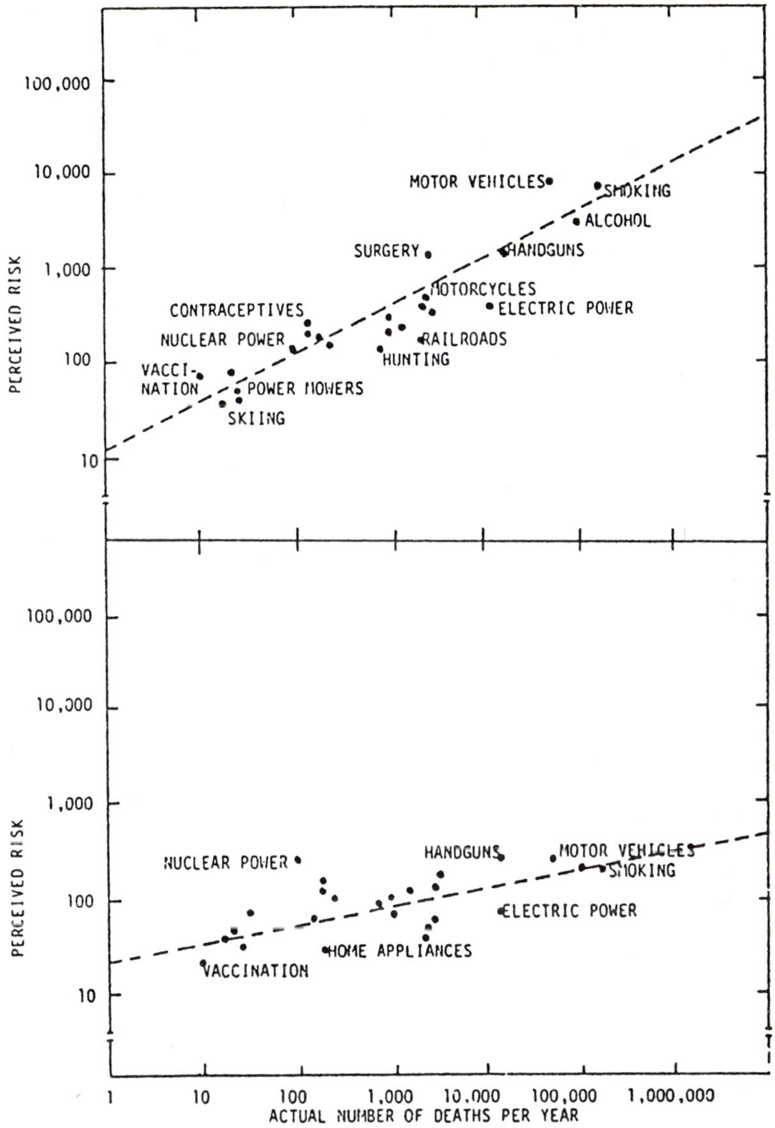

Figure 10.3. Judgments of perceived risk for experts (top) and lay people (bottom) plotted against the best technical estimates of annual fatalities for 25 technologies and activities. Each point represents the average responses of the participants. The dashed lines are the straight lines that best fit the points. The experts' risk judgments are seen to be more closely associated with annual fatality rates than are the lay judgments.

Table 10.3. Fatality Estimates and Disaster Multipliers for 30 Activities and Technologies

Activity or Technology	Technical Fatality Estimates	Geometric Mean Fatality Estimates Average Year		Geometric Mean Multiplier Disastrous Year	
		LOWV	Students	LOWV	Students
Smoking	150,000	6,900	2,400	1.9	2.0
Alcoholic beverages	100,000	12,000	2,600	1.9	1.4
Motor vehicles	50,000	28,000	10,500	1.6	1.8
Handguns	17,000	3,000	1,900	2.6	2.0
Electric power	14,000	660	500	1.9	2.4
Motorcycles	3,000	1,600	1,600	1.8	1.6
Swimming	3,000	930	370	1.6	1.7
Surgery	2,800	2,500	900	1.5	1.6
X-rays	2,300	90	40	2.7	1.6
Railroads	1,950	190	210	3.2	1.6
General (private) aviation	1,300	550	650	2.8	2.0
Large construction	1,000	400	370	2.1	1.4
Bicycles	1,000	910	420	1.8	1.4
Hunting	800	380	410	1.8	1.7
Home appliances	200	200	240	1.6	1.3
Fire fighting	195	220	390	2.3	2.2
Police work	160	460	390	2.1	1.9
Contraceptives	150	180	120	2.1	1.4
Commercial aviation	130	280	650	3.0	1.8
Nuclear power	100[a]	20	27	107.1	87.6
Mountain climbing	30	50	70	1.9	1.4
Power mowers	24	40	33	1.6	1.3
High school & college football	23	39	40	1.9	1.4
Skiing	18	55	72	1.9	1.6
Vaccinations	10	65	52	2.1	1.6
Food coloring	--[b]	38	33	3.5	1.4
Food preservatives	--[b]	61	63	3.9	1.7
Pesticides	--[b]	140	84	9.3	2.4
Prescription antibiotics	--[b]	160	290	2.3	1.6
Spray cans	--[b]	56	38	3.7	2.4

[a] Technical estimates for nuclear power were found to range between 16 and 600 annual fatalities. The geometric mean of these estimates was used here.

[b] Estimates were unavailable.

the most fatalities that was apparent in Fig. 10.2 for a different collection of hazards. Also as in Fig. 10.2, the moderate overall relationship between lay and technical estimates was marred by specific biases (e.g., the underestimation of fatalities associated with railroads, X-rays, electric power, and smoking).

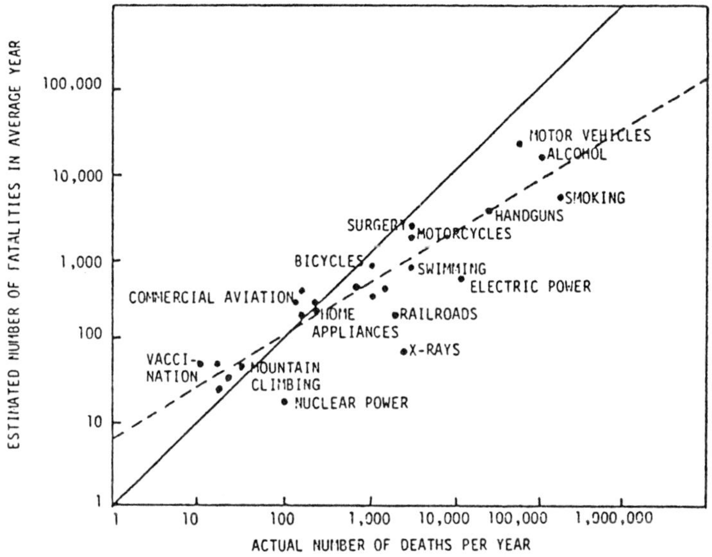

Figure 10.4. Lay people's judgments of the number of fatalities in an average year plotted against the best estimates of annual fatalities for 25 activities and technologies. The solid line indicates accurate judgment, while the dashed line best fits the data points. These results have much the same character as those shown in Figure 10.2 for a different collection of hazards. Low frequencies were overestimated and high ones were underestimated. The overall relationship is marred by specific biases (e.g., the underestimation of fatalities associated with railroads, X-rays, electric power, and smoking).

Disaster Potential: The fact that the LOWV members and students assigned very high risk values to nuclear power along with very low estimates of its annual fatality rates is an apparent contradiction. One possible explanation is that LOWV members expected nuclear power to have a low death rate in an average year but considered it to be a high-risk technology because of its potential for disaster.

In order to understand the role played by expectations of disaster in determining lay people's risk judgments, we asked these same respondents to give for each activity and technology a number indicating how many times more deaths would occur if next year were "particularly disastrous" rather than average. The averages of these multipliers are shown in Table 10.3. For most activities, people saw little potential for disaster. For the LOWV sample all but five of the multipliers were less than three, and for the student sample all but six were less than two. The striking exception in both cases is nuclear power, with a geometric mean disaster multiplier in the neighborhood of 100.

For any individual an estimate of the expected number of fatalities in a disastrous year could be obtained by applying the disaster multiplier to the estimated fatalities for an average year. When this was done for nuclear power, almost 40 percent of the respondents expected more than 10,000 fatalities if next year were a disastrous year. More than 25 percent expected 100,000 or more fatalities. These extreme estimates can be contrasted with the Reactor Safety Study's conclusion that the maximum credible nuclear accident, coincident with the most unfavorable combination of weather and population density, would cause only 3,300 prompt fatalities (9). Furthermore, that study estimated the odds against an accident of this magnitude occurring during the next year (assuming 100 operating reactors) to be about 2,000,000:1.

Apparently, disaster potential explains much or all of the discrepancy between the perceived risk and frequency of death values for nuclear power. Yet, because disaster plays only a small role in most of the 30 activities and technologies we have studied, it provides only a partial explanation of the perceived risk data.

Qualitative Characteristics: Are there other determinants of risk perceptions besides frequency estimates? We asked experts, students, LOWV members, and Active Club members to rate the 30 technologies and activities on nine qualitative characteristics that have been hypothesized to be important (18). These ratings scales are described in Table 10.4.

Mean ratings were quite similar for all four groups. Particularly interesting was the characterization of nuclear power, which had the dubious distinction of scoring at or near the extreme on all of the undesirable characteristics. Its risks were seen as involuntary, delayed, unknown, uncontrollable, unfamiliar, catastrophic, dread, and fatal. This contrasted sharply with the characterizations of non-nuclear electric power and another radiation technology, X-rays. Electric power and X-rays were both judged more voluntary, less certain to be fatal, less catastrophic, less dreaded, more familiar, and less risky than nuclear power (see Fig. 10.5).

Table 10.4. Risk characteristics rated by LOWV members, Active Club members, students, and experts

---

Voluntariness of risk
Do people face this risk voluntarily? If some of the risks are voluntarily undertaken and some are not, mark an appropriate spot towards the center of the scale.

risk assumed voluntarily	1	2	3	4	5	6	7	*risk assumed involuntarily*

Immediacy of effect
To what extent is the risk of death immediate--or is death likely to occur at some later time?

effect immediate	1	2	3	4	5	6	7	*effect delayed*

Knowledge about risk
To what extent are the risks known precisely by the persons who are exposed to those risks?

risk level known precisely	1	2	3	4	5	6	7	*risk level not known*

To what extent are the risks known to science?

risk level known precisely	1	2	3	4	5	6	7	*risk level not known*

Control over risk
If you are exposed to the risk, to what extent can you, by personal skill or diligence, avoid death?

personal risk can't be controlled	1	2	3	4	5	6	7	*personal risk can be controlled*

Newness
Is this risk new and novel or old and familiar?

new	1	2	3	4	5	6	7	*old*

Chronic/Catastrophic
Is this a risk that people have learned to live with and can think about reasonably calmly, or is it one that people have great dread for--on the level of a gut reaction?

common	1	2	3	4	5	6	7	*dread*

Severity of consequences
When the risk from the activity is realized in the form of a mishap or illness, how likely is it that the consequence will be fatal?

certain not to be fatal	1	2	3	4	5	6	7	*certain to be fatal*

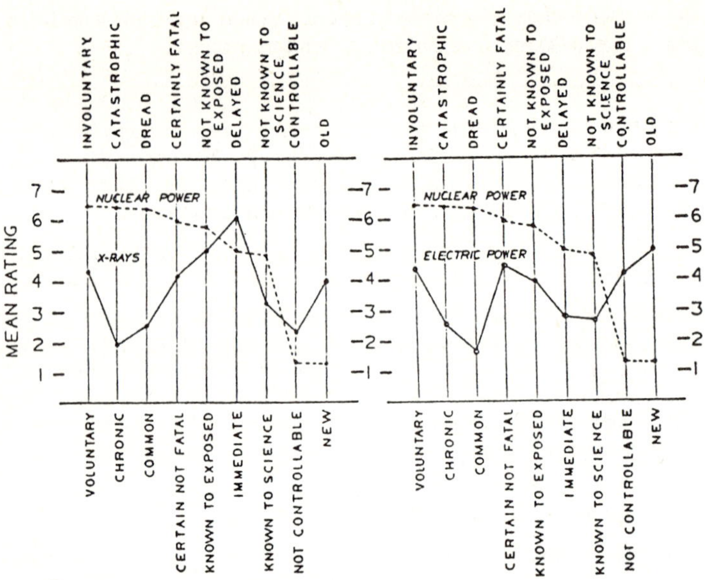

Figure 10.5. Qualitative characteristics of perceived risk for nuclear power and related technologies. In the right-hand diagram, risk profiles for nuclear power and non-nuclear electric power are compared. In the left-hand diagram, nuclear power and X-rays are compared. Each profile consists of nine dimensions rated on a seven-point scale. The instructions that elicited these responses are reproduced in Table 10.4. The perceived qualities of nuclear power are dramatically different from the comparison technologies. The source of this data is the LOWV sample studied by Fischhoff et al.[16]

Across all 30 hazards, ratings of dread and of the likelihood of a mishap's being fatal were closely related to lay judgments of risk. In fact, the risk judgments of the LOWV and student groups could be predicted almost perfectly from ratings of dread and lethality and the subjective fatality estimates for normal and disastrous years (19). Experts' judgments of risk were not related to any of the nine risk characteristics (20).

Many pairs of risk characteristics tended to be correlated with each other across the 30 activities and technologies. For example, risks faced voluntarily were typically judged to be well known and controllable. These interrelations were sufficiently high to suggest that all the ratings could be explained in terms of a few basic dimensions of risk. In order to identify such dimensions, we conducted a factor analysis of the correlations from each group (principal components analysis with varimax rotation to simple structure). We found that the nine characteristics could be represented by two underlying factors which appeared to be the same for each group. Figure 10.6 illustrates the factor scores for each hazard within the common space. Hazards at the high end of the vertical dimension or factor (e.g., food coloring, pesticides) tended to be new, unknown, involuntary, and delayed in their effects. Hazards at the other extreme of this factor (e.g., mountain climbing, swimming) had the opposite characteristics. High (right-hand) scores on the horizontal factor (e.g., nuclear power, commercial aviation) were associated with events whose consequences were seen as certain to be fatal, often for large numbers of people, should something go wrong. Hazards low on this factor (e.g., power mowers, football) were seen as causing injuries, rather than fatalities, to single individuals. We have labeled the vertical factor as "Unknown Risk" and the horizontal factor as "Dread Risk." In sum, even though the four groups had somewhat different perceptions of the riskiness of the various hazards (Table 10.2), they tended to characterize these hazards similarly.

Judged Seriousness of Death: In a further attempt to improve our understanding of perceived risk, we examined the hypothesis that some hazards are feared more than others because the deaths they produce are much "worse" than deaths from other activities. We thought for example, that deaths from risks imposed involuntarily, from risks not under one's control, or from hazards that are particularly dreaded might be given greater weight in determining people's perceptions of risk.

When we asked students and LOWV members to judge the relative "seriousness" of a death from each of the 30 activities and technologies, however, the differences were slight. The most serious forms of death (from nuclear power and handguns) were judged to be only about two to four times worse than the least serious forms of death (from alcoholic beverages and smoking). Furthermore, across all 30 activities, judged seriousness of death was not closely related to perceived risk of death.

Reconciling Divergent Opinions

Our data show that experts and lay people have quite different perceptions about how risky certain technologies are. It would be comforting to believe that these divergent risk judgments would be responsive to new evidence so that, as information accumulates, perceptions would converge towards one "appropriate" view. Unfortunately, this is not likely to be the case. As noted earlier in our discussion of availability, risk

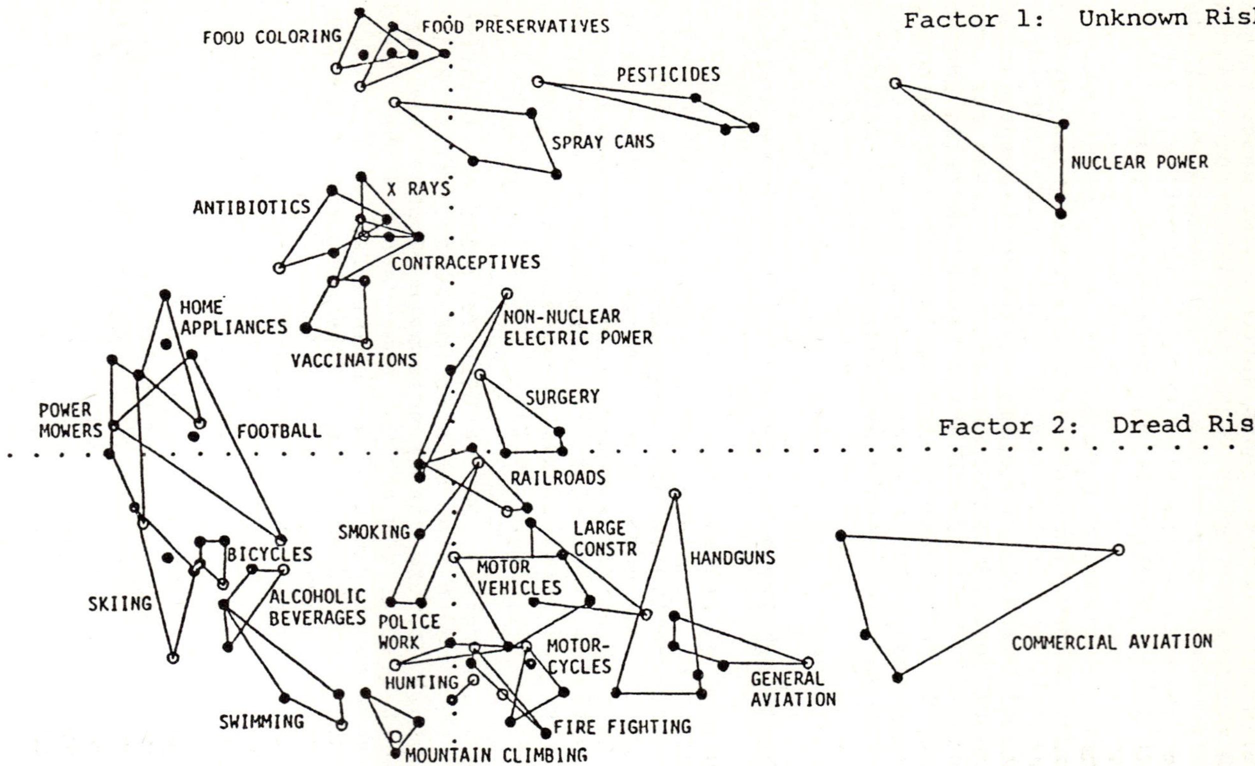

Figure 10.6. Location of 30 hazards within the two-factor space obtained from LOWV, student, Active Club and expert groups. Connected lines join or enclose the loci of four group points for each hazard. Open circles represent data from the expert group. Unattached points represent groups that fall within the triangle created by the other three groups.

perception is derived in part from fundamental modes of thought that lead people to rely on fallible indicators such as memorability and imaginability.

Furthermore, a great deal of research indicates that people's beliefs change slowly and are extraordinarily persistent in the face of contrary evidence (21). Once formed, initial impressions tend to structure the way that subsequent evidence is interpreted. New evidence appears reliable and informative if it is consistent with one's initial belief; contrary evidence is dismissed as unreliable, erroneous, or unrepresentative. Thus, depending on one's predispositions, intense effort to reduce a hazard may be interpreted to mean either that the risks are great or that the technologists are responsive to the public's concerns. Likewise, opponents of a technology may view minor mishaps as near catastrophes and dismiss the contrary opinions of experts as biased by vested interests.

From a statistical standpoint, convincing people that the catastrophe they fear is extremely unlikely is difficult under the best conditions. Any mishap could be seen as proof of high risk, whereas demonstrating safety would require a massive amount of evidence (22). Nelkin's case history of a nuclear siting controversy (23) provides a good example of the inability of technical arguments to change opinions. In that debate each side capitalized on technical ambiguities in ways that reinforced its own position.

THE FALLIBILITY OF JUDGMENT

Our examination of risk perception leads us to the following conclusions:

(1) Cognitive limitations, coupled with the anxieties generated by facing life as a gamble, cause uncertainty to be denied, risks to be distorted, and statements of fact to be believed with unwarranted confidence.

(2) Perceived risk is influenced (and sometimes biased) by the imaginability and memorability of the hazard. People may, therefore, not have valid perceptions even for familiar risks.

(3) Our experts' risk perceptions correspond closely to statistical frequencies of death. Lay people's risk perceptions were based in part upon frequencies of death, but there were some striking discrepancies. It appears that for lay people, the concept of risk includes qualitative aspects such as dread and the likelihood of a mishap's being fatal. Lay people's risk perceptions were also affected by catastrophic potential.

Disagreements about risk should not be expected to evaporate in the presence of "evidence." Definitive evidence, particularly about rare hazards, is difficult to obtain. Weaker information is likely to be interpreted in a way that reinforces existing beliefs.

The significance of these results hinges upon one's acceptance of our assumption that subjective judgments are central to the hazard management process. Our conclusions mean little if one can assume that there are analytical tools which can be used to assess most risks in a mechanical fashion and that all decisionmakers have perfect information and the know-how to use it properly. These results gain in importance to the extent that one believes, as we do, that expertise involves a large component of judgment, that the facts are not all in (or obtainable) regarding many important hazards, that people are often poorly informed or misinformed, and that they respond not just to numbers but also to qualitative aspects of hazards.

Whatever role judgment plays, its products should be treated with caution. Research not only demonstrates that judgment is fallible, but it shows that the degree of fallibility is often surprisingly great and that faulty beliefs may be held with great confidence.

When it can be shown that even well-informed lay people have difficulty judging risks accurately, it is tempting to conclude that the public should be removed from the hazard-management process. The political ramifications of such a transfer of power to a technical elite are obvious. Indeed, it seems doubtful that such a massive disenfranchisement is feasible in any democratic society.

Furthermore, this transfer of decisionmaking would seem to be misguided. For one thing, we have no assurance that experts' judgments are immune to biases once they are forced to go beyond their precise knowledge and rely upon their judgment. Although judgmental biases have most often been demonstrated with lay people, there is evidence that the cognitive functioning of experts is basically like that of everyone else.

In addition, in many if not most cases, effective hazard management requires the cooperation of a large body of lay people. These people must agree to do without some things and accept substitutes for others; they must vote sensibly on ballot measures and for legislators who will serve them as surrogate hazard managers; they must obey safety rules and use the legal system responsibly. Even if the experts were much better judges of risk than lay people, giving experts an exclusive franchise on hazard management would involve substituting short-term efficiency for the long-term effort needed to create an informed citizenry.

For those of us who are not experts, these findings pose an important series of challenges: to be better informed, to rely less on unexamined or unsupported judgments, to be aware of the qualitative aspects that strongly condition risk judgments, and to be open to new evidence that may alter our current risk perceptions.

For the experts, our findings pose what may be a more difficult challenge: to recognize their own cognitive limitations, to temper their assessments of risk with the important qualitative aspects of risk that influence the responses of lay people, and somehow to create ways in which these considerations can find expression in hazard management without, in the process, creating more heat than light.

ACKNOWLEDGMENT

The authors wish to express their appreciation to Christoph Hohenemser, Roger Kasperson, and Robert Kates for their many helpful comments and suggestions. This work was supported by the National Science Foundation under Grant ENV77-15332 to Perceptronics, Inc. Any opinions, findings and conclusions, or recommendations expressed herein are those of the authors and do not necessarily reflect the views of the National Science Foundation.

REFERENCES

1. We have not attempted here to review all the important research in this area. Interested readers should see C.H. Green, "Risk: Attitudes and Beliefs," in Behavior in Fires, ed. D.V. Canter (Wiley, New York, in press); R.W. Kates, Risk Assessment of Environmental Hazard (Wiley, New York, 1978); and H.J. Otway, D. Maurer, and K. Thomas, "Nuclear Power: The Question of Public Acceptance," Futures, 10 (April, 1978), 109-118.

2. A. Tversky and D. Kahneman, "Judgment under Uncertainty: Heuristics and Biases," Science, 185 (1974), 1124-1131.

3. A. Tversky and D. Kahneman, "Availability: A Heuristic for Judging Frequency and Probability," Cognitive Psychology, 4 (1973), 207-232.

4. S. Lichtenstein, P. Slovic, B. Fischhoff, M. Layman, and B. Combs, "Judged Frequency of Lethal Events." Journal of Experimental Psychology: Human Learning and Memory, 4 (1978), 551-578.

5. B. Fischhoff, P. Slovic, and S. Lichtenstein, "Fault Trees: Sensitivity of Estimated Failure Probabilities to Problem Representation," Journal of Experimental Psychology: Human Perception and Performance, 4 (1978), 342-355.

6. B. Fischhoff, P. Slovic, and S. Lichtenstein, "Knowing with Certainty: The Appropriateness of Extreme Confidence," Journal of Experimental Psychology: Human Perception and Performance, 3 (1977), 552-564.

7. S. Lichtenstein, B. Fischhoff, and L.D. Phillips, "Calibration of Probabilities: The State of the Art," Decision Making and Change in Human Affairs, ed. H. Jungermann and G. de Zeeuw (D. Reidel, Dordrecht, The Netherlands, 1977).

8. M. Hynes and E. Vanmarcke, "Reliability of Embankment Performance Predictions," Proceedings of the ASCE Engineering Mechanics Division Specialty Conference (University of Waterloo Press, Waterloo, Ont., 1976).

9. U.S. Nuclear Regulatory Commission, Reactor Safety Study: An Assessment of Accident Risks in U.S. Commercial Nuclear Power Plants (WASH-1400, NUREG-75/014); Nuclear Regulatory Commission, Washington, DC, 1975.

10. U.S. Nuclear Regulatory Commission, Risk Assessment Review Group, Risk Assessment Review Group Report to the U.S. Nuclear Regulatory Commission (NUREG/CR-0400; Nuclear Regulatory Commission, Washington, DC, 1978).

11. U.S. Congress, House, Committee on Government Operations, Subcommittee on Conservation, Energy, and Natural Resources, Teton Dam Disaster: Hearings... (94th Cong., 2d sess.; Government Printing Office, Washington, DC, 1976).

12. R.W. Kates, Hazard and Choice Perception in Flood Plain Management (Research Paper, 78; Department of Geography, University of Chicago, Chicago, IL, 1962).

13. K. Borch, The Economics of Uncertainty (Princeton University Press, Princeton, NJ, 1968).

14. "Doubts Linger on Cyclamate Risks," Eugene Register-Guard, 14 January 1976.

15. E.E. David, "One-Armed Scientists?," Science, 189 (1975) 891.

16. B. Fischhoff, P. Slovic, S. Lichtenstein, S. Read, and B. Combs, "How Safe is Safe Enough? A Psychometric Study of Attitudes towards Technological Risks and Benefits," Policy Sciences, 8 (1978), 127-152; P. Slovic, B. Fischhoff, and S. Lichtenstein, "Expressed Preferences," unpublished manuscript (Decision Research, Eugene, OR, 1978).

17. The correlations between perceived risk and the annual frequencies of death were .92 for the experts and .62, .50, and .56 for the League of Women Voters, students, and Active Club samples, respectively.

18. W. Lowrance, Of Acceptable Risk (William Kaufman, Los Altos, CA, 1976).

19. The multiple correlation between the risk judgments of the LOWV members and students and a linear combination of their fatality estimates, disaster multipliers, dread ratings, and severity ratings was .95.

20. A secondary finding was that both experts and lay persons believed that the risks from most of the activities were better known to science than to the individuals at risk. The experts believed that the discrepancy in knowledge was particularly great for vaccinations, X-rays, antibiotics, alcohol, and home appliances. The only activities whose

risks were judged better known to those exposed were mountain climbing, fire fighting, hunting, skiing, and police work.

21. L. Ross, "The Intuitive Psychologist and His Shortcomings," Advances in Social Psychology, ed. L. Berkowitz (Academic Press, New York, NY, 1977).

22. A.E. Green and A.J. Bourne, Reliability Technology (Wiley Interscience, New York, NY, 1972).

23. D. Nelkin, "The Role of Experts on a Nuclear Siting Controversy," Bulletin of the Atomic Scientists, 30 (1974), 29-36.

Large-Scale Risk As a Problem Of Technological, Psychological and Political Judgment[1]

Charles Vlek, Hans Kuyper, Henk Boer

Institute for Experimental Psychology, Social Psychology Division, University of Groningen, the Netherlands

INTRODUCTION

The purpose of this paper is to identify, order, and explain basic concepts, problems, and issues in what may be called the societal debate on large-scale risks. In Section 2 the problem is introduced with a discussion of large-scale risk-taking by analogy with a small-scale example. A number of policy questions are listed and six different kinds of "risk research" are identified. Various problems characterizing large-scale risky decisionmaking are enumerated. In Section 3 a theoretical overview is given of various concepts of risk, both from a statistical-economical point of view and as the result of a psychological analysis. Section 4 starts outlining a multidisciplinary field research project serving as a source of ideas, findings, and conclusions treated in this and later parts of the paper. It then deals with a number of problems lying at the theoretical basis of methods and procedures for risk analysis. In Section 5 the perception and judgment of

1 The writing of this paper has been possible under research grants from the Netherlands Ministries of Housing, Physical Planning and Environmental Hygiene, and of Education and Sciences. The author is indebted to B. Ale, K. Altena, G. Bartels, R. Geerts, M. Molag, K. Sikkema, P.J. Stallen, and especially S. Stemerding, without whose partnership in research and discussions the present set of ideas, conclusions and suggestions would not have developed this far. Thanks are also due to the directors of the Advanced Studies Institute in Les Arcs (1983), V. Covello, J. Mumpower, P.J. Stallen and R. Uppuluri, who created and monitored the opportunity that eventually resulted in the composition of this paper. Mrs. W. Toet-Hollander deserves gratitude for her faithful assistance in its production.

NATO ASI Series, Vol. G4
Environmental Impact Assessment, Technology Assessment, and Risk Analysis. Edited by V.T. Covello et al.

risk are discussed in connection with an empirical study about the evaluation of several hazardous materials' uses by different interest groups. Possible evaluation and decision procedures concerning large-scale risky activities are briefly surveyed in Section 6, together with the design and some results of a decision-theoretic study on the use of hazardous materials. Finally, Section 7 gives a brief summary and touches upon a number of broader issues to put the paper in its proper context.

The paper is meant to give a written account of a series of lectures held in the Summer of 1983 at an Advanced Studies Institute sponsored by NATO and the U.S. National Science Foundation in Les Arcs, Haute-Savoie, France. The author has been working for quite some time as an experimental researcher in both cognitive and social psychology, with a strong interest in Bayesian decision theory and problems of human judgment and choice behavior.

THE PROBLEM

To understand societal debates on large-scale risks it is instructive to consider a small-scale example. Suppose that, to attend a Summer seminar you have to travel from abroad to the mountain resort at Les Arcs in the French Alps not far from Chamonix. By some rapid mode of transportation you arrive in Geneva, from where you have decided to cover the remaining 150 kilometers using your own 15-year old sports bicycle (shipped in advance). The trip will bring you along scenic roads and villages; your final goal can only be reached via a number of fairly high mountain passes.

To contemplate and discuss the risks of an unusual tour like this there may be four relevant parties involved: you yourself, your partner--wife or friend--, your bicycle shopkeeper and repairman, and your 78-year old mother. Let us briefly consider each in turn, with their backgrounds, views, and preferences.

You yourself are having to make it, both safely and in time. You are an amateur cyclist who is certainly not in top condition for long-distance mountain cycling. You do have a fair amount of experience, though, and you are well aware of your own as well as your bike's physical limitations.

Your bicycle shopkeeper has recently done a maintenance job on your bike. He knows it quite well. "The bike is O.K. but it is not new. Don't demand too much of it. Your rear wheel is slightly out of balance. It'll be wise to keep your brakes well adjusted. Do take some spare parts and vital tools along."

Your partner in life who will stay home sympathizes with your love of bicycling but tends to be a bit worried about the physical strains exerted both on the bike and yourself. "It must be fascinating, but you don't have the condition for it. Please try not to crash into the side of a down-hill hair-pin. I'd like to see you back the way you are."

Your old mother lives in a remote countryside village which she rarely leaves. She once tried to learn riding a bicycle, but her clumsiness and lack of opportunity left her with an intuitive fear for this "two-wheeled piece of circus equipment." She tends to consider bicycle riding as unwise and unnecessary. "But if you must...."

It is clear that the four parties involved are entitled to their own opinions all of which have some relevance to your own attitudes and behavior. The major question involved--"What is the probability of a serious accident?"--does, however, not have an obvious answer.

Now let us look at some of the difficulties of mountain bicycling. Suppose that, after having proudly enjoyed the majestic view at the top, you are coming down a 1600-meters pass. The road is narrow and winding with some scattered surface irregularities. For every kilometer it descends by about 100 meters, so you can't let the bike go free. Thus the demands on your front and rear brakes are heavy, and you have to balance body-and-bike very carefully to follow the bends in the road. Other road users may be coming on or trying to overtake you. The high mountain air is cold and the speed of descending makes your eyes water, which occasionally blurs a clear view of things ahead. The same swift, cold air stiffens up your muscles, causing incidental tremblings that preclude fine motor adjustments at crucial moments.

The risks are inherent in a subtle and rapidly changing combination of road lay-out, the condition of your bike (especially the brakes), and your own condition -- or what is left of it after you pedaled up the 1600-meters pass. How could you estimate the level of risk? Will you want to assess any probabilities? Could anybody else realize what hazards are involved at the detailed operational level of this trip? Had you yourself anticipated these hazards (or some of the particular joys and benefits)?

Obviously, you yourself, your life-partner back home, your bicycle shopkeeper and your old mother are bound to have radically different images, expectations, and evaluations concerning the nature and possible consequences of this bicycle trip. Given that the others involved allow you the freedom that goes with peculiar preferences, there are aspects that are differently judged in connection with differences in knowledge and experience and the inferences based upon them. It is a very personal experience at last, and an assessment of skilled controllability, that form the pivots of <u>your</u> decision to bike or not to bike. And then--when you actually do--many things appear to deviate from what you had been able to anticipate. Thus improvisation is the last of your personal safety strategies.

In the above example, the separate but concurrent aspects of risk-taking behavior stand out clearly. Technical assessments of the bike's condition are essential, as well as an (amateur) medical/psychological judgment of your own skills, possibilities, and limitations. Judgmental -- representational as well as evaluational -- tasks have to be performed. And some decision rule or scheme is followed when outcomes of assessments and judgments -- including those by others -- are to be combined into an ultimate preference and optimal action.

When it comes to large-scale risk-taking, similar components are involved. In evaluating the risks and benefits of the bulk storage and transportation of LPG as a car-fuel, for example, a technical risk analysis is called for. Part of it should consist of a "psycho-technical" analysis of possible man-machine or man-equipment interactions in order to assess human failure scenarios and probabilities. It would be useful to know how different interest groups judge and evaluate the LPG activity and comparable options with respect to a number of relevant utility attributes. And one would not be able to decide on what to do best, without the use (at least implicitly) of a decision scheme in which the outcomes of the various analyses are to be combined into a final choice of policy.

There are, of course, a number of characteristic differences between large- and small-scale risk taking. The problem of human control over large-scale complexity can be succinctly formulated as follows. Complex socio-technical systems serving the population at large, are widely suspected (and sometimes experienced) to be insufficiently controllable both with respect to their safety and the nature and distribution of their benefits. Such systems imply large-scale activities which are seen by many as incomprehensible, and therefore essentially undecidable. Possible consequences are thought to be unimagineable, irreversible, and potentially catastrophic.

Even experts have proven that they cannot understand and handle all aspects of these systems appropriately. And the resulting benefit-costs/risks ratio is inequitable for certain subgroups of the population. Due to the massive and wide-spread nature of their benefits such systems pose a number of dilemmas-of-control which we are only beginning to realize. Some of these will be briefly discussed in the final section of this paper.

The complexities of large-scale technological developments have led policy makers and others to pose a number of crucial questions related to safety issues; ten such questions are presented in Table 1. In this collection, risk analysts and decision theorists will recognize fundamental issues related to definitions, assessment, and measurement techniques; judgmental procedures and decision rules; and not the least, seemingly unsolvable problems of social decisionmaking in fairness to those most concerned. The ten questions also reveal that the issues appeal to various scientific disciplines.

In the past ten or so years three globally different approaches have developed: (1) risk analysis, originating in the technical and engineering sciences; (2) the study of the perception and judgment of risks, having its roots in the social (especially the psychological) sciences; and (3) decision analysis, a practical methodology stemming from Bayesian decision theory. Further differentiations and amplifications in recent years allow us to distinguish the six different kings of "risk research" mentioned in Table 2. Notoriously missing in Table 2 is the topic of socio-economic benefits, a blind spot in the eye of many a risk researcher. We should also realize that models of human failure have received comparatively little attention in the study of large-scale risks. Just these points might lead us to wonder whether the various topics could be integrated in a global framework to serve as a basis for the design of full-scope, risk-benefit studies of large-scale technological developments. Each line of research indicated in Table 2 may provide results that by themselves are necessary but insufficient to formulate reasonable answers to pertinent policy questions.

Ultimately, every safety problem is a decision problem. Hence it is meaningful to look upon the various kinds of risk research from the perspective that, after all, a choice has to be made. Are we going to invest more into safety measures? Will we run the risks and eventually do our best to live with the after-effects of an accident? Do we opt for a risk-sharing strategy?

Table 1: Ten policy questions about large-scale risks

1. Why do certain government decisions about large-scale activities turn out to be unacceptable for segments of the population?
2. How could it be understood that risks associated to certain kinds of activity draw much public attention while others don't?
3. Can risks and risky activities be categorized systematically, so that (in)comparabilities become apparent and differentiated safety policies can be formulated?
4. What might explain apparent discrepancies between "calculated" and "experienced" risks; on what grounds could the one be called "objective" and the other "subjective"?
5. In which way can potential effects of possible serious accidents be estimated and charted with sufficient validity, and how could their degree of seriousness (or disutility) be acceptably assessed?
6. When it comes to assessing probabilities of accidents in complex systems, to what extent can assumptions and conditions be met that are inherent to the mathematical definition of 'probability' with its frequentistic, logical and personalistic interpretations?
7. When 'risk' is pragmatically defined as some function of both the probability and the seriousness of a possible (set of) accident(s), then how should these two factors be weighted and combined?
8. By what method could a valid causal model be formulated about the structure and succession of both technical and human factors involved in a risky activity, so as to allow for the design of effective safety measures?
9. Are there any good (scenario) methods to help us anticipate broadly and timely enough the social and political meaning and consequences of new technological developments?
10. What could a policy making body best undertake when views and preferences concerning large-scale risky activities consistently differ among different interest groups in the population?

Table 2: Six major topics of current risk research

1. Technical systems reliability
2. The human operator as a source of failure probabilities
3. Human cognitive representations of risky activities, as related to judgments and preferences
4. Human expressions and consequences of anxiety and fear
5. Policy design and (social) decision analysis
6. Effects on judgments and preferences of information and communication about risks

Complex decisionmaking can be modelled in terms of a number of successive stages. Table 3, adapted from Vlek and Wagenaar (1979), gives an idea of what an analytical decision process might consist of. The nine stages should not be seen as strictly successive; there may, of course, be feedback loops that enable a decisionmaker to re-think what she or he did at an earlier stage. The model as a whole should function as an interface between someone's "internal representation" of the world and her or his actual interactions and experiences with that world. Almost any stage therefore implies information-gathering activities -- either from "within" or from the outside task environment -- that should support an adequate run through that stage. Stage I is put in parentheses as it would not belong to a decision process proper, but rather to some long-term rational decision strategy.

Table 3: A model of successive stages in complex decision making

A. recognition of a problem (threat, opportunity, desire for change)
B. hierarchical goal analysis (to specify objectives)
C. design of, and search for feasible options
D. description of relevant possible consequences
E. estimation of probabilities of consequences
F. assessment of value or utility of consequences
G. application of decision rule (weights, trade-offs, combination rule)
H. choice and commitment
I. (observation of consequences and learning from experience)

Obviously now, the six kinds of risk research displayed in Table 2 pertain to different stages of the decision process modelled in Table 3.

- o Social decision rules are hard to apply, if only because not a single one exists that is justifiable from a choice-theoretical point of view.

At the individual level the problems of large-scale decisionmaking have to do with cognitive representations of status quo, sets of goals, policy options, and of their possible consequences; they are also inherent in the making of value judgments and trade-offs as well as the final combination of a multitude of relevant considerations. There can be no social decisionmaking without problem representations, assessments, and overall judgments in the heads of individuals participating in the decision procedure.

At the social level there are problems of deliberation and negotiation, of the organization of information and communication processes, and of designing and selecting suitable arrangements for the participation of relevant interest groups and persons. The foremost problem of social choice is the integration of discrepant views and the combination of diverging preferences.

It would be quite an advantage for any system of government and public administration if these social and individual problems of large-scale decisionmaking concerning risky technological developments could be approached from a coherent general framework. Would that lead to a revised set of major topics for risk research? Let us for the moment answer this question by noting that for an adequate evaluation and control of largescale technological developments, we need at least the following:

- o Adequate descriptions of technical installations, procedures, and processes

- o Valid models of human skilled behavior and probable failure in demanding tasks

- o Useful and accepted definitions of "risk," "benefit" and "acceptability," and the implied key-concept of probability

- o Reliable knowledge about how representative interest groups perceive and judge costs, risks, and expected benefits of an activity

Table 4: Basic characteristics of large-scale activities

1. The authority to decide upon, and to control the implementation of the activity rests with few people, whereas possible consequences fall to many.
2. Possible negative as well as positive consequences are difficult to assess and evaluate beforehand. They are (a) massive, (b) far-reaching in time, (c) replete with unintended, creeping side effects, and (d) hard to reverse.
3. Possible consequences are different for different groups of people, both in nature and degree. Some profit more than others, and there are those who definitely lose.
4. Goals and values underlying judgments and preferences may differ widely among different interest groups.
5. Knowledge about the nature of the activity tends to be unevenly distributed; there are experts, educated lay-people as well as totally uninformed persons concerned; among all these there are small and great pretenders.

o Practical rules and criteria as pivots of socially-acceptable decision procedures

In the sections to follow these nuclear points of interest will be considered in more detail. Obviously, the definitional questions come first.

WHAT IS "RISK"?

As a preamble to a large field-research project on the personal judgment of risks (Vlek and Stallen, 1979, 1981), thorough theoretical explorations and discussions have resulted in an expose of various rational and personal conceptions of risk (Vlek and Stallen, 1980). The main points will be recapitulated below, whereby some revisions may occur in light of recent publications by others. With regard to rational conceptions of risk the work of Coombs (1972), Libby and Fishburn (1977) and Schaefer (1978) are to be acknowledged. As far as personal and social dimensions of risk are concerned the influence of Lowrance (1976), Otway and Fishbein (1977; see also Otway and Von Winterfeldt, 1982), Rowe (1977) and Fischhoff, Slovic, Lichtenstein, Read and Combs 1978; see also Slovic et al., this volume) should be recognized.

A rational decision theorist would conceptualize any problem of acceptable risk as a choice problem. Formally speaking we would define a risky choice problem as a probability distribution over a set of possible outcomes to each of which may be associated a consequence utility. Such a distribution could be characterized by its various moments, among which are the mean or expected value (or utility), the variance of valued outcomes, and the skewness as an index of asymmetry around the mean. Distinguishing between relative gains and losses as possible outcomes of available options, we may list six different formal definitions of risk, as in Table 5.

Table 5: Six formal definitions of risk

* the probability of a loss;
* the size of a (credible) loss;
* expected loss, i.e. the product of the probability and the size (or negative value) of a possible loss;
* the variance of the probability distribution over the values (or utilities) of all possible relevant outcomes;
* a semi-variance, for example: the variance of the distribution of all negatively valued outcomes about the average expected loss (their common mean);
* a linear function of the expected value and the variance of the total distribution of possible outcomes;

The first three definitions have the advantage of being applicable to separate possible losses; when there are several of these one would simply take their sum total. The standard risk parameter from decision theory would be the variance; a greater spreading of both gains and losses around their expected value means greater risk: betting for a pound is more risky than betting for 10 pennies. The last two definitions in Table 5 are specifications that one may or may not find useful to apply. Especially Number 6 is more of an open definition designed for the purpose of describing the choice behavior of laboratory subjects.

There are several problems associated with these definitions of risk. First of all, each and every one of them would be insufficient as a basis for making decisions. Indeed, one might ask whether one needs to know how "risky" a particular option is in order to be able to make an adequate decision about it. Secondly, each definition is an abstract and strongly reductionist parameter whose meaning is all the more difficult to grasp the greater the number of possible different outcomes and the dimensionality of their consequences. Third, the six definitions are not all independent of one another; when it is judged useful to apply several of them simultaneously one would not know how to weigh their relative importance.

The natural transition to a discussion of personal and social dimensions of risk lies in the statement that formal definitions of risk are too abstract, global, and one-sided to be suitable as a basis for risky choice. Admittedly, there are recent developments in risk analysis whereby risk is rather loosely determined as a collection of points in a graphical probability X size of accident space (e.g., Kaplan and Garrick, 1981). A similar conceptualization is used by the Groningen Provinciale Waterstaat (1979) authorities, who have ventured to add some numerical criteria enabling them to identify "acceptable," "unacceptable" and "to be thoroughly analyzed" sets of risk. But these and other "norms for acceptable risk" cannot be considered to result from an explicit consideration of costs, risks, and benefits of the particular activity being regulated. Thus, in a number of respects, they are inevitably arbitrary.

In section 6 we will deal with the question of whether in such cases a full-scale decision analysis ("socialized" or not) would be conceptually and methodologically feasible. At this moment it is relevant enough to go into the types of dimensions, variables, and aspects that seem to need a decision-maker's attention when it comes to judging large-scale risks. Table 6, taken from Vlek and Stallen (1980), summarizes the first 19 of 32 relevant "aspects of risky decision problems" organized into 11 major categories. The four categories not shown in Table 6 are labelled: (H) knowledge and experience about the risky activity, (I) personal condition of the subject, (J) social considerations (norms, habits, possible inequities), and (K) confidence in experts/regulators. The heading of Table 6 suggests how the various aspects are related to the evaluation of "acceptability."

Table 6: First 7 of 11 categories of fundamental aspects of risky decision problems. The implied relationships with "acceptability" are hypothetical.

IT IS HYPOTHESIZED, THAT THE RISKS OF AN ACTIVITY ARE (UNTHINKINGLY, TACITLY, VOLUNTARILY, OR PERFORCE) MORE READILY ACCEPTED TO THE EXTENT THAT:

(A. VOLUNTARINESS OF EXPOSURE)

(1) there are fewer comparable options available;

(2) someone's personal influence on the decision is larger;

(3) correction or change of the selected action is easier;

(4) the importance of intended benefits is larger;

(B. CONTROLLABILITY OF CONSEQUENCES)

(5) accident probabilities can be influenced by safety measures;

(6) there is less room for human error, recklessness, and sabotage;

(7) accident-effects can be contained by rescue operations;

(8) accident-effects are reversible;

(C. DISTRIBUTION OF CONSEQUENCES IN TIME)

(9) intended benefits are obtained at a shorter term;

(10) undesired consequences have delayed (a.o.t. immediate) effects;

(D. DISTRIBUTION OF CONSEQUENCES IN SPACE)

(11) intended benefits materialize at a shorter (social) distance;

(12) undesired consequences occur at a greater (social) distance;

(E. CONTEXT OF PROBABILITY ASSESSMENT)

(13) judged (relative) accident-frequency is lower;

(14) imaginability of accident-causation is less;

(15) final assessment of accident-probability is less ambiguous;

(F. CONTEXT OF ACCIDENT EVALUATION)

(16) an accident violates less central values or less basic needs;

(17) imaginability of final accident-effects is less;

(18) final assessment of accident-seriousness is less ambiguous;

(G. COMBINATION OF ACCIDENT-PROBABILITY AND 'SERIOUSNESS')

(19) risks are of the type 'relatively high-probability/small-loss';

The list of aspects and their categories elicits a number of comments. First, it is apparent that qualitative aspects of both expected benefits and losses (such as, economic growth, new job development, somatic and health effects, and environmental damage) are largely folded up in aspects Numbers 4 and 16, respectively. Their number and kind are dependent upon the particular problem being analyzed. Secondly, the list of aspects can be divided into evaluative aspects (which have to be weighted) and informative aspects (which may be believed or not). Third, Table 6 is suggestive of a number of psychological mechanisms (e.g., numbers 9 and 19) likely to operate in risky situations. Fourth, it also may make the reader think about a number of assessment problems associated with the use of certain variables. Fifth, one may easily imagine how controversial debates might arise over a number of variables, when it comes to judging large-scale activities.

A crucial question about Table 6 is whether a scheme like this be compatible with the idea of a formal (social) decision analysis. From a decision theoretic point of view we may then note that several aspects (e.g., numbers 4, 9-12, 16) can be considered as attributes in a multi-attribute utility analysis. Some of these may be further decomposed; others may keep their status of "proxy attribute" (Keeney and Raiffa, 1976, especially chapter 2). Others (numbers 5-8, 13-15) specify the conditions characterizing probability assessment; (we may note in passing that controllability considerations thus far seem to play an unimportant role in the operationalization of the probability concept). Finally, some aspects concern the structure and context of the decision situation someone might be in: are there few or many options; how flexible can I still be after cutting the knot; how big, or how limited, is my decision power? Thus a number of aspects given in Table 6 could be accommodated in a carefully conducted social decision analysis. Other aspects, however, seem to indicate that certain decision problems easily exceed the limits of any procedure that could be theoretically justified.

From a dynamic-psychological point of view it is attractive to define risk globally as a "lack of sufficient control" over an intended course of action, or more generally about the course of one's daily life. It would seem that the notion of controllability lies at the heart of expert and public controversies about large-scale risks. The multidimensional meaning of this concept is, however, difficult to specify. In addition, we don't know how to indicate ideal levels of controllability so as to live comfortably forever after.

The concept of perceived or expected control might be operationalized in terms of the decision process stages given in Table 3. For example, a risky choice problem might appear insufficiently controllable to the extent that one has only vague ideas about the sort of goal that one would like to pursue (stage B). Or the lack of controllability may be associated with the stage of probability estimation (E) for which one lacks essential information. Table 6, too, suggests a number of aspects in terms of which a risky situation may impress one as being short of controllability.

The psychological dynamics and consequences of variations in perceived controllability of a demanding (stressful or risky) situation are roughly sketched in the model of Fig. 1 (distilled from recent research on stress, anxiety, and human performance; see for example Lazarus, 1980; Eysenck, 1982; and Stallen and Tomas, 1982). Keeping in mind the example of mountain bicycling sketched in Section 2, we might interpret this model as saying that a general feeling or appraisal of controllability of a demanding situation lies at the basis of a person's level of "activation." The latter, functionally working out as a pattern of four different processes, is qualitatively different depending upon whether the situation is seen as a "challenge" or as a "threat." The nature and intensity of the various activation components determine the effectiveness of a subject's processing of relevant information (perceptual and otherwise). Thus there may be either positive or negative effects on the subject's main task performance as well as any contextual (subsidiary) performance.

According to the model of Fig. 1, a lack of sufficient controllability leads to a situation being perceived as threatening. The immediate effects may be inefficient or even deficient task performance whereby major catastrophes might result from an accumulation of small errors. Repeated or chronic exposure to insufficiently controllable situations may eventually result in psychosomatic complaints and withdrawal syndromes such as a state of "learned helplesness" (Garber and Seligman, 1980), or in patterns of "reactance" (Brehm, 1972) manifested as vandalism, public agression, and the like. The question of what degree of situational controllability would be optimal for different groups of people can not be easily answered on the basis of available theories and evidence.

An important perspective has been developed by Wilde (1978, 1982) in his theory of risk homeostasis. The basic idea is almost as simple as it is convincing. A risk-taker, Wilde says, continuously adapts his or her behavior so as to maintain a situation of exposure to a roughly constant level of risk. The latter depends upon an implicit evaluation and weighting of the costs and benefits of safe behavior versus those of risky behavior. By consequence, when a particular activity is made safer through external factors, the individual will manifest a safer conduct only for a limited period of re-adaptation.

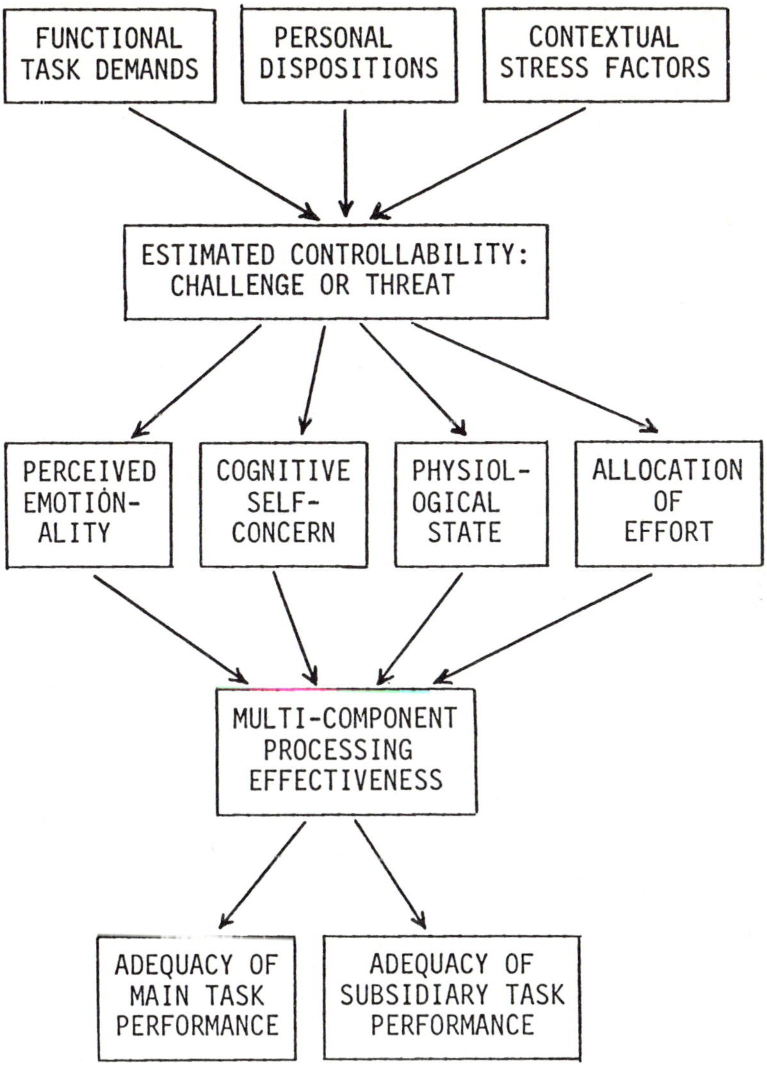

Figure 1: A model of process components characterizing human behavior in demanding (stressful) situations.

Thereafter a return will be observed to his or her original, "personal" level of risk. Risky behavior, resulting in a certain accident rate, can thus be made safer only by increasing the relative costs (material and other) of an accident, which would bring down someone's target level of risk.

The notion of risk homeostasis is intuitively appealing, and it has far-reaching implications for personal and public safety policies. Although the theory has hardly been put to the test at the level of the individual risk-taker, Wilde's (1982) ideas are supported by, e.g., Starr's (1976) observations that accident rates associated with the use of agricultural machinery and medical anesthetics tend to remain rather constant in time, despite improvements and refinements in safety respects. Users just take on more challenging jobs when their tools have become more reliable. There are, however, several critical questions to be answered before the idea of risk homeostasis can be taken into account in the practice of safety policies.

In Section 7 below, an attempt is made to put the theoretical problems of risk in the context of various broader issues. At this point we may see answers emerge to some of the policy questions listed in Table 1. Consider, for example, questions 1 and 2. A certain government decision may be unacceptable for segments of the public to the extent that it reduces their degree of (perceived) control over important aspects of their daily life. For this same reason certain types of risky activity may get more attention than others, even when the latter manifest a higher accident rate. Question 4 in Table 1 may by now receive the obvious answer that, in comparison to "experienced risk," "calculated risk" follows from an entirely different conception. In an attempt to integrate the two--psychological and mathematical--concepts, the former might be considered to be the flesh and blood of the latter which can be no more than a formal abstraction. An emerging question for policy makers might be: to what extent should "personal control over the course of one's life" be a prominent political value in the design and implementation of coherent safety policies?

LIMITATIONS OF PROBABILITY ASSESSMENT IN RISK ANALYSES

In Sections 4, 5, and 6, specific attention will be given to three particular areas of risk research, namely risk estimation, the perception and judgment of risks, and decision procedures concerning large-scale risky activities. Theoretical comments as well as empirical conclusions will be presented.

These are based on, or have been sharpened by, a multidisciplinary research project on the description and judgment of risks associated with the use of hazardous materials, notably LPG (a car-fuel) and SO_2 (or sulphur dioxyde as used to facilitate processes in the beet sugar industry). The project as a whole is outlined in Table 7. In this section we will look into some problems of technical-statistical risk analysis (or what Rowe, 1977, would call "risk estimation").

To be practical, any theory of risk should be a reductionist abstraction of a class of realistic cases to which it applies. The drawbacks of reduction and abstraction, however, are particularly felt in the field of technological risk analysis. There, essential components of basic parameters such as those defined in Table 5 are estimated using different methods and techniques. For example, through reliability analysis the operational limitations of a piece of material or equipment under given circumstances are estimated. In a failure-mode-and-effect analysis specific disturbances are imposed on a system, and their separate or combined effects are recorded. Over a longer period of time, accident statistics may be kept of systems under normal operation. To study the nature of particularly feared accidents, historical case studies may yield the required information about antecedent conditions, the coincidence that made it happen, and the immediately occurring loss and damage with their aftereffects on the social and physical environment. The nature and intensity of accident effects may be more specifically studied using theoretical models of, for instance, sudden heavy gas dispersion in free air, or penetration into the earth's surface of a certain toxic material.

These various approaches to chart risks all necessarily rely on the existence (or the creation) of a database from which inferences are made concerning the value of key parameters. Theoretically, the ideal database for a risk analyst would be a stochastic process having a fixed set of stationary characteristics, which generates observable mishaps and accidents in a regular manner. In practice, of course, databases often fall short of this model situation. One frequent deficiency is that necessary data have not been recorded. For example, one cannot properly estimate the relative frequency of a particular accident if one is unable to specify the number of occasions whereby the accident could have happened but did not. Or one may have only very global recordings of past accidents so that relative frequencies of accidents occurring under specific circumstances can not be estimated precisely enough.

Table 7: A multidisciplinary project on hazardous materials

Title:	Description and judgment of risks associated to the use of hazardous materials.
Granting agency:	Netherlands Ministries of Housing, Physical Planning and Environmental Hygiene, and of Education and Sciences.
Executing institute:	Environmental Studies Center of the University of Groningen, with support from the Social Psychology Department and the Department of Technical Chemistry.
Particip- ating staff:	K. Altena, M. Boekhoudt, H. Boer, R. Geerts, H. Kuyper, M. Molag, K. Sikkema, W. Sprong, S. Stemerding, G. Tiemessen, W. Tiemessen, Ch. Vlek, and several others.
Duration:	The project runs from February 1980 until April 1984.
Goals:	Clarify concepts, (further) develop and test assessment methods, draw substantive conclusions about risks and risk judgments. List, develop and test alternative procedures for improved regulatory decision making. Demonstrate possible relevance of risk analysis and risk perception research.
Basic design:	Three subprojects, each carried out once specifically for LPG, and once for (relatively unknown) SO_2.
1. Risk analysis:	In subproject 1, a technical-statistical analysis was carried out in an attempt to inventory and objectify the risks of LPG and SO_2, and to indicate possible ways for further increasing their safe application.
2. Risk evaluation:	In subproject 2, psychometric analyses were carried out on judgments of risks and benefits of various hazardous materials including LPG and SO_2. Raw data had been collected from several hundred respondents belonging to six basically different interest groups.
3.Decision making:	In subproject 3 a number of decision situations were analyzed concerning the use of LPG and SO_2 both in terms of existing procedures and with regard to possible improvements therein. An important objective was the design of 'better' decision procedures that would reduce at least some of the grounds for controversy.
Reports:	There will be six reports: on the technical risk analyses of LPG (2) and SO_2 (3), on the psychological studies specific to LPG (4) and SO_2 (5), and on the combined decision-analytic studies of LPG and SO_2 (6), respectively; there will be a separate "Overview and Summary" report (1). All reports are in Dutch; only the latter (no.1) will also become available in English, later in 1984.

A principal drawback of virtually any database is that it pertains to the past. In reality, ever-changing conditions of life (e.g., through improved safety measures or increasing human neglect) may jeopardize probability estimates that one might wish to make in view of observed relative accident frequencies.

The problems of assessing risks on the basis of observed data, be it from laboratory tests, experiments, accident-statistics or case studies, are crucially related to the concept of probability. To recapitulate briefly some essentials: probability is a mathematical model parameter characterizing a well-defined sample space (e.g., a die, set of playing-cards or wheel of fortune) with mutually-exclusive and exhaustive classes of elements (an odd number, an ace, a zero). The probability of a future draw of an element is specifiable on the assumption that the conditions of drawing remain substantially the same.

Technological risk analysis is virtually married to the probability concept, but almost exclusively so under its frequentistic and logical interpretations. Thus technological probability estimates tend to suffer from deficient statistics on past (opportunities for) accidents, and from nonstationary "sampling conditions" that foreclose extrapolations into the future. In addition, logically inferred probability estimates (e.g., from fault trees) are heavily dependent upon the particular combination model formulated to characterize the system under consideration. Thus probability in its frequentistic and logical interpretations seems useful for risk analysts only to the extent that the required frequencies and the logic are there, while the axioms and assumptions of probability theory are reasonably fulfilled. In many realistic cases this is extremely hard to demonstrate.

In its personalistic interpretation, probability is a degree of belief that can be expressed as a certain willingness to bet on the outcome of an uncertain event. "I'll bet you 10 to 1 that I'll be home before 7:00," implies that the speaker is about 90 percent sure that (s)he can make it. When properly measured by means of axiomatic methods (see several contributions in De Zeeuw, Vlek and Wagenaar, 1970; Spetzler and Staël von Holstein, 1975), well-considered intuitive judgment may be numerically expressed and tested for coherence. In most realistic risk and decision analyses such assessments are inevitably called for, especially so at a decomposed level. A sensitivity analysis should reveal what degree of exactness is required.

The assessment of personal probabilities, too, is not free of problems of measurement and interpretation (see, for instance, the practical review by Wallsten and Budescu, 1980). But its basic tenet is irresistible: when there is an acute shortage of reliable data your creative intuition can only go by some generalized knowledge and experience (as when making a voyage of discovery); then you had better express your "true" hypotheses and beliefs as validly and as orderly as possible, using methods that are theoretically justified. To the extent that there exists such a thing as "generalized expert knowledge and experience," the personal probability estimates of different experts should tend to converge. Otherwise we are left with a situation where the long-run average best probability assessor tends to win rather than lose his bets.

A last (but not the least) factor complicating any probability analysis of risks concerns the way a theoretical sample space is constructed from what in fact is a variety of possible different accident processes. Here, probabilistic reasoning demands an artificial separation between "cause" and "effect," which in practice often runs parallel to the earlier and later parts of a whole accident process with reference to a critical point in time (often called the "coincidence"). Any way of structuring a probability sample space, therefore, implies a particular model of the process whereby a given set of consequences comes about.

It has already been indicated that there is a dearth of valid models of possible human failure in the operation of complex technical systems. At the same time the relative frequency of serious accidents due to human error is alarmingly high. From a technical-engineering background Haastrup (1983) estimates this number to be 35 percent for the process industries. The psychologist Wagenaar (1983) reviews work showing this percentage to be very much higher for, e.g., aviation (Hartman, 1979), road traffic (Sabey and Taylor, 1980), the shipping-trade (Karlsen and Kristansen, 1980), and the nuclear power industry (Lloyd et al., 1979). In view of these figures, modeling the reliability of the human operator in complex systems seems at least as urgent as the technical reliability analysis of those systems themselves. The notion of an exhaustive sample space of independent possible outcomes may not be that suitable as a basis for modeling accident processes that are in fact multifactorially determined.

The various limitations of probability assessment in practical risk analyses should reduce the belief that probability as such is a sensitive variable in the solution of ill-defined, complex decision problems. It is not. What may be (partially) true in the probability theorist's laboratory may be impossible to find in reality. Other factors, like knowing what choice there are or what consequences should be considered important, may be more useful and relevant. They may well render an optimal decision virtually insensitive to anything more than a fairly global expression of uncertainty.

From what has been noted above the following empirical conclusions could almost have been predicted. In subproject 1 (cf. Table 7) it appeared that by far most historical accidents related to the use of LPG had been caused by external factors that were not part of pre-designed LPG systems. It was also concluded that case studies did not provide a sufficient basis for the assignment of probabilities to specific future accidents, neither would they allow for valid damage expectations to be formulated. Available models of accident effects and possible damage to the environment did provide global indications of possible negative effects under given conditions. The risks of LPG appeared to be further reducible through measures that would decrease the probability and/or the negative effects of particular incidents. Finally, it is worth noting that the investigators considered "risk" to be quantifiable only when they adopted a subjective attitude toward both probability and effect. In their view, expert personal estimates on many points are inevitable, due to the great amount of factual uncertainty involved.

PERCEPTION AND JUDGMENT OF RISK

In section 3 a number of theoretical definitions of risk were discussed (Table 5). Their formal, abstract, and necessarily one-sided character led us to consider a number of informal aspects of risky decision problems (Table 6). The section was concluded with a brief discussion of dynamic-psychological models of risk-taking in demanding situations (e.g., Fig. 1). In the present section psychological research on large-scale risk will be looked at in more detail, both from a theoretical and an empirical point of view.

The Meaning of Risk Perception Research

Classical psychological research on the role of risk in decisionmaking was primarily rooted in a descriptive interest: given such and such well-defined risky situations, how much risk (mostly, probability of a loss) would the subject be willing to take? Slovic (1964) and Kogan and Wallach (1967) give reviews of a literature revealing vigorous attempts to discover consistent situational, group, and personality influences on the amount of risk preferred.

Despite much research it appeared especially impossible to demonstrate the existence of personal "risk-taking propensities" that would generalize over a variety of risky situations (Brichácek, 1970; Slovic, 1971; Kozielecki, 1974)[2]. Willingness to take well-specified risks was found to be mainly a function of situational variables and group influences. The nature of the former ("What makes for situational risk?") has only recently begun to receive systematic attention. Group influences on risk preferences were given special attention in studies on the "risky shift" that could occur in group decisionmaking through polarization of attitudes (Myers and Lamm, 1976).

Under the increasing influence of formal decision models in psychology, risk research has gradually taken a different course. The work of Slovic and Lichtenstein (e.g., 1968; and, joined by Fischhoff, 1977) and Coombs (1972, 1975; Coombs and Avrunin, 1977) has concentrated on definition questions regarding risk and whether it was at all distinguished by decisionmakers and played a significant role in determining their preferences.

Somewhat surprisingly, a rather new type of research on the "perception of risks"[3] from large-scale technological developments seemed to develop rather independently from these existing lines with the introduction of amateur-psychological "dimensions of subjective risk" proposed by -- among others -- Lowrance (1976), Rowe (1977), and Ravetz (1977). This

---

2 At a more fundamental level, however, we should note that empirical evidence points at the existence of consistent personal levels of "trait-anxiety." Upon confrontation with a stressful situation these would trigger an individual into an actual state of anxiety over possible negative outcomes or risk (e.g., Eysenck, 1982). Thus high trait-anxious persons might judge a given situation as more risky than low trait-anxious persons. Remarkably, there is not much research connecting ideas about personal levels of trait-anxiety to people's willingness to take risks or their judgments of risk.

3 Of course, only a small part of what we tend to call risk is actually "perceived." For the remainder, risk is being inferred, structured, reasoned and judged against the background of human knowledge, experience and values. The terms "risk perception" and "perceived risk" are used here as general labels.

obviously has to do with the ill-defined nature of large-scale risks, and it strengthens the point made in Section 3 that formal definitions of risk are too abstract to account for a variety of situational factors that seem to influence personal judgments.

Thus risk perception research arose in response to questions about the way different groups of persons conceptualize risks and risky activities such that it became clear why they assigned to them particular overall judgments of riskiness and acceptability. Rather than asking what characteristics of choice alternatives with explicit probability distributions of outcomes would be useful as definitions of risk, risk perception researchers have been trying to uncover the cognitive structure by which risks and risky activities were represented in the minds of those worried and/or responsible. Compared to small-scale, immediate risks which are at most "emotionally reasoned" (cf. the example of mountain bicycling in Section 2), large-scale risks are supposed to be "cognitively reasoned" since their very nature (cf. Table 4) makes it nearly impossible to judge them immediately by more than vague intuition.

The methods used in research on perceived risk are more generally used in the social sciences to describe cognitive structures of perceived sets of stimulus materials that are ill-defined *a priori*. Slovic, Fischhoff, and Lichtenstein (1980) had several groups of subjects explicitly judge 90 different "risk stimuli" (e.g., car driving, bicycle riding, use of nuclear weapons, and coal mining) with respect to some 18 risk characteristics such as "voluntariness of exposure," familiarity to scientists," and "catastrophic potential." An overall factor analysis of the collected ratings revealed that the 18 risk characteristics could be condensed into three factors. These were interpreted as (1) "innocent" vs. dread risk, (2) known vs. unknown risk, and (3) number of people exposed and degree of personal exposure.

Vlek and Stallen (1979, 1981) conducted a psychometric field study with a large number of respondents from different age and sex groups, and living at systematically different distances from a heavily industrialized area in the Rotterdam Harbor region. In one task, subjects had to order 26 risky activities -- such as "drunken driving", "transportation of chlorine gas," and "mass use of pesticides", each briefly described on a card -- according to the "seriousness of their risks and dangers." An individual differences scaling procedure (carried out on the recorded, categorized rank orders) resulted in a two-dimensional configuration of the 26 risk stimuli. Its primary dimension could be interpreted as "size of a potential accident," and most subjects appeared to agree that "riskiness" increased with the value of an activity on that dimension. The secondary dimension was labelled "degree of organized safety." About half of the subjects judged riskiness to increase with decreases in "organized safety;" the other subjects roughly seemed to indicate that riskiness increased with increases of organized safety. This research further revealed that independent personal judgments of riskiness, beneficiality, and acceptability of the same set of 26 activities were related to one another in a way that would help to understand why certain dimensions of risk are the subject of social controversies, while others are not.

A third descriptive strategy starts from general similarity judgments on pairs of risky activities. When all possible different pairs of a sufficient number of risk stimuli have thus been rated, a multi-dimensional scaling analysis may yield a configuration of the stimuli from which the original similarities may be approximately reproduced. At an individual level this was done for respondents in the various interest groups of subproject 2 (Table 7), yielding a two-dimensional solution for each person separately. Thereafter a subject's judgments of the same set of stimuli on a number of explicit "risk aspects" (as in Table 6) were linearly fitted into his or her own stimulus configuration. Thus it became possible to estimate -- by the squared multiple correlation coefficient -- the "saliency" that an explicit risk aspect must have had during the pairwise similarities judgment task. Kuyper and Vlek (1984, in press) give more details and some results.

These and similar descriptive analyses on personal judgments of risky activities do, however, suffer from a number of weak points summarized in Table 8. First, the underlying research strategy is utterly descriptive. Raw judgments are taken for granted while the question is being answered as to what fundamental dimensions might help us explain their structure and variation. The analyses are often carried out on aggregate group data so that little can be learned about individual processes of risk judgment. Secondly, these raw judgments are collected in a relatively quick-and-dirty manner by way of written or verbal responses. This implies that they are sensitive to a number of factors obstructing a clear correspondence between expressed preferences and actual behavioral choices.

Third, due to the lack of basic (large-scale) risk theory it has been too easily believed that experimental respondents could be asked, and would be able to compare, risky activities that rarely had to be compared in practice (for example: skiing and the use of pesticides, or car driving and the consumption of contraceptive pills). Decision-theoretically, risks should only be compared to the extent that the activities

Table 8: Weaknesses of risk-benefit perception research

1. research strategy utterly descriptive; little is learned about processes of risk judgment
2. raw judgments taken for granted and subject to peculiarities in the attitude-behavior relationship
3. tasks require comparisons among basically incomparable activities (e.g., skiing and the use of pesticides)
4. risks to be judged independently of benefits and other contextual decision factors; seems unjustified
5. little attention for "risk-taking can be fun" (if you can manage)
6. studying relationship between subjective and "factual" annual fatality rates does not seem to be relevant for safety policies

to which they are associated can be incorporated into one and the same decision problem (e.g., "would you rather have your electricity from a nuclear power station or from a wind turbine park?").

A fourth difficulty resides in the fact that risks are to be judged independently of any benefits. This completely ignores the desperation of experimental decision researchers over the elusive nature of the risk concept and its connection with expected value (e.g., Coombs, Donnell and Kirk, 1977; Schaefer, 1978). From a strict decision-theoretic point of view, risk is a by-product of structuring and numerically characterizing a decision problem, whose formal solution is entirely feasible by way of a probability and utility analysis over possible consequences of any alternative. Psychologically, it is as yet unclear whether observable patterns of fear, anxiety or worry are directly related to any available definition of risk as it applies to a particular situation. We should hasten to realize, of course, that these patterns are related to several informal and specific variables characterizing such situations. But the combined effects and possible interactions of the latter have not been documented with respect to large-scale risky activities (see also footnote 2). For one thing, it would be worthwhile to investigate whether risks, as formally defined, can at all be judged independently of benefits and other, possibly contextual aspects of risky decision problems.

Fifth, the idea that "risk-taking can be fun" is hardly deliberated in recent risk perception research (e.g., Allen, 1983). To what extent can (and do) large-scale technological risks per se fulfill human needs for arousal and sensation? Do they underlie a positive "utility of gambling," or do they constitute a challenge (to some ...) rather than a threat (to others ...)? It would seem necessary to relate this point to the concept of perceived controllability as discussed in Section 3. Risk can indeed be fun, if you think you can manage.

One peculiar issue in the judgment of risk is the relationship and the possible discrepancies between subject-estimated frequencies of annual fatalities due to a variety of causes, and known figures available from statistical databases (Lichtenstein et al., 1978; Vlek and Stallen, 1979; Green and Brown, 1980; Renn, 1981; Von Winterfeldt, John and Borcherding, 1981). Why should it be troubling us, and what indeed could we undertake, when it is noted that most people underestimate the annual number of deaths through the medical use of anesthetics whereas they tend to overestimate fatalities due to riding on motorbikes? To begin with, the "objective" figures themselves would be a meager basis for making any general safety decision, for reasons enumerated in section 4 on the limitations of probability assessment. In addition, it can not be assumed that particular individuals base their own risk-taking behavior upon intuitive statistics about a large number of anonymous cases, that are unlikely to contain the particular sets of circumstances under which they are likely to confront a given kind of risk.

Personal estimates of annual fatality rates due to varous causes do provide interesting grounds for interpretation and further questions. This type of study, however, should give way to more fundamental research on the kind of scripts, scenarios, or generalized hypotheses that different people tend to use when asked how risky an unfamiliar activity, product or profession is for the average citizen or for a particular group of citizens.

Recent research on the perception and judgment of risks has been the focus of other criticisms as well (e.g., Otway and Thomas, 1982). Reflecting upon the somewhat naive assumption that the public's perception of risks from nuclear power could be "corrected" in view of expert assessments, Thomas and Bailley (1982, p. 21) note:

> But as subjectivism has gained strength, not least among the experts and risk analysts themselves, social scientists have tended to suspend judgment on the usefulness and applications of their research, and pressed on with data collection in the service of "understanding" but with no clear idea of how this understanding can and should be used. (...) there is a danger that the research objectives are confused to avoid acknowledgement of the political consequences that all knowledge of this kind implies.

Despite these various reservations and criticisms we must conclude that psychological research on the perception and judgment of risks and benefits seems urgently needed and potentially very useful. Several arguments for this are summarized in Table 9. Related, and other arguments, are discussed by Slovic, Fischhoff and Lichtenstein (1982). It should also be

Table 9: Relevance of risk-benefit perception studies

* clarify concepts, variables and methods to constitute an efficient language for communicating about risky activities;
* bring to light implicit assumptions, tacit suppositions and personal hypotheses that may influence risk debates improperly and unintendedly;
* suggest experimental research to test theoretical hypotheses about (large-scale) risk-taking mechanisms, either individual or collective;
* provide theoretically sound and methodologically rigorous descriptions of substantive points of view about the nature and value of risks and benefits;
* provide operational dimensions to be used as (dis-) utility attributes in formal policy decision analyses;
* allow for characterization of decision situations of different parties or persons concerned;
* raise question of optimal social decision procedures in view of problem scope and tensions among experts, administrators and the general public.

clear by now that the theoretical basis of this research can and should be strengthened and elaborated, as was already suggested in Section 3.

Evaluation of Hazardous Materials' Uses by Different Interest Groups

A number of issues discussed above have been the focus of attention in subproject 2 of the study described in Table 7. The original questions underlying this psychological investigation were:

1) What do relevant interest groups know about various risky activities (such as using different hazardous materials)?

2) How do such groups judge various attributes of a risky activity, and how do these judgments combine into an overall evaluation of an activity's "acceptability"?

3) What is the influence of pertinent information on individual and group judgments?

4) How do some groups think that other groups evaluate certain risky activities?

In addition to these questions there were several hypotheses about effects on judgments of different kinds of interest in LPG and degrees of decisional power associated with various groups. Altogether these led to the following empirical approach.

Method: Interest groups relevant for questions around LPG were identified and heuristically ordered along two factors -- Kind of Interest and Decisional Power -- following the design given in Table 10. Decisional power should be understood in terms of large-scale decisions about LPG. Each of the 220 subjects was carefully approached and taken through a psychometric interview led by a trained interviewer. Interviews were conducted individually in a subject's home or office, and lasted for about two and a half hours in two episodes separated by a break (which could be between 15 minutes and one day). Interviewers performed on the basis of a standard text consisting of questions and instructions accompanying appropriate stimulus items. Subjects gave their answers orally as well as in writing, mostly in multiple-choice response format, whereby specific response sheets were used throughout. The major part of the interview consisted of the following five tasks:

Table 10: Subject design of field experiment.

KIND OF INTEREST IN USE OF LPG	DECISIONAL POWER (AND EXPERTISE) ABOUT LPG high	low
direct	20 company managers	40 gas station attendants
indirect	20 civil servants Economic Affairs	40 private car owners using LPG as fuel
conflicting	20 civil servants Environmental Affairs	40 male neighbours / 40 female neighbours } of LPG gas station

1) Rating the similarity of 45 pairs of the 10 hazardous activities (stimuli), given in table 12

2) Judging the same 10 stimuli on a number of standard attributes (Table 13)

3) Responding to 20 statements concerning different basic attitudes

4) Responding to 36 items testing one's knowledge about hazardous materials

5) Indicating the position of "an acceptable LPG gas-station" on nine different criteria

Here, the main results will be discussed as obtained through Tasks 2, 3, and 5. An analysis of pairwise similarity judgments and the "saliency" (cf. Section 5.1) of standard attributes is presented, among other things, in Kuyper and Vlek (1984, in press).

Results on Attitude Ratings: In Task 3 all respondents in the LPG-study were asked to respond to 20 attitude questions cast in multiple-choice format. The set of 20 were originally designed to be an operationalization of five global attitudes. A typical statement contributing to a general score on "feelings of insecurity toward the use of hazardous materials" was, for example:

"Are you ever worried yourself about the possibility of getting involved in a serious accident with hazardous materials?"

1. I am rarely or never worried about it,
2. I am sometimes worried about it,
3. I am regularly worried about it,
4. I am continuously worried about it.

(please mark one)

Under general "appreciation for technological developments" a typical statement was:

"Science and technology stimulate developments that I can approve of":

1. I very much agree with this,
2. I tend to agree with this,
3. I tend to disagree with this,
4. I very much disagree with this.

(please mark one)

A reliability analysis over the five original attitude constructs and an accompanying factor analysis over the 20 statements led to a condensation into three basic attitude scales: (1) "feelings of insecurity toward hazardous materials," (2) "feelings of powerlesness with regard to hazardous materials policies," and (3) "general appreciation for technological developments." Table 11 presents the average scores on these reliable attitude scales of the six groups of respondents in Table 10. It is apparent from Table 11 that the lowest general feeling of insecurity is noted with the group of gas station attendants who are supposed to be both directly interested in the use of LPG and have a low decisional power. LPG-managers, too, have low feelings of insecurity. The highest insecurity seems to occur with the two groups, both high and low in decision power, who are supposed to have a conflicting interest regarding the use of LPG.

With respect to "feelings of powerlesness with regard to hazardous materials policies" there is an overall significant "effect" of the degree of decisional power: those three groups low in decisional power score higher on the powerlesness-scale than their relatively powerful counterparts. This result just validates the heuristic definition and selection of the groups of respondents.

"Appreciation for technological developments" appears to increase with the degree to which one has a direct interest in the use of LPG, regardless of a group's supposed degree of decisional power. Conflicting interest groups definitely score lowest on this attitude scale.

Table 11: Average scores of the six respondent groups on three general attitude scales (see text).

KIND OF INTEREST IN USE OF LPG	DECISIONAL POWER (AND EXPERTISE) ABOUT LPG high		low
direct	3.5	insecurity	2.4
	4.3	powerlesness	5.7
	8.5	technology	7.2
indirect	3.4	insecurity	3.3
	4.2	powerlesness	4.8
	6.4	technology	6.6
conflicting	5.1	insecurity	5.0
	4.1	powerlesness	5.0
	6.0	technology	5.6

Results on Attribute Ratings: In Task 2 every respondent had to judge the ten one-sentence stimuli listed in Table 12 with respect to each of the 11 standard attributes given in Table 13. Upon request they could receive a short description and information on each stimulus. Please note, that in the present terminology, numbers 1-5 in Table 12 would be called small-scale activities, while Numbers 6-10 would be labeled large-scale. The attributes had been formulated in terms of five categories.

Table 12: Risky activities used as stimuli for similarity judgment and explicit attribute rating

1. Using bleaching-liquid for household purposes.
2. Using pain-killers for an extended period of time.
3. Carrying out radiation therapy for medical purposes.
4. Using calor gas at home (for cooking, heating).
5. Driving a car running on LPG.
6. Storage of LPG at a gas station close to a built-up area.
7. Transporting LPG in tank trucks.
8. Storage of suphur dioxyde (SO_2) close to a built-up area.
9. Transporting sulphur dioxyde (SO_2) in tank trucks.
10. Operating nuclear power stations for electricity generation.

Table 13: Set of 11 standard attributes of risky activities. Italicized are brief labels to be used in text.

1. *Imaginability* of a serious accident.
2. *Voluntariness* of exposure to risks.
3. *Social (dis) advantages.*
4. *Personal (dis) advantages.*
5. *Probability* of a serious accident.
6. Probability of an accident in which I am involved *(probability self).*
7. *Avoidability* of accidents.
8. Adequacy of *safety measures.*
9. Size of a serious accident (*catastrophic potential).*
10. Overall *riskiness.*
11. Overall *acceptability.*

Attribute 7, for example, was presented as follows.

"I believe that possible accidents due to this activity are:

* totally unavoidable (1),
* somewhat unavoidable (2),
* reasonably avoidable (3), (please mark one)
* quite well avoidable (4),
* perfectly avoidable (5)".

By exception, attribute 11 was formulated in terms of a three-category scale, as follows.

"Everything taken together I believe this activity to be:

* unconditionally acceptable (1),
* acceptable only if additional safety measures are taken (2), (please mark one)
* absolutely unacceptable (3).

After the necessary checks on the raw data had been carried out a two-mode factor analysis was conducted, with respondents x stimuli as replications equaling a total of 2200. The overall solution appeared to have two factors which could be interpreted as "dangerousness" and "advantageousness," respectively. Separate factor analyses for each stimulus of Table 12 justify the conclusion that the two-factor structure harboring the 11 standard attributes is not stimulus-specific. For example, for all ten stimuli the attribute "riskiness" appeared to have the highest positive loading on factor 1, followed by "probability," "probability self," "catastrophic potential" and "personal (dis)advantages." The two bundles of attributes loading highest on the respective factors appeared to be nonorthogonal to one another. This led us to the observation that nine of the standard attributes of Table 13 (excepting "imaginability" and "voluntariness") could each be interpreted in a favorable as well as an unfavorable direction. For instance, an activity might be considered favorable to the extent that accidents with it would seem "avoidable." Or a judgment of unfavorableness would be assigned to an activity for which "inadequate safety measures" had been taken.

Therefore, a separate reliability analysis was carried out to test whether there were sufficient grounds for the construction of one summary scale called "favorability of the activity," based on the nine attributes indicated before. The reliability coefficient (α) turned out to be moderately high (0.66-0.74) for the "small-scale" activities in Table 12, with the exception of car driving on LPG (0.80). The favorability scales for the "large-scale" activities (numbers 5-10) had a reliability coefficient of between 0.78 and 0.87. These coefficients would not increase when one or more attributes were ommitted from the scale.

Consequently, each respondent's ratings on the nine relevant attributes (numbers 3-11 in Table 13) have been averaged to obtain, for each activity, a "favorability" score. Table 14 presents average favorability scores for the six groups of respondents with respect to the three LPG stimuli of Table 12.

For all three activities the average favorability score is highest for the direct-interest groups and lowest for the conflicting-interest groups. This relationship between kind of interest and height of the favorability score is most pronounced for "LPG transport," but it is statistically highly significant for all three cases. As one would expect there is no "effect" of differences in decisional power. Not surprisingly, the five large-scale stimuli of Table 12 turned out to reveal most of the intergroup differences in favorability, thus manifesting their (potential) controversialism.

Table 14: Average 'favorability' scores for six groups of respondents and three LPG-stimuli (nos. 5-7 in table 12).

KIND OF INTEREST IN USE OF LPG	DECISIONAL POWER (AND EXPERTISE) high		low
direct	4.1	LPG-driving	3.8
	3.6	LPG-storage	3.3
	3.6	LPG-transport	3.3
indirect	3.3	LPG-driving	3.7
	2.2	LPG-storage	2.5
	2.5	LPG-transport	2.7
conflicting	3.2	LPG-driving	3.2
	2.2	LPG-storage	2.0
	2.2	LPG-transport	2.4

Results on Setting Safety Criteria: Task 5 presented the respondents with a number of variables useful to characterize LPG gas stations along the road. Respondents had to indicate a position on each variable that would make the station "acceptable" for them in that particular respect. The four most sensitive variables have been dichotomized as follows: (1) the station could lie either inside or outside urban districts, towns or villages, (2) it could be located at either less or more than 150 ms. from a reasonable collection of occupied houses or buildings, (3) the LPG storage tank could be situated either above or under the ground, and (4) client motorists would be either allowed or not allowed to fill up their LPG gas tanks themselves. The underlined words outside, more, under and not indicate the response category that would correspond to a strict attitude with respect to the safety of the LPG gas station.

Table 15 gives the percentage of respondents in each of the six groups who had chosen to respond in the "strict" direction of each of the four variables. The last row in each division of Table 15 contains an average overall "strictness" score that was composed of each respondent's pattern of acceptability criteria.

Table 15: Percentages of respondents within each group setting 'strict' criteria of outside, more, under and not (see text) for LPG gas stations. The last row in each subdivision represents an average over-all score indicating 'strictness of safety criteria'.

KIND OF INTEREST IN USE OF LPG	DECISIONAL POWER (AND EXPERTISE) high		low
direct	5	outside	25
	0	more	28
	70	under	43
	35	not	63
	2.3	STRICT	3.2
indirect	70	outside	55
	50	more	63
	70	under	70
	95	not	50
	4.9	STRICT	4.5
conflicting	80	outside	80
	25	more	63
	85	under	78
	70	not	85
	4.2	STRICT	5.2

Again, there is a highly significant relationship between the kind of interest and the strictness of a group's safety criteria. Groups with a direct interest in LPG are most lenient, those with a conflicting interest tend to set the strictest safety criteria. The exception is the group of civil servants Environmental Affairs (Table 10) yielding an unexpected low percentage of people responding in the "more than 150 ms." category. Had they adhered to an interim rule decreed in 1978 by the Netherlands Head Inspector for Environmental Hygiene, saying that 150 ms. should be considered a critical distance, their average strictness score would have been higher. Of course, the greater technical knowledge of this particular group may have caused them to be somewhat more lenient of the distance variable than their Head Inspector.

Some Conclusions: Above, only a limited set of results could be presented from subproject 2 as described in Table 7. Herein, however, one general set of conclusions from that project is sufficiently illustrated. Basic attitudes, attribute ratings of risky activities, and specific safety criteria for LPG gas stations tend to follow a similar pattern of differences among groups differing in the kind of interest and their degree of decisional power with respect to the use of LPG.

A comparatively high appreciation for technological developments and low feelings of insecurity concerning hazardous materials (Table 11) go together with relatively favorable attribute ratings of various LPG activities (Table 14); they also correlate with a low degree of strictness in setting safety criteria for LPG gas stations (Table 15). The main factor mediating these group differences seems to be the kind of interest (direct, indirect or conflicting) a group has in the production, distribution and further use of LPG as a car fuel. This is supported by a subtle, but rather consistent difference between the two indirect-interest groups in Tables 11, 14, and 15: Obviously LPG car drivers (indirect interest, low decisional power) have a somewhat more direct interest in LPG than civil servants responsible for Economic Affairs (which include the topic of energy in the Netherlands).

Theoretically, this pattern of empirical results may be taken to manifest the existence of broad and well-organized cognitive structures that respondents address when asked to give judgments concerning the use of hazardous materials in general and of LPG in particular. Specific judgments about hazardous materials and LPG gas stations seem to be thoroughly embedded into a more general set of basic attitudes, which are in turn related to personal role characteristics. This may explain at least one negative finding from this study: The prior provision of a carefully balanced brochure, giving technical as well as economical information about LPG, did not appear to affect either the structure or the substance of judgments of those subgroups of respondents who had carefully read it before participating in the psychometric interview. This would seem to indicate that public information about particular large-scale activities whose risks are being debated, should be designed and communicated in a broad and much-encompassing fashion, in line with their special characteristics and the corresponding problems of large-scale decisionmaking as discussed in Section 2 of this paper.

DECISION-ANALYTIC APPROACHES TO RISK

It was noted in Section 3 that decision theorists consider safety problems to be decision problems. In their view, which level of risk is acceptable depends upon which alternative course of action in a risky choice problem is most attractive, or least rejectable (Vlek and Stallen, 1980; Von Winterfeldt, 1980; Derby and Keeney, 1981). Thus two critical questions are (1) What sort of decisions are constituting the basis of comprehensive safety policies? and (2) Which approach may be followed to arrive at "good decisions"? In this section these questions will be discussed as a preamble to a presentation of some results from subproject 3 of the study described in Table 7.

Types of Decisions and Different Approaches

Safety policies may be fruitfully considered as resting upon three different types of decision. First, there are the strategic decisions pertaining, for example, to the introduction of a new enterprise such as scale oil production from tar sands. In fact, such decisions are infrequently taken explicitly. Rather they occur as an outgrowth of an ongoing technological development. Past investment patterns sometimes make them hard to avoid. Second, tactical decisions deal with problems of, for example, the siting of a particular plant as a base for the implementation of an accepted general strategy. Third, operational decisions may deal with the conditions under which the plant management is given permission to put it into operation.

One problem in understanding public and expert controversies over technological risk is to specify the type of decision that is being debated either implicitly or explicitly. For example, public protests against a siting proposal for a particular industrial plant may be easily met if a change in the siting plan would have no consequences for the implementation of a higher-level strategy (e.g., about total production volume). Or, formal government objections against specific plant operations may, when met, eventually lead to a close-down for reasons of uneconomically high costs of safety measures. When the different parties in a public debate differentially incorporate such considerations in their views of what the problem is, they will, in fact, find themselves talking about different problems.

Another problem in risk acceptance debates often is that government regulators are pressed to take decisions that they are not lawfully empowered to take. For example, at the level of strategic decisions the Netherlands government has little to say in the way of weighing the risks against the expected benefits of a new large-scale activity.[4] Instead, government safety policies are restricted to setting the standards for, for example emissions of toxic wastes or the routing of hazardous materials transports, which would prevent certain accidents from happening or would limit their catastrophic potential. If we accept that safety problems are decision problems, then it seems recommendable that regulatory agencies be given the authority to advise on how to optimally implement strategic goals adopted by their constituents.

The distinction among strategic, tactical, and operational decisions naturally suggests an elaboration into a hierarchy of decisions taken in a time-ordered sequence to implement an adopted general strategy. This would help clarify the process whereby operational risks arise and might be reduced. For it is often in past decisions and their follow-ups that the roots of current risks are to be sought. Such a causal-decision analysis would be related to the causal-structure framework proposed by Hohenemser, Kasperson and Kates (1981).

The second question: "How should optimal safety decisions be taken?" may be given various answers. Large-scale safety problems not being identical to small-scale laboratory gambles, formal models for risky decision have their limitations of practical application. Table 16 presents a number of ways to arrive at an acceptable solution to a complex decision problem. A thorough discussion of these various possibilities lies beyond the scope of this paper. We should note, however, that methods numbers 2, 3 and 4 correspond to the "revealed," "institutional," and "expressed" preferences approaches distinguished by a.o. Kates (1977). Numbers 5-7 are based on formal methods and rules for complex decisionmaking, while numbers 8-10 represent more pragmatic approaches that seem to concentrate on deliberation, negotiation, and the outcomes of cognitive-emotional processes. Evidently, some ways of decisionmaking raise a lot of questions about the who and how of the procedure to be followed. Table 16 deals with questions related to substantive decision criteria, methods for structuring and judging choice alternatives, and procedures for interpersonal debate and social judgment.

Table 16 also suggests that the choice of a judgment method, a set of decision criteria and a procedure for social debate rests upon a meta-decision. In light of a number of require-

---

4 One exception is the Public Inquiry into Future Energy Policies mentioned in Section 7, which precedes a parliamentary decision on future electricity generation in the Netherlands. Table 1: Ten policy questions about large-scale risks.

ments, possibilities, and limitations, present-day decision theorists would opt for a full-scale decision analysis (number 7 in Table 16) as the best practical means towards complex safety decisions. For example, Von Winterfeldt (1980) claims that:

> Decision analysis is the only comprehensive and practical methodology for aiding complex decisions. However, decision analysis needs to be adapted to the political and institutional realities of decision making on problems of technological risks. (p. 37)

Table 16: Possible ways to arrive at acceptable decisions

1. check amount or size of critical consequences against natural levels or dosages
2. accept an activity when its risks resemble and do not exceed those of activities apparently accepted thus far
3. accept an activity when preferences for its main consequences can be inferred from institutionalized laws, regulations and ways of living
4. let relevant interest groups express their current preferences and select the activity that appears most preferred, or least rejected
5. perform a cost-benefit analysis and go for the option having the highest benefit-cost ratio
6. choose the option with the lowest risk among the set of goal- (or cost-) effective options
7. perform a full-scale decision analysis and choose the course of action having the highest expected utility
8. give the problem to an experienced professional decision maker together with the necessary expert advice, and follow his or her recommendations
9. let representatives of a broad spectrum of interest groups debate under a well-trained and neutral chairperson, and follow this group's recommendation to the extent that it is coherent
10. announce the problem widely and declare a trial period in which you carefully observe the reactions - verbal, emotional and otherwise - of a variety of different interest groups; select the course of action that promises to cause the least social turmoil

A detailed discussion of, and further references about, decision analysis are given by Fischhoff, Lichtenstein, Slovic, Derby and Keeney (1981), who also review a number of its drawbacks and shortcomings (see also Kates, 1977; and Schwarz, 1979). Most of these fall in the category of individual-level problems of structuring and evaluating alternative courses of action. For example, "knowing the facts" may be very difficult and this goes for "assessing labile values" and "considering equity criteria" as well.

In the course of the empirical study described in Table 7 we have come to realize, however, that some of the more nasty problems of large-scale decisionmaking reside in the presence of different interest groups. These tend to have their own views of the basic problem, as well as their own sets of goals and values, and they tend to weigh certain sources of information as more important than others. They might even have different preferences in connection with the meta-decision about which way of decisionmaking to adopt. In view of these social- and individual-level problems of formal decision analysis it is questionable whether any practical application to complex safety problems could, in the end, be very much more than a compromise based on possibilities 7, 8 and 9 in Table 16.

Some Results of an Empirical Study[5]

Although subproject 3 of the study on hazardous materials' uses (Table 17) was started with the good intention to apply and test methods based on "rational decision theory," our basic strategy has gradually taken a different tack. Differences in views and preferences of relevant interest groups as well as practical limitations in invited respondents' motivation, curiosity and available time, have led us to the conclusion that the meta-decision about acceptable methods of large-scale decisionmaking should get more attention than some formal evaluation of alternative hazardous materials' uses by itself. The latter, therefore, is taking up only a modest position in the results to be reported. We will discuss the part of the subproject that deals with the use of SO_2 (sulphur dioxyde as used to facilitate beet sugar production) rather than LPG.

Method: Three different interest groups were carefully approached to yield 24 civil servants responsible for hazardous

---

5 The principal investigator and data-analyst of subproject 2 has been Hans Kuyper, whereas Henk Boer has played a central role in subproject 3. Both are associated with the Environmental Studies Center of the University of Groningen.

materials regulation (employed by the department of traffic and waterways), 24 beet sugar company managers, and 24 people living in the neighborhood of two large beet sugar factories. These 96 participants were each individually invited and taken through a carefully designed interview held at the respondent's home or office. The interview lasted for about two hours and consisted of standard instructions and questions to be responded to verbally or through filling in pre-designed response sheets.

There were three basic tasks. First, respondents had to consider eight different ways of producing beet sugar, either with or without the use of more or less sulphur dioxyde to facilitate parts of the process. As part of this task they also had to indicate which aspects of the eight alternatives they found most important to take into account when one option had to be selected.

In their second task respondents considered 10 different interest groups of people who (might want to) play a role in regulatory decisionmaking about the use of SO_2. They then had to rank order the 10 groups according to the "decisional influence" each group should, in the respondent's view, be given in the social regulatory process. In a later part of this task respondents also had to rank order eight relevant group characteristics (such as technical expertise or being exposed to risks) following the degree to which each characteristic would entitle a group to play a significant role. Finally, respondents judged each of the 10 groups with respect to their position on each of the eight group characteristics employed (e.g. "How technically expert do you judge labor union representatives to be?").

The third task demanded of respondents to rank order six different ways of deciding about the optimal use of SO_2, with respect to their attractiveness and feasibility. Each procedure was briefly described on a card. In addition they were asked to rate the desirability of a number of different properties that might characterize a procedure for regulatory decisionmaking.

Some Results: Table 17 gives average importance ranks assigned by the 24 respondents in each group to six relevant aspects of possible ways to use (or not use) SO_2 in the production of beet sugar. Statistically, the italicized figures are significantly different from the other two figures in a particular row. Evidently, managers of beet sugar companies express a fairly distinct view: Their favorite aspects are quality and price of the sugar, and they assign less weight to the amount of SO_2 used and the manner in which it is handled and stored. There is not much difference between the views of civil servants and neighbors who are both paying more attention to the SO_2.

Table 17: Average importance ranks of six characteristics of different ways to produce beet sugar with or without SO_2, as obtained from three different interest groups.

ATTRIBUTES	GROUPS OF RESPONDENTS civ.serv.	managers	neighbors
SO_2 is/is not produced at the plant	3.7	2.9	3.8
sugar quality	3.2	5.0	3.7
manner of transport. and storage of SO_2	4.3	3.3	3.9
amount of SO_2 used	3.9	2.9	4.0
number of SO_2 transports required	3.5	2.5	3.5
final sugar price	2.3	4.4	2.1

In connection with their most important utility attributes, company managers express a strong preference for the "current process" in which a fair amount of SO_2 is used in two critical stages of sugar production. In contrast, the groups of civil servants and neighbors most prefer the so-called German process without SO_2, which results in a type of sugar that is slightly less pure-white and slightly more expensive.

Table 18 presents the same three groups' average ranks concerning the decisional influence each of 10 different possible interest groups would deserve in the regulatory process. All three groups appear to allow the greatest decisional influence to government safety inspectors (1) and civil servants responsible for environmental affairs (4). The three groups also seem to agree that civil servants responsible for economic affairs (3) and relevant labor unions (6) deserve the least decisional influence. For the other interest groups considered there are significant differences in the degree of decisional influence the three groups are willing to assign.

Remarkably, both the civil servants and the company managers assign the highest average rank to their own groups, numbers 2 and 8, respectively. Neighbors of beet sugar factories seem to be less arrogant: They recognize the decisional weight of interest groups 1 and 4, followed by 8, 9 and 5, thereby putting their own group at a modest average rank. Again, company managers have an outspoken view: They would rely most on their own group's preferences, and they tend to assign the least decisional influence to citizens for environmental protection (7), relevant labor unions (6) and people living in the neighborhood of a beet sugar factory (5).

Table 18: Average decisional-influence ranks due to 10 possible different interest groups, as assigned by three groups of respondents.

INTEREST GROUP CONSIDERED	GROUPS OF RESPONDENTS		
	civ.serv.	managers	neighbors
1. government safety inspectors	7.5	7.9	7.7
2. civ.serv. traffic/waterways	8.3	6.3	5.2
3. civ.serv. economic affairs	3.0	4.1	3.7
4. civ.serv. environm. affairs	7.5	7.9	7.7
5. neighbors of sugar factory	5.6	3.5	5.8
6. relevant labor unions	3.5	3.3	4.6
7. citizens for environm. protect.	5.0	2.4	5.3
8. managers beet sugar factories	6.5	9.1	5.9
9. lower personell sugar company	5.0	5.4	5.9
10. sugar processing industries	3.0	5.2	3.2

In line with these rankings according to decisional influence are respondents' judgments about which group characteristics most entitle a group to play a significant role. For example, all three groups agree that "technical expertise," "being exposed to risks and nuisance," and "having a well-considered opinion" are most important bases for an interest group's decisional influence. Other aspects, like "defending consumer interests" and "constituting an organization" are considered relatively unimportant.

Also in line with rankings of decisional influence are the three groups of respondents' ratings of each of the 10 interest groups with respect to various group characteristics. The relatively low position assigned by the managers to the citizens for environmental protection (7) in Table 18, for example, is related to the managers' low average judgment of this particular interest group's degree of technical expertise, which significantly deviates from the corresponding judgment by the civil servants and the neighbors.

As a final result to be reported, Table 19 presents the average "desirability" ranks assigned by the three groups to each of six alternative decision procedures (cf. Table 16). Somewhat surprisingly, the three groups are largely in agreement concerning the most preferred decision approaches: "professional judgment" and "decision analysis" score highest, followed by "revealed preferences" (a safe historical basis ..?). Cost-benefit analysis scores lowest in two of the groups, and it is significantly less looked down at by company managers. The latter definitely see the least use for a reliance on expressed preferences, an approach that is intermediately preferred by civil servants as well as neighbors.

Table 19: Average desirability-ranks of six different decision procedures as assigned by three groups of respondents. Numbers in parentheses refer to table 16.

DECISION PROCEDURES	GROUPS OF RESPONDENTS		
	civ.serv.	managers	neighbors
natural levels or dosage (1)	2.9	2.7	3.0
revealed preferences (2)	3.4	3.7	3.4
expressed preferences (4)	2.7	1.8	3.3
cost-benefit analysis (5)	2.0	2.8	1.9
decision analysis (7)	4.8	4.7	4.1
expert prof. judgment (8)	5.2	5.3	5.3

Preliminary Conclusions: Subproject 3 of the study described in Table 17 is still being written up. Therefore, only preliminary conclusions can be presented here. First, respondents from different interest groups appear to assign different importance weights to essential utility attributes of alternative processes of beet sugar production. This may help explain differences among preference orders concerning production processes as given by the same groups. There is no easy way to combine such consistent differences into an overall preference order acceptable by all groups.

Second, if one seeks to resolve the problems of social choice by assigning different degrees of decisional influence to different interest groups, then it should be realized that each group tends to judge the others in its own way with respect to relevant group characteristics. This may explain why interest groups differ in the way they rank various groups with respect to the decisional influence these would be entitled to get in the regulatory process.

Third, an attempt to reach agreement on a meta-decision about which decision approach to select for the resolution of a particular safety problem might not easily meet with success. Different interest groups tend to weight different properties of decision procedures in divergent ways so that, in consequence, their preference orders over a set of possible different procedures may deviate from one another. Empirically, the present results are not disheartening on this point. From the preference ranks given in Table 19 one might conclude that some wise combination of professional judgment and explicit decision analysis may well be both a workable and an acceptable procedure for tackling complex safety decision problems.

SUMMARY AND SOME BROADER ISSUES

In this paper a psychological decision theorist's view is given of the so-called societal debate on large-scale risks from technological developments. Ten policy questions about this general topic led to a charac- terization of large-scale activities and to a listing of cor- responding problems of policy decisionmaking. It was concluded that formal definitions of risk and "acceptability" badly needed the flesh and blood of a more specific collection of psychological and social aspects of risky decision problems. Separate attention was given to limitations of probability assessment in risk analyses, to the perception and judgment of risk, and to decision-analytic approaches toward safety prob- lems, respectively. Intermittent empirical illustrations were derived from a multidisciplinary field-research project on hazardous materials' uses (Table 17). Along the way, a number of suggestions for further theorizing and empirical research have been given. To round off, we will consider a few general issues that may put the paper in its proper context.

The Real Problem

The essence of the problem of large-scale risks seems to lie in threats of loss of human control over the course of individual as well as social life. Increasing sophistication and differentiation of technological applications continuously call for a strengthening and elaboration of control systems and regulations by industry, government and other social institutions. The individual and social benefits stemming from large-scale facilities pose a number of dilemmas-of-control. One dilemma is that increased control tends to go along with abstract and global conceptualizations of a system, at the expense of concrete knowledge of its specific operations. A second dilemma resides in the stability required for expert system-managers to operate on the basis of an accurate and reliable model representation of the system at hand. With highly complex systems, however, stability may easily turn into a degree of rigidity which obstructs any flexibility needed to adapt to changing circumstances. A third dilemma-of-control occurs with the introduction of new technological applications. Their social meaning and consequences could not have been tested under controlled experimental conditions. But when they become apparent in full scope it may be too costly or plainly impossible to change or halt unwanted developments (Collingridge, 1980). Remarkably enough, these and other dilemmas are often "solved" by means of further extensions in the techno-bureaucratic control apparatus. This development is in sharp contrast with emancipatory movements and tendencies within modern industrial societies.

The Meaning of Risk: Large-scale risks from technological developments constitute one focus of attention when modern society's real problem is being debated. Formal models of risky decision offer a set of theoretical concepts, which is perhaps good enough to describe the surface structure of a safety problem. It seems rather necessary, however, to understand the deep and dynamic structure of large-scale risks as they may afflict feelings and emotions as well as intended-rational analyses and judgments concerning policy options. "Risk," like stress, can be given a stimulus-definition, as when described in terms of the characteristics of an external environment. It may also be given a response-definition in terms of observable behavior patterns such as sweating, screaming, running away, or meeting the challenge head-on. Most useful, however, seems a transactional definition of risk, in terms of observational and appraisal processes, emotional and activation patterns, and behavioral or performance processes. Although the latter definition is tuned to the level of the individual person, it may well be given an operational meaning in terms of collective reaction and response processes.

Levels of Theorizing: The societal debate on large-scale risks looks different from different theoretical perspectives. Desires for setting practical standards compete with pleas for understanding the deeper causes of safety problems. Acknowledging the hierarchical structure of safety policies in terms of strategic, tactical and operational decisions (cf. section 6) may help to improve the setting of agendas for future research. Wynne (1983) has argued that there is a vast alienation between the innovation and commitment phase of a new technological application and its social implementation. This, he claims, frequently leads to organizational incoherence and thus unviability of the technology. Adequate multilevel modeling of what it means to develop, test, and implement a large-scale facility might limit the full extent of such multifaceted problems.

Demands for an Interdisciplinary Approach: Together, the various sections of this paper demonstrate both the need for, and the possibilities of an interdisciplinary approach to the management of large-scale risks. There is a special appeal to representatives of the engineering, psychological and political sciences. However, researchers from different backgrounds and devotions may easily find themselves constructing a true Tower of Babel when attempting, from their own perspective, to deal with what seems to be one and the same problem. At the end of a comprehensive review of recent multidisciplinary research on human judgment and decisionmaking Pitz and Sachs (1984) say:

> As soon as one tries to use theoretical insights to solve significant social problems, the impossibility of disciplinary compartmentalization becomes apparent.

This statement is highly applicable to the societal debate on large-scale risk. Its main issues may prove unresolvable unless people from widely different inclinations and training sit together and talk while attempting to demolish the blinding walls of scientific disciplines.

Multiparty Approaches and Public Participation: At the end of Section 6 it was concluded that some wise combination of expert professional judgment and explicit decision analysis seemed most acceptable for dealing with complex decision problems. Such an approach needs to be elaborated and "socialized." When socio-technical systems are meant to serve, and potentially harm, ever larger segments of a country's population, the development of pluralistic social decision schemes becomes all the more urgent. Kunreuther and Linnerooth (1983, ch. 3) have modeled a multiattribute multiparty (MAMP) approach to complex decisionmaking, which they and others have illustrated to be useful in understanding policy formation concerning the siting of liquid energy gas facilities. In various countries, methodologically less rigorous approaches have been followed, for example, the Windscale Inquiry in the U.K. (see Wynne, 1982), the Berger Inquiry (Gamble, 1978) and the activities of the Ontario Royal Commission on Electric Power Planning (Dooley and Robinson, 1982) in Canada; and very recently, the Public Inquiry into Future Energy Policies (or "Broad Societal Debate") in the Netherlands (1981-1983). The theoretical and methodological basis of such multiparty inquiries and debates can and should be improved.

Rationality, Objectivity and the Role of Experts: This issue is nasty and complicated because it boomerangs at the researcher's own position. In a thorough discussion of "the two camps on rationality" Jungermann (1983) concludes that rationality is conditional upon the model formulated to represent a complex problem. Only within a well-defined model may one attempt to be rational, and in a number of different ways at that. Perfect, comprehensive rationality is a utopian dream. Therefore, whoever claims to be rational should specify the model in terms of which this would be so, and in which particular regard (s)he means to so behave. A similar argument pertains to the concept of "objectivity." Modern philosophers of science and cognitive psychologists would be among the first to proclaim the downfall of methodological neopositivism.

They would maintain that facts and values cannot be separated, and that even scientists are dependent upon socio-historical and (sub-)cultural schemas of concepts and their interrelationships in order to model parts of the world and interpret empirical data about them. In line with this, experts are easily suspected these days, because they are being denied the objectivity and value-neutrality traditionally assigned to them. Thus, we might conclude that scientific rationalism is gradually giving way to some sort of democratic relativism (Feyerabend, 1980), in which complex safety problems are resolved after a decent amount of consensus has been reached through a procedure that different interest groups have been able to subscribe to in advance.

REFERENCES

1. Allen, V.L., Eutropistic reactions to risks and hazards. In B. Berglund and C. Lévy-Leboyer (Eds.), New trends in environmental psychology. Beverly Hills, CA: Sage, 1983.

2. Brehm, J.W., Responses to loss of freedom: a theory of psychological reactance. Morristown, NJ: General Learning Press, 1972.

3. Brichácek, V., Use of subjective probability in decision-making. In G. De Zeeuw et. al. (Eds.), Acta Psychologica, 1970, vol. 34, no. 2/3 (see reference below).

4. Collingridge, D., The social control of technology. London: F. Pinter, 1980.

5. Coombs, C.H., A review of the mathematical psychology of risk and risk-taking. Michigan Mathematical Psychology Program Report MMPP 72-6, 1972.

6. Coombs, C.H., Portfolio theory and the measurement of risk. In M.F. Kaplan and S. Schwartz (Eds.), Human judgment and decision processes. New York: Academic Press, 1975.

7. Coombs, C.H. and Avrunin, G., Single-peaked functions and the theory of preference. Psychological Review, 1977, 84, 216-230.

8. Coombs, C.H., Donnell, M.L., and Kirk, D.B., An experimental study of risk preferences in lotteries. Michigan Mathematical Psychology Program Report MMPP 77-6, University of Michigan, 1977.

9. Derby, S.L. and Keeney, R.L., Risk analysis: understanding "How safe is safe enough?" Risk Analysis, 1981, 1, 217-224.

10. De Zeeuw, G., Vlek, C.A.J., and Wagenaar, W., Subjective probability; theory, experiments, applications. Acta Psychologica, 1970, special volume 2/3.

11. Dooley, J.E. and Robinson, J.B., Nuclear risk management in Canada: an examination of the Ontario Royal Commission on Electric Power Planning. Final Report prepared for the Beyer Institute of the Royal Swedish Academy of Sciences and the Swedish Energy Research and Development Commission. University of Toronto Institute for Environmental Studies, July 1982.

12. Eysenck, M.W., Attention and arousal: cognition and performance. Berlin: Springer Verlag, 1982.

13. Feyerabend, P., Democracy, elitism, and scientific method. Inquiry, 1980, 23, 3-18.

14. Fischhoff, B., Slovic, P., Lichtenstein, S., Read, S., and Combs, B., How safe is safe enough? A psychometric study of attitudes towards technological risks and benefits. Policy Sciences, 1978, 9, 127-152.

15. Fischhoff, B., Lichtenstein, Slovic, P., Derby, S.L., and Keeney, R.L., Acceptable risk. London: Cambridge University Press, 1981.

16. Gamble, D., The Berger Inquiry: an impact assessment process. Science, 1978, 199, 946-952.

17. Garber, J. and Seligman, M.E.P. (Eds.), Human helplessness. New York: Academic Press, 1980.

18. Green, C.H. and Brown, D., Through a glass darkly: perceiving perceived risks to health and safety. Paper prepared for Workshop on Perceived Risk (Eugene, OR), Dundee: Duncan of Jordanstone College of Art, 1980.

19. Groningen Provinciale Waterstaat, Normen op het gebied van de veiligheid met betrekking tot gevaarlijke stoffen. Groningen (Neth.), 1979.

20. Haastrup, P., Design error in the chemical industry. In Proceedings of the 4th International Symposium on Loss Prevention and Safety Promotion in the Process Industries. E.F.C.E. Publications Series No. 33, volume 1, 1983.

21. Hartman, B., Human factors aspects of aircraft accidents and incidents. AGARD Conference Proceedings, 1979, No. 254.

22. Hohenemser, C., Kasperson, R.E., and Kates, R.W., Causal structure: a framework for policy formulation. In C. Hohenemser and J.X. Kasperson (Eds.), Risk in the technological society. American Association for the Advancement of Science, Boulder, CO: Westview Press, 1981.

23. Jungermann, H., The two camps on rationality. In R. Scholz (Ed.), Decisionmaking under uncertainty. Amsterdam: North-Holland Publishing Company, 1983.

24. Karlsen, J.E. and Kristansen, S., Statistical survey of collisions and groundings of Norwegian ships for the period 1970-1978. Det Norske Veritas Report No. 80-0199, Oslo 1980.

25. Kaplan, S. and Garrick, B.J., On the quantitative definition of risk. Risk Analysis, 1981, 1, 11-27.

26. Kates, R.W., Summary report. In R.W. Kates (Ed.), Managing technological hazard: research needs and opportunities. University of Colorado Institute of Behavioral Science, Program on Technology, Environment and Man, Monograph No. 25, 1977.

27. Keeney, R.L. and Raiffa, H., Decisions with multiple objectives: preferences and value tradeoffs. New York: Wiley, 1976.

28. Kogan, N. and Wallach, M.A., Risk taking as a function of the situation, the person and the group. In Th. Newcomb (Ed.), New directions in psychology, volume 3. New York: Holt, Rinehart and Winston, 1967.

29. Kozielecki, J., Environment and personality as determinants in decisionmaking. Polish Psychological Bulletin, 1974, 5, 3-11.

30. Kunreuther, H. and Linnerooth, J. (Eds.), Risk analysis and decision processes. Berlin: Springer Verlag, 1983.

31. Kuyper, H. and Vlek, Ch., Contrasting risk perceptions among interest groups. Acta Psychologica, 1984 (in press).

32. Lazarus, R.S., The stress and coping paradigm. In C. Eisdorfer, D. Cohen and A. Kleinman (Eds.), Theoretical bases for psychopathology. New York: Spectrum, 1980.

33. Libby, R. and Fishburn, P.C., Behavioral models of risk taking in business decisions: a survey and evaluation. Journal of Accounting Research, 1977 (Autumn), 272-292.

34. Lichtenstein, S., Slovic. P., Fischhoff, B., Layman, M., and Combs, B., Judged frequency of lethal events. Journal of Experimental Psychology (Human Learning and Memory), 1978, 4, 551-578.

35. Lloyd, R.C., Haeberlin, S.W., Clayton, E.D., and Carter, R.D., Assessment of criticality safety. Nuclear Technology, 1979, 42, 13-22.

36. Lowrance, W.W., Of acceptable risk: science and the determination of safety. Los Altos, CA: Kaufmann, 1976.

37. Myers, D.G. and Lamm, H., The group polarization phenomenon. Psychological Bulletin, 1976, 83, 602-627.

38. Otway, H. and Fishbein, M., Public attitudes and decision-making. International Institute for Applied Systems Analysis, RM-77-54, 1977.

39. Otway, H. and Thomas, K., Reflections on risk perception and policy. Risk Analysis, 1982, 2, 69-82.

40. Otway, H. and von Winterfeldt, D., On the social acceptability of technologies. Policy Sciences, 1982, 14, 247-256 (also, with different title, in H. Kunreuther (Ed.), Risk: A seminar series. International Institute for Applied Systems Analysis, CP-82-S2, Laxenburg, Austria, 1982).

41. Pitz, G.F. and Sachs, N.J., Judgment and decision: theory and application. Annual Review of Psychology, 1984 (in press).

42. Ravetz, J. (Ed.), The acceptability of risks. London: Barry Rose Ltd., in association with the Council for Science and Society, 1977.

43. Renn, O., Man, technology and risk. A study on intuitive risk assessment and attitudes towards nuclear energy. Jülich (GFR): Kernforschungsanlage, Programmgruppe Kernenergie and Umwelt, June 1981.

44. Rowe, W.D., An anatomy of risk. New York: Wiley, 1977.

45. Sabey, B.E. and Taylor, H., The known risks we run: the highway. In R.C. Schwing and W.A. Albers, Jr. (Eds.), Societal risk assessment. New York: Plenum Press, 1980.

46. Schaefer, R., What are we talking about when we talk about "risk"? A critical survey of risk and risk preference theories. International Institute for Applied Systems Analysis, RM-78-69, Laxenburg, Austria, 1978.

47. Schwarz, W.B., Decision analysis, a look at the chief complaints. New England Journal of Medicine, 1979, 300, 556-559.

48. Slovic, P., Assessment of risk taking behavior. Psychological Bulletin, 1964, 61, 220-233.

49. Slovic, P., Information processing, situation specificity, and the generality of risk-taking behavior. Oregon Research Institute ORI Research Bulletin, Vol. 11, no. 3, Eugene, OR, 1971.

50. Slovic, P. and Lichtenstein, S., Relative importance of probabilities and payoffs in risk taking. Journal of Experimental Psychology Monograph, 1968, 78 (3) part 2.

51. Slovic, P., Fischhoff, B., and Lichtenstein, S., Cognitive processes and societal risk taking. In H. Jungermann and G. de Zeeuw (Eds.), Decisionmaking and change in human affairs. Dordrecht (Neth.): Reidel, 1977 (also in J.S. Carroll and J.W. Payne, Eds., Cognition and social behavior. Potomac, MD: Erlbaum, 1976).

52. Slovic, P., Fischhoff, B., and Lichtenstein, S., Facts and fears: understanding perceived risk. In R.C. Schwing and W.A. Albers, Jr. (Eds.), Societal risk taking: how safe is safe enough? New York: Plenum Press, 1980.

53. Slovic, P., Fischhoff, B., and Lichtenstein, S., Why study risk perception? Risk Analysis, 1982, 2, 83-93.

54. Spetzler, C.S. and Staël von Holstein, C.-A.S., Probability encoding in decision analysis. Management Science, 1975, 22, 340-358.

55. Stallen, P.J. M. and Tomas, A., Psychological aspects of risk: the assessment of threat and control. In H. Kunreuther (Ed.), Risk: a seminar series. International Institute of Applied Systems Analysis, CP-82-S2, Laxenburg, Austria, 1982.

56. Starr, Ch., General philosophy of risk-benefit analysis. In H. Ashley, R.L. Rudman and C. Whipple (Eds.), Energy and the environment: a risk-benefit approach. New York: Pergamon, 1976.

57. Thomas, K. and Baillie, A., Public attitudes to the risks, costs and benefits of nuclear power. Paper prepared for a Joint SERC/SSRC Seminar on Research into Nuclear Power Development Policies in Britain, June 1982.

58. Vlek, C.A.J. and Stallen, P.J.M., Persoonlijke beoordeling van risico's; over risico's, voordeligheid en aanvaardbaarheid van individuele, maatschappelijke en industriële aktiviteiten. University of Groningen: Institute for Experimental Psychology, 1979. (Shorter English report in Vlek and Stallen, 1981, see below; results on subject-estimated

annual fatality rates briefly dealt with in Stallen and Tomas, 1982; see above.)

59. Vlek, Ch. and Stallen, P.J., Rational and personal aspects of risk. Acta Psychologica, 1980, 45, 273-300.

60. Vlek, Ch. and Stallen, P.J., Judging risks and benefits in the small and in the large. Organizational Behavior and Human Performance, 1981, 28, 235-271.

61. Vlek, Ch. and Wagenaar, W.A., Judgment and decision under uncertainty. In J.A. Michon, E.G. Eijkman and L.F.W. De Klerk (Eds.), Handbook of Psychonomics. Volume 2. Amsterdam: North-Holland Publishing Company, 1979.

62. von Winterfeldt, D., Four theses on the application of risk assessment methods. In J. Conrad (Ed.), Society, technology and risk assessment. London: Academic Press, 1980.

63. von Winterfeldt, D., John, R.S., and Borcherding, K., Cognitive components of risk ratings. Risk Analysis, 1981, 1, 277-287.

64. Wagenaar, W.A., Menselijk falen. Nederlands Tijdschrift voor de Psychologie, 1983, 38, 209-222.

65. Wallsten, T.S. and Budescu, D.V., Encoding subjective probabilities: a psychological and psychometric review. Chapel Hill: University of North Carolina, L.L. Thurstone Psychometric Laboratory, 1980.

66. Wilde, G.J.S., Theorie der Risikokompensation der Unfallverursachung und praktische Folgerungen für die Unfallverhütung. Hefte zur Unfallheilkunde, 1978, 130, 134-156.

67. Wilde, G.J.S., The theory of risk homeostasis: implications for safety and health. Risk Analysis, 1982, 2, 209-225.

68. Wynne, B., Rationality and ritual: The Windscale Inquiry and Nuclear Decisions in Britain. Chalfont St. Giles (Bucks.): British Society for the History of Science, 1982.

69. Wynne, B., Redefining the issues of risk and public acceptance; the social viability of technology. Futures, February 1983.

Psychological Perspectives on Technology as Societal Option, Source of Hazard and Generator of Environmental Impacts

Kenneth H. Craik

Institute of Personality Assessment and Research, University of California, Berkeley, CA

INTRODUCTION

Technology assessment (TA), risk assessment (RA), and environmental impact assessment (EIA) constitute distinctive but related mechanisms for promoting a proper relation of technology to society and the environment. Respectively, they treat technology as societal option (TA), as source of hazard (RA), and as generator of environmental impacts (EIA). Each of these activities raises psychological issues that are being usefully addressed in research programs.

In a broader sense, the ordinary citizen of a pluralistic, industrial society comes into possible contact with technology in at least three ways: 1) at the level of public policy discourse concerning issues of technology, 2) within the context of decisionmaking about specific technologies and their possible consequences, and 3) as part of the everyday, here-and-now, directly experienced environment.

A societal decisionmaking model for the socio-technological environment will provide a framework for locating technology assessment, risk assessment, and environmental impact assessment within the sequence of policy discourse, decision processes, and environmental outcomes. Psychological issues will be reviewed within this context. (1)

1 This chapter draws in part upon ideas and results from a research program being conducted with David M. Buss and Karl M. Dake, and supported by National Science Foundation Grant PRA-8020017. The author takes sole responsibility for the present formulation, however.

NATO ASI Series, Vol. G4
Environmental Impact Assessment, Technology Assessment, and Risk Analysis. Edited by V.T. Covello et al.

DECISIONMAKING MODEL FOR THE SOCIO-TECHNOLOGICAL ENVIRONMENT: OVERVIEW

Before discussing psychological issues arising within the activities of technology assessment, risk assessment, and environmental impact assessment, these activities must be placed within a conceptual framework of societal decisionmaking.

Flow charts of the decision process can be found within discussions of technology assessment (1, 2, 3, 4), risk assessment (5, 6, 7, 8, 9), and environmental impact assessment (10, 11, 12, 13, 14, 15). However, a unified model incorporating all three endeavors explicitly is rarely encountered.

A unified model of societal decisionmaking regarding the socio-technological environment is presented in Fig. 1. The model has three major components: policy analysis, planning, and environmental monitoring. Policy analysis and planning are considered as technical endeavors that contribute to broader, ongoing societal activities of policy discourse and decision processes, respectively. Technology assessment, risk assessment, and environmental impact assessment are treated as specialties within planning, and are ordered according to the closer affinity of technology assessment to policy analysis and environmental impact assessment to the implemented environment.

The environment represents the everyday, here-and-now physical manifestations of implemented socio-technological decision outcomes. The environmental component of the model encompasses various states, including the pre-implementation environment and the ordinary environment, in which implemented technological processes and facilities function according to expectations. It also includes the malfunctioning environment, which ranges from minor mishaps and failures in technological functioning to disastrous events, which in turn lead to environmental states of emergency, damage assessment, and two phases of reconstruction. Various feedback loops in the model convey the role of policy analysis and planning in seeking to foster adaptive socio-cultural evolution. Feedback from the post-decision environment to planning and policy analysis requires some form of informal or formal environmental monitoring.

POLICY ANALYSIS AND POLICY DISCOURSE

An adaptive relation to its environment is one of the functional prerequisites of a viable society (16). Technology, in

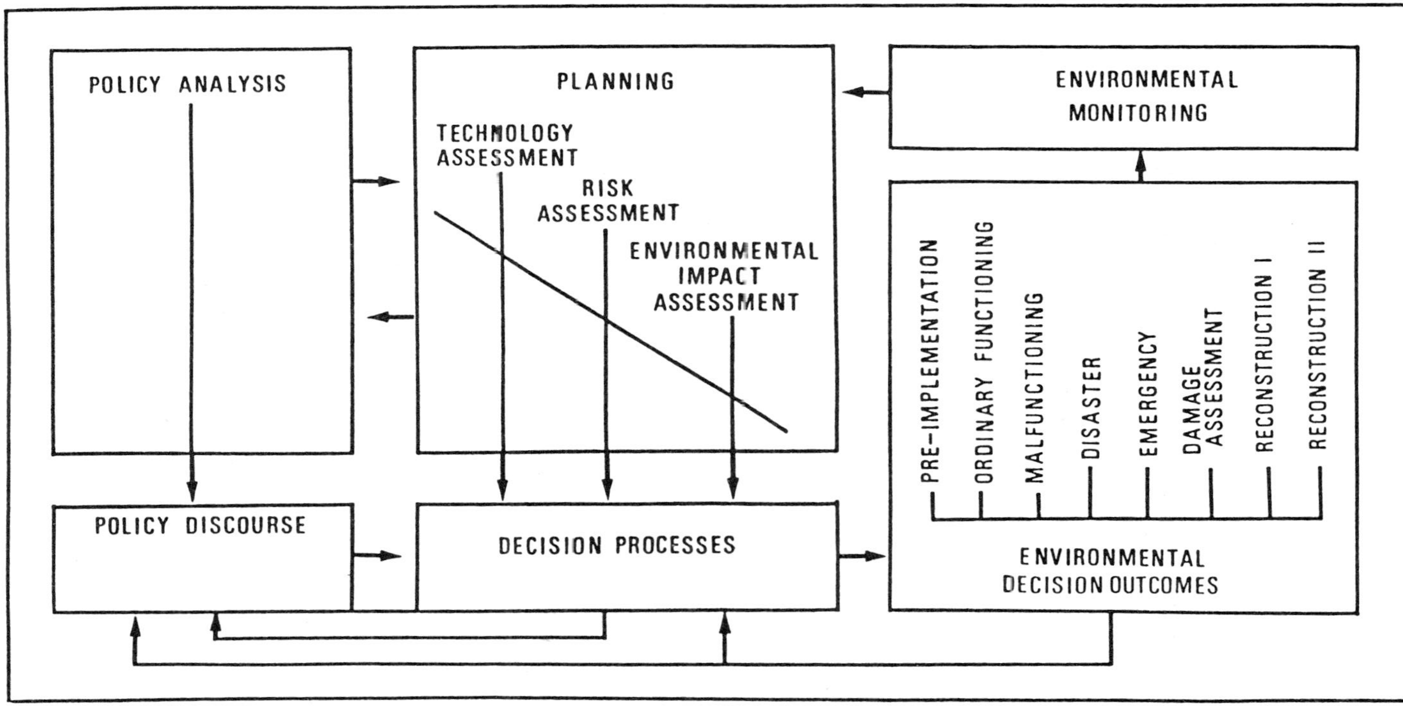

DECISION-MAKING MODEL FOR THE SOCIO-TECHNOLOGICAL ENVIRONMENT

FIGURE 1

its generic "tools use" as opposed to "hi-tech" sense, is a primary linkage between a society and its environment. An increasingly important public policy issue in modern industrial societies is the proper relation of society to the environment and the role to be played by technology in establishing it. Policy analysis seeks to contribute to a clarification of the ongoing public discourse about technology and its feasible uses (Fig. 1).

Within contemporary democratic societies, discourse on policy issues regarding technology and related matters is maintained through a continuing flow of communication in the public arena: within agencies, legislative bodies, the mass media and informal discussions among citizens. Policy advocates, experts and political leaders formulate and structure discourse about technological matters but the responses of the general public are also monitored through several channels, including social science surveys, public opinion polls, and formal voting. Systematic monitoring of the general public's views is certainly preferable to the use of implicit, unchecked, and possibly inaccurate assumptions as a guide to policy and planning (17). Few studies have gauged the accuracy of decisionmakers' working assumptions about public perceptions (18, 19, 20) and more are warranted.

Once available, decisionmakers and policy advocates often become keenly interested consumers of the findings from surveys, polls and ballots (21, 22). When the views of the general public regarding technological policy are gathered, they must be interpreted as well as recorded.

Psychological research can contribute to the understanding of the structure of contemporary public discourse about technology and its management. Within the United States and other industrial nations, discourse among experts and public leaders over the past two decades has generated an evolving and expanding issues complex in which policies regarding technology are interrelated with issues concerning resources, population, environment, economics, energy, decision strategies, and broad societal purposes (23, 24, 25, 26, 27, 28).

Distinctive orientations toward this issues complex can be analyzed as contemporary worldviews. The implications of individual variations among persons in the endorsement of various contemporary worldviews are being investigated in psychological research (23, 29, 30). Guided in particular by an analysis of expert discourse on issues of technology and society (31), two perspectives have been studied: Contemporary Worldview A embodies the notion of a high-growth, high-technology, free enterprise society, with a pro-business stance on regulation, materialistic goals, and centralized, rational, quantified decisionmaking processes, while Contemporary Worldview B stresses leveling off material and technological growth, concern about social and environmental impacts of growth, lower levels of consumption, decentralization of both control and population, goals of human self-actualization, and participative decisionmaking guided by non-materialistic values.

Variations among persons in their endorsement of these contemporary worldviews are systematically related to psychological issues arising within the technical activities of the planning process, as well as to preferences for decision processes in planning (29, 30).

ASSESSMENT ACTIVITIES WITHIN THE PLANNING PHASE

The activities of technology assessment, risk assessment, and environmental impact assessment are located primarily within the planning, design, and implementation phase of decisionmaking (Fig. 1). They seek to provide the planning function with objective forecasts about technological options and projects, as well as follow-up evaluations of decision outcomes.

Technology assessment is a specialized activity within the process of technological development and application. It examines specific potential technological applications as societal options and yields a broad analysis of possible environmental, economic, demographic, institutional, and social impacts, identifies impacted groups, and gauges the feasibility of implementation of the specific technological application (1, 3, 4, 32, 33, 34). Although TA attends to possible consequences across a set of domains, it can be considered <u>technology-centered</u>, in that the entities it tracts and monitors are specific technologies and their possible fates.

Risk assessment is a specialized activity within the process of hazard or risk management (35, 36, 37, 38). Hazard management seeks to achieve acceptable levels of safety from threats to life, health, and well-being arising from technologies and natural events (8, 39, 40, 41). Although several other forms of adverse impacts can be considered (economic, political, international), as in the case of energy technologies (42), risk management has been largely <u>safety-centered</u>, in terms of striving to insure that threats to life and health are kept at acceptable levels. Within this context, RA identifies adverse outcomes and offers probabilistic estimations of their likelihood.

Environmental impact assessment is a specialized activity within the process of environmental planning, design, and management and in its broadest sense encompasses both the natural and built environments. Its purpose is to identify the range of possible impacts of proposed projects upon existing environments (12, 43, 44, 45). In the case of the environmental impact statements mandated by the U. S. National Environmental Policy Act of 1969 (NEPA), the purpose is full disclosure of impacts for consideration by governmental agencies, private and public interest groups and the general public. In the fields of environmental planning and design, impact assessment and post-construction evaluations are viewed as forming components of a feedback cycle (13, 15). The proposed projects reviewed for their environmental impacts are typically landuse changes that entail technological applications. Although the source of impacts may be technological in kind, EIA can be considered as clearly <u>environment-centered</u>, in that monitoring the quality of the environment is its abiding concern.

Of the three forms of assessment, TA is closer to policy analysis, in that it considers possible technological applications at an early stage and broad level. EIA is closer to decision outcomes in that it considers the impacts of specific implemented forms of technological applications.

Within the United States, each of the three specialities has grown out of distinctive legislative histories (15, 32, 46), and they tend to be involved with different governmental units (e.g., for TA, the Office of Technology Assessment and the Office of Science and Technology; for RA, the Federal Drug Agency and the Nuclear Regulatory Commission; for EIA, the Council on Environmental Quality and the Environmental Protection Agency). However, much overlap on the latter score occurs, and the assessment activities share fundamental aspects of rationale and function.

Each activity can be viewed as a contribution to the ongoing process of socio-cultural evolution, as conceptualized by Campbell (47). He has offered a general model for processes that tend toward adaptive fit through variation and selective retention. Table 1 presents the model and three illustrative processes: biological evolution, learning, and socio-cultural evolution. In each case, the process involves three components: 1) a source of variation in the units of analysis, 2) consistent selection criteria and mechanisms for retention and elimination of units, and 3) mechanisms for the preservation of selected variants. In biological evolution, genes are the unit of analysis, with selection criteria applied through differential reproductive opportunities retaining certain random variants that are preserved through duplication via the chromo-

Table 1

The Concept of Variation and Selective Retention, with Three Applications

Context:	Biological Evolution	Learning	Socio-cultural Evolution
Units	genes	responses, actions	skills, customs recipes for living, policies, plans, designs reform programs
Variations:	random variation in gene frequency	exploratory trial and error responses	innovations deviants
Consistent selection criteria:	adaptive function	adaptive function	adaptive function
Retention climination mechanism:	differential reproduction opportunities	reinforcement (drive reduction)	imitation diffusion legitimacy
Mechanisms for preservation of selected variants:	duplication via chromosome-gene system	memory	socialization training sanctions

Adapted: Campbell, D.T.-Variation and selective retention in socio-cultural evolution. In H.R. Barringer, G.I. Blankston, and R.W. Macks (Eds.), <u>Social change in developing area</u>. Cambridge: Schenkman, 1965.

(author's table - no permission required)

some-gene system. In learning, selection criteria are manifested in differential reinforcement of exploratory trial-and-error responses, with the selected variants preserved via the memory system of the organism.

In socio-cultural evolution, the units are societal and institutional forms such as skills, customs, plans, designs, and procedures. When confronted with a challenge to its equilibrium, a viable society or social entity emits a wide variety of alterations, innovations, and other deviations from customary practice in an effort to cope with its new situation. Consistent selection criteria are then applied to identify among those new forms the alternatives that display functional adaptation. Societal mechanisms such as legitimacy, imitation, and diffusion come into play to retain the selected variants. Additional mechanisms, such as sanctions, policies, legislation, training programs, and socialization are mobilized for the preservation and duplication of the selected variants.

Within this perspective, the activities of technology assessment, risk assessment, and environmental impact assessment can be viewed as efforts to enhance this adaptive process, to make it more explicit and objective, and to promote its continuing effectiveness. In their capacity for systematically providing information, analysis, and follow-up monitoring of the processes of technological development, hazard management, and environmental planning, the activities of TA, RA, and EIA do not yield decisions but instead provide intelligence and foresight to the decisionmaking process and establish a clear notion of accountability regarding the process and outcomes of decisionmaking.

Thus, with reference to Fig. 1, the use of systematic assessment is itself a policy stance. Feedback from both decision processes and decision outcomes to policy analysis and discourse is essential to the mechanism of retention-elimination based upon consistent selection criteria. Furthermore, feedback from environmental monitoring of decision outcomes is a basis for gauging and improving the validity of TAs, RAs, and EIAs.

Psychological Research Issues in TA, RA, and EIA

Psychological research can play at least three roles in the technical assessment functions of TA, RA, and EIA. First, psychological analysis of the basic constructs and processes involved in these assessment functions can contribute to conceptual clarification and improved effectiveness. Second, psychological research can provide primary input to the causal analysis of technological impacts and consequences. Third, the communication of the products of technical TAs, RAs, and EIAs to users, including other experts and decisionmakers as well as the general public, raises psychological issues that may bear upon their reception, use, and acceptability.

Psychological Research and TA: Technology assessment can take two forms: problem-initiated analysis and technology-initiated analysis (32). The former tracks the specific impacts of actual or possible technological applications, (e.g., subsonic aircraft noise, air pollution generated by automobiles). Cause and effect systems models can readily formulate and address these problems. Technology-initiated analysis is more distinctive, more open-ended, and more ambitious, seeking to identify the full range of possible societal consequences of new technologies. This task of TA can be viewed as requiring the effective deployment of the creative imagination of persons knowledgeable about potential impacts. Psychological research on the creative process is pertinent to this endeavor (48). For example, a modification of the Guilford Consequences Test (49) can be used to elicit a broad range of societal consequences of specific technological developments from the general public or selected elite samples of respondents (50, 51). The second mode of psychological contribution to TA is the more straightforward use of research findings in the analysis of technological impacts. Thus, for example, in a retrospective TA of mass television, results on the effects of TV violence upon the behavior of viewers would be relevant. The third mode of contribution takes the form of research on how information generated by TAs is used by policy analysts and others, and on what modifications in format and content would increase the effective use of products of TA (52).

Psychological Research and RA: Psychological research bearing upon risk assessment has examined some of its basic constructs. Laypersons' judgments of technological risk appear to be only moderately related, or even unrelated, to their estimations of annual societal fatalities (53, 54, 55). In contrast, annual societal fatalities figure prominently as an index of risk in many RAs and also related positively to judgments of risk made by technical experts (54, 56). More generally, the technical construct of risk entails identified adverse consequences and their estimated probabilities (57), although the precise formulation of probability can elicit disagreement among experts (58, 59, 60). In communicating quantitative technical RAs to the general public, laypersons' readiness to think of potential hazardous events in probabilistic terms (8, 54) and their grasp of probabilistic concepts become central issues (56, 61).

Ways to incorporate these public concerns about risk into the technical exercise of risk assessment are not at all obvious, but presumably would entail attention 1) to the qualitative nature of the hazards involved, that is, whether they are perceived as dreadful, unknown, etc. (53, 54), and 2) to the prevailing level of public confidence in the management of the hazard (41). Actually, the recognition of public conceptions of risk may be more appropriately considered in the post-RA decision processes for establishing the acceptability of technologies and the management of hazards associated with them (Fig. 1) (35, 39, 62, 63, 64).

Psychological Research and EIA: Within the United States, the impetus to assess environmental quality derives directly from the National Environmental Policy Act of 1969 (NEPA), which established the Council on Environmental Quality (CEQ) and the Environmental Protection Agency (EPA), called for monitoring and reporting the quality of the environment, and mandated the preparation of environmental impact statements that fully disclose the consequences of proposed projects, including non-economic considerations.

Many subsequent legislative acts and agency decisions have given clear standing to such non-economic goals as aesthetic and scenic quality, visibility and air quality, minimization of noise, and so on. The criteria for assessing these conditions have been formulated generically as observer-based, or perceived, environmental quality indices (PEQIs) (65).

A substantial amount of research has been devoted to the use of psychological assessment techniques for gauging scenic quality and the effects of project impacts upon it (66, 67, 68), as well as the assessment of the perceived quality of residential and institutional environments (13, 69). A beginning is being made to incorporate these assessment techniques institutionally within federal agency functioning and to establish standards for their use (70).

Many of the projects that entail environmental impact statements regarding scenic quality derive from the implementation of energy technologies: surface mining, power plants, electric power transmission lines, off-shore oil drilling, and so on. And of course technology is impressively used in the construction of highways, bridges, high-rise buildings, and other consequential transformations of landuse that entail EIAs. The siting of technological facilities and impacts of both kind occur well along the process extending from policy analysis to implementation (Fig. 1). EIAs deal with the here-and-now physical manifestations of technology as it occurs within the everyday environment.

The potential impact of a technological project upon the pre-implementation environment is often difficult for both experts and laypersons to envision (11, 71). As a means to increase foresight about impacts gauged in EIAs and to facilitate public participation in the process, various forms of environmental simulation have been employed. One dynamic environmental simulation facility enables observers to "walk" or "drive" through scale models of urban, suburban, or natural environments. A remotely guided periscope and tiny lens, controlled manually or by computer, moves through the scale model, projecting an eye-level view on television monitors. In addition, stopframe color films and videotapes can be made of the simulated tours. Thus, the technique can provide previews of alternative environments for feedback in the planning and design process, for use in public hearings and the preparation of environmental impact statements, and for experimental man ipulations in environmenatal psychological research (72, 73).

Research has been directed to appraising the psychological adequacy of the environmental simulation technique (74) and to its use in improving the effectiveness of general public hearings and public participation (11, 71, 73). The mandate for including the general public more fully in the EIA process is widely accepted and has engaged a range of social science approaches (15, 75).

Psychological Research Related to Choice of Decision Process

The technical activities of TA, RA, and EIA provide information, analyses, and forecasts that contribute to the decision process but must be weighed along side other considerations (Fig. 1). Thus, for example, identification of adverse consequences and estimation of their probabilities in an RA do not establish what is an acceptable option or decision (41, 56, 61, 62, 76). In a general sense, the decision process leads to a limited array of end results: 1) accept and implement an option, 2) modify an option or mitigate its impact, and 3) reject an option. How complex information and multifarious factors are to be considered in moving toward such decisions is itself a meta-decisional problem and a wide array of approaches to such societal decisions is available (39).

Current research is examining the psychological implications of laypersons' views on three sets of issues concerning the societal process of risk management: 1) who should make decisions about technology and technological hazards 2) how these societal decisions should be approached, and 3) what attributes are desirable criteria for the various formal and informal approaches to societal decisions (23, 29, 30).

As an example of this line of inquiry, the endorsement of Contemporary Worldviews A and B have been related to preferences regarding societal decision processes. Contemporary Worldview A is associated with a preference for business leaders, professionals, and regulatory agencies, while Contemporary Worldview B is associated with a preference for the general public, the courts, and individual consumers as the loci of societal decisions. Contemporary Worldview A is related to a positive evaluation of cost benefit analysis, political judgment, revealed preference, professional judgment, and implied preference as decision approaches, while Contemporary Worldview B is related to a positive evaluation of expressed preference, societal emergent, and societal scenario as decision approaches. As criteria for appraising societal decision approaches, Contemporary Worldview A is related to positive judgment of these attributes: economically-oriented, compatible with institutions, logically sound, decisive, past-oriented, and politically acceptable. In contrast, Contemporary Worldview B is associated with positive judgments of these criteria: accountable to the public, creative, intuitive, future-oriented, and conducive to learning (29, 30).

The processes of TA, RA, and EIA have in practice displayed distinctively salient features which nevertheless possess important cross-assessment implications. Technology-initiated analyses in TA have stressed the imaginative generation of wide-ranging possible consequences of technological options, yet this task is also shared by RA and EIA. RA has focused upon systematic estimation of the probabilities to be associated with various consequences of technological options, yet this kind of judgment or information is relevant to the impacts considered in TA and EIA as well. EIA has been pursued with an explicit charge of facilitating public participation in the process as much as possible, yet in a pluralistic democratic society this aspiration is comparably appropriate to TA and RA. A final point is that all three forms of assessment deal with forecasts and thus follow-up monitoring of the environmental outcomes of decisions offers necessary feedback for gauging and improving their validity.

ENVIRONMENTAL MONITORING OF DECISION OUTCOMES

Most decisionmaking models for the socio-technological environment proceed through the planning phase only, or else continue to implementation and follow-up monitoring in an abbreviated and abstract way. The model presented in Fig. 1 undertakes a more detailed consideration of the implemented environment as an outcome of the decision process. This component recognizes that implemented technologies take on site-contextual embodiments within the everyday environment. The model deals with the implemented environment as an array of possible states. The pre-implementation environment represents the state of affairs prior to the introduction of the new technology as process and facility. The environment of ordinary functioning represents the post-implementation environment in which the technology is operating more or less as planned and intended. The environment of malfunctioning represents a set of states ranging from minor problems (e.g., mechanical failure in a kitchen appliance) to disaster (e.g., failure of a dam; nuclear accident). The post-disaster environment has been particularly neglected in this literature. However, a useful post-disaster model (based upon earthquake and flood events) is offered by Kates and Pijawka (7). Their environmental phases include periods of emergency (search and rescue, damage assessment), restoration (debris removal, resumption of retail trade, return of refugees), reconstruction I (rebuilding, resumption of predisaster functioning) and reconstruction II (major construction of commemorative and betterment developments).

Several points deserve attention in this consideration of the implemented environment as an outcome of the decision process. First, the model does not imply an inevitable sequence from ordinary functioning to malfunctioning to disaster; indeed, the likelihood of these states of the environment is a central concern of risk assessment. Second, environmental impact assessment typically deals with the change from the pre-implementation environment to the environment of ordinary technological functioning. Third, presumably generalizable scientific or professional knowledge of the implemented environment of specific technologies is the basis for the implicit and explicit forecasts entailed in TA, RA, and EIA and employed during the planning phase of decisionmaking. Fourth, the location and spatial distribution of sites for potentially hazardous technological facilities is an important topic upon which TA, RA, and EIA may usefully converge. And fifth, the term environmental monitoring within the model (Fig. 1) refers to various systematic processes for describing and evaluating the implemented environment, as a means for providing feedback to the planning and policy phases of decisionmaking.

Psychological Aspects of the Environment as Decision Outcome

Implemented technologies are manifested as processes and facilities taking their place within the everyday, here-and-now environment. The field of environmental psychology deals with persons' impressions of the everyday physical environment (77,

78). It encompasses cognitive, affective, and behavioral factors; that is, how persons categorize and describe the environment, their emotional reactions to it, and their conduct in and towards it.

A useful distinction can be made between the perceptual-cognitive and the affective components of environmental meaning (79, 80). A pertinent but unexplored issue regarding perceptual-cognitive meaning is the way in which laypersons construe technology and what they judge to be prototypical instances of it. Analysis of the differentiation between "natural" versus "artificial" settings is pertinent to this issue (81), but technology as a cognitive category (82) warrants specific detailed examination.

Environmental impact assessment and risk assessment deal with matters that bear differentially upon the affective meaning of the environment. EIA tends to focus upon perceived environmental quality and scenic values (65). Thus, for example, environmental aesthetics can be formulated in terms of the environment as a source of affect, primarily of a pleasurable kind (83). On the other hand, analyses of the qualitative characteristics of perceived risk have found "dread" to be an important attribute (53, 54). Other characteristics examined thus far have been cognitive rather than affective attributes (e.g., certain to be fatal, involuntary). To highlight the affective implications of the dread characteristic, synonyms might be noted: fright, terror, horror, fearfulness, trepidation, aversion, dismay, anxiety, worry, misgiving, apprehension, and suspicion (84).

Russell and Pratt (85) have conceptualized the affective meaning of environments as a two-dimensional bipolar space defined by eight variables falling in the following circular order: pleasant, exciting, arousing, distressing, unpleasant, gloomy, sleepy, and relaxing (Fig. 2). In clockwise order, EIA appears to be associated with matters related to the 2 to 4 o'clock zone, while RA is associated with the 10 to 11 o'clock zone. More generally, environmental perception research has tended to deal with the 1 to 9 o'clock range and risk perception with the 9 to 12 o'clock range of affects. A fuller application of concepts and methods from the psychological study of emotion offers a promising avenue to research linking risk perception to environmental perception more generally.

Although unintended at the point of implementation, malfunction of technologies is a prospect considered by both experts and laypersons. Thus, persons respond to both the actual and possible environment. Risk perception deals with psychological inquiry concerning impressions of potential hazard from both

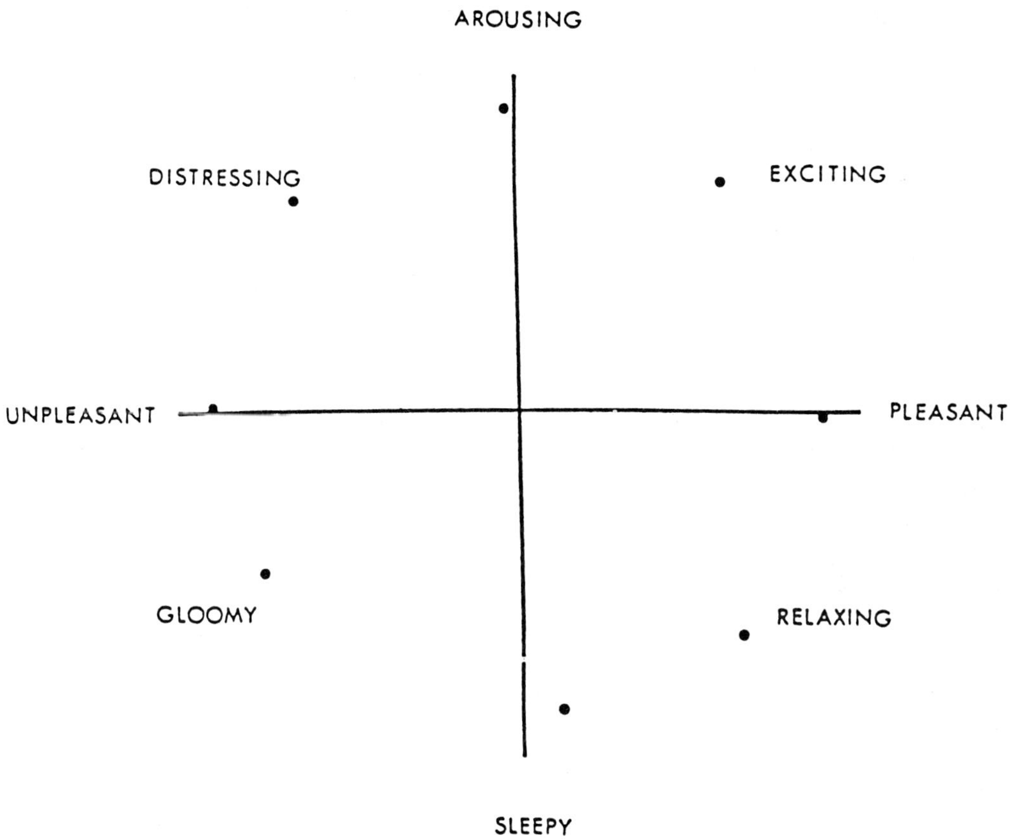

Figure 2

Circular ordering of eight affective descriptors

(From Russell & Pratt, 1980)

technological and natural sources in the environment (56). Thus, risk perception addresses issues regarding the environment of the possible. Ordinary environmental perception can be considered as an active process, importantly influenced by psychological sets and other cognitive factors. Leff and Gordon view persons as encountering a setting with something like a plan to attend to and process certain kinds of information about it. Thus, they have instructed research participants to take certain cognitive sets intentionally when presented with an environmental scene (e.g., to attend to the shapes, lines, colors and textures of the scene; to attend to imagining changes that would make the scene better). Participants then rate these cognitive sets on their difficulty, interest, pleasure and also indicate how the sets affect the general experience of the scene. Research comparing the cognitive set to imagine the possible hazardous potentials in the setting could be usefully compared to the array of other cognitive sets that Leff and Gordon have explored (86, 87, 88). In everyday life, of course, these psychological sets are rarely so explicit or intentional and are typically organized into larger units, often around the role the person is enacting at the time, or more general personal dispositions (89).

The possible also occasionally becomes the actual, and the fields of psychological stress and disaster research deal with persons' response to the event of malfunctioning and the sequence of environmental states and decisions that follow upon it. The environment of malfunctioning can entail events ranging from minor mishaps to major disasters. Minor failures of technological functioning are stressors capable of contributing to the kind of daily hassles that generate important psychological costs (90). These affective reactions may entail frustration rather than fright but their possibly persistent and chronic nature also has to be taken into account and studied.

The response to life-threatening disastrous events and the post-disaster environment has been studied much more often from a sociological than psychological perspective (91). Recent social psychological research (92) has offered evidence in support of an affiliative model of escape behavior in buildings during fires. This model predicts that individuals with close psychological ties will attempt to escape in groups of two or more; flight is assumed to involve escape to familiar persons and places. In contrast, the panic model assumes that escape involves a homogenous population of persons concerned with self-preservation and competing with each other for limited exits. These two models serve to illustrate the importance of psychological assumptions that are unavoidable in planning emergency arrangements and facilities for conditions of disaster (93). The results also raise the issue that hazards can be viewed from an individual point of view (55), from a small group perspective (92), or from a societal vantage point (56).

By incorporating the sequence of environmental states following upon a disastrous event, the model (Fig. 1) suggests that at least some of the qualitative characteristics of perceived risk (53, 54) may be considered as attributes of laypersons' conceptions of disastrous events and the states of post-disaster environment (e.g., certain to be fatal, effects immediate, effects observable, globally catastrophic, threatens future generations). Research is warranted on the extent to which perceived risk is related specifically to imaginative representations of the disaster and post-disaster environments.

The decisionmaking model for the socio-technological environment (Fig. 1) points to psychological research on the siting of hazardous technological facilities as a focus of inquiry that would pull together the concerns of TA, RA, and EIA (94, 95, 96, 97). From the perspective of EIA, issues raised by the siting of a nuclear power plant, for example, focus on the impact upon environmental quality found in comparing the pre-implementation setting to the ordinarily functioning implemented environment. RA and risk perception deal not with the here-and-now environment but with the possible environment and perceptions of it. TA addresses the task of forecasting the fate of new technologies and, in part, the ways in which perceptions of the implemented technologies may create issues concerning their feasibility and public acceptance.

Finally, from a broad policy perspective, the model helps to identify the many forms which the issue of public acceptability can take. First, public acceptability may be an issue regarding policy analysis and the place of technology, considered at the broadest level of societal policy. Second, public acceptability may be an issue regarding the reception of specific technical products of TA, RA, or EIA. Third, public acceptability can focus upon confidence in or approval of the choice of decision processes used in the management of technologies and their associated hazards. Finally, public acceptability may bear upon attitudes toward specific implemented technologies and their positive and negative impacts upon everyday life and the here-and-now environment within which it takes place.

REFERENCES

1. Jones, M.V., A technology assessment methodology: Project summary. Springfield, VA: National Technical Information Service, 1971.

2. Mayo, L.H., The management of technology assessment. In R.G. Kasper (Ed.) Technology assessment: Understanding the social consequences of technological applications. New York: Praeger, 1972, pp. 71-122.

3. Vlachos, E. & Hendricks, D.W., Technology assessment for water supplies. Fort Collins, CO.: Water Resources Publications, 1977.

4. Armstrong, J.E. & Harman, W.W., Strategies for conducting technology assessments. Boulder, CO.: Westview Press, 1980.

5. Kates, R.W., Experiencing the environment as hazard. In S. Wapner, S.B. Cohen & B. Kaplan (Eds.), Experiencing the environment. Plenum Press, 1976, pp. 133-156.

6. Kasperson, R.E., Societal management of technological hazards. In R.W. Kates (Ed.), Managing technological hazard: Research needs and opportunities. Boulder, CO.: Institute of Behavioral Sciences, University of Colorado, 1977. pp. 49-80.

7. Kates, R.W. & Pijawka, D., From rubble to monument: The pace of reconstruction. In J.E. Haas, R.W. Kates & M.J. Bowden (Eds.) Reconstruction following disaster. Cambridge, MA: MIT Press, 1977. pp. 1-23.

8. Sorensen, J.H. & White, G.F. Natural hazards: A cross-cultural perspective. In I. Altman, A. Rapoport & J.F. Wohlwill (Eds.) Human behavior and environment, Vol. 4, New York: Plenum Press, 1980. pp. 279-313.

9. O'Riordan, T., Risk perception studies and policy priorities. Risk Analysis, 1982, 2, pp. 95-100.

10. O'Riordan, T., Public opinion and environmental quality: A reappraisal. Environment and Behavior, 1971, 3, pp. 191-214.

11. Appleyard, D., Understanding professional media: Issues, theory and a research agenda. In I. Altman & J.F. Wohlwill (Eds.) Human behavior and environment, Volume 2. New York: Plenum Press, 1977. pp. 47-89.

12. Corwin, R., Heffernan, P.H., Johnston, R.A., Remy, M., Roberts, J.A. & Tyler, D.B., Environmental impact assessment. San Francisco, CA: Freeman, Cooper & Co., 1975.

13. Friedmann, A., Zimring, C. & Zube, E.H., Environmental design evaluation. New York: Plenum Press, 1978.

14. Mitchell, B., Geography and resource analysis. New York: Longman, 1979.

15. Zube, E.H., Environmental evaluation: Perception and public policy. Belmont, CA: Brooks/Cole, 1980.

16. Aberle, A.K., Cohen, A.K., Davis, M.J., Jr. & Sutton, F.X., The functional prerequisites of a society. Ethics, 1950, 60, pp. 100-110.

17. White, G.F., Formation and role of public attitudes. In H. Jarrett (Eds.) Environmental quality in a growing economy: Essays from the Sixth RFF Forum. Baltimore, MD: Johns Hopkins University Press, 1966, pp. 105-127.

18. Hendee, J.C. & Harris, R.W., Foresters' perception of wilderness-user attitudes and preferences. Journal of Forestry, 1970, 68, pp. 759-762.

19. Buhyoff, G.J., Wellman, J.D., Harvey, H. & Fraser, R.A., Landscape architects' interpretations of people's landscape preferences. Journal of Environmental Management, 1978, 6, pp. 255-262.

20. Thomas, K., Swaton, E., Fishbein, M., & Otway, H.J., Nuclear energy: The accuracy of policy makers' perceptions of public beliefs. Behavioral Science, 1980, 25, pp. 332-344.

21. Converse, P.E., Public opinion and voting behavior. In F.I. Greenstein & N. W. Polsby (Eds.), Handbook of political science, Volume 4, Reading, MA: Addison-Wesley, 1975.

22. Spangler, M.B., The role of interdisciplinary analysis in bridging the gap between the technical and human sides of risk assessment. Risk Analysis, 1982, 2, pp. 101-114.

23. Buss, D.M. & Craik, K.H., Contemporary worldviews: Personal and policy implications. Journal of Applied Social Psychology, in press.

24. Cotgrove, S., Catastrophe or cornucopia: The environment, politics and the future. New York: Wiley, 1982.

25. Dunlap, R.E., Ecologist versus exemptionalist: The Ehrlich-Simon debate. Social Science Quarterly, 1983, 64, pp. 200-203.

26. Milbrath, L.W., Environmental beliefs and values. In M. Hermann (Ed.) Handbook of political psychology. San Francisco, CA: Jossey-Bass, in press.

27. O'Riordan, T., Environmentalism. London: Pion, 1976.

28. Thompson, M., Among the energy tribes: The anthropology of the current policy debate. Laxenburg, Austria: International Institute for Applied Systems Analysis, unpublished report, 1981.

29. Buss, D.M., Craik, K.H., & Dake, K.M., Contemporary worldviews and perception of the technological system. Contemporary issues in risk analysis: The social and behavioral sciences. New York: Plenum Press, in press.

30. Buss, D.M., Craik, K.H., & Dake, K.M., Perceptions of technological hazards and their management. In B. Berglund & C. Levy-Leboyer (Eds.) Applications of environmental psychology: Recent research on environmental hazard and unfavorable environments. New York: Plenum Press, in press.

31. Reuyl, J.S., Harman, W.W., Carlson, R.C., Levin, M.B., & Witwer, J.C., Solar energy in America's future: Preliminary assessment. Stanford, CA: Stanford Research Institute, 1977.

32. National Academy of Engineering. A study of technology assessment. Washington, D.C.: U.S. Government Printing Office, 1969.

33. Kasper, R.G., (Ed.) Technology assessment: Understanding the social consequences of technological applications. New York: Praeger, 1972.

34. Tarr, J.A. (Ed.), Retrospective technology assessment. 1976. San Francisco, CA: San Francisco Press, 1977.

35. Kates, R.W. (Ed.) Managing technological hazard: Research needs and opportunities. Boulder, CO: Institute of Behavioral Sciences, University of Colorado, 1977.

36. Schwing, R.C. & Albers, W.A., Jr., (Eds.) Societal risk assessment: How safe is safe enough? New York: Plenum Press, 1980.

37. Hohenemser, C. & Kasperson, J.X. (Eds.) Risk in the technological society. Boulder, CO: Westview Press, 1982.

38. Covello, V.T., Menkes, J. & Nehnevajsa, J., Risk analsysis, philosophy, and the social and behavioral sciences: Reflections on the socpe of risk analysis research. Risk Analysis, 1982, 2, pp. 53-58.

39. Fischhoff, B., Lichtenstein, S., Slovic, P., Derby, S.L. & Keeney, R.L., Acceptable risk. New York: Cambridge University Press, 1981.

40. Weinberg, A.M., Reflections on risk Assessment. Risk Analysis, 1981, 1, pp. 5-8.

41. Otway, H.J. & Thomas, K., Reflections on risk perception and policy. Risk Analysis, 1982, 2, pp. 69-82.

42. Buehring, W.A., Foell, W.K. & Keeney, R.L., Examining energy/environment policy using decision analysis. Energy Systems and Policy, 1978, 2, pp. 341-367.

43. Dickert, T.G. & Domeny, K.R. (Eds), Environmental impact assessment: Gudelines and commentary. Berkeley, CA: University Extension, University of California, Berkeley, CA, 1974.

44. Wandesforde-Smith, G., Environmental impact assessment and the politics of development in Europe. In T. O'Riordan & R.K. Turner (Eds), Progress in resource management and environmental planning, Volume 2. New York: Wiley, 1980. pp. 205-237.

45. O'Riordan, T. & Sewell, W.R.D., (Eds.), Project appraisal and policy review. New York: Wiley, 1981.

46. Covello, V.T. & Menkes, J., Issues in risk analysis. In C. Hohenemser & J.X. Kasperson (Eds.) Risk in the technological society. Boulder, CO: Westview Press, 1982. pp. 287-301.

47. Campbell, D.T., Variation and selective retention in socio-cultural evolution. In H.R. Barringer, G.I. Blanksten & R.W. Mack (Eds.), Social change in developing areas. Cambridge: Schenkman, 1965.

48. Barron, F. & Harrington, D.M., Creativity, intelligence and personality. Annual Review of Psychology, 1981, 32, 439-476.

49. Christensen, P.R., Merrifield, P.R. & Guilford, J.P., Consequences Test. Beverly Hills, CA: Sheridan Supply Co., 1958.

50. Barron, F. Creative person and creative process. New York: Holt, Rinehart & Winston, 1969.

51. Craik, K.H. Environmental psychology. In Craik, K.H., Kleinmuntz, B., Rosnow, R.L., Rosenthal, R., Cheyne, J.A., & Walters, R.H., New directions in psychology, 4. New York: Holt, Rinehart & Winston, 1970. pp. 1-122.

52. Wilton, P.C. & Myers, J. Research on the utilization of technology assessment information. Berkeley, CA: Department of Political Science, University of California, unpublished report, 1980.

53. Fischhoff, B., Slovic, P. Lichtenstein, S., Read, S., & Combs, B. How safe is safe enough? A psychometric study of attitudes towards technological risks and benefits. Policy Sciences, 1978, 9, pp. 127-152.

54. Slovic, P., Fischhoff, B. & Lichtenstein, B. Facts and fears: Understanding perceived risks. In R.C. Schwing & W.A. Albers, Jr. (Eds.) Societal risk assessment: How safe is safe enough? New York, Plenum Press, 1980. pp. 181-216.

55. von Winterfeldt, D., John, R.S. & Borcherding, K. Cognitive components of risk ratings. Risk Analysis, 1981, 1, pp. 227-288.

56. Slovic, P., Fischhoff, B. & Lichtenstein, S. Why study risk perception? Risk Analysis, 1982, 2, pp. 83-93.

57. Kaplan, S. & Garrick, B.J. On the quantitative definition of risk. Risk Analysis, 1981, 1, pp. 11-27.

58. Abramson, L.R. Some misconceptions about the foundations of risk analysis. Risk Analysis, 1981, 1, pp. 229-230.

59. Kaplan, S. & Garrick, B.J. Some misconceptions about misconceptions: A response to Abramson. Risk Analysis, 1981, 1, pp. 231-233.

60. Kasper, R.G., Perceptions of risk and their effects on decision making. In R.C. Schwing & W.A. Albers, Jr. (Eds.) Societal risk assessment: How safe is safe enough? New York: Plenum Press, 1980, pp. 71-84.

61. Ruckelshaus, W.D. Science, risk and public policy. Washington, D.C.: U.S. Environmental Protection Agency, 1983. 20 pp.

62. Cole, G.A. & Withey, S.B. Perspectives on risk perceptions. Risk Analysis, 1981, 1, pp. 143-163.

63. Gardner, G.T., Tiemann, A.R., Gould, L.C., DeLuca, D.R., Doob, L.W. & Stolwijk, J.A.J. Risk and benefit perceptions, acceptability judgments and self-reported actions toward nuclear power. Journal of Social Psychology, 1982, 116, pp. 179-197.

64. Vlek, C. & Stallen, P. Rational and personal aspects of risk. Acta Psychologica, 1980, 45, pp. 272-300.

65. Craik, K.H. & Zube, E.H. (Eds.) Perceiving environmental quality: Research and applications. New York: Plenum Press, 1976.

66. Zube, E.H., Brush, R.O. & Fabos, J.G. (Eds.) Landscape assessment: Values, perceptions and resources. Stroudsburg, PA: Dowden, Hutchinson & Ross, 1975.

67. Elsner, G. & Smardon, R.C. (Eds.) Our national landscape: A conference on applied techniques for analysis and management of the visual resource. Berkeley, CA: USDA Pacific Southwest Forest and Range Experiment Station, 1979.

68. Rowe, R.D. & Chestnut, L.G. (Eds.) Managing air quality and scenic resources at national parks and wilderness areas. Boulder, CO: Westview Press, 1983.

69. Carp, F.M. & Carp, A. Perceived environmental quality of neighborhoods: Development of assessment scales and their relation to age and gender. Journal of Environmental Psychology, 1982, 2, pp. 295-312.

70. Craik, K.H. & Feimer, N.R. Setting technical standards for visual assessment procedures. In G. Elsner & R.C. Smardon (Eds.) Our national landscape: A conference on applied techniques for analysis and management of the visual resource. Berkeley, CA: USDA Pacific Southwest Forest and Range Experiment Station, 1979.

71. McKechnie, G.E. Simulation techniques in environmental psychology. In D. Stokols (Ed.) Perspectives on environment and behavior: Theory, research and applications. New York: Plenum Press, 1977. pp. 169-189.

72. Appleyard, D. & Craik, K.H. The Berkeley Environmental Simulation Project: Its use in environmental impact assessment. In T.G. Dickert & K.R. Domeny (Eds.) Environmental impact assessment: Guidelines and commentary. Berkeley, CA: University Extension, University of California, 1974. pp. 121-126.

73. Appleyard, D. & Craik, K.H. The Berkeley Environmental Simulation Laboratory and its research program. International Review of Applied Psychology, 1978, 27, pp. 53-55.

74. Craik, K.H. The psychology of the large scale environment. In N.R. Feimer & E.S. Geller (Eds.) Environmental psychology: Directions and perspectives. New York: Praeger, 1983. pp. 67-105.

75. Andrews, R.N.L. & Waits, M.J. Environmental values in public decisions: A research agenda. Ann Arbor, MI: School of Natural Resources, University of Michigan, 1978.

76. Douglas, M. & Wildavsky, A. Risk and culture: An essay on the selection of technological and environmental dangers. Berkeley, CA: University of California Press, 1982.

77. Craik, K.H. Multiple scientific paradigms in environmental psychology. International Journal of Psychology, 1977, 12, pp. 147-157.

78. Holahan, C.J. Environmental psychology. New York: Random House, 1982.

79. Ward, L.M. & Russell, J.A. The psychological representation of molar physical environments. Journal of Experimental Psychology: General, 1981, 110, pp. 121-152.

80. Craik, K.H. Comments on "The psychological representation of molar physical environments," by Ward & Russell. Journal of Experimental Psychology: General, 1981, 110, 158-162.

81. Wohlwill, J.F. The concept of nature: A psychologist's view. In I. Altman & J.F. Wohlwill (Eds.) Human behavior and environment, Volume 6, New York: Plenum, 1983. pp. 5-37.

82. Rosch, E. Natural categories. Cognitive Psychology, 1973, 4, 328-350.

83. Wohlwill, J.F. Environmental aesthetics: The environment as a source of affect. In I. Altman & J.F. Wohlwill (Eds.) Human behavior and environment, Volume 1. New York: Plenum Press, 1976. pp. 37-86.

84. Landau, S.L. & Bogus, R.J. (Eds.) The Doubleday Roget's thesaurus in dictionary form. Garden City, NY: Doubleday, 1977.

85. Russell, J.A. & Pratt, G. A description of the affective quality attributed to environments. Journal of Personality and Social Psychology, 1980, 38, 311-322.

86. Leff, H.L., Gordon, L.R. & Ferguson, J.G. Cognitive set and environmental awareness. Environment and Behavior, 1974, 6, pp. 395-447.

87. Leff, H.L. & Gordon, L.R. Environmental cognitive sets: A longitudinal study. Environment and Behavior, 1980, 12, pp. 291-328.

88. Leff, H.L. Experience, environment and human potentials. New York: Oxford University Press, 1978.

89. Craik, K.H. A role theorectic analysis of scenic quality judgments. In R.D. Rowe and L.G. Chestnut (Eds.) Managing air quality and scenic quality resources at national parks and wilderness areas. Boulder, CO: Westview Press, 1983. pp. 117-126.

90. Lazarus, R.S. & Cohen, J.B. Environmental stress. In I. Altman & J.G. Wohlwill (Eds.) Human behavior and environment, Volume 2. New York: Plenum Press, 1977. pp. 90-127.

91. Quarantelli, E.L. (Ed.) Disasters: Theory and research. Beverly Hills, CA: Sage, 1978.

92. Sime, J.D. Affiliative behavior during escape to building exits. Journal of Environmental Psychology, 1983, 3, pp. 21-42.

93. Canter, D. (Ed.) Fires and human behavior. New York: Wiley, 1980.

94. Van der Pligt, J., Van der Linden, J. & Ester, P. Attitudes to nuclear energy: beliefs, values and false consensus. Journal of Environmental Psychology, 1982, 2, 221-231.

95. Baker, E.J., West, S.G., Moss, D.J. & Weyant, J.M. Impact of offshore nuclear power plants: Forecasting visits to nearby beaches. Environment and Behavior, 1980, 12, pp. 367-407.

96. Woo, T.O. & Castore, C.H. Expectancy-value and selective exposure as determinants of attitudes toward a nuclear power plant. Journal of Applied Social Psychology, 1980, 10, pp. 224-234.

97. Kunreuther, H.C. & Ley, E.V. (Eds.) The risk analysis controversy: An institutional perspective. New York: Springer-Verlag, 1982.

Prediction, Diagnosis, and Causal Thinking in Forecasting*

Hillel J. Einhorn and Robin M. Hogarth

Center for Decision Research, Graduate School of Business, University of Chicago, 1101 East 58th Street, Chicago, Illinois 60637**

INTRODUCTION

Imagine that you lived several thousand years ago and belonged to a tribe of methodologically sophisticated cave-dwellers. Your methodological sophistication is such that you have available to you all present day means of the methodological arsenal--details of the principles of deductive logic, probability theory, access to computational equipment, etc. However, your level of substantive knowledge lags several thousand years behind your methodological sophistication. In particular, you have little knowledge about physics, chemistry, or biology. In recent years, your tribe has noted an alarming decrease in its birth-rate. Furthermore, the tribe's statistician estimates that unless the trend is shortly reversed, extinction is a real possibility. The tribe's chief has accordingly launched an urgent project to determine the cause of birth. You are a member of the project team and have been assured that all means, including various forms of experimentation with human subjects, will be permitted to resolve this crucial problem.

The above story illustrates the following points:

(1) The ultimate goal of forecasting is to provide guidance for taking action. Therefore, forecasting is intimately tied to decisionmaking and should be evaluated within this context. For example, although the tribe's statistician may use a time series model of past birth rates to predict the downward trend, this is of limited usefulness to the tribe's leaders. Indeed,

---

** Support for this work was provided by a contract from the Engineering Psychology Program, Office of Naval Research.

NATO ASI Series, Vol. G4
Environmental Impact Assessment, Technology Assessment, and Risk Analysis. Edited by V.T. Covello et al.

one might find small consolation in the fact that the date of the tribe's extinction can be precisely predicted. Clearly, the need to take appropriate action to change the process is of paramount concern.

(2) While much attention in the forecasting literature is devoted to forecast accuracy (see Hogarth and Makridakis (1981) for a review), this focus may be misplaced. In particular, when viewed within the broader context of how forecasts affect actions, and how both affect outcomes, the criterion of forecast accuracy is problematic. In order to illustrate this, consider the following three situations and the relevance of forecast accuracy in each: (a) A weather forecaster predicts that a hurricane will come ashore at a certain time and place and actions contingent on the forecast are taken (e.g., boarding up property, evacuation). Since these actions do not affect the hurricane itself, the relevance of forecast accuracy as a criterion is obvious and its evaluation relatively simple. However, even in those cases where one cannot typically exert control over the process (as when large physical systems are involved), actions can sometimes be taken to affect the process; consider, for example, "seeding" hurricanes (Howard, Matheson, and North, 1972). (b) Economic forecasters predict a recession next year and the government reacts by taking action to stimulate the economy. Clearly, action is taken to intervene in the process and the resulting outcomes are a joint function of the factors on which the prediction is based and the action taken. Under these circumstances, the meaning of forecast accuracy is ambiguous; in fact, one can imagine that forecast accuracy is not desirable. For example, the more effective the action is in changing the process, the less accurate the initial forecast will be. (c) People in a small town hear a rumour that the banks are about to fail. They think that if this forecast is accurate, they had better withdraw their money as fast as possible. Accordingly, they go to the banks to close their accounts (those skeptical of the forecast see many people withdrawing money and either take this as a sign that the rumour is true or foresee the consequences of waiting too long, thus joining the crowd in either case). By the end of the day the banks have failed, thereby confirming the rumour. This case differs from (b) in that the actions taken lead to confirmation of the predictions (demonstrating the so-called "self-fulfilling" prophecy effect). The perniciousness of these cases is that awareness of the effect of actions on outcomes is low and can lead to overconfidence on forecasts that are of low or even zero accuracy (see Einhorn and Hogarth, 1978; Einhorn, 1980).

(3) In order to understand the relations between predictions, actions, and outcomes, one needs a causal model of the process. Such a model must be developed through the use of what we call diagnostic or backward inference. That is, past observations, events, and data are used as evidence to infer the process(es) that produced them. Such inferences are called diagnostic since they involve going from visible effects such as symptoms, signs, and the like, to their prior causes. For example, one might consider a particularly large decline in last month's sales as symptomatic of a more fundamental malady (such as an incompetent sales force). We consider diagnostic inference to be based on causal thinking, although in doing diagnosis one has to mentally reverse the time order in which events were thought to have occurred (hence the term "backward inference"). On the other hand, predictions involve forward inference, i.e., one goes forward in time from present causes to future effects. However, it is important to recognize the dependence of forward inference/prediction on backward inference/diagnosis. In particular, it seems likely that success in predicting the future depends to a considerable degree on making sense of the past. Therefore, people are continually engaged in shifting between forward and backward inference in both making and evaluating forecasts. Indeed, this can be eloquently summarized by Kierkegaard's observation, "Life can only be understood backwards; but it must be lived forward."

A second important aspect of diagnostic inference concerns the process by which hypotheses are formed and relevant variables found. For example, why are certain variables chosen as "relevant" to the phenomenon in question while others are concidered irrelevant or of lesser importance? This issue goes to the heart of all prediction problems although it has not received the attention it deserves. For example, consider the clinical versus statistical prediction controversy in psychology, which has raged off and on for over 25 years (Meehl, 1954; Sawyer, 1966; Dawes, 1979). One succinct conclusion of this literature (and one that we subscribe to) has been made by Dawes and Corrigan; namely, "...the whole trick is to decide what variables to look at and then to know how to add." (1974, p. 105). Assuming that we can add, how do we decide what variables to look at? On this crucial point we have little guidance other than some ill-defined notions of experience, intuition, gut feeling, hunch, and so forth. From our perspective, one must have some hypothesis or theory for selecting relevant from irrelevant variables. Indeed, relevance can only be understood in relation to some model (usually implicit) of what generates the variable to be predicted. For example, imagine being asked to predict the result of mixing two chemicals with known atomic structures.

Without some knowledge of chemistry and physics, picking relevant from irrelevant variables is meaningless. Therefore prediction depends on backward inference which involves both the forming of hypotheses to interpret the past and the choosing of relevant from irrelevant variables in that interpretation.

We now turn to consider the details of the diagnostic process with particular emphasis on how people seek relevant variables and test for causal significance. These issues are examined in the following way: (1) we discuss the role of similarity and differences in directing attention to variables; (2) we posit the existence of probabilistic signs used for inferring causal relations and call these "cues to causality;" (3) we identify the following cues and discuss their roles in causal thinking: temporal order, constant conjunction, contiguity in time and space, number of alternative explanations, similarity, predictive validity, and robustness; (4) the role of the cues in inferring causality is emphasized by discussing the nature of spurious correlation and the existence of causal relations in the absence of correlation; (5) we conclude by considering implications for forecasting.

THE DIAGNOSTIC PROCESS

We conceive of diagnosis as consisting of three interrelated phases: (1) finding relevant variables; (2) linking variables into causal chains; and (3) testing the causal significance of the links in those chains. In this paper we limit ourselves to how two, or at most, three variables become linked in a causal manner. We begin by first discussing the general issues of what directs attention to variables and what is meant by "causal relevance."

Much psychological research indicates that processes of perception and judgement are sensitive to differences or deviations from present states, adaptation levels, and reference points (e.g., Helson, 1964; Kahneman and Tversky, 1979). Therefore, we propose that in searching for a cause of some effect, which is itself a deviation from the normal or average, attention is directed toward prior deviations or abnormal states of comparable length and strength. For example, our cave-dwellers probably asked themselves what unusual event (or events) preceded the decline in births. Moreover, if the effect of interest is large (i.e., is of substantial duration and/or strength), we expect that the suspected causes are judged to be of comparable size. These conjectures follow from what Kahneman and Tversky call the "representativeness" heuristic (1972), which in this case implies that similar causes have similar effects. Indeed, Mill noted that this is a deeply rooted belief that "not only reigned supreme in the ancient world, but still possesses almost undisputed dominion over many of the most cultivated minds" (cited in Nisbett and Ross, 1980, p. 115). Mill thought that such a belief was erroneous and many cases exist in which similarity has been misleading (see, e.g., Shapiro, 1960; Shweder, 1977). On the other hand, it is difficult to imagine how one could search for variables without using some notion of similarity. Of particular importance is the non-literal use of similarity through analogy and metaphor (e.g., Ortony, 1979). Indeed, analogies and metaphors provide models of phenomena and thus direct attention to specific aspects and variables. Moreover, their use engages one's prior knowledge so that the explanation for a new or poorly understood phenomenon can be integrated with what is already known or believed. For example, in trying to understand how the brain works, one could consider it as a computer, a muscle, or a sponge. Note how the metaphors direct attention to different, but known, variables and processes. Thus, a computer model suggests informational input, retrieval, and computational processes; a muscle model suggests that processes are strengthened and weakened with amount of use, that thinking can be a "strain" (cf. Shugan, 1980), and so on; a sponge model suggests a more passive "soaking up of information," etc. As should be clear, the choice of a particular metaphor is crucial since it directs attention to a limited set of variables, thereby excluding others. However, this is precisely the function it should serve (we consider this further in the "Implications" section).

While analogies and metaphors help direct attention to variables, not all variables considered are seen as causally relevant. Therefore, it is important to delineate how this occurs. In order to do so, we discuss the concept of a "causal field" and a causal variable as a difference-in-that-field (Mackie, 1974). The following example illustrates both concepts. Imagine a worker in a chemical plant who contracts cancer and sues the company for causing his disease. His lawyer argues that the cancer rate of workers in this factory is nine times the national average for workers in comparable industries. Note that the "field" in this argument is industries of a certain type and the causal argument rests on a difference (higher cancer rates) in this field. However, the defense lawyer asks: "Would the worker have gotten cancer if he didn't work in the chemical plant?" In order to answer this, he then shows that there is a history of cancer in the worker's family (none of whom worked in chemical factories); the worker has smoked cigarettes for 20 years prior to getting cancer; and so on. The important point to note is that the field has now shifted to people who have certain family histories and habits; and in this field, cancer is not unusual and thus not a difference in the field. Hence, this particular case of cancer is not causally related to working in chemical factories. Of course, one could strengthen the prosecution case if the field were again changed--e.g., showing that the particular chemical plant had nine times the cancer rate of other chemical plants making the same products. The reason that this information strengthens the prosecution's case is that by narrowing the field to chemical plants making the same products, the possibility of alternative explanations is reduced, thereby making the difference in the narrowed field more causally relevant.

Although people are sensitive to differences-in-a-field, there are many differences to which one could attend. Therefore, what mechanisms guide the allocation of attention to those variables of greatest causal significance? An important consideration in this regard concerns the factors that distinguish between differences that are causal versus those that are called "conditions." For example, imagine a house fire in which a gasoline can was found. While the flammable material is a difference in the field (since most people do not have gasoline cans in their homes), it is not likely to be seen as the cause of the fire. Indeed, one is more likely to consider the throwing of a lighted match or cigarette near the can as the cause. As Mackie (1974) points out, there are no general rules for distinguishing causes from conditions, although the following are useful: (1) events are more causal than standing conditions (sparks rather than flammable material cause fires); (2) events that are intrusive are more causal than those that generally occur; (3) something abnormal or wrong is more causal than what is normal and right (e.g., the accident was caused by the person veering to the left--not by the other person who drove straight ahead). It is important to stress that there are exceptions to all three rules; e.g., poverty, which is a standing condition, may be seen as the cause of social unrest. Therefore, each of these rules only provides probabilistic evidence for distinguishing causes from conditions.

The mechanism by which causal links are formed and tested is now considered by introducing the concept of "cues to causality." We consider the term "cues" as having a specific meaning that corresponds with its use in Brunswik's psychology (1952; see also Hammond, 1955; Campbell, 1966). Specifically: (1) the relation between each cue and causality is probabilistic; that is, each cue is only a fallible sign of a causal relation; (2) people learn to use multiple cues in making inferences in order to mitigate the potential errors arising from the use of single cues; (3) the use of multiple cues is facilitated by the intercorrelation (redundancy) between cues in the environment. Such intercorrelation reduces the negative effects of omitting cues and aids in directing attention to the presence of other cues; and (4) although multiple cues reduce uncertainty inference, they do not entirely eliminate it. This is as true for backward as it is for forward inference.

In order to provide a conceptual framework for discussing the cues, imagine that one has a causal candidiate in mind (X) for explaining some effect, Y. The causal strength of X will depend on its being a difference-in-the-field, which can be assessed by considering it relative to Y and $\bar{Y}$; i.e., does X discriminate between the occurrence or absence of Y. For example, in trying to determine why some people got sick after eating in a particular restaurant (Y), we immediately want to know if those who didn't get sick ($\bar{Y}$), ate the same food (X). If the number of cases of $X \cap \bar{Y}$ and $X \cap Y$ are comparable, the strength of the connection is clearly diminished. However, this is not the only difference that can be considered. One could further ask whether Y is a difference-in-the-field of X versus $\bar{X}$; i.e., the field is reversed so that what is figure becomes ground, and vice versa. In this case we wish to compare the number of people who got sick without eating the suspected food ($\bar{X}$), to those who didn't get sick ($\bar{Y} \cap \bar{X}$). These differences, as well as those above, can be summarized in a standard 2 x 2 table, as shown in Table 1. The frequencies in each of the four cells can be thought of as individual components that are combined in various ways to judge the relation between X and Y. The cues to causality come into play by directing attention to different parts of the 2 x 2 table. We now consider the cues.

Table 1. Contingency table between X and Y

	Y	$\bar{Y}$
X	Cell a = $f(X \cap Y)$	Cell b = $f(X \cap \bar{Y})$
$\bar{X}$	Cell c = $f(\bar{X} \cap Y)$	Cell d = $f(\bar{X} \cap \bar{Y})$

Note: f stands for frequency

(1) The cue of temporal order has two related functions: (a) since causes are assumed to precede effects, the search for relevant causal variables focuses on only those events that preceded the effect and thereby reduces the number of variables to be considered; (b) in terms of Table 1, temporal order is

used to label the axes of the 2 x 2 table as being either X (independent variable) or Y (dependent variable).

(2) The cue of constant conjunction is defined as the degree to which two variables occur together, holding contiguity in time and space constant. For example, consider a specific case of classical conditioning where a bell is repeatedly paired with food such that an animal learns to salivate at the sound of the bell. However, when the bell is not always followed by food, the relation is more difficult to learn. Note that temporal order is also involved in this example; in fact, attempts at "backward conditioning" (giving food and the bell) have generally been unsuccessful (however, see Spetch, Wilkie, and Pinel, 1981). In terms of Table 1, we consider that constant conjunction directs attention to the joint frequency of X and Y (cell a).

(3) We treat the cues of contiguity in time and space together although they are conceptually distinct. The reason for considering them together is that they both influence the relevance of constant conjunction as a causal cue and thus concern cell a in Table 1; i.e., contiguity is viewed as a cue to causality via its effects on constant conjunction. This occurs in two ways: (a) Together with similarity, attending to differences-in-a-field, and temporal order, contiguity in time and space aids in focusing attention on what variables occurred close in time to, and in the vicinity of, some effect Y. Indeed, Siegler has shown that for young children (5-6 years old, temporal contiguity is a very strong cue for inferring causality (Siegler and Liebert, 1974; Siegler, 1976). Moreover, these studies show that older children are less dependent on contiguity alone, being able to make use of multiple cues. Nevertheless, in the absence of high contiguity, variables that are causally related may not be noticed as important, as in the low temporal contiguity between pregnancy and intercourse. Additionally, the importance of contiguity cues can be seen by noting that certain causal distinctions are made when contiguity conflicts with other cues to causality. For example, consider the distinction between a "precipitating" and "underlying" cause. The former is generally some action or event that is high in temporal and spatial contiguity but low in similarity of length or strength with the effect. Thus, the precipitating cause of World War I was an assassination in Sarajevo but the underlying causes were economic upheaval, German nationalism, and so on. (b) Once variables have been identified, we hypothesize that constant conjunction is weighted by the degree of temporal and spatial contiguity between the variables. Therefore, although two variables may have a high joint frequency of occurrence, we would expect their judged causal strength to be lowered as

temporal and spatial contiguity were decreased. Similarly, we expect that increased contiguity in time and space will strengthen the causal relevance of variables (including those with low joint frequency).

(4) The number of alternative explanations, or competing variables as causes of Y, is an important negative cue to causality (i.e., the greater the number, the less the causal relevance of some X for Y). As stressed by Campbell and colleagues (Campbell and Stanley, 1963; Campbell, 1975; Cook and Campbell, 1979), causal strength should be evaluated by the ruling out of alternative explanations. Indeed, Mackie (1974) states that the primitive notion of a cause involves asking oneself the question: "Would Y have occurred if X had not?" The greater the number of alternative explanations underlying a "yes" answer, the lower the causal relevance of X for Y. Note that the posing and answering of the above question (which is called a "counterfactual conditional") may either involve doing a real experiment or what is called a "thought" experiment. In the former, one compares the effect of X on Y with that of $\bar{X}$ on Y (the control group condition). In this way, the conterfactual question is easily answered. Moreover, note that a control group allows one to infer that X is the only difference in the field that can affect Y (since the control group is the field). However, even in real or quasi-experiments, alternative explanations can exist if other cues to causality are ambiguous. For example, Campbell (1969) points out that (a) the gradual introduction of some experimental change makes the determination of its causal impact more difficult than a sharp introduction, as, for example, when a remedial program is phased in over a long time period rather than implemented all at once; (b) unless replication of the $X \cap Y$ relation is accomplished by the introduction of X over multiple time periods or over different units within the same time period, evidence of constant conjunction is weak; (c) the determination of causal relations when effects are not contiguous in time and space with manipulated variables is problematic. For example, variables that have "lagged" effects or cumulative effects over time are difficult to isolate.

When real or quasi-experiments are not possible, one can nevertheless engage in the following thought experiment in order to answer the question, "Would Y have occurred if X hadn't?" Imagine the world before X, go forward to where X would occur, and then delete it from the scenario. Now run the scenario forward from that point to see if Y happens or not. Clearly, in such thought experiments the construction of "possible worlds" and imaginary scenarios is crucial for judging causal significance. In fact, according to Mackie (1974):

> The key item is a picture of what would have happened if things had been otherwise, and this is borrowed from some experience where things were otherwise. It is a contrast case rather than the repetition of like instances that contributes most to our primitive concept of causation. (p. 57)

The use of alternative explanations as a cue to causality has important implications for making inferences in general and for interpreting experiments in particular. As recognized by Hume, "not only in philosophy, but even in common life, we may attain the knowledge of a particular cause merely by one experiment, and after a careful removal of all foreign and superfluous circumstances" (as quoted in Mackie, 1974, p. 25). As a case in point, consider the following one-shot case study with a single datum: the occurrence of a huge explosion near Los Alamos, New Mexico, in July 1945 which no one doubted to be the effect of detonating an atomic bomb. Clearly, inferring causality in this poorly designed experiment was not difficult, whereas assessing causality in the most meticulously designed experiments in social science is often problematic at best. When one considers why the causal inference is so strong in the bomb example, ask yourself the following question: "Would an explosion of such magnitude have occurred if an atomic bomb had not gone off?" While it is possible to think of alternative explanations for the explosion, they are so unlikely as to be virtually non-existent. Moreover, note how the other cues to causality point to a causal relation (temporal order, contiguity in time and space, similarity of the "unusualness" of effects and cause, and so on). Therefore, even in one-shot case studies with no control group, causality can be inferred (see Campbell (1975) for an illuminating discussion of this issue).

The use of counterfactual questions for assessing alternative explanations can be conceptualized by referring back to Table 1. Note that in backward or diagnostic inference, the question, "Would Y have occurred if X hadn't?" focuses attention on cell c, which contains the $Y \cap \overline{X}$ instances. In forward inference, on the other hand, one can ask the question, "Would not-Y occur if X had?" In this case, attention is focused on cell b, containing the $\overline{Y} \cap X$ cases. In either situation, one moves from considering only cell a in Table 1 to other components of the X, Y relation.

(5) As noted earlier, similarity plays a crucial role in the finding of relevant variables. Furthermore, analogy and metaphor can be used to understand new phenomena by linking them to one's prior knowledge. However, one can ask how good a

certain analogy might be, as, for example, between the human brain and the computer. Whereas both possess many common features, they also contain distinctive aspects. Tversky (1977) has proposed a theory of how such common and distinctive aspects are combined to form similarity judgments. Specifically, he posits that the judgment of similarity between two objects is a weighted linear function of the features they have in common less the distinctive features of each. The parameters of this function reflect differential attention paid to common and distinctive components. Tversky's formulation parallels the judgment of causal strength in the following way. Consider Table 1 and note that the frequencies in cells a and d can be seen as representing the constant conjunction of the variables (analogous to common features), whereas the frequencies in cells b and c can be seen as instances that disconfirm constant conjunction or affirm alternative explanations (analogous to the distinctive features of two objects). If one considers how attention can be shifted to these different aspects by such factors as whether the task involves forward or backward inference, or what other cues to causality are present (temporal order, contiguity), then judged causal strength can be seen as a weighted linear function of confirming and disconfirming evidence (for experimental results, see Schustack & Sternberg, 1981).

(6) The degree to which one variable can predict another, denoted as predictive validity, is an important cue to causality. Predictive validity is commonly measured by the correlation coefficient and it is noteworthy that this involves all four cells of Table 1. That is, the correlation between X and Y, denoted by r_{xy}, is given by the formula $r_{xy} = (ad-bc)/[(a+b)(c+d)(a+c)(b+d)]^{1/2}$. Other things being equal, the greater the predictive validity of some variable, the greater its causal relevance. However, when other causal cues are involved, things are not equal. This makes the use of statistical correlation as a causal cue more uncertain than is generally realized. We discuss this in some detail by considering the meaning of "spurious correlation." Thereafter, we consider what we call "causalation," which results when variables that are causally related show low or no statistical correlation.

On the Psychology of Spurious Correlation

Every student who has taken an introductory statistics course containing a section on correlation is told: Correlation does not imply causation. Unfortunately, the factors that do imply causation are rarely if ever discussed and students are left without further guidance. To make matters worse, several days later the same students are warned against "spurious correlation," i.e., the correlation between two variables due to the commmon causal influence of some third factor. Since the concept of spurious correlation suggests that some correlations are more (or less) causally related than others, it is natural to ask how one can tell the difference (cf. Simon, 1954). However, one is disappointed to learn that there are no simple rules for doing this and that one must exercise one's judgment in this matter. This is not to say that such judgments will always be difficult to make or that they will vary according to the person doing the judging. For example, consider the correlation between the number of pigs and the amount of pig-iron (Ehrenberg, 1975). Such a correlation does seem spurious when the common causal factor, "economic activity," is considered. On the other hand, consider the correlation between amount of rain and number of auto traffic accidents in a city, over the course of a year. Such a correlation does not seem spurious. What is the difference between these two cases?

If we make use of the cues to causality, the spuriousness of the correlation between pigs and pig-iron becomes apparent. First, consider the cues that do point to a causal relation; namely, constant conjunction, contiguity in time, and predictive validity. However, the cues that point away from a causal relation are: temporal order (which cannot be used to specify which variable is cause or effect); low contiguity in space (it being unlikely that farms and factories are in close physical proximity); many alternative explanations for either variable (for example, the answer to the question, "Would pig production have gone up if the production of pig-iron hadn't?" has many yes answers); the similarity of the variables only with respect to their names, all else being quite dissimilar; and the robustness of predictive validity seeming low when, for example, one considers the lack of relation in non-industrialized countries. Taken together, the negative evidence regarding a causal relation seems much stronger than the positive evidence. Indeed, the judgment that the relation is spurious is made easily and quickly.

Now consider the second case: the temporal order of rain and accidents is clear; constant conjunction, contiguity in time and space, and predictive validity are all high; similarity, via the use of prior knowledge about the effects of slippery roads, is high; robustness is also high since predictive validity holds in many cities over widely dispersed geographical locations. The only negative cue is the number of alternative explanations since the answer to the question "Would accidents have increased if it hadn't rained?" might be yes. However, even in this case, there may be few competing alternatives. Therefore, taken together, the cues strongly point to a causal relation and a judgment that the correlation is non-spurious seems warranted.

The generalization that can be made from the above is the following: The judged strength between variables is a joint function of their correlation and the causal cues that are implicit in the labelling of the variables. This statement has several important implications: (1) The labelling of variables should have a major effect on their judged relational strength, holding statistical correlation constant. Evidence for this comes from a study by Jennings, Amabile, and Ross (in press). They found that when people viewed scatterplots of variables labelled X and Y, the statistical correlation had to be quite high for people to see a relationship. However, when the variables were given labels that engaged prior knowledge (making use of the cues to causality), the statistical correlation needed to see a relationship was much lower. (2) The effects of labelling have also been studied in probability learning studies and confirm the importance of cues to causality in learning the relations between variables. For example, Adelman (1981) showed that subjects in a multiple cue probability task learned quite well when variable labels were congruent with statistical predictiveness, but not otherwise. Camerer (1981) showed the subjects were able to learn a disordinal interaction only when the variables were labelled in accord with prior beliefs (this involved factors that were thought to affect the price of wheat futures in a commodity market). When the same task was given as an abstract problem with variables labelled as X_1, X_2, and Y, no learning occurred. (3) When statistical correlation and cues to causality conflict, spurious correlation is not the only outcome; e.g., a low or zero statistical correlation could mask a true causal relation. In order to illustrate this, let us return to our cave-dwellers, who have hit upon the hypothesis that sexual intercourse is related to pregnancy. Accordingly, they designed and carried out the following experiment: One hundred females were allocated at random to an intercourse condition, and 100 to a non-intercourse condition. As indicated in Table 2, 25 females became pregnant, and 175 did not. Given our present world knowledge we can state that the five people in the no-intercourse/yes-pregnancy cell represent "measurement error," i.e., faulty memory in reporting, lying, etc. Is intercourse important for pregnancy? The statistical correlation is small ($r=0.34$) and our cave-dwellers, in their methodological sophistication, might well question whether the hypothesis is worth pursuing. Indeed, if the sample size were smaller, the correlation might not even be statistically "significant." Moreover, even with a significant correlation,

r^2=0.12, which is hardly a compelling percentage of the Y variance accounted for by X.

Table 2. Data matrix for hypothetical intercourse–pregnancy experiment

		Pregnancy Yes	Pregnancy No	
Intercourse	Yes	20	80	100
	No	5	95	100
		25	175	200

The important implication of the above hypothetical experiment is the following: Although correlation doesn't necessarily imply causation, causation doesn't necessarily imply correlation. We have somewhat facetiously labelled examples of the latter as "causalations," giving them equal standing with the better-known and opposite concept of spurious correlation. The importance of causalation is that it demonstrates that sole reliance on statistical measures for understanding and interpreting data is insufficient (see also Simon, 1954). This conclusion is not surprising to those who have always maintained that judgment be a part of the evaluation of evidence. However, the delineation of the cues to causality gives one some hint as to the components of such judgments.

(7) We have used the conceptual device of Table 1 to show how the various cues to causality direct attention to different aspects of causal strength. However, the cue of robustness goes beyond a single 2 x 2 table and explicitly raises the question as to whether, and to what degree, the predictive validity between X and Y varies as a function of other variables (see Toda, 1977). For example, imagine that there is a positive correlation between smoking (X and $\overline{X}$) and lung cancer (Y and $\overline{Y}$). Now consider that the correlation is computed separately for men and women with the following result: the correlation is positive for men but negative for women (i.e., women who don't smoke are more likely than smokers to get lung cancer). Note that by sub-dividing the original sample into sub-groups, one now considers several 2 x 2 tables and asks whether the overall correlation holds in each. If it doesn't, the relation is not robust and the causal relevance of X for Y is decreased. In the above example, one is likely to be suspicious of any causal relation if the sign of the correlation changes in one of the sub-groups. On the other hand, if predictive validity is robust, it points more strongly to a causal relation.

IMPLICATIONS

We now consider three implications of the above analysis for forecasting: (1) the selection of variables and the building of forecasting models; (2) the evaluation of forecast accuracy; and (3) learning to improve forecasting ability.

(1) We have emphasized the role of similarities and differences as well as the cues to causality in directing attention to variables in the diagnostic process. However, a significant feature of causal/diagnostic thinking is the remarkable speed and fluency which people seem to have for generating explanations and accommodating discrepant fact into expanded hypotheses (Fischhoff, 1975; Slovic and Fischhoff, 1977; Tversky and Kahneman, 1980). A useful analogy of this process is provided by multiple regression analysis. In deriving a model (backward inference) one seeks the combination of variables and parameters that maximizes some measure of fit, e.g., R^2. Furthermore, by increasing variables, R^2 approaches 1. However, and often to the user's surprise, the measure of fit typically "shrinks" on prediction. Thus, the power of post hoc explanations is unfortunately matched by the paucity of predictive validity.

Since diagnostic thinking is so fluent, one must guard against the way cues to causality quickly restrict our interpretation of the past. Of particular significance in this regard is the use of metaphors in guiding attention and providing models of complex phenomena. Since metaphors, like other cues to causality, are of imperfect validity, errors can be made by relying on single images. "The map," as the general semanticists remind us, "is not the territory" and thus we specifically recommend the method of multiple metaphors as a way of guarding against the premature adoption of a single model. That is, instead of focusing on a single metaphor, experiment with several. For example, consider how one might view forecasting. In this paper we have adopted a medical metaphor by focusing on the diagnostic process and the effects of treatments (actions) on outcomes. However, consider a ballistics metaphor in which forecasting is likened to aiming and shooting at a target (see Hogarth, 1981). Such a metaphor leads to consideration of quite different variables and issues. Therefore, while no single model is correct, each directs attention to different factors, thereby providing a more complete picture of the phenomenon.

Whereas the cues to causality play an important role in structuring and stabilizing our perceptions of reality, such stability may be purchased at the cost of novelty and originality. That is, the cues direct attention to what is "obvious," thereby reducing innovation and creativity. This suggests that one way to facilitate creative thinking might be to go specifically against the cues. For example, when dealing with a complex outcome, one might search for a dissimilar and simple causal candidate rather than a similar and complex one.

(2) As emphasized at the beginning of this paper, the evaluation of forecast accuracy is problematic without a causal understanding of the factors that influence outcomes. In particular, we stressed the fact that forecasts are made to aid decisionmaking and this makes outcomes a function of actions as well as predictions. The difficulties of evaluating forecast accuracy under these conditions can be conceptualized by considering two factors: (a) to what extent are people aware that their actions can affect outcomes? and (b) to what extent do their actions actually affect outcomes? If, for the sake of simplicity, we only consider two levels of each factor (i.e., aware-not aware, and actions affect outcomes-don't affect outcomes), Table 3 presents a convenient way to summarize the various possibilities.

Table 3. Categorization of forecasting situations by awareness and efficacy of actions

		Actions affect outcomes: Yes	Actions affect outcomes: No
Awareness of actions affecting outcomes	Yes	(1) Forecast accuracy conditioned on action; $p(Y\|X, A) \neq p(Y\|X)$	(3) Illusions of control; superstitions.
	No	(2) Self-fulfilling and self-defeating prophecies	(4) Simple forecast accuracy is adequate.

Cells, 1, 2, and 4 were discussed earlier in our examples of economic forecasting, rumours of bank failure, and hurricane prediction, respectively. We simply note that in the first two cases (cells 1 and 2), the evaluation of forecast accuracy should be conditioned on the action taken. That is, let Y be some outcome, X a forecast, and A an action based on the forecast; one is then interested in p(Y/X, A) rather than p(Y/X). As we have noted elsewhere (Einhorn and Hogarth, 1978), when actions affect outcomes, forecasts and outcomes are not conditionally independent of action; i.e., $p(Y/X, A) \neq p(Y/X)$.

The situation represented in cell 3 deserves special attention since it involves taking action to affect outcomes when such actions are useless. When action involves physical manipulation or direct intervention, Langer(1975) has shown that people are prey to an "illusion of control;" i.e., actions are seen to affect outcomes that are generated by random processes (cf. Lopes, in press). For example, people tend to believe that the lottery tickets they personally select have a greater chance of winning than those selected for them by a lottery administrator. Similarly, the cognitive activities of planning and forecasting in organizations can lead to illusions of control and overconfidence by restricting attention to those consequences one can imagine while diverting attention from those not considered (Hogarth and Makridakis, 1981). Moreover, it should be noted that illusions of control are conceptually identical to certain types of superstitious behaviour. Indeed, the joint presence of randomness and the cues to causality (especially temporal order, constant conjunction, and high contiguity in time and space) inevitably lead to some superstitious behaviour (Skinner, 1966). Thus, one can consider superstition as the cost one pays to gain causal knowledge. The issue remains, of course, as to whether the cost outweighs the associated benefits (cf. Killeen, 1978).

(3) Given the difficulties of evaluating forecast accuracy, what can be done to improve forecasting ability? We consider two related approaches: (a) experimentation to untangle the forecast-action-outcome sequence; and (b) systematizing thought experiments.

(a) Opportunities for experimentation are often overlooked. For example, consider personnel selection or advertising. In both cases one could carry out full or partial experiments by randomly selecting employees or stopping advertising completely. While such experiments are typically unfeasible, partial experiments could provide much useful information. For example, one could randomly admit a small percentage of candidates and advertising could be stopped in selected time periods or areas (see Cook and Campbell (1979) for relevant experimental designs).

(b) To systematize thought experiments, we suggest the following procedure. Imagine that sales have increased and you wish to determine the causal significance of a prior advertising campaign. A real experiment would involve using a control group to test whether no campaign would have led to the same level of sales. Using the symbols of Table 1, the control group outcome would focus on the $\bar{X} \cap Y$ cell (i.e., c). Moreover, in the absence of a real experiment this datum would be supplied by the answer to the counterfactual question "Would Y have occurred if X hadn't?" However, answering only this question would be a weak form of inference. We also suggest that people ask themselves questions that illuminate the $X \cap \bar{Y}$ and $\bar{X} \cap \bar{Y}$ cells. The former involves imagining whether no sales increase ($\bar{Y}$) would have followed the advertising campaign (X); the latter asks whether no campaign ($\bar{X}$) would have been followed by no increase in sales ($\overline{Y}$). Systematically posing these different counterfactual questions can increase the power of thought experiments by generating information analogous to that available from real experiments.

CONCLUSION

Forecasting depends on the use of data and the exercise of judgment. Indeed, a colleague has succinctly summarized the theme of this paper by saying that one cannot understand statistics without psychology (see also Hogarth, 1975). While agreeing, we also note that the meaning and effects of uncertainty are central to the psychology of thinking and thus call for statistical expertise. In this paper we have emphasized the probabilistic nature of cues to causality and the uncertainties associated with backward and forward inference. Moreover, we have stressed the need to understand how judgment affects the generation and testing of formal statistical models. We believe that the development and use of statistical techniques will benefit from understanding judgmental processes. Without such understanding, which depends crucially on causal thinking, simply forecasting the future will not ensure the future of forecasting.

REFERENCES

Adelman, Leonard, "The influence of formal, substantive, and contextual task properties on the relative effectiveness of different forms of feedback in multiple-cue probability learning tasks," Organizational Behavior and Human Performance, 27 (1981), 423-442.

Brunswik, Egon, The Conceptual Framework of Psychology, Chicago: University of Chicago Press, 1952. Camerer, Colin F., "The validity and utility of expert judgment," unpublished Ph.D. dissertation, University of Chicago, 1981.

Campbell, Donald T., "Pattern matching as an essential in distal knowing," in Hammond, K.R. (ed.), The Psychology of Egon Brunswik, New York: Holt, Rinehart and Winston, 1966.

Campbell, Donald T., "Reforms as experiments," American Psychologist, 24 (1969), 409-429.

Campbell, Donald T., "Degrees of freedom and the case study," Comparative Political Studies, 8 (1975), 178-193.

Campbell, Donald T., and Stanley, Julian C., Experimental and Quasi-experimental Designs for Research, Chicago: Rand-McNally, 1963.

Cook, Thomas D., and Campbell, Donald T., Quasi-experimentation: Design & Analysis for Field Settings, Chicago: Rand-McNally, 1979.

Dawes, Robyn M., "The robust beauty of improper linear models in decisionmaking," American Psychologist, 34 (1979), 571-582.

Dawes, Robyn M., and Corrigan, Bernard, "Linear models in decisionmaking," Psychological Bulletin, 81 (1974), 95-106.

Ehrenberg, Andrew S.C., Data reduction: Analyzing and Interpreting Statistical Data, New York: Wiley, 1975.

Einhorn, Hillel J., "Learning from experience and suboptimal rules in decisionmaking," in Walsten, T.S. (ed.), Cognitive Processes in Choice and Decision Behavior, Hillsdale: Erlbaum, 1980.

Einhorn, Hillel J., and Hogarth, Robin M., "Confidence in judgment: persistence of the illusion of validity," Psychological Review, 85 (1978), 395-416.

Fischhoff, Baruch, "Hindsight≠foresight: the effect of outcome knowledge on judgment under uncertainty," Journal of Experimental Psychology: Human Perception and Performance, 1 (1975), 288-299.

Hammond, Kenneth R., "Probabilistic functionalism and the clinical method," Psychological Review, 62 (1955), 255-262.

Helson, Harry, Adaptation-Level Theory, New York: Harper, 1964.

Hogarth, Robin M., "Cognitive processes and the assessment of subjective probability distributions," Journal of the American Statistical Association, 70 (1975), 271-289.

Hogarth, Robin M., "Beyond discrete biases: functional and dysfunctional aspects of judgmental heuristics," Psychological Bulletin, 90 (1981), 197-217.

Hogarth, Robin M., and Makridakis, Spyros, "Forecasting and planning: an evaluation," Management Science, 27 (1981), 115-138.

Howard, Ronald A., Matheson, James E., and North, D. Warner, "The decision to seed hurricanes," Science, 176 (1972), 1191-1202.

Jennings, D., Amabile, T. M., and Ross, L., "Informal covariation assessment: data-based versus theory-based judgments," in Kahneman, D., Slovic, P., and Tversky, A. (eds), Judgment Under Uncertainty: Heuristics and Biases, New York: Cambridge University Press, in press.

Kahneman, Daniel, and Tversky, Amos, "Subjective probability: a judgment of representativeness," Cognitive Psychology, 3 (1972), 430-454.

Kahneman, Daniel, and Tversky, Amos, "Prospect theory: an analysis of decision under risk," Econometrica, 47 (1979), 263-291.

Killeen, Peter T., "Superstition: a matter of bias, not detectability," Science, 199 (1978), 88-90.

Langer, Ellen J., "The illusion of control," Journal of Personality and Social Psychology, 32 (1975), 311-328.

Lopes, Lola L., "Doing the impossible: a note on induction and the experience of randomness," Journal of Experimental Psychology: Human Learning and Memory, in press.

Mackie, John L., The Cement of the Universe: A Study of Causation, Oxford: Clarendon Press, 1974.

Meehl, Paul E., Clinical versus Statistical Prediction: A Theoretical Analysis and Review of the Literature, Minneapolis: University of Minnesota Press, 1954.

Nisbett, Richard E., and Ross, Lee D., Human Inference: Strategies and Shortcomings of Social Judgment, Englewood Cliffs: Prentice-Hall, 1980.

Ortony, Andrew, "Beyond literal similarity," Psychological Review, 86 (1979), 161-180.

Sawyer, Jack, "Measurement and prediction, clinical and statistical," Psychological Bulletin, 66 (1966), 178-200.

Schustack, Miriam W., and Sternberg, Robert J., "Evaluation of evidence in causal inference," Journal of Experimental Psychology: General, 110 (1981), 101-120.

Shapiro, Arthur K., "A contribution to the history of the placebo effect," Behavioral Science, 5 (1960), 109-135.

Shugan, Steven M., "The cost of thinking," Journal of Consumer Research, 7 (1980), 99-111.

Shweder, Richard A., "Likeness and likelihood in everyday thought: Magical thinking in judgments about personality," Current Anthropology, 18 (1977), 637-658.

Siegler, Robert S., "The effects of simple necessity and sufficiency relationships on children's causal inferences," Child Development, 47 (1976), 1058-1063.

Siegler, Robert S., and Liebert, Robert M., "Effects of contiguity, regularity, and age on children's causal inferences," Developmental Psychology, 10 (1974), 574-579.

Simon, Herbert A., "Spurious correlation: a causal interpretation," Journal of the American Statistical Association, 49 (1954), 467-479.

Skinner, Burrhus F., "The phylogeny and ontogeny of behavior," Science, 153 (1966), 1205-1213.

Slovic, Paul, and Fischhoff, Baruch, "On the psychology of experimental surprises," Journal of Experimental Psychology: Human Perception and Performance, 3 (1977), 544-551.

Spetch, Marcia L., Wilkie, Donald M., and Pinel, J. P. J., "Backward conditioning: a reevaluation of the empirical evidence," Psychological Bulletin, 89 (1981), 163-175.

Toda, Masanao, "Causality, conditional probability and control," in Aykac, A., and Brumat, C. (eds), New Developments in the Applications of Bayesian Methods, Amsterdam: North Holland, 1977.

Tversky, Amos, "Features of similarity," Psychological Review, 84 (1977), 327-352.

Tversky, Amos, and Kahneman, Daniel, "Causal schemas in judgments under uncertainty," in Fishbein, M. (ed.), Progress in Social Psychology (Vol. 1), Hillsdale: Erlbaum, 1980.

Risk Judgment and the Communication of Hazard Information: Toward a New Look in the Study of Risk Perception

Timothy C. Earle* and George Cvetovich**

*Battelle Human Affairs Research Centers, Seattle, WA

**Western Washington University, Bellingham, WA

INTRODUCTION

During the years immediately following World War II, a group of American psychologists developed a body of work in the study of perception which became known as the "New Look" (Bruner and Goodman, 1947; Krech and Crutchfield, 1948; Bruner and Klein, 1960). The New Look was part of a larger movement at that time which aimed "to bring psychology out of the laboratory and into the area of improving human society" (Hochberg, 1964, p. 101). While the New Look was predated by related work in perception (the work of Brunswik discussed in Hammond, 1966) and social psychology (Sherif, 1935) what was new was the sharp focus on the effects of social factors such as economic status on the perception of certain attributes of objects.

Our paper is subtitled "Toward a New Look in the Study of Risk Perception." This updated reference to the New Look is not meant to call to mind the controversial empirical results of that work but rather to evoke and acknowledge a certain spirit and direction of research. Of central significance to both the New Look and to us are the study of the effects of social factors on cognitive processes and the production of socially useful research. Rather than the perception of object attributes, however, we have chosen to study judgments of risk. And we have attempted to enhance the social utility of our work by studying the risk judgments of real people in real hazard situations.

The study of the risk judgments of individuals dealing with hazards in their everyday lives suggests a complex definition of risk that takes into account the wide variety of personal experiences and concerns among the public. A significant recent contribution to our understanding of risk by Fischhoff and his colleagues (Fischhoff, Slovic, and Lichtenstein, 1983;

NATO ASI Series, Vol. G4
Environmental Impact Assessment, Technology Assessment, and Risk Analysis. Edited by V.T. Covello et al.

Fischhoff, Watson, and Hope, in press) has illuminated this complexity by closely examining the subjective definition of risk. There are two basic points (Fischoff, Watson, and Hope, 1983): (a) There is no single definition of risk suitable for all problems (and this is true for both scientists and the public); (b) The choice of definition in any individual case is a political one, in the sense that it expresses someone's values about the importance of different adverse effects. Risk is thus multidimensional and variably defined both within and among individuals. Furthermore, risk is just one of several factors considered by individuals when making risk decisions. People don't ordinarily judge abstract levels of risk. Instead, they engage in risk decisions, choosing among several options (among modes of travel, for example). In these decisions "each option has a variety of relevant features, one of which is its level of risk" (Fischhoff, Watson, and Hope, 1983, pg. 6). The other features include costs and benefits each of which is also, of course, subjectively defined. All of these risk-decision features, risks, costs, and benefits, are products of personal, social, and cultural forces which together produce a rich diversity of risk-decision outcomes (Douglas and Wildavsky, 1982; Thompson, 1983).

In a recent exploratory study designed to provide a preliminary look at the characteristics of real-world risk decisions, we interviewed 68 persons living in a moderate-sized city in Washington state. The interviews were conducted by telephone and inquired about the significant risk decisions concerning health or safety that individuals had made in the past six months. Each respondent was asked to select a single risk decision from his experience, and further detailed information about that decision was collected. One series of items focused on the consequences considered by individuals in their risk decisions. The results given in Table 1 show that all of the consequences suggested in the questionnaire were reported by a majority of the respondents to have been considered in their risk decisions. Benefits, significantly, were considered by 91.2 percent of the respondents. Respondents were also asked to indicate who was affected by the hazards involved in their risk decisions. The data in Table 1 show that the most significant risk decisions for most respondents (57.4 percent) affected only themselves or themselves and known others.

These preliminary results provide some insight into the complex variety of everyday risk decisions. As would be expected, individuals are primarily concerned with risks that affect themselves. These risks are multidimensional, however, and the decisions that involve those risks also include other consequences, particularly benefits and costs.

Table 1. CONSEQUENCES CONSIDERED AND PERSONS AFFECTED IN SELECTED RISK DECISIONS

- CONSEQUENCES CONSIDERED IN RISK DECISIONS

CONSEQUENCE	% CONSIDERED
BENEFITS	91.2
COSTS	73.1
ENVIRONMENTAL EFFECTS	51.5
ILLNESS	71.6
DEATH	71.6
RISK TO OTHERS	69.1
UNCERTAINTY ABOUT OUTCOMES	71.2
DREAD & ANXIETY ABOUT OUTCOMES	74.2

- PERSONS AFFECTED BY HAZARDS INVOLVED IN RISK DECISIONS

SELF ONLY	39.7
SELF & KNOWN OTHERS	17.7
GENERAL POPULATION	42.6

Our approach to a New Look in the study of risk perception is centered on the examination of risk judgments as aspects of risk decisions made by real people in real hazard situations. The use of the term "risk judgment" rather than the more familiar "risk perception" is deliberate. There are two basic reasons for our preference for risk judgment. First, judgment is a more general term than perception. True, perception has come to be used in a very broad, all-encompassing way (for a discussion of the variety of uses of the term perception in social psychology see Tajfel, 1969). In the study of risk perception, perception has come to be synonymous with "subjective" (see the discussion of "actual" vs. "perceived" risks in Fischhoff, Slovic, and Lichtenstein, in press); the use of the term risk perception owes more to historical accident and methodological analogy than it does to analysis of the phenomena being studied. These free-wheeling uses of perception seem to us to sacrifice too much to popular usage. Inappropriate labeling of the phenomena we study may, for example, establish barriers to alternative conceptions and methods, thereby limiting the development of our understanding. In the present case, we choose to view perception as a particular type of cognitive process which is useful to distinguish it from others.

Our general term, judgment, refers to a continuum of cognitive processes which range from perception-like intuitive processes at one end to analytical reasoning at the other. The distinction between perception and analytical reasoning is rooted in the work of Brunswik (1952; 1956) who described perception as a probabilistic, intuitive, continuous, and rapid process, as opposed to analytical reasoning which he saw as a deterministic, discontinuous, and slow process. Hammond has elaborated the pioneering work of Brunswik through his development of Social Judgment Theory (Hammond, Stewart, Brehmer, and Steinmann, 1975) and his explication of quasi-rational thought which consists of both intuitive and analytical components (Hammond and Brehmer, 1973; Hammond, 1983; Hammond, Hamm, Grassia, and Pearson, 1983; Earle, 1972). In our view, risk judgments can be intuitive and perception-like, analytical, or quasi-rational. It is likely that most risk judgments are quasi-rational, composed of varying mixtures of intuitive (implicit, nonsequential, and non-recoverable) and analytical (explicit, sequential, and recoverable) elements.

The second reason for our use of risk judgment rather than risk perception builds on the first. The more explicit identification of cognitive processes inherent in the use of risk judgment and its associated cognitive continuum (as opposed to the loose use of risk perception) provides a framework for the exploration of the relations between task environments and cognitive processes. This point is illustrated by examples drawn from two fields related to the study of risk judgment. First, in the behavioral approach to economics, Cyert and Simon relate task characteristics to the use of various behavioral rules:

> "Behavioral rules (rules of thumb) are modes of behavior that the firm (or individual) develops as guides for making decisions in a complex environment with uncertainty and incomplete information. These rules generally incorporate the decisionmakers' assumptions about the nature of the environment and the nature of the firm itself. In general the "thing" we refer to as judgment is capable of being decomposed into a set of behavioral rules. As uncertainty gives way to certainty and as knowledge increases, the behavioral rules change and move closer to those derived by a priori reasoning about a simplified situation." (Cyert and Simon, 1983, pg. 105).

In our second example, Tversky and Kahneman connect task characteristics to the use of judgment heuristics:

> "Judgments of probability vary in the degree to which they follow a decompositional or a holistic approach and in the degree to which the assessment and the aggregation of probabilities are analytic or intuitive ... at one extreme there are questions (e.g., What are the chances of beating a given hand in poker?) that can be answered by calculating the relative frequency of "favorable" outcomes. Such an analysis possesses all the features associated with an extentional approach: It is decompositional, frequentistic, and algorithmic. At the other extreme, there are questions (e.g., What is the probability that the witness is telling the truth?) that are normally evaluated in a holistic, singular, and intuitive manner . . . Decomposition and calculation provide some protection against conjunction errors and other biases, but the intuitive element cannot be entirely eliminated from probability judgments outside the domain of random sampling." (Tversky and Kahneman, 1983, pg. 310).

In the study of risk judgments, as in these related examples, it may be possible to make connections between cognitive processes and environmental characteristics. And further, the cognitive processes used in risk judgments may eventually be related to subsequent behavior in hazardous environments.

Although no studies exploring the relations among environmental characteristics, cognitive processes, and behavior in hazardous environments have yet been completed, such studies are a goal of our approach. The aspect of the environment of central interest is the information environment inhabited by the hazard: we would expect the hazard information available to, and used by, individuals to affect their risk-judgment cognitive processes, and the risk judgments in turn to affect future information-seeking and risk-mitigation behaviors. Studies of hazard information environments would therefore seem a good starting place for the development of our New Look.

In the first main section of this paper we describe three studies of hazard information environments (all carried out in collaboration with Michael K. Lindell of Battelle/HARC in Seattle). Two of the studies examine the content of mass media coverage of hazards, one technological (the gasoline crisis of 1979 in the U.S.) and one natural (the Mount St. Helens eruptions of 1980). The third study explores the suitability of science writers for their role as communicators of hazard information to the public. Together these studies provide some understanding of how the news media attempt to identify and meet the hazard information needs of their users.

The work on hazard information environments has led to the development of conceptual models as guides for future research. One model describes our view of risk judgment as a dynamic process in which an individual adapts to the hazardous environment through a series of steps over time. The second model describes the hazard communications system, consisting of the hazard, technical experts, professional decisionmakers, communications media, and the public. Of particular interest in the hazard communications system is the role of scientific information. The descriptions of these two conceptual models constitute the second main section of the paper. The third and final section is devoted to our view of future risk judgment research.

STUDIES OF HAZARD INFORMATION

The purpose of these studies is to examine aspects of the information environment in which individuals make their judgments of risk. All three studies look at the relations between the news media and hazards. The two media content studies explore these relations at the level of individual news media and within the context of particular hazards. The study of science writers is more abstract, inquiring into the general beliefs of these important media workers across a broad spectrum of hazards.

The Gas Crisis Of 1979

This study was designed by Earle and Lindell (1980) to investigate the role of the mass media in the development of long waiting lines at gasoline stations during the 1979 gasoline shortage. Most of the data used in this study were derived from the news media--four newspapers and three television networks. The newspapers were the Los Angeles Times, the New York Times, the Denver Post, and the Portland Oregonian. These newspapers were chosen on the basis of two criteria: (1) They represent geographically diverse areas of the country, and (2) they are the leading newspapers in two cities in which significant gas lines occurred in 1979 (Los Angeles and New York) and in two cities which managed to avoid significant gas lines in 1979 (Denver and Portland). The three television networks, CBS, NBC, and ABC, are those American networks with nightly news programs. The total time period during which these news media were studied begins on January 1, 1979 and extends to June 30, 1979, a period encompassing all of the primary events leading up to the development (in some locations) of waiting lines at gas stations. The television analysis was designed to explore the early events, while the newspaper analysis was designed to focus on the events immediately preceding and during the development of gas lines. The national scope of the network news made it a suitable medium for the primarily national early events; the more local scope of subsequent events of the gas shortage requires the use of a local medium. For this study, local newspapers were used. Data from the newspapers and the television networks were used to provide three categories of information: (1) Information about the events that preceded the lines, (2) information about public response, and (3) information about the content of the media. Additional data included public opinion polls and American Automobile Association surveys of gasoline station closings. These sources provided information on public concerns about shortages and about availability of opportunities to purchase gasoline, respectively (for details on the content analysis procedures, see the original report.)

Results of the analysis of media coverage can be summarized in several points:

1) Network television news probably affected public opinion on the gas crisis by contributing to the increased salience of the general energy issue during the early months of 1979, before the issue became obtrusive in the lives of many people. Although there was extensive television coverage of events related to the gasoline shortage, there is no evidence that this coverage produced gas lines.

2) Substantially restricted opportunities to purchase gas (as indicated by levels of Sunday gas station closings in excess of 80 percent) preceded the development of long gas lines.

3) No consistent relations were found between newspaper content and the development of long gas lines.

4) The general relationship between newspapers and the gas crisis was shown to have been a complex interaction based in part on events responded to by the newspapers and in part on content initiated by the newspapers. Newspapers not only reported the gas crisis events, but in some cases helped to explain them and may have contributed toward their solution.

5) The most prominent newspaper content category in each of the four cities studied reflected the individual nature of each city's crisis. The newspapers in gas-line cities gave relatively great prominence to responses to events, while newspapers in non-gas-line cities gave relatively great prominence to suggested ways of avoiding the worst manifestations of the crisis.

Of these results, points 1, 2, and 3 are directly related to questions concerning the effects of the news media on the gas lines: Did the news media cause, trigger, or lengthen the gas lines? First, it is clear that the lines were not caused by the news media; if that were the case, lines would have developed in all of the cities studied. Second, the results indicate that newspaper content alone was not sufficient to trigger nor lengthen the gas lines. Assessment of the effects of newspaper content is not a simple matter, as points 4 and 5 of the results suggest. The general relationship between the gas crisis and newspaper contents was complex, consisting of a variety of functions only one of which was the reporting of events occurring at gas stations. But the notions of triggering and lengthening are based soleley on that one function, the reporting of events known at least as well to the readers (i.e., those who observed or participated in the gas lines) as to the reporters. In addition to reports of gas station events, however, newspapers carried reports of government actions, effects of the shortage and improvements in supply, and reports of occurrences that many members of the public would have been totally unaware of had they not read about them in their newspapers. These latter reports, then, were categories of newspaper content that could have, and quite possibly did, affect public beliefs and behaviors.

During the gas crisis (and at other times as well) the newspapers did not simply passively respond to occurrences that took place beyond their control. The newspapers of course actively selected their contents, but in two groups of content categories, ways and means of coping and beliefs about causes of the shortage, the newspapers also actively initiated some of the occurrences for their contents. In this way the newspapers responded to the gas station events by attempting to provide useful services to their readers, information for understanding the lines, and techniques for dealing with them. This may well have been the most important way in which the newspapers affected the lines. Indeed it is possible that the newspapers could have contributed to the shortening of the lines by providing drivers with information on transportation alternatives together with some encouragement to use them. Less obviously, newspaper coverage could have shortened the lines by providing a community-wide view of the problem. The newspapers allowed the public to see "the big picture" or at least "a big picture." Individuals could thus compare their own personal situations with those of other persons in other places and with events that occurred in the past as well as possible future events. The newspapers, in short, could help drivers to reduce the uncertainty they were experiencing by enabling them to make sense of the lines and to cope with them.

At a more general level, these results are consistent with, and represent an extension of, Zucker's (1978) theory of variable news media influence. According to Zucker's theory, the content of the news media can be expected to influence public beliefs and behavior when the issue involved is non-obtrusive and new. Thus, network television news would be expected to affect public opinion on the salience of the energy issue during the early weeks of the crisis. When the gas crisis events became obtrusive in the lives of people, when gas station closings became common, no media effects would be expected and none were found. Thus far, the results were consistent with Zucker's. The results went beyond Zucker's by noting that the media coverage of the gas crisis consisted of much more than the reporting of events occurring at gas stations. Newspapers also contained stories focusing on responses to the gas lines. These responses were new, non-obtrusive issues for many readers, contents capable of affecting their beliefs and behaviors. Of particular interest were the responses of the newspapers themselves, providing background information on the crisis and suggesting techniques to deal with it. The most significant effects of the news media on the gas lines may well have been ones that led to the shortening and ending of the lines.

Mount St. Helens

An opportunity to extend our investigations of newspapers and hazards from the technological to the natural arose in the spring of 1980 with the renewal of volcanic activity at Mount St. Helens (Lepman and Mullineaux, 1982; Decker and Decker, 1981). The Mount St. Helens eruptions generated a complex set of social interactions characterized by these factors:

1) Involvement of many diverse actors--The four most significant groups were the scientific community, the government, the public, and the news media.

2) All were forced to deal with volcanic events that could be predicted, if at all, only with a high degree of uncertainty. Available historical data indicated, moreover, that volcanic activity could take place over an extended period of time, as much as twenty years.

3) Significant decisions had to be made by all persons involved regarding the nature of the hazard and its possible impacts as well as the likelihood of specific impacts affecting themselves and others. Actions to mitigate these impacts needed to be taken and their consequences carefully considered.

Central to each of these factors and to the entire set of interactions related to Mount St. Helens was the basic dichotomy between the scientists and non-scientists. As is the case with many natural and technological hazards, the events at Mount St. Helens were best known to and understood by the scientists. Scientists were asked by others for predictions of the future behavior of the mountain. They were thus obligated to communicate with the public, and they did so through the mass communications media.

In covering the Mount St. Helens events, newspapers were confronted with this problem: how can highly technical scientific information be usefully conveyed to their readers who lack scientific training? This problem, common to many hazard situations, was especially salient with regard to Mount

St. Helens because of (a) the relatively low level of scientific understanding of volcanic phenomena by the experts (Lepman and Mullineaux, 1982) and (b) the relatively low level of scientific sophistication of the predominantly rural population most exposed to the effects of the mountain (Perry and Greene, 1982). The situation at Mount St. Helens became one of newspapers having to communicate scientists' probabilistic and guarded descriptions and predictions to readers who likely viewed scientists as professionals whose task it was to produce practical solutions to pressing problems (Pion and Lipsey, 1981; Garfield, 1982). Mount St. Helens thus presented a conflict between the highly uncertain knowledge available from experts and the public's expectations and demands for clear, unambiguous, and immediately useful information (Nigg, 1982). The primary focus of the Mount St. Helens media analysis was on how several newspapers dealt with this conflict.

Results from this study (Earle, Southwick, and Lindell, 1981) suggest that both the amount and type of coverage of the 1980 period of volcanic activity were related to two characteristics of the newspapers studied: frequency of publication and proximity to the mountain.

1) Amount of coverage--Coverage decreased with distance from the mountain, and coverage was greater in weeklies than in dailies. These results were in accord with previous studies and discussions. Tichenor, Donohue, and Olien (1980) pointed out that weeklies tend to be dominated by local news. The Mount St. Helens story was local news for the two weeklies in the present study. Since dailies carry both local and non-local news, even a daily for which the volcano was a local story (Longview Daily News) would be expected to contain proportionally less coverage than the local weeklies. Within the daily newspapers, on the other hand, proximity to Mount St. Helens greatly affected the amount of coverage. As suggested by Larson (1980), the local newspaper contained vastly more coverage than the national newspaper; the regional newspaper contained moderately more coverage than the national paper. This effect of proximity on amount of coverage can be attributed to a number of factors. Most significant perhaps is the fact that, in general, the potential effects of the volcanic activity decreased rapidly with distance. The information needs of readers thus similarly declined from the local to the regional to the national newspaper. In addition to greater information needs, the volcano story also was of

greater interest to local readers because it was a globally-significant event happening to them. Local readers were participants in a gripping national drama with a world-wide audience. Of course they wanted to read about it and about themselves. Finally, for local reporters and editors, Mount St. Helens was the biggest, most dramatic story of their professional careers. They knew it, and they worked overtime to generate the extensive coverage they judged the story demanded.

2) Use of scientific information--Dailies used more scientific information than weeklies, more scientific sources and, overall, more sources per article. These results strongly support the previous studies of Tichenor, Donohue, and Olien (1980), particularly with regard to differences in styles of conflict-resolution between weeklies and dailies. Weeklies, according to Tichenor and his co-workers, tend to suppress differences of opinion and work toward consensus. Dailies, on the other hand, strive to maintain communication among persons and groups with diverse opinions and interests. Due to the highly uncertain nature of the events, the large number of diverse actors and the serious consequences of the decisions to be made, the Mount St. Helens story was a particularly rich context in which to study the styles of conflict resolution used by newspapers. Though the responsible government agencies, particularly the United States Forest Service and the United States Geological Survey, quickly established a system for coordinating information dissemination, there remained ample opportunity for the publication of conflicting or at least confusingly dissimilar opinions from a variety of presumably equally well-qualified scientific observers.

   The weeklies resolved these conflicts by ignoring them. Both weeklies tended to rely on local sources to provide information from and for affected or potentially affected people. The weekly in the high risk area provided more practical, concrete information. The two weeklies did not differ in source usage: both cited few sources. Thus, rather than seek out diverse opinions from the wide variety of available experts, scientific and otherwise, the editors and reporters for the weeklies tended to present their own personal views of the events or rely on local sources. These editors and reporters may be understood as acting as buffers between a potentially conflict-producing set of events and opinions, on the one hand, and their readers' desires for harmony, predictability, and control, on the other. The dailies, in contrast, encouraged the expression of divergent opinions. The Longview Daily News, for example, cited more than sixty sources of background information during the coverage period; the neighboring Cowlitz County Advocate cited only three.

The low use of scientific information by the weeklies may have been related to a view of science as the final arbiter or the purveyor of truth. If the scientists had been able to provide a single, simple, concrete explanation of what was happening together with a clear, definite and non-disruptive prescription of what to do, then the weeklies would probably have been as open to them as the dailies. Or, alternatively, the level of interest of the dailies would have declined to that of the weeklies. But such was not the case at Mount St. Helens. There, the scientists could provide little in the way of immediately useful information; they could offer only general guidelines. To the individual living in the risk area, general guidelines may be worse than none at all. Based on the scientists' highly uncertain prognoses, government officials may establish restricted areas, thereby preventing some local persons from leading their normal lives. Faced with the opinions of scientists who seemed no better informed than they themselves, but whose opinions could affect their lives, local residents experienced confusion and some hostility (see, for example, Volcano Cabin Owners, 1980). The weekly newspapers did not handle this conflict between scientists and local residents by becoming involved in the controversial scientific issues. Instead, they presented the views of the local residents. As events subsequently evolved, the scientists were as surprised as anyone by what happened.

3) <u>Local and national perspectives on coverage</u>--The national daily (New York Times) used more human interest items derived from local residents and more items without cited source. The local (Longview Daily News) and regional (Portland Oregonian) dailies used more information about government activities and more government and local sources. The local daily presented more detail about volcanic effects on the mountain but used fewer scientific sources. These results support Larson's (1980) general characterizations of the differences between local and

national coverage: the local daily provides more detailed coverage, the natioanal more human interest. The results are similarly in accord with Quarantelli's (1978) finding that local newspapers report more from a "command post" point of view. Thus, the local and regional dailies in the present study used more government information and sources; the national daily expressed its independence through its frequent use of no source at all. Underlying all of these differences between local and national dailies would seem to be the differences in the uses that the newspaper's readers would have for the information. Local readers need to know more details than national readers because the details may affect them. Similarly, local readers need to know about the activities of government. The "command post" point of view is a useful one for local dailies, at least on an initial basis, because that is where the action is. The national daily can afford to explore other points of view because of the lack of direct consequences of the event for its readers.

A pervasive theme underlying all three of these categories of results is the newspapers' response to the needs of the readers. This is consistent with the conclusion drawn by Earle and Lindell (1980). The sensitivity of newspapers to readers' needs is particularly striking in the case of the weeklies. The readers of each of the two weeklies consisted of relatively homogeneous groups, each with specific needs brought on by the volcanic activity at Mount St. Helens. The weeklies seemed fully aware of these differing sets of needs, and each responded appropriately within the limits of its format and style.

In attempting to meet the needs of its readers, any newspaper must interact with a structured information environment. As a consequence, a newspaper covering a particular aspect of a story seeks information, not in a random fashion, but from specific sources. This explains why newspapers did not differ in the types of sources used within each of the content categories. The critical factor differentiating the coverage by each of the newspapers, then, was the selection of types of content: Earthquake Activity, Background Information, Government Action, and so forth. Once the type of content was selected, the sources--and to a significant degree the specific content--followed.

Science Writers

The individuals whose job it is to communicate complex technical information in terms comprehensible to the public occupy (along with their editors) a critical position in the hazard information system. In a story like Mount St. Helens where there may be general agreement within the community about the information needs of readers, science writers may exercise relatively little discretion in writing their stories: readers' needs lead to content categories, which lead to certain sources, which in turn produce the primary content. In other hazard situations, however, there may be little agreement within the community about information needs. In many cases, the public may be entirely ignorant of the issue. In these cases, nuclear power for an obvious example, science writers can have a strong impact on the salience of an issue and also on the public's knowledge about that issue. Because of the critical importance of science writers to the communication of hazard information, we have attempted to learn more about their views on scientific and technological issues (Earle, 1981; Earle, Lindell, and Rankin, 1981; Lindell and Earle, 1981).

In one task, a group of science writers randomly selected from the membership of a professional organization were asked to judge the salience of nine technological and scientific issues: Recombinanat DNA (genetic) research; suspected cancer-causing agents; cloning research; nuclear wastes; computers and privacy; use of animals in medical research; toxic chemicals and toxic metals such as DDT and mercury; psychological testing; and pollution from coal-burning power plants. A discriminant function analysis was used to compare the judgments produced by the science writers with those from other respondent groups. The other respondents were sampled from these groups: Nuclear Engineers; Chemical Engineers; Environmentalists; Hazardous Facility Communities [residents of six communities in which potentially hazardous industrial facilities are located: Red Wing, Minnesota; Waterford, Connecticut; Everett, Massachusetts; Savannah, Georgia; Forsyth/Colstrip, Montana; and Pueblo, Colorado]; and General Communities [residents of six communities chosen without regard to hazardous industrial facilities: San Francisco, Houston, Chicago, Detroit, Boston, and Atlanta]. The analysis yielded three functions accounting for 90 percent of the variance among groups (Fig. 1):

1) The first discriminant function contrasts the two public groups with the remaining groups. The most significant difference contrasts the hazardous facility communities with the science writers and the nuclear engineers. The issues given high importance

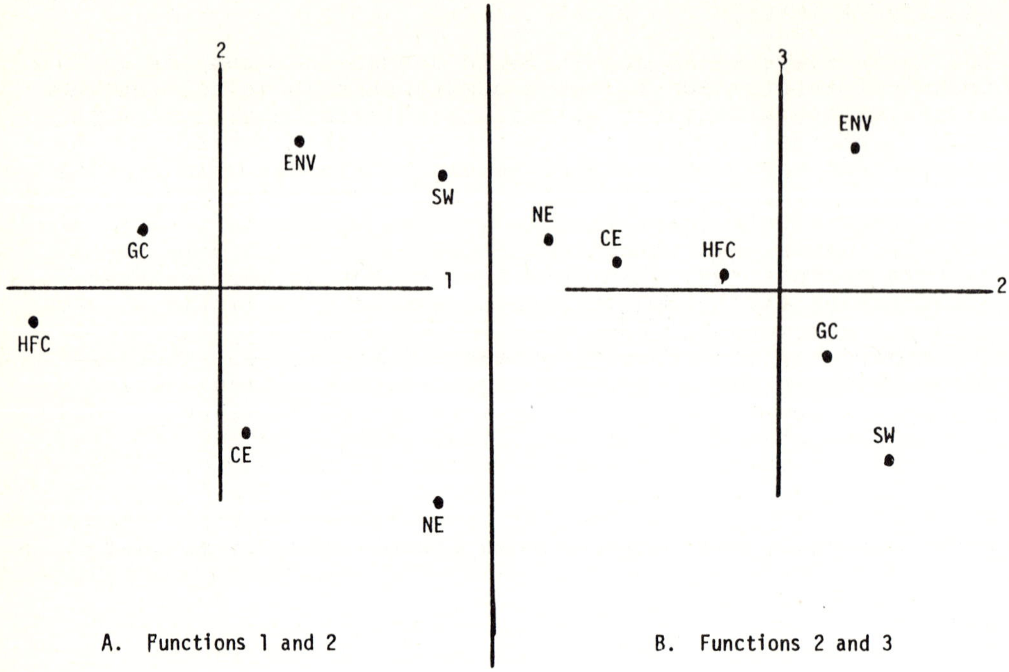

Figure 1. RELATIVE POSITIONS OF SIX GROUPS OF RESPONDENTS ON THREE DISCRIMINANT FUNCTIONS.

GC = GENERAL COMMUNITIES
HFC = HAZARDOUS FACILITY COMMUNITIES
ENV = ENVIRONMENTALISTS
SW = SCIENCE WRITERS
CE = CHEMICAL ENGINEERS
NE = NUCLEAR ENGINEERS

ratings by the public groups were psychological testing, suspected cancer-causing agents, use of animals in medical research, and nuclear wastes. The remaining groups differed from the public groups due to their relatively high importance judgments for pollution from coal-burning power plants, recombinant DNA research, cloning research, and toxic chemicals and toxic metals. This function thus roughly contrasts issues of general public interest with issues of interest to those with greater knowledge of science and technology. This contrast accounts for 46 percent of the variance among groups.

2) The second function contrasts the two groups of engineers, nuclear and chemical, with the informed observers of technology--the environmentalists and the science writers. The latter two groups differed from the engineers by their relatively strong beliefs in the importance of suspected cancer-causing agents, cloning research, toxic chemicals and toxic metals such as DDT and mercury, computers and privacy, and nuclear wastes. The contrast here is between the relatively strong interest in the issues by environmentalists and science writers and the relatively low interest in them by the engineers. This contrast accounted for 29 percent of the variance among groups.

3) The third function, accounting for 15 percent of the variance, contrasts the science writers with the environmentalists. The science writers more strongly emphasized recombinant DNA research, computers and privacy, suspected cancer-causing agents, psychological testing, use of animals in medical research, and cloning research. The environmentalists give higher ratings to the importance of nuclear wastes, toxic chemicals and metals, and pollution from coal-burning power plants. While both these groups have been described as informed observers of technology, this contrast shows that the science writers were more interested than the environmentalists in issues outside the sphere of environmentally-related issues of health and safety.

This discriminant function analysis demonstrates that science writers are distinguished by their concern about a variety of scientific and technological issues, both environmentally-related and others. These results suggest that science writers, unlike the other groups of respondents, are not a special interest group for any single issue or group of issues. Instead, the special interest of the science writers is the entire field of scientific and technological issues. This general high level of interest on the part of the science writers is further illustrated by the fact that they gave the highest average importance ratings; these were followed by the environmentalists, the general communities, the hazardous facility communities, and the nuclear engineers. The chemical engineers gave the lowest importance ratings to this set of scientific and technological issues.

Another task included in the same survey asked respondents to judge--considering only health and safety concerns--the minimum distances from several industrial facilities where they would be willing to live or work. Eight industrial facilities were included: natural gas, coal, oil, and nuclear power plants, oil refinery, nuclear waste disposal facility, storage area for liquified natural gas, and toxic chemical disposal facility. The assumption underlying this task is that a person's beliefs about the health and safety hazards at a facility will be highly related to how close to the facility the person would locate herself or himself. Of course a person's willingness to live or work near a potentially hazardous industrial facility may be affected by many other factors as well. For that reason, respondents were specifically requested to base their judgments only on "risk to health and safety."

The distance judgments were first cross-tabulated in a group by distance-categories matrix. This information was then simplified first by rank-ordering the cumulative percentages of respondents within each group willing to live or work within each distance category for each hazardous facility and, second, by averaging these rankings across the eight facilities. These data are shown in Table 2. Taking the range of the rankings across facilities as a measure of variation within each group, the six respondent groups form three categories, those with low, medium, or high variablility. The general tendency of the environmentalists was to make relatively small, if any, distinctions among the various facilities; they indicated a desire to stay away from any hazardous industrial facility. The science writers, like the environmentalists, saw little difference among the facilities. Unlike the environmentalists, however, the science writers showed a moderate willingness to live or work close to hazardous facilities. The final low-variability group, the chemical engineers, were somewhat less willing to accept either a nuclear power plant or a nuclear waste disposal facility than the remaining facilities, but they nonetheless were strongly accepting of all of the hazardous facilities. The three low-variability groups, then, each showed a different consistent policy: the environmentalists were relatively strong in their rejection of

Table 2.

HAZARDOUS FACILITY DISTANCE JUDGMENT RANKINGS AVERAGED OVER DISTANCE CATEGORIES, BY GROUP AND FACILITY

FACILITY	GROUP					
	General Communities	Hazardous Facility Communities	Environ-mentalists	Science Writers	Chemical Engineers	Nuclear Engineers
Natural-Gas-Burning Power Plant	5.2	3.8	4.8	3.2	1.0	3.0
Toxic Chemical Disposal Facility	4.4	3.4	4.6	2.6	1.0	5.0
Coal-Burning Power Plant	3.4	2.0	5.4	3.8	1.0	5.4
Nuclear Power Plant	5.2	3.6	5.2	3.8	2.2	1.0
Oil Refinery	3.4	2.5	4.9	3.8	1.2	5.2
Nuclear Waste Disposal Facility	5.5	4.7	4.0	3.4	2.0	1.4
Oil-Burning Power Plant	4.8	3.0	4.6	2.6	1.0	5.0
Storage Area for Liquified Natural Gas	4.1	2.4	5.2	3.8	1.0	4.5
$\bar{X}$	4.5	3.2	4.8	3.4	1.3	3.8

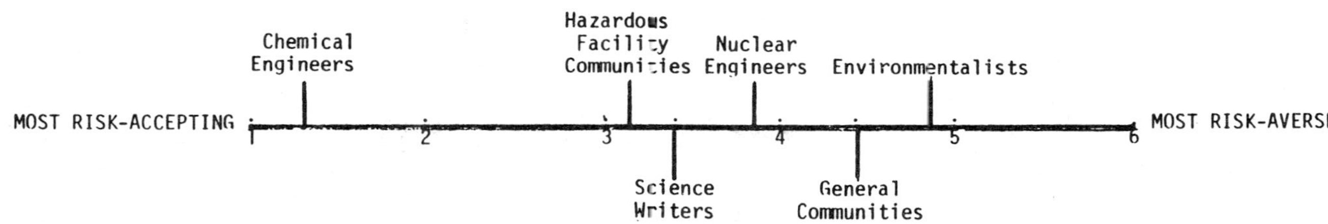

the idea of living close to all hazardous industrial facilities; the science writers showed a moderate willingness to live next to hazardous facilities; and the chemical engineers were relatively strong in their willingness to live or work near hazardous facilities.

The two public groups had medium variability in their rankings across facilities. The two groups expressed similar sentiments on the relative riskiness of the facilities: the three lower risk facilities for both groups were the coal-burning power plant, the oil refinery, and the storage area for liquified natural gas; the three higher risk facilities were the nuclear waste disposal facility, the natural-gas-burning power plant, and the nuclear power plant. What separated the general communities from the hazardous facility communities was the expressed reluctance of the former to live or work as close to any of the hazardous facilities. For each of the eight facilities, the hazardous facilities group was more willing than the general communities group to accept small distances. This result indicates a general tendency toward risk acceptance by the hazardous facility communities. Although each of these communities was selected because of its proximity to one of these facilities (coal-burning power plant, nuclear power plant or storage area for liquified natural gas) respondents in these communities were relatively more accepting of all of the facilities named in this task. The probability of such a result occurring by chance is extremely small ($p < .005$, sign test).

Only one group, the nuclear engineers, had high variability in rankings across facilities. The nuclear engineers had an extremely large range of 4.4, ranking 5.4 (tied for least willing to live close) on the coal burning power plant and 1.0 (most willing to live close) on the nuclear power plant. While they were, as might be expected, accepting of the risks presented by a nuclear power plant and a nuclear waste disposal facility, the nuclear engineers were moderate in their position toward a natural-gas-burning power plant and strongly rejecting of the risks associated with the remaining facilities. The nuclear engineers thus were very different from the chemical engineers. While the chemical engineers were relatively strongly accepting of the risks presented by all of the facilities, the nuclear engineers accepted only those connected with the work in which they, themselves, were engaged.

It is admittedly somewhat artificial to speak of an overall tendency toward risk acceptance or aversion on the part of groups to respondents since such a tendency may offer little insight into the actual risk judgments made by individuals in real situations. Nonetheless, such results offer some insight into the differences in risk judgments among groups. The main effect for groups on the distance judgment task is illustrated in the graph at the bottom of Table 2. The six respondent groups are located on a single risk-judgment dimension labeled "Most Risk-Accepting" at the lower end and "Most Risk-Averse" at the upper end. The locations of the groups correspond to their mean rankings across all eight hazardous facilities. These mean rankings thus provide a good summary picture of the general relative risk judgments of the six respondent groups. The most risk-accepting group was the chemical engineers; the most risk-averse group was the environmentalists. Several comparisons are of particular interest: (a) the above-average risk-aversion of the nuclear engineers compared with the extreme risk-acceptance of the chemical engineers; (b) the greater risk-acceptance of the hazardous facility communities compared with the general communities; and (c) the centrist position of the science writers. This last result is of particular significance for our study of the hazard information environment since it is evidence in support of the view of science writers as relatively objective observers of controversial issues.

These two sets of results indicate that science writers have a general high level of interest in scientific and technical issues and are neither risk accepting nor risk averse with regard to hazardous industrial facilities. These attributes appear to be appropriate to the critical communications role occupied by science writers: Their strong concern should enable them to uncover new issues and follow up the latest developments in old ones. And their middle-of-the-road position on industrial hazards should lead to relatively impartial coverage and a public perception of fairness.

In summary, our studies of the mass media within the hazard information environment system suggest that the news media generally attempt to identify and meet the information needs of their users and that, among the many attributes required of them, science writers appear to be well equipped at least in terms of level of interest in their subject matter and the ability to approach it impartially.

A NEW LOOK CONCEPTUAL FRAMEWORK

As indicated by the foregoing studies of the communication of hazard information, one of the major concerns of our New Look is the role of information in the development of risk judgment. Our other major concern is the conceptualization of risk judgment as a dynamic process. In this section we present conceptual models of both of these central areas of interest.

Risk Judgment

Our model of risk judgment is illustrated in Fig. 2. Risk judgment is conceived of as a dynamic process in which an individual adapts to the hazardous environment through a series of steps over time. Each step consists of a judgment, followed by an action (behavior of some sort), followed finally by an outcome which serves as feedback for the individual. This is a simplification of course; judgments, for example, may be followed by other judgments before action is taken and feedback recieved. The individual uses the feedback as he or she initiates another judgment--action--outcome loop, and learning occurs over time (Fig. 2a). The overall adaptation process between an individual's risk judgment process and the natural or technological hazard process is shown in Fig. 2b.

This conception of risk judgment as a dynamic process differs markedly from typical static, discrete conceptions (e.g. Starr and Whipple, 1980) and is based on Hogarth's (1981) discussion of continuous judgment processes. "In continuous processing ...," Hogarth points out, "judgment can be thought of as providing a temporal background of mental activity that is punctuated by particular choices." (Hogarth, 1981, p. 201). If cognitive processes, such as risk judgment, are studied only as discrete incidents, then a distorted psychology of risk will evolve which minimizes the importance of feedback and which identifies as "biases" heuristics that may be functional in continuous environments (see also Ebbeson and Konecni, 1980). In addition to Hogarth, several other researchers have recently called for increased attention to the continuous nature of human experience (Vlek and Stallen, 1981, p. 269; Carroll, 1980, p. 75).

The elements which together form the dynamic risk judgment process are presented in Fig. 2c. These elements are similar to those identified by Vlek and Stallen (1981). As discussed earlier in this paper, risk judgments are only one of several ingredients of risk decisions. Also included are judgments of benefits and costs. For simplicity, the conceptual model presented here deals only with risk judgments. A more complete description of risk decisions would include all of the relevant factors.

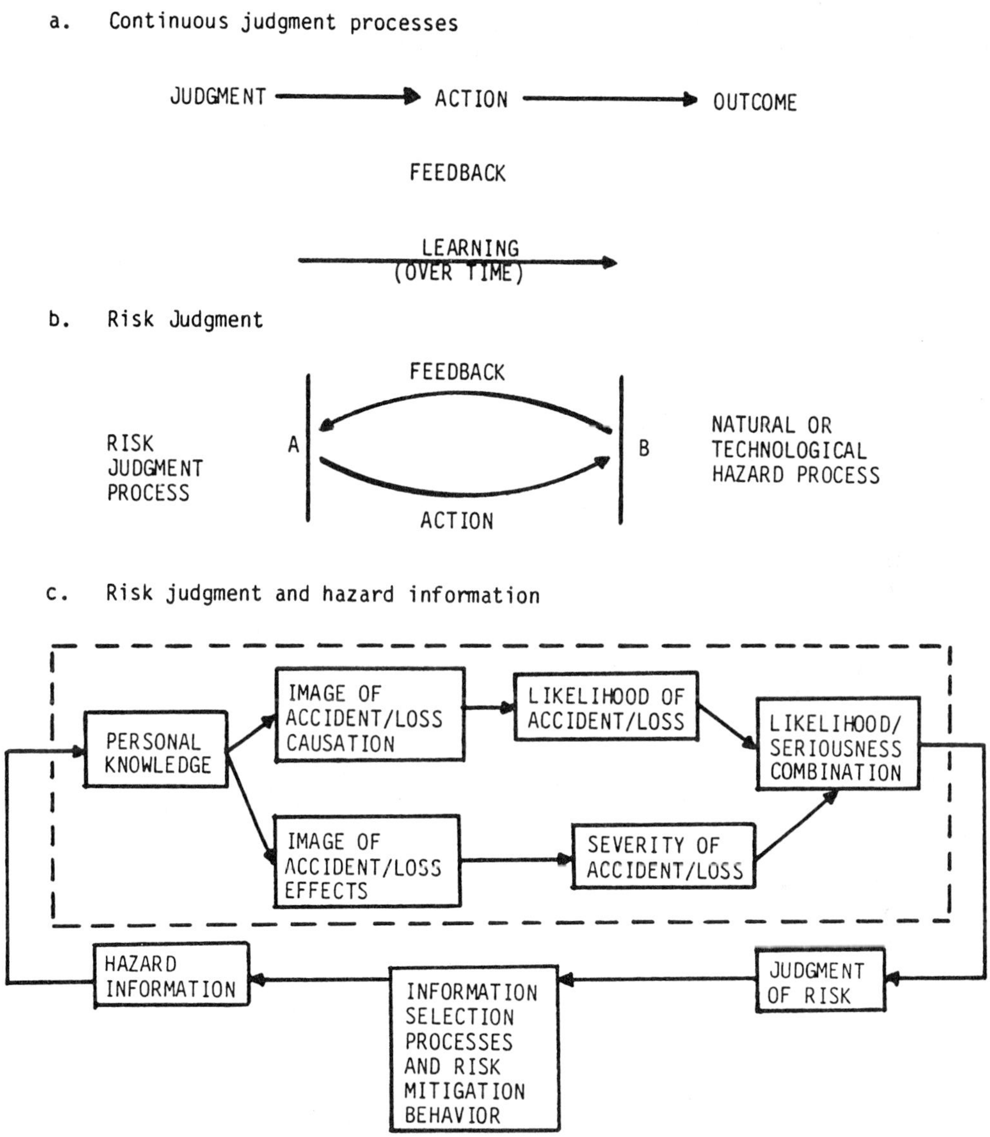

Figure 2. RISK JUDGMENT: A DYNAMIC PROCESS

By risk judgment we mean a compound of likelihood and consequences generated in varying ways by individuals. The process, in outline, is this: an individual's personal knowledge of a hazard enables him to (a) imagine sequences of accident causality which lead to a conception of the likelihood of a loss or accident, (b) imagine accident effects which are evaluated and lead to a conception of the seriousness of a loss or accident, and (c) form a likelihood-seriousness combination which produces the overall conception of riskiness.

The judgment of a risk is a difficult task. Most individuals do not calculate a risk by explicitly combining (multiplying) probabilities and consequences (see Earle, Lindell, and Rankin, 1981, for an analysis of this point; see also Starr and Whipple, 1980). Instead, individuals often must judge risks on the basis of what they remember hearing or observing about the hazard. In situations of this sort, people tend to simplify their judgment tasks by using heuristics (Kahneman and Tversky, 1972; Tversky and Kahneman, 1973; and Tversky and Kahneman, 1974). The implications of heuristic usage in judgment processes have been elucidated in a series of studies by Slovic and his colleagues (see, for example, Slovic, Kunreuther, and White, 1974; Slovic, Fischhoff, and Lichtenstein, 1977; Slovic, Fischhoff, and Lichtenstein, 1980). These researchers have found that single, static judgments of risk, as a result of the use of heuristics, are highly variable among individuals and often fallible when compared with risks calculated according to certain normative principles. Hogarth (1981), however, has suggested that heuristic usage may be seen in a more positive light when studied as part of a continuous process rather than as part of a single discrete incident. Prominent in our dynamic model of risk judgment is the modification of judgment over time as the result of the selection and use of new hazard information. These individual risk-judgment processes occur within the larger context of the hazard communications system. A model of this complex system is presented next.

A Model of the Hazard Communications System

In the dynamic risk judgment process depicted in Fig. 2, the internal cognitive processes (enclosed in the dashed line) operate on the hazard information to which the individual has been exposed. Hazard information can be obtained through direct personal experience or through various communications media. A simplified version of this hazard communication system is presented in Fig. 3. Four primary groups of individuals are identified:

1) <u>Technical experts</u> who generate scientific and technical information about the hazard.

2) <u>Decisionmakers</u> who manage the public response to the hazard.

3) <u>Mass media</u> representatives who communicate information about the hazard to the general public.

4) <u>Public</u> individuals who learn about the hazard through the hazard communications system, and who, on that basis, formulate their risk judgments.

The arrows in Fig. 3 indicate the direction of information flow. Individuals in each of the groups thus can learn about the hazard through direct personal experience. In addition, inter-group communication is possible within all pairs of groups. Communication between certain groups is more likely than between others. The public is more likely to receive information from the mass media than from technical experts, for example. Finally, while not indicated in the figure, there is also extensive communication within groups. Technical experts communicate with one another, either directly or through professional journals, for example, and interpersonal communication among members of the public can be very important in the development of their risk judgments.

As illustrated in Fig. 4, the information communicated within the hazard communications system can be characterized by two dimensions: communications medium and content. Content is dichotomized into "scientific information" and "non-scientific information." Scientific information is information that is characterized by the general methods of science i.e. measurement and logic (Boulding, 1980). Scientific information is generated by and frequently attributed in the news media to scientific or technical sources (scientists or other researchers). The essence of scientific information rests in its logical structure and quantitative content. Non-scientific information is defined by the absence of the primary characteristics of scientific information, i.e. lack of quantification and lack of logical structure. The distinction between scientific and non-scientific information is of course not as sharp as we have drawn it here: there is clearly a continuum of information, ranging from the strictly scientific (a mathematical argument, for example) to the totally non-scientific (qualitative personal impressions). We use a dichotomy for simplicity of presentation. One possible advantage to considering a hazard information continuum is the opportunity to relate it to Hammond's cognitive continuum

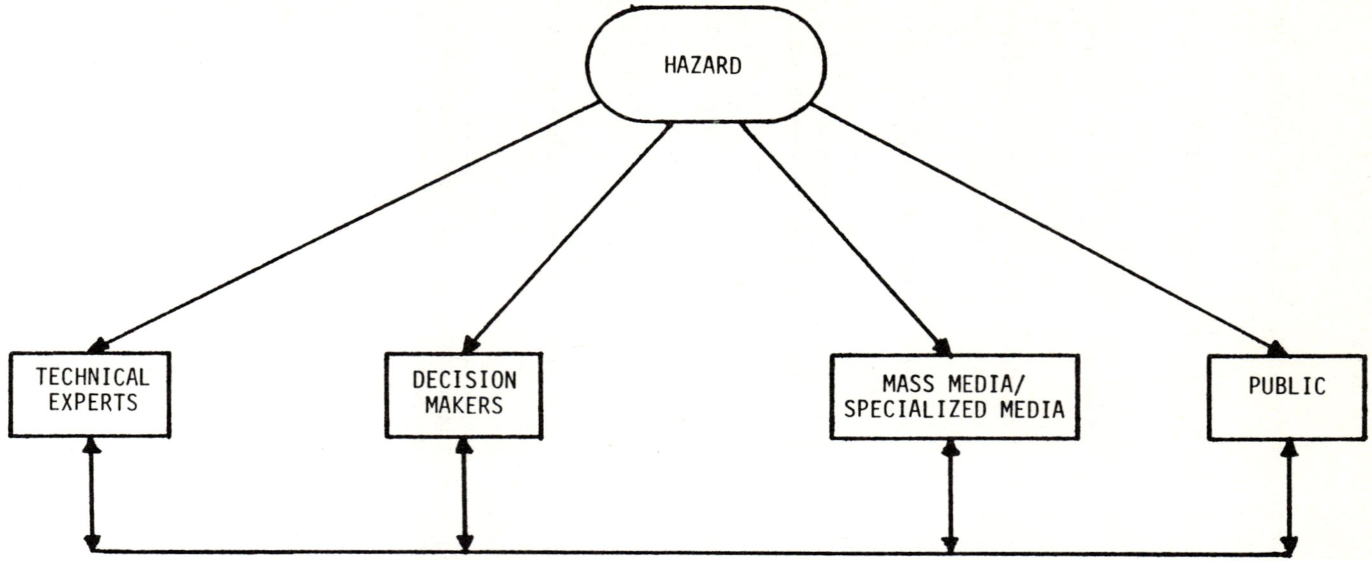

Note: The arrows indicate the direction of information flow.

Figure 3. THE HAZARD COMMUNICATIONS SYSTEM

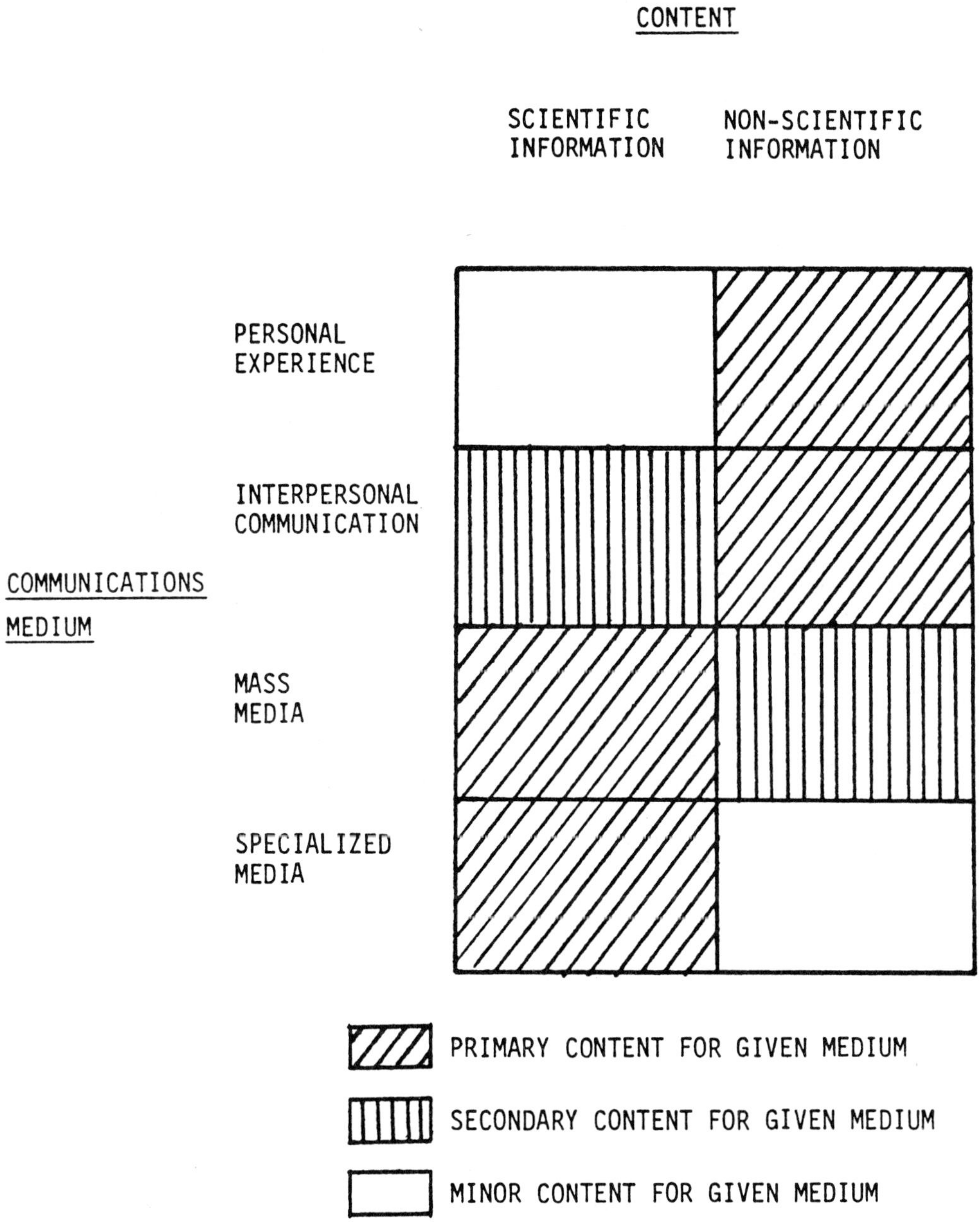

Figure 4. CONCEPTUAL DIMENSIONS OF HAZARD INFORMATION

(Hammond, Hamm, Grassia and Pearson, 1983). Studies investigating these relations are suggested in the final section of this paper.

Four communications media are identified in Fig. 4. For each medium the hypothesized relative significance of the two types of content, scientific and non-scientific, is designated. The reasoning behind these characterizations of the media are, briefly:

1) Personal experience--This medium includes the direct experience of a hazard by an individual. A resident of a house that is damaged by a flood receives information about that hazard through personal experience. Whatever the primary content of personal experience, it is clearly something other than scientific information (with the possible exception of certain personal experiences of scientists).

2) Interpersonal communication--This includes informal, unstructured communication between individuals. Two neighbors discussing the potential ill effects of a nearby waste disposal facility are exchanging information through interpersonal communication. The content of interpersonal communication may consist of both scientific and other types of information; since most individuals are not knowledgeable about scientific issues (Miller and Barrington, 1981), the primary content of interpersonal communication is likely to be non-scientific information (see, for instance, Colligan, Pennebaker, and Murphy, 1982). Exceptions would include discussions, for example, between a public official and a technical expert.

3) Mass media--Included in the mass media are newspapers, radio, and television. An individual reading about a local earthquake hazard in a newspaper is receiving information through a mass medium. In their coverage of technological and natural hazards, the mass media include both scientific and non-scientific information. The primary hazard information sources in the mass media tend, however, to be scientific (see Earle, Southwick, and Lindell, 1981, for an analysis of newspaper information sources for Mount St. Helens).

4) Specialized media--This communications medium is more complex than the others in that it includes a diverse set of specific media. These media have in common relatively narrow ranges of informational content that appeal to relatively small, homogeneous audiences with

similar interests and backgrounds. Beyond these common features, the specialized media diverge: scientific magazines, technical journals, newsletters, workshops, seminars, colloquia, public meetings and hearings, etc. A member of the general public reading about low-level radiation in a scientific magazine and a local government official attending a workshop on waste disposal are examples of the communication of hazard information through specialized media. The content of the specialized media consists almost entirely of scientific and technical information.

In summary, our model of the hazard communications system identifies the source and receiver of the information, the medium used, and the type of information involved. In the remainder of this section, we discuss several topics concerning the operations of the hazard communications system. We first take a brief look at the relations between risk judgment and hazard information needs. This is followed by a discussion of information seeking. The final and major topic is the role of scientific information in risk judgments.

The Relations Between Risk Judgment and Hazard Information Needs

The hazard communications system described above contains four communications media: personal experience, interpersonal communication, mass media, and specialized media. Each of these media, it was argued, is associated with different types of information which induce a wide variety of cognitive processes. A further question concerning these communications media is whether they are differentially associated with a variety of risks. As an additional part of the exploratory study described earlier in this report, respondents were asked whether or not they used each of the four communications media in their selected risk decision. Table 3 shows that personal experience and interpersonal communication were used at equally high rates, followed by mass media and, finally, specialized media. This measure of media use was combined with a measure of risk decision characteristics composed of a combination of the risk decision consequences and the persons affected (see Table 1). The results given in Table 3 indicate that the different media are associated with different risks. The various media thus seem to fill different information needs.

The appropriateness of the communications medium/risk decision-characteristics associations presented in Table 3 seems clear. Personal experience, for example, is favored in decisions affecting only oneself or oneself and known others.

Table 3. HAZARD COMMUNICATIONS MEDIA AND RISK DECISION CHARACTERISTICS

MEDIUM	% USE	RISK DECISION CHARACTERISTICS
PERSONAL EXPERIENCE	85.3	RISKS DUE TO UNCERTAINTY ABOUT OUTCOMES CONSIDERATION OF BENEFITS EVENT INVOLVES SELF OR SELF AND KNOWN OTHERS
INTERPERSONAL COMMUNICATION	85.3	(WEAKLY RELATED TO THE SAME CHARACTERISTICS AS PERSONAL EXPERIENCE)
MASS MEDIA	58.8	RISKS INVOLVING THE GENERAL PUBLIC CONSIDERATION OF ENVIRONMENTAL EFFECTS
SPECIALIZED MEDIA	45.6	CONCERN WITH PERSONAL RISKS INVOLVING UNCERTAINTY AND/ OR DEATH

Personal experience would seem to be particularly suitable for "non-technical" decisions affecting oneself, as is the case here. "Technical" decisions (decisions that can be illuminated by scientific or technical information) affecting only oneself are associated with specialized media. Examples of this latter type of decision frequently arise in the treatment of medical problems. The mass media, finally, are certainly suitable sources of information about risks affecting the general public and the environment.

Information Seeking

One approach to the problems of the communication of scientific information is to investigate the conditions under which individuals go beyond the mass media and actively seek out additional information. Taking this approach, one of the earliest systematic investigations of the mass media, a now-classic field study by Hyman and Sheatsley (1947), explored "some reasons why information campaigns fail." Results of a national survey on whether particular news items had been heard were interpreted to indicate that individuals actively seek out, or select from among different messages, information that supports their existing attitudes and avoid information that contradicts them. This selective exposure hypothesis inspired much research in the 1950s and the 1960s (Freedman and Sears, 1965; Katz, 1968; Mills, 1968; McGuire, 1968; Sears, 1968). A number of investigations reported evidence supporting the existence of psychologically-motivated selection of information (Mills, 1968). Several considered reviews of the literature, however, conclude that support comes mostly from studies with spurious results that can be interpreted in alternate ways (Freedman and Sears, 1965; Lowin, 1968; McGuire, 1968).

One "good" feature of the selective exposure hypothesis, and perhaps an important reason for its popularity, is that it characterizes people as being actively involved in making judgments about being exposed to information. Rather than strictly defensive, as suggested by the selective exposure hypothesis, however, the active interaction of the "media target" and media content is often positively adaptive--an effort to make better risk judgments and decisions about risk mitigation behavior. Recent research suggests that information seeking or interest may be strongly motivated by interest in a general topic, the belief that current information is inadequate and that attending to new messages will yield useful information (Janis and Mann, 1977; Janis and Rausch, 1970; Grunig, 1974). Weinstein (1979) studied information seeking about the threat of environmentally-caused cancer. He found that more people were interested in learning what the hazard might be than in receiving information that minimized the danger. Interest in the general topic differentiated the information seekers from the nonseekers.

Similar results were obtained by a recent unpublished study examining the effects of an information campaign against driving while intoxicated (DWI). Early in 1983 the state of Washington sponsored a public campaign consisting of short radio and television "spots" and newspaper "ads" indicating the seriousness of the DWI problem, basic facts about it (e.g., number of DWI deaths), and the consequences if convicted of DWI. Near the end of the campaign a random sample of 190 people over the age of 18 years with numbers listed in the Bellingham, Washington directory were questioned via telephone interviews. Almost all of those surveyed (93 percent) had seen or heard at least one of the state-sponsored public announcements. The study participants were given the titles of two pamphlets dealing with DWI and asked to rate interest in reading each. The results showed that interest in new information is related to personal experience with DWI incidents. People for whom DWI was a personally relevant issue because they or a member of their family had been a driver, passenger, or a pedestrian victim in a DWI incident, judged the risks of DWI as higher and were more interested in new information than those without a personal involvement in the issue. Judgment of high risk was also related to a high frequency of exposure to messages about DWI in the media and to a pattern of either light consumption of alcohol or complete abstinence. An assessment that personal knowledge is inadequate was related to infrequent exposure to DWI messages in the media and an interest in new information. Heavy drinking patterns, defined as the consumption of more than two alcoholic drinks per day, were associated with a judgment that DWI is not a serious problem, little interest in the informational pamphlets, and a belief that the blood level required for a DWI conviction be raised and the minimum prison sentence be lowered. Those with a heavy drinking pattern seemed to be more concerned with reducing the dangers of being arrested for DWI than with the hazard of injury and death that DWI poses to drivers and others.

It is appealing to assume that the simple step of disseminating good-quality, exact information can improve public risk assessment processes. This assumption, based frequently on the erroneous belief that hazard information can be interpreted unequivocally, makes the job of risk management appear to be much easier than it really is. In the real world of hazard managers, matters are more complicated. The dissemination of hazard information, for example, can be counter-productive to the goals and best intentions of those distributing it. Public education programs may increase confusion over complicated issues (e.g., Nelkin, 1977) or, rather than allaying possibly unfounded fears, produce new apprenhensions. People clearly are not passive media targets who absorb information indiscriminately. Instead, they are often active seekers and selectors of information. Improvements in societal risk assessment, management, and policy development require an understanding of the psychosocial forces that influence this selection.

The Role Of Scientific Information In Risk Judgments

Depending on the nature of the hazard, the general public could have access to information through one or more of the communications media discussed above. Due to relatively low levels of scientific education and technical training, however, few members of the general public utilize the specialized media. Most members of the public thus must rely on personal experience, interpersonal communication, and the mass media for hazard information. And, as indicated in Fig. 3, it is through the mass media that most members of the general public receive scientific information about a hazard (see Nealey and Rankin, 1978, for a study of public media use and nuclear power). To better understand the communication of scientific information through the mass media to the general public we must first look briefly at the general problem of media effects and then more closely examine specific aspects of scientific information.

Mass Media Effects. Theoretical perspectives on the effects of the mass media on public attitudes and behavior are reviewed in Earle and Lindell (1980). These authors concluded that Zucker's (1978) theory of variable news media influence provides valuable insights into media effects because it clearly distinguishes situations in which effects would be expected, from those in which effects would be unlikely. Zucker identifies two key principles in his theory--obtrusiveness and duration. On an obtrusive issue, i.e. one with which members of the general public have had considerable personal experience, the mass media would be expected to exercise minimal influence. The duration principle assumes that attitudes change in response to new information and that the public eventually loses interest in non-obtrusive issues. Media influence thus would be expected to be greatest with new, emergent issues and least with old, familiar issues. We would therefore expect scientific information contained in the mass media to maximally influence public hazard judgments when the hazard is new and the public has had little personal experience with it.

Scientific information and the mass media: The bulk of the evidence on the efficacy of the communication of scientific information through the mass media points to three major problems: (a) scientific information regarding significant public issues is frequently conflicting, confusing, and inconclusive; (b) scientific information often does not reach those who may need it most; (c) science writing and writing techniques are frequently inadequate and ineffective. The research findings supporting each of these conclusions are briefly reviewed.

a. Scientific information regarding significant public issues is frequently conflicting, confusing, and inconclusive.

- On important issues, scientific evidence is often inconclusive, leading by default to decisions based on ideology and political and economic concerns (Douglas and Wildavsky, 1982; Farago, 1976; Holden, 1980; Holdren, 1976; Kolata, 1980; Mazur, 1973; Meltsner, 1979; Nelkin, 1977; Nelkin, 1979; Sjoberg, 1980; Trachtman, 1981; Wildavsky and Tennenbaum, 1981). In his critique on the communication of scientific information, Trachtman argues that the Yes/No, A or B type of decisions faced by consumers and citizens are ill served by the equivocal evidence typically communicated by scientists through the media. And the more significant the issue, the more likely there is to be conflict among the scientific experts.

- The uncertainty generated by scientific controversies can produce stress and other negative effects among the public (Holden, 1980; Holdren, 1976; Kolata, 1980; Kraus, Mehling, and El-Assal, 1963; Meltsner, 1979).

- Scientific information in the mass media tends to be ineffective in altering public behavior or leads to inaction (Novic and Sandman, 1974; Robinson, 1963; Trachtman, 1981; Wiebe, 1973).

- o Scientific information in the mass media is often inaccurate (Sjoberg, 1980; Tankard and Ryan, 1974; Trachtman, 1981).

The basic problem underlying much of the inaccuracy of mass media science communication is what Meltsner calls "the fundamental inadequacy of scientific knowledge," (Meltsner, 1979, p. 354). This central difficulty is highlighted also by Trachtman:

> "...To try to communicate to the general public something of the real feel of scientific research--tentative, sometimes seeming to take two steps backwards for every one forward, ultimately imprecise in technique and measurement, full of experimental error and, therefore, not at all like the idealized reports of utterly rational and totally inerrant experiments published in the scientific journals--is enormously difficult and extremely unsatisfying to the casual reader..." (Trachtman, 1981, p. 14)

The basic uncertainty of scientific information clashes with the public's perceived need for clear, unambiguous direction.

b. Scientific information often does not reach those who may need it most.

- o The audience for mass media scientific information tends to consist of those members of the public who are most scientifically sophisticated (Davis, 1958; Donohue, Tichenor, and Olien, 1974; Donohue, Tichenor, and Olien, 1975; Knott and Wildawsky, 1980; Miller and Barrington, 1981; Miller, Suchner, and Voelker, 1980; Tichenor, Olien, and Donohue, 1976; Weinstein, 1979; Witt, 1976). As a consequence, scientific information communicated in the mass media tends to be ineffective in informing the uninformed.

c. Science writing and writing techniques are frequently inadequate and ineffective.

- o Rather than focus on the intrinsic difficulties of scientific information, a number of investigators have identified problems stemming from the writers and the techniques they use (Altman, 1979; Cole, 1975; Dunwoody, 1980; Fishlock, 1974; Friedman, 1981; Funkhauser and Maccoby, 1974; Greenberg, 1979; Grunig, 1974; Jensen, 1977; Krieghbaum, 1967; Perl, 1979; Perlman, 1974; Pfund and Hofstadter, 1981; Randell, 1979; Rubin and Sachs, 1973; Ryan and Dunwoody, 1975; Sachsman, 1976; Shepherd, 1981; Swan, 1979; Tannenbaum, 1963). The common focus of these studies is on the development of writing techniques that will facilitate the communication of scientific information to the lay public.

Arguments and evidence endorsing the efficacy of mass media communication of scientific information are relatively scarce. The advocates include those who argue that: (a) the value of scientific information is clear to all and such information need only be well-communicated (Shen, 1975; Schloen and Day, 1975); (b) unbiased, value-free scientific evidence on policy controversies can be communicated to the public (Mazur, 1981; for examples of unsuccessful attempts see Johns, 1974; Livingston, 1974; Morrison, 1974; Patterson, 1974); and (c) the problem of scientific information communication needs to be reframed with the focus and communications from the point of view of the receiver (Dervin, 1980; Genova and Greenberg, 1979; Grunig, 1980; Slovic, Fischhoff, and Lichtenstein, 1980).

The factors involved in the communication of hazard information to the general public can be summarized best by reference to Fig. 4. Hazard information is often scientific and technical information; such information is communicated primarily through the mass media and the specialized media. Previous research provides the following picture of the effects of scientific information communicated through these media:

- o Scientific information in the mass media has maximal effects on public risk judgment when the issue is new and the public has had little personal experience with the hazard.

- o Scientific information in the mass media is often conflicting and inconclusive, and it tends not to reach those who need it most (i.e. the uninformed).

- Scientific information in the specialized media is subject to many of the limitations of scientific information in the mass media (i.e. access to and use of scientific information are related to socioeconomic status and education, etc.). It is possible, however, that if general interest in a new issue, judged risk regarding an old issue, or stakes in the community are individually or collectively strong enough, that scientific information in the specialized media may be sought out and used.

It seems likely also that if an individual is motivated to seek hazard information in the specialized media that he or she would previously have been motivated to seek it in the mass media. Since the mass media are easier to access than the specialized media, a failure to find desired information in the mass media is likely to precede information seeking in the specialized media. Finally, an individual may attempt to locate needed hazard information through interpersonal communication with a person who presumably has access to scientific information through the mass media and/or the specialized media.

<u>Public use of scientific information</u>: Our exploratory study of risk decisionmaking provides a preliminary view of the use of scientific hazard information by the general public. In respondents' selected risk decisions, scientific information was associated with the use of specialized media and with the following risk decision characteristics: concern with personal risks involving uncertainty, consideration of benefits, anxiety about risks to others, seeking help from others, and personal control. These results support our notion that individuals seeking scientific hazard information will make use of specialized media, and give some indication of the types of risk decisions which can prompt such an information search.

In addition to where scientific information is obtained and for what types of decisions, we also inquired about the amount and judged consistency of that information. The results presented in Table 4 show that use of scientific information was uniform across the first three quarters of the distribution, with a dropoff of about 10 percent in the fourth quarter. A surprising 45.5 percent of the respondents reported that more than half the information they used in making their selected

Table 4. REPORTED USE AND CONSISTENCY OF SCIENTIFIC INFORMATION

- Use of Scientific Information

Scientific Information (%)	Reported Use (%)
0 - 25	27.9
26 - 50	26.5
51 - 75	27.9
76 - 100	17.6

- Consistency of Scientific Information

	Scientific Information	All Information
Judged Consistent (%)	72.6	48.4
Judged Inconsistent (%)	27.3	51.6

risk decisions was scientific. Even more surprising, given our discussion of the "fundamental inadequacy" of scientific information, is the finding that a large majority (72.6 percent) of the respondents who used scientific information judged it to be consistent. In contrast, slightly over half (51.6 percent) of the respondents found some inconsistencies among all of the information they used. Rather than avoid scientific information because it is inconsistent and therefore difficult to use in risk decisionmaking, our respondents appear to have sought out scientific information because it is more consistent (and more informative and helpful) than other types of hazard information.

These results provide a refreshing antidote to the pessimism of the preceding analysis: scientific information need not necessarily be seen as complex and contradictory. Indeed, in situations in which the scientific information used suggests a single course of action in easy-to-understand lay language (as in many health-related problems), then scientific information may be most welcome and useful. On the other hand, when scientific information suggests multiple courses of action and contains technical concepts and arguments, then scientific information may be rejected in favor of other simpler guides to decisionmaking. Public use of scientific hazard information thus may be related to the complexity of the logical structure and quantification in the information.

<u>Scientific information and professional decisionmakers</u>: While the general public may be the ultimate group for whose benefit scientific hazard information is generated, the initial target audience for that information frequently is groups of professional decisionmakers. The communication of hazard information to professional groups differs in several ways from the communication of hazard information to the general public. Like the general public, of course, professional groups make use of the mass media. Professional groups, however, are much better able to use the scientific and technical information contained in the mass media. The greatest difference between professional groups and the general public is in the use of specialized media. Many studies have discussed the heavy use of specialized media by professional groups (Greene and Gori, 1982; Bates, 1979; Bingham, Hawkins, Frendreis, and Leblanc, 1978; Bingham and McNaught, 1976; Lambright, 1979; Rogers and Shoemaker, 1971; Sabatier, 1978; Szanton, 1981; and Utech and Utech, 1974). In defining "use of hazard information" we follow recent conceptual developments that have produced a broad and diffuse interpretation. Weiss, for example, (Weiss and Bucuvalas, 1980; Weiss, 1980; Weiss, 1979) defines "using research" as "taking research into account in the work of one's office." Weiss and Bucuvalas (1980) identified three major

factors affecting knowledge utilization: (1) relevance, (2) truth test (research quality and conformity to user expectations), and (3) utility test (action orientation and challenge to the status quo). Other researchers have also attempted to specify general sets of significant factors (Rothman, 1980; Sabatier, 1978). Still others have concluded that the number of significant factors seems to be endless and overwhelming (Dickman, 1981; Larson, 1980).

Though professional groups clearly make greater use of scientific and technical information than the general public, it is also clear that the use of such information by some groups of professionals is limited by a number of factors. As with the general public, a central constraint on the use of scientific information by professionals is its "fundamental inadequacy." Crandall and Lave identified this problem in the decisions of regulators in federal agencies: "The difficulty is not that science has nothing to offer; rather it is that science is not perceived as helpful because it is inconclusive. Since regulators do not see scientific data as helpful, they assign research a lower priority and a lesser role in policy formulation" (Crandall and Lave, 1981; p. 17). Although professional decisionmakers are better equipped by training to use scientific information, the fact that they have to deal with more of it, and at a level closer to the original technical jargon, may lead them, like the public, to simplifying decision strategies.

The ways in which both the public and professional decisionmakers deal with scientific hazard information in their risk decisions is a prime area for further research in the New Look approach to risk judgment. These, along with other new directions, are discussed next.

SUMMARY AND CONCLUSIONS

We have attempted in this paper to outline the elements of a New Look in the study of risk perception. In the first main section of the paper we described three studies of hazard information environments. The second main section was devoted to a description of our conceptual framework, including conceptual models of risk judgment and the hazard communications system. In this third and final section, we briefly review the key points of the earlier sections and describe our view of future risk judgment research.

Hazard Information

Studies of two hazard situations, the 1979 gas crisis and the 1980 volcanic activity at Mount St. Helens, demonstrated that individual news media work independently to identify the particular information needs of their users and to develop the content to meet those needs. Science writers, the news media workers frequently responsible for hazard information, were shown to be highly concerned and relatively impartial observers of scientific and technological issues. These studies suggest that the news media, despite their shortcomings (Scanlon, Morton, and Luuko, 1979), are basically well suited to the job of communicating hazard information to the public.

Inquiries into the actual risk decisions made by real people in everyday situations showed those decisions to be based on complex, subjectively-defined conceptions of risk as well as on other consequences such as benefits and costs. The types of information sought, and communications media used, were reliably related to the types of risks involved in everyday risk decisions: information from personal experience, for example, was most heavily used in "non-technical" decisions affecting oneself, while the mass media were used as sources of information about risks affecting the general public. The scientific hazard information used in everyday risk decisions was judged to be more consistent than other types of information.

New Look Conceptual Framework

The first step in the development of our conceptual framework was to move from the notion of risk perception to that of risk judgment. Risk judgments were conceptualized as:

1) Comprising Complex, multidimensional elements in the risk decisions made by real people in real hazard situations

2) Involving cognitive processes ranging along "a continuum demarcated by intuitive cognition at one pole and analytical cognition at the other" (Hammond, 1983; p. 15)

3) Comprising a dynamic process in which an individual adapts to a hazardous environment through a series of steps over time.

Each of these adaptive steps may include behavior related to information selection and risk mitigation. This hazard information was discussed within the context of a hazard communications system which included the source and receiver of the information, the medium used and the type of information. Hazard information was dichotomized into scientific and non-scientific, with the former being characterized by logical structure and quantification. It was suggested that public use of scientific hazard information may be related to the complexity of these two factors.

The View Ahead

As is evident from the preceding review, the primary focus of our New Look is on the interaction of risk judgments and hazard information. Our principal research objective therefore is clarification of the role of scientific hazard information in the development of public (and professional) risk judgments. The appropriate design for such a study is the longitudinal field study: the study over time of real people making risk decisions about actual hazards.

Longitudinal field studies of course are not the only paths to risk judgment enlightenment: a thorough understanding of risk judgments clearly must be based on research results generated in a variety of ways. One dimension of this variety, however, is that of laboratory studies conducted in artificial places and time vs. field studies carried out in real places and time. By real places and time we mean the ordinary places of everyday life--homes and workplaces--and the normal flow of time over days and weeks. These are the places and times in which people develop their risk judgments, and certain aspects of that development, such as the effects of hazard information, can only be studied there. We emphasize longitudinal field studies for this reason. Also, because of their difficulty and expense, they need all the encouragement they can get (see O'Riordan, 1982, for further arguments supporting field studies of risk judgment).

Our New Look can also benefit greatly from controlled laboratory studies. Ideally suited for such an approach, for example, would be investigations of the relations between the hazard information environment and risk judgment cognitive processes: scientific hazard information is an information environment in which risk judgments are made, and the characteristics of this environment very likely affect the cognitive processes used in making risk judgments. The techniques used by Hammond and his colleagues to study the relations between environmental characteristics and cognitive processes may be well suited for adaptation to the study of risk judgments (Hammond, Hamm, Grassia, and Pearson, 1983).

With regard to scientific hazard information, it seems clear that the type of cognitive process elicited would depend both on the training of the individual judge and on the complexity of the information. Complex scientific information (complex in logical structure and quantification), for example, may lead to analytical cognitive processes in a trained scientist and to intuitive processes in an untrained lay person. Further insight into the development of these judgment processes may be gained through application of process tracing techniques (Svenson, 1979).

It is clear to us that an integrated program of research based on our New Look conceptual framework (or any other) requires both laboratory and field studies interacting in mutually beneficial ways. The amount of insight into risk judgments produced by any research approach will depend greatly on the extent and quality of that interaction.

REFERENCES

Altman, L. K., Ethos of medical news reportage. Annals of the New York Academy of Sciences, 1979, 329, 342-345.

Bates, T. F., Transferring Earth Science Information to Decisionmakers. Problems and Opportunities as Experienced by the U.S. Geological Survey. Geological Survey Circular 813. Arlington, Virginia: U.S. Geological Survey, 1979.

Bingham, R. D., Hawkins, B., Frendreis, J. P., and LeBlanc, M. P., Professional Associations as Intermediaries in Transferring Technology to City Governments. Milwaukee: The University of Wisconsin (prepared under a grant from the National Science Foundation), 1978.

Bingham, R. (assisted by T. P. McNaught), The Adoption of Innovation by Local Government. Lexington, Massachusetts: Lexington Books, 1976.

Boulding, K. E., Science: Our common heritage. Science, 1980, 207 #831-836.

Bruner, J. S. and Goodman, C. C., Value and need as organizing factors in perception. Journal of abnormal and social psychology, 1947, 42, 33-44.

Bruner, J.S. and Klein, G. S., The functions of perceiving: New Look retrospect. In B. Kaplan and S. Wapner (Eds.), Perspectives in Psychological Theory. New York: International University Press, 1960.

Brunswik, E., The conceptual framework of psychology. In International Encyclopedia of Unified Science, (Vol. 1, No. 10). Chicago: University of Chicago Press, 1952.

Brunswik, E., Perception and the Representative Design of Psychological Experiments, (2nd edition). Berkeley: University of California Press, 1956.

Carroll, J.S., Analyzing decision behavior: The magician's audience. In T.S. Wallsten (Ed.), Cognitive Processes in Choice and Decision Behavior. Hillsdale, New Jersey: Lawrence Erlbaum Associates, 1980.

Cole, B.J., Trends in science and conflict coverage in four metropolitan newspapers. Journalism Quarterly, 1975, 52, 465-471.

Colligan, M. J., Pennebaker, J. W., and Murphy, L. R., Mass Psychogenic Illness: A Social Psychological Analysis. Hillsdale, New Jersey: Lawrence Erlbaum Associates, 1982.

Crandall, R. W. and Lave, L. B. (Eds.), The Scientific Basis of Health and Safety Regulation. Washington, D.C.: Brookings Institution, 1981.

Cyert, R. M. and Simon, H. A., The behavioral approach: With emphasis on economics. Behavioral Science, 1983, 28, 95-108.

Davis, R. C., A Report on a Nation-Wide Survey for the National Association of Science Writers. Ann Arbor: Survey Research Center, Institute for Social Research, University of Michigan, 1958.

Decker, R. and Decker, B., The eruptions of Mount St. Helens. Scientific American, 1981, 244, 68-80.

Dervin, B., Communication gaps and inequities: Moving toward a reconceptualization. In B. Dervin and M. J. Voigt (Eds.), Progress in Communication Sciences, Volume 11. Norwood, New Jersey, Ablex Publishing Corporation, 1980.

Dickman, F. B., Work activities, settings, methodologies, and perceptions: Correlates of evaluative research utilization. Knowledge: Creation, Diffusion, Utilization, 1981, 2 (3), 375-387.

Donohue, G. A., Tichenor, P. J., and Olien, C. N., Mass media functions, knowledge and social control. Journalism Quarterly, 1974, 50 (4), 652-659.

Donohue, G. A., Tichenor, P. J., and Olien, C. N., Mass media and the knowledge gap: A hypothesis reconsidered. Communication Research, 1975, 2 (1), 3-23.

Douglas, M. and Wildavsky, A., Risk and Culture. Berkeley: University of California Press, 1982.

Dunwoody, S., The science writing inner club: A communication link between science and the lay public. Science, Technology and Human Values, 1980, 5 (3), 14-22.

Earle, T. C., Intuitive and Analytical Thinking in Consistent and Inconsistent Multiple-Cue Learning Tasks. Unpublished doctoral dissertation, University of Oregon, Eugene, 1972.

Earle, T. C., Public Perception of Industrial Risks: The Context of Public Attitudes Toward Radioactive Waste. BHARC (411/81/003), Battelle Human Affairs Research Centers, Seattle, Washington, 1981.

Earle, T. C. and Lindell, M. K. The Role of the News Media in the 1979 Gasoline Crisis. BHARC (411/80/002), Battelle Human Affairs Research Centers, Seattle, Washington, 1980.

Earle, T. C., Lindell, M. K. and Rankin, W. L., Risk Perception, Risk Evaluation and Human Values Concerning Nuclear Waste Management. BHARC (411/81/007), Battelle Human Affairs Research Centers, Seattle, Washington, 1981.

Earle, T. C., Southwick, L. L., and Lindell, M. K., Newspaper Coverage of Mount St. Helens: Patterns of Content and Information Sources. BHARC (411/81/035), Battelle Human Affairs Research Centers, Seattle, Washington, 1981.

Ebbson, E. B. and Konecni, V. J., On the external validity of decisionmaking research: What do we know about decisions in the real world? In T. S. Wallsten (Ed.), Cognitive Processes in Choice and Decision Behavior. Hillsdale, New Jersey: Lawrence Erlbaum Associates, 1980.

Farago, P., Science and the Media. London: Oxford University Press, 1976.

Fischhoff, B., Slovic, P., and Lichtenstein, S., The "public" vs. the "experts": Perceived vs. actual disagreements about the risks of nuclear power. In V. Covello, G. Flamm, J. Rodericks, and R. Tardiff (Eds.), Analysis of Actual Vs. Perceived Risks. New York: Plenum, in press.

Fischhoff, B., Watson, S. R., and Hope, C., Defining Risk. Eugene, Oregon: Decision Research, 1983.

Fishlock, D., What makes a good science writer? Nature, 1974, 250 (5469), 747-749.

Freedman, J. L. and Sears, D. O., Selective exposure. Advances in Experimental Social Psychology, 1965, 2, 57-97.

Friedman, S. M., Blueprint for breakdown: Three Mile Island and the media before the accident. Journal of Communication, 1981, 31 (2), 116-128.

Funkhauser, G. R. and Maccoby, N., An experimental study on communicating specialized science information to a lay audience. Communication Research, 1974, 1 (1), 110-128.

Garfield, E., Is public confidence in science declining? Current Contents, 1982, 45, 5-13.

Genova, B.K.L. and Greenberg, B. S., Interests in news and the knowledge gap. Public Opinion Quarterly, 1979, 43, 79-91.

Greenberg, D., An overview of science journalism in the American press. Annals of the New York Academy of Sciences, 1979, 329, 346-347.

Greene, M. R. and Gori, P. L., Earthquake Hazards Information Dissemination: A Study of Charleston, South Carolina. U.S.G.S. Open-File Report, 1982, in preparation.

Grunig, J. E., Three stopping experiments on the communication of science. Journalism Quarterly, 1974, 51, 387-399.

Grunig, J. E., Communication of scientific information to nonscientists. In B. Dervin and M. J. Voigt (Eds.), Progress in Communication Sciences, Volume 11. Norwood, New Jersey, Ablex Publishing Corporation, 1980.

Hammond, K. R., The Psychology of Egon Brunswik. New York: Holt, Rinehart and Winston, 1966.

Hammond, K. R., Unification of Theory and Research in Judgment and Decision Making. Boulder: Center for Research on Judgment and Policy, Institute of Behavioral Science, University of Colorado, 1983.

Hammond, K. R. and Brehmer, B., Quasi-rationality and distrust: Implications for international conflict. In L. Rappoport and D. Summers (Eds.), Human Judgment and Social Interaction. New York: Holt, Rinehart and Winston, 1973.

Hammond, K. R., Hamm, R. M., Grassia, J., and Pearson, T., A Second Direct Comparison of Intuitive, Quasi-Rational and Analytical Cognition: Use of the Multitrait-Multimethod Matrix. Boulder: Center for Research on Judgment and Policy, Institute of Behavioral Science, University of Colorado, 1983.

Hammond, K. R., Stewart, T. R., Brehmer, B., and Steinmann, D. O., Social judgment theory. In M. F. Kaplan and S. Schwartz (Eds.), Human Judgment and Decision Processes. New York: Academic Press, 1975.

Hochberg, J. E., Perception. Englewood Cliffs, New Jersey: Prentice-Hall, 1964.

Hogarth, R. M., Beyond discrete biases: Functional and dysfunctional aspects of judgmental heuristics. Psychological Bulletin, 1981, 90, 197-217.

Holden, C., Love canal residents under stress. Science, 1980, 208, 1242-1244.

Holdren, J. P., The nuclear controversy and the limitations of decisionmaking by experts. Bulletin of the Atomic Scientists, 1976, 32, 20-22.

Hyman, H. H. and Sheatsley, P. B., Some reasons why information campaigns fail. Public Opinion Quarterly, 1947, 11, 412-423.

Janis, I. L. and Mann, L., Decision Making: A Psychological Analysis of Conflict, Choice and Commitment. New York: Free Press, 1977.

Janis, I. L. and Rausch, C. N., Selective interest in communication that could arouse decisional conflict. Journal of Personality and Social Psychology, 1970, 14, 46-54.

Jensen, D., The loneliness of the environmental reporter. Columbia Journalism Review, 1977, 15, 40-42.

Johns, M. W., "Nuclear Reactors": A diatribe for political purposes. Science Forum, 1974, 7 (3), 5-6.

Kahneman, D. and Tversky, A., Subjective probability: A judgment of representativeness. Cognitive Psychology, 1972, 3, 430-454.

Katz, E., On re-opening the question of selective exposure to mass communications. In R. P. Abelson, E. Aronson, W. J. McGuire, T. M. Newcomb, M. J. Rosenberg, and P. H. Tannenbaum (Eds.), Theories of Cognitive Consistency: A Sourcebook. Chicago: Rand McNally, 1968.

Knott, J. and Wildavsky, A., If dissemination is the solution, what is the problem? Knowledge: Creation, Diffusion, Utilization, 1979, 1, 537-577.

Kolata, G. B., Love canal: False alarm caused by botched study. Science, 1980, 208, 1239-1242.

Kraus, S., Mehling, R., and El-Assal, E., Mass media and the fallout controversy. Public Opinion Quarterly, 1963, 27, 191-205.

Krech, D. and Crutchfield, R. S., Theory and Problems of Social Psychology. New York: McGraw-Hill, 1948.

Krieghbaum, H., Science and the Mass Media. New York: New York University Press, 1967.

Lambright, W. H., Technology Transfer to Cities: Processes of Choice at the Local Level. Boulder, Colorado: Westview Press, 1979.

Larson, J. F., A review of the state of the art in mass media disaster reporting. In Committee of Disasters and the Mass Media, National Research Council (Eds.), Disaster and the Mass Media. Washington, D.C.: National Academy of Sciences, 1980.

Lepman, W. and Mullineaux, D. R. (Eds.)., 1980 Eruptions of Mount St. Helens, Washington. Washington, D.C.: United States Geological Survey, 1982.

Lindell, M. K. and Earle, T. C., Comparative Analysis of Risk Characteristics of Nuclear Waste Repositories and Other Disposal Facilities. BHARC (411/81/005), Battelle Human Affairs Research Centers, Seattle, Washington, 1981.

Livingston, J., "Open Pit Mining": Precisely what the public needs. Science Forum, 1974, 7 (3), 6-7.

Lowin, A., Further evidence for an approach-avoidance interpretation of selective exposure. Journal of Experimental Social Psychology, 1969, 5, 265-271.

Mazur, A., Disputes between experts. Minerva, 1973, 11, 243-262.

Mazur, A., The Dynamics of Technical Controversy. Washington, D.C.: Communications Press, 1981.

McGuire, W. J., Selective exposure: A summing up. In R. P. Abelson, E. Aronson, W. J. McGuire, T. M. Newcomb, M. J. Rosenberg, and P. H. Tannenbaum (Eds.), Theories of Cognitive Consistency: A Sourcebook. Chicago: Rand McNally, 1968.

Meltsner, A. J., The communciation of scientific information to the wider public: The case of seismology in California. Minerva, 1979, 17 (3), 331-354.

Miller, J.D. and Barrington, T.M., The acquisition and retention of scientific information. Journal of Communication, 1981, 31 (2), 178-189.

Miller, J. D., Suchner, R. W., and Voelker, A. M., Citizenship in an Age of Science. New York: Pergamon Press, 1980.

Mills, J., Interest in supporting and discrepant information. In R. P. Abelson, E. Aronson, W. J. McGuire, T. M. Newcomb, M. J. Rosenberg, and P. H. Tannenbaum (Eds.), Theories of Cognitive Consistency: A Sourcebook. Chicago: Rand McNally, 1968.

Morrison, R. W., Canadians should produce their own such publications. Science Forum, 1974, 7, 7-8.

Nealey, S. M. and Rankin, W. L., Nuclear Knowledge and Nuclear Attitudes: Is Ignorance Bliss? BHARC (411/002), Battelle Human Affairs Research Centers, Seattle, Washington, 1978.

Nelkin, D., Technological Decisions and Democracy: European Experiments in Public Participation. Beverly Hills, California: Sage Publications, 1977.

Nelkin, D., Scientific knowledge, public policy and democracy. Knowledge: Creation, Diffusion and Utilization, 1979, 1 (1), 106-122.

Nigg, J. M., Communication under conditions of uncertainty: Understanding earthquake forecasting. Journal of Communication, 1982, 32, 27-36.

Novic, K. and Sandman, P. M., How use of mass media affects views on solutions to environmental problems. Journalism Quarterly, 1974, 50 (4), 680-684.

O'Riordan, T., Risk-perception studies and policy priorities. Risk Analysis, 1982, 2, 95-100.

Patterson, W. C., Helping the public judge contentions issues involving science. Science Forum, 1974, 7 (3), 3-4.

Perl, P., The underreported story of shipyard hazards. Annals of the New York Academy of Sciences, 1979, 354-356.

Perlman, D., Science and the mass media. Daedalus, 1974, 103, 207-222.

Perry, R. W. and Greene, M. R., Citizen Response to Volcanic Eruptions: The Case of Mount St. Helens. New York: Irvington Publishers, 1982.

Pfund, N. and Hofstadter, L., Biomedical innovation and the press. Journal of Communication, 1981, 31 (2), 138-154.

Pion, G. M. and Lipsey, M. W., Public attitudes toward science and technology: What have the surveys told us? Public Opinion Quarterly, 1981, 45, 303-315.

Quarantelli, E. L., The Command Post Point of View in Local Mass Communications Systems. Columbus, Ohio: Disaster Research Center, The Ohio State University, 1978.

Randell, W. S., Compilation of the story concerning bischloromethyl ether fatalities. Annals of the New York Academy of Sciences, 1979, 329, 357-359.

Robinson, E. J., Analyzing the impact of science reporting. Journalism Quarterly, 1963, 40, 306-314.

Rogers, E. and Shoemaker, F. F., Communication of Innovations: A Cross-Cultural Approach. New York: Free Press, 1971.

Rothman, J., Using Research in Organizations. Beverly Hills, California: Sage Publications, 1980.

Rubin, D. M. and Sachs, D. P., Mass Media and the Environment. New York: Praeger, 1973.

Ryan, M. and Dunwoody, S. L., Academic and professional training patterns of science writers. Journalism Quarterly, 1975, 52, 239-246, 290.

Sabatier, P., The acquisition and utilization of technical information by administrative agencies. Administrative Science Quarterly, 1978, 23, 396-417.

Sachsman, D. B., Public relations influence on coverage of environment in the San Francisco area. Journalism Quarterly, 1976, 53, 54-60.

Scanlon, J., Morton, G., and Luuko, R., Media coverage of crisis: Better than reported, worse than necessary. Journalism Quarterly, 1978, 55, 68-72.

Schloen, L. H. and Day, S. B., The role of the communications fellow in the conceptual development of biosciences communications. In S. B. Day (Ed.), Communication of Scientific Information. Basel: Karger, 1975.

Sears, D. O., The paradox of de facto selective exposure without preferences for supportive information. In R. P. Abelson, E. Aronson, W. J. McGuire, T. M. Newcomb, M. J. Rosenberg, and P. H. Tannenbaum (Eds.), Theories of Cognitive Consistency: A Sourcebook. Chicago: Rand McNally, 1968.

Shen, B.S.P., Science literacy and the public understanding of science. In S. B. Day (Ed.), Communication of Scientific Information. Basel: Karger, 1975.

Shepherd, R. G., Selectivity of sources: Reporting the marijuana controversy. Journal of Communication, 1981, 31 (2), 129-137.

Sherif, M., A study of some factors in perception. Archives of Psychology, No. 187, 1-60.

Sjoberg, L., The risks of analysis. Acta Psychologica, 1980, 45, 301-321.

Slovic, P., Fischhoff, B., and Lichtenstein, S., Behavioral decision theory. Annual Review of Psychology, 1977, 28, 1-39.

Slovic, P., Fischhoff, B., and Lichtenstein, S., Informing people about risk. In L. Morris, M. Mazis, and B. Barofsky (Eds.), Product Labeling and Health Risks, Banbury Report 6. Cold Spring Harbor, New York, 1980.

Slovic, P., Kunreuther, H., and White, G. F., Decision processes, rationality and adjustment to natural hazards. In G. F. White (Ed.), Natural Hazards, Local, National and Global. New York: Oxford University Press, 1974.

Starr, C. and Whipple, C., Risks of risk decisions. Science, 1980, 208, 1114-1119.

Swan, J., Occupational health: Press priorities and responsibilities. Annals of the New York Academy of Sciences, 1979, 329, 348-353.

Szanton, P., Not Well Advised. New York: Russell Sage Foundation and the Ford Foundation, 1981.

Tajfel, H., Social and cultural factors in perception. In G. Lindzey and E. Aronson (Eds.), The Handbook of Social Psychology, (2nd Edition), Volume 3. Reading, Massachusetts: Addison-Wesley, 1969.

Tankard, J.W., Jr. and Ryan, M., News source perceptions of accuracy of science coverage. Journalism Quarterly, 1974, 51, 219-225, 334.

Tannenbaum, P. H., Communication of science information. Science, 1963, 140 (3567), 579-583.

Thompson, M., To Hell with the Turkeys! A Diatribe Directed at the Pernicious Trepidity of the Currrent Intellectual Debate on Risk. Working paper RC-5. College Park, Maryland: Center for Philosophy and Public Policy, University of Maryland, 1983.

Tichenor, P. J., Donohue, G. A., and Olien, C. N., Community Conflict and the Press. Beverly Hills, California: Sage Publications, 1980.

Tichenor, P. J., Olien, C. N., and Donohue, G. A., Community control and care of scientific information. Communication Research, 1976, 3 (4), 403-424.

Trachtman, L. E., The public understanding of science effort: A critique. Science, Technology and Human Values, 1981, 6 (36), 10-15.

Tversky, A. and Kahneman, D., Availability: A heuristic for judging frequency and probability. Cognitive Psychology, 1973, 5, 207-232.

Tversky, A. and Kahneman, D., Judgment under uncertainty: Heuristics and biases. Science, 1974, 185, 1124-1131.

Tversky, A. and Kahneman, D., Extensional versus intuitive reasoning: The conjunction fallacy in probability judgment. Psychological Review, 1983, 90, 293-315.

Utech, H. P. and Utech, I., The Communication of Innovations Between Local Government Departments. Pilot Study for the National Science Foundation, Office of National Research and Development Association. Washington, D.C.: National Science Foundation, 1974.

Vlek, C. and Stallen, P.J., Judging risks and benefits in the small and in the large. Organizational Behavior and Human Performance, 1981, 28, 235-271.

"Volcano cabin owners to defy 'stay out' order." Seattle Post Intelligencer, May 17, 1980, p. A5.

Weinstein, N. D., Seeking reassuring or threatening information about environmental cancer. Journal of Behavioral Medicine, 1979, 2, 125-139.

Weiss, C. H., The many meanings of research utilization. Public Administration Review, 1979, 39, 426-431.

Weiss, C. H., Knowledge creep and decision accretion. Knowledge: Creation, Diffusion, Utilization, 1980, 1.

Weiss, C. H., and Bucuvalas, M. J., Truth tests and utility tests: Decisionmakers' frames of reference for social science research. American Sociological Review, 1980, 45, 302-313.

Wiebe, G.D., Mass media and man's relationship to his environment. Journalism Quarterly, 1973, 50, 426-432, 446.

Wildavsky, A. and Tenenbaum, E., The Politics of Mistrust: Estimating American Oil and Gas Resources. Beverly Hills, California: Sage Publications, 1981.

Witt, W., Effects of quantification in scientific writing. Journal of Communication, 1976, 26 (1), 67-69.

Zucker, H. G., The variable nature of news media influence. In B. D. Ruben (Ed.), Communications Yearbook 2. New Brunswick, New Jersey: Transactions Books, 1978.

"Who's Afraid of Technological Risk?": An Attempt to Model Reasonable Emotions

P.J.M. Stallen

Centre for Technology and Policy Studies, Netherlands Organization for Applied Scientific Research, P.O. Box 541, 7300 AM APELDOORN - The Netherlands

INTRODUCTION

Toxicologists typically distinguish three levels of exposure - acute, chronic and subchronic. Toxicological knowledge of these levels differs considerably. We know quite a lot about acute exposures but much less about chronic poisoning, particularly in cases not connected with occupational hazards. We know virtually nothing about subchronic exposures.

It should not come as a surprise that we have even less knowledge of the psychic and social effects of exposure to environmental hazards. For example, social scientists still know practically nothing about injuries to mental public health, such as fear or the social effects of the loss of communality (e.g., Erikson, 1976) caused by environmental pollution.

In a recent personal communication, a Hoffman La Roche physician closely involved in the Seveso disaster--a situation in which one might speak of acute exposure--noted that there had been no psychological examination of persons living in the contaminated area. The accident at Three Mile Island (Harrisburg) was one of the first cases in which such an examination did take place (see Dohrenwend et. al., 1981). In cases of chronic exposure--if we may describe the cases of Lekkerkerk and Gouderak (Holland) or Love Canal (USA) as such--social scientists have, to date, carried out research in scattered instances only (e.g. Levine 1982, Edelstein 1982). Perhaps one could call the ambient presence of various hazardous technological activities or processes a situation of subchronic exposure, but virtually no study has been undertaken to assess their separate effects on psychological health. Why?

In this article I will present the argument that the study of human behaviour is more complex than the study of physical

NATO ASI Series, Vol. G4
Environmental Impact Assessment, Technology Assessment, and Risk Analysis. Edited by V.T. Covello et al.

processes (sections 1 and 2). I will also attempt to model reasonable emotions about technological risk (Sections 3-5). My aim is to show that psychological effects can be discussed sensibly.

We end with Robert Pirsig's frustrating observation (Zen and the Art of Motorcycle Maintenance) that, if an attempt is successful, it will only leave us with more interesting questions.

THE HUMAN CAUSE OF THE HAZARD SEQUENCE

Figure 1 shows the hazard causal sequence developed by researchers at Clark University, Worcester (Mass.). Physicians and toxicologists usually deal with the end of the sequence, with the relation between exposure and first or higher order consequences. The power of man to look and think ahead, however, entails that he may also worry about events that have not yet led to a certain exposure or outcome. He may sometimes consider certain situations threatening, whereas there is only a small possibility that harmful releases will occur. As such expectations and worry highly depend on the presumed effectiveness of the corresponding control strategies and, in many cases, these expectations cannot be quantified better than as a "certain degree of belief," this will bring about great differences of opinion in the appraisal of the risk.

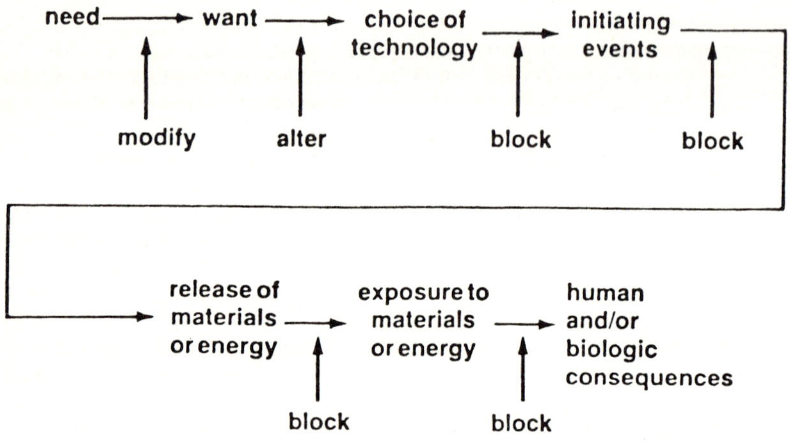

Figure 1. Hazard Causal Sequence as Developed by Hohenemser, Kasperson, and Kates (1982).

In a recent address, Wagenaar (1983) observed how important, on the one hand, human error is as a cause of many accidents, but how little is known about how to counteract such error (for instance by better system designs). As a result, trust in human innovative power may make some remain unworried about situations that others feel highly threatening. Human factors are, in principle, present everywhere in the chain of Fig. 1, because every stage has its own possibilities of human control. The farther one goes back in the sequence, the less consensus there will be on the eventual risk of an undesirable consequence. In their recently published book (Risk and Culture), social anthropologist Mary Douglas and political scientist Aaron Wildavsky (1982) point to the earlier stages of control in this chain. They argue that we cannot fully understand why we worry about some risks and not about others as long as we do not realize how the social structure (control stage 1) in our society causes us to pursue certain values. "Common values lead to common fears." These different human "inputs" to the hazard causal sequence make the study of possible undesirable psychological effects of exposure to technological risks more complicated than the study of medical and biological effects, however complex the latter may be.

RISK PERCEPTION OR RISK CONCEPTION?

In the past few years studies have been conducted in the Netherlands, Germany, Great Britain, and the United States about the meaning of risk. One of the general conclusions that can be drawn is that the term risk, and the reasons why a certain activity is called risky, depends highly on the person and situation in question. It appears that among the general public riskiness is considered to be a combination of the catastrophic potential of an activity or the impact of its effects, if any, on the one hand; and the controllability of any undesirable events on the other.

By comparison, risk analysts, and probably policymakers as well, define risk in terms of the estimated number of fatal accidents per unit of exposure. We will not go into detail as to whether such a definition of riskiness sufficiently covers all kinds of other impacts on man, environment, and society. It does not, as Hohenemser et al. (1982) have shown. Restricting ourselves to the limited conception of estimated risk, this estimate inevitably contains elements of personal judgement and feelings. Sometimes these elements are present to a minimum extent, such as in cases where observations can be done within an established framework or paradigm. However, the subjective element will increase strongly when the results of various studies are to be integrated or when data are to be extrapolated. The majority of full-fledged, policy-relevant risk analyses are of this complex nature.

More important perhaps, there is also a difference in decisionmaking perspectives between the public and experts. Expert assessments typically pertain to regulatory policy. Here, it is quite reasonable to assume that one (statistical) death is just as bad as another. However, expert data concerning the harmful effects of an event on the average citizen are only one fraction of the data relevant to those exposed to the event (although it should be emphasized that people often pay insufficient attention to such base-rate information). No wonder there is often a communication gap between the expert or government official and the general public in cases where public safety is concerned. And perhaps we may even say this: no wonder that somebody who considers one element of the definition of the term risk as the only relevant one ("ten to the minus umpteenth power") will be suspected of unwillingness to participate constructively in a discussion, and thus, of attempts to preserve the status quo. On these grounds we dispute the correctness of the frequently heard statement that there is a gap between the manner in which the public judges the safety of a technology and the factual, objectively determined risks thereof--suggesting that the public should rely on the risk estimates of experts.

A SENSE OF FEELINGS OF INSECURITY WITH RESPECT TO ENVIRONMENTAL THREAT

Let us now turn to a specific psychological effect--a feeling of insecurity, such as the feeling of being threatened or feeling anxious. This is one of the possible consequences of exposure to a dangerous technological activity which a public authority may be willing to consider in its decisionmaking or standard-setting processes. A good example is the accident at the Three Mile Island-2 nuclear reactor. A local action group submitted a request to the Court saying in effect that the Nuclear Regulatory Commission should, in conformity with the National Environmental Policy Act, investigate whether and to what extent the restart of the TMI-1 power plant would cause psychological damage to the population similar to that which occurred at the TMI-2 reactor. The action group won the case and the Commission called in a number of social psychologists to assist them in preparing an environmental impact assessment report.[1] In view of the government bill on Environmental Impact Assessment, it is quite conceivable that the need for such research may also be felt in the Netherlands. The "Openbaar Lichaam Rijnmond" has initiated an investigation on this matter and in the remaining part of this paper I will discuss this research, which is being carried out by A. Tomas and myself. Our study focuses on the following questions: 1) To what extent do the people of Rijnmond feel insecure as a result of technological and industrial activity in the Rijnmond area? 2) What is the relative importance of such feelings of anxiety amidst other problems with which one is faced in daily life? 3) How do people cope with their feelings of insecurity?

Let me first make a more philosophical remark about scientific answers to such questions. It is my belief that a scientific approach can be applied to such matters as feelings of anxiety and insecurity. Emotions are all too easily called irrational, and thus, irrelevant. It is claimed that they are liable to change and therefore should not be referred to when advice is needed. However, they are always there, sometimes as cool expert knowledge, sometimes as heated disagreement. Moreover, psychoanalysis has shown that emotions are not so disordered as to be unpredictable.

I do not mean to say that people are fearful only when a psychologist has stated that they are fearful. Every science, and social science in particular, is based on a construction of reality in thoughts, however convincing and inspiring that construction may be. Consequently, observations of feelings of insecurity because of technological threat, or observations of psychological health effects in general, cannot be made without due consideration of this creative and subjective element.

The comprehensive definition of the term "health" by Zielhuis (1980) also exemplifies the importance of the subjective experience of human functioning: "Health is a condition, in principle not static, of the organism which, in the opinion of both the individual and the experts functions entirely satisfactorily, seen in the light of the given physical and mental faculties, measured by age, sex, the condition of the population to which the individual belongs and checked against

---

1 The Court's decision has meanwhile been appealed and the Supreme Court of the United States has rejected the claim of the action group. However, they did not preclude the possibility of such claims being allowed in the case of factual or real (as opposed to potential or expected) changes in the natural environment, for instance by waste disposal.

views, such as these exist dependent on the state of science and on the health care objectives based thereon, time in which one lives, cultural pattern and social ideas." (p. 382). One of the things on which health is based is a "sound" field of tension between individual and social judgement. Health is also based on a field of tension between personal desires on the one hand and individual and social faculties on the other. It is this field of tension to which we have focused our attention in our study of the Rijnmond Population. I will try to describe briefly the psychological model underlying this study.

THE "SITE" MODEL: (COPING) WITH A SENSE OF INSECURITY WITH RESPECT TO THREAT FROM THE (TECHNOLOGICAL) ENVIRONMENT.

Following the approach taken by an American research group headed by Richard Lazarus, we distinguish between appraisal processes, which may lead to feelings of insecurity (stress) and the process of coping with such feelings. Figure 2 shows the two processes. There are two types of appraisal processes: the process of assessing or evaluating a potential threat ("primary appraisal") and the process of examining the available means to stand up to a threat ("secondary appraisal"). There are also different types of coping with a feeling of insecurity. I will return to this point later. Figure 2 illustrates the assumed presence of an effective, emotional reservoir from which the individual derives his specific values and motives, and a cognitive reservoir in which (for convenience) all intellectual elements are localized. For the moment let us assume that every appraisal or type of coping process has an emotional and an intellectual component. The bowtie symbol in Fig. 2 indicates mechanisms of selection. By presenting two consecutive internal psychological processes we can illustrate how the individual learns: he can do so by comparing his judgment of the situation at 5 with that he had thought the situation would be at 3 (see dotted arrow). Figure 3 details the appraisal portion of the internal reaction process to a threat. In each situation one first asks oneself whether the activity to which the threat is directed should nevertheless be continued, given the envisaged benefits.

APPRAISAL

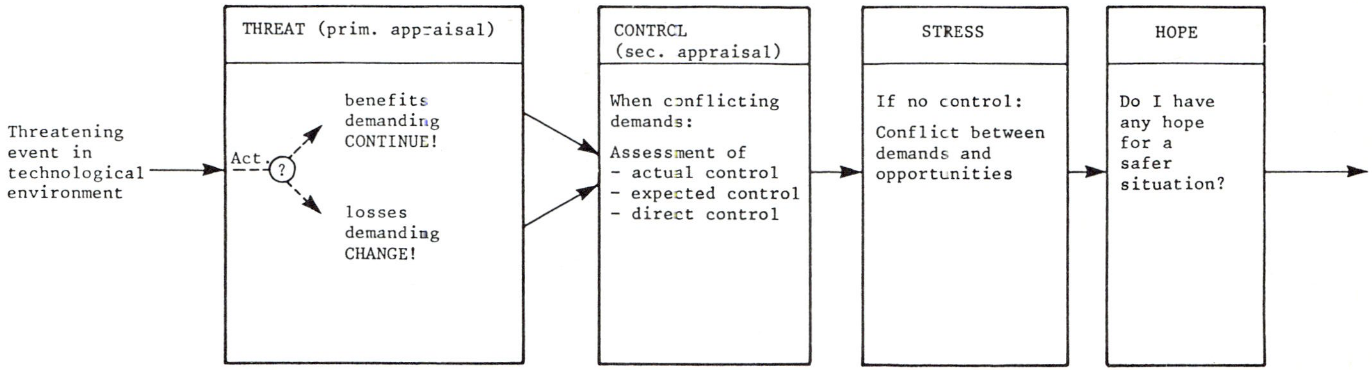

Figure 2. Representation of individual response to threat from the (technological) environment (Source: Stallen & Tomas, 1983.

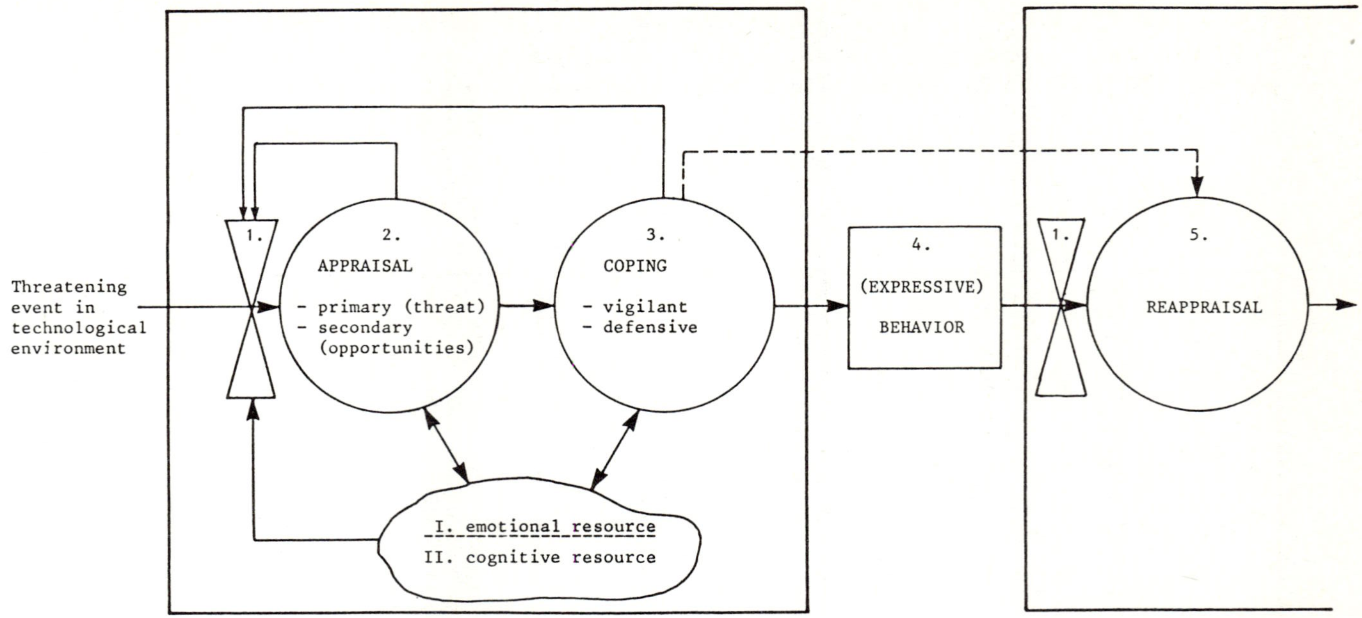

Figure 3. Detailed representatives of appraisal part of individual response to technological threat (Source: Stallen & Tomas, 1983).

If the expected benefits do not distinctly outweigh any possible disadvantages or damage, one will be inclined, particularly when it is a matter of great importance, to examine whether one is in a position, or has the power, to do something about the threatening cost-benefit ratio. Here again, the aspect of control is involved, which also played an important part in Fig. 1. Perhaps one will, without much trouble, directly find a way to change one's activity, so as to escape the threat (actual control). If not, one will attempt to discover alternatives (expected control). If that, too, seems to be impossible, there will be only one way left to avoid a serious conflict or a serious feeling of insecurity between desired benefits and the desirability of giving these benefits up. That is by hoping that the threat may be reduced before the situation develops unfavorably (cf. Janis & Mann, 1977).

Such hope may be based on confidence in finding alternatives, and also to some extent on confidence in the action of others (for instance public authorities). In our model, hope is the ultimate resource for coping with a feeling of insecurity in an alert, problem-focused manner. In situations where hope of finding an alternative is absent, the individual will be faced with unbearable conflict. Therefore, he will have to find ways of artificially reducing his sense of insecurity by: 1) underestimating or even denying the threat, 2) overestimating the envisaged benefits, 3) thinking that there is sufficient time left, 4) blaming others for events that are actually due to his own lack of power, or 5) simply not thinking about the threat and the problem. These are called "possibilities of repression" in psychoanalysis and there are thought to be no fewer than 30 types.

By and large current risk perception studies concentrate on what we called primary appraisal of threat, or they deal with some overall assessment with no explicit awareness of its underlying dynamic. It is our hope that the model presented in Fig. 3 offers a tool with which to make finer observations about how people respond to, or behave, with respect to technological risks.

HYPOTHESES

In our study of the Rijnmond population we will be using this model for mapping out the various ways in which one can feel insecure. We expect to find persons who have no feeling of concern (there is no threat), persons who feel that the dangers typically belong to the situation (the risks are accepted because of the benefits), people who will behave defensively (the situation is too dangerous and there is no hope of change for the better), and, finally, persons who experience feelings of insecurity because they hope that by expressing these feelings in some way or another they may arrive at a safer situation. We are also testing other hypotheses, some of which are opposite to the above. For instance, we will test the hypotheses that those who feel threatened are persons who already "by nature" are afraid of any threat or uncertainty. In plain speech, they are cowards. If this appears to be true, one should direct one's attention to the threatened person rather than to the threatening situation. Another hypotheses is that people are venting their emotions on industry. In reality, the argument goes, their feelings of insecurity have other causes, such as unemployment or a disappointing economic situation. Since they hold industry accountable for this, they are merely using industrial safety as a surrogate means for showing their discomfort.

CONCLUSION

Let us assume that this study produces good theory and good measuring tools. What we would then like to do is examine situations of acute exposure to dangers caused by industrial activity or industrial products. By doing so, we will gain a better insight into the external causes or conditions underlying feelings of threat or insecurity. In our study we pay little attention to external factors. For example, what is the significance of social supports or social ties in a group or community to feelings of insecurity and to ways of dealing with these feelings? (The views put forward by Douglas and Wildavsky might be relevant here.) And if we find answers to such questions, the next question is" How can we reduce or prevent feelings of insecurity?" Other questions will also emerge. How can people be compensated for additional exposure to risks? What principles of social justice should be used when allocating risks? What safety philosophy should govern or justify such allotment? This is, of course, not only the province of the psychologist, but of all those interested in social decisionmaking. But I will not address this subject, as Brian Wynne and others have written on this and related issues elsewhere in this volume.

REFERENCES

Dohrenwend, B. et al. (1981), "Stress in the community: A Report to the Presidents Commission on the Accident at Three Mile Island," Ann. New York Acad. Sci., 365, 159-174.

Douglas, M., & Wildavsky, A. (1982), Risk and culture, Berkeley: University of California Press.

Edelstein, M.R. (1982), The Social and Psychological Impacts of Groundwater Contamination in the Legler section of Jackson. Ramapo College, New Jersey.

Erikson, K., (1976), Loss of Communality at Buffalo Creek, Amer. J. Psychiatr., 133, 302-305.

Hohenemser, C., Kasperson, K.E., Kates, R.W. (1982), "Causal Structure: A Framework for Policy Analysis," in Hohenemser, C. and Kasperson, J.X. (eds.), Risk in the technological society, Boulder, Col.: Westview Press.

Janis, I. and Mann, L. (1977), Decision Making, New York: Free Press.

Levine, A., (1982), Love canal: Science, Politics, and people. Lexington Books: New York.

Stallen, P.J.M. & Tomas, A. (1983), "Psychological risk: The Assessment of Threat and Control," in Ricci, P. Sagan, P. and Whipple, C. (eds.). Technological Risk Assessment, Sijthoff: Den Haag.

Vlek, C.A.J. & Stallen P.J.M. (1981), "Judging Risks and Benefits in the Small and in the Large," Org. Beh. and Hum. Perf., 28, 235-271

Wagenaar, W.A. (1983), Menselijk Falen, Rijksuniversiteit, Leiden (Inaugural Address, Feb. 11, 1983).

Zielhuis, R.L., (1980) "Van represseif naar preventief onderzoek," T. Soc. Geneesk., 58, 381-388.

Psychological Aspects of Scenarios*

Helmut Jungermann

Institut fur Psychologie, Technische Universitat Berlin, Dovestr.1-5, D-1000 Berlin 10

INTRODUCTION

A major step in the evaluation of technologies, and policies concerning technologies, is often the development of scenarios describing possible futures under various conditions. The notion of scenario was introduced for military planning by Herman Kahn while at RAND Corporation in the 1950s. Since then, the scenario method has been applied in many areas of long-range policymaking. Typical examples are scenarios for U.S. agricultural policy (1), environmental policy (2), world energy supply strategy (3), helium storage policy (4), climate change (5), or aviation communications technology (6). A recent example are the four scenarios for energy policy in West Germany, developed by a Temporary Committee of the German Parliament. These scenarios describe alternative energy policy options for the next 50 years and represent a wide range of expectations, assumptions, goals, and values with respect to, for instance, population growth, lifestyle changes, economic growth, change of economic structure, and availability of energy resources and technologies for their transformation (7).

The increasing importance of the scenario method is documented by the increasing number of publications and conferences. The reason for the obvious rise in interest in scenarios is the long-time horizon of sound and comprehensive technology assessment and policy planning studies. Such studies often have to anticipate possible futures of more than 10 years, sometimes up to more than 50 years, because it takes such a

*A number of ideas expressed in this paper originated in discussions with Detlof von Winterfeldt. I would also like to thank Friedrich Fleischer, Lutz Hausmann, and Michael Lugtenburg for their contributions and comments.

NATO ASI Series, Vol. G4
Environmental Impact Assessment, Technology Assessment, and Risk Analysis. Edited by V. T. Covello et al.

long time for new technologies to develop and diffuse and for their impacts on environment and society to manifest themselves. The extreme uncertain, dynamic, and complex character of the futures to anticipate, however, makes most traditional procedures of futures research unfeasible. The purpose of scenarios is to provide descriptions of possible futures in terms of their basic and major events, trends, and activities.

A look into the literature shows that the discussion on scenarios is dominated by economists, engineers, and planners. Contributions from the social sciences, and particularly from psychology, are rare. The experience with TA studies, environmental and social impact assessment studies, and policy planning studies has shown, however, that two (necessary, not sufficient) requirements must be met if such studies are to have a chance in the political process. First, non-technical elements must be considered explicitly, e.g., change in society's value system, political controversies, demand shifts, and international relations. Second, experts must be able to explain and justify their judgments about possible futures, i.e., the assumptions about developments and the preferences with respect to values and goals that guide their judgments. Psychology offers some insights into how people--and experts are also people--use their knowledge and their intellectual abilities when confronted with tasks such as scenario construction. Hopefully, one can derive some consequences for the construction process that might improve presently available techniques.

The paper will be structured as follows: first, features and types of scenarios will be described briefly and a definition will be provided as a basis for the following discussion. Second, characteristics of presently used techniques for constructing scenarios will be described and evaluated to demonstrate the potential contribution of psychology. Third, I will present some research results from various areas of experimental cognitive psychology relevant to scenario construction. I will focus on the organization of knowledge, the salience of information, the outset of information search, and the orientation of information search. Finally, some conclusions will be drawn and potential research tasks will be suggested.

THE SCENARIO CONCEPT

The term "scenario" is not a very precisely defined concept. A number of interpretations exist, but most of them share the following features:

- Scenarios are <u>hypothetical</u>, i.e., they describe possible or potential futures

- Scenarios describe <u>processes</u>, i.e., they represent sequences of events over some period of time

- Scenarios consist of <u>states</u>, <u>events</u>, <u>actions</u>, and <u>consequences</u> which are conditionally or casually related

- Scenario elements are judged with respect to their <u>importance</u>, <u>desirability</u>, and/or probability

- Scenarios include the depiction of an <u>initial state</u> which usually but not necessarily lies in the present, and of a <u>final state</u> at a fixed time horizon

The purpose of developing scenarios is usually to provide a context in which alternative developments, e.g., the implications of a set of policies, can be explored. Scenarios define the parameters and assumptions which then can be studied in detail with more specific methods. In this sense, scenarios provide the framework for an entire technology assessment or policy evaluation study, embedding the other parts of the study within them. For instance, if population growth has been defined in a scenario as a major variable to be considered, a computer model may be used to predict population growth over some given time period with some given parameters.

It might be useful to distinguish the scenario concept from other related terms, such as forecasting and planning. Forecasting aims at predicting what will happen, i.e., at predicting the actual occurrence of future events, whereas scenarios describe potential or possible events and actions. Planning asks how we can make things happen, i.e., which set of actions would solve a given problem or achieve a certain goal, whereas scenarios describe which actions one could take and to which consequences these actions might lead. The distinguishing feature of scenarios, it appears, is their "hypotheticality" (8), and it illuminates the power of the scenario concept. Forecasting has turned out to be extremely difficult--some might say impossible--in technology assessment or in environmental and social impact assessment due to the number of variables and the complexity of their interactions. Planning requires that the goals to achieve or the problems to solve are given and well-defined, but this is mostly not the case. Scenarios are developed as perspectives over a long time range that helps structure the problem and decide about goals.

The features described above still leave room for a wide variety of scenario types. Ducot and Lubben (9), for example, have classified scenarios along three axes:

- Exploratory-Anticipatory: In exploratory scenarios, the question is what the effects are when certain causes are given, while in anticipatory scenarios one asks for the possible causes given certain effects

- Descriptive-Normative: Descriptive scenarios simply state an ordered set of possible events irrespective of their desirability or undesirability, while normative scenarios take the policymaker's values and interests explicitly into account

- Trend-Peripheral: A trend scenario represents the normal surprise-free forward operation of the present main trends, while a peripheral scenario is produced when the policymaker allows some parameters to take unlikely and extreme values

The resulting three-dimensional classification shows various different types of scenarios. For instance, the scenario for an LNG accident (10) could be considered as an exploratory, descriptive, and peripheral type of scenario. The energy policy scenarios developed in West Germany (7), on the other hand, could be classified as exploratory, normative, and trend scenarios. The typology is useful in that it highlights salient features of available scenarios as well as possibilities of constructing scenarios for problems not yet, or inappropriately, studied with the method.

Of course, the analytic structure for representing the problem to be investigated is in no way determined by this characterization of scenarios. One analytic structure, however, is a <u>decision tree</u>. A decision tree consists of acts-events-sequences in which the final events represent the outcomes to be considered. The logic of the decision tree is that the branches of the tree fan out without cross-connections and loops. Energy policy, for example, could in principle be described with a decision tree; i.e., many long, complicated sequences, ending up in all possible states of the system, depending on alternative policies and realizations of uncertain variables. In practice, however, the complexity of the problem makes such an approach unfeasible. But scenarios can be conceived as <u>paths in a decision tree</u>, as a subset of acts-events-sequences which appear reasonable or desirable under certain assumptions. The development of scenarios may be seen as a tool to handle the bushy-mess problem (11) by identifying a limited number of more or less possible and desirable sequences. An example illustrating this function very explicitly and formally is given by Keeney & Raiffa (12) in their Mexico Airport case in which they developed detailed scenarios for each of a class of alternatives for governmental action. The conceptualization of scenarios as paths in a decision tree does not imply that the sequences of elements linking the present to the future will always be explicit. Snapshot scenarios, for example, describe explicitly only possible future states, not the events and actions leading up to these states. However, such a description of future states is probably impossible without implicitly making assumptions about such sequences. Otherwise, they might come close to science fiction (which is not my subject in this paper).

We can now define a scenario as a consistent representation of the hypothetical development of a system from an initial state to some expected and/or desired final state; analytically, it may be conceived as a path in a decision tree.

THE CONSTRUCTION OF SCENARIOS

Techniques for constructing scenarios range from "soft," speculative and qualitative descriptions to "hard," rigorous and quantitative modeling. While the former often lead to results which are too general to help the policymaker identify and evaluate options, the latter often lead to results too specific to provide the policymaker with a broad picture of the possible future. From a superficial look at the literature, one can gain the impression, however, that there exists a set of efficient, sound, and standardized techniques for constructing scenarios from which one can choose, depending on the problem and the purpose. But a closer check reveals that most proposed techniques are only loosely defined procedures with little theoretical justification and empirical validation (13, 14).

A number of procedures have been proposed for constructing scenarios which have not been developed specifically for that purpose but represent more general tecniques to aid creative thinking. Examples are Brainstorming and Brainwriting (15), Synectics (16), Morphological Analysis (17), or the Delphi Method (18). The Delphi method has been discussed explicitly for scenario building (19); the principle remains the same, namely, to ask experts to generate events or express opinions about the events in an iterative group session until some consensus is achieved. Another technique to mention is GERT, a combination of network theory, probability theory, and simulation (20). It is a graphical evaluation and review technique which policy analysts can use to develop alternative scenarios and likely outcomes by changing assumptions, and see what effects result in the problem structure. The main problem with many techniques is that the specifics of the construction process are rarely ever spelled out; for instance, which questions open the process and how it should be framed, which elements are generated first and which second, how the time horizon is set, etc.

Mitchell, Tydeman, and Georgiades (21) reviewed 18 variants of scenario construction and came to distinguish nine characteristics in which these techniques may differ. One such feature is the degree to which relations between scenario elements are newly developed or taken from precanned structures. In an inductive, bottom-up approach, the elements are first generated and then, using matrix and graph theoretical procedures, their interrelations are gradually established. In a deductive, top-down approach, on the other hand, clusters of interrelated elements are used as input for the scenario development. Another feature is the handling of time: Single-period models specify one discrete possible state of the system at the end of a defined time interval, while multi-period models specify several time intervals and thus lead to scenarios with several stages. Other features are the character of the scenario elements (events or trends), the role of probabilities, the request of subjective information search and processing, and the purpose for which the scenario is developed. Mitchell et al. (21) suggest an approach that tries to incorporate the advantages of the other techniques. The main feature of their procedure is the explicit recognition of various levels of detail in the description of the future and the use of scenario-generation techniques within a process that generates structure at each level. I will not describe the entire procedure here but only mention the first step of it because it illustrates the crucial point of all procedures. The step is: "Search the environment for items relevant to the problem context," and various approaches to this step are suggested (e.g., Brainstorming). The rationale for this search is never made explicit; for instance, how the criterion for relevance is defined.

Several authors have tried to schematize the process of generating scenarios. A comparison of the suggested schemata shows a wide range in the number of steps, their specification, their order, etc., but also a number of commonalities, in particular feedback loops, iterative proceeding, and the general principle of first decomposing and then recomposing the issue at stake. However, a general schema for the construction process does not exist. That many experts have little confidence in available techniques for constructing scenarios is evidenced by the observation that in most technology assessment, environmental and social impact assessment, or policy evaluation studies in which scenarios were developed, the process of scenario construction itself is either not mentioned at all or is only given little consideration. I know of no study in which the process is explicitly documented. A reasonable assumption is that in these cases the experts did not apply any particular technique at all; this, for example, was true for the German Enquete Commission.

We can conclude that no generally accepted, theoretically and empirically sound scenario construction methodology exists. The existing techniques rely to a large extent on what is often called the disiplined intuition of the experts developing scenarios.

COGNITIVE FACTORS IN SCENARIO CONSTRUCTION

When people develop scenarios they have to use the knowledge that is stored in their cognitive system--even if it is the knowledge that they have to look for information elsewhere. What is of interest, then, is which characteristics of the human information processing system--also called memory--control this use of knowledge, or, more specifically, which factors and rules determine the choice of events and actions that are included in a scenario. To some degree, this depends on the nature of the task. Two features in particular distinguish scenario construction from other cognitive activities: first, the exploration of alternative futures over a long period of time is an extremely <u>complex</u> task. From the enormous number of possibly relevant elements, a reasonable set of events, actions, goals, and values must be selected. The cognitive problem is to retrieve "old"--but relevant--knowledge. Second, the exploration of the future requires <u>productive</u>, rather than reproductive thinking. People must go beyond the information in memory and be creative in order to generate events, trends, actions; or to imagine consequences that might not have been considered previously. The cognitive problem is to form "new"--and relevant--knowledge. Now, exploration of the future is not a completely unusual task for the cognitive system; whenever people consider new options they explore the potential implications of these options in the future. We therefore can learn something about the process of scenario construction--which necessarily and inherently is a subjective process in the beginning--by looking at people's cognitive functioning in similar situations.

Over the last twenty years, research in cognitive psychology has provided considerable insights into the human information processing system which we might well use to better understand, and possibly improve, scenario constructing processes. I will describe some research results that are particularly interesting. For a general presentation of relevant research see Hogarth (27).

Organization of Knowledge

We still do not understand very well how knowledge is represented in memory. Human knowledge is extremely extensive and everything seems to be related to everything else. However, one concept has become generally accepted as catching an important characteristic of the internal structured organization of knowledge--the schema concept. A cognitive schema is a complex unit of knowledge, consisting of generalized experiences--e.g., we have a (rather simple) "table" schema that represents the general features of all tables that we have seen or heard of, i.e., legs and a horizontal plate attached to them. A schema combines the information relevant to our experiences with tables, and whenever this schema is retrieved, all the information within it is made available. Schemata can also represent stereotypic sequences of events and actions, such as "dining in a restaurant;" this type of schemata is called "scripts" (22). Cognitive scripts work as bundles of expectations about the order as well as the occurrence of events and actions. For instance, the restaurant script consists, at the very minimum, of "entering, ordering, eating, paying, and leaving"--and we would be very surprised if somebody would pay before he or she ordered a menu.

Schemata are developed through our private and professional experience and represent a very efficient organization of knowledge, facilitating storing, processing, and retrieving information. But schematic thought is also subject to distortions and errors, i.e., it may result in incorrect temporal or causal assumptions about occurrences of relations within event sequences. For example, when Europeans come to the U.S. for the first time, they are usually puzzled when the hotel manager asks them to present a credit card or pay their bill when they check in; their temporal assumption about the order of events is different, and the causal assumption resulting from their new experience might be that they somehow do not look very trustworthy to the manager. More relevant to scenario construction is the finding by Bower, Black, and Turner (23) that subjects who have read a stereotypic story and are then asked to recall what they have read, tend toward false recognition of non-mentioned events; they apparently fill their memory gaps with elements that are part of their schema and thus should have occurred in the story. Another example illustrating the role of schemata is the study by Gettys, Maning, Mehle, and Fisher (24) that was concerned with factors influencing hypothesis generation. In one of their experiments, subjects had to generate hypotheses about the relations between a set of data presented to them. Subjects who had been informed about some possible hypotheses beforehand were strongly influenced in the subsequent hypothesis generation task with respect to the quantity and quality of hypotheses generated, as compared to subjects who had not received such prior information. When people have to generate hypotheses (or events), they are not able to retrieve all the relevant knowledge from memory that should be considered. The phenomena described are essential characteristics of what is called "deformation professionelle," i.e., the observation that in their professional life, people develop schematic ways of thinking which, although usually very efficient, can hinder them from seeing, understanding, or imagining new situations. A loss in cognitive flexibility is the price we pay for the economic principle of schematic thinking.

The organization of our knowledge in schemata is crucial for scenario construction because in this particular task we have no criterion for applying schemata. Usually, we apply schemata according to the similarity between schema and situation, i.e., how many features they share. However, when we develop scenarios we want to imagine sequences of events and actions that we have not and cannot have experienced before, i.e., dissimilarity is a criterion rather than similarity. Similarity of the imagined futures to past experiences is a criterion only insofar as it keeps us within the boundaries of plausibility and prevents us from fantasizing about unrealistic futures. But within these boundaries, scenarios should represent a wide range of possible futures which are dissimilar rather than similar to past situations. This is the whole idea of the method. But dissimilarity is at best a criterion ruling out the direct application of schemata, but does not provide a positive rule of using our knowledge. In such situations of uncertainty, we might expect that people will tend to either apply inappropriate schemata, i.e., use apparently similar existing schemata, or try to assimilate the task to available schemata, i.e., modify the task in some features.

Considering research on cognitive schemata, then, a distinction between an inductive, bottom-up approach and deductive, top-down approach to scenario construction (21) seems problematic. There is nothing like an unstructured retrieval and generation of elements which afterwards can be structured, because the information elements are always embedded in cognitive schemata. The retrieval of one element will lead to the activation of other elements of the same schema and preclude an independent sampling of other elements. Thus, scenario construction must be conceived as an iterative process of generating and structuring that strongly depends on existing schemata.

Saliency of Information

The obvious question is which factors determine where we start looking for information in our memory. This depends on which parts of our knowledge--i.e., which cells or cell networks--in the system are activated and thus salient. But when and why becomes information salient to a person, and thus a candidate for inclusion in a scenario or a cue that triggers off the search for other information to be included. Various factors have been studied of which I will mention only some that are particularly relevant for scenario construction. Maybe the most important factor is the frequency with which an event has been directly or indirectly experienced. For instance, events that are frequently well-publicized, such as large-scale accidents or crimes, become salient in our memory, and instances for such events can easily be retrieved. For example, airplane accidents, the discharge of toxic chemicals, a LNG transportation accident, receive wide and extensive media coverage making them highly salient--at least for a while. This cognitive availability of information can result in biased judgments of the probability of events: the more salient the event is, the more probable it is judged (25, 26). But this increases the chance of this event being included in a scenario. It is important to note, though not surprising, that the frequency with which an event is publicized is not necessarily highly correlated with the actual occurrence of that event.

Saliency also depends on the concreteness of information. Information that is more vivid, describing specific features or instances, is more easily activated (and therefore retrieved), than abstract information such as statistical summary data. For example, a picture of lungs devastated by tar impresses many smokers much more than cancer statistics. That is why environmental groups concerned about acid rain publish pictures of dead trees if they want to catch the attention of the public. Information coded in memory by vivid, concrete, and specific associations can become--sometimes disproportionately--salient and thus again have a better chance for inclusion in a scenario (27). Unless, of course, we assume that the experts who develop scenarios are free from such influences, but my experiences with expert groups does not support this assumption very strongly.

Sometimes, however, it is numbers that are more specific and concrete than images. For example, the energy policy scenarios developed by the German Enquête Commission focus on technical and economic variables such as economic growth, changes in economic structure, growth of basic materials industries, energy saving, energy use, etc. These are familiar terms that can be defined relatively easily and, even more important, quantitatively; these variables can thus be explained, discussed, and defended. Variables such as life style, private and social needs, esthetic values (and their changes over 50 years) are relatively abstract and hard to define such that agreement can be reached about the meaning of the terms and their measurement. Most assessment studies concentrate, for understandable reasons, on "hard" variables and data, but the danger of this orientation along salient information has become obvious in many cases. The controversies in the public, but also among experts, often were not about the technical or economic aspects of the problem but rather about the social and political implications of a new technology or project. I have the impression, however, that technology and policy assessment have learned from this experience (28).

Outset of Information Search

Closely related to the first two aspects I have discussed is the dependence of knowledge retrieval on the outset of information search in memory. Information in memory can be conceived as being organized in the form of a network of associations. In an associative network, pieces of information are represented as nodes connected by arcs. According to one theory, if a node is activated, activation spreads along paths through the network from the currently active node to other nodes, depending on the strength of activation and on the strength of associative relations (29). This implies for scenario construction that what will be generated in this process depends also on where the search for relevant information is started, i.e., which information is salient in the beginning--intentionally or unintentionally.

There is some research on the role of the goal as starting point in a problem structuring situation. In an experimental study, Pitz, Sachs, and Heerboth (30) investigated the effect of various techniques on the generation of options and objectives, based on means-ends-analysis and script theory. Groups were given a decision problem and a list of objectives related to this problem, and were then asked to generate as many reasonable options as possible. Three groups, for example, were given the list of objectives and asked to generate options that satisfied the objectives one at a time, two at a time, or all simultaneously. Although the differences were small, there was a tendency for the group focusing on one objective at a time only to produce more options than the group who tried to meet all objectives simultaneously. Jungermann, von Ulardt, and Hausmann (31) proposed a network model for the representation of goals and actions with several specific assumptions regarding the spread of activation through the network. In their experiment, they studied the effect of different degrees of goal explication on subjects' creativity and found that the number of actions generated increased with the degree of goal explication, thus supporting the subjects' information search process.

The relation to scenario construction is evident because an essential feature of scenarios is that they are implicitly or explicitly goal-oriented. Goals play a major role as general assumptions about values or as final states possibly occurring. They determine to a large degree how the alternative futures will be conceptualized. An example is the conflict about nuclear power. One side defines acceptability of this technology by referring to its safety in terms of the expected number of fatalities directly attributable to the technology, whereas the other side is more concerned with (un)acceptability in terms of increased centralization of state power and of future hazards from radioactive waste. Both sides, starting from quite different goals, conceptualize the problem very differently and thus explore quite different futures; there is no agreement about the scenarios which need to be developed in determining the acceptability/unacceptability of nuclear power (32).

One can generally observe that differences between scenarios developed by groups opposing each other differ not so much in detail but in their values and goals. It is interesting to note, however, that it is mostly the groups opposed to technological projects who formulate their goals and values explicitly, while the other side tends to assume certain goals and values more implicitly. But the consideration of possibilities is never, and probably can never be, independent of values and goals. Sometimes, goal orientation is an intentional process, sometimes not. While the intentional case is less difficult to identify, because the groups mostly make their values and goals quite explicit, the case where values and goals unintentionally set the bounds for exploring the future is often less obvious. In the latter case there might either exist a broad societal consensus, or society might have no particular interest in the problem (which, of course, may change very quickly). Of course, if one aims at optimizing scenario construction in the sense that the generated scenarios will be most comprehensive with respect to facts as well as values, the best strategy is to make the goals explicit for the scientific and for the social community, and to look at the problem from different goal perspectives.

Orientation of Information Search

We can look at knowledge retrieval from still another angle. We can ask for the orientation of information search, i.e., what kind of elements and relations people are actually looking for when engaging in scenario developing. They do not want to think up just any possible futures. For instance, it makes little sense to generate only scenarios which favor the same policy. Rather, one wants to generate scenarios such that their final states favor different policies. It might also be useful to generate scenarios that separate, perhaps for the same policy, final states that favor different values. And, of course, one wants to catch in the construction the relevant variables which matter for policymaking; scenarios should describe significant possible developments that offer the policymaker feasible alternative options.

It has been hypothesized by Kahneman and Tversky that the "joints" of a scenario (i.e., the nodes of the path in the decision tree that are generated) will be events that are low in redundancy and high in causal significance. They define a non-redundant event as representing a local minimum in the predictability of the sequence, a point at which significant alternatives might arise; and a causally significant event as one whose occurrence alters the values that are considered normal for other events in the chain that eventually leads to the target of the scenario (33). If this rule should govern the selection of variables for a scenario, it would help achieve one of the goals stated above, i.e., identifying relevant and significant components of possible futures. However, as Kahneman and Tversky point out, the rule might also lead to a bias for scenarios in which dramatic events mark causal transitions. Events that are produced by slow and incremental changes might be neglected. An illustration for that danger is perhaps the focus in the risk assessment literature on risks from hazardous technologies leading potentially to catastrophes and the relative neglect of risks from technologies producing slow and creeping damages.

The notion of causality is evidently of interest for the orientation of information search and generation. Variables that are considered causally relevant for future events will be looked for because they offer an opportunity for action. But what makes variables appear causally relevant? According to Einhorn and Hogarth (34) causal relevance of variables is assessed, for example, by the following cues:

- The temporal order of events--because causes are assumed to precede effects, the search for relevant causal variables focuses on only those events that preceded some effect and thereby reduces the number of variables considered

- Predictive validity--the higher one variable can predict another, measured by the correlation coefficient, the more causally related to this variable does it appear

- Similarity, and particular metaphors and analogies are also cues to judge causal relevance--the use of a metaphor or an analogy can induce a causal interpretation of otherwise apparently irrelevant phenomena

- The number of alternative explanations--the greater the number of competing causal variables, the less the causal relevance of some particular variable for the effect

The last idea is closely related to the notion of causal schemata, as discussed by Tversky and Kahneman (35). These authors examine the role of causal schemata for the judgment of probability of some future event. The easier it is to construct a scenario (as a causal schema) leading to the event, the higher is its estimated likelihood. But it is difficult in complex situations to construct sequences of variations of interacting factors over a long period of time. People therefore consider only very simple and available scenarios, in which many factors do not vary at all and only the most obvious variations take place, and interacting changes are rare (35). The implication is that events that can be produced by a number of unlikely ways will often not be included in scenarios. Ten years ago most Americans would probably have had great difficulty in imagining a plausible causal sequence that would lead to the impeachment of a President; a few years later this possibility was highly available in the minds of almost everyone.

These findings and ideas are relevant for a discussion of two techniques often used to develop scenarios, the forward and the backward approach. The <u>forward</u>, exploratory construction process focuses on separating crucial events and actions that will produce more or less desirable outcomes. It is usually fact-oriented, while goals and values are only implicitly defined as boundaries of the whole range of scenarios. The backward anticipatory process focuses on the goals to be achieved and asks for the policies that would lead to these states. It is value-oriented, because the goals are explicitly defined. The forward approach makes the prognostic construction of sequences easier than the backward approach because the latter asks to consider a multitude of sequences that could lead to the final state. Both techniques will thus usually not produce the same scenarios. The discrepancies, however, can provide valuable insight, and the best strategy will often be to combine both techniques in an iterative procedure (36).

What Is a Good Scenario?

I have described a number of cognitive factors that inevitably play a role in the scenario construction process. On the one hand, the cognitive processes involved provide us with extraordinary abilities for exploring complex possible future developments; on the other hand, they are subject to influences that might impair the quality of the scenarios developed. But how can we find out how good a scenario or a set of scenarios actually is? This is a crucial point for the scenario methodology, because it cannot use a falsification criterion: One cannot make an error in developing scenarios because there are no right or wrong scenarios. Forecasts can turn out to have been wrong, and the forecasting model can then be changed; plans can turn out to be unfeasible, and better plans can then be made. But scenarios cannot fail; by definition they represent possible futures, none of which must actually occur. There is no logically correct solution and no environmental criterion for validating a scenario.

The two criteria most often mentioned for judging the quality of scenarios are consistency and plausibility. However, I have never seen them explicitly applied in the literature, and there are indeed no rigorous tests for consistency (leaving aside gross logical errors) and plausibility. We have here the same problem as in the scenario construction process itself. We can judge consistency only on the basis of standards and rules proven by experience and accepted in the present, and we can judge plausibility only on the basis of our present theoretical and empirical knowledge. But because in scenarios we want to describe futures that transcend our present experience and knowledge, these are shaky grounds. That is to say, the judgment of consistency and plausibility is influenced by the same cognitive factors as is the construction of scenarios itself.

The effects described in this paper can only be partially mitigated by letting scenarios be developed in groups who directly or indirectly work together, as, for example, in a Delphi group. It is true that in groups the construction process can be improved. Members can--and should--correct each other and bring in their own viewpoints, based on their individual personal experience and professional expertise. In this sense, heterogeneity of a group is important for stimulating a creative discussion. On the other hand, groups only work effectively if there is a minimum of homogeneity, i.e., if the group members share to some degree experiences and attitudes, expertise, and perspective. This supports the construction process but may also restrict the range and direction of thinking in much the same way as cognitive factors influence the individual construction process. Actually, rather than correct each other, people might disregard their personal opinion and knowledge to agree with the group, as exemplified in Janis' analysis of President Kennedy's staff handling the Bay-of-Pigs affair when almost everybody became a "victim of group-think" (37). And, second, as soon as certain schemata, perspectives, and approaches of thinking have been introduced into a group it is cognitively difficult (although of course not impossible) for other group members to adopt other views. For example, framing options about some medical technologies in terms of lives lost might make it difficult to consider the problem in terms of lives saved (38). The German Enquête Commission that had to develop energy policy scenarios seems to have represented an almost ideal composition, covering a wide range of values and expertise without being too heterogenous. But still, an environmentalist group of experts criticized the scenarios developed and suggested other scenarios, putting more emphasis on ecological aspects. In general, constructing scenarios in a group can, but must not, mitigate the effects of individual cognitive factors. A group guarantees in no way the quality of the scenarios developed.

The judgment of the quality of a scenario, however, does not rest only on its consistency and plausibility. It also depends on the context in which, and for which, it is developed. Scenarios might be primarily constructed to identify policy options, or to provide a basis for reaching a consensus in major aspects among conflicting groups, or to facilitate the communication of a complex problem to a non-expert public. And a set of scenarios might serve one of these functions very well and the others very poorly.

The literature generally neglects the problem of how to evaluate scenarios, and this probably reflects that the problem is rarely tackled in practice. Considering the increasing use of the method, this is really surprising. Some kind of evaluation of the method will be required if it is to survive. One possibility would be a multi-attribute approach which catches the multitude of relevant evaluative criteria (36, 39).

SUMMARY AND CONCLUSION

Scenarios are thought to provide a basis for thinking, discussing, and deciding about technology policies, and about environmental or social developments, by offering a framework within which options, activities, trends, tasks, and goals can be made explicit and coherent. They represent simplified models of possible futures and help focus the attention on crucial variables and decision points.

There is no established scenario construction methodology. The techniques proposed have little theoretical justification and empirical validation. One rarely finds a detailed description of any technique, and even less of a technique's actual application. The available techniques leave wide room for expert judgment and intuition. Because the task of scenario construction requires heavy reduction of complexity and highly creative thinking and because there are no rigorous criteria for the quality of scenarios, the cognitive processes involved operate under particular uncertainty. Organization of knowledge, saliency of information, and outset and orientation of information search are important aspects to take into account.

To better control the potentially constructive as well as destructive effects of important characteristics of the human information processing system, we ought to focus on the _process_ of scenario construction. Without confining the intuition of experts when developing scenarios, we should operationalize what is meant by _disciplined_ intuition. This requires, first, studying more systematically and thoroughly the ways in which cognitive factors can influence scenario construction. Experimental as well as case studies are needed. Second, it requires the development and testing of techniques that use optimally cognitive competence and counteract cognitive biases. The available knowledge from cognitive psychology already provides some ideas in this respect. For example, one could design various types of sensitivity analysis (e.g., assuming different goals or framing problems differently), procedures to elicit the underlying value structures (e.g., value trees), or approaches to build acts-events-sequences (e.g., combining forward and backward construction).

REFERENCES

(1) Stanford Research Institute, Center for the Study of Social Policy (SRI/CSSP), Handbook of Forecasting Techniques. Prepared for Institute for Water Resources, U.S. Army Corps of Engineers, Fort Belvoir, VA (1975).

(2) Environmental Protection Agency (EPA), Alternative Futures for Environmental Policy Planning: 1975-2000 (Washington, DC, U.S. Government Printing Office, 1975).

(3) Häfele, W., IIASA's World regional energy modelling, Futures, 12, 18-34 (1980).

(4) Epple, D. and Lave, L.B., Helium: Investments in the future, The Bell Journal of Economics (1980).

(5) Epple, D. and Lave, L.B., Planning for climate change, Methodology of Climate Impact Assessment (Scope, 1981).

(6) Chen, K., Jarboe, K., and Wolfe, J., Long range scenario construction for technology assessment, Technological Forecasting and Social Change, 20, 27-40 (1981).

(7) Deutscher Bundestag, Bericht der Enquête Kommission Zukünftige Kernenergiepolitik: Drucksache 8/4341 (Bonn, Presse- und Informations-zentrum, 1980).

(8) Häfele, W., Hypotheticality and the new challenges: The pathfinder role of nuclear energy, Minerva, 12, 303-322 (1974).

(9) Ducot, C. and Lubben, H.J., A typology for scenarios, Futures, 12, 51-57 (1980).

(10) Keeney, R.L., Siting Energy Facilities (New York, Academic Press, 1980).

(11) Raiffa, H., Decision Analysis (Reading, MA, Addison-Wesley, 1968).

(12) Keeney, R.L. and Raiffa, H., Decisions with Multiple Objectives: Preferences and Value Tradeoffs (New York, Wiley, 1976).

(13) Mitchell, R.B., Tydeman, J., and Georgiades, J., Structuring the future - application of a scenario-generation procedure, Technological Forecasting and Social Change, 14, 409-428 (1979).

(14) Lippold, H. and Welters, K., Szenario-Technik (Berlin, Zentrum Berlin für Zukunftsforschung, 1976).

(15) Osborn, A.F., Applied Imagination (New York, Scribner, 1953).

(16) Gordon, W.J.J., Synectics (New York, Harper & Row, 1961).

(17) Stein, M.I., Stimulating Creativity, Vol.I,II (New York, Academic Press, 1974, 1975)

(18) Linstone, H.A. and Turoff, M., The Delphi-Method: Techniques and Applications (Reading, MA, Addison-Wesley, 1975).

(19) Sutherland, J.W., Architecting the future: A Delphi-based paradigm for normative system building, in H.A. Linstone and M. Turoff, eds., The Delphi-Method: Techniques and Applications (London, Addison-Wesley, 1975).

(20) Moore, L. and Clayton, E., GERT Modeling and Simulation (New York, Petrocelli, 1976).

(21) Mitchell, R.B., Tydeman, J., and Georgiades, J., Structuring the future - application of a scenario-generation procedure, Technological Forecasting and Social Change, 14, 409-428 (1979).

(22) Schank, R.C. and Abelson, R.P., Scripts, Goals, Plans and Understanding (Hillsdale, NJ, Lawrence Erlbaum, 1977).

(23) Bower, G.H., Black, J., and Turner, T., Scripts in text comprehension and memory, Cognitive Psychology, 11, 177-220 (1979).

(24) Gettys, C.F., Manning, C., Mehle, T., and Fisher, S., Hypothesis generation: A final report of three years of research, Technical Report 15-10-80 (Norman, OK, University of Oklahoma, 1980).

(25) Tversky, A. and Kahneman, D., Availability: A heuristic for judging frequency and probability, Cognitive Psychology, 5, 207-232 (1973).

(26) Combs, B. and Slovic, P., Causes of Death: Biased newspaper coverage and biased judgments, Journalism Quarterly, 56, 837-843 (1979).

(27) Hogarth, R., Judgment and Choice: The Psychology of Decision (Chichester, etc., Wiley, 1980).

(28) Paschen, H., Summary and evaluation of the results of the symposium "Internationales Symposium über die Rolle der Technologiefolgenabschätzung im Entscheidungsprozeß", 19.-21. Oktober 1982 in Bonn (Berlin, Umweltbundesamt, 1983).

(29) Collins, A.M. and Loftus, E.F., A spreading-activation theory of semantic memory, Psychological Review, 82, 407-428 (1975).

(30) Pitz, G.F., Sachs, N.J., and Heerboth, J., Procedures for eliciting choices in the analysis of individual decisions, Organizational Behavior and Human Performance, 26, 396-408 (1980).

(31) Jungermann, H., von Ulardt, I., and Hausmann, L., The role of the goal for generating actions, in P.C. Humphreys, O. Svenson, and A. Vari, eds., Analyzing and Aiding Decision Processes (Amsterdam, Budapest, North-Holland/Hungarian Academic Press, 1983).

(32) Humphreys, P., Value structures underlying risk assessment, in H. Kunreuther, ed., Risk: A Seminar Series (Laxenburg, International Institute for Applied Systems Analysis, 1981).

(33) Kahneman, D. and Tversky, A., The simulation heuristic, in D. Kahneman, P. Slovic, and A. Tversky, Judgment under Uncertainty: Heuristics and Biases (Cambridge, MA, Cambridge University Press, 1982).

(34) Einhorn, H.J. and Hogarth, R.M., Prediction, diagnosis, and causal thinking in forecasting, Journal of Forecasting, 1, 1-14 (1982).

(35) Tversky, A. and Kahneman, D., Causal schemas in judgments under uncertainty, in M. Fishbein, ed., Progress in Social Psychology (Hillsdale, NJ, Lawrence Erlbaum, 1980).

(36) Jungermann, H. and von Winterfeldt, D., Cognitive aspects and analytical methods of risk identification: The construction of scenarios, in H. Jungermann, D. von Winterfeldt and R. Coppock, Analysis, Evaluation and Acceptability of Hazardous Technologies and their Risks: A Workshop Report, IIES-dp 82-2 (Berlin, Internationales Institut für Umwelt und Gesellschaft, 1982).

(37) Janis, I.L., Victims of Groupthink (Boston, Houghton Mifflin, 1972).

(38) Tversky, A. and Kahneman, D., The framing of decisions and the psychology of choice, Science, 211, 1453-1458 (1981).

(39) Edwards, W., Guttentag, M., and Snapper, K., Effective evaluation: A decision theoretic approach. In M. Guttentag (ed.), Handbook of evaluation research. Beverly Hills, CA: Sage Publications, 1975.

SECTION II: Psychological Perspectives

Part II: Case Studies

The Empirical Investigation of Citizens' Preferences with Respect to Four Energy Scenarios

O. Renn, H.U. Stegelmann, G. Albrecht, U. Kotte

Program Group: Technology and Society, Nuclear Research Centre, Post Office Box 1913, D-5170 Julich-1, Federal Republic of Germany

INTRODUCTION

"How can polarized social conflict involving risk be reduced and replaced by an atmosphere of trust and mutual respect among the opposing parties? What research is needed to design an environment in which effective multiway communication, constructive debate, and compromise can take place?" (1)

This quote taken from a list of 10 relevant research questions in the field of risk and technology assessment refers to one of the most promising and demanded, but at the same time most ambitious tasks of social science research: the promotion of social integration. This task is based on the assumption that science could not only provide data for clarifying relationships between given phenomena, but could also offer procedural advice on how to accomplish means of conflict resolution. But can science, and in particular social science, contribute to conflict management? Is science the appropriate instrument for interfering with political debate and facilitating the formation of compromises? Is the role ot science overstressed by aiming toward a mediating function between social interest groups?

Most sub-systems of society perceive themselves as central agents for the well-functioning of society in general. They consider their contribution an essential effort to support the present coherence of social forces and a decisive means for shaping the future performance of society. The sub-system of science might also overstate its social influence by demanding a superior position as an arbiter in social conflict.

What kind of services can science offer to the policymaker in order to facilitate conflict resolution? There are basically

NATO ASI Series, Vol. G4
Environmental Impact Assessment, Technology Assessment, and Risk Analysis. Edited by V.T. Covello et al.

four possible contributions of utilizing scientific expertise for political consulting. 1) Discussion of false inferences: Using the methods of scientific inquiry it is possible, at least in principle, to investigate the likelihood of various consequences of proposed policy actions and to exclude statements which can be proven wrong; 2) Modeling of causal structures: Since reality is characterized by a high degree of complexity, any policymaker is unable to overlook the consequences of his political action. With the aid of an abstract model, scientists can create a simplified picture of causal interdependencies and provide some insights into the probable consequences of policy options; 3) Forecasting of future developments and consequences: In addition to causal relationships between policy actions and social consequences, scientists can describe the direction and probability of technological and social change by combining data from past experiences and revealing general trends of social and economic development. They might also provide information by which politically desirable states of society can be achieved; and, 4) Providing arenas for conflict resolution: Scientists can propose procedures and formal methods for facilitating conflict management. The process of decisionmaking can be designed in such a way that the affected interest groups perceive a fair chance to participate in the final decisionmaking process and the general public can gain the feeling that their concerns have been taken seriously. Such a mediating role of science can only be accomplished if the arena is being shaped by scientists without interfering in the arbitration process by means of their own values and preferences (2).

These four principal contributions of science to policymaking have been challenged by the actual performance of scientific expertise in the political arena. First, experts themselves are bound into the network of social interests and often act as advocates of policies or ideologies. Second, the separation between values and facts which underly the cooperative function of policymakers and scientists is rather artificial and dubious. As a result, either the scientists come up with only those facts that support their own preferences (technocratic planning) or the politicians use factual expertise merely as a justification of preformulated political decisions. Third, interest groups are more interested in scientists as advocates (of their own case) and less interested in scientists as mediators. By utilizing scientific expertise to legitimize one's own claims, it is more probable to gain a higher proportion of social influence at the end. It can also be expected that one-sided scientific support reinforces the motivation and mobilization of all group members (a vital interest for survival, if the interest group is not yet institutionalized). Only when groups involved in the conflict realize that polarization does not help any of them to proceed and when policymaking is paralyzed might scientists be called in to initiate a first move.

Overlooking the past experience of policy consulting, the results are rather discouraging--even simple facts are disputed among scientists. They construct models to back-up their own political point of view; the scope of forecasting ranges from predicting extinction of all life on earth to a coming age of paradise; and, no one seems to show any interest in the integrative role that science is able to play. On the other hand, a German official of the Federal ministry for Research and Technology stated recently: "We knew in advance what was right or wrong. But since we faced public opposition, we spent millions of dollars for scientific consulting. But what we got is even a bigger mess. Now we are in real trouble!" Instead of reconciliation, a lot of conflicts in society were aggravated by engaging scientists as consultants.

There is no easy way out. On one hand, the potential of scientific expertise has not been utilized in a proper way; on the other hand, there is an increasing demand for scientific expertise in the policymaking process. This is due to four reasons: 1) Modern industrial societies have developed such a high level of complexity that merely "guessing" remains as the only viable alternative to scientific inquiry. Common sense or political experience are not sufficient anymore to foresee the consequences of political actions; 2) Modern industrial societies are characterized by a high degree of interdependencies between economic, social, and political parameters resulting in a high sensitivity of the system for minor changes. Policy actions in one field might affect the outcomes of a completely different social segment. This requires more scientific research and advice; 3) Because of the high degree of industrialization (connected with overpopulation), the potential of modern technologies and the interaction between political decisionmaking and the physical and social environment, the magnitude of possible outcomes has reached a level where global catastrophes can be caused by human activities. Ecological collapse, atomic war, famine, or systems breakdown are the most serious problems in the modern world. Since major political decisions might create irreversible consequences which make such catastrophes more probable, it is vital that any social system has access to at least a rough scientific estimation of the magnitude of consequences which are connected with one or the other policy option; and 4) As the public in industrialized countries is better educated than in former times and is thus aware of the complexity, interdependencies, and increase in the potential magnitude of consequences, they demand more scientific reasoning in the policymaking process and show a high degree of awareness to arguments which are put forward to policymakers by scientists. At the same time, they are not willing to accept all the decisions that are formed by the legitimate policymaking bodies. They claim that their interests have to be considered in the process of decisionmaking and hence they demand more means of public participation. The incorporation of public concerns and the implementation of participation requires scientific consulting to design procedures and arenas which guarantee high quality of the decision output as well as social stability and public transparency.

Hence, modern societies are in urgent need of all four contributions that science can offer in principle. But we have not yet developed viable ways of structuring the necessary cooperation between science and politics. There have been excellent records in small-scale applications of scientific consulting, when defined problems have to be solved within local communities. But on a national or even international level, the past experiences are less promising. In particular, models are missing of how to make use of the potentials of scientific expertise, without being caught in the trap of ideology, technocracy, or post-rationalization.

One attempt to structure a model is the study of social compatibility of energy supply systems which is carried out by the research group "Technology and Society" of the Nuclear Research Center, in Julich (Federal Republic of Germany). The following article and the contribution by H. Jungermann, D. von Winterfeldt, and R. Keeney (all in this volume) are devoted to this approach.

ENERGY POLICIES: A SERIOUS CHALLENGE FOR SOCIAL SCIENCE CONSULTING

The "Julich Social Compatibility Study" is based on the general assumption that the contribution of Social Sciences to facilitating conflict resolution can only be effective as long as the role of the scientist is confined to the process of catalysis--a term borrowed from chemistry meaning, in our context, "aiding or speeding up the social process of finding compromises without interfering with one's own values or preferences." Based on this general hypothesis five more refined assumptions were formulated which directed the empirical conceptualization of our approach: 1) Science and politics pursue different interests when cooperating with each other. The ideal scientist has the ambition to help the politician to find the best solution to a given problem, hence he looks for an optimization rule to choose between options according to expected outcomes. The ideal politician, however, wants to maximize public support and to retain the general trust in the principal ability of the political elite to handle complex issues and to master social conflicts. Hence, in his sense, the justification of a decision is more important than the quality of its outcomes. This divergence of interests is inherently based on the specific functions that both subsystems perform within the overall societal system, regardless of whether more personal goals, like gaining reputation, prestige, power, or money are involved. So the first requirement of our approach was to provide not only facts and models to assist the process of choosing the right option, but also to address the problem of legitimizing and implementing the best choice. If this problem turns out to be too difficult, the process has to be iterated until a second best solution is to be created with both aims in mind. This could serve a dual purpose: improve both political efficiency and progress in scientific rationality. 2) Following our general requirement that a scientist should not insert personal values and preferences into the consulting task, some kind of a distinction between facts and values has to be made. At least, we assume that scientists belonging to different stake-holder groups can reach agreement on basic properties of objects or policies, if those properties can be measured physically (amount of emissions, cost, technical performance, reliability, and so forth). Furthermore, we are convinced that even the spectrum of possible conclusions referring to uncertain events of future effects can be restricted to a limited range, provided that scientists with different value backgrounds are willing to cooperate and that the situation in which the judgments have to be formulated facilitates agreement. It should be noted that we are not demanding value-free research or neutral scientists. Instead, we are pursuing a procedure in which factual knowledge is based on the agreement among scientists with various value commitments. 3) With respect to the value side, we consider value-based preferences of any citizen as important as any preferential judgment expressed by politicans or professionals. According to the democratic goal "one man-one vote," the sum of revealed public preferences should have the highest legitimative power. 4) In order to measure preferences of citizens in a valid way, two assumptions must be made: first, that the amount of knowledge which is necessary to express national preferences (referring to the expected consequences and their probabilities) can be gathered and transferred to the public; and second, that a common set of evaluative criteria can be established. The knowledge has to come from the experts in this field (c.f. assumption 2). The criteria can best be formulated by those who are regarded as stakeholders in the respective issue. 5) Any compromise aiming towards conflict resolutions can only be implemented, if at least the moderate stakeholder groups find a common agreement and the general public gains the feeling that their concerns were handled in a fair and serious manner.

Based on these five assumptions, our complex model of consulting in energy policies consists of three basic elements:

Step 1: Identification and selection of concerns and evaluative criteria.
For this step, the technique of value-free analysis was used to elicit the general concerns and values present in contemporary German society. We interviewed representatives of nine leading social interest groups and asked for their values and concerns with respect to energy policy. The objective of this exercise was to collect all relevant values and concerns and to construct a common catalogue of criteria which all German stakeholder groups were able to agree on. This catalogue was used as a general yardstick for the assessment of probable consequences of each energy option (Step 2) and as a reference list to elicit the general preferences of the public (Step 3). (The concept, procedure, and the results of this analysis are described in more detail by R. Keeney and D. von Winterfeldt in their articles.)

Step 2: The identification and measurement of impacts and consequences related to different energy options.
The criteria derived from the value tree were operationalized and transferred into indicators. Those indicators were formulated in such a way that they could be applied to energy systems as well as energy scenarios. Experts with different value preferences were asked to judge various energy systems and energy scenarios according to each indicator. In a joint venture, the experts who had strongly disagreed on one or the other indicator were urged to find either a compromise or to define a range of possible solutions. This task was facilitated by an informal situation of face-to-face interaction. For the social and political impacts, a delphi method was used to reach final agreement. In addition, some highly distinguished scientists of well-known insti-

tutions, like the Prognos Consulting Corporation in Switzerland, brought in their expertise to validate our ratings. The indicators and the final measurements were listed in a report without constructing indices or assigning weight to any of the 71 indicators (3). Step 2 is also described by Jungermann in his article on the two social compatibility studies.

Step 3: Aggregation and weighting of the imppact profiles by randomly selected citizens:
In a four-day seminar, randomly selected citizens received all the necessary information to understand and handle the impact profile. They were also confronted with scientific and political arguments (via hearings) in support for each of the possible energy options. In order to restrict the scope of options, three to five local energy supply scenarios and four national scenarios (constructed by the Enquete Commission of the German Federal Parliament) had to be evaluated. It was the task of the participants to assign weights to each subcriterion and criterion of the indicator list (knowing the range of all scientifically derived ratings for each option) and to evaluate the overall performance of each scenario. Finally, they had to give recommendations with respect to the demanded priorities in future energy policymaking.

The following chapters will only deal with the third element: the elicitation of public preferences. This part of our study is not only the most ambitious one, but encompasses a very new method of preference measurement: the concept of planning cells. As a matter of fact, this concept was used in our study for the first time on a national level. The instrument of planning cells claims to serve two basic purposes: measurement of public preferences and direct citizen participation.

"THE PLANNING CELL" AS POLICYMAKING INSTRUMENT IN THE DECISIONMAKING PROCESS

The procedures for participation of the public in complex planning tasks which have been introduced and practiced up to now remain unsatisfactory in many respects. Peter C. Dienel has developed a new instrument for citizens' participation, the so-called "planning cell." The research unit "Citizens' Participation and Planning Procedures" at the University of Wuppertal has up to now tested this procedure in the field of siting, town planning, and urban renewal. This model claims to open up new possibilities of participation for the private citizen and to improve the efficiency of decisionmaking in the planning process.

Dienel defines the planning cell as "a group of citizens who are selected by a random process and are given paid leave from their workday obligations for a limited period of time in order to work out solutions for given, soluble planning problems with the assistance of advisors on procedure." (4)

The group, random selection, time limit, and the advisors on procedure are to be regarded as the significant structural elements in the planning cell. A group of the citizens actually means a small group of about 25 people who work on the predefined task in a group process. Since the citizens involved have been selected by a random procedure, they are not individually concerned with the planning problems to be solved. To encourage them to participate, they are assigned the socially highly esteemed role of a "consultant" in the public planning process. The seriousness of the planning task to be solved is also made clear by the honorarium which the citizen receives for his function as a "consultant." The limited participation period prevents the citizen from being alienated from his real social role; he only changes his perspectives for a brief period. Moreover, this limited period prevents personal interests specific to the proceedings from arising and possibly hindering planning appropriate to the task. The essential task of the advisors on procedure consists of ensuring that the necessary specialized information is presented and processed, and making an effective group process possible.

Figure 1 illustrates the course of a planning cell's work. The flow-chart shows that evaluation procedures take place in all phases of implementation. Individual evaluations, evaluations by the working groups and plenary sessions are collected with the aid of standardized evaluation questionnaires. The completed evaluation questionnaires represent in a certain way the skeleton output relevant to the planning task. The evaluation and processing of these results according to statistical methods is the main concern of the final working phase. The so-called "citizens' report" is compiled and published as the last step from the insights and recommendations obtained with respect to solving the problems.

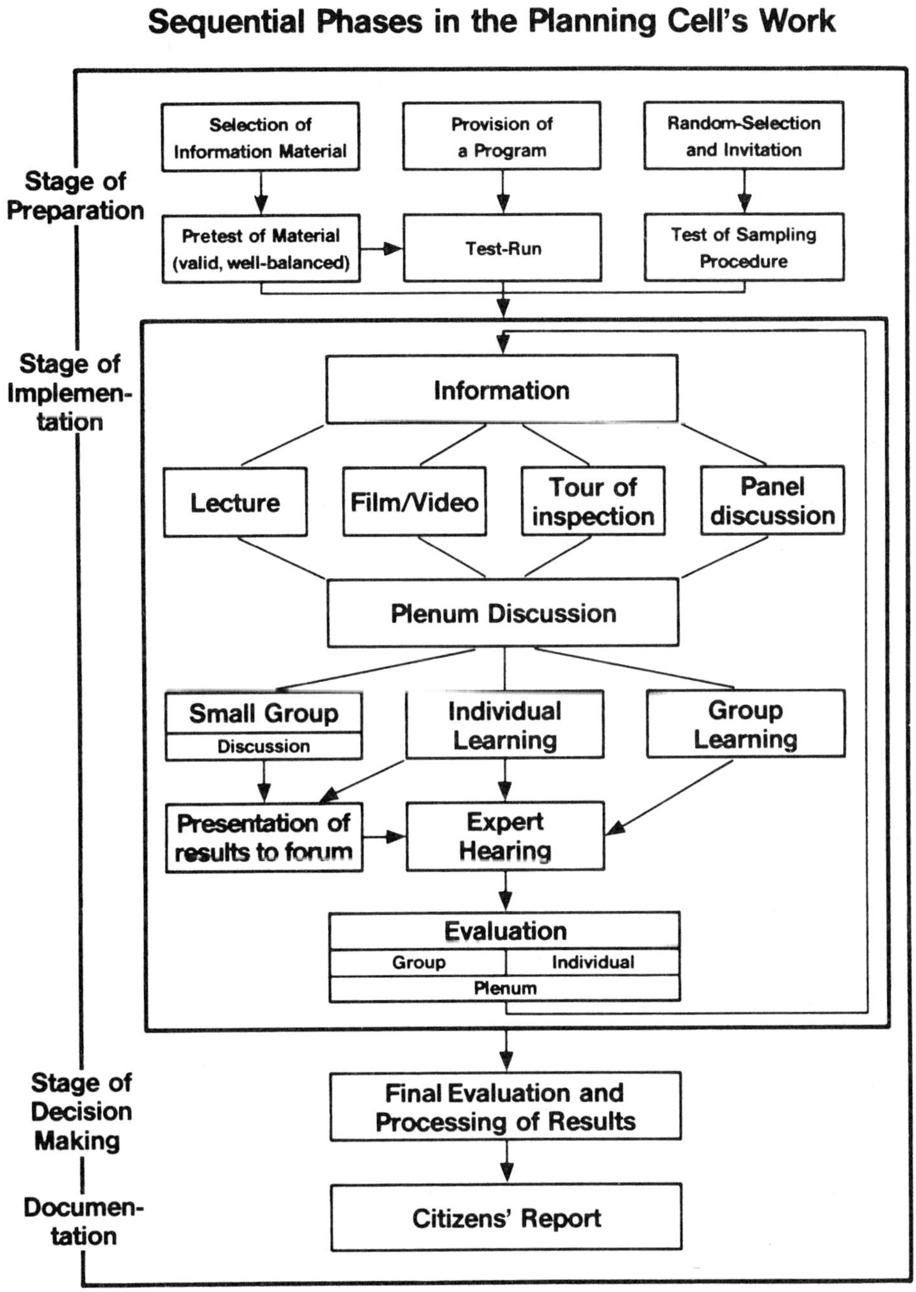

Fig. 1: A Flow-Chart of the Planning Cell Procedure

An advantage of the planning cell is the random distribution of participants, for this selection procedure distributes the changes of participation evenly. The social selectivity (which in other procedures leads to a preponderance of members of the middle and upper classes) is thus offset; citizens of all classes are involved in a planning cell procedure. It is even more significant that by means of random selection a panel results in which the factor of direct interest is largely ruled out. In contrast, other technocratic and popular planning procedures are very strongly influenced by pressure groups; they are thus susceptible to strategic maneuvers and are often unable to form a consensus. The planning cell specifically offers those not concerned and thus not directly interested a possibility of mediation between polarized interest groups. In this way the unproportionate influence of pressure groups is avoided and the production of solutions to the problem is also encouraged.

An increase in the efficiency of complex planning processes is derived from the following points: 1) It is founded on a broad database due to the inclusion of heterogenous social positions; 2) The tendency of traditional policymaking bodies to no longer fulfill the planning tasks in a way appropriate to the tasks themselves but rather to the benefit of their own institutions, increased by bureaucratization and technocratic planning structures, is largely mitigated by the random selection and limited duration of the planning; 3) New, unpredictable solution approaches can be achieved on the basis of immediate practical experience and special needs of the citizens; and, 4) the inclusion of social values, which takes place by introducing the subjective values of the individual citizen, contributes to the ability to form a consensus on the solutions discovered.

In order to guarantee comparability of results, the information and evaluation procedures and the whole sequential situation must be standardized. This parallel connection neutralizes randomly occurring biases and distortions and thus increases the representativeness of the planning output. This then provides the planning process with increased immunization against external pressure. The implementation of several planning cell runs with similar results improves their validity, since even in the case of different constellations of interests, the same solutions might be worked out. Finally, this procedure is intended to render complex planning processes more controllable.

This last advantage seems to be doubtful at least, because the planning cell depends to a large extent on the amount of given information and the preliminary work by the sponsoring agency.

Moreover, it can hardly control the process of decisionmaking itself.

The complex of information is in general a problem for this procedure. The quality of the output relevant to planning is decisively dependent upon the quality of the factual information fed in. There is a great danger of manipulating the results via the selection and provision of information (5).

Another negative aspect should be mentioned. Probably the citizens will tend to prefer traditional solutions since as representatives of the "silent majority" they will hardly be capable of suggesting innovative approaches. The extensive integration of the procedure into the official planning processes is also criticized because this could lead to a rapid adaptation to existing schemes of decisionmaking (6).

The random selection of the citizen is at the same time an exclusion procedure. This results in the disregard of an important principle of citizens' participation, namely, that every interested party can participate in the planning.

Even more important are the constitutional misgivings which imply that a type of "ancillary parliament" could be constituted by the planning cells (7). Even though it is clear that the citizens' report only makes recommendations and that parliamentary competence will not be encroached upon, it cannot eliminate the emerging problem of the double legitimation of political decisions.

Although the weakness of the procedure mentioned above should not be ignored, it still seems that the "planning cell" model offers specific advantages which justify its application in our project. In our opinion these are particularly to be found in the acquisition of public preferences, the representation of social values, and in the fact that those involved are not committed to any interest group.

This procedure can assist in gathering insights which can contribute to improving the legitimation and also the efficiency of planning processes.

THE BASIC FUNCTIONS AND TASKS OF PLANNING CELLS IN THE JULICH "SOCIAL COMPATIBILITY" STUDY

In addition to the assumptions and requirements which were stated in the first chapter, it is necessary for the application of the planning cell concept to consider the conditions and characteristics of the political arena in which energy policies have to be formulated and implemented. In contrast to some other political arenas, the energy scene is characterized by the following three major features: 1) lack of unanimity among the scientific experts about facts; 2) lack of credibility with respect to scientists and policymakers on the side of the public; and 3) lack of readiness of the stakeholders to engage themselves in the process of forming a compromise.

Any participation model aiming toward improving the decisionmaking process has to take these problems into account. A functional solution is only to be achieved if the following three requirements are met: 1) the scientific dissent must be represented within the participation process without frustrating people; 2) the legitimative power of the political decisionmaking process must be strengthened by introducing participation; and 3) a mediating force between the polarized stakeholder groups has to be fulfilled by the participatory agents.

The planning cell model seems to serve these requirements. By compiling the necessary information profiles of performances for each energy option--constructed by scientists with different points of view--a legitimate way of conveying the basic facts to the participants was found without oppressing dissent, but also without creating frustration caused by confusion.

In addition, experts with different opinions were invited to present their arguments. Small working groups within the planning cells were formed to process all the given information and to extract the factual input. The purpose of improving the legitimacy of the political decisionmaking process and the depolarization of the conflict was pursued by a random selection procedure. Since everybody was eligible to serve as a participant in the planning cell, the uninvolved citizen's opportunity of identifying with members of the planning cell was guaranteed to a greater extent than in the case of traditional decisionmakers. Without interest groups being represented, the effect of polarization could be reduced, because a large number of participants with "in between" points of view were able to act as mediating agents.

Most specifically, we expected three different types of outcomes as a result of the planning cell procedure:

1) Task-oriented outcome: In order to avoid the mistake of overtaxing the ordinary citizen by filling up his schedule with numerous sophisticated tasks, we reduced the amount of evaluative task to the following essential questions:

- Evaluation of various energy systems and conservation technologies via multiple choice questions (simple attitude measurement)

- Evaluation of one's own energy situation and description of future plans referring to energy installations and insulation

- Weighting of the criteria and sub-criteria to evaluate energy systems and scenarios

- Using the criteria-set to rate the overall performance of large-scale power plants and to rate the proposed consequences of four energy scenarios developed by the Enquete-Commission of the German Parliament

- Making a final decision on the four energy scenarios and comparing this decision with the calculated result of the former ratings (using the algorithm Σ criteria weight times score of rating)

- Evaluating the local energy supply situation and suggesting new solutions to meet local energy problems

- Trying to establish a compromise between centralized and decentralized energy systems (group task)

2) Knowledge-oriented outcome: The classical data collection via survey techniques provide more or less random opinion profiles. In case of the planning cell, however, the gathering of opinions is proceeded by an information process. Thus, it is possible to measure opinions or beliefs, after they have gone through the cognitive stress of information and controversial discussions. In particular, it is interesting to investigate the changes in opinion after information and discussions have taken place and to study the attitude-forming process as a function of social

positions, general values, and demographic variables.

3) Policy-oriented outcome: As a combination of the results referring to the evaluative tasks and the revealed social structure of preferences, general inferences should be possible dealing with:

   - Suggestion of means to overcome polarization and to form compromises

   - Identification of elements in energy policies which will probably meet serious public opposition

   - The range of possible elements for constructing acceptable compromises

THE AGENDA OF THE PLANNING CELLS

The program of the planning cell's seminar was arranged in such a way that information, discussions, small group working periods and evaluations via questionnaires alternated with each other.

The first day of the program started with a general introduction into the purpose and the procedure of the seminar. This introduction was followed by an inspection tour to one of the close-by energy facilities. By visiting a power generation station or a central heating system, some of the key-problems of energy supply and demand were discussed. The local context was further described by the first lecture focussing on the possibilities and options to improve the local energy situation. After lunch this discussion was broadened to include the national perspective. Based on a short introduction of energy policies in Germany, the participants-divided into small working groups-had to deal with some of the major problems of today's energy debate (e.g. waste heat utilization, oil supply, energy consumption, or new energy sources). The purpose of these small group-arrangements was to impart familiarity with the present energy situation in Germany.

At the end of the day, the criteria to evaluate energy systems were introduced and explained. Again, working groups were formed (by random assignment) to simulate some practical applications of the criteria and subcriteria. The list of criteria and subcriteria are summarized in Table I. The first day was completed by handing out a questionnaire asking for a rank ordering of the criteria by each participant individually. The four days' schedule of each planning cell procedure is summarized in Fig. 2.

The second day was characterized by lectures containing the necessary information about energy technologies and their impacts. After each lesson the participants, gathered in small groups, discussed the important topics and directed specific questions to the lecturer. At the end of each session a questionnaire was handed out to all participants asking for the general feeling of favorableness or uneasiness with respect to the discussed topic. The lectures dealt with the following subjects: fossil fuel, heating systems, energy conservation, renewable energy sources, power generation, and nuclear energy. For each of these subjects, an expert from universities or from other uncommitted research institutes was invited to present the basic information.

In the preliminary test runs of the planning cells, the controversial issues of nuclear and solar energy were presented by two speakers holding opposite points of view. But this procedure turned out to be very time-consuming and frustrating with respect to the confusion caused by the controversial live debate. Instead, we prepared two video films containing precise statements of the nation's leading pro- and anti-nuclear, as well as pro- and anti-solar advocates. These statements were ordered according to the criteria lists which we had already introduced to the participants at the first day of the seminar. In addition, we invited two technical experts (one with a moderate pro-nuclear, one with a moderate pro-solar point of view) to present the technical information and to answer questions. With the exception of the planning cell run at Berlin where the energy seminar had been conducted for the first time, this procedure was regarded as neutral and well-balanced by all participants.

At the end of the second day we arranged a first trial for the participants to work with the given set of criteria. The task was to evaluate different power plant systems according to the eight criteria and the 31 subcriteria. Small groups of three to five participants were asked to assign ratings ranging from -2 to +2 to each subscriterion. The rating referred to the evaluation of nuclear, coal, hydro, oil, solar, and wind power. Then an index was constructed summarizing the overall intentions of all subcriteria and reflecting the final evaluation of a given power plant with respect to each main criterion. This complex task was not entirely performed by

Social Compatibility of Energy Systems: Agenda of the Planning Cells

	Tuesday	Wednesday	Thursday	Friday
Session 1	Introduction: The Energy Situation	Fossil Energy Heating Systems	City Planning and Energy Supply or Continuation of Session 4 of Wednesday	Generation of Goals and Aims for Energy Policies
Session 2	Inspection Tour of Local Energy Facility	Energy Conservation Renewable Energy Sources	Introduction to 4 Energy Options* Group Evaluation of Options	Final Evaluation: Local Energy Situation (Recommendations)
	Lunch	Lunch	Lunch	Lunch
Session 3	Local Energy Supply and Demand: Situation and Options	Nuclear Energy	Political Hearing	Individual Evaluation and Decision on the four Energy Options
Session 4	Introduction of Evaluation Criteria	Electrical Power Generation Evaluation of Power Generation Plants		Continuation of Session 3 Evaluation of the Seminar and the Participatory Process

*The four energy options have been developed by the German Parliamentary Enquete-Commission on nuclear energy

Fig. 2: Agenda of the Planning Cells

Table I: List of Criteria and Subcriteria to Evaluate Energy Systems or Scenarios

1.	Financial and Material Requirements
1.1	Cost of Today (Consumer, Investor)
1.2	Cost Development
1.3	Technical Material and Efficiency
2.	Security of Supply
2.1	Availability
2.2	Susceptibility to Technical Failures
2.3	Reserves and Resources of Energy
2.4	Potential to Meet Energy Demand
2.5	Flexibility
3.	Economic Effects
3.1	Labor Market
3.2	National Competitiveness
3.3	Structural Development
4.	Environmental Impacts
4.1	Local Consequences
4.2	National Consequences
4.3	Global Effects
5.	Health and Safety
5.1	Risks for the Employees
5.2	Risks for the Public
5.3	Potential Threat of Catastrophes
5.4	Problems for Future Generations
6.	Social Impacts
6.1	Standard of Living
6.2	Social Justice
6.3	Social Security
6.4	Impacts for the Working Place
6.5	Flexibility for Life-Styles
6.6	Personal Well-Being
7.	Political Impacts
7.1	Impacts on Constitutional Rights
7.2	Impacts on Decisionmaking Process
7.3	Level of Self-Determination
7.4	Impacts on Political Stability
8.	International Effects
8.1	Impacts of International Trade
8.2	Peace
8.3	International Justice (Distribution of Wealth)

each small working group. Only two criteria and the corresponding subcriteria were assigned to each group for evaluation. After the rating procedure, the speakers of the small groups had to report on their rating and to explain the reasons, why they had chosen a certain rating to represent their view. No instructions were given on how to come to an agreement within one working group. Except for two cases, unanimity could be achieved without interference by the chairman.

In order to facilitate the rating procedure and to make the evaluations more accurate, the results of our impact analysis were handed out to all participants and explained to them. The impact analysis contained only the scores of the selected 71 indicators and left open the spaces for the evaluation of the subcriteria and criteria; thus factual information was given without prejudicing the subjective ratings of each group. As already mentioned, the impact analysis contained only those indicator scores which were not contested by one side or the other. If there was a dissent, this was indicated by reporting both estimations.

The evaluation of the five options to generate electricity consumed more time than we initially anticipated. On an average, each group took around one to two hours to make the evaluation of one criterion (summarizing up to four subcriteria). Although we occasionally observed strategic response patterns assigning only positive evaluations to the most favorite candidate and negative ones to the least favorite, most participants were highly motivated to fulfill their task as accurately as possible. Since our main goal was to end up with a well-balanced judgement by the participants, we tried not to push the citizens to hurry up with their ratings, but extended this session to the morning of the next day, skipping the planned session on city planning and energy demand.

Accordingly, the third day started with the completion of the questionnaire for the power plant evaluation. In the second session, two members of the scientific secretariat of the German Enquete Commission of the German Parliament on Nuclear Energy presented the four energy scenarios dealing with possible energy futures in the Federal Republic of Germany. The scenarios were constructed in such a way that different political options were operationalized in terms of consistent energy supply and demand models for the year 2000 and the year 2030. One of the two speakers explained the four scenarios to the audience and pointed out the main differences between them (the four scenarios are illustrated in Fig. 3).

In particular, the role of nuclear energy (options 1 and 2 utilizing this technology to a large extent, options 3 and 4 rejecting it) and the role of energy conservation and solar systems (options 1 and 2 with a moderate amount of conservational and solar technologies, options 3 and 4 concentrating on these two) were discussed in detail. A working-group session was organized after the lecture to make the participants more familiar with the fundamental structure of the scenarios. They could also raise questions and direct them to the two speakers of the scientific secretariat. One speaker was more in favor of options 1 or 2 and the other of options 3 or 4. Before lunch the participants went through the same exercise as the day before. They were asked to evaluate the four scenarios according to the subcriteria and to form indices for each criterion. As an additional task, they had to perform a provisional weighting of the subcriteria in order to make the construction of the indices more valid. Again, this task was done in small groups.

After lunch a hearing was organized between the audience and politicians from the German Parliament representing the four parties (Christian Democrats, Social Democrats, Liberal Party, and Green Party). The objective of this hearing was to make the participants familiar with the political arguments related to each of the four options. Also, the hearing served the function of strengthening the level of self-confidence with regard to the participant's own judgment. After two and a half days of intensive learning, most citizens had gained a degree of expertise which sometimes turned out to be superior compared to the knowledge of the politicians represented at the podium. The reactions of the politicians varied from astonishment to anger after realizing that ordinary citizens were able to detect technical mistakes or biased argumentation.

Supported by this experience, the participants were asked to give their final judgment on the fourth day. In a first session they went through an exercise of generating their own evaluative criteria by defining social goals and aims which should be met by energy systems. The new set of criteria was later used as a second cluster of independent variables to predict the intuitive judgment of each individual. Afterwards the local energy situation was discussed again. Some recommendations were formulated to improve the work of the regional planning agency.

Both afternoon sessions were reserved for the individual evaluation of the four energy options of the German Enquete Commission. The evaluation process was divided into seven steps: 1) Rank ordering of the criteria; 2) Assignment of cardinal weights to each criterion; 3) Rating each of the four

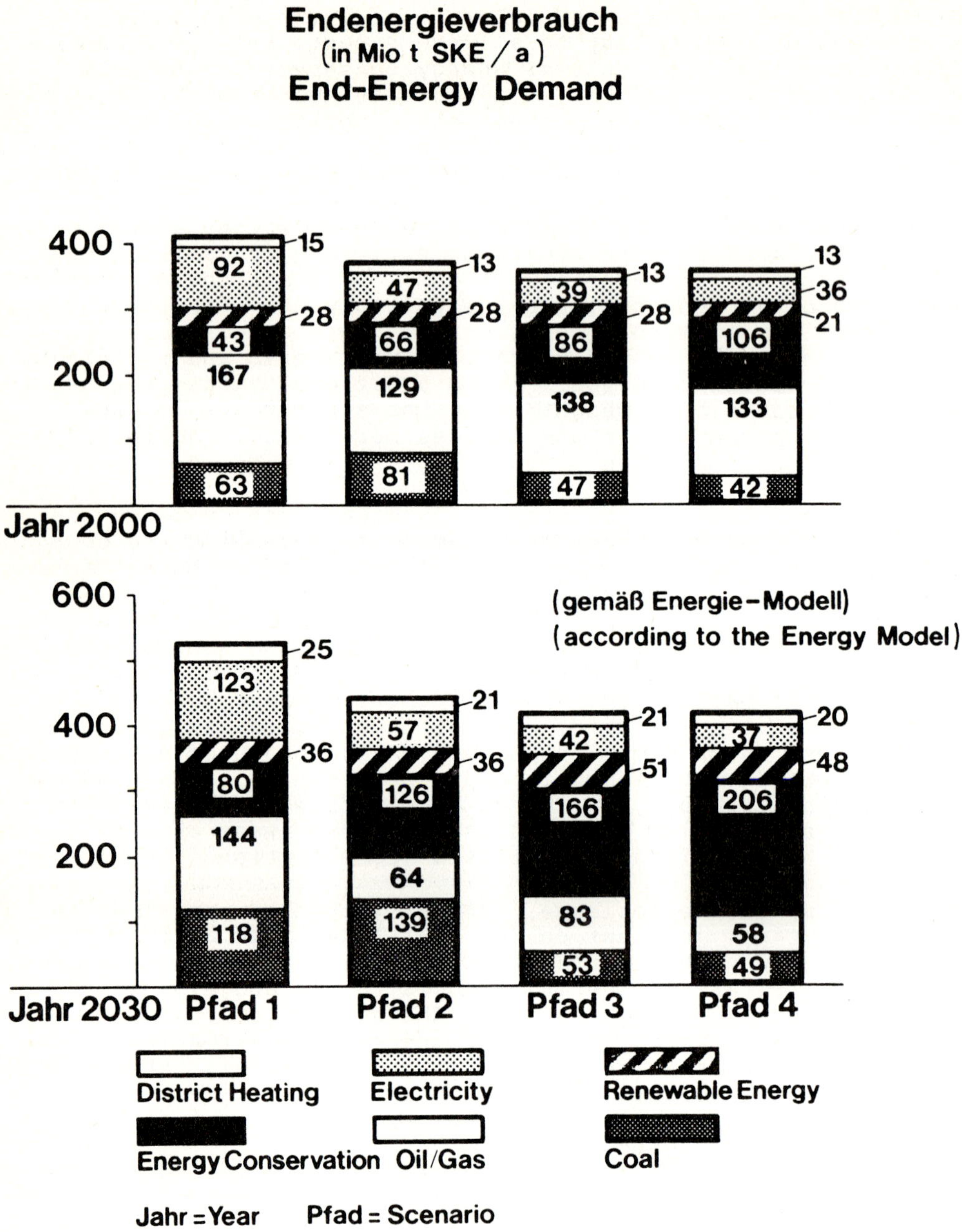

Fig. 3: The End-Energy Demand according to the Four Energy Scenarios of the German Parliamentary Enquete Commission

scenarios according to each of the eight criteria (scores ranging from -2 to +2) on the basis of our impact profiles; 4) Intuitive judgment about the best and second best scenario; 5) Contrast of the intuitive judgment with the calculated result of the rating and weighting process (using the formula criteria weights multiplied by scenario ratings); 6) Subjective explanation of resulting discrepancies between the intuitive and the calculated results; and 7) Collection of political arguments to "sell" one's own decision to the general public.

Whereas steps 1 through 4 and 7 are self-explanatory, steps 5 and 6 need a little more elaboration. After the participants had indicated their intuitive preferences with respect to the four options, they could relax during a coffee break. During this time, the individually assigned weights for each criterion and the ratings for the evaluation of the four options were multiplied and summed up for each of the four scenarios. The procedure, the mathematics, and the results were then presented to the participants. If a discrepancy between the intuitive and the calculated preference order occurred, the respondent was handed a questionnaire to write down the possible reasons for this gap. Specifically, we asked if the list of criteria was regarded as incomplete, if there were interactional effects between the criteria (non-linear relationships) or if the respondents had objections concerning our model. These questions came close to the border-line of the mental ability of many participants and caused some frustration and demotivation. But by explaining the task to each individual separately and giving him the feeling of doing extremely well, we were able to regain his attention and to reveal some of the underlying causes that probably initiated the gap between the intuitive and the calculated results.

Finally, each participant evaluated the whole seminar as well as his own role as a lay consultant. Before departing, participants were also asked to fill out a questionnaire with regard to their personal data (age, sex, income, education, profession, party affiliation, opinion leader scale, religion, family status, and others). A glass of champagne and a check for the honorarium of being a consultant marked the end of the seminar.

THE IMPLEMENTATION OF THE PLANNING CELLS

In the time period from June to April 1983, we organized 24 planning cells at seven German locations. Four test runs had been carried out at two German towns to develop valid and reliable scaling techniques and to test our program agenda. Substantial changes were made after the pretests. Therefore, fully comparable data is only available for 20 planning cells.

Since the number of citizens who participated at each planning cell seminar varied from 12 to 32, we did not end up with 500 persons, but only with 427. Including the participants of the pretest, 482 citizens took part in our study.

The selection of the locations to run planning cells was made according to the following specifications: 1) Representation of all major regions in Germany; 2) Variation in size of population; 3) Variation of building structures and housing development; 4) Variation of environmental quality; and 5) Representation of familiar and unfamiliar energy supply situations (including renewable energy sources).

As a further selection rule, we considered only those locations where investigations of the local energy structure had been carried out before, so that we could use existing reports or written material that could provide us with the necessary information to prepare the local options and to describe possibilities to improve the regional energy situation. By following these rules to choose the appropriate towns and cities, we found seven locations which seemed to fit into our list of requirements: 1) Berlin-Neukolln - part of the city of Berlin, 300,000 inhabitants of predominantly low social stratus, mainly old buildings, reconstruction measures underway (sanitation), large amount of old-fashioned heating systems, high pollution; 2) Gelsenkirchen - larger city in the Ruhr-district of about 300,000 inhabitants, 100,000 living in the center, predominantly members of the working class, densely populated and highly industrialized area, high proportion of foreign workers, mainly old, some renovated buildings, high amount of coal-fired systems, extremely high pollution; 3) Offenbach - small city of about 110,000 inhabitants of predominantly medium social stratus, in the center of the southwest, densely populated industrial areas, mixture of old and new buildings, high potential for district heating because of industrial waste heat, high pollution; 4) Norderstedt - typical affluent suburban area at the outskirts of Hamburg, about 66,000 inhabitants of medium and high stratus, predominantly new buildings, single-family homes with backyards, some applications of solar energy, medium pollution; 5) Viersen - medium-sized town of around 80,000 inhabitants of predominantly high-medium stratus, mixture of rural and industrial structure, more new than old buildings, high amount of electrical heating systems, low to medium pollution; 6) Soest - medium-sized town in the neighborhood of a densely populated industrialized area (Ruhr-district), 40,000 inhabitants of predominantly medium stratus, mixture of old and new buildings, mainly gas and oil heating, resort area for the industrial center nearby, medium pollution; and 7) Neustadt an der Waldnaab - small rural town of 5,600 inhabitants of predominantly low and medium stratus, small center with old buildings, new single-family homes in the development areas, mainly oil or wood heating systems, good chances for renewable energy (in particular biogas), low pollution.

After the seven towns had been chosen, the random selection of participants for our seminar took place. For this purpose we used the central register at the town hall in which the addresses of all inhabitants are listed. (In Germany any resident has to register in his home town.) According to previous experiences with planning cells, we needed about 100 addresses to end up with 25 participants. Thus, we counted on 75 percent failures or refusals. This assumption was even too optimistic. On an average, only 20 percent of all persons that we invited to participate actually attended our seminar. The invitation process itself underwent three stages: 1) Formal invitation letter containing the purpose of the seminar, the general idea of the planning cell method and some details of the contract that had to be signed by the participants (for example concerning the honorarium); 2) Personal visit to each potential participant by one member of the research team to explain personally the details of the planning cell and to prepare the contract; and 3) If necessary, personal or telephone requests to the employer of each potential participant asking for the permission to grant his employee four days of vacation.

The organization of planning cells is a highly complex task. Babysitting services have to be provided, adequate rooming has to be found, the local authorities have to be consulted and the local press informed. All this necessary preparation was carried out by the University of Wappertal (Prof. P. Dienel and his colleagues). Also, the contract was signed only by Prof. Dienel or his substitute to avoid any commitment or compliance effects which might occur if the name "Nuclear Research Center" appeared on the contract. (Working out the program, setting up the agenda, preparing the evaluations and running the scenarios were all a joint venture of the Nuclear Research Center and the University of Wuppertal).

The high amount of refusals to participate (about 80 percent) was mainly due to job or family ties which prevented many potential participants from accepting the invitation. Some persons were not satisfied with the honorarium that we offered, some expressed their lack of interest in the subject, some were opposed to the procedure in general or were just irritated, some had other commitments, and so on. It was surprising, though, that few respondents refused because of self-perceived incompetence. Although some women insisted that their husbands or sons would have been a much better choice, they were frequently persuaded to attend the seminar. It was sufficient to explain the random selection technique and its relevance for the representativeness of the results. Of course, it is possible that the feeling of incompetence was a latent cause for refusal, but for social status reasons it was not explicitly expressed to the research team.

Provided the case that the invitation was accepted, there were only four persons out of 487 who did not show up at the first day of the seminar, and only one person who dropped out of the seminar before it ended. In spite of this encouraging experience, a success rate of 20 out of 100 makes it difficult to draw statistical inferences and to predict the general public attitudes or beliefs. Since family or professional commitments are not distributed equally, we expected some systematic distortions. Surprisingly, the dispersion of demographic characteristics came very close to the general distribution reported for the population of Germany. As can be seen in Table II and III, there were only minor differences in the distribution of sex and age classes (considering the fact that we restricted our samples to persons older than 18 years).

There was also a fairly good representation of all social classes in our samples, with a slight over-representation of middle classes and a slight under-representation of high classes. The strongest bias that we were able to detect referred to professional affiliations. Compared with the national average, we had attracted far too many white collar workers (in particular, government employees) and not enough members of the working class. This is indicated in Table IV.

We also were confronted with an over-representation of students, both from high school and college. Since students have more time to spare, this bias is not astonishing. The percentage of housewives, unemployed, or retired persons did not deviate from the general average in spite of the expectation that these groups are likely to be over-represented when paid for attendance.

Looking over these results, we may conclude that the samples cannot be regarded as representative for the whole German population nor for the locations that we had selected. But the dispersion of demographic characteristics and professions within the samples justifies the use of inferential statistics with respect to relations between defined social groups. By comparing the results of the planning cells with similar

Table II:

SEX DISTRIBUTION

(Comparison of Planning Cells with German Population)

SEX	Planning Cells (n = 482)	Total* Population (n≈47 Mio.)
male	53.1 %	47.7 %
female	46.9 %	52.3 %

* West-German Population ≥ 18 years

Source: Statistical Yearbook of the Federal Republic of Germany 1981, edited by Statistisches Bundesamt Wiesbaden 1981

Table III:

DISTRIBUTION OF AGE CLASSES
(Comparison of Planning Cells with German Population)

Age	Planning Cell	Total Population*
18-25	16.6%	13.9%
26-35	22.2%	17.3%
36-45	21.0%	20.0%
46-55	17.2%	16.0%
56-65	16.0%	12.5%
> 65	6.0%	20.3%
Missing	1.0%	
	100.0%	100.0%

* West-German Population ≥ 18 years

Source: Statistical Yearbook of the Federal Republic of Germany 1981, edited by Statistisches Bundesamt Wiesbaden 1981

Table IV:

DISTRIBUTION OF PROFESSIONAL AFFILIATIONS

Self-Employed	3.73 %
Employees	31.33 %
Government and Public Employees	9.34 %
Workers	10.79 %
Housewives	19.09 %
Students	9.75 %
Retired Persons	13.90 %
Missing	2.10 %
n = 482	100.00 %

results of national surveys, we might also be able to validate some general conclusions. These can be found by combining specific results of the planning cells with similar tendencies revealed in representative surveys.

THE RESULTS OF THE PLANNING CELLS

The last run of a planning cell seminar took place in April, 1983. In the following months the phase of data processing was started, but has not been completed up to now. Thus, we are only able to report on a selection of results which have been analyzed so far. Specifically, we concentrated our efforts to study the preference structure of the citizens with respect to the four energy scenarios. But we also looked into the general attitude and belief structure. For this article, we selected the following topics: 1) Beliefs about different energy systems; 2) Weights and rank order of criteria; 3) Preference structure with respect to the four energy options of the Enquete Commission; and 4) Predictive power of the utility and factor models reflecting the intuitive decision.

After the presentation of the results, we will draw some general conclusions about the lessons we learned about the merits and pitfalls of the planning cell instrument.

Beliefs About Different Energy Systems

In all planning cells we found rather clear-cut beliefs concerning the different energy systems. Coal energy was perceived as a native energy source and thus beneficial to society, but also as environmentally harmful. There was an overwhelming majority who favored strict antipollution laws with respect to coal-fired power stations, even if the consumers had to pay high prices for their electricity. Many participants refused to make tradeoffs between environmental quality and cost when deciding about the impacts of coal energy. In order to prevent acid rain problems and the deterioration of the German forest--the "hottest" political issue in Germany in the year 1983, at least with respect to the environment--they demanded immediate action to be taken regardless of the cost.

With respect to oil and gas, most people were afraid of dependencies from the supplier nations, but this topic did not have the same saliency or emotional power as the issue of environmental quality. The reliance on more native energy sources was unanimously requested, but not for the price of major economic disadvantages or consumers' reduction of utilities. Since coal was regarded as environmentally harmful and oil and gas as politically dangerous, the future hope focused on solar energy and conservation. There were high expectations concerning the potentials of the alternative technologies or solar systems, but a very low personal interest in purchasing one. Most people believed that solar energy was not yet technically feasible and that the systems would be improved over time. Most participants agreed with the statement that the Federal Government of Germany should spend more money for research and development of solar systems. These programs should be financed by reducing the research grants for nuclear energy, by raising the price of electricity (but only modestly), or by transferring money from other parts of the federal budget. As expected, tax increase was the least favorite method to finance solar research.

Technologies for energy conservation (insulation, heat pump, passive solar utilization) were also highly approved by most of the participants. There were hardly any objections against state interventions into the market economy in order to speed up the market penetration of conservation technologies. Also, stricter laws or regulations concerning insulation or building permits were demanded from government, even if the participants were aware that it would increase the cost of building new homes. However, there were clear indications that governmental regulations restricting consumption or controlling individual heating behavior were unanimously rejected.

In spite of the general favorableness toward conservation, each individual perceived only a limited scope of possibilities to save more energy in his home. A vast majority (over 70 percent) of all participants indicated that they either perceived no chance of implementing new conservational installations or that they were convinced that they had already done everything possible. Institutional, economic, structural, and social barriers restraining an effective conservational program were frequently mentioned, but only few generic ideas were expressed about how to overcome these difficulties. There was high support for the suggestion that the government should finance conservational installations in advance and should then be paid back by the customer out of the surplus money he had saved by consuming less energy.

We experienced a very ambivalent reaction toward nuclear energy. Around 30 percent of all participants were fundamentally opposed to any use of nuclear power. This group perceived nuclear energy as very expensive, environmentally harmful, dangerous, and socially unacceptable. Information which contained positive arguments was rejected either as interest-driven manipulation or as short-term thinking. This specific group also criticized the ratings or our impact analysis as soon as they discovered any positive scoring with respect to nuclear energy. The distrust concerning the nuclear ratings was not extended to the scores of the other energy systems.

The vast majority of all participants showed a strong degree of ambiguity: on the one hand, they perceived nuclear power as necessary, economical, and promising, but on the other hand they expressed a strong degree of discomfort with this type of generating electricity. Most people supported the recommendation to confine the use of nuclear energy to that amount that all other energy sources together could not meet.

The difficulties encountered in this limitation of nuclear growth were well understood, particularly the question of who determines the amount of nuclear energy that cannot be substituted for. In spite of this problem, most participants expressed their feeling that the nuclear program should be restricted to the minimum of what should be regarded as necessary and that nuclear installations should be replaced, if other technically or economically feasible solutions were available. However, almost everyone of this majority group voted against a complete shutdown of nuclear power plants. They were convinced that after all nuclear energy might play a major role in the future, provided that the safety problems, the reprocessing and waste disposal problems, and the negative social impacts (like police state methods) could be managed in a satisfactory way. Also most people believed that in the long run nuclear energy had the potential to be the most important energy source for the Federal Republic of Germany, but the appropriate technology for this purpose was still to be developed.

Weights and Rank Order of Criteria

For the purpose of evaluating the four options of the Enquete Commission of the German Parliament, eight criteria and 31 subcriteria had been handed out to the participants. The task of the participants was to assign weights to each subcriterion and criterion and later on to evaluate the four options according to each criterion. The theoretical approach was based on a modified MAU-model by simplifying the utility measurement. Since we deliberately provided extensive information about the four options and the criteria, we had to control the effects of our information input on the participants' scoring behaviour. We assumed that the rank order of criteria was derived from personal values and should therefore not be altered by the information process; for our information was meant to focus only on facts and their (controversial) interpretation. In order to test the influence of the information process, we asked the participants to make a rank order of the main criteria on the first and the last day of the seminar.

Figure 4 illustrates the medians of the rank order for all eight criteria comparing first and second measurement. Evidently all observed changes are only of minor magnitude and the sequential order remains the same. Based on the individual scores, however, there is a high degree of change between the first and the second measurement. But these differences vary in general between one or two places up or down the rank order. The reason for these alterations is probably based on random effects (lack of memory) and not on real shifts in value preferences. The contingency coefficient[1] exceeded the amount of 0.60 for all criteria. We were also able to prove that the preference structure of our samples was rather similar to the preference structure which has been measured in previous commercial survey studies (for instance, Infratest Poll 1981). We are justified in concluding that our samples represent the general populations, view fairly well and that our information process has obviously not changed the preference order.

Looking at the priorities revealed by the weighting procedure, it does not seem surprising (knowing the general beliefs) that health/safety and environmental quality form the top of the hierarchy. The general economic concerns, in particular security of supply, are rated higher than the more specific concerns of financial and material requirements. This criterion gained more importance over the four days' information period, whereas the relevance of the environmental effects was rated slightly lower at the last day compared with the rating of the first day. Political, social, and international aspects were regarded as less important for the evaluation of energy systems.

We performed a subdivision of the cardinal weighting scores (ranging from 0 to 100) according to the intuitive preference for one of the four energy scenarios in order to analyze how the importances of criteria vary when individuals prefer a more nuclear or a more solar-oriented energy option. In Table V the criteria weights are listed according to the scenario preferences. The table conveys the immediate impression that the weighting scores do not differ dramatically, even in the most prominent cases of scenarios 1 and 4. The means of the weights for economic impacts, health and safety, political impacts, and international effects are not significantly different between the respondents in favor of the pro-nuclear scenario 2 and the respondents in favor of the non-nuclear scenario 3.

It is probably no surprise that proponents of a nuclear scenario assign more weight to cost and less to environment and social impacts. More astonishing is the unexpected homogeneity between the different preferential groups. The saliency of social values connot be the decisive factor in forming energy preferences.

---

1 Pearson contingency coefficient:

$$C = \sqrt{\frac{\chi^2}{N + \chi^2}}$$

Comparison of rank order of criteria at the beginning and at the end of the planning cell run (medians)

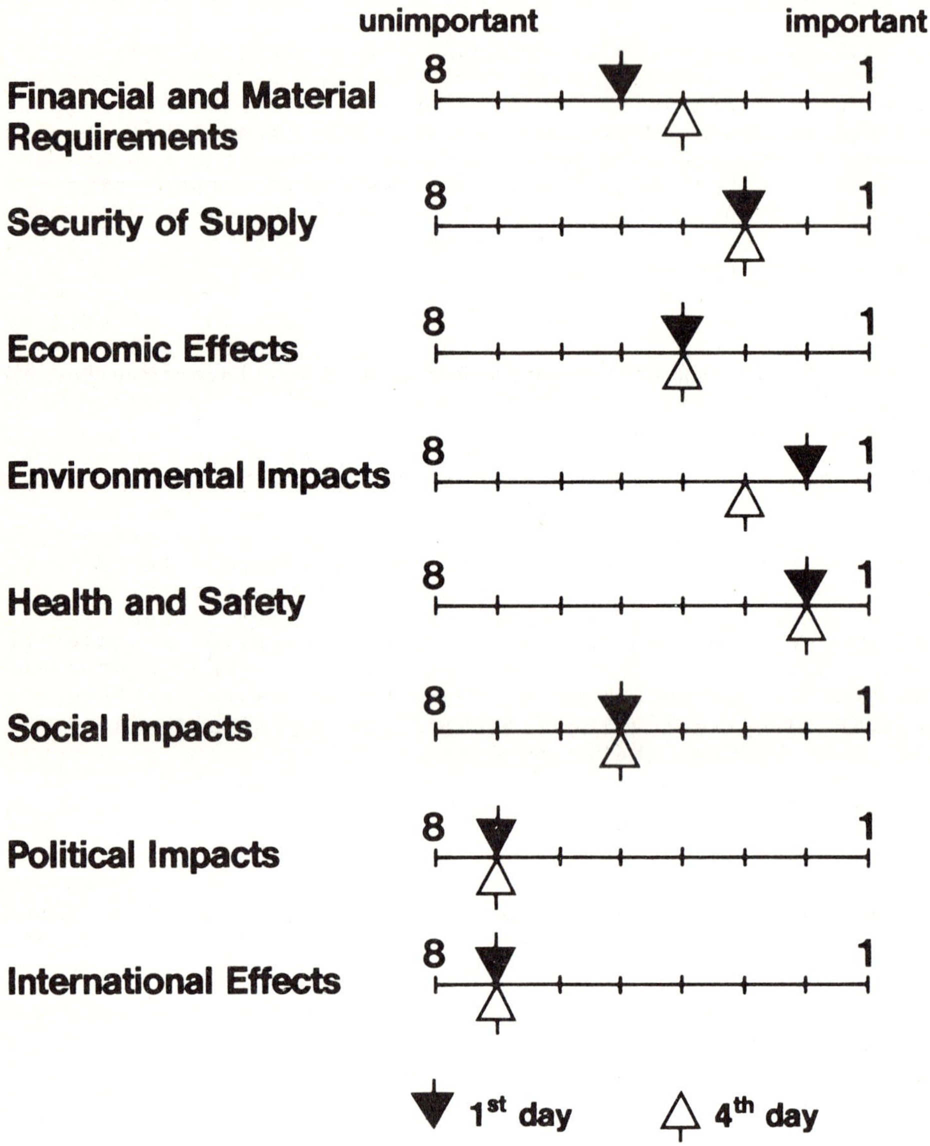

The preferences for one of the four energy scenarios are much more determined by the perceived performance of the four energy scenarios with respect to the eight criteria. This is illustrated in Table VI. This table lists some selected ratings of the four energy scenarios again according to the intuitive scenario preference. The range of possible evaluations is restricted from -2 to +2.

In contrast to the weights given to each criterion, the ratings for the performance of each scenario differ considerably. All respondents believe that environmental quality and health/safety impacts should be regarded as the most important criteria for different energy systems. But the judgment--whether nuclear or solar scenarios should be evaluated as superior when those criteria are used as yardsticks--reveals the gap between the different preferential groups quite clearly.

Thus, there is much more agreement on the weights to be assigned to various criteria than on the perceived properties of each evaluated option with respect to this criterion. This conclusion can be confirmed by a variety of other studies in Germany and the United States (8).

The Intuitive and Calculated Preference Structure with Respect to the Four Energy Scenarios

Figure 5 shows the results of the intuitive preference measurement and the calculated preferential order. The moderate pro-nuclear option 2 gained the highest approval, followed by the most moderate non-nuclear option 3. Most of the respondents who gave first priority to option 2 or 3 also assigned the second priority to the other moderate option (either scenario 2 or 3). Thus, there is a clear indication that more moderate scenarios are preferred.

The two pro-nuclear options together were chosen less frequently than the two non-nuclear options. Around 16 percent of participants preferred the extreme solar and conservational scenario 4 as opposed to only three percent who preferred the extreme pro-nuclear scenario 1. Thus, there is a considerable group of highly motivated and convinced citizens with a strong anti-nuclear commitment, but there is not an equal sized pro-nuclear fraction. Also, more than 70 percent of the persons who preferred option 2 (moderate pro-nuclear) moved to the moderate non-nuclear scenario 3 when asked for the second priority. The proponents of scenario 3, however, were equally divided: 50 percent assigned their second priority to scenario 2, the other 50 percent to scenario 4. So, in spite

Table V: The weights of the eight criteria as a function of the preference for one of the four energy scenarios (means)

	1	2 ↔	3	4	t-Test p (2vs3)
1. Financial and Material Requirements	16.25	12.65	9.49	8.48	0.00
2. Security of Supply	20.42	17.72	15.23	11.83	0.00
3. Economic Effects	18.33	13.26	13.04	10.0	0.71
4. Environmental Impacts	12.25	16.20	19.10	22.66	0.00
5. Health and Safety	13.08	18.89	20.14	20.60	0.08
6. Social Impacts	8.75	9.57	11.52	13.98	0.00
7. Political Impacts	7.67	6.51	7.36	8.0	0.05
8. International Effects	8.5	6.71	7.12	9.30	0.37

Table VI: The unweighted ratings of the four energy scenarios as a function of the preference for one of the four scenarios (means)

Preference for:		1	2	3	4	t-Test p (2vs3)
Economic Effects	Scenario 2	0.58	0.82	0.26	-0.37	0.00
	Scenario 3	-0.42	0.15	0.55	0.51	0.00
Environmental Impacts	Scenario 2	-0.50	-0.16	-0.65	-0.95	0.00
	Scenario 3	0.85	0.40	0.69	0.26	0.00
Political Impacts	Scenario 2	0.33	0.3	-0.31	-0.93	0.00
	Scenario 3	-0.58	0.25	0.68	0.70	0.00
International Effects	Scenario 2	0.08	0.16	-0.28	-0.95	0.00
	Scenario 3	-0.08	0.27	0.59	0.74	0.00

Holistic and calculated judgment on the preference for one of the four scenarios

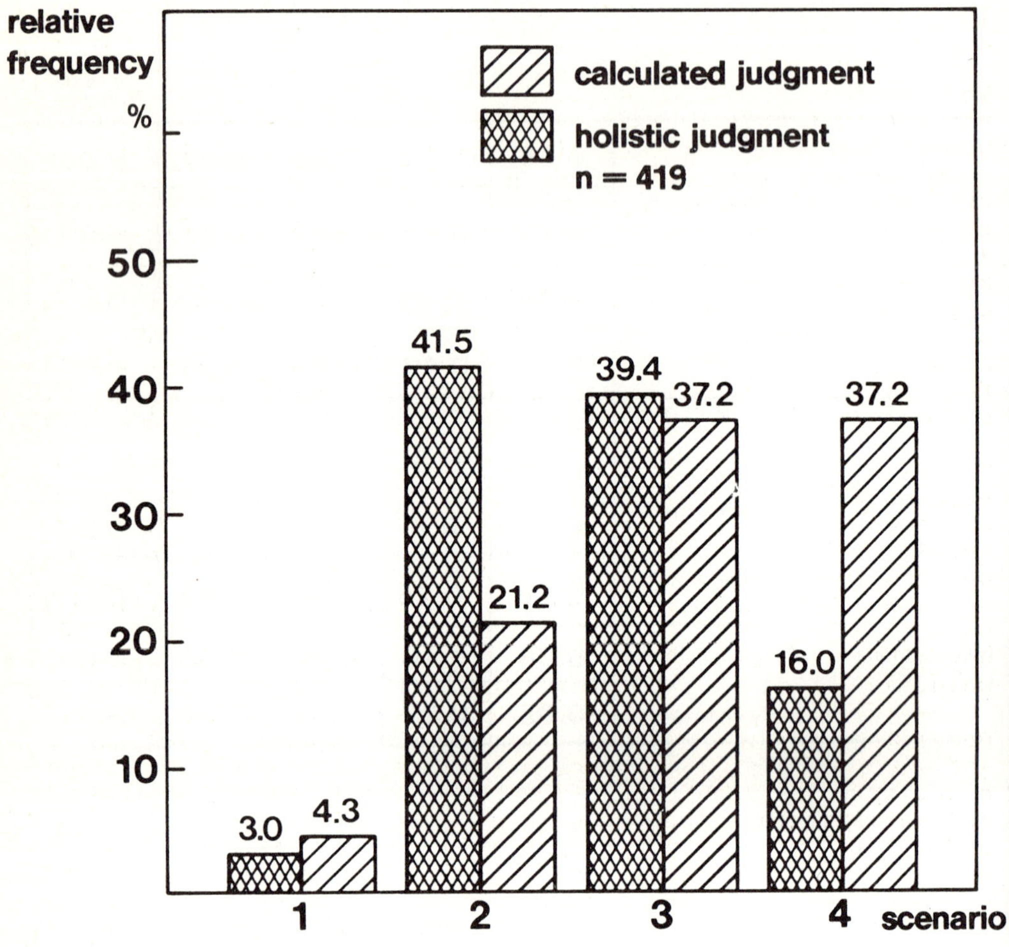

(replacement of missing data by means)

of the highest score for the moderate pro-nuclear option 2, there is a tendency to perceive the share of nuclear energy as a burden which almost half of the respondents are ready to accept for mainly economic reasons, whereas the other half would prefer this burden to be replaced by conservation or solar systems. The later group is less willing to sacrifice its anti-nuclear option as the former group is willing to abandon its nuclear preference.

This tendency is also clearly indicated by the calculated preference structure which was constructed by multiplying the weights of each criterion with the perceived performance of each scenario and summing up all the products for the eight criteria. According to the calculated results, three quarters of all participants gave the highest priority to options 3 and 4. There is quite a good correspondence between the calculated and the intuitive judgment in case of scenarios 1 and 3, but a big gap in the case of scenarios 2 and 4. In particular, the calculated preference of 37.2 percent in favor of scenario 4 strongly contrasts with the intuitive preference of 16 percent for this scenario. In the last paragraph of this chapter we offer some hypothetical explanations for this deviation.

The high degree of ambiguity revealed by this comparison between the two different preference measurements can also be studied by looking at the two components of the calculated index: weights and ratings. We have already shown that the weights themselves do not highly discriminate between the different preferential groups; we will concentrate our analysis on the ratings. Table VII provides an overview of the unweighted ratings (arithmetic means), Table VIII of the weighted ratings (means of weights times ratings).

Comparing scenario 2 with 3, it is easy to detect that the anti-nuclear scenario 3 receives better ratings in four out of eight criteria (environment, health/safety, political effects, and international consequences). Only in one case (security of supply) is the pro-nuclear scenario regarded as superior. When we take a look at the weighted profiles, the evaluations are somewhat more balanced. According to the three economic criteria, option 2 has been associated with slightly higher advantages as opposed to the ratings of option 3. According to the social, political, and international consequences, scenario 3 is still superior compared to scenario 2. With respect to health/safety and environmental quality, all participants rated the non-nuclear scenarios much more positively than the nuclear ones. Thus, the question arises: Why is there a strong tendency toward a positive evaluation of options 3 and 4, when the intuitive decision turned out to be in favor of option 2, at least for the relative majority? This quesion can only be

Table VII: The mean values of the perceived performance of the four energy scenarios with respect to eight evaluative criteria (unweighted ratings)

Criterion	SCENARIO PREFERENCE 1		2		3		4	
	Mean	Median	Mean	Median	Mean	Median	Mean	Median
Financial and Material Requirements	-0.75	-1	0.25	0	0.32	0	-0.53	-1
Security of Supply	-0.2	0	0.41	1	0.46	1	0.12	0
Economic Effects	0.12	0	0.5	1	0.29	0	-0.1	0
Environmental Impacts	-1.2	-1	-0.5	-1	0.48	1	1.0	1
Health/Safety	-1.0	-1	-0.32	0	0.63	1	1.2	1
Social Impacts	-0.4	0	0.25	0	0.35	0	-0.1	0
Political Impacts	-0.6	-1	-0.1	0	0.44	1	0.1	0
International Effects	-0.6	-1	-0.1	0	0.45	1	0.4	1
Summed Mean	-0.58		0.05		0.43		0.26	

Table VIII: Mean values of the perceived performance of the four energy scenarios with respect to eight evaluative criteria (weighted ratings)

Criterion	SCENARIO PREFERENCE							
	1		2		3		4	
	Mean	Median	Mean	Median	Mean	Median	Mean	Median
Financial and Material Requirements	-7.9	-10	4.2	10	3.7	0	-6.3	-5
Security of Supply	-3.8	0	8.3	10	7.1	5	-0.2	0
Economic Effects	2.2	0	7.1	5	2.8	0	-3.1	0
Environmental Impacts	-23.5	-20	-9.6	-10	9.2	10	-18.9	20
Health/Safety	-22.1	-20	-6.1	0	12.4	15	22.6	20
Social Impacts	-3.2	0	2.5	0	5.0	0	-1.2	0
Political Impacts	-4.8	-5	-0.3	0	3.3	0	0.4	0
International Effects	-4.6	-5	-0.7	0	2.9	0	3.0	0
Summed Mean	-58.2		4.7		43.1		28.5	

answered by investigating the predictive power of our evaluation matrix and the statistical analysis of the resulting errors in prediction.

Predictive Power of the Utility and Factor Models

Before elaborating on the different models of predicting the intuitive judgment with respect to the four energy scenarios, it seems worthwhile to take a closer look at the distribution of MAU-scores which characterize the overall performance of the four scenarios. Figure 6 shows the frequency curves for all four scenarios.

There is a clear indication that scenario 1 (highly pro-nuclear) has been evaluated as inferior compared to the three other options. The more moderate pro-nuclear scenario 2 has a peak at the zero-point. This supports our impression that option 2 is associated with a high degree of ambiguity. In contrast, scenario 3 has been evaluated much more positively (see the two peaks, one at +50 and the other one at +85). Also, the variance of evaluations is smaller compared to scenario 2. This indicates that scenario 3 has never been rated as totally negative; scenario 2 has been rated occasionally and very rarely as totally positive (which is also true for scenario 2). Scenario 4 (the most pro-conservation, pro-solar, anti-nuclear scenario) has a particularly interesting shape, because there are two peaks of around equal size, one at the zero point and one at 50. If we only consider this distribution of evaluations, we are able to describe more precisely the scoring behavior of the respondents.

If a participant of our seminar evaluated the extreme scenario 4 as ambivalent in its outcomes, and the scenarios 3 and 2 as positive, he would probably prefer scenario 2 over all other options. If a participant perceived scenarios 3 and 4 as rather positive and scenario 2 as more ambivalent, he chose scenario 3 when asked for his intuitive preference. An intuitive priority for the two extreme scenarios 1 and 4 was always accompanied by an extreme positive evaluation of the chosen alternative and a clear cut negative perception of the three remaining options. This conclusion can be further reaffirmed when the ranges of the overall evaluations (calculated preferences) are subdivided according to the intuitive preferences. Figure 7 illustrates this relationship.

Most difficult to explain is the scoring behavior of the relative majority preferring the more moderate pro-nuclear option 2. This group assigns almost equal scores to scenarios 2 and 3. Also, the most anti-nuclear scenario 4 gains a high

Verteilung der summierten Pfadbewertung für die 4 Pfade der Enquete-Kommission.
Distribution of evaluation scores with respect to four energy scenarios.

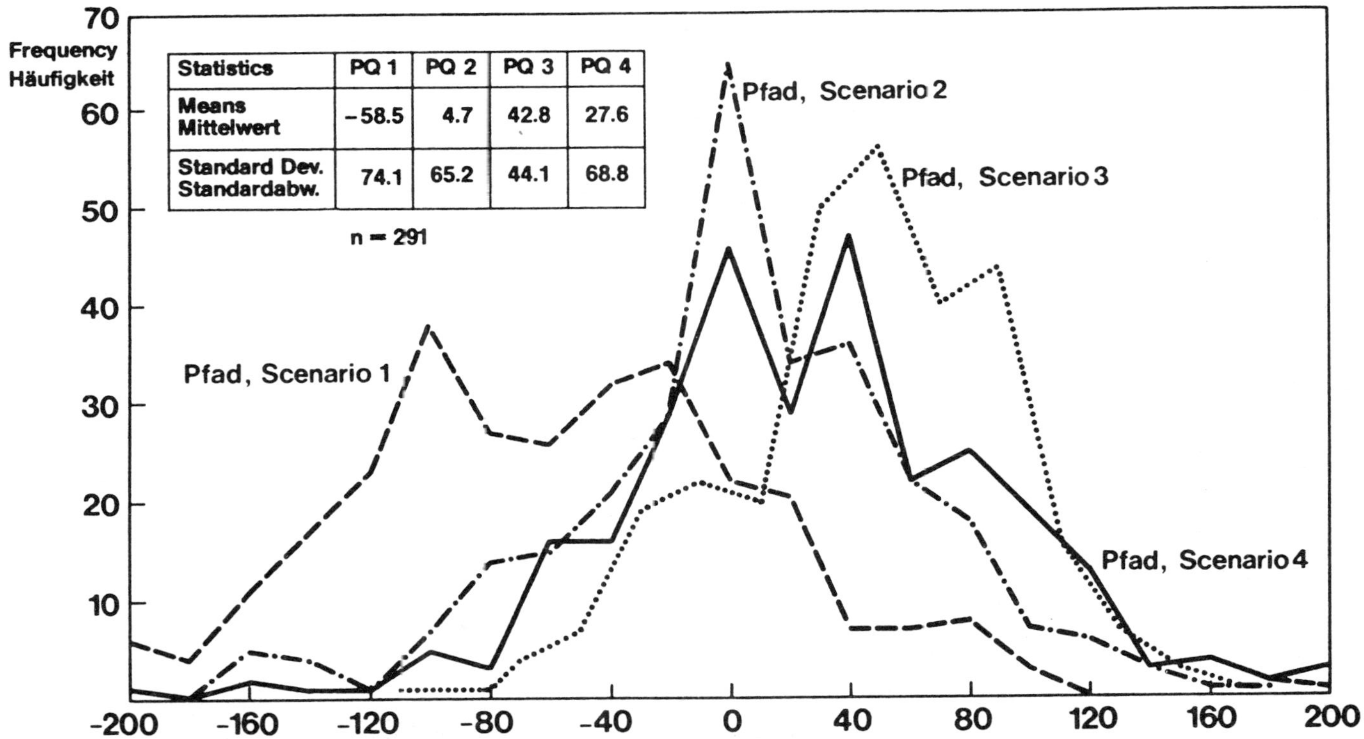

Fig. 6: The frequency distribution of respondents's MAU-scores to evaluate the four energy scenarios (Pfad = scenario)

degree of appreciation. Participants who voted in favor of scenario 3 evaluated scenario 4 as at least equally positive. But why then did they decide to give priority to scenario 3?

In order to explain the difference between the intuitive and the caluclated preferences, it is necessary to compare inconsistencies on an individual level. For this purpose, we considered two basic models, one based on the sum of the unweighted ratings for each scenario and the other one based on the sum of the weighted ratings. We performed a cross-tabulation between the intuitive preferences and the model-based calculated preferences. Those preferences were simply derived by choosing the scenario which reached the highest score on the summed ratings (weighted and unweighted). Figures 8 and 9 illustrate the results of this comparison for both models.

As can be expected from the analysis so far, most errors in prediction occurred for the group preferring scenario 2. For both models, the proportion of errors amounts to around 56 percent. According to the MAU model, 25 percent belonging to this group scored highest on scenario 3 and the other 25 percent on scenario 4. Looking at the group of respondents who intuitively favored scenario 3, both models reveal that only a marginal percentage assigned a higher score to options 1 or 2, but that virtually all errors were due to a calculated preference toward option 4. The calculated preference is much more likely to be one step further toward a "softer" scenario compared to the intuitive preference. If we look at the diagonal line where the perfect fits are described, we can detect hardly any error below that line (meaning that few participants rated the more nuclear option better on an analytical base than on an intuitive base). The vast majority of all errors are above the line.

Another interesting result can be found by looking at the two models. There is hardly any difference between the weighted and unweighted version of the model. The total percentage of correct prediction is 50.13 percent for the weighted and 51.07 percent for the unweighted model. We can even detect that the weights bias our model even more toward a soft preference structure. It is clearly reaffirmed, though, that the weights do not play a decisive role in the formation of preferences.

This result appears to be counter-intuitive. We had expected that the scaling process based on arguments and rational reasoning would reveal a stronger priority for options 1 and 2, whereas the intuitive judgment influenced by evaluations, visions, and affects would focus on options 3 and 4. But just the opposite was the case.

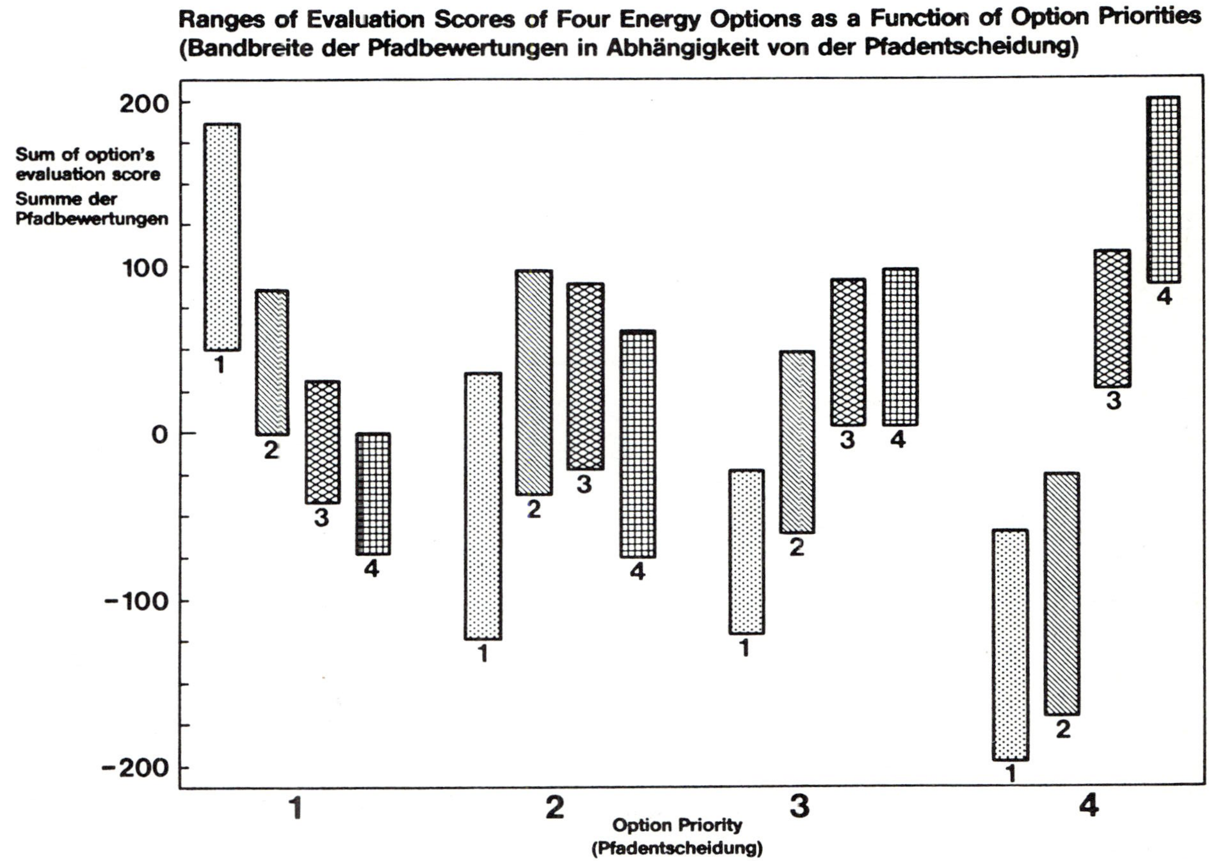

Fig. 7: The ranges of scores for each scenario evaluation as a function of the intuitive preference for one of the four scenarios

Contrast between the holistic judgment and the calculated preference via the sum over the multicriteria ratings (unweighted) with respect to the four energy scenarios*

Scenario Priority	Model: max $[\sum_{1}^{8} \text{ratings}]$ 1	2	3	4	total
1	61.54	23.08	15.38	0	13 (=100%)
2	6.32	44.25	29.31	20.11	174 (=100%)
3	1.21	4.85	57.58	36.36	165 (=100%)
4	1.49	2.99	44.78	50.75	67 (=100%)
total	5.25	21.48	42.48	30.79	419* (=100%)

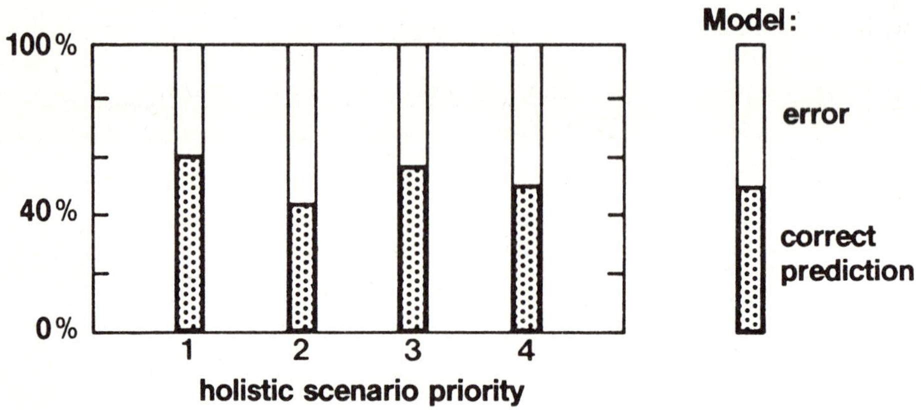

Fig. 8:

Percentage of correct prediction for the overall model: 51.07%

*replacement of missing data by means

Contrast between the holistic judgment and the calculated preference via a simplified MAU-model (weights x ratings) with respect to four energy scenarios*

Scenario Priority	Model: max $[\sum_{1}^{8}$ weights x ratings$]$ 1	2	3	4	total
1	53.85	38.46	7.96	0	13 (=100%)
2	5.75	43.86	25.86	24.71	174 (=100%)
3	0.61	3.64	52.12	43.64	165 (=100%)
4	0	2.99	35.82	61.19	67 (=100%)
total	4.3	21.24	37.23	37.23	419* (=100%)

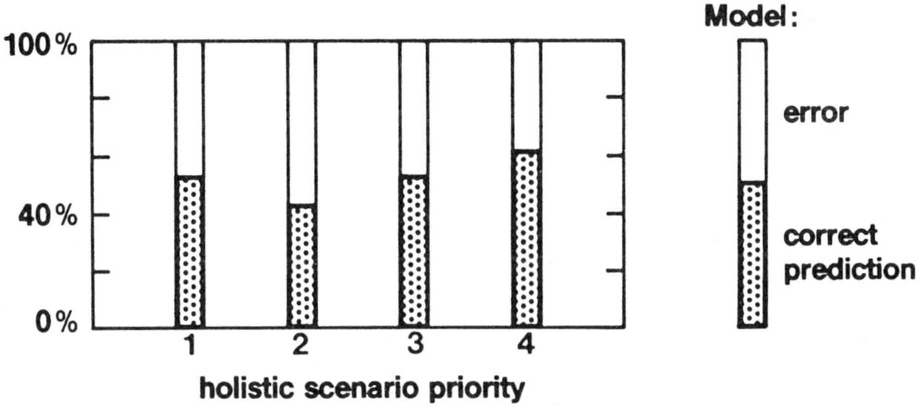

Fig. 9:

Percentage of correct prediction for the overall model: 50.13%

*replacement of missing data by means

What kind of explanation could we find to understand this scoring behavior? First, we tried other models to construct a meaningful index of summarizing the ratings of the participants. In order to rule out the strong ties between health/safety and environmental aspects, we combined the ratings for those two criteria into one and based our calculation on a seven-criteria set. But the results did not differ from those of the two original models. After this exercise we used some statistical procedures to construct appropriate models based on multiple regression and factor analysis. Neither model showed any improvement in the percentage of correct prediction. For the purpose of comparison, Fig. 10 illustrates the relationships between the intuitive judgment and the calculated results of a factor model. All ratings were multiplied by the factor score coefficient (standing for an indirect measurement of importance for each criterion with regard to the overall intention of all criteria).

Since the factor and the regression models were characterized by the same biases as the previous models, there is little support for the assumption that the discrepancy between the calculated and the intuitive judgment is due to a latent structure of values which were not overtly expressed by most of the participants. For example, because of social desirability reasons most proponents of scenarios 1 and 2 did not admit that in their view economic values were superior to health and safety values. The weights expressed by the advocates of scenarios 3 and 4, however, came close to the real preference structure of the group. We are still testing the hypothesis of latent value biases by constructing more refined models. But to draw a preliminary conclusion, there is no doubt that the ratings themselves are extremely biased toward a positive evaluation of scenarios 3 and 4, and even latent weights assigned to each criterion cannot compensate for this effect. Of course, there is also the possibility that we chose the wrong criteria or omitted decisive ones which could be regarded as the underlying causes for the superiority of option 2 in the intuitive rating process. Since we had asked all persons to give explanations for resulting discrepancies between the intuitive and the calculated judgment, we were able to test this hypothesis, at least to the extent that the participants could consciously recall their judgments. Indeed, we found some evidence that the list of criteria induced three biases toward scenarios 3 and 4: 1) Many respondents indicated that the three economic criteria (cost, security of supply, and economic consequences) should have more weight for the decisionmaking process, if all three of them received negative evaluations. They thought it would be inadequate to add up the weights for each single economic criterion; extra weight should

Contrast between the holistic judgment and the calculated preference via factor analysis and the unweighted ratings (factor score coefficients times ratings) with respect to four energy scenarios*

Scenario Priority	Model: max $[\sum_{1}^{8}$ coeff. x ratings]				total
	1	2	3	4	
1	61.54	15.38	15.38	7.69	13 (=100%)
2	6.90	42.53	30.46	20.11	174 (=100%)
3	1.82	4.24	58.18	35.76	165 (=100%)
4	1.49	0	46.27	52.24	67 (=100%)
total	5.73	19.81	43.44	31.03	419* (=100%)

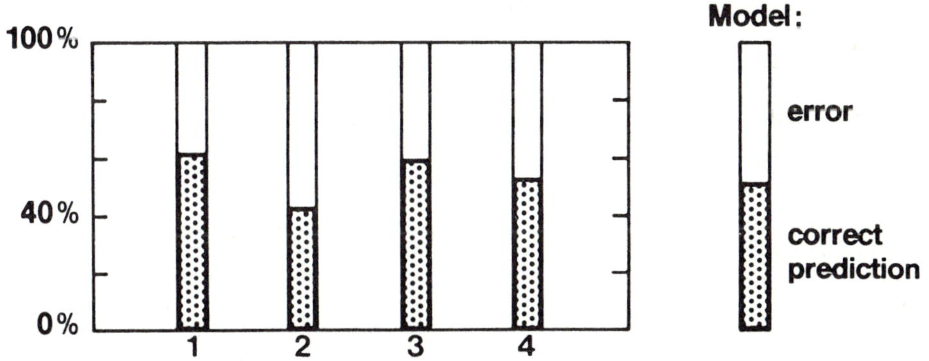

Fig. 10:

Percentage of correct prediction for the overall model: 50.83%

*replacement of missing data by means

Table IX: The preference for energy scenarios 1 or 2 versus 3 or 4 as a function of age (n = 419)

Age	Energy Scenarios 1/2	Energy Scenarios 3/4	
18 - 40	32.7 %	67.4 %	191 (= 100 %)
≥41	55.2 %	44.8 %	223 (= 100 %)
	185	229	414

be given to the joint effect of all three criteria; 2) Many respondents expressed some objections to the placement of technical feasibility as a subcriterion of the criterion security of supply. They suggested that feasibility should be a separate criterion independent from economic security; and 3) Some respondents missed the criterion "political implementability." They expressed the opinion that option 4 could not be implemented because of pressure group influence. This reason was most frequently expressed by those participants who voted for option 3, but ranked highest on option 4.

Apart from these explanations there were only three participants who suggested completely new criteria or modified the criteria list. Some could not give any explanation for the discrepancy. But the most striking result was that during the discussion most of the respondents who had to explain the deviation between intuitive and calculated preference referred to non-cognitive reasons. They admitted that most of the arguments would lead to a more favorable view of options 3 or 4, but they trusted the official policymakers in Germany who were believed to have a better insight view on the subject and were thus more competent on this issue. Since most German institutions are in favor of options 1 or 2, the participants accepted this judgment and acted accordingly. Reference groups, such as parties, unions, or professional societies, were also cited.

There was no sharp distinction between cognitive and noncognitive explanations. Since most of the respondents gave more than one answer, the criticism of the criteria list and the reference to social group judgments were mixed. By reviewing the results, we gained the impression that the cognitive aspects reflected the degree of rationalization of reference group influence to a large extent.

There are some indications to back up this impression. Figures 11, 12, and 13 provide an overview on the distribution of option priorities as a function of sex, age, and party preference. Whereas sex had no impact on the intuitive preference for one of the four options, we detected quite intensive relationships between age, party preference, and option priority. The more conservatively people voted in national elections and the older they were (in particular over 40 years old), the more they preferred options 1 or 2.

Older people and conservative voters tend to express more trust in established institutions and assign a higher degree of credibility to politicians and scientists. Indeed, we could prove that the majority of the proponents for option 2 having higher scores for option 3 or 4 belonged to the group of conservative and older people. We are still in the process of analyzing the motivational roots of the discrepancies, but right now it already seems justified to conclude that the discrepancies between intuitive and calculated preference is mainly due to loyalty vis-a-vis reference group judgments.

CONCLUSIONS

The use of planning cells as a participatory element in the political decisionmaking context turned out to be a highly complex but rewarding effort. From the experience with more than 20 planning cells, we have gained a better understanding of the chances and the limits of this model. Also, we encountered some difficulties which were not visible when we started our program. Our experiences justify the further development of the planning cell concept, but at the same time revealed some of the serious problems and drawbacks related to the idea of picking random samples for policy evaluation. These problems have to be solved before this instrument is used in wider scope to facilitate decisionmaking under conflict.

The chances and advantages of the planning cell concept refer to the following purposes: 1) The elicitation of ordinary citizens' preferences concerning complex decisions (with input of appropriate information); 2) The understanding of underlying reasons for specific attitudes or beliefs; 3) The measurement of "popularity" or hostility towards proposed governmental policies (after relevant information has been presented); and 4) The prediction of opposition against certain objects or policies.

The preference for one of the four energy scenarios as a function of sex

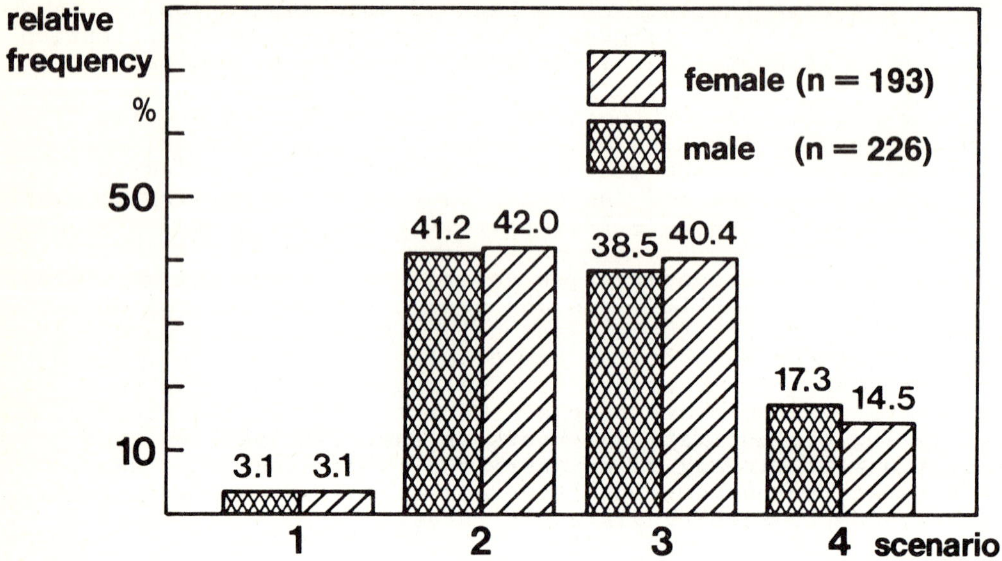

We believe that these four purposes of the planning cell have been fulfilled during our study and could even be improved. Since the participants of planning cells cannot represent the whole population, inferential statistics can only be applied with special care. By comparison with the result of national surveys, there was a good chance of validating common trends and describing more specific results from the planning cell sample. In contrast to these positive aspects, we faced quite a few limitations which should also be mentioned. These refer to: 1) The incapability of ordinary citizens to create new or generic solutions to conflict management or resolution; 2) The difficulty to elicit tradeoffs between different attributes; 3) The problems of aggregating individual wieghts or ratings and interpreting mean values because of differences between groups, different perceptions of what the weights and rankings mean, response sets, distortions because of social desirability and other latent factors; and 4) The limits of the legitimation effect in polarized disputes (results will be interpreted as manipulated as soon as they deviate from one's own opinion).

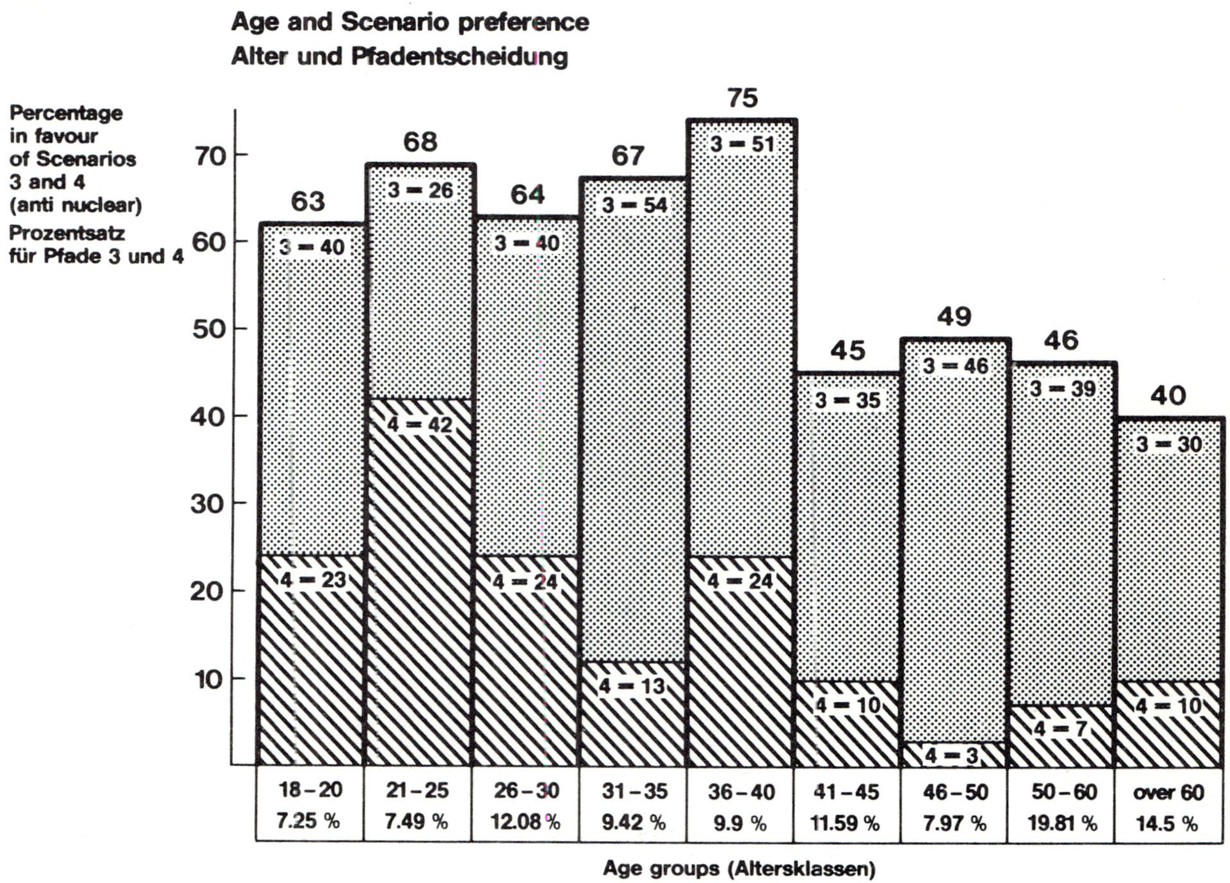

Fig. 12: The preference for one of the four energy scenarios as a function of age

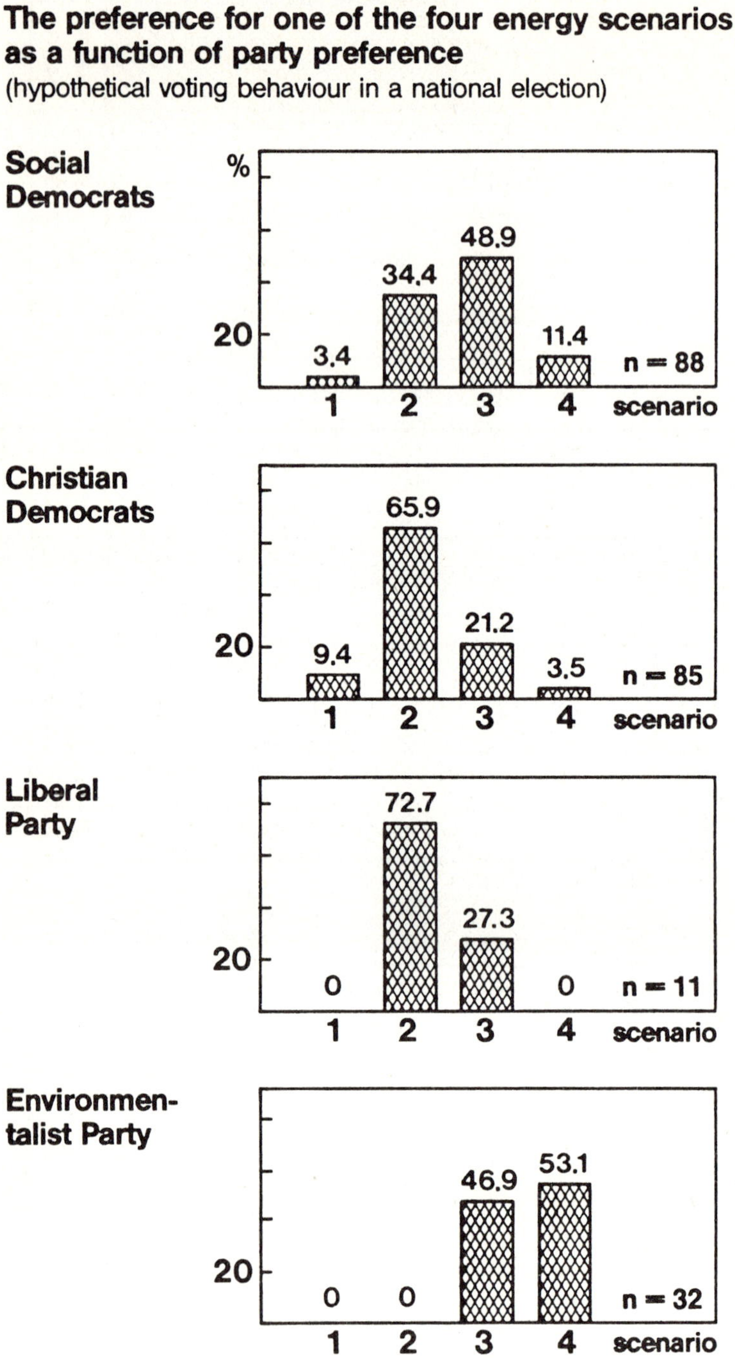
The preference for one of the four energy scenarios as a function of party preference
(hypothetical voting behaviour in a national election)
Social Democrats
%
20
3.4
34.4
48.9
11.4
n = 88
1 2 3 4 scenario
Christian Democrats
20
9.4
65.9
21.2
3.5
n = 85
1 2 3 4 scenario
Liberal Party
20
0
72.7
27.3
0
n = 11
1 2 3 4 scenario
Environmentalist Party
20
0
0
46.9
53.1
n = 32
1 2 3 4 scenario

Fig. 13:

Considering these limitations, it is evident that planning cells can provide information about the acceptance of objects or policies (after all necessary information is given), but hardly any information about the acceptability of objects or policies, at least when highly complicated, national issues are at stake. Since our project is mainly directed toward testing the acceptance of energy policies--if all the impacts were known--we obviously had chosen the right instrument. Since we contented ourselves with the measurement of acceptable procedures, one question of course immediately comes up--what about participation? Was it not the goal of the whole exercise to introduce public participation into the decisionmaking process?

This question was one of the key problems within the planning cells, too. The participants expressed their discomfort with the double role they had to play. They were consultants and test persons at the same time. If they were regarded merely as a jury to make judgments with respect to given options, we should not be interested in the underlying motives or reasons. If they are regarded as test persons for improving our knowledge of people's preferences, we should not give them the impression of being our "consultants." In order to avoid this conflict, we pursued a double strategy: we promised the participants that their expressed preferences would be documented and handed over to the Federal Ministry of Research and Technology (Citizens Report) without speculating over motives or underlying reasons. But we also informed them that the policymakers had a viable and legitimate interest in receiving information about why people express special preferences and what kind of values might be involved. This interest can only be met by using survey methods and statistical data processing techniques.

Although this double strategy led to some irritation on the side of the participants, we had the feeling that we succeeded in conveying the purpose for creating such a double role of planning cell participants. The whole procedure reflects our main assumption that participation in complex matters cannot substitute for political decisionmaking. Rather, it should be regarded as a decision aid to form and shape political judgments according to the latent and overt value structure of the concerned public. If this assumption is accepted, the planning cell might be the ideal instrument to collect the relevant feedback from society and to reveal the intuitive preferences and values which should be the guideline of democratic policymaking.

REFERENCES

1. P. Slovic, B. Fischhoff, S. Lichtenstein, and D. MacGregor, "Perceived Risk: Psychological Factors and Social Implications," Paper presented at the NATO Advanced Study Institute, Les Arcs, France, August 21-31, 1983 (Manuscript from Decision Research, A Branch of Perceptronics, 1201 Oak Street, Eugene, Oregon 97401, USA).

2. O. Renn, "Wissenschaftliche Politikberatung in Spannungsfeld von Wertwandel and Legitimationskrise," in Perspektiven angewandter Sozialforschung, H. Klages, ed. (Westdeutscher Verlag, Opladen), in press.

3. G. Albrecht, H. U. Stegelmann (eds): Bilanz der Energieversorgung (High-Tech Verlag, Munchen), in press.

4. P. C. Dienel, Die Planungszelle (Westdeutscher Verlag, Opladen 1978).

5. U. von Alemann, Partizipation - Demokratisierung - Mitbestimmung (2 edition, Westdeutscher Verlag, Opladen 1978).

6. Ibid.

7. D. Siebenborn and H. Nehm, Die Planungszelle - Eine Form der Bürgerpartizipation, Der Städtetag, 7 (1980), pp. 398-402.

8. W. G. Stillwell, D. von Winterfeldt, and R. S. John, Value Tree Analysis of Energy Supply Alternatives (Technical Report SSR1 81-2, Social Science Research Centre, University of Southern California, Los Angeles (1981) or O. Renn, "Man, Technology and Risk," (Jül-Spez-115, Report of the Nuclear Research Centre, Julich, 1981).

Attitude Studies by the IAEA/IIASA Risk Assessment Group

O. Renn, E. Swaton*

Program Group: Technology and Society, Nuclear Research Centre Julich, Post Office Box 1913, D-5170 Julich-1, Federal Republic of Germany

INTRODUCTION

Since the beginning of our history, technology and man have been in a state of tension: man has learned to make use of instruments to refine and perfect his non-specialized organic nature in any desired direction. Technology helps him to move faster than any animal, to see, hear, and smell better, to obtain food faster and in greater quantities, to protect himself more effectively against hazards and natural risks, to prolong his span of life, and to ensure continuous propagation of his species. However, a price must be paid for this progress: increasing specialization and differentiation in society result in anonymous and incomprehensible social structures which the individual can no longer understand, technological systems of ever greater perfection and cost increase the risk of being killed by the machine which man himself has created. The acceleration of production efforts creates more and more environmental pollution which, in the final analysis, can destroy the foundations that support human life. Last but not least, technology can also amplify the negative developments in human society: the more efficient our technology, the greater is the potential for catastrophic events when aggression takes place, expressed in terms of crime, terrorism, civil unrest, or war.

There is no doubt that ambivalence towards technology exists. Many types of societies have deliberately forgone any forced advancement of technological progress because the social dynamics associated with the development of new instruments

* IAEA/IIASA Risk Assessment Group, International Atomic Energy Agency, Vienna International Centre, Post Office Box 100, A-1400, Vienna, Austria

NATO ASI Series, Vol. G4
Environmental Impact Assessment, Technology Assessment, and Risk Analysis. Edited by V.T. Covello et al.

would have threatened the static conditions in these societies. (e.g., ancient China). Therefore, in the 19th century, technological progress meant overcoming rigid hierarchies and postfeudal power conditions. Until the past decade, the advancement of science and technology was considered a priority task of society, associated with a highly positive value and progressive image. However, within a short period of time the problems created by major industrial facilities, an increasing awareness of the environment, disillusionment as to the limitations of scientific research and technology in important areas (such as cancer research), employment problems due to automation, and saturation phenomena in the consumer area have initiated a reinterpretation process which has directed the spotlight of perception to the ambivalence of technology. In addition, the concepts for soft or alternative "countertechnologies" have stimulated the increasingly critical discussion on technology and its consequences (1).

In this situation, where cognitive contradictions and a successive change of value orientations and ethics both introduce uncertainty into man's thinking, studies on the perception of technology achieve their special significance. The future development of technologies depends to a large extent on the readiness of society to cope with the new skepticism of its citizens. In particular, the public debate has been focused on energy systems. Should we proceed with building nuclear power reactors? Is it worthwhile to invest into the development of renewable energy sources? How much primary energy can we save, if all means of employing efficient techniques for energy conversion are utilized? Is energy conservation feasible?

These questions cannot be answered exclusively by the expert community. Value judgments, concerns about the future situation of mankind, and political considerations play a major role in the energy controversy. If those parameters are not taken into account, the mere technocratic decisionmaking process will ensure a rise of social opposition. The other extreme, the refusal of factual expertise, leads to a polarization of social groups caused by the inability of the political institutions to play a mediating role between the different factions, since mediation must be based on some solid ground of factual knowledge.

In this dilemma, social studies of energy systems have to serve three functions: 1) They should provide knowledge about the motives and mechanisms of reasoning which shape the public perception of technology; 2) They should evaluate existing or propose new means and techniques to resolve conflicts without pre-determining their potential outcome; and 3) They should provide a political framework which allows a proper incorporation of facts, values, and ambiguities in the decisionmaking process.

The following article covers only the first aspect. We will report on the attitude surveys regarding various energy technologies which have been conducted under the auspicies of the IAEA/IIASA Risk Assessment Group in various countries. The main purpose of these studies has been the detection of latent patterns which govern the mental process of assimilating and evaluating energy systems. Also, a cross national comparison regarding the structure of salient belief factors is one of the central objectives of the IAEA/IIASA investigations.

A CLASSIFICATION OF RESEARCH APPROACHES TO STUDY THE PERCEPTION OF TECHNOLOGIES

The questions of how individuals perceive technolgies and how they evaluate information about technologies is part of quite a variety of psychological, social psychological, and sociological theories. Psychological theory has emphasized the individual process of common sense reasoning, incorporating the social environment and specific motivational factors. The purpose of this research is directed towards the individual process of understanding the representation and assimilation of technological information and the formation of an overall judgment.

Psychological theory assumes that human beings have specific value clusters which influence the weighting of attributes connected with the perception of a given object. Specific psychological mechanisms of common sense reasoning combine the perceived properties of an object with the relevant attributes which have been weighted by individual values prior to the intuitive assessment process. Depending on the psychological model used, the resulting judgment consists of a linear combination of attributes and perceived consequences or, in addition, is influenced by common sense heuristics to cope with uncertainty (3).

Social psychological research concentrates on the interaction between social environment (social value, norms, and roles) and personal judgment. The perception of technologies is being understood as a process of deriving attributes about specific objects from general social values and personal attitudes and linking these attributes to the perceived properties of the object or technology (4).

According to social psychological theory, the perceived properties of an object are transferred into a belief system in which not only object-related cognitions but also situation-related factors (such as the perception of those people who favor the object) and symbolic attributes (such as national pride) are combined to an overall image of the object. After each of these beliefs has been evaluated by the degree of emotional saliency (good - bad), an overall judgment can be derived (5). Sociological research finally addresses the problem of group responses to technologies, concentrating on the influence of social values, institutional constraints, reference group judgments, communication, and power interchange (6).

The means of combining the beliefs into an overall judgment are of minor interest. The research focuses on the roots of the belief-forming process. If social groups are convinced that they will increase their power or gain more resources (money, status, social influence) they will collect or even create positive beliefs and neglect negative ones. Via reference groups influence, all individuals who feel themselves represented by these groups will adopt this view and will shape their belief-system according to the desired result. The cognitions or attributes which are used to justify the judgment are rationalization-strategies to back up personal or group-related interests. General value commitments and perceptions of the society (general attitude systems) are the intervening factors which, on one hand, determine the definition of gains and losses in the social game of acquiring influence and, on the other hand, structure the arguments to legitimate self-interest vis-a-vis all other groups in society (7).

The sociological approach does not apply to the investigation of the individual reaction towards technologies. Yet it is often said that the study of personal beliefs and evaluations --both domains of psychological or social-psychological research--disguises the real causes of the social perception of technologies because the individual beliefs are more justifications of preformulated social judgments made by reference or interest groups. But the fact that interests, power play, and value conflicts play a role in the acceptance of new risk sources does not contradict the fact that people will "mentally" absorb and assess risk sources with the aid of innate or acquired cognitive processing patterns. Only the consideration of both facts will provide a uniform overall picture of the situation. Thus, psychological research is in a far better position to clarify why it is primarily nuclear energy which is confronted with acceptance problems and not, for example, chemical plants, refineries, or automobiles, while sociological research is better able to provide information on the problems regarding the organization of protest behavior and the emergence of general resistance (8).

These preliminary remarks seem to be necessary to elucidate the expressive power and also the limitations of psychological and social psychological research. If the process of absorption and the processing of judgment formation with respect to technologies is successfully traced and typical patterns of people's reaction to new risk sources are identified, a first and important step towards classification of relationships between man, technology, and risk will have been taken.

SEU-THEORY VERSUS ATTITUDE THEORY

Among the psychological and social psychological approaches, two main theories have been applied to study human responses towards technologies: the theory of the subjectively expected utility (SEU-theory) and the attitude theory. According to SEU-theory, the relationship between the expected benefit and loss determines the images of technologies which people shape in the course of assimilating information about new objects. This rationalistic approach can be interpreted as a variant of the value-expectation-concept within the framework of the psychological theory (9).

Attitude theory relies more on psychological learning theories (stimulus-response-models). People memorize specific beliefs about objects through communication and personal experience. Those beliefs are summarized to an overall judgment indicating the degree of favorableness towards the object (10).

For the purpose of describing the differences of SEU and attitude theory more accurately, it is necessary to define some key terms which are needed to understand the different approaches.

> <u>Object Perception</u> - Object perception describes the process of mentally representing and assimilating information and experience with respect to a physical object or entity (11).
>
> <u>Values</u> - A value is a conception, explicit or implicit, distinctive of an individual or characteristic of a group, of the desirable which influences the selection from available modes, means, and ends of action (12).

Belief - Belief represents the cognitive images a person has about a given object, i.e., it is a probability judgment whether an attribute is or is not, and to which degree it is associated with the perception of an object. The subjective feeling of goodness or badness which is linked with each attribute refers to the affect a person might have and is called subjective evaluation (13).

Attitude - Attitude is a mental and neutral state of readiness, organized through experience, exerting a directive or dynamic influence upon the individual's response to all objects and situations with which it is related (14).

Concerns - A concern refers to a state of positive or negative responsiveness of individuals to become aware of and process any information or personal experience regarding salient areas of interest on that matter.

In Fig. 1 we have made an attempt to depict the interconnections between beliefs, concerns, values, attitudes, and perceptions. We have started with five categories: the physical environment, the social environment, the cultural environment, the psychological motives, and the socialized world view. Any individual is confronted with a specific object which is embedded in a social situation and a cultural context (symbolic meaning). The physical properties of the specific object as well as the specific characteristics of the situation are perceived by the individual. The perceived properties are not identical with the real properties. Limited access to information, specialized selection filters, and general concerns govern the perception process. Parallel to the perception of properties, a subjective assimilation of social characteristics takes place: both processes are combined in the subjective assessment of consequences that are seen as associated with the object. Also, at this stage, associations derived from the cultural context or from personal emotion are included.

The next step is the processing of the perceived object properties, situational characteristics, predicted consequences, and associations into a belief system. The selection of what enters into the belief systems, the mode of abstraction from personal experience or concrete information to generalized convictions, and the way of ordering the respective items into salient clusters are influenced by the values,

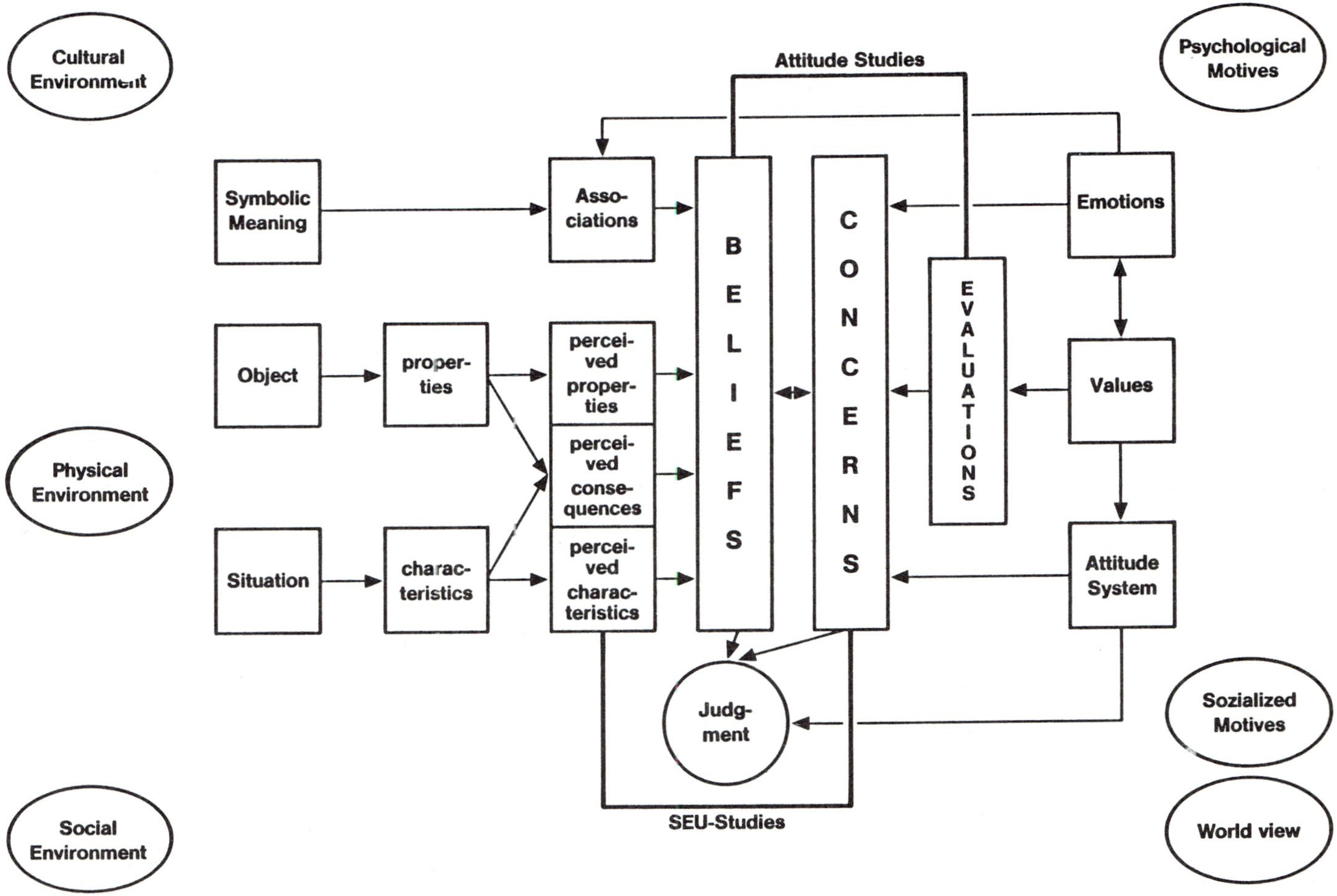

Figure 1: The Mental Process of Technology Perception and the Explanatory Value of SEU and Attitude Theory

emotions, and attitudes towards similar objects via evaluations and concerns.

The last step refers to the process of balancing the positive and negative beliefs aiming towards a general evaluative judgment with respect to the object. For this purpose, the beliefs are ordered according to their subjective importance, the judgments of reference groups are incorporated, the personal consequences of each possible judgment are assessed, and the possible outcome is compared to earlier experience with similar objects.

This outline is, of course, just an analytical tool to understand the process of attitude formation. The different stages are interlinked in the real world and proceed much more unconsciously, as we have pointed out. Yet, it represents a theorectical framework which helps us to direct our research efforts and techniques to the key stations of people's reasoning.

Coming back to the original question regarding the difference between SEU- and attitude theory, we can use the above outline to illustrate the two approaches. SEU-theory combines the level of concerns with the level of perceived properties, characteristics, and consequences. This is done in three steps: 1) We investigate what matters to people, what they are concerned about. The selection of concerns roots in the value orientations and the general attitude system (symbolizing the real world for each individual); 2) we introduce a specific object and ask for the perceived properties with respect to the salient concerns and the probability of their occurrence; and 3) we combine the information via a simple linear or a more sophisticated nonlinear model to predict the overall judgment of a specific person.

The attitude model relies on a very similar procedure but uses a different input. In attitude theory, researchers investigate the beliefs about an object first, and then try to elicit the motivational saliency of each belief by asking for the general evaluations which underly the intention of this belief. Finally, they use different algorithms to combine beliefs and evaluations to determine the direction of attitudinal commitment.

SEU-theory has the advantage that it can be used with objects that are fairly new, because it is based on primary perceptions of properties and general concerns which are not specifically related to the object. It is also more appropriate if the psychological mechanism of perceiving, assimilating, and processing information about general characteristics of objects such as risk, uncertainty, personal benefits) is to be studied.

Attitude theory applies more to research about the understanding of motives and cognitive reasons which underlie the particular responses of people towards a concrete technology. Studies based on this approach are in general more reliable in predicting an overall judgment with regard to a specific technology. This is the reason why we chose an approach based on attitude theory to investigate the motives and beliefs that determine the mental responses of people towards energy systems.

THE ATTITUDE CONCEPT

After the classification of the different psychological and social psychological approaches to study responses of people towards technologies, it is necessary to give a brief introduction into the various attitude concepts. Otherwise our reasons for choosing the Fishbein approach for our own investigations will not be understandable.

The first differentiation among attitude concepts was made between defining the term attitude as a physiological readiness of the body to react as opposed to conceptualizing the same term as a mental state. Since somatic reactions turned out to be undecisive with respect to their psychological meaning, only the mental concept of attitude survived. Measuring finger pulse, galvanic skin responses, or heartbeat rates does not indicate any specific response towards an object, but merely reveals the existence of psychic tensions which might have different reasons and call for various interpretations. Therefore, attitude research at present relies mainly on oral responses or observations of behavioral reactions.

After this issue has been settled, a new controversy came about between the mediative and behavioristic schools. According to the mediative approach, attitude refers to a mental state of favorableness or unfavorableness towards an object exerting a dynamic incentive to behave accordingly (15). The behavioristic school defines attitude as the probability of similar behavior towards an object in similar situations (16). Whereas the behavioristic approach has only been interested in the consistency of overt behavior, leaving aside the underlying mental causes and reasons (black box), the mediative approach intends to investigate the mental process of constructing a relationship between a subject and an object regardless of whether this relationship will be transferred into an observable action (17). Over time the behavioristic approach has been contested because 1) Attitudes towards objects may exist in spite of any overt behavior ever having taken place (e.g. prejudices against foreigners, etc.); 2) Behavior is not only determined by attitudes toward an object or an action, but also is influenced by a set of other aspects which range from personal characteristics to social roles; and 3) The actual behavior is influenced by external parameters (situaton, object, time).

These obstacles cannot be compensated by the advantage to include an operationalized measurement of attitudes within the theoretical definition.

In contrast to the behavioral concept, the mediative refers to the process of mental processing of external stimuli. Shaw and Wright hold that "Attitude refers to a relatively enduring system of evaluative concepts of beliefs which have been learned about the characteristics of a social object or a class of social objects" (18). The problem here is the difficulty to operationalize a mental state.

Within the general framework of mediative attitude definitions, there are different concepts depending on the number and structure of components which determine attitude. The classical consistency theory distinguishes three equally important components: cognitive, affective, and conative.

The cognitive component refers to all the beliefs that a person connects with a certain object; the affective component describes all the emotions and evaluations that are interlinked with the beliefs; and the conative component summarizes all the behavioral intentions which will probably occur if the person faces a situation in which these intentions are a socially accepted way of reaction. Classical consistency theory assumes that individuals try to order beliefs, emotions, and intentions in such a way that a consistent system of positive and negative feeling is developed which can be comprised into an overall judgment. This judgment is a decisiive factor for one's own action (19).

This classical approach was altered by other theories. Rosenberg, for instance, divided attitudes into two segments: cognitive beliefs and affects. Those two components form the central inputs for the overall attitude, but at the same time the more peripheral determinants of the readiness to act according to the overall feeling are relevant. Behavior is also influenced by innate and situational factors, which are unrelated to one's own perception of the object (20).

It is to Rokeach's credit to have pointed out the distinction between beliefs referring to an object and those referring to the situation in which the object is introduced. When comparing the SEU and the attitude models we already mentioned the difference between the perception of the object itself and its social circumstances--both perceptions constitute the subjective assessment of consequences. According to the model of Rokeach, the beliefs connected with the object and the beliefs connected with the situation are combined to a holistic judgment (21).

A multidimensional attitude theory has been developed by Triandis. The starting point of this model is a complex reality which shapes the cognitive structure and the affective evaluations by learning processes. Behavioral intentions are loosely intercorrelated with these socialized patterns of perceptions, but they are more thoroughly influenced by other factors, such as personal properties, behavior of reference groups, or social norms. The actual behavior is a function of the strength of the attitude (not its direction), the behavior intentions and other situational variables (22).

Finally, we should mention the theorectical work by Vroom who set up a rather similar approach to SEU-theory. According to his model, beliefs can be substructured into instrumental, probabilistic, and consequential aspects. These aspects are evaluated by the valence of values and emotions and transferred into an attitude (23). An illustration of the basic features of all consistency models can be seen in Fig. 2.

As a viable alternative to the consistency models, other concepts have been proposed which regard attitudes as a composition of emotionally evaluated beliefs (24) based on the following definition: "A person's attitude toward any object is a function of his beliefs about the object and the implicit evaluative responses associated with these beliefs" (25). The attitude theory developed relies on the following assumptions (26): 1) Any given object is related to various attributes; 2) Associated with each of the attributes is an implicit evaluative response, i.e., an attitude; 3) Through learning and experience, these evaluative responses are associated with the attitude object; 4) These evaluative responses summate; and 5) On future occasions the attitude object will elicit this summated evaluated response, i.e., the overall attitude.

If these assumptions are regarded as valid, attitudes can easily be elicited by collecting data on beliefs about the respective attitude object and the evaluative weight given to each attribute. Other unidimensional attitude models rely

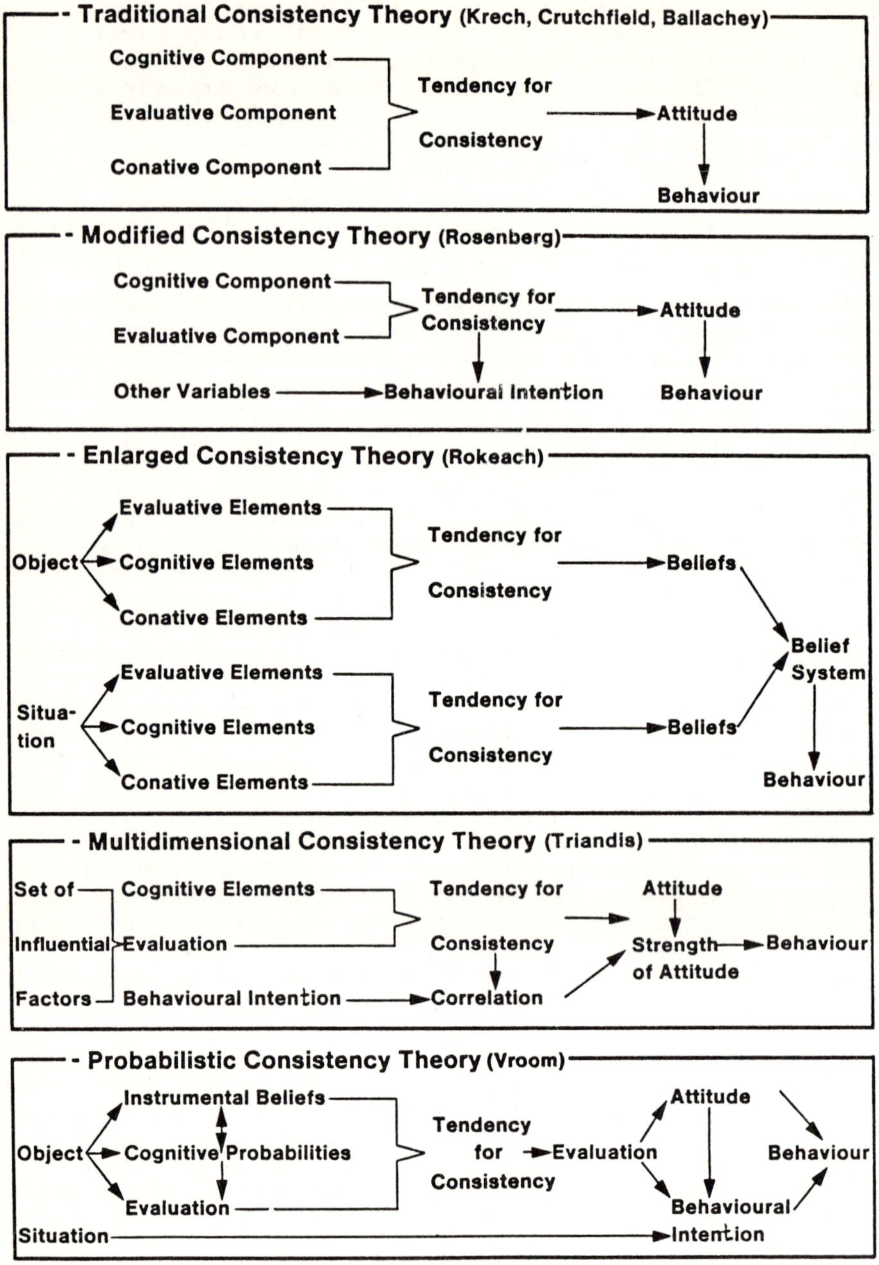

Figure 2: A Review on the Basic Assumptions of Consistency Theories on Attitudes and Behaviour

solely on the composition of affective statements about the object in question.

Figure 3 summarizes the relations among beliefs, attitudes, intentions to perform specific behaviors, and behavior. A good explanation of this figure has been given by H. Otway in an article on risk perception:

> "We can see that a person holds many beliefs about each attitude object, i.e., he has "learned" a number of "facts" about it. But this attitude only pre-disposes him to behave in a consistent way with respect to that object when a large number of behaviors are observed. Attitude does not pre-dispose him to perform any specific behavior with respect to that object, and thus attitude would not be expected to show any relation to some specific behavior. However, to repeat, the attitude toward the object is of interest because it gives us an indication of the overall pattern of behavior with respect to the object." (27)

By comparing the different attitude models, it seems evident that the multidimensional approaches have a more complex and realistic theorectical base, but encounter enormous problems in establishing an appropriate measurement technique. In particular, the composition rule for combining the different dimensions into one holistic judgment cannot be derived by theorectical assumptions and is open to subjective variations. However, the simple affective scales to measure attitudes cover only partially the complexity of attitudes and have empirically been proven as bad predictors for general behavior (28).

As a good compromise between the theoretical complex multidimensional attitude concepts and the simple affective concepts based upon a single-scale measurement, we decided to choose the in-between model of Fishbein which covers at least two dimensions of attitudinal patterns and provides for a precise and well interpretable measurement procedure. The general drawback of the Fishbein model is the assumption that no response biases exist among the beliefs and that no interaction takes place between the evaluation and the beliefs. It is evident that in reality this cannot be accomplished. In practical research, however, a feasible solution is to state a general thematic frame prior to the

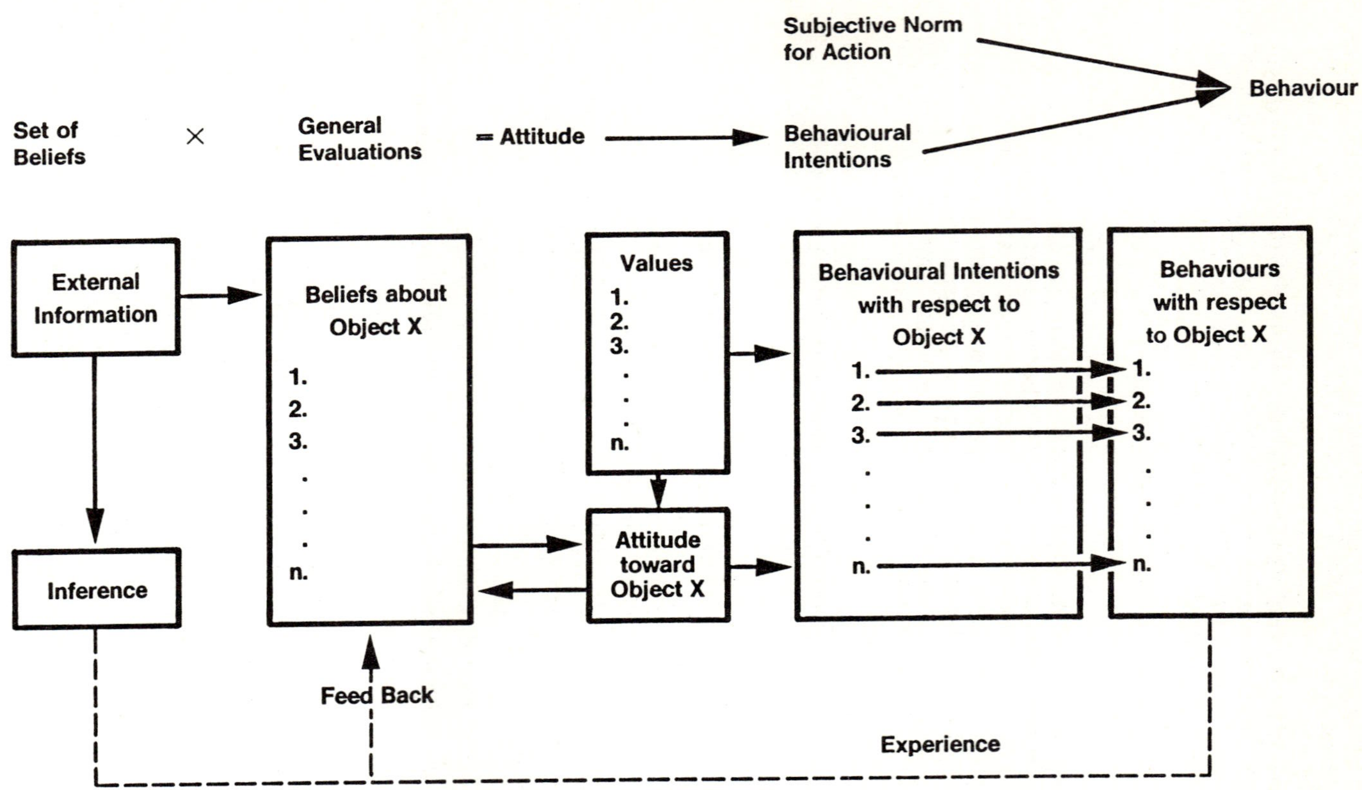

Figure 3: The Relations between Attitudes, Beliefs, Behavioural Intentions and Behaviour According to Fishbein

measurement of evaluations and to introduce the concrete attitude object only when the beliefs are presented. The intercorrelations among beliefs can be detected by using special statistical techniques like multiple stepwise regression analysis. Another advantage of the Fishbein model is its applicability for comparisons between samples from different nations since the formalized concept can be transferred to different culture contexts without opressing cultural variations in reasoning and object perceptions.

GENERAL METHODOLOGY OF THE FISHBEIN TECHNIQUE

According to the attitude theory of Fishbein and Ajzen, a person's attitude toward any object is a function of his beliefs about that object and the implicit evaluative responses associated with those beliefs.

Formally, beliefs and attribute-evaluations are combined as follows:

$$A_o = \sum_{i=1}^{n} b_i e_i \text{ where}$$

A_o is the attitude toward the given object,

b_i is the belief about this object, expressed as subjective probability that the object is related to attribute i;

e_i is the evaluation of attribute i; and

n is the number of beliefs (units of information) a person holds about the attitude object in question.

This formal representation of the model suggests a twofold measure of attitude. Since the sum of the belief-evaluation products represents an indirect measure of attitude, this measurement can be validated by a direct approach to measure attitude. The technique selected for this purpose is the "Semantic Differential" developed by Osgood, et al. (29).

With this method, the respondent is presented with a series of seven-place bipolar adjective scales. The end-points are defined by adjectives which have generally been shown to have evaluative connotations; e.g., good-bad, harmful-beneficial. The respondent is asked to rate the object by placing a check mark at the point on the scale which he feels indicates the appropriate description. A direct measure of attitude is obtained through summating over the responses to this set of bipolar adjectives which have an evaluative connotation with regard to the given attitude object.

The correlation coefficient between the direct and indirect measurement demonstrates the reliability of a chosen set of attributes as relevant units of information.

A further test of the validity of the Fishbein scaling technique can be carried out by asking the respondents directly what their feelings are considering the perceived properties of a given object. This direct measurement of the attitudinal commitment can also be correlated with the indirect measurement by the Fishbein algorithm.

Thus, several separate and independent measures of attitude are available: the sum of the eb-products, interpreted as an indirect representation of attitude; the sum over the ratings of the adjectives of the semantic differential interpreted as a direct representation of attitude, and finally, attitude by the direct scale (PRO/CON scale) where respondents simply indicate their degree of liking or disliking of the attitude object.

THE APPLICATION OF THE FISHBEIN TECHNIQUE TO ENERGY SYSTEMS

Design of a questionnaire pertaining to the Fishbein technique requires a careful selection of the attributes. The development of the questionnaire for the IAEA/IIASA studies underwent the following stages: a first version was put together after extensive screening of mass media and relevant literature. In addition, some 100 persons in Vienna were interviewed about what came to their mind when they thought of "nuclear power." Particularly this interview procedure is of considerable importance because it permits the researcher to select those concerns which are forwarded most often (in Fishbein-terminology, overall salient beliefs). Although a person's attitude will be determined by only a few salient beliefs, it is necessary to use a larger set of relevant beliefs in order to detect particular clusters of beliefs among various social groups. The identification of underlying determinants of attitudes via statistical procedures, like factor analysis, also requires a larger amount of items for the deduction of valid results. A number of 30 to 40 different beliefs seems to be sufficient to meet both requirements.

Accordingly, a pilot questionnaire was designed and applied in Austria to a stratified sample (30). The next step implied some modifications of the original version of the quesionnaire so that it could be applied in other countries as well which meant the inclusion of issues which has been repeatedly raised in different countries. Furthermore, the questionnaire was extended by another form of evaluating the beliefs. All items were again to be judged, but now from the point of view of how important these issues were perceived to be in the ongoing energy debate. This, we felt, would permit not only a better understanding (cross-validation) of the relevance of the selected items and the role these issues have in acceptance or rejection of energy technologies, but also provide an interesting means of comparison between different samples in cross national comparison.

After the necessary pretests, the questionnaire used in later studies consisted of five parts, each having separate instructions. All instructions were written on the questionnaire, permitting application to groups of respondents instead of single face-to-face interviewing. Running through the whole interview takes between 30 to 90 minutes, depending on the speed of the respondent's decisionmaking and the number of energy systems included.

The first part of the questionnaire is devoted to elicit the attribute evaluations, where respondents express their subjectively felt goodness or badness of each of the 30 attributes. The verbal anchoring of the scale for this part is "good-bad." For example:

Promoting my nation's industrial development

BAD: ____: ____: ____: ____: ____: ____: ____: GOOD
very quite slightly don't know slightly quite very

It should be noted that here the attributes are presented without reference to any specific energy source.

In the second part of the questionnaire, the same 30 attributes are presented, but now linked with the attitude objects for each of the selected energy systems. This belief strength is measured again on a bipolar seven-place scale with the end points now labelled "unlikely-likely." The respondent is asked whether and how strong the attribute is related to the attitude object. For example:

The use of nuclear energy promotes my nation's industrial development

UNLIKELY: ___: ___: ___: ___: ___: ___: ___: LIKELY

This part is repeated separately for each energy source under investigation.

The third part is designed using the techniques of the Semantic Differential, as described above. Here each bipolar seven-place scale has a different verbal anchoring, namely pairs of adjectives found meaningful to describe an energy source. Sixteen pairs of adjectives were selected and again each energy source was rated separately.

The next section of the questionnaire (part 4) was designed to explore the awareness of respondents about those issues which are prevalent in the discussions about acceptance of nuclear energy. Thus, emphasis in this part is given to the level of information of respondents rather than their personal attitudes, and can be used as a means for cross-checking the salience of the attributes selected. The instruction given for this section asks the respondents to indicate their opinion about the importance of the 30 statements with regard to the ongoing discussions about the use of nuclear energy. Again, a seven-place scale is used whose end points are labelled "very unimportant-very important." For example,
Being harmful to future generations

VERY UNIMPORTANT ___: ___: ___: ___: ___: ___: ___: VERY IMPORTANT

Results derived from these importance ratings lend themselves to interesting comparisons with samples from other countries because the level of awareness and information is indicated. Furthermore, a lack of information (e.g., about particular benefits) can be detected.

Subsequently, a straight forward measure of the personal opinion about nuclear energy (PRO/CON scale) is elicited on a 7-point scale with the labels "very unfavorable-very favorable." Part 5 of the questionnaire is devoted to demographic information such as sex, age, years of schooling, occupation, etc.

CROSS NATIONAL APPLICATION

During the 1970s the questionnaire was translated into German, Spanish, French, Portuguese, Finnish, and Japanese. Psychologists and social scientists from various countries have been trained to use the Fishbein technique and to conduct surveys in their countries. So far data on attitude towards the use of nuclear energy have been collected from Brazil,

Colombia, Japan, the Philippines, Austria, Finland, and the Federal Republic of Germany. Similar questionnaires have been used in France and Canada. The main purpose of all these studies is to gain a better understanding for controversial issues and to identify underlying basic social values which guide the evaluation of various beliefs. In particular, national or cultural differences in perceiving the same object might give some insight information about the structure of the individual perception process. Some unique features seem to play a decisive role in almost all countries (such as safety, long-term risks, and indirect economic benefits), wereas others turn out to be rather specific with respect to national particularities.

Cross-national surveys always run into the difficulty of finding appropriate samples which allow comparisons between countries. Any comparison relies on at least one common denominator for which differences can be identified and measured. Because countries differ in their cultural heritage, their social structure, and their economic system, it is essential to base comparative research on a fixed and constant parameter that holds true for each of the observed countries and can be used as a yardstick to interpret deviations between countries in a meaningful way. Such a common parameter might be an equal degree of knowledge, a share of common values, or a homogeneous structure of social positions. Because attitudes on energy systems are not independent from the level of object-related knowledge (which varies from country to country) and because there is no indication that nuclear energy is perceived in terms of indentical values, it is necessary to restrict the scope of social positions in order to create a homogeneous background for comparisons.

As long as rather similar countries are selected for the analysis of attitudes, a survey based on random sampling of the general public can be used. For the purpose of cross national survey between nations of different cultural background, the range of social positions has to be predefined to end up with comparable samples. In this case we confined our samples to students of technical and natural sciences, opinion leaders, politicians, or other educated classes. Nevertheless, the attitudes of the general public in each participating country were also measured to compare the results of these surveys with the finds of the elite samples. This comparison--if restricted to one country or culture only--is a valid and useful instrument to investigate the gap between the perceptions of the general public as opposed to those of technical or cultural elites.

STATISTICAL METHODS OF DATA PROCESSING

Apart from the normal statistical procedures such as frequency distribution analysis and correlations, three more sophisticated methods have been applied to reduce the large amount of variables to a smaller number of salient elements. The techniques involved are factor analysis, multiple regress, and discriminant analysis.

Factor analysis is a method used to determine the underlying cognitive structure in a given set of attributes by comparing the similarities of variance distributions resulting in a combination of items with high intercorrelations. The main purpose of a factor analysis is to derive dimensions which are inherent in a larger set of items.

Multiple regression indicates the strength of a relationship between one dependent variable and a set of independent variables whereby the intercorrelations between the items of the independent variable set are excluded from the analysis. This procedure explains for each independent variable (e.g., attribute) the additional amount of the declared variance of the dependent variable (e.g., semantic differential as a direct measurement).

Discriminant analysis provides a viable yardstick for evaluating the relative distance between various sub-groups of a given sample. A whole set of possible discriminative variables can be investigated and the result of the analysis reflects the relative significance of items as explanations for group differences.

A BRIEF ILLUSTRATION OF THE FISHBEIN TECHNIQUE: THE AUSTRIAN STUDY ON NUCLEAR ENERGY

In the earlier studies of the Austrian public and Austrian decisionmakers (31), the original 39 belief items were subjected to factor analysis to explore the underlying dimensions which characterize the salient determinants of people's feelings towards nuclear energy. This factor analysis produced a factor structure consisting of four belief clusters: Factor 1, psychological risk; Factor 2, economic and technical benefits; Factor 3, social-political risks; and; Factor 4, environmental and physical risks.

Two groups were subsequently drawn from the total sample representing the most pro-nuclear and anti-nuclear respondents. Table I shows the contributions made to the formation of the opposite attitudes by each of these dimensions. The pro-nuclear attitude is largely due to the contributions of factors 2 and 4 and thus rely on the technical, economic, and environmental considerations. The anti-nuclear commitment can be best explained by factors 1 and 3, concentrating on the psychological and sociopolitical aspects of nuclear energy. The strongest differentiation between the two groups is the psychological risk dimension (30).

Table I: Contributions Made to Attitude by Each Belief Dimension (Public Groups)

Dimension	Attitude contribution (Range ± 9)	
	Pubpro (n = 50)	Pubcon (n = 50)
Psychological risk	− 1.6**	− 6.4
Economic-technical benefits	+ 3.2**	− 0.2
Socio-political risk	− 1.2*	− 3.2
Environmental risk	+ 1.9**	− 1.4

* Difference significant at 0.05 level
** Difference significant at 0.01 level

In order to validate these finds, the sum of all evaluated beliefs were correlated with the semantic differential measure. The correlation coefficient of v = 0.63 is highly significant. Also, the sum over the factor items was correlated with the semantic differential measure (for the purpose of demonstrating the validity of the factor structure), resulting in a coefficient of r = 0.66.

Thus, we feel justified to conclude that the main reason for people in Austria to oppose or favor nuclear energy is the saliency of psychological and sociopolitical aspects. If persons are convinced that nuclear energy poses "psychological" risk on their life (such as delayed effects, nonsensible dangers, no personal control), and that at the same time sociopolitical impacts of this technology exist, a negative attitude is likely to be formed regardless of how the physical and environmental risks are perceived (31). This application of the Fishbein model illustrates its usefulness in explaining different responses of people towards energy technologies.

Next, we will describe in more detail a comprehensive study on attitudes towards nuclear power, comparing students in three different countries. This description is intended to outline the distinct steps of a case study and to provide an understanding for the method, its application, and its potential.

A CASE STUDY: A COMPARISON OF ATTITUDES IN THREE NATIONS TOWARDS NUCLEAR POWER

During the time from 1978-1982, three surveys of students from Technical Universities in the Federal Republic of Germany (FRG), Japan, and the Philippines were conducted using the questionnaire which was developed by the IAEA/IIASA Risk Assessment Group (32). All necessary translations were made by bilingual social scientists; several pretests proved the validity of the translation.

As previously indicated, international comparison demands a certain degree of homogeneity with respect to the selected samples if the issue of the survey is related to either experience or knowledge. Since attitudes on energy systems rely partly on the level of knowledge, we decided to confine our samples to students of technical and natural sciences. We assumed that students of these disciplines have at least a basic understanding of the functions and purposes of different energy systems regardless of their country of origin.

COMPOSITION OF THE SAMPLES

Three student samples were taken from Technical Universities. The German students were enrolled in Aachen and Cologne (N = 150), the Philippine students in Manila (N = 174), and the Japanese students in Tokyo (N = 36) and Osaka (N = 84).

With regard to sex distribution, the sample from FRG consisted of 63 percent males and 37 percent females, from the Philippine 26 percent males and 69 percent females (5 percent of the sample did not indicate their sex in the questionnaire); and in Japan 96 percent males and four percent females responded to the questionnaire. The age distribution for the three samples is given in Table II.

Table II: Age Distribution of the Three Samples

	FRG		PHILIPPINES		JAPAN	
Age Categ.	Absolute Values	Percent	Absolute Values	Percent	Absolute Values	Percent
18-29	94	62.7%	124	71.3%	107	89.9%
30-45	55	36.7%	18	10.3%	11	9.2%
46-59	1	0.7%	2	1.1%	1	0.8%
Missing Cases			30	17.2%	1	0.8%

VALIDATION OF THE MODEL

As indicated in the description of the Fishbein technique, the significance of the belief items can be tested by correlating the indirect measurement represented by the sum over the evaluated belief items with the direct measurement which is defined as the sum over the relevant adjectives of the semantic differential. The second direct measurement, the pro/con scale, is also included. This was done for the three samples separately. Table III shows the correlation coefficients between the two direct attitude scores (PC, ΣSD) and the indirect measure (Σeb).

Table III: The Correlation Coefficients between Direct and Indirect Measurements of Attitudes (Model Validation)

Attitude Measures	JAPAN	PHILIPPINES	FRG
PC -Σeb	0.59684	0.47485	0.73599
PC -ΣSD	0.71027	0.65471	0.86473
Σeb -ΣSD	0.68520	0.52331	0.81172

All correlations are significant on the one percent probability level, indicating that the design of the questionnaire represents a valid instrument for investigating attitudes. The differences of the correlation coefficients between the three samples can be attributed to the specific variance distribution of each sample.

Examination of Fig. 4 where the respondent frequencies are given in percentages for each response category for each sample, shows that the Japanese students were predominantly for the use of nuclear energy (categories 2 and 3 account for 68 percent of the sample); the Philippine student sample includes two groups, one very opposed (category -3 accounts for 25 percent) and one slightly in favor (category 1 represents 20 percent). The FRG students are also composed of two groups, with 45 percent of the sample being for the use of nuclear power (categories 2 and 3), and a smaller group of 25 percent (categories -3 and -2) being against.

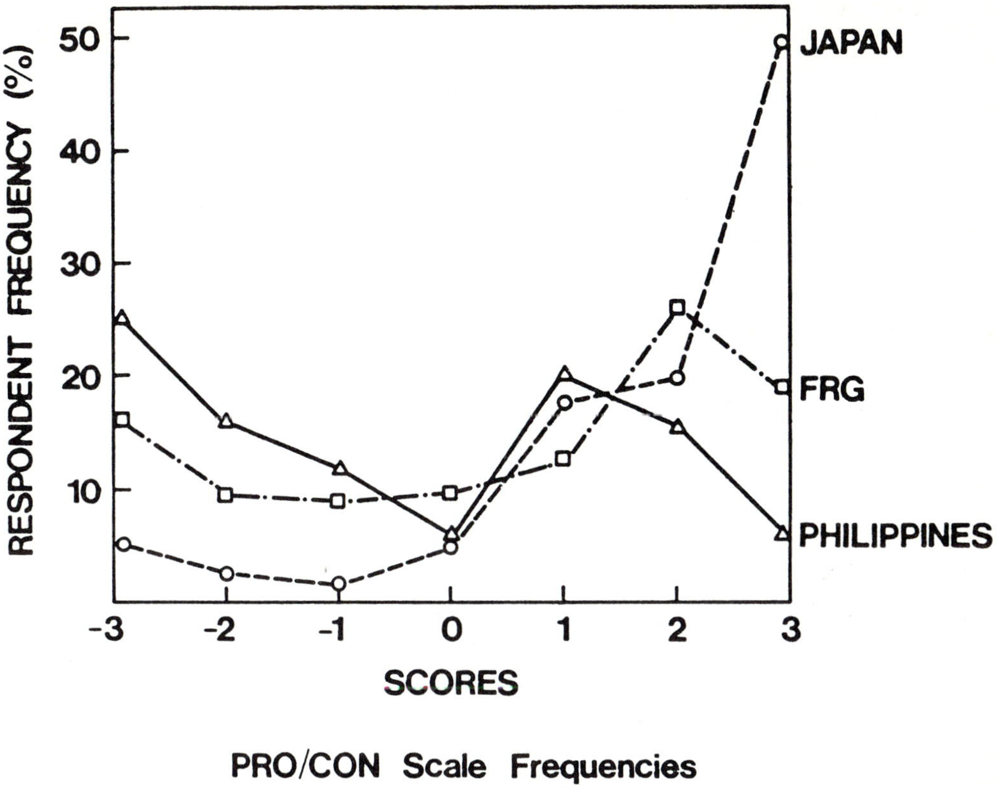

Figure 4: The Frequency Distribution of the PRO/CON Scale Referring to Nuclear Energy

As a general remark it can be noted that except for the Japanese students who are predominantly in favor of the use of nuclear power, the other two student samples from FRG and the Philippines appear to include both interest groups, proponents and opponents.

The second direct attitude measure available is the adjectives of the semantic differential. Figure 5 shows the mean scores for the three samples. The favorable Japanese resondents see the use of nuclear energy as very important and useful, quite good, modern, and worthwhile, but also realize that it is quite controversial and slightly dangerous. In contrast to this relatively clear perception, Philippine students view the same concept as quite wrong, oppressing, and useless. FRG students have the least strong feelings about the use of nuclear energy, perceiving it as quite modern and useful, slightly important, and worthwhile, but slightly dangerous.

Taking the three samples together, the students agree that nuclear power is important, worthwhile, and good, but also that it is dangerous, oppressing, and controversial. Some clue for interpretation of the attitudes of the respective samples might be the fact that there is no nuclear power plant in operation and only one in construction in the Philippines, whereas there are about 15 nuclear power plants operating in the FRG and 24 in Japan. Respondents from these two countries seem to have a more positive attitude towards nuclear energy.

THE BELIEF STRUCTURES

After having established the composition of the various samples with regard to their orientation PRO or CON the use of nuclear energy, the following analyses concentrate on identification of determinants for these attitudes.

Application of factor analysis to the belief scores is expected to reveal the cognitive structure of the respondents concerning their perception of the issues pertaining to the use of nuclear energy. The method used is principal component analysis with subsequent Varimax rotation. This technique produces underlying dimensions which are independent, i.e., orthogonal factors. Since the three samples differed in their attitudes towards nuclear power, it is not anticipated that the factor structures will be identical. Nevertheless, the clustering of items with high intercorrelations could be an informative indicator for general issues of interest in the respective societies.

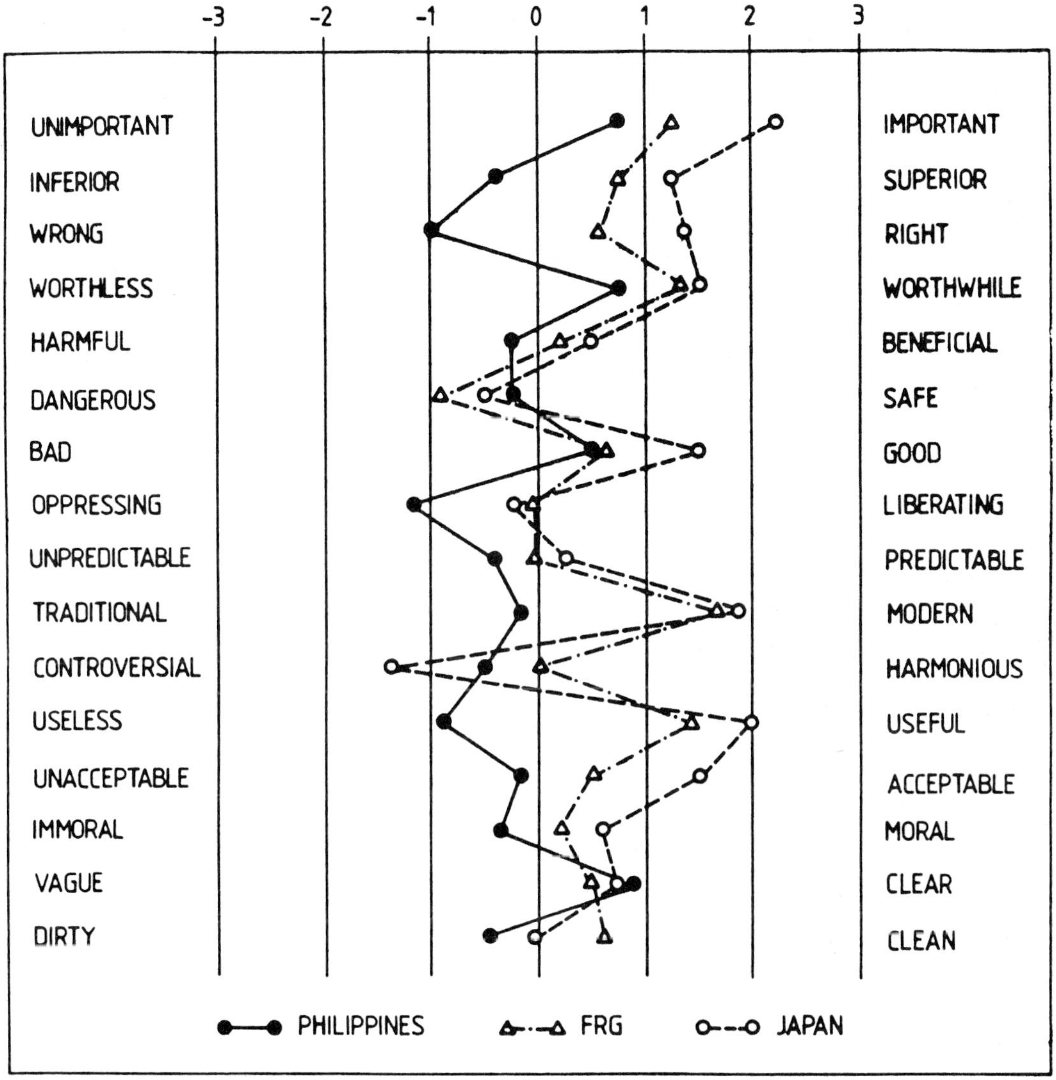

Figure 5: Means of the Semantic Differential to Characterize Nuclear Energy

Philippines

The result of factor analysis of the belief scores on nuclear energy are given in Table IV. Items with high intercorrelations cluster around two aspects, the risk considerations and the benefits, whereby the latter are distinguished by having an effect on the national level or a rather personal level.

Factor I represents the strongest concern with an eigenvalue (i.e., explanatory power for overall variance) of 7.15 and has been labelled: "negative impacts of large-scale technology." The issues involved include the lack of active control over the hazard and the involuntary exposure to it, the concern about large accidents, and the health hazards created, whether directly or indirectly through burdening the environment.

Factor II comprises issues of national interest such as progress and industrial development, prestige, and stimulation of research and has been termed "progress in national development." This factor has an eigenvalue of 3.93.

Factor III seems to represent the more personal benefits respondents attribute to the use of nuclear energy. Thus, the leading items for this factor are referring to the economic production of electricity and the capacity to cover the energy needs on a long-term basis. Another aspect of personal concern, the provision of jobs, is also included in this factor, labelled "fringe-benefits." The eigenvalue of this factor is 3.27.

In order to determine the influence of the cognitive structure on the attitudinal commitments, a multiple regression analysis was undertaken with the semantic differential as dependent and the factors of the belief systems as independent variables. The purpose of the regression analysis was to detect the order of influential strength of each independent variable for the explanation of the dependent criterion. The stepwise procedure of the regression assures that only then will an independent variable be included in the analysis if it significantly adds to the amount of variance already explained by those variables which have been selected through the analysis so far.

For the samples from the Philippines it can be demonstrated that the best predictor for the sum of the semantic differential is Factor III, named "fringe-benefits" ($r = 0.54$; $p \leq 0.01$). This factor is followed by the other benefit-related factor, "progress in national development" ($r = 0.51$, change in R^2) = 0.12; $P \leq 0.01$).

Table IV: Factor Structure of Beliefs (PHILIPPINES)

FACTOR LOADING	BELIEF STATEMENT: THE USE OF NUCLEAR ENERGY...
	FACTOR I: Negative impacts of large scale technology
.87	...exposes people to hazards which they cannot influence by any actions of their own
.78	...involves hazardous agents which cannot be detected by man's senses
.78	...leads to accidents which affect large numbers of people at the same time
.74	...leads to environmental pollution
.71	...has an impact on people's health
.71	...leads to dependency on small groups of specialists
.70	...is harmful to future generations
	FACTOR II: Progress in national development
.84	...leads to technological progress
.81	...promotes my nation's industrial development
.78	...increases my nation's prestige
.64	...prevents brown-outs*
.57	...stimulates scientific and technological research
	FACTOR III: Fringe-benefits
.78	...provides a cheap energy source
.67	...is a long-term solution to energy needs
.60	...leads to a more even distribution of income among nations
.55	...helps to conserve natural resources
.53	...leads to increased employment

*item only included in Philippine survey

Factor I "negative impacts of large-scale technologies," seems to play only a minor role with regard to their attitudes, since the single correlation amounts to only $r = 0.31$ ($P \leq 0.01$) and the additional amount of declared variance is below one percent ($P = 0.12$). Considering the composition of the sample (about 50 percent pro and 50 percent con, see Fig. 4) it is interesting to note that it is the two benefit factors which seem to be the predominant aspects in the attitude formation of the Philippine sample.

Federal Republic of Germany

Factor analysis of belief scores on nuclear energy of FRG students yielded four factors as shown in Table V. Only one of these factors emphasizes the benefits of nuclear energy, whereas three express concern about the various negative consequences of this energy source. The beneficial aspects cluster in Factor I, "economic progress," whose eigenvalue of 6.33 indicates that a relatively large part of the total variance is explained by this factor. Items refer to cheap electricity production, improvement of standard of living, and various benefits on a national level.

The second factor with an eigenvalue of 4.78 is very similar to Factor I of the Philippine students. Thus, the same labelling, "negative impacts of large-scale technology," was used.

Factor III represents a collection of concerns which could develop into a threat, probably not tomorrow but in the near future. Therefore, it is described as "potential for threat," including issues such as terrorist activities, passive exposure, international conflicts, and the long-term radioactivity of wastes. The eigenvalue of this factor is relatively low, 2.48.

The possible impact of nuclear power on society is expressed in Factor IV, "restriction of social flexibility," with an eigenvalue of 2.42. Items loading high on this factor refer to the possibility of a restricted societal development and the dependency on big industry and its highly specialized professionals.

Computation of multiple regression coefficients as indicators for the attitudinal commitment revealed that in contrast to the Philippine sample, the perception of "negative impacts of large-scale technologies" turned out to be the predominant factor attitude as determined by the direct measurement. The simple correlation was $r = 0.83$ (P 0.01). "Economic

Table V: Factor Structure of Beliefs (FRG)

FACTOR LOADING	BELIEF STATEMENT: THE USE OF NUCLEAR ENERGY...
	FACTOR I: Economic progress
.75	...provides a cheap energy source
.75	...improves our standard of living
.71	...assures the economic independence of my country
.68	...promotes my nation's industrial development
.62	...leads to technological progress
	FACTOR II: Negative impacts of large scale technology
.74	...leads to environmental pollution
.68	...has an impact on people's health
.64	...leads to accidents which affect large numbers of people at the same time
.63	...has a long-term impact on climate
.60	...is harmful to future generations
	FACTOR III: Potential for threat
.71	...provides a source of threats from terrorists
.59	...exposes people to hazards which they cannot influence by any actions of their own
.44	...involves a technology which is usable as a tool in international politics
.42	...postpones the development of alternative energy sources
.41	...requires management of dangerous wastes
.40	...restricts personal freedom through rigorous security measures
	FACTOR IV: Restriction of social flexibility
.67	...leads to consumption-oriented society
.56	...leads to dependency on small groups of specialists
.51	...restricts options for future societal development
.47	...concentrates power in big industrial enterprises

progress," was the second best predictor, adding a surplus amount of 6 percent to the declared variance ($P = 0.02$). Although the two other risk factors, "potential for threat" and "restriction of social flexibility," correlate highly with the semantic differential ($r = 0.80$ and 0.81 respectively) they do not add any explanatory power to the multiple correlation coefficient, after Factor II (impacts of large-scale technology) has been included as first predictor.

The predominant concerns in Factors III and IV are thus covered by Factor II, at least in those areas which play a significant role in the formation of this sample's attitudes. It seems evident that potential threat and social risks as highly complex beliefs do discriminate between different attitudinal positions, but are not the underlying reason for this initial formation of positions towards nuclear energy. Rather, both factors are seen as necessary consequences of the perceived negative impacts of large-scale technologies. Respondents who related highly negative impacts with the use of nuclear energy were just as well convinced that the threat for political misuse and for social flexibility was connected with the general impacts of large-scale technologies.

Japan

The factor structure of the Japanese student sample also comprises four factors, again with only one of them referring to beneficial aspects of nuclear energy but three dealing with threats, hazards, and negative impacts (Table VI). However, the factor combining the beneficial aspects occupies first place among all factors. With an eigenvalue of 4.69, it refers mainly to economic advantages such as industrial development, standard of living, and increased employment; thus, Factor I was labelled "economic prosperity." Factor II deals with the "impact on society" from the use of nuclear energy and has an eigenvalue of 3.41. Items with this factor include concerns about the national and international power distribution and health considerations. The third factor, whih combines possible impacts on society in the future has been termed "long-term hazards" and has a relatively low eigenvalue of 2.79.

Factor IV consists of mainly the same items as Factor II of the FRG sample, thus, it has been labelled "potential for threat." This factor also has a quite low eigenvalue (2.57).

Regarding the relevance of these four factors for the attitudinal commitment of this student sample from Japan, "economic prosperity" has the highest correlation ($r = 0.64$; P 0.01) and therefore takes the first position in the stepwise

Table VI: Factor Structure of Beliefs (JAPAN)

FACTOR LOADING	BELIEF STATEMENT: THE USE OF NUCLEAR ENERGY...
	FACTOR I: Impact on society
.64	...concentrates power in big industrial enterprises
.61	...leads to diffusion of knowledge for construction of weapons
.58	...has an impact on people's health
.55	...leads to environmental pollution
.50	...involves a technology which is a tool in international politics
	FACTOR II: Economic prosperity
.71	...promotes my nation's industrial development
.70	...improves our standard of living
.69	...leads to technological progress
.63	...leads to increased employment
.62	...leads to a more even distribution of income among nations
.61	...increases my nation's prestige
	FACTOR III: Long-term hazards
.76	...restricts options for future societal development
.62	...has a long-term impact on climate
.61	...is harmful to future generations
	FACTOR IV: Potential for threat
.72	...requires management of dangerous wastes
.62	...provides a source of threats from terrorists
.57	...exposes people to hazards which they cannot influence by any actions of their own
.57	...leads to accidents which affect large numbers of people at the same time

multiple regression. The third factor, "long-term hazards," has the second best predictive power, immediately followed by Factor II, "impact on society." Both factors increase the share in the explanation of the dependent variable with 3 percent and 2 percent, respectively, whereby the additional explanatory power of Factor III can be regarded as significant ($p = 0.07$), and Factor II does not add significant ($p = 0.07$) to the overall declared variance of the semantic differential. Similar to the Philippine sample, economic considerations are the best discrimination for different attitudinal positions. Both risk factors, "impacts on society" (with a simple $r = 0.40$) and "long-term hazards" (with a simple $r = 0.45$) are definitely important for the formation of attitudes, but they are by far not as decisive as the benefit factor. The multiple regression analysis has demonstrated that--after the benefit factor has been accounted for--the two risk factors do not improve the prediction of attitude. The "potential for threat" factor barely contributes as attitude determinant with a simple correlation of $r = 0.27$, $p = 0.05$). In the multiple regression analysis, this factor has no predictive power at all.

Comparative Remarks

As expected, the factor structures derived from the three samples were not identical. While analysis of the beliefs from the German and Japanese students yielded three risk-orientated and one benefit factor, Philippine students differentiated between two benefit dimensions, but perceived only one risk dimension. In spite of this general difference in the factor structure, the benefit factor played a major role in all the samples regarding their relative influence on the explanation of the variance of the attitudinal direction. For the Japanese and the Philippine students, the economic factor turned out to be the best predictor for their attitudinal position; for the German sample, it was the second best. For the German students, the negative impacts of large-scale technology discriminated best between favorable or unfavorable attitudinal positions towards nuclear energy. In contrast, the Philippine students did not share this concern. Another factor, "potential for threat," was also represented in two samples, but made a relatively low contribution to attitude for both German and Japanese students.

CROSS-NATIONAL ATTITUDE DETERMINANTS

To investigate the differences in the responses between the three student samples in more detail, two statistical procedures were applied. First, the weighted belief items were used separately as independent variables for a stepwise multiple regression, again with the sum over the adjectives of the semantic differential as dependent criterion. In order to have a larger range within the variance of the independent predictors and to include evaluated beliefs, the scores of the Exb scale were used instead of the belief-scores only. Second, a discriminant analysis was performed to yield an order of those belief items which discriminate most highly between the three samples. Also, it seems interesting to investigate how precisely the membership of each individual case to one of the three samples can be predicted only knowing the scores of some salient beliefs.

The Salient Beliefs for the Formation of Attitudes

The five predictors which were found to be most relevant for the explanation of the attitude scores towards nuclear energy as derived from the semantic differential are presented in Table VII. As expected, each sample has its own specific pattern of issues relevant for their attitude formation. In line with the findings of the previous chapter, a benefit-related argument occupied first place with the Philippine and the Japanese sample, whereas the German students chose a risk argument as most decisive for their attitude towards nuclear energy. For them, health consideration and a potential impact on the environment are prime issues, together with the realization of the importance to produce energy in an economic way. Japanese students put most weight on the accomplishment of technical progress, but are considerably influenced by the threat of large accidents, the hazards for future generations, and the dependency on experts. Improvement of the standard of living is the key issue for the Philippine students, followed by the related reflections about economic energy production and the enhancement of research. However, concern about the environment is also an important part of their attitudinal commitment.

In general, it can be noted that, considering the five most important predictors for all three samples, benefit-related issues have the largest influence on the formation of attitudes, with the sample from the Philippines occupying the

Table VII: Multiple Regression of eb Scores

(Most Important Predictors in Descending Order)

		R(changed)
GERMAN STUDENTS		
13.	Having an impact on people's health	.73**
17.	Providing a cheap energy source	.79**
20.	Leading to environmental pollution	.82**
14.	Postponing the development of alternative energy sources	.84
5.	Leading to technological progress	.85
JAPANESE STUDENTS		
5.	Leading to technological progress	.50**
18.	Leading to accidents which affect a large number of people at the same time	.62**
4.	Being harmful to future generations	.69**
10.	Leading to dependency on small groups of specialists	.73**
25.	Leading to increased employment	.75
PHILIPPINE STUDENTS		
1.	Improving our standard of living	.46**
17.	Providing a cheap energy source	.55**
26.	Stimulating scientific and technological	.60**
20.	Leading to environmental pollution	.64**
21.	Restricting options for future societal Development	.65

** Changes in correlation significant (p 0.01)

three top positions. German and Japanese students consider both benefits and risks as decisive for their attitudes.

The Discriminative Power of Weighted Beliefs

The most important question in cross-national surveys is the distinction between typical belief and evaluation patterns which form the basic skeleton of each cultural identity. Identification of attitudes towards energy is certainly too confined as a concept to permit in depth interpretation of cultural and social properties which influence the response patterns to new technologies. But at least the concerns which are predominant in one country compared to the others can be revealed.

In order to detect the main differences between the three samples, discriminant analysis was used. Depending on the parameters used for the statistical calculation, all variables can be ordered according to the degree to which their variance discriminates between the samples. Table VIII shows the results for each item. In the first line the most discriminative item is listed followed by the second most discriminative and so on. In total, 15 items out of 30 proved to be significantly different. This rather large number is a good indication for the relevance of national particularities.

Inspection of Table VIII demonstrates that there are distinct differences with regard to the expected benefits of nuclear energy. Whereas German students emphasize the advantages of cheap energy supply and of conserving natural resources, the Philippine students apparently disregard these two benefits, but are convinced that nuclear energy can increase the industrial development of their country and the national prestige. Those two benefits are considered of no avail to the German students, though. The Japanese respondents lie in between. Similar to the Germans, they regard nuclear energy as an inexpensive way of generating electricity; like the Philippine students, they believe in the incentive role of nuclear energy for the development of the national industry.

Regarding the question of conserving natural resources and increasing national prestige, the Japanese respondents relate both issues to the utilization of nuclear energy, although not as strongly as the German or the Philippine respondents. In contrast to the German and Philippine sample, the Japanese regard nuclear power as a long-term solution to their energy problems. All three samples react more homogeneously on the risk side. But there are still some distinct patterns which are worthwhile mentioning. The Japanese respondents perceive

Table VIII: Discriminant Coefficients and Mean Differences Between the Three Samples

	EB	Lambda	Sig.	Mean German Students	Mean Japanese Students	Mean Philippine Students
1	Conserve natural resources (7)	0.86	0.00	4.42	2.09	-0.37
2	Increase nation's prestige (22)	0.77	0.00	0.09	1.34	2.64
3	Assuring economic independence (12)	0.69	0.00	1.92	1.33	-2.16
4	Technology as tool in international politics(15)	0.65	0.00	-1.28	-1.78	0.29
5	Management of dangerous wastes (6)	0.60	0.00	-0.08	1.73	-2.71
6	Leads to equal income distribution among nations (30)	0.57	0.00	-2.45	-1.14	-0.94
7	Promotes industrial development (3)	0.54	0.00	1.93	3.14	3.58
8	Restricts options for future societal development (21)	0.52	0.00	+2.27	+0.92	+0.31
9	Postpones the development of alternative energy sources (14)	0.49	0.00	-0.22	2.04	-1.15
10	Provides cheap energy (17)	0.48	0.00	2.45	2.59	-1.38
11	Long-term solution to energy needs (19)	0.48	0.00	0.35	2.21	-0.13
12	Restricts personal freedom (2)	0.46	0.00	1.9	1.3	-2.16
13	Leads to environmental pollution (20)	0.45	0.00	-1.95	-2.37	-4.84
14	Leads to proliferation (28)	0.44	0.00	-1.46	-1.35	-0.89
15	Uses up valuable land (9)	0.43	0.00	-1.28	-1.77	0.29

hardly any risk in connection with radioactive wastes; the Germans show medium concern, whereas the Philippine students have a rather negative view on the waste problem. This negative evaluation is also predominant with regard to environmental pollution and, rather unexpectedly, about the restriction of personal freedom. German and Japanese students are less concerned about environmental pollution as a consequence of nuclear power and do not believe that personal freedom might be endangered by the implementation of nuclear power. Only the international threats seem to be more decisive for the German and Japanese negative view of nuclear power. Proliferation is seen as a high probability risk factor by these two samples, whereas the Philippine students are not as much concerned with this possible menace to world peace.

Two more differences should be mentioned. First, the Japanese students do not perceive nuclear energy as a competitor for alternative energy sources but as a complement. Both the German and Philippine respondents believe that the use of nuclear power will restrain the development of alternative energy sources. Second, while the Germans feel that nuclear energy has the potential to increase scientific research, the Philippine respondents reject this possibility, and Japanese respondents are somewhat undecided.

Most of the results fit into a consistent mosaic characteristic for each country. German and Japanese students reflect their industrial heritage by ascribing the role of a promoter of nuclear energy which helps to provide inexpensive electricity, encourage economic progress, and increase national independency. The Philippine sample perceives nuclear energy as an imported technology with rather doubtful economic advantages and high risks. But they do link nuclear energy with some positive symbolic attributes: Increase of prestige as well as encouragement for modernization and industrial development. The motivation to go nuclear is more functional on the German side, more symbolic on the Philippine side. The Japanese respondents react more like the German students; however, in some aspects concerning national prestige and economic development they agree with the Philippine statements. Functional attitudes are typical for highly industrialized western cultures; the Japanese are still partly influenced by traditional value systems, but at the same time highly motivated by modern functional evaluations. If the risk aspects had not had such a strong impact on the Japanese attitudes, their overall judgment combining functional and symbolic aspects would tend to a rather well-balanced and stable attitude towards nuclear energy.

A further interpretation of the results could lead to substantial errors, since proponents and opponents of nuclear energy are not equally distributed in each sample. Thus, some of the differences revealed in the discriminant analysis are due to differences in the overall attitude distribution within the samples rather than to national differences. If one keeps the attitude distribution constant, some of the results would have to be modified. In particular, the social risks of restricting personal freedom and of adding more restraints to a flexible development of society are only related to negative attitudes towards nuclear energy. Therefore, the emphasis of the Philippine sample towards societal risk is not originated by national differences, but caused by the higher frequency of opponents within the Philippine sample. If the distributional effect is eliminated by statistical precedures, there is no significant difference between the three samples with regard to societal and social risks, while all other differences between German, Philippine, and Japanese students were still existent.

The importance of national properties in the attitude formation can further be demonstrated by the predictive power of the two discriminant functions. Using two discriminant functions which represent the differences between the three samples, 71 percent of all cases could be correctly classified. It is indicated that the belief system not only discriminates between proponents and opponents of nuclear energy, but between different national samples.

THE INVESTIGATION OF IMPORTANCES

The objective of this part of the study was to determine the awareness of respondents about the current issues in the nuclear debate. The information derived is intended to give an indication about the informational background of the three samples rather than attitudinal aspects. Table IX gives the mean values of the importance ratings of each attribute for the three samples and the ranks assigned on the basis of the mean values.

Generally, it can be noted that the German students tend to have lower importance ratings than the remaining two samples, who appear to consider a substantial part of the issues presented as relevant in the debate about the use of nuclear energy. Inspection of the ranks as expressions of priorities, however, shows that there is an overall agreement between the three nations. This is also reflected in the rank correlation

TABLE IX: Importance Ratings and Respective Ranks

		PHILIPPINES		JAPAN		FRG	
		means	rank	means	rank	means	rank
1	Improving our standard of living	4.57	13	4.55	13	3.03	20
2	Restricting personal freedom through rigorous security measures	3.89	25	3.27	28	3.06	19
3	Promoting my nation's industrial development	4.64	8	4.24	16	3.54	14
4	Being harmful to future generations	4.62	12	4.73	10	3.58	13
5	Leading to technological progress	4.74	4	4.79	8	4.19	7
6	Requiring management of dangerous wastes	4.56	14	5.44	1	4.54	6
7	Helping to preserve natural resources	4.69	5	4.66	11	5.14	1
8	Providing a source for threats from terrorists	4.06	22	4.08	19	2.79	26
9	Using up valuable land	4.12	23	3.63	25	2.53	28
10	Leading to dependency on small groups of specialists	4.19	21	3.69	24	2.86	22
11	Exposing people to hazards which they cannot influence by any actions of their own	4.66	7	4.55	13	3.43	16
12	Assuring the economic independence of my country	4.56	14	3.80	23	4.07	8
13	Having an impact on people's health	5.13	1	4.92	6	3.73	11
14	Postponing the development of alternative energy sources	3.76	28	3.20	29	2.85	23
15	Involving a technology which is a tool in international politics	4.25	20	4.12	18	2.83	25
16	Having a long-term impact on climate	4.30	19	4.49	15	3.22	18
17	Providing a cheap energy source	4.41	17	5.01	2	4.60	5
18	Leading to accidents which affect large numbers of people at the same time	4.83	3	5.00	3	3.71	12
19	Being a long-term solution to energy needs	4.63	9	4.81	7	5.07	2
20	Leading to environmental pollution	4.63	9	4.76	9	3.79	9
21	Restricting options for future societal development	4.83	18	3.57	26	3.03	20
22	Increasing my nation's prestige	3.53	29	2.51	27	2.11	30
23	Leading to a consumption-oriented society	3.46	30	3.10	30	2.27	29
24	Concentrating power in big industrial enterprises	3.99	24	3.86	22	2.65	27
25	Leading to increased employment	4.68	6	4.66	11	4.81	3
26	Stimulating scientific and technological research	4.98	2	5.00	3	4.79	4
27	Reducing the need to conserve energy	3.89	25	3.89	21	3.46	15
28	Leading to diffusion of knowledge for construction of weapons	3.77	27	4.18	17	2.84	24
29	Involving hazardous agents which cannot be detected by man's senses	4.63	9	4.98	5	3.74	10
30	Leading to a more even distribution of income among nations	4.45	16	3.99	20	3.74	10

coefficients, which demonstrate a highly significant concordance of priorities (Philippines - Japan: $r = 0.79$, Philippines - FRG: $r = 0.77$, Japan - FRG: $r = 0.72$).

Taking the first four top ranking issues for each sample, it can be seen that Philippine and Japanese respondents want risks and benefits of nuclear energy to be discussed whereas FRG respondents mainly are interested in the benefits. With regard to the priorities expressed by Philippine students, they have focused their attention on potential health impacts and large accidents on the risk side, and on stimulation of scientific and technological research and progress on the benefit side. Japanese are also attentive to the potential of large accidents and to research in science and technology, but furthermore deem more elaborate discussions on waste management and on economic ways to produce necessary energy. In the FRG, students appear to be concentrating their attention on the beneficial aspects of generating energy with nuclear power, with conservation of natural resources, long-term solutions to energy needs, increased employment, and stimulation of research amongst their top priorities. This might be interpreted as an indication of their interest to hear more about the benefits operational nuclear power plants will ensure rather than being overwhelmed with information about risks being reduced.

Regarding the lowest priorities assigned by the three samples, they all agree that a concern about consumption-oriented society is negligible and that their nation's prestige is not at stake in the debate about nuclear power. Furthermore, the often heard argument that advancement of nuclear energy might lead to a shortage of funds and interest for development of alternative energy sources does not appear to be a relevant issue.

SUMMARY AND CONCLUSIONS

We presented a case study of an empirical investigation using the Fishbein attitude model to reveal the underlying dimensions of people's perception of nuclear energy. We compared the belief structure of each sample, tried to reveal the most significant factors that influence the general attitudinal commitment, and searched for major differences between the three nations. What results did we find and how can they be interpreted? Let us first summarize the most important findings:

1) The three student samples differ in their composition of pro- and anti-nuclear points of view. The Japanese students were predominantly in favor of the use of nuclear energy; the German students were divided in their attitudinal structure, but with the majority on the pro-nuclear side, whereas the Philippine student sample showed the most anti-nuclear resentments. Since the size of the three samples does not permit any inference about how representative the distribution of attitude is compared to the attitude of all students in each country, the difference between the three samples might be due to sampling distortions as to real national discrepancies.

2) Looking at the scores of the semantic differential, the response pattern of the three samples is supported by the primary analysis of the attitude distribution. The more favorable Japanese respondents see nuclear energy as very important, useful, modern, and worthwhile, assigning only a few negative attributes (controversial and slightly dangerous) to this energy technology. In contrast, Philippine students view nuclear energy as quite wrong, oppressing and useless, conceding, though, that this energy source might be important for the future. The German students perceive nuclear energy as quite modern and useful, but also as dangerous.

3) A closer examination of the belief stucture using factor analysis revealed three to four basic dimensions which account for the variance of most of the 30 single items. The Japanese students have some concerns about indirect impact on society (such as power concentration or proliferation), long-term hazards (such as harm for future generations), and potential for threat (such as accidents with large numbers of affected people), but perceive at the same time quite a potential for economic prosperity. The German students are also concerned about the negative impacts of large-scale technologies and the potential for threat. Furthermore, they are concerned about the restrictions of social flexibility (such as dependency on small groups of experts). Like the Japanese students, they also emphasize the benefit of nuclear power

concentrating on its positive effect towards economic progress. The Philippine sample put more emphasis on the negative impacts of large-scale technologies than the other respondent groups, but perceived also some progress in national development and indirect fringe benefits which might or might not be associated with the extension of nuclear power.

4) The division into various underlying dimensions does not necessarily provide any conclusions about the importance of those factors with respect to the overall attitude of the respondents. Regression analysis was used to investigate the salient beliefs which are decisive for attitudinal commitment. The regression analysis revealed a rather ambiguous picture: the attitudes of all these groups were influenced by positive and negative items simultaneously. Direct advantages, such as cheap energy, and disadvantages, such as health effects or pollution, turned out to be the most important items for the attitude of the German students. More indirect advantages, such as technological progress and employment, and disadvantages (such as being harmful to future generations or affecting a large number of people) are underlying determinants for the attitude structure of the Japanese sample. It was surprising that almost only positive benefit items govern the direction of attitudes of the Philippine sample, in spite of their more negative view towards nuclear power, since there was a universal agreement among all Philippine students that nuclear energy is dangerous and risky. For this reason only the perception of benefits discriminates between positive and negative attitudes.

5) Discriminant analysis was used to detect the main differences between the three national samples. Concerning the benefits of nuclear power, the German students emphasized the advantages of cheap energy supply and of conserving natural resources--two aspects which are less relevant for the Philippines. Rather, the Philippine students are convinced that nuclear energy increases the industrial development of their country and the national prestige--these two aspects have no relevance for

the German students. The Japanese respondents give credit to all four aspects. Concerning the risk side, the samples reacted more homogeneously. All respondents are worried about health effects and the waste problem; the Philippine students were most concerned. Political risks such as proliferation are more typical for the two industrialized countries (Germany and Japan), whereas social risks, such as the potential threat to personal freedom, were quite predominant for the Philippine sample.

6) The importance of national properties could be demonstrated by the predictive power of the two discriminate functions which represent the differences between the three samples. The discriminative power of the belief items was higher between the three national samples than between the pro and con groups of the combined samples. How economic advantages and personal and social risks are perceived with respect to nuclear power seems to depend much more strongly on national properties than on favorable or unfavorable positions. Although the protest against nuclear energy is international, the arguments and basic evaluation appear to considerably differ between various contries. Thus, it seems necessary to study these national properties more closely in order to investigate the social factors which influence the national perception of nuclear power.

7) Considering the importances of each belief item, there are only four differences among the three samples. With regard to the priorities, the Philippine students focused their attention on potential health impacts, large accidents (risk side), and on the stimulation of research and progress (benefit side). Japanese students were also attentive to the potential of large accidents and to research in science and technology, but were concerned about waste management and energy supply. German students concentrated their attention on the beneficial aspects of nuclear power, ranging from economic advantages to conservation of natural resources.

What kind of general conclusions can be drawn from these results? First, it seems apparent that all respondents agree that nuclear power can be associated with direct health risks, but also with political and social threats which are due to the perceived impacts of large-scale technologies. Whereas in the German sample both risk levels are highly intercorrelated, the two other samples differentiated between the two levels. There was hardly any argument between the pro- and anti-nuclear groups about the direct risks, including the long-term hazards and the waste problem. But if the indirect risks, like social threats for personal freedom or political risks of proliferation, were seen as a major consequence of nuclear energy, a negative attitude was likely to be formed.

But the most salient beliefs mainly cluster around the benefit items. Again, with the exception of the German students --exposing a predominantly technical point of view--the economical or financial aspects were less important as opposed to the indirect incentives to improve the employment rates, the development of industry, and the long-term supply situation. Thus, the purely instrumental advantages proved to be insufficient to compensate for the perceived risks. If the respondents were not convinced that in the long run the economy and the native society would benefit from the use of nuclear power, their attitudes were at least ambiguous, if not negative.

There were clear cut differences between the three nations. The German students had the most pragmatic approach to the nuclear issue. They were concerned about health risks, waste disposal and pollution, somewhat worried about proliferation, and influenced by economic aspects, such as energy prices or technological progress. The Japanese did take these direct impacts of nuclear energy into account, but concentrated on the risks of large-scale technologies and their indirect benefits. In particular, elements of progress and national prestige were relevant, assigning some symbolic value to the nuclear technology. The students from the Philippines focused their attention on the indirect and symbolic aspects of nuclear power. Being dependent on the delivery of power stations from abroad, nuclear energy was not perceived as an instrument for national independence or self-reliance. Positive symbolic attributes refer to the improvement of the standard of living, national prestige, or the stimulation of scientific progress. If these benefits were not regarded as probable, the negative view on the risk aspects, ranging from the waste problem to the potential threat to personal freedom, outweighed the positive beliefs. Thus, according to our analysis it does not seem surprising that the Philippine sample was characterized by the percentage of negative attitudes.

A lot of questions remain to be answered. What are basic values which determine the genesis of specific beliefs? Why do people differ in various countries? How is the attitude of nuclear power connected with perceptions of other technologies or technological change in general? One survey cannot answer all these questions. We hope that our investigation will encourage new research in this area and that these studies will serve as a mosaic stone in the evaluation of scientific knowledge about man and technology.

REFERENCES

1. O. Renn, Man, Technology, and Risk, (Jul-Spez-115, Reports of the Nuclear Research Centre Julich, 1981), or M. Douglas and A. Wildavsky, Risk and Culture (University of California Press, Berkeley, San Francisco, 1982).

2. K.R. Hammond, G.H. McCleeland, and J. Mumpower, Human Judgment and Decision Making (Praeger Special Studies, Hemisphere Publishing Corp., Boulder, Col. 1978).

3. A. Tvevsky and D. Kahneman, The Framing of Decision and the Psychology of Choice, Science, 211, 453-458 (1981).

4. J.L. Janis and L. Mann, Decision Making (The Free Press, New York, 1977).

5. H.J. Otway, Perception and Acceptance of Risks, Zeitschrift fur Umweltpolitik, 2 (1980) pp. 593-616.

6. O. Rammstedt, Was heibt soziologisch Risiko? Manuscript for the Workshop: Analysis, Evaluation and Acceptability of Hazardous Technologies and Their Risks. (Berlin, Dec. 14-15, 1981).

7. J. Hoos, "Risk Assessment in Social Perspective", in Perception of Risk, Proceedings of the 15th Annual Meeting, National Council on Radiation Protection and Measurements, eds. (Washington D.C. 1980), pp. 57-84. For a special application on energy systems, see: R. Caputo, World in Collisions: Is a Rational Policy Possible for Countries in West Europe? (Manuscript, California Institute of Technology, Pasadena (1983) prepared for Energy Policy.

8. O. Renn and E. Swaton, Psychological and Sociological Approaches to Study Risk Perception (Manuscript, Invited Paper for the Workshop ENVIRISK 83, San Miniato, June 22-24, 1983).

9. H. Jungermann, "Introduction: Decision Making in Theory" in Psychologische Entscheidungstheorie, Wayne Lee, author (Beltz, Weinheim, Basel 1977) pp. 9-37.

10. M. Rokeach, Beliefs, Attitudes and Values (San Francisco 1969).

11. O. Renn and H.P. Peters, "Intuitive Risk Perception: Research Results of Attitude Survey Towards Risk and Technology" in Proceedings of the International ANS/ENS Topical Meeting on Probabilistic Risk Assesment, American Nuclear Society ed. (Port Chester, Sept. 20-24, 1981), pp. 1464-1480.

12. C. Kluckhorn, "Values and Value Orientations in the Theory of Action", in Toward a General Theory of Action, T. Parsons, E. Shils eds., (New York and Evanston 1962) pp. 388-433.

13. M. Fishbein and I. Ajzen, Belief, Attitude, Intention and Behaviour: An Introduction to Theory and Research (Addison-Wesley, Reading, Mass., 1975).

14. G.W. Allport, "Attitudes", in Handbook of Social Psychology, C. Murchinson ed. (Clark University Press, Worcestar 1935) pp. 798-844.

15. W.J. McGuire, "The Nature of Attitudes and Attitude Change", in Handbook of Social Psychology, G. Lindzey and E. Avanson, eds. (Vol. III, Addison-Wesley, Reading, Mass. 1969) pp. 136-314.

16. M.L. De Fleur and F.R. Westie, Verbal attitudes and event acts: An experiment on the salience of attitudes. American Sociological Review 23 (1958) pp. 667-673.

17. W. Meinefeld, Einstellung und soziales Handeln, (UTB, Reinbek bei Hamburg 1977).

18. M.E. Shaw and J.M. Wright, Scales for the Measurement of Attitudes (McGraw-Hill, New York 1967).

19. D. Krech, R.S. Crutchfield and E.L. Ballachey, Individual in Society (McGraw-Hill, New York 1962).

20. M.J. Rosenberg, Cognitive structure and attitudinal affect, Journal of Abnormal and Social Psychology 53 (1956) pp. 367-372.

21. M. Rokeach, "The nature of attitudes", in International Encyclopedia of the Social Sciences, vol. 1 D.L. Sills, ed. (New York, 1968) pp. 449-458.

22. H.C. Triandis, "Toward an Analysis of the Components of Interpersonal Attitudes", in Attitude, Ego-Involvement and Change, C.W. Sherif and M. Sherif, eds. (Wiley, New York 1967) pp. 227-270.

23. V.H. Vroom, Work and Motivation (New York 1964).

24. Reference No. 18 and C. Murphy, L. Murphy and T. Newcomb, Experimental Social Psychology (Harper & Row, New York, 1937).

25. M. Fishbein, An Investigation of the Relationships between Beliefs about an Object and the Attitude toward that Object. Human Relations 16 (1963) pp. 233-240.

26. M. Fishbein and I. Ajzen, Belief, Attitude, Intention and Behaviour. An introduction to theory and research (Addison, Wiley, Reading, Mass. 1975).

27. H. Otway, Perception and Acceptance of Risks, Zeitschrift fur Umweltoplitik 2 (1980) p. 97.

28. A.W. Wicker, Attitudes vs. Actions: The relationship of verbal and event behavioural responses to attitude objects. Journal of Social Issues 22 (1969) pp. 41-78.

29. C. Osgood, G. Suci, and P. Tannebaum, The Measurement of Meaning (University of Illinois Press, Urbana 1957).

30. J.J. Otway and M. Fishbein, Public Attitudes and Decisionmaking (RM-77-54, Laxenburg, Austria, International Institute for Applied Systems Analysis, 1977).
or:
K. Thomas et al. Comparative Study of Public Beliefs about Five Energy Systems (RR-80-15, Laxenburg, Austria: International Institute for Applied Systems Analysis, 1980).

31. Reference on No. 30 and K. Thomas et al., Nuclear Energy: The Accuracy of Policy-Makers Perception of Public Beliefs (RM-80-18, Laxenburg, Austria: International Institute for Applied Systems Analysis, 1980).

32. E. Swaton and O. Renn, Attitudes towards Nuclear Power. A Comparison between Three Nations (RM-X-83, Laxenburg, Austria, International Institute for Applied Systems Analysis, in press).

Structure of Public Opinion, Nuclear Debate and Risk Assessment: A French Survey

E. Stemmelen,* F. Fagnani,** J.L. Brenot,* J.P. Pages,*
J. Fagnani***

CONTEXT OF THE FRENCH ENERGY POLICY AND THE NUCLEAR DEBATE

An Old And Ambitious Program

France, and to a lesser degree, Japan are the only industrial countries to keep to their schedules for a stepped-up nuclear program (1). By January 1, 1981, gross installed nuclear power was 15 GWe, of which 15,900 were MWe light water reactors. Another 27 PWR reactors, including ten 1,300 MWe reactors, were under construction. When completed, they will provide an additional 35.1 GWe. Since 1981, one out of three Kwh has been nuclear; by 1984, the proportion is expected to be one out of two. Starting in 1985, nuclear power will furnish 60% of electricity production, which at that point will no longer need to rely on oil. (See Fig. 1).

France is a country which began very early to develop a nuclear research and development policy. One of the first political decisions on the development of civilian uses of nuclear energy, contained in the National Assembly vote in July 24, 1952, was to implement a five-year program calling for the construction of two generators (50 and 100 to 150 MW), a treatment plant, and the choice of a site. In 1956, EDF, the national utility guaranteeing the majority of electricity production with the monopoly of its distribution, launched its first nuclear program involving three reactors of the natural uranium graphite gas type to be set up at Chinon on the Loire. A second program of the same type was launched by EDF in 1963, comprised of three power plants (St. Laurent in 1963, Bugey 1 in 1965, and St. Laurent 2 in 1966). During the period of

* Commissariat à l'Energie Atomique, IPSN, BP 6, 92260, Fontenay-aux-Roses, France
** C.E.P.N, BP 48, 92260, Fontenay-aux-Roses, France
*** CNRS, Laboratoire de Géographie Humaine, Paris I.

NATO ASI Series, Vol. G4
Environmental Impact Assessment, Technology Assessment, and Risk Analysis. Edited by V.T. Covello et al.

1965-70, the relative decrease in the crude oil price, as well as strained relations between the two principal institutions concerned (CEA and EDF), contributed to a delay in decisions to be taken. Despite the provisions of the fifth plan (1965-70), envisaging the setting up of one or two reactors each year, no new nuclear project was initiated between 1967 and 1969. Political changes in 1969 finally led to the decision to aban-

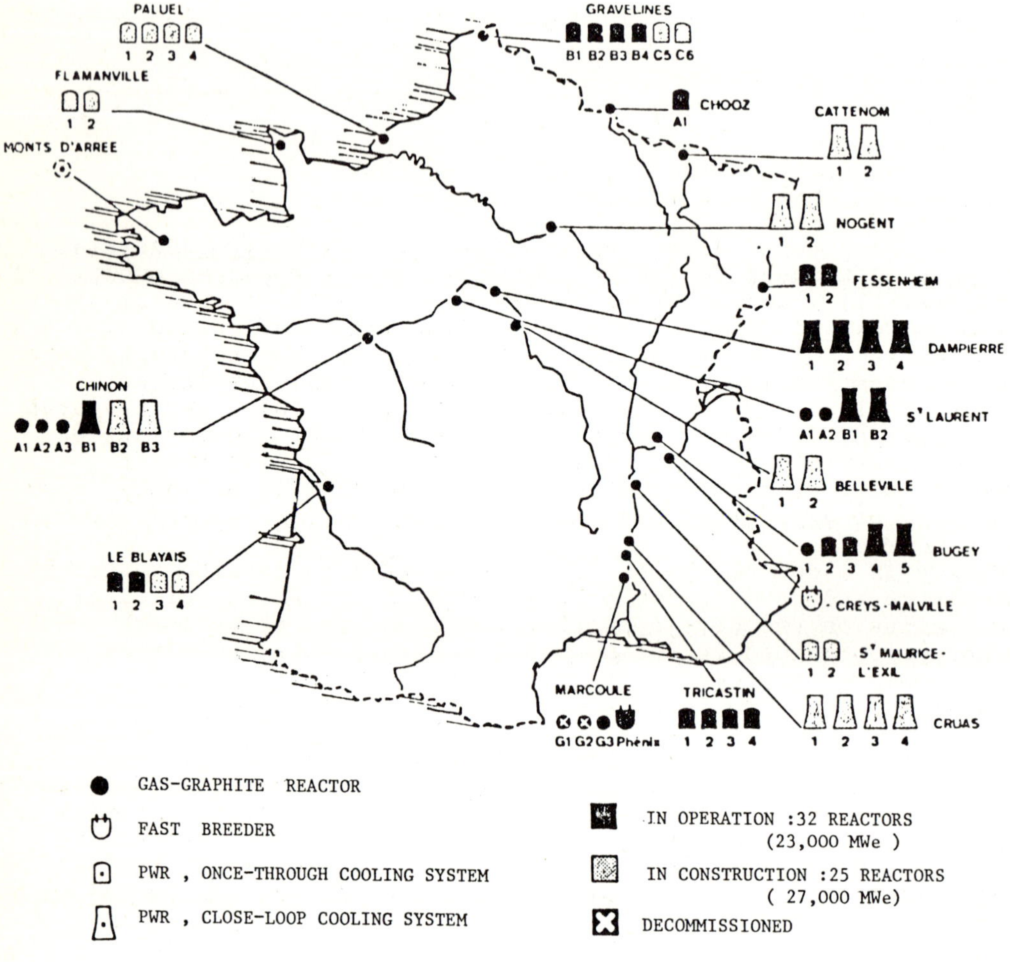

Figure 1 : French nuclear power plants (situation 01.01.1983)

(CEA/DPG MAP)

don the French system in favor of those developed by the USA. It was within this new framework of development that the French electro-nuclear program became more ambitious from 1971 onwards and was intensified even more after the oil crisis in 1974 (Messmer Plan of 1974) (2). In 1970, capacity in France amounted to 1600 MWe.

An Energy Policy Based on Continuity (Despite the Political Changes)

In October 1981, for the first time in the history of the Fifth Republic, Parliament actually discussed and voted on the government's energy and nuclear policy. This debate was a significant change in comparison with the pre-Mitterand period. Under President Valery Giscard d'Estaing, from 1974 to 1981, the French nuclear program was considered the sine qua non of national independence, and the government showed extreme firmness in the face of anti-nuclear protests (3). For example, in 1980 the clash between the inhabitants of the Breton village of Plogoff and the national riot police was very violent. All nuclear-related decisions were concentrated in the hands of the executive branch (the President and the government) and the publicly-owned companies concerned: Electricité de France (EDF) and the Commissariat à l'Energie Atomique (CEA). Moreover, the administration kept tight control over the choice of sites, which perfectly illustrated the way France's highly centralized system worked (4).

Once elected, President Francois Mitterand was expected to fulfill campaign promises made to a constituency critical of the French nuclear program. The new president immediately took some highly symbolical decisions: he called a halt to the Plogoff project, as well as to the extension of a military camp at Larzac, which had mobilized ecologists ever since 1973. Finally, in July 1981, President Mitterrand suspended work on five nuclear sites until Parliament could discuss these projects. All these measures were favorably received by the opposition groups.

As a matter of fact, the Confédération Francaise du Travail (CFDT), the second most important trade-union confederation in France, and in particular members of the nuclear union at the La Hague reprocessing plant, had an active part in formulating criticisms of the nuclear program. Although the Confederation never actually took an anti-nuclear stand, it did serve as a link between anti-nuclear activists and the French Left Wing. Following the accident at Three Mile Island in 1980, a joint petition for "another energy policy" was circulated by the CFDT, a number of left-wing organizations (among them, the Socialist Party) and by anti-nuclear groups. The petition was a compromise solution based on calls for more democratic decision-making, a demand for a national referendum and for regional energy plans, and a proposal for a nuclear moratorium until a democratic energy debate could be held.

The main ecological and anti-nuclear groups joined forces under the banner "Ecology Today" to support the candidacy of Brice Lalonde for the 1981 presidential election. Lalonde was defeated on the first round of voting. On the second and final round, spokesmen for the movement refused to come out openly for Francois Mitterand. But the Left benefitted from the fact that it was "favorably inclined."

However, despite electoral promises of the Socialist Party to the contrary, the reappraisal of France's energy policy brought about singularly few changes in the nuclear program itself. Following the October 1981 debate, the government decided to get the construction of six additional reactors (five 1,300 MWe and one 900 MWe) under way during 1982 and 1983. This is only three less than the previous government had planned for. The government also decided to carry on with the expansion of the reprocessing plant at La Hague, whose annual capacity will gradually be raised from 400 to 1,600 tons, and the building of the fast breeder reactor "Superphenix."

Which changes in the energy policy program did the socialist government promote at the opening of the October 1981 Parliamentary session?

Table 1 is a comparison of the energy forecasts of the preceding government based on the Eighth Plan with those used in October 1981. The latter were based on the "Hugon" Report, requested by Prime Minister Pierre Mauroy, to prepare for the parliamentary debate. The lower figure for the 1990 forecast is based on a hypothesis of a 2.5% average annual growth rate of gross domestic product (GDP); the higher figure, which the government was hoping to meet, corresponds to a 5% growth. (Forecasts of the Eighth Plan were based on an estimated annual growth rate of 3.5%.)

Actually, the margin of uncertainty for 1990 reflects the uncertainty surrounding the evolution of the general economic situation.

The main lines of the new government's policy, as they were developed during the October 1981 debate, are summarized below.

Energy-saving measures, particularly those aimed at moderate consumption, have been stepped up; they were presented as the major innovation of the governmental team. A new body, the "Agence pour la Maîtrise de l'Energie," was created in April 1982. The creation of this agency could prove to be a major

Table 1 Energy forcasts for 1990 in France

Consumption of primary energy in Quadrillion BTU'S	1981		1990 (Eighth Plan)	1990 (Hugon Report) (2,5%)	(5%)
Coal	1.44		1.41	1.35	1.62
Oil	3.99		2.91	2.74	3.06
Gas	1.05		1.80	1.16	1.32
Hydraulic	0.64		0.60	0.62	0.62
Renewable energies	0.15	(+)	0.51	0.39	0.47
Nuclear	0.84		3.13	2.57	2.84
TOTAL	8.1		10.36	8.83	9.93
Total consumption of electricity in BTU'S	2.37		4.28	3.45	3.96

(+) Figure for 1980

factor in broadening the spectrum of expertise on energy matters, which had previously been the exclusive domain of the administration and both public and private companies directly involved in the main energy-producing chains. It goes without saying that the new agency is not to the liking of part of the French energy Establishment.

Though the 1990 forecasts for renewable energies are similar, the present government has considerably increased allocations, so as to reach its objective of some 0.4 quads. In addition, funds for research have tripled.

Though coal-consumption figures are roughly the same, the present government has provided bigger guarantees for national coal production, one of the long-standing concerns of the Communist Party, which is now part of the government coalition. In 1982, subsidies were raised to 11.1 centimes per

British thermal unit (about 1.6 cents U.S),[1] which guaranteed a 1982 production of 18.4 million tons. The objective of 30 million tons by 1990 requires so great an investment that it is highly unlikely that it will be met.

Lastly, the curtailed nuclear program -- 6 instead of 9 units to be under way during 1982-1983, representing 55 GWe instead of 59 GWe -- appears to be more realistic, given the slump in the growth rate of electricity consumption (3.8% in 1981 compared with 12.5% in 1978).

Which are the Socialist government's justifications for continuing the nuclear program? The arguments put forward by the Mauroy government to justify the fact that its nuclear policy is a departure from the Socialist Party's election platform -- only 39 GWe and 40 nuclear reactors by 1990 -- reveal the underlying motivations of the new regime. The government has continued to stress the need for national energy independence (in 1980, 72% of primary energy was imported, which represented 26% of all imports) and to call attention to the highly competitive cost of nuclear energy. According to EDF, the cost of a nuclear kwh during base production in 1990 will be 16 centimes (2.3 U.S. cents) against 26 (3.7 U.S. cents) for coal and 56 (8.5 U.S. cents) for oil. More important, it has been using three types of arguments to show that its policy is consistent with its overall plan to fight the recession:

- Employment is a top priority
- An energy base must be ensured for the day when economic growth picks up
- The country must preserve its technological advances.

These arguments are also consistent with the Left's penchant for public intervention in the national economy and its commitment to preserving the State's decisive role in the nuclear fuel cycle.

The main changes introduced by the Socialist government in energy policy have been the reform of decisionmaking procedures with a marked intention to increase public participation. Yet (until now) these attempts to "democratize" the process have contributed more to increasing the socio-political consensus in

---

1 One U.S. dollar equals 7 francs.

favor of the nuclear program than to creating new avenues for action for opponents of the program.

Towards a Broader Consensus: Reforms in the Decisionmaking Process

The theme of "defending local democracy" had often helped to unite anti-nuclear groups with local officials, especially among the Socialists. Indeed, a number of local and regional officials had been short-circuited by EDF, which exercised control over the worksites. They also had to contend with the highly centralized administration. EDF chose to work with what could euphemistically be called its "clients" who were in fact only those officials willing to cooperate. Of course, the "fiscal" fall-out of a nuclear installation was considerable. In 1979, for example, the Fessenheim plant (two 900 MWe units) brought in 4.8 million francs (about $0.7 million) to the Haut-Rhin department, while the municipality itself shared another 6.6 million ($0.94 million) with neighboring and other localities. What is more, since 1979, inhabitants living near nuclear plants pay less for their electricity than the going rate.

Although the Public Inquiry procedure was modified in 1976 to let local inhabitants express their opinions on major installations, it was more often than not a mere formality, and this invariably set off anti-nuclear demonstrations. The Socialist government therefore tried to make up institutional changes in these proceedings.

The bill proposing the democratization of public inquiry and environmental impact studies was ratified by a vast majority in Parliament on May 17, 1983.

The law proclaims that the public inquiry cannot last less than a month. The Commissary in charge of the inquiry or the President of the Commission for inquiry are now endowed with more powers and will be able to call for a public meeting, if they deem necessary, to be held with the plant contractor and with the consent of the "Commissaire de la Republique," formerly known as "Prefect." These public meetings will resemble the Public Hearings held in the United States. As soon as the public inquiry has begun, the file must be made available to the public for an unlimited amount of time, upon request at the Prefecture.

This reform undoubtedly introduces procedures that will improve public information in different points of view. It indicates a genuine attempt by the government to break away from the formality and arbitrariness of past procedures.

Alongside the global decentralization reform put through by the new government, "Regional Energy Agencies" have been created. They are under the authority of Regional Councils. Municipal and Regional Councils were consulted to determine the localities for the six new nuclear sites that the government decided to pursue.

These agencies have a three-fold purpose: to provide general information on energy questions, to bring their regional situation to the attention of national policymakers, and to take decentralized initiatives, especially in such fields as energy-saving measures and renewable energy sources. They are also designed to elaborate and supervise the implementation of regional energy plans.

"Information Committees" open to local officials, unions, and associations as well as scientists (among others) have been set up in a number of localities where nuclear plants are to be installed. A similar national inquiry committee is looking after the La Hague plant.

For anti-nuclear activists, however, this is still only a "caricature of consultation," inasmuch as the government has always said that it alone will take the final decision. The idea of local referendums, as well as of national referendum, has gone by the boards. But the point is now the following: is the anti-nuclear movement able to participate in the energy debate and to influence the energy choices?

The Marginalization of the Anti-Nuclear Movement

The election of a Socialist government has not slowed down the marginalization of the anti-nuclear movement which had already grown under the former administration (5). Furthermore, the institutional reforms introduced since May 1981 have reinforced the isolation of anti-nuclear groups. Unable to significantly influence energy policy, these groups can only draw attention to a potentially dangerous consequence of decentralization. As locally elected public officials find their prerogative reinforced, there is a risk that they will favor short-term employment options to the detriment of evaluating the long-term social cost of industrial development.

Of all the "social movements" that arose in the wake of the May 1968 student revolt, the anti-nuclear movement is without a doubt the one having the greatest difficulty adapting to the new political situation (6). The feminist and consumer movements apparently find themselves able to take advantage of the new possibilities for institutional action and experimentation offered by the present government. On the contrary, within the ecological movement, centrifugal tendencies are increasingly strong.

Attempts at unification of the two biggest groups in the "radical" wing of the ecology movement--the "Network of Friends of the Earth" and the Political Ecology Movement, failed during their respective annual meetings in June 1982. Yet unification was probably the only way they could have attracted such regional groups as the Ecological Federation of the Midi or that of Alsace. According to sociologist Serge Moscovici, who is one of the movement's theoreticians: "It is presently much less dynamic and is incapable of creating a structure for itself. This lack of direction is perturbing the ecology movement and is leading to its demise."

We believe that a number of parameters specific to the French situation go a long way to explain the characteristics of the anti-nuclear movement:

First, the discretionary procedures used by the previous government to impose its nuclear program polarized the debate even more, enabling the most radical elements of the anti-nuclear movement not only to attract some of the less radical members but also to make broader alliances on the local level. However, it also made for a basic weakness: opposition groups were incapable of influencing the decisionmaking process on energy issues.

The political and electoral polarization between the Left and the Right, which has characterized the French scene since 1973, also helped to politicize the ecology movement and local protest groups. But at the same time, it made it all the more difficult to structure the movement and to take independent political action.

Inquiries in fact show that voting for ecology candidates (and of course approving some of their ideas) is, and always has been, quite distinct from participating in an anti-nuclear movement. The score of ecology candidates has not been systematicaly higher in districts where anti-nuclear demonstrations have taken place. There is a big difference between voters and hard-core anti-nuclear activists. Whereas the vast majority of activists are politically to the left, those who vote the ecology ticket are just as often to the right as to the left. Political scientists interpret the ecology vote in the presidential election as a partial shift of middle-of-the-roaders, who were forced to choose between the

Right and the Left on the final round of voting. The surveys described later on conflicting issues in French society show that whereas activists and ecology voters are equally in favor of such issues as liberalized abortions, tolerance of homosexuals, or doing away with the draft, they generally differ on ideological grounds. Whereas activists are progressive, voters tend to be more conservative. The former believe in State intervention on social questions, which is in sharp contrast to the deliberately "modern" individualism of voters.

If the changes that have come about since May 1981 led to the socio-political isolation of the anti-nuclear groups, this is because they were deprived of a number of trump cards. The new context revealed the opposition movement's inability to adapt on these three levels:

- Inability to carve out a positive role in the newly institutionalized negotiations created by the decentralization reform

- Inability to actively intervene in the reformulated energy debate

- Inability to adapt to the evolving social debate on technology

The October 1981 decisions sanctioned the perhaps irreversible evolution of the energy debate. The question is no longer of being for or against nuclear energy, or even of a moratorium. The debate now centers on the relation between economic growth and energy consumption and on the different forms of energy that would best meet the energy crisis.

Anti-nuclear groups have taken little part in this debate. They have let the CFDT become the only alternate source of expertise. In our opinion, the political and ideological structures which the ecology and anti-nuclear movement had to confront have changed. This makes it very difficult, if not impossible, for the movement to adapt without undergoing major upheavals.

Following internal discussions between public officials and industrial representatives, the government decided in July of 1983 to moderately reduce the program for 1983-1984. Two reactors, instead of three, will be built each year because of a decrease of electricity demand combined with France's low economic growth rate (0.2% in 1981, 1.5% in 1982 and an estimated 2% in 1983).

PUBLIC OPINION AND NUCLEAR ENERGY: EMPIRICAL ANALYSIS

The Different Kinds of Investigation

Without question, the nuclear program and its strong resulting controversy have impressed upon the average citizen's life emotions of interest, anxiety, confidence, and refusal. In general, the study of public opinion is limited to evaluating the rating of nuclear energy in terms of "pros and cons." This information does not help very much for understanding public opinion as too many factors may interfere; but obtained in surveys at regular time intervals, it enables us to evaluate the trends shown in Fig. 2.

The observed discrepancies between these three surveys are more apparent than real; they are implied by the formulation of each question and the mode of response which is respectively a four, five, or six point-scale. In our surveys, the EDF-SECED ones, the percentage sum is relative only to people who declare acceptance or strong acceptance and the "maybe yes" group which rises 25% is not included. In the two others, this group is part of pro-nuclear people.

The rate during these last 10 years is not really changing due to any controversy or big event such as Three Mile Island. We cannot see where nuclear arguments, advantages, or drawbacks intervene in this trend which is, we recognize, a very rough indicator.

There are numerous actors who play in the social game associated with the nuclear energy subject, outside the political arena. These include the following:

- Technicians and scientists who express themselves either as individuals or through commissions, working groups, or academies
- Decisionmakers in governmental agencies and companies who define directions and push the issues
- Associations and social groups which are identified as having a public interest position or which are looking for it
- The media

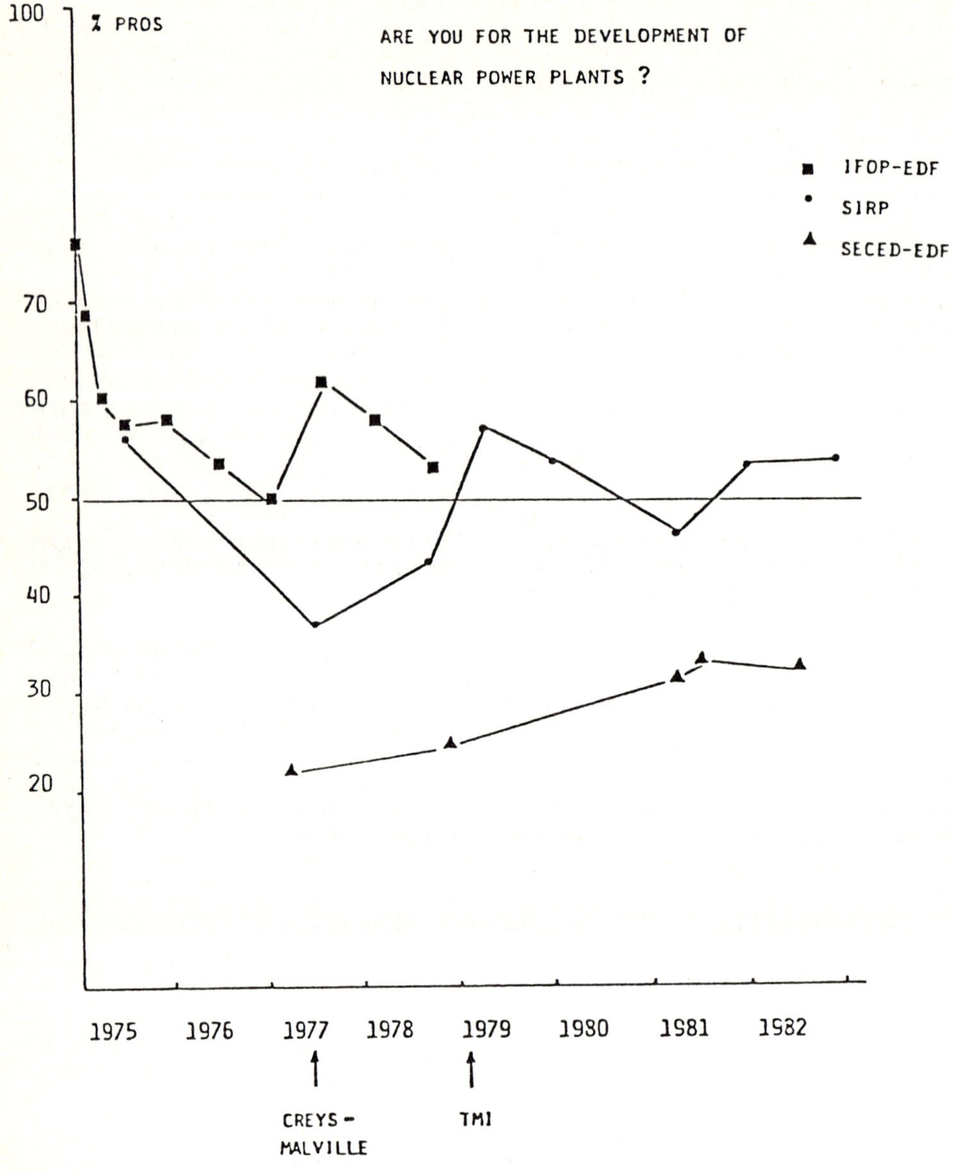

Figure 2 : Evolution of public opinion between 75 and 82 through different surveys

These actors produce declarations which are very different both in form and in content. When the debate becomes hard, social values considerations join the technological issues and the discourse which began as a technical one, with its true-false scale, calls in the values with its good-bad scale. The nuclear energy which was a technical and economical response to a definite need is now a conflictual issue, that is, a field where different social forces communicate and clash.

Often, the average citizen does not perceive clearly these conflicting interests. Apart from some fears or anxieties he may feel, he would appear to be scarcely involved, for he is too far from decision centers to have any immediate influence on the course of events. At this stage, there are two possibilities: focus on the individual and his personality or consider him as a social actor.

One can work on actors either by describing their history (7,8,9,10), by using ethnology of small local communities (11), by analyzing local conflicts (7), and by seeing if the anti-nuclear movement is able to transform itself into a national social movement (4,5,12). The study of messages is also possible in the field of semiology, restricted to lexicology (13,14) or extended to connected significations, secondary meanings (14). To understand anti-nuclear positions, psychiatrists and psychologists take an interest in atom anxiety and personality influence (16,17). Other researchers (18) prefer to explain attitudes and their resulting behavior by analyzing the belief system about energy. For us, technical arguments allow the individual to "rationalize" his position, a position which is already defined if not always expressed, and which is consistent with ideological positions.

Public Opinion Structures and Nuclear Energy: Goal and Method

Based on the principle of non-autonomy of the nuclear energy issue, our research tries to see its connection with other conflictual problems. For any of these, the individual takes a position and states where and how he stands in relation to other members of society. At any time, current conflicts on which people express themselves define the present status of public opinion. A priori, diversity seems to be its main character; in fact, structures can be found -- they are the major axes of communication and therefore of social differentiation. Interest lies in the study of the link between the nuclear issue and the obtained structures.

To do this, five national surveys, based on several thousand people, have been carried out in France between March 1977 and June 1982 (19,20,21). These are shown in Table 2.

Table 2: Public opinion structures

	MARCH 1977	NOV. 1978	MARCH 1981	JUNE 1981	JUNE 1982
SAMPLE SIZE	1000	1054	1622	528	1653
PARTICULARS	31	22	37	35	39
THEMES	56	66	76	84	80
OTHERS	45	44	8	2	20

The objective of these surveys was to investigate the social significance underlying the public attitude toward nuclear power plants. What are the perceptions, how can they be associated with attitudes, how are attitudes related to economic and social conditions? The wider public (randomly chosen individuals) were asked to give their viewpoint on more than 50 conflictual topics (abortion, death penalty, nuclear power, immigration, and atomic weapons among others). Simultaneously, we paid attention to socio-demographic, cultural, and political characteristics of the people interviewed. Table 3 presents some of these topics.

TABLE 3: Examples of questions

Liberalization of the abortion law is a good thing
Price rises can be avoided
Construction of nuclear power plants must be continued
God exists
One may have confidence in the judicial system
There are too many immigrant workers
Pollution is extremely worrisome
The death penalty must be abolished
Conventions must be respected
One should no longer get married
Unemployment is very worrisome
One no longer feels secure
Nuclear weapons are indispensable for France
Pornography must be strongly fought against

Data are submitted for statistical analysis and interpretation. Standard statistical processing (rates and cross-tabulations) gives a short and incomplete description of the national sample and an appraisal of the degree to which any topic draws a consensus. To go further, data are submitted to a standard multidimensional technique, that is, factor analysis; correlations between the issues are used to discover the main axes of social differentiation and a detailed analysis is made of the relative positions of the numerous social groups considered. Subsequently these analyses are used to follow trends of public opinion; if many small changes occur in some issues, are the major dimensions unchanged? Does the significance of propositions gradually alter? If so, in what direction? An interesting feature -- a reverse movement of the nuclear energy theme -- arose, especially after the French presidential election of 1981.

In Fig. 3, the themes are displayed on a circular target. Each one is a point.

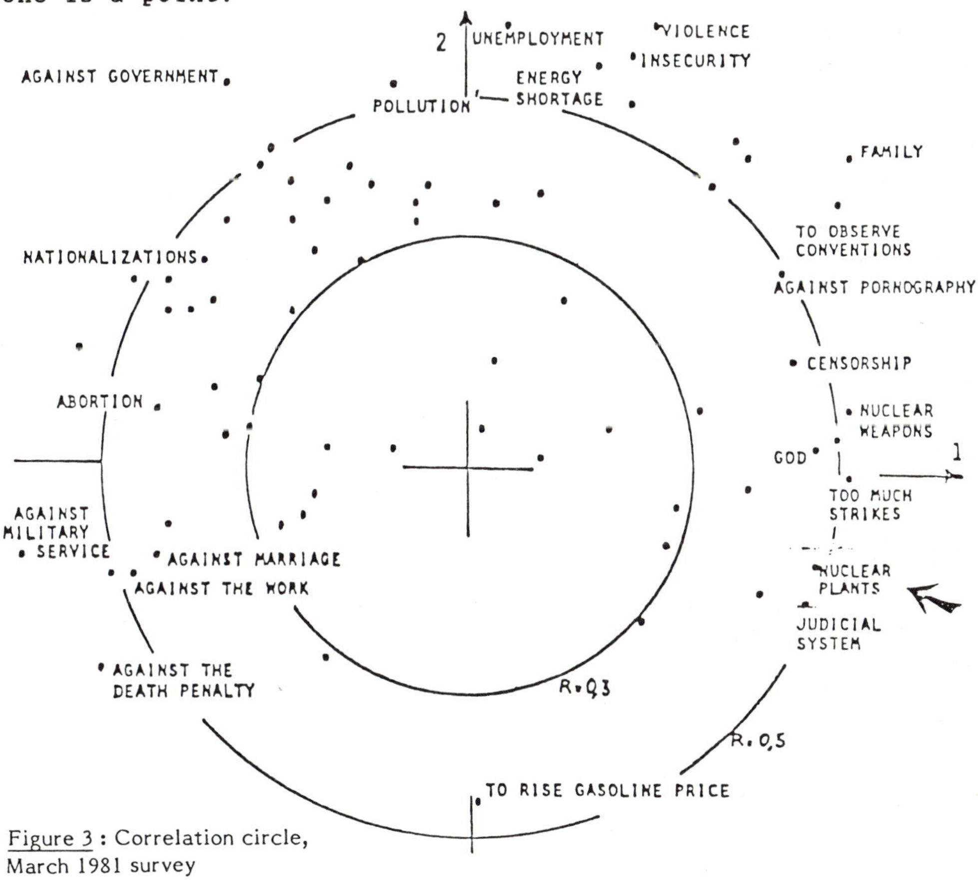

Figure 3 : Correlation circle,
March 1981 survey

The rule is as follows: the further the point from the center, the stronger the importance of the issue to the public, and the closer the relationship to the neighboring issues. The basic dimensions of our target are characterized by labels. These are already an interpretation of the factors obtained by our multidimensional analysis; they appear in Fig. 4.

Let's take the issue "We must continue to build nuclear power plants." Firstly, it lies in periphery, thus it is important to public opinion. To the same extent, issues like unemployment, insecurity, family unity, pollution, the death penalty, nuclear power plants (and atomic weapons) allow the citizen to take a clear-cut position. But for nuclear plants -- not as some people think - this position is not from the standpoint of violence and anxiety, but rather from a "sacred" and "stability" standpoint.

The horizontal axis of our "target" corresponds to themes linked both to "eternal" values and to the recent achievements of our society. On one hand, we have the ideal synthesis of "taboo and totem," often linked to patriotism. On the other hand, we have a double rejection of such ideas in the labor organization and even in the work ethic. In less symbolic and more sociological terms, the horizontal axis expresses the conflict in society and has the ideological dimension in the doctrinal sense. The meaning behind this becomes clear: the atomic weapons and the "God exists" theme are close because the rationality involved here is of an ideological nature. In another reference field such as "defense of life,"they would have been opposed to each other, as "abortion free" and "suppression of death penalty" would be.

If we emphasize the significance to be given to this axis, it is with a view to clarify the controversy surrounding nuclear power plants. It is essential to understand that this conflict is closely connected with sacred symbols and with doctrinal choices. The issues are not simply considered from the technical and economic standpoint. In 1981, opposition to nuclear power (civilian and military) meant supporting a movement which denounced the technico-economic system and rejected the traditional values.

Decisionmakers would certainly like to see a change in attitude toward a wider consensus and even toward serenity (see bottom of target). Is such a shift still possible?

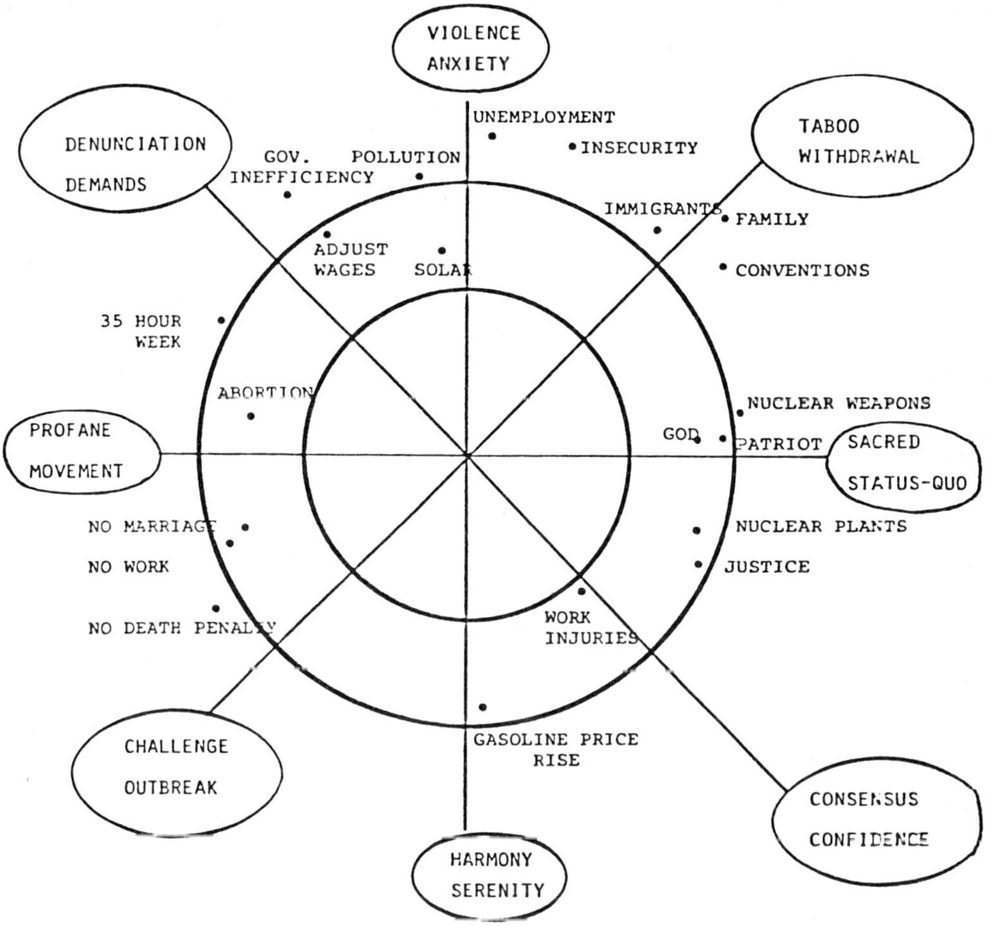

Figure 4 : Some of the 76 issues in the march 1981 survey

Essay of Interpretation of Public Opinion Evolution

In the beginning, the nuclear issue was not conflictual and was perceived in the serenity-anxiety field. Nuclear supporters were characterized by their attitudes of challenge, by their reliance on science (see Fig. 5).

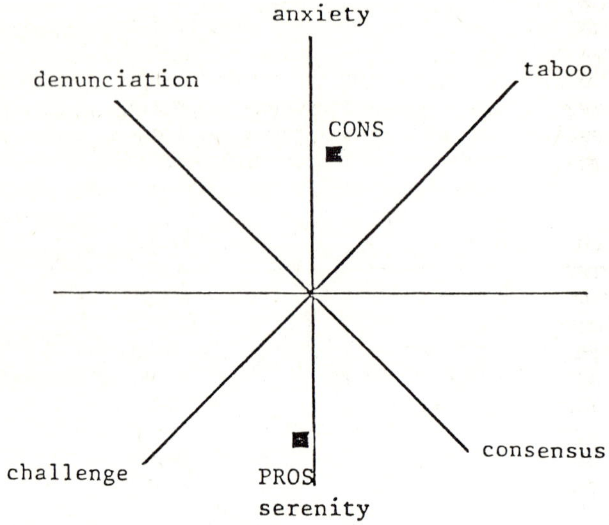

Figure 5: Opinion on nuclear power plants: the 1950s

Technical and economical aspects were neglected. A review of texts of this period, especially in the media, leads one to think that an image of challenge-serenity and also of prestige was associated with the civil nuclear industry. To the contrary, opponents to nuclear power plants (and of the bombs?) were the defenders of taboos and criticized the knowledge, an attitude linked to anxiety and violence fears. Here, the anxiety referred to has nothing to do with psychology and even less with metaphysical "angst"; it is the psycho-social concept of the anxious attitude.

It is true that anxiety was for a time the decisive factor behind the shy opposition to nuclear plants, but only up to 1960-1965. It seems curious that this analysis is still widespread even these days among the Nuclear Establishment. Therefore, the traditional studies of anti-nuclear people are too often limited to looking for anxiety and even distress created by the nuclear plant.

In 1958 the World Health Organization said "anxiety exists and pursues to a remarkable degree. It threatens to have pathological effects on mental health" (22). But 10 or 20 years later, this explanation is no longer sufficient. This is widely demonstrated by the changes in attitude: if anxiety was the reason for anti-nuclear movements, why did they originate around 1970 and not as early as the 1950s? It is sometimes said, "Inhibitions come back." This explains nothing. It is the come-back itself which needs understanding. Also, the problem is more one of finding out why anxiety resulted around the 1950s in passivity and later in opposition and struggle. Or in other terms, why the average anti-nuclear citizen shifted from conventions' respect to rejection.

The second stage started in the 1960s and began to develop as illustrated in Fig 6.

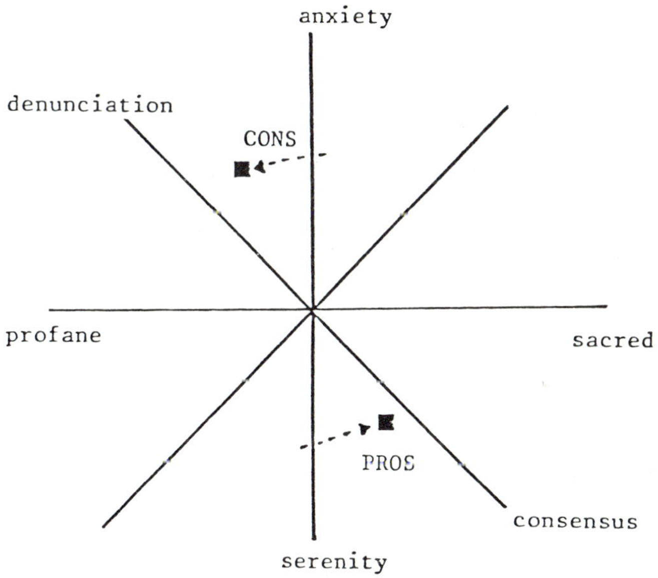

Figure 6: Opinion on nuclear power plants: the 1960s

The essential point is the evolution along the horizontal axis. The challenge party will soon not only "drop" their support for the nuclear issue, but will even turn against it. In fact, it wasn't individuals who changed their minds. With a few exceptions, the contestation of the 1960s and 1970s was not supported by those who advocated it during the 1940s and

1950s. It is the general change in values and widespread innovations in everyday life that make what seemed revolutionary yesterday look conservative today. Let us consider the consensus/rejection split which appeared for nuclear power around 1965, traces of which still remain in 1980. This aspect is related to issues which uphold technical progress and either avoid or underline its possible negative consequences. In our surveys, we reference themes such as "honest oil companies," "animal experimentation," and "increase gasoline prices." Some people show a total confidence in each area: security, unemployment, workplace injuries, justice. The whole industrial society is accepted and even defended and this is why we can talk about a consensus or considered approval. It is an almost promethean challenge to nature. It is clear that at the other side one finds opposing issues voicing questions raised by the technological development. These themes criticize the modern world and, more specifically, the techno-structure which is typical of developed societies. It is a challenge to culture in response to the challenge to nature mentioned above. In this way the opposition to nuclear power rests on a culture/nature guide line. In the first step, it is a debate on ways of action, organization, and development modalities. In the next step, a global controversy expands and intensifies from the day when nuclear power becomes an essential element in the survival of society. It then acquires a quasi-sacred character; it lies not simply on technical/practical considerations but on the idea of obligation and therefore of taboo.

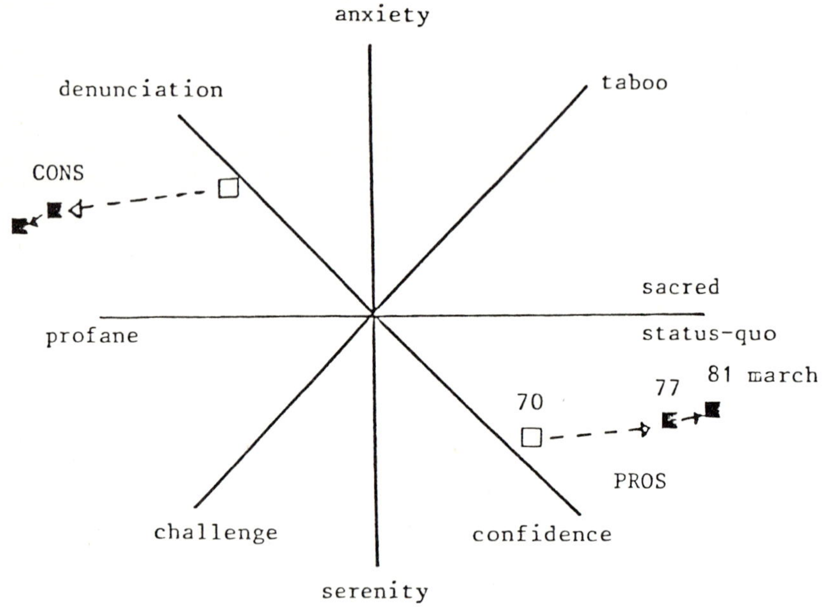

Figure 7: Opinion on nuclear power plants: the 1970s

When we said "We can no longer do without nuclear power," we were in that context. The 1970s could be illustrated as shown in Fig. 7.

"What might have been limited to technical disputes in a fair evaluation of the risks of the new technology and to a disagreement over determining levels of acceptable risks to populations, gradually changed in several Western countries, to a real passionate battle with acceptance or rejection of nuclear power plants at stake." (23,p.446).

Supporters of nuclear power played on factors such as prestige and necessity. So we repeat: the anti-nuclear movement arose from this "sacralisation" of civil nuclear power which was largely due to scientists and political decisionmakers, the former for the prestige associated with nuclear power, the latter for the "necessity" that they gave to it. Obviously groups with varying degrees of organization and differing concerns caught this opportunity. It is true that nuclear power has been used as a scapegoat, but to have this privilege, some conditions had to be present, all related to the idea of "sacralization." Nuclear power met these conditions because some people had given it almost "sacred" significance (see Fig. 8).

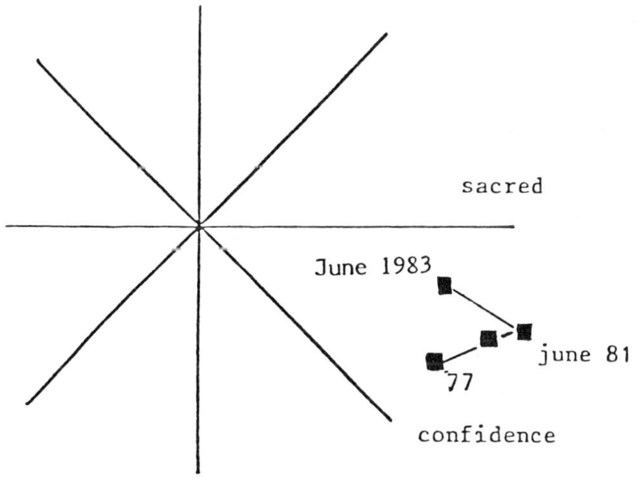

Figure 8 : Opinion on nuclear power plants: recent evolutions of pros

It is fruitful to observe in public surveys the most recent developments. Many things happened in France between March 1981 and June 1983. It seems that with the new administration, which does not claim to be pro-nuclear, a "desacralization" process has begun (backwards on the horizontal axis) and we are coming back to a more technical and economical debate and a less controversial one (the point moves away from the target periphery). This development is more interesting to us than a few more or less percentage points in favor of nuclear power.

PUBLIC ATTITUDES TOWARD RISKS

The different studies developed on public attitudes toward nuclear energy show that fear and anxiety are far from being a unique factor of explanation. Ideological dimensions particularly play a fundamental role. Is it possible to draw similar conclusions in the case of risk perception in general? Is it relevant, furthermore, to compare public attitudes toward different risks? These are the types of questions that we now envisage handling.

A Psycho-Social Approach of Perceptions

It is likely that the layman is, in front on a so-called "risk," in a similar position as the "expert" in following the sequences of cognitive and evaluative operation -- identification, evaluation, and control. But, we know very little of this whole mental process. The "feeling of insecurity" that we study here can be considered broadly as the result of a discrepancy: the individual compares, implicitly or not, the resources that he thinks are available and those that seem to him necessary to overcome the situation. The fear is stronger as this balance seems to be negative for him. The approach might be parallel to the one of the "expert" or the decisionmaker involved in risk management; but it is obvious that the estimates of the magnitude of the risk and of the capacity to respond adequately are very different from one individual to the other. In fact, we have to distinguish strongly between two kinds of situation: whether or not the danger is actually present; whether it represents an immediate, effective threat or only a potential one. The psychology is referring here to two different concepts: fear and anxiety. We are dealing more ourselves with the second aspect of potential, perceived threat where the term "anxiety" is too strong to be the right reference. In fact, we are not interested in investigating pathological individual situations but rather social perceptions of average "normal" individuals. It is the reason why we speak of "feelings of insecurity" related to different social objects including technologies.

The links with the previous analysis can now be outlined. The perception of risk related to an object expresses the negative part of the perception of this object itself and, in this perspective, the preceding method of investigation may be applied. This approach leads to an analysis of the socio-cultural dimensions of the perception of risks, in the same manner as we have investigated nuclear energy. The method is also inductive and empirical; it follows an old statistical tradition in existence since Karl Spearman (24), who was trying to find a unique dimension to express the concept of intelligence.

Here, we are going to ask a representative sample of individuals to react to a whole set of risky situations and objects. The score given by the individuals will then be analyzed by a statistical method which is able to extract the essential dimensions of the answers (principal component analysis).

The preliminary survey carried out in January 1982 was based on a questionnaire which can be divided into three main parts:

- o Socio-demographic characteristics
- o Attitudes toward a series of 40 activities in terms (respectively) of the magnitude of their risks, of their utility, and of the level of priority of safety promotion
- o Attitudes toward a series of controversial themes appearing at the moment in the French scene. (See supra page 19).

Preliminary Results of Survey

The questionnaire proposes a list of 40 activities as diverse as possible: alcoholic beverages, motorcycles, mountain climbing, nuclear energy, hunting, X-rays, planes, and food additives among others. The individual is asked to characterize each activity on a scale of risk which has seven levels. A first analysis can then be carried out on the statistical means. The activities perceived as the more dangerous are, for example, the following: narcotics (mean: 6.4), tobacco, alcoholic beverages, motorcycles (mean: 5.2), mountain climbing, nuclear energy (4.4). The less dangerous are: bicycles, and trains (1.8). The statistical tests show that age and socio-professional group of the respondents are factors of differentiation of risk perception. Women are also more sensitive than men, especially in the case of pharmaceuticals. These results are very similar to those published by P. Slovic, who inspired this approach (26).

A second scale (with seven levels) is applied to the score activities concerning their perceived utility. An analysis, in terms of comparison of the mean scores, can also be performed. In that case, the medical activities (surgery, vaccination, radiology) and the industrial activities (civil engineering, oil, chemical industry, iron and steel) are judged the more useful (mean: 6). On the opposite side (2.5) are the following: narcotics, tobacco, food coloring, alcoholic beverages, hunting, automobile races, and others. In that, the influence of factors such as age, sex, socio-professional group, and level of instruction is not very significant.

Lastly, a third scale rates the demand for more safety. It is mostly in the case of narcotics (mean: 6) and nuclear energy (5.7) that this demand appears. The emphasis put on the prevention of risks linked to these two activities is quite obvious and is fairly different than in the case of the other activities (races, downhill skiing, mountain climbing, and hunting). Women are more insistent than men on the level of effort which ought to be devoted to protection. This is particularly true for those activities which present corporal or vital dimensions such as medicine, food, fertilizers, and nuclear energy.

Looking for Indicators

The 40 scales of risk associated with each activity can be considered as ways of expression of individual risk perception. Principal component analysis allows these 40 variables to be produced on a unique scheme (correlation circle), corresponding to the nature of their statistical relationships framework (linear correlation). The graph obtained (Fig. 9) can be simply interpreted as shown below.

On the horizontal axis, all activities are located on the same side. This axis constitutes a global indicator of anxiety. The individuals using important scores of risk are doing so for all activities and the same appears, on the contrary, for those people using rather low scores.

On the vertical axis, the individual and collective activities are contrasted. Among collective activities, those related to energy provision are particularly differentiated. So, beyond the variability observed in the distribution of anxiety and concern which appears independent of the object of projection, a fundamental distinction is found between risk-related activities. The ones that are seen as corresponding to individual free choice contrast sharply with those which are seen as resulting from collective and social choices.

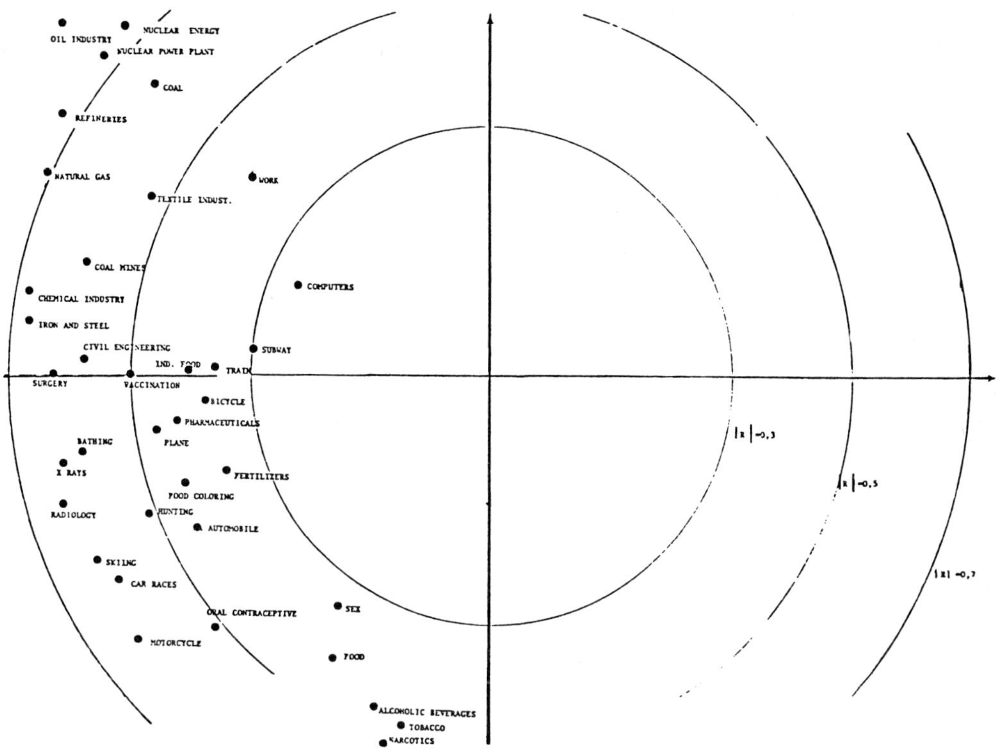

Figure 9 : Perceived risk associated with 40 activities.

Correlation circle (plane 1,2)

Similar analysis can be undertaken on perceptions of utility and on security and safety demand. But utility appears to be a non-univocal concept. In fact, it has to be split in two different dimensions: social utility on the one hand has to be differentiated from what is perceived as individual, selfish interest, and reward. As far as the security demand is concerned, the results show no difference with those obtained from the data on risk perception itself (see Fig. 9).

A Psycho-Social Analysis

The survey on the structure of public opinion was devoted, at the onset, to understanding public attitudes toward nuclear energy. It actually produces a framework to permit a follow-up of public opinion and of its evolution in the short-term. As we have mentioned supra, the questionnaire used in the survey on risk perception contained 20 items in common with the preceding general survey (Part II). Through this set of questions, it is possible to redefine, by using appropriate statistical calculation, the axes defining the original "principal plane." The method permits, through a limited number of questions in common, to make a junction or "grafting" between different surveys (in terms of population interviewed as well as themes explored).

We have to be reminded of the interpretation of the principal axes of the general survey. Horizontally, the first axis opposes "status-quo" and "movement," and vertically, "anxiety" to "serenity" (see Fig. 10). Then it is possible to plot the points representing the answers to the 40 risk scales of activities on the same plane. Technically, we have to calculate the linear correlation coefficient between these 40 variables and the principal components. We observe that most of the activities appear at the top of the graph, in the region labeled "anxiety," which seems quite visible. However, one activity seems an exception to that rule, "Work." Expressing the idea that work is dangerous is not the result of a general feeling of insecurity but, rather, the challenge of a value. Such an attitude is indeed characteristic of that area in the graph. Through the expression of the existence of occupational risks, it is more a denial of the moral value system linked to work that appears. At the opposite end is the variable "narcotics" on the side of "taboo" and withdrawal on traditional values. It is probably the same for oral contraceptives, sex, and food, where the moral arguments appear directly underlying the projections in terms of "risk." In that case, the ethical criteria constitute the immediate basis of what is perceived as a societal risk (see Fig. 10).

Some other activities are conceived as equally dangerous for "hidden" reasons that the analysis makes patent: food additives, industrial food, and nuclear energy appear in the area of "denunciation" and "demand." We can infer that judging these activities dangerous expresses a critical reactive attitude toward society and its socio-political organization.

We can ascertain that it is not so for traditional industries or energy systems: the ideological dimension of risk perception is especially true in the case of nuclear energy, which is one kind of explanation of its controversial potential.

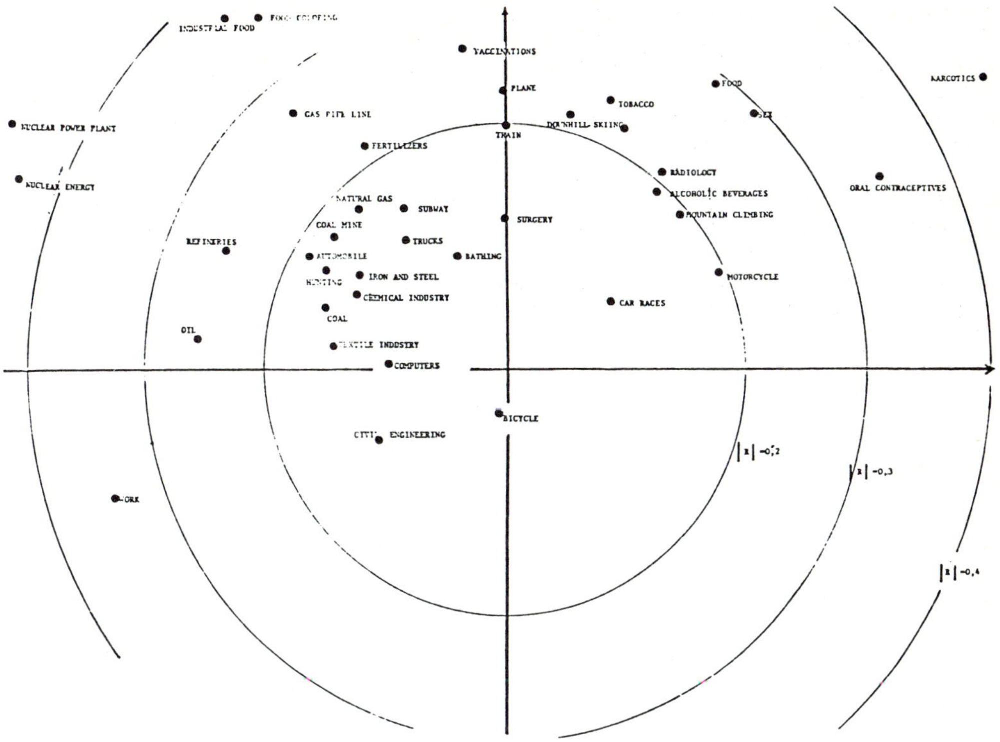

Figure 10 : Perceived risk associated with 40 activities.

Projection on reference plane 1,2

(SECED) - 82 Survey.

These examples illustrate the interest and limitations of the statistical approach of public perception of risk. From a general point of view, it demonstrates the multi-dimensional aspects of risk perception taken at its psycho-social level. For certain activities, if a synthetic indicator of fear and concern is conceivable and meaningful, the social signification of declaring that they are risky is far more ambiguous. Expressing the fact that occupation and nuclear energy are dangerous activities is not actually stating a concern (vertical axis) but criticizing some social norms (horizontal axis). Conversely, if the denunciation of the dangers of narcotics is certainly a mark of anxiety, it also expresses a feature of conservatism. In conclusion, it seems difficult to deal with risk perception in restricting oneself only at a technical level. The necessity of embedding the matter of risk in the broader context of values and ideology was one of our basic assumptions. The proposed method is one among the empirical solutions to such an objective.

REFERENCES

1. J. Fagnani, J.P. Moatti
"The french socialist government's energy policy: continuity and changes." To be published in "Journal of policy analysis and management." Cambridge U.S.A.

2. F. Fagnani, J.P. Pages
Nuclear energy and public opinion
European coordination centre for research and documentation in social sciences, 1981, occasional paper no. 1, p. 25-39, Vienna.

3. J. Fagnani, J.P. Moatti
"Environmental issues in France: the institutional and political context and social actors in the opposition." Conference "Environmental policies and politics." E.C.P.R. Lancaster, U.K., April 1981.

4. J. Fagnani, J.P. Moatti
Environmental questions: an overview of conflicts in France. Conference "Environmental politics and policies: from problem awareness to implementation assessment," European Consortium for Policy Research (E.C.P.R). Florence, Italy, March 1980.

5. J. Fagnani, J.P. Moatti
"L'opposition anti-nucléaire en France: enjeux et acteurs du débat." Conférence "Bürgerinitiativ-bewegung und die Kernenergiefrage:ein internationales Vergleich." University of Bielefeld, West Germany, November 1979.

6. J. Fagnani, J.P. Moatti
"The French socialist government's energy policy and the decline of the anti-nuclear movement." Conference "Comparative research on nuclear energy policy: the state of art." E.C.P.R. Berlin, West Germany, December 1982.

7. F. Fagnani, A. Nicolon
"Nucleopolis - Matériaux pour l'analyse d'une société nucléaire" P.U. Grenoble, 1979.

8. L. Timbal-Duclaux
"Ecologie, énergie nucléaire et opinion publique," Rapport EDF, 1979.

9. P. Simonnot
"Les nucléocrates" P.U. Grenoble, 1978.

10. F. de Gravelaine, S. O'dy
"L'état EDF." A. Moreau Editeur, 1978.

11. I. Chiva
"Perception du risque autour de quelques centrales nucléaires." Colloquium on the risks of different energy sources, PARIS, 24-26/01/1980.

12. A. Touraine
"La prophétie anti-nucléaire," Seuil, 1980.

13. J.F. Picard et J.M. Fourgous
"L'émergence du débat nucléaire dans la grande presse de 1968 à 1976." Colloque sur les implications psycho-sociologiques du développement de l'industrie nucléaire, PARIS, 13-15/01/1977.

14. J.P. Delziani et C. Carde
"Approche pscychologique de l'évolution du thème nucléaire à partir d'un quotidien populaire de 1950 à 1976." opus cité cf. ref. 13

15. H. Stemmelen
"La presse et le nucléaire," Thèse 3è cycle, Paris VII, 1980.

16. Y. Pelicier
"Exposé des recherches de l'équipe de psychologie (service de psychiatrie des adultes, Hopital Necker)". opus cité. ref. 13.

17. G. Guedeney et G. Mendel
"L'angoisse atomique et les centrales nucléaires," Payot, Paris, 1973.

18. F. Niehaus et E. Swaton
"Public risk perception of nuclear power." Opus cité cf. ref. 11.

19. F. Fagnani et J.P. Pages
"Nuclear energy and public opinion," Occasional Papers, pp. 25-38, European Coordination Center for Research and Documentation in Social Sciences, Vienna, 1981.

20. J.P. Pages, G. Morlat, E. Stemmelen
"Structure de l'opinion publique et débat nucléaire dans la société française contemporaine," Revue générale Nucléaire, p. 140-149, no. 2, 1982.

21. J.P. Pages et coll.,
"Les structures de l'opinion publique en 1982," Tl et 2, Note LSEES 82/07, CEA, 1982.

22. OMS
"Questions de santé mentale que pose l'utilisation de l'énergie atomique à des fins pacifiques," Genève, 1958.

23. B. Goldschmidt
"Le complexe atomique," Fayard, Paris, 1980.

24. K. Spearman
"General intelligence objectively determined and measured." Am. j. of psychology, 15, 1904, 201, 293.

25. P. Slovic
Rating the risks, Environment, 117, 1979, 14, 20.

Improving Scientists' Judgments Of Risk

Kenneth R. Hammond, Barry F. Anderson,*
Jeffrey Sutherland**, Barbara Marvin***

Institute of Behavioral Science, Center for Research on Judgment and Policy, University of Colorado

INTRODUCTION

Despite the high concern about the risks from scientific and technological developments, little progress has been made toward providing the public with usable information about such risks. As a result, doubt, confusion, and uncertainty hamper formation of public policies concerning the further development of nuclear power, the disposal of toxic waste, the use of food additives and insecticides, the causes and consequences of acid rain and other scientific and technological issues. Moreover, policy makers who rely on scientists for information about such issues are often frustrated by conflicting counsel. For example, commenting on the report from the Ford Foundation Energy Project, Brooks (1975) complained that: "The whole thing presents a divergence of values and political prejudices which is truly depressing. If experts are so at loggerheads, how can the public and the politicians fail to be confused? The expert views seem to leave almost unlimited scope for all possible national energy policies."

Growing awareness of the confusion surrounding risk assessment resulted on 5 May 1981 in the introduction in Congress of H.R.3441, "a bill to establish a coordinated program under the Office of Science and Technology Policy for improving and facilitating the use of risk analysis." This bill is the latest of roughly fifteen attempts at "improving the quality of

* Portland State University
** Health Sciences Center, University of Colorado
*** Center for Research on Judgment and Policy, University of Colorado

NATO ASI Series, Vol. G4
Environmental Impact Assessment, Technology Assessment, and Risk Analysis. Edited by V.T. Covello et al.

the scientific component" of estimates of "risks to human life, health, and environment" (Nehnevajsa & Menkes, 1981).[1]

Disagreement about values, or what ought to be, is inevitable and deserves more analytical treatment than it currently receives.[2] In the present article, however, we consider only scientific dispute about facts, when such dispute directly bears on the public interest. We do so for three reasons: agreement on scientific facts is the scientists' primary responsibility, disagreement among scientists frustrates the development of public policy regarding risk, confuses members of the public, and reduces their confidence in science, and disagreement about the facts, we believe, can be reduced through cooperation between cognitive psychologists and the scientists in dispute. The first two points are hardly

---

1 Persistent need for improving risk assessment can be seen from the following: "When it comes to the effect of chemical exposure on hormones, chromosomes, and the like, the experts acknowledge how little they agree, and how little they know. They disagree about acceptable measurement techniques and about the reliability of raw data. They disagree even more about the inferences to be drawn from the facts." Bazelon (1980, p. 661). Also, "Any conclusions (scientists) present that are not based firmly on scientific evidence are subject to question by the general public, and the revelation of biases and logical weaknesses can quickly lead to an emotional criticism of science generally. It does not take many of these incidents to turn a large fraction of the population away from science, especially when public opinion is sharply divided" Spilhaus (1980, p. 43). "Important to the future of science and technology is the fact that the public has somewhat lost confidence in the ultimate value of the scientific endeavor . . . . they are less certain that scientific research will inevitably yield public benefit," Handler (1980). Confidence in science will almost certainly be decreased as a result of the widely publicized results of a study, (Cole, Cole, and Simon, 1981) indicating that scientists differed considerably in their judgments of the quality of research projects and that chance played a definite role in determining which projects would receive funding.

2 There is a growing literature on the role of values in science; for reviews see Science, Technology and Human Values, a quarterly journal published by MIT Press; for an example of the integration of science and social values in policy making, see Hammond and Adelman (1976).

debatable, but the third invites test. Therefore, we analyzed a situation in which scientists' disagreement about facts had for several years effectively blocked the efforts of a quasi-official citizens' committee to inform its constituents about threats to their health and safety from toxic emissions from a nearby, federally-owned industrial plant. The scientists' dispute has also prevented formation of public policy regarding the plant by state and federal agencies directly involved. In what follows, we first identify the primary source of persistent disagreement among scientists. Second, we indicate a remedy that requires the interaction between cognitive psychologists and disputing scientists. Third, we test that remedy in the context of a worked-out example.

THE PRIMARY SOURCE OF DISAGREEMENT AMONG SCIENTISTS

When citizens and their representatives demand to know "How much risk does a given installation or action impose on the population likely to be affected?" scientists and engineers can seldom immediately perform the experiments, or carry out the statistical studies, that will provide rigorously derived answers. Rather, they must almost invariably extrapolate and generalize from the pure science of the laboratory to circumstances not yet studied. To do so, they must resort to "judgment." Without the cognitive safeguards provided by the fully analytical methods of logic and experimentation, scientists who exercise judgment are reduced to employing the same quasi-rational cognitive processes that laypersons must use, with much the same consequences, uncertainty, misunderstanding, and persistent dispute that leads to personal animosity--epiphenomena which frustrate, delay and often misdirect the development of public policy. (For comment regarding generalizaton beyond controlled conditions, see Brooks, 1975; see also Hammond, Mumpower, Dennis, Fitch, and Crumpacker, in press, for a description of fundamental obstacles to the use of scientific information in public policy making).

Scientific Method, Cognitive Safeguards, and Judgment

The scientific method incorporates certain cognitive safeguards, conventions involving definition, measurement, causal logic, explicitness, and criticism that guard against mistakes and misunderstanding in constructing experiments and in drawing inferences from data. Broadly speaking, the principal function of these cognitive safeguards is to prevent both unsystematic errors (illogic) and systematic errors (bias). There are, however, various degrees to which it is feasible to apply these safeguards, once the cognitive security provided by the laboratory experiment is given up. As Fig. 1 illustrates, reductions in analytical cognition are accompanied by increases in quasi-rational judgment (Hammond, 1978).

Mode 1 represents the experiments in physics and chemistry where scientists fully control the circumstances in which they work. Every effort is made to eliminate quasi-rational judgment. As a result, scientific work benefits from all the advantages of analytical cognition: high confidence in the logic underlying the experiment and in the validity of the results, together with the cognitive security provided by the retraceability of the process. In Mode 2, where full experimental control is impossible, as is often the case in the macro-biological sciences and usually the case in the social sciences, scientists must employ such cognitive safeguards provided by the statistical method as random assignment, use of control groups, and high reliance on mathematical logic and computation. The statistical method provides a gain in feasiblity of application over pure experimentation in the area of public policy. For example, it is the method used in clinical trials of drugs to provide the facts upon which a social policy to permit or prohibit use of a drug will be based. Analytically powerful as the statistical method may be, the validity of the results obtained are more vulnerable to challenge than results produced by pure experimentation. Statistically-based studies gain in feasibility at the cost of increased uncertainty, unequivocally indicated by the numerical probability value attached to each result. Thus a door is opened to potential conflict over the correct conclusion to be drawn regarding causality, or even mere association, as well as to uncertainty regarding the correct application of the statistical method itself (see Chalmers, Smith, Blackburn, Silverman, Schroeder, Reitman, and Ambroz, 1981, for a quantitative system "constructed to evaluate the design, implementation and analysis of randomized controlled [clinical] trials").

One of the safeguards which statistical experimentation demands, the random assignment of subjects to conditions, has become increasingly difficult to achieve, particularly in situations involving societal risk. In order to acquire information in as rigorous a manner as possible in those circumstances where random assignment cannot be achieved, as is often the case in epidemiological research, it has been necessary to resort to quasi-experiments. (see, for example, Lilienfeld & Lilienfeld, 1980, pp. 10-13; Campbell, 1963, pp. 212-242; Campbell & Stanley 1963, pp. 171-246; Cook & Campbell,

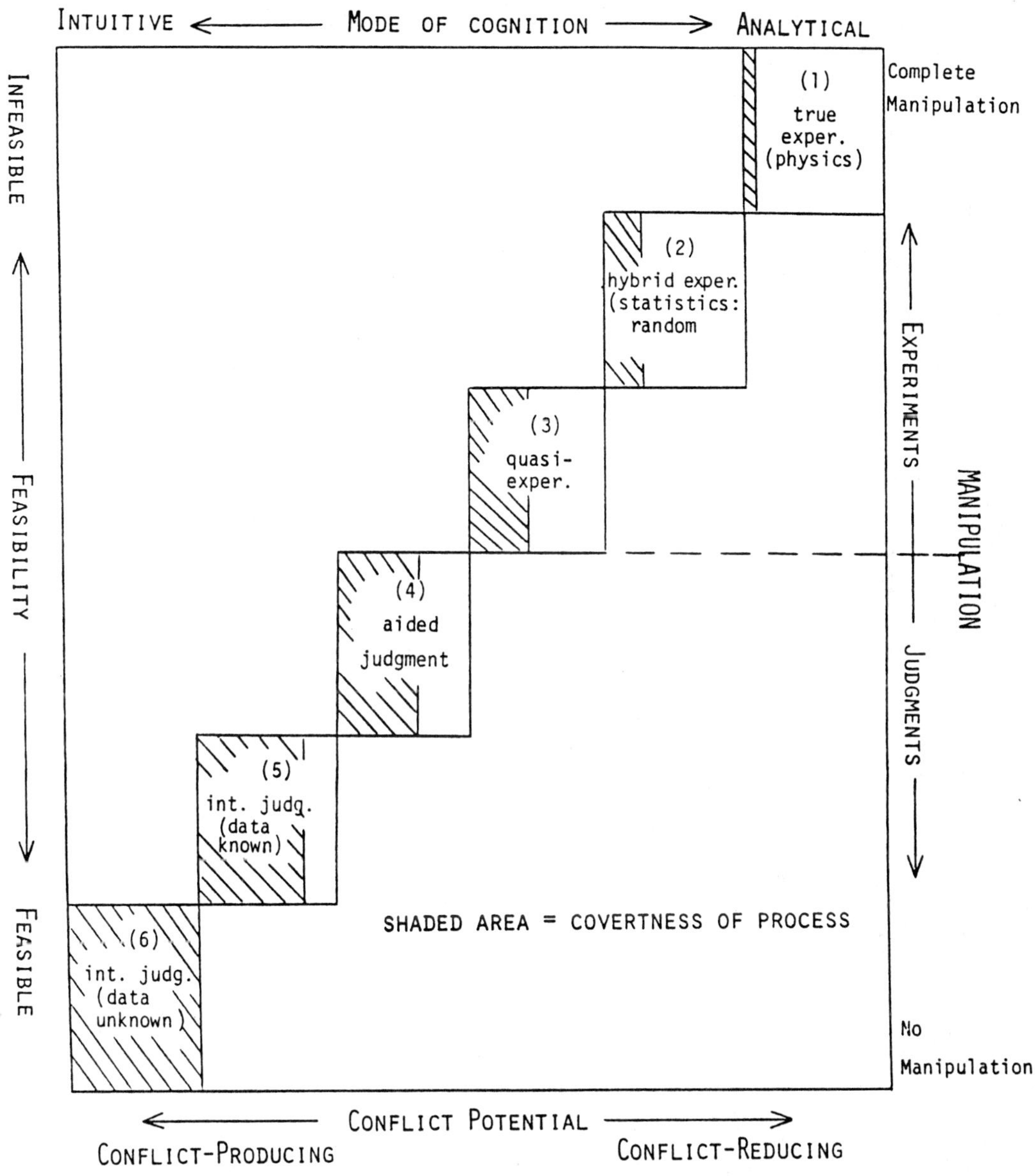
INTUITIVE
MODE OF COGNITION
ANALYTICAL
INFEASIBLE
FEASIBILITY
FEASIBLE
Complete Manipulation
(1) true exper. (physics)
(2) hybrid exper. (statistics: random
(3) quasi-exper.
(4) aided judgment
(5) int. judg. (data known)
(6) int. judg. (data unknown)
EXPERIMENTS
MANIPULATION
JUDGMENTS
SHADED AREA = COVERTNESS OF PROCESS
No Manipulation
CONFLICT POTENTIAL
CONFLICT-PRODUCING
CONFLICT-REDUCING

Figure 1.

1979.) Quasi-experiments (Mode 3) are more feasible to apply than are conventional statistical experiments and have been very useful in their investigation of public policy questions (Cook & Campbell, 1979; see also Riecken, Boruch, Campbell, Coplan, Glennan, Pratt, Rees, & Williams, 1974). But the lack of statistical safeguards in Mode 3 increases the need for quasi-rational judgment in the evaluation of the validity of the results produced by quasi-experimental methods (see Anderson, Deane, Hammond, McClelland, and Shanteau, 1981, for definitions of commonly used terms in this field). The critical role of judgment in epidemiological research, for example, may be seen in the report to the Surgeon General on the effects of smoking and health in which the authors state that: "The causal significance of an association is a matter of judgment which goes beyond any statement of statistical probability." The results of quasi-experiments are much more susceptible to dispute than are the results of pure experiments (Mode 1) or statistical studies (Mode 2) because many of the safeguards provided by conventional random assignment and other features of statistically-based experimental design have been lost. (Epidemiological studies can usually be characterized as quasi-experiments because they often involve some manipulation but seldom include random assignment. "Smoking and Health: Report of the Advisory Committee to the Surgeon General of the Public Health Service" (1979, p. 20) sets forth "criteria for judgment" of causality.)

The distinction between the quasi-experiments of Mode 3 and the aided judgments of Mode 4 is significant. Once active manipulation of conditions is no longer feasible, inquiry is restricted to passive cognitive operations. Judgment emerges from its previously minor role to become the major form of cognitive activity employed. An advantage is thereby gained, for it is much more feasible to apply the passive mode of cognition to social problems, particularly to matters of risk, than to attempt to manipulate conditions experimentally. The great disadvantage of the passive cognitive mode is that all the important cognitive safeguards associated with experimentation are lost. For although judgment can be aided by propositional logic (e.g., modus tollens, modus ponens), and the logic of the statistical method, these safeguards against illogic and bias are seldom sufficient guarantees against errors. Indeed, such safeguards may be used incorrectly or even ignored (see Tweny, Doherty, and Mynatt, 1981, especially pp. 115-128). As the methods of manipulation, test, and replication for drawing conclusions about cause and effect become infeasible, imagination, metaphor, and analogy are used instead. Extrapolation and generalization from results obtained in Modes 1, 2, and 3 thus become vulnerable to the illogic and bias that the scientific method was designed to prevent.

In Mode 4, aids to judgment are provided in order to reduce the covert cognitive activity of unaided judgments. Although such aids take a variety of forms they have the common aim of making evident the information relied upon, and the process by which the information is organized into a judgment. Thus, implicit theories become explicit and subject to criticism. Such externalization is accomplished through the quantitative analysis of the judgment process, rather than through reliance on subjective descriptions, which often omit reference to essential parameters of the process. In short, the aim of aids to judgment is to externalize as much of the judgment process as possible in order to retain as many cognitive safeguards against illogic and bias as possible (see Hammond, McClelland, & Mumpower, 1980, for a review; see also Einhorn & Hogarth, 1981).[3]

In those circumstances when judgment aids are not used (Mode 5), when only (a) the data brought before the policy maker, citizen, or scientist and (b) the judgment made in response to the data are known, feasibility is greater than in Mode 4, but even those cognitive safeguards associated with aided judgment are lost. For in Mode 5 the processes of organizing the data into a judgment are largely private and can only be ascertained by unverifiable self-reports on how the judgment was reached. As a result, those who utilize Mode 5 are defenseless against the charge that their reports of how they arrived at a judgment are false; that although they claim facts "a" and "b" to be more important than fact "c", the "truth" is that they considered self-serving fact "c" to be most important. The only remaining safeguard is than an explicit set of data has been agreed upon. (An example of how Modes 5 and 6 produce suspicion may be seen in the attack on the integrity of scientists and the National Academy of Sciences by Boffey, 1975; for a review of Boffey see Polsby, 1975.)

In those circumstances where persons are free to select the data to which they will attend, as well as free to organize the data into a judgment as they see fit (Mode 6), even this last cognitive safeguard is eliminated. The cognitive activity of Mode 6 is very easy for anyone to use, and therefore, is the most feasible and frequently employed mode of cognition.

---

3 Too often those commenting on risk assessment ignore the costs of intuitive judgments while extolling their possible benefits. Thus Ashby (1979) believes "hunch" to be superior to aided judgment: "All attempts to rely on quantification . . . are likely to make the decisions worse, not better" (p. 23). Such sweeping condemnations of attempts to establish cognitive safeguards typically overlook the need for cognitive safeguards. See also Ashby (1976) for similar views.

The Pervasive Role of Intuitive Judgment in Ascertaining Risk

Scientists prefer to base their judgments on a common, delimited and well-defined set of data, as many National Research Council Committee reports have demonstrated. Therefore many scientific judgments of risk are carried out in Mode 5.[4] But scientists are frequently required to judge the relative risk of various alternatives without reference to a specific, delimited set of data (e.g., "Dr., what is your judgment of the risk posed by...?"). In such situations judgment takes place in Mode 6. The pervasive reliance on intuitive judgment which dominates Modes 5 and 6 often results in conflicting and self-contradictory results, induces nonproductive dispute, leads people to converge on mutually destructive strategies, and blocks the formation of a rational public policy (Hammond & Brehmer, 1973, pp. 338-391; see also Brehmer & Hammond, 1977, pp. 79-103). Lowrance's (1976) widely-cited remark that "A thing is safe if its risks are judged to be acceptable" indicates the key role of intuitive judgment in risk assessment.

Many examples of the pervasive role of intuitive judgment in scientific assessment of risk can be cited. Okrent, for example, documents the use of Mode 5 during assessment of risk from a chemical plant on Canvey Island in the Thames River (1980, p. 373):

> The Health and Safety Executive recommended that improvements be made that would reduce the likelihood of each of the risk estimates by a factor of 2 or 3. With these improvements, it was judged, the risks would be acceptable... The Health and Safety Executive will have the responsibility for evaluating the risk assessment and deciding on the acceptability of the risk.

In this case the Health and Safety Executive recognized the highly intuitive nature of such judgments and decisions. As Kletz pointed out: "The report admitted on the last page that it may have exaggerated the risks, as it stated, 'Practical

---

4 The Report of the National Research Council's Committee on the Biological Effects of Ionizing Radiation (1980) provides an example of the consequences of the cognitive activity of Mode 5, namely, irreducible dispute that frustrates the development of social policy. The report contains a prefatory letter from Philip Handler (then President of the National Academy of Sciences) to D. Castle (then the Administrator of the Environmental Protection Agency) that documented irreducible dispute by noting that two members of the committee refused to endorse the report.

people dealing with industrial hazards tend to "feel in their bones" that something is wrong with risk estimates developed in the body of the report'." (1981, p. 43)

In another case Okrent (1980, p. 373) reports risk assessment under Mode 6: "the study... did not establish quantitative criteria against which to judge the acceptability of the risk at each of the proposed sites," Gilinsky, a member of the Nuclear Regulatory Commission, was protesting forced participation in Mode 6, when he stated:

> We are confronted with this legitimate plea for a broad standard of safety because the reordering of priorities between safety and licensing has left a vacuum which must be filled. What we have been using in place of a standard for safety was an answer: The answer was provided by the Rasmussen Reactor Safety Study, which was taken to say, in effect, there was little to worry about, and by the political imperative allowing plants to continue in operation, willy-nilly.
>
> Why does this question arise here when it does not in automobile or airline safety, where we also lack an explicit safety goal? It is because we have a significant statistical sample for making estimates of the safety of airplanes--anyone can look it up in the world almanac. This is not true for nuclear reactors. (1980, p.5)

The societal consequences of failing to specify criteria are indicated by Lowrance (1981):

> I'll use the hypothetical example of a waste-site problem. If that high authority took it upon itself to review the waste site and only considered the cancer-causing properties of the chemicals and did not look at possible reproductive health threats, even by setting its own rules for review, it already has made a social judgment: Cancer is what is important and reproductive issues are not so important.

REMEDY

The bill, H.R.3441, concerning risk assessment now before Congress, states: "It is desirable to better delineate the scientific basis for agency decisions from their more subjective, judgmental aspects" (p. 3). But if scientific judgments of risk are to be improved, it will be necessary to subject scientists' judgment processes to the same critical examination that scientists apply to other natural phenomena. Science, in short, can and should provide its own remedy for the absence of cognitive safeguards in scientists' judgments of risk. (For some examples, see Hammond and Adelman, 1976; Mumpower, Veirs, & Hammond, 1979; Fisch, Hammond, & Joyce, in press; Stewart, Joyce, & Lindell, 1975, pp. 249-262.)

Judgment Analysis

Providing information for citizens and public policy makers ordinarily forces scientists to give up the analytical mode of cognition, and requires them to use the modes of cognition in which their analytical skills are no longer applicable. Without the cognitive safeguards of Modes 1, 2, and 3, scientists perform little better in Modes 4, 5, and 6 than a person who knows nothing of the cognitive requirements of science (a point we shall demonstrate below). When experimentation is not feasible, scientific judgment must be used to select the data to be considered in forming a judgment of risk, decide how the criteria should be weighed, determine the functional relations between criteria and risk, and most important, decide how such material should be aggregated, or organized into a judgment of risk. The latter parameter is of particular importance in scientific judgments because principles that implicitly organize data into a judgment (as in Modes 5 and 6) are, in fact, covert, untested scientific theories.

Techniques developed from empirical research in judgment can be, and have been, successfully used to achieve greater awareness of the four parameters of judgment processes and thus to help scientists to move from Modes 5 and 6 to Mode 4. Externalizing these parameters and their values makes it possible to discover and compare the source of differences in scientists' judgments, and thus to aid in their revision when appropriate.

The following example illustrates this process.

AN EXAMPLE

The Rocky Flats Plant (RFP), is a federally-owned industrial plant near Denver, Colorado that has been under suspicion since 1953 of discharging plutonium into the atmosphere and thus creating a risk to the health and safety of thousands of nearby citizens. The uncertainty regarding the risk from the operation of the RFP led the governor of Colorado and the congressman representing the district in which the RFP is located to form in June, 1978, the Rocky Flats Monitoring Committee (RFMC). Its stated purposes included the following:

> to provide state and local government officials with continuous advice (regarding) . . . the welfare of local citizens; to conduct an extensive information program to educate the public about the Rocky Flats Plant; to study and evaluate reports, accidents and various health standards applicable to the Rocky Flats Plant and to educate the public about the same. (Governor's Executive Order, 8 June 1978.)

Fulfillment of these objectives clearly rests on the assumption that scientists can and will provide the RFMC with information regarding the possible risk from the RFP; advice, information, evaluation, and education are wholly dependent upon scientific sources.

The RFMC encountered a serious obstacle in its effort to carry out its mission. Scientists, local and national, provided different and thus confusing judgments regarding the risk to health and safety from the RFP. Several years of increasingly acrimonius dispute among scientists made it impossible for the RFMC to inform itself or its constituents about the risk from the RFP. Consequently, in February, 1981, the RFMC approached the Center for Research on Judgment and Policy (CRJP) at the University of Colorado and requested assistance in resolving the dispute.

STEPS TO SOLUTION

The effort to reduce the dispute among scientists was carried out in three Phases.

Phase I

The specific aim of Phase I was to determine whether differences among scientists did in fact exist, and, if so, whether such differences were due simply to differences in disciplinary background (e.g., epidemiology vs. radiology).

Eleven scientists were chosen for study on the basis of their knowledge about and interest in risks to health from plutonium emissions from the RFP. The scientists agreed that the proper method for studying the effect of their respective disciplines on their judgments was to provide relevant information about a number of persons with differential exposure to plutonium radiation, ask each scientist to judge the risk of cancer for each person described, and to compare the judgments by discipline. This work was undertaken; a report (Hammond & Marvin) to the RFMC concluded that the eleven scientists did in fact:

> make different judgments of cancer risk from identical information, that the differences among scientists were stable, and were not due to differences in professional background and training, but due to differences in the way in which the judgments were made. Specifically, in a study of scientists' judgments of 40 (hypothetical) individuals exposed to cancer risk, large differences were found in the relative weights (importance) they assigned to the six factors (e.g., smoking, diet, etc.) deemed causative of cancer. Additionally, differences in implicit theories of risk, i.e., differences in principles for organizing information from several causative factors, may also be a source of differences in judgments of cancer risk.
>
> It is therefore concluded that the principal sources of disagreement are locatable, if not already located by the present study, and that it is feasible to reduce disagreement and thus to remove this impasse from the work of the RFMC. Because these tentative conclusions are based on a brief pilot study, limitations are indicated and recommendations regarding future work are included. (1981, p. i)

Thus the utility of applying the above framework to the analysis of scientists' judgments of risk was demonstrated. First, it was discovered that agreement on the factors to be considered could be achieved and that these factors could then be made explicit. Second, once the factors were defined, a straightforward application of judgment analysis using a form of multiple regression statistics, aided by an interactive computer program designed specifically for judgment analysis could be employed (Hammond, Stewart, Brehmer, & Steinmann, 1975, pp. 271-312; Hammond, Rohrbaugh, Mumpower, & Adelman, 1977, pp. 1-30). This analysis made it possible to ascertain quickly (a) the relative weights each scientist attached to each factor, (b) functional relations between factors and judgments, (c) self-consistency of judgments, and (d) inter-judge agreement. It was therefore possible to provide a report for the RFMC and the eleven participating scientists within a work-period of one month. Third, it became obvious that a nonretraceable process of organizing information into a judgment of risk for the hypothetical persons exposed to radiation was being employed instead of an explicit principle derived from research findings. The scientists' judgments in this situation could therefore be described as quasi-rational (Anderson, Deane, Hammond, McClelland, & Shanteau, 1981). That is, their public disagreements were produced not by clear, fully-documented differences in methods of aggregating the common information but by cognitively implicit methods of aggregating information.

Phase II

The purpose of Phase II was to determine whether disagreement could be removed by assisting the scientists to construct an agreed-upon, defensible, organizing principle for aggregating data into a prediction of risk. That is, an attempt would be made to develop an explicit theory of risk of cancer from plutonium emissions from the RFP.

The CRJP research group asked five of the most widely disagreeing scientists among the 11 studied in Phase I to participate in Phase II. The aim was to find a wholly analytical method of defining predictor variables, their functional relations with cancer, interactions among the variables, and, most important, a defensible and explicit method of aggregating information over the predictor variables. Such a method would not include subjective weighting of variables. Instead, each variable to be included would have to be defended by a specific reference to directly related studies. Furthermore, each scientist's suggestion for each step of the process would be checked by the other four scientists; differences would be settled by reference to the relevant literature.[5]

Phase II thus began with an analytic approach to the structuring of the problem itself. First, different kinds of cancers were distinguished, and it was agreed that different cancers should not be grouped together, but should be considered separately. Second, the scientists agreed that the most important cancer to be considered in relation to the RFP should be lung cancer. Third, predictor variables were then considered; the goal was to arrive at a set of predictors that would maximize validity and minimize redundancy. The set of predictor variables used in Phase I was reexamined and revised from this perspective.

Factors Considered

Plutonium was included as an important predictor variable, because it is clearly relevant to both the presence of the RFP and lung cancer. Medical x-radiation was included because the scientists agreed that it is the most important of the lung-cancer-producing agents involving radiation, other than plutonium. Smoking was included because of its causal relevance to lung cancer. It was also agreed that, except for smoking, the most important lung-cancer-producing agent that does not involve radiation is asbestos. Asbestos was not included because it was learned that there is no appreciable exposure to asbestos in the Rocky Flats area. Age of exposure to radiation was also included in the set of predictor variables because of the long latency period between exposure and the appearance of the disease.

The final step in the problem-structuring portion of Phase II was to separate epidemiological problems from problems in health physics. The question of how best to measure the level of respirable plutonium was therefore reserved for the final phase of the project.

Phase II thus differed from Phase I in the more analytic approach taken to structuring the prediction problem, and in selection of the set of predictor variables. In Phase I, hypothetical individuals were presented for judgment, each individual having been characterized by a measurement on each variable deemed to be associated with cancer. In Phase II,

---

5 Squire's (1981) development of a procedure for ranking animal carcinogens fails to specify the parameters of the judging process with the result that the weighting (scoring) system and other parameters constitute an implicit additive theory of risk that is proposed for regulatory purposes. Dall, on the other hand, was explicit about the arbitrary use of an additive organizing principle: "Because we can't generalize about the toxic effects of multiple chemical exposure, it is almost impossible to estimate the health impacts of such exposure other than to assume that they are simply additive." In Ember, (1980).

however, variables, not individuals, were presented one at a time for evaluation, and the functional relationship between probability of lung cancer and each of the predictor variables was specified directly by the scientists. The predictor variables were then presented in pairs, and principles for combining them were put forward, criticized and revised on the basis of the research literature.

Results

The approach taken in Phase II produced a much more satisfactory level of agreement than had been achieved in Phase I. The five scientists agreed on the factors to be considered, the general forms of the functional relations between plutonium and lung cancer (close to linear), smoking and lung cancer (convex downward but close to linear over the range of concern), and x-radiation and lung cancer (convex downward but close to linear over the range of concern). The scientists also agreed that the latency period for the development of lung cancer should be taken to be 25 years. Agreement on these conditions made it possible to construct an analytical principle for aggregating the data into a prediction.

The principle was derived from the following information:

1. The annual probability of lung cancer in nonsmokers is taken to be 7/100,000 (Doll & Hill, 1964; Hammond, E. C., 1966; Doll & Peto, 1978).

2. The effects of smoking are predicted by the equation in Doll and Peto (1978).

3. The effects of exposure to plutonium and smoking are obtained by modifying these probabilities by ratios taken from a study by Archer, Gillam, Wagoner, and James (1978). (See Fig. 1.)

4. The effects of x-radiation are determined by relating the dosage in question to the equivalent plutonium dosage in terms of REM's (*Final Environmental Impact Statement: Rocky Flats Plant Site, Golden, Jefferson County, Colorado*, 1980).

5. The effects of age of exposure to plutonium are based on mortality figures taken from the U.S. Department of Health and Human Services Life Tables, p. 25.

Agreement on this analytical principle established an explicit, defensible method for organizing the relevant facts into a judgment of risk. For example, if the amount of respirable plutonium in the atmosphere downwind from RFP is known, then persons living in that area who know the number of cigarettes they smoke each day, the amount of radiation they have received and are receiving from medical or occupational sources, and their age at initial exposure, could readily find in a table the effect of measured plutonium in the atmosphere on their personal, annual risk of lung cancer.

In addition, policy makers who must judge the acceptability of the increase in the amount of risk placed on their constituents as a result of atmospheric plutonium can now base their decision on an agreed-upon scientific analytical principle (Anderson, Hammond, Berg, Hamman, Johnson, Marine, & Sutherland, 1981). The advantage of such an analytical principle is that, unlike pooled intuitive judgments, its rationale and empirical base is clear. Therefore, it is readily subject to modification if it should be found to be in error, and as new scientific findings make modification appropriate.

Phase III: Development of Tables of Estimated Risk

The purpose of this phase of the study was to use the above principle, together with the results of previous empirical research, to develop risk tables which could be easily understood by the general public. These tables would be entered by level of plutonium exposure, current age, and smoking habits, to yield the relative risk of lung cancer during the next year for a specific individual compared to a person of the same age who is a nonsmoker and has never been exposed to Rocky Flats plutonium.

Use of the analytic principle developed in Phase II requires converting an ambient air level of plutonium to dose in REMs to the lung. The ambient air level of Pu239+240, in microcuries per cubic meter of air, is measured, reported, and published by the Rocky Flats Plant (RFP), the Department of Energy (DOE), and the Colorado Department of Health (CDH). A panel of four experts in health physics was recruited to develop a valid dose estimation procedure (Sutherland, 1982).

Tables of risk were then developed using the following procedure:

1. Specify individual age, average number of cigarettes smoked per day, average ambient air level of Pu239+240

in microcuries per cubic meter of air, and age at initial exposure to this ambient air level.[6]

2. Calculate the expected incidence of lung cancer for persons of the same age and smoking status using the equation of Doll and Peto (1978). This equation is valid for persons who began smoking at ages 16-25.

3. Determine the cumulative dose in REMs to the lung from Pu239+240 inhalation. Assume six REMs of plutonium inhalation exposure is equivalent to one working level month of radon exposure as estimated in the BEIR III report (1979). (Doses from x-radiation should be added.)

4. Adjust expected incidence of lung cancer for persons of same age and smoking status for working level months equivalent exposure using Fig. 1 of Archer, Gillam, Wagoner, and James (1978). The method of adjustment is specified in detail by Sutherland (1982).

5. Calculate relative risk of lung cancer by dividing the incidence rate calculated in procedure #4 by the expected incidence of lung cancer in nonsmokers with no Rocky Flats plutonium exposure.

The equation for determining the probability of lung cancer during the next year for a person of specified age 'a' and who smoked 'c' cigarettes per day starting at age 16-25 is:

$$I = (0.273 \times 10^{-7})\ (C+6)^2\ (A-22.5)^{4.5}\ \exp\ (0.00018\ (DOSE))$$

The first three terms of the equation represent risk from smoking alone (Doll & Peto, 1978). The last (exp(0.00018 (DOSE))) represents the increased risk from plutonium exposure and the interaction effect between cigarette smoking and plutonium inhalation. The method of dose calculation for a person exposed to a continuous ambient air level of Rocky Flats plutonium is specified in detail by Sutherland (1982).

Eight ambient air levels of Rocky Flats Pu239+240 were chosen for analysis (see Table 1). The first level presented (level

---

6 In the tables exposure to Rocky Flats plutonium since birth was assumed. For periods of exposure less than the entire life span, risk may be extrapolated from values in the tables by multiplying the fraction of life span at risk by the difference between risk for level of exposure of interest and risk for zero plutonium exposure and adding this value to risk at zero plutonium exposure.

(0)) is for no Rocky Flats plutonium exposure. Relative risks in Tables 2-4 for this level reflect age and smoking risk only. Level (1), 0.1E-9 microcures of Pu239-240 per cubic meter of air, is a typical background level recorded at CDH in metropolitan Denver (A Risk Evaluation for the Colorado Plutonium-in-Soil Standard, 1976). No increase in risk is noted at background fallout levels. (Compare levels (0) and (1) in Tables 2-4.)

TABLE 1

Ambient Air Levels of Pu239+240

DESCRIPTION	LEVEL	MICROCURIES/CU M
No exposure	(0)	0.0
Typical Background Fallout	(1)	0.1X10-9
Low RFP Sampler East Gate	(2)	0.5X10-9
Typical RFP Sample East Gate	(3)	0.1X10-8
Typical EML Sample East Gate	(4)	0.2X10-8
Typical CDH Sample East Gate	(5)	0.5X10-8
DOE/CDH Guideline	(6)	0.6X10-7
Less than 1/50th of the 1957 release at the stack	(7)	0.1X10-5

RISK AND LIVING NEAR THE RFP DURING 1974-1978

Level (2), 0.5E-9 microcuries per cubic meter, was the lowest average reading reported from RFP east gate samples during the years 1974-1978 (Barker, 1981). Levels (3), (4), and (5) are typical levels reported from RFP, EML (DOE Environmental Measurements Laboratory) and CDH samplers respectively during the same period (see Fig. 2).

No increase in radiation is noticeable at these levels of plutonium exposure in any age group at any smoking level. Under normal operating conditions during 1974-1978, the plant presented no noticeable radiation hazard to persons living in its vicinity. (Compare levels (2), (3), and (4) in Tables 2-4.)

TABLE 2

Lung Cancer Relative Risk for Nonsmoking White Males

Plutonium Ambient Air Levels

	(0)	(1)	(2)	(3)	(4)	(5)	(6)	(7)
40	1.00	1.00	1.00	1.00	1.00	1.00	1.00	1.06
45	1.00	1.00	1.00	1.00	1.00	1.00	1.00	1.06
50	1.00	1.00	1.00	1.00	1.00	1.00	1.00	1.07
ge 55	1.00	1.00	1.00	1.00	1.00	1.00	1.00	1.08
60	1.00	1.00	1.00	1.00	1.00	1.00	1.01	1.09
65	1.00	1.00	1.00	1.00	1.00	1.00	1.01	1.10
70	1.00	1.00	1.00	1.00	1.00	1.00	1.01	1.10
75	1.00	1.00	1.00	1.00	1.00	1.00	1.01	1.11

RISK OF PLUTONIUM EXPOSURE AT DOE/CDH GUIDELINES

Level (6), 0.6E-7 microcuries per cubic meter, is the DOE/CDH guideline for release of soluble plutonium aerosols in controlled areas. At this level of exposure, a one percent increase in relative risk is noted for nonsmokers. A similar percentage increase is noted for smokers. Since no noticeable increase in risk should be observable at DOE/CDH guideline exposures, this guideline needs more careful evaluation by federal and state regulatory authorities.

TABLE 3

Lung Cancer Relative Risk for White Male Smokers

20 Cigarettes per Day

Plutonium Ambient Air Levels

Age	(0)	(1)	(2)	(3)	(4)	(5)	(6)	(7)
40	2.82	2.82	2.82	2.82	2.82	2.82	2.83	2.98
45	5.34	5.34	5.34	5.34	5.35	5.35	5.36	5.69
50	8.50	8.50	8.50	8.50	8.50	8.50	8.53	9.11
55	12.11	12.11	12.11	12.11	12.11	12.12	12.17	13.08
60	16.04	16.04	16.04	16.04	16.05	16.05	16.12	17.46
65	20.18	20.18	20.18	20.18	20.18	20.19	20.29	22.12
70	24.44	24.44	24.44	24.44	24.45	24.45	24.59	26.99
75	28.76	28.76	28.76	28.76	28.76	28.77	28.94	31.99

During 1957, a fire and explosion at the plant produced unmonitored plutonium releases during the fire and for six days thereafter due to power outage. Monitoring resumed seven days after the fire and a release as recorded equal to 50 times level (7) (Johnson, 1981). Actual release may have been an order of magnitude or more greater.

Therefore, some individuals downwind from the plant may have received the cumulative dose in 1957 equivalent to more than 50 years exposure at level (7) due to the 1957 fire. As Tables 2-4 indicate, the increase in risk for lifetime exposure varies from four percent for 40-year-old males to 11 percent for 75-year-old males. It is essential, therefore, that the Rocky Flats Emergency Response Plan allow for evacuation of persons living near the plant in the event of an accident. Careful consideration should be given to zoning regulations by public officials for development near the plant, since it is much

TABLE 4

Lung Cancer Relative Risk for White Male Smokers

40 Cigarettes per Day

Plutonium Ambient Air Levels

	(0)	(1)	(2)	(3)	(4)	(5)	(6)	(7)
40	8.82	8.82	8.82	8.82	8.82	8.83	8.85	9.33
45	16.73	16.73	16.73	16.73	16.73	16.73	16.79	17.81
50	26.60	26.60	26.60	26.60	26.60	26.61	26.71	28.53
55	37.91	37.91	37.91	37.91	37.92	37.92	38.09	40.95
60	50.22	50.22	50.22	50.22	50.23	50.24	50.47	50.64
65	63.17	63.17	63.17	63.18	63.18	63.20	63.52	69.24
70	76.50	76.50	76.51	76.51	76.52	76.54	76.96	84.47
75	90.02	90.02	90.02	90.03	90.04	90.07	90.60	100.12

easier to evacuate a business office than families and small children from residential dwellings and schools.

Relationship Between Plutonium Risk And Smoking Habits

While radiation risk percentage increases are similar for smokers and nonsmokers, smoking alone increases the risk of lung cancer far more than possible increases in risk from past Rocky Flats Plant operations. Figure 2 illustrates this phenomenon for a 75-year-old male. Relative risk is shown for various

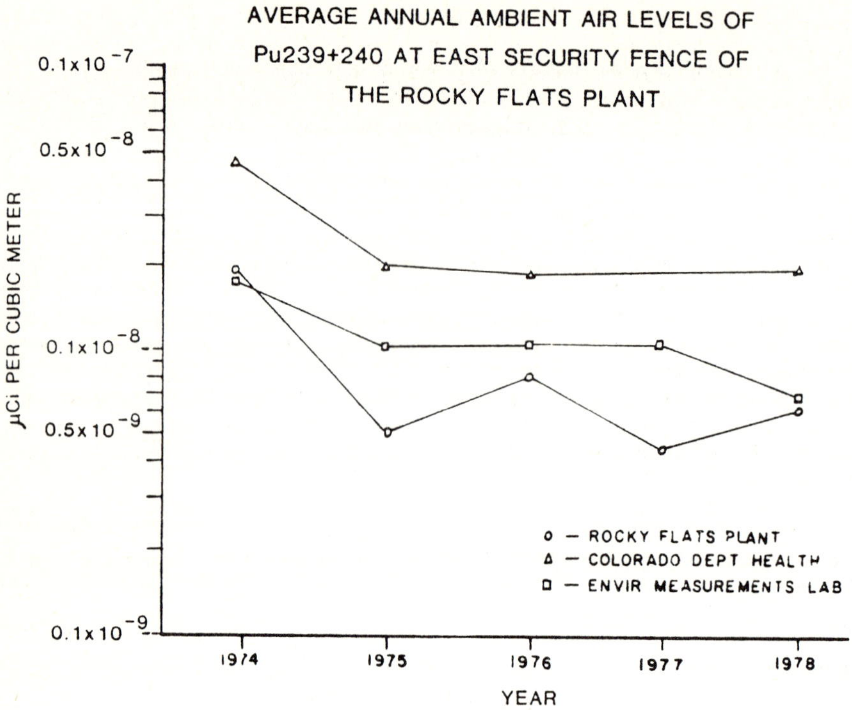

Figure 2

levels of plutonium exposure and cigarette consumption. Only levels (5), (6), and (7) are labeled on the figure because the first four levels produce no noticeable increase in risk.

For nonsmokers, the relative risk of lung cancer varies from 1.0 at level (5) to 1.11 at level (7). For persons smoking 20 cigarettes per day, relative risk increased from 28.77 at level (5) to 31.99 at level (7). For persons smoking 40 cigarettes per day, the relative risk increased from 90.07 at level (5) to 100.12 at level (7). In all cases, for a 75-year-old male, the increase was 11 percent over this range of plutonium exposures.

From a public health perspective, an 11 percent increase in lung cancer risk to nonsmokers will produce less than one additional lung cancer case per 100,000 population per year. For smokers, that increase will be three to 10 per 100,000 additional lung cancer cases per year. From an individual smoker's perspective, then, the first precautionary action to take if working or living near Rocky Flats is to stop smoking.

The achievement of this study is that, using Tables 2-4, individuals can now determine their annual relative risk from living or working near the plant for ambient air plutonium levels measured and published by the CDH. The tables will also be important to public officials responsible for overseeing development of the Rocky Flats environs and for protecting the health of workers and the public. These officials can now see that there is a small possible increase in risk at the DOE/CDH guideline level and a larger increase in risk from accidents. In addition, they must grapple with the dilemma posed by risk to smokers. Unlike nonsmokers, smokers incur significant risk of lung cancer from an accident at the RFP. But that risk is a direct result of their voluntary acceptance of an even greater absolute risk of lung cancer from cigarette smoking. The question is whether the public should bear the huge costs of reducing the involuntary risk of lung cancer for smokers, when smokers could reduce that risk themselves by giving up cigarettes.

RISK OF LUNG CANCER IN A 75-YEAR-OLD MALE

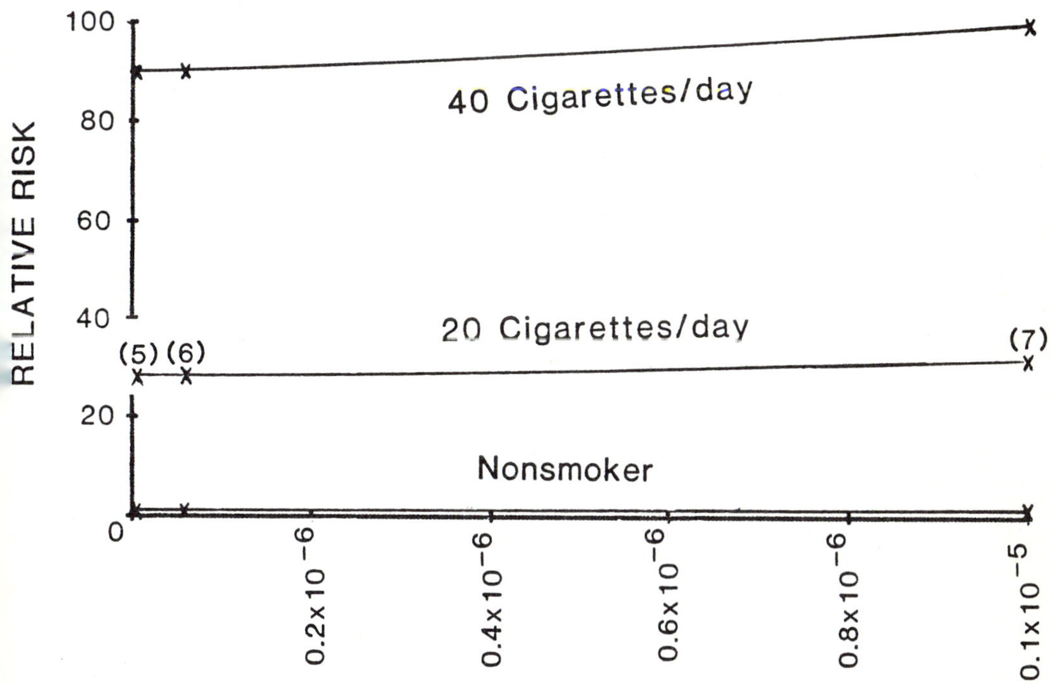

Pu AMBIENT AIR LEVEL (μCi/cu m)

Figure 3

Errors of Estimate

The error of estimate in Table 1 is a function of the following:

Error Induced By the Dose Estimation Procedure: This depends primarily on the relative biological effectiveness (RBE) of inhaled plutonium in producing cancer of the lung. The BEIR III report states that this RBE is in the range 8-15 based on review of historical data. Table 1 was calculated with an assumed RBE of 10.

Error Due to Population Sampling: (in the data of Doll and Peto (1978) and Archer, Gillam, Wagoner, and James (1978)). Cancer incidence rates are based on counts of cancers occuring in specified population which follow a Poisson distribution. Confidence limits on the curves from these studies vary from 0.871-1.154 times the dose-response rates in smokers estimated by Doll and Peto to 0.459-2.73 times the dose-response rates for nonsmoking uranium miners estimated by Archer, Gillam, Wagoner, and James where the sample size is smallest (Bailar and Ederer, 1964).

Error of Unknown Magnitude Introduced By Using Data from Health Effects of Human Radon Exposure to Estimate Effects of Plutonium Exposure: Both radon daughters and plutonium particles are alpha emitters. However, the distribution of particles in the lung is slightly different for different isotopes. No data are available in the literature to allow estimation of resulting differences in the rate of lung cancer induction in humans.

The latter source of error is of unknown magnitude because it is introduced by the judgment of a panel of epidemiologists that the data on human radon exposure is the best available data for use in estimation of plutonium inhalation effects. This is precisely where the mode of inquiry moves from Mode 3 (quasi-experiments) to Mode 4 (aided judgments). While data from a Mode 3 epidemiologic study of plutonium exposure is highly desirable and may be available in the future, the estimates of risk produced by Mode 4 are a substantial improvement over the statements produced by judgments based on unknown data of unknown bias and covert assumptions (Modes 5 & 6).

CONCLUSION

Scientists' judgments of risk should be improved through the use of judgment analysis because it describes scientists' cognitive activity rather than their motives, externalizes parameters of the scientists' judgment process that otherwise remain obscure, and assists in the development of explicit, rather than implicit, theories that are open to criticism and modifications based on the scientific literature. All of these advantages are part of the scientific tradition of employing cognitive safeguards against systematic and unsystematic errors.

Judgment analysis cannot, however, provide the cognitive safeguards that are offered by experimentation or even quasi-experimentation--although judgment analysis necessarily supplements the latter. Nor can it drive out every aspect of covertness, subjectivity, flawed logic, or bias--unintentional or deliberate. Nevertheless, as the present example illustrates, scientists who bitterly disagreed over the risk to citizens' health from plutonium emissions achieved agreement through the use of techniques produced by research on human judgment. The generality of the value of such efforts remains to be seen; but we must learn to make the best use of our present knowledge. For at present, scientific judgments are the only scientifically unexamined part of the risk assessment process.

The results of eliminating disagreement among scientists in the Rocky Flats case are clear. The citizens committee was able to carry out its charge: educating it constituents about the risks to health posed by emissions from the Rocky Flats Plant. Public officials found that risks must be evaluated in the context of other risks; that an involuntary risk (accidental emissions of plutonium) can achieve significance in the context of a voluntary risk (smoking).

ACKNOWLEDGEMENT

We thank J. Healy, W. Hendy, J. Berg, R. Hamman, C. Johnson, and W. Marine for their contribution. We thank J. Grassia, M. Luhring, and D. Victor for assistance in preparation of this article. Supported by funds from the Rocky Flats Monitoring Committee, the University of Colorado, and the Engineering Psychology Programs, Office of Naval Research, Contract No. 00014-77-C-0336, Work Unit Number NR 197-038 and NIH Grant CA 31682.

REFERENCES

Anderson, B. F., Deane, D. H., Hammond, K. R., McClelland, G. H., & Shanteau, J. C. Concepts in judgment and decision research: Definitions, sources, interrelations, comments. New York: Praeger, 1981.

Anderson, B. F., Hammond, K. R., Berg, J., Hamman, R., Johnson, C., Marine, W., & Sutherland, J. Second report to the Rocky Flats Monitoring Committee concerning scientists' judgments of cancer risk (Center for Research on Judgment and Policy, Report No. 233). Unpublished manuscript, University of Colorado, Institute of Behavioral Science, 1981.

Archer, V. E., Gillam, J., Wagoner, J., & James, L. Factors in exposure-response relationships of radon daughter injury. Conference/workshop on lung cancer, epidemiology, and industrial applications of sputum cytology. Golden, CO: Colorado School of Mines Press, 1978.

Ashby, E. Protection of the environment: The human dimension. Proceeding of the Royal Society of Medicine, 1976, 69, 721-730.

Ashby, E. Reflections on the costs and benefits of environmental pollution. Perspectives in Biology and Medicine, 1979, 23, 7-24.

Bailar, J. C., III, & Ederer, F. Significance factors for the ratio of a Poisson variable to its expectation. Biometrics, 1964, 20, 639-643.

Bazelone, D. L. Science, technology, and the court. Science, 1980, 208(4445), 661.

BEIR III: The effects on populations of exposure to low levels of ionizing radiation. Washington, DC: National Academy of Sciences, Advisory Committee on Biological Effects of Ionizing Radiation, 1979.

Boffey, P. M. The brain bank of America: An inquiry into the politics of science. New York: McGraw-Hill, 1975.

Brehmer, B. & Hammond, K. R. Cognitive factors in interpersonal conflict. In D. Druckman (Ed.), Negotiations: Social-psychological perspectives. Beverly Hills, CA: Sage Publications, 1977.

Brooks, H. Expertise and politics: Problems and tensions. American Philosophical Society Proceedings, 1975, 119(4), 257-261.

Campbell, D. T. From description to experimentation: Interpreting trends as quasi-experiments. In C. Harris (Ed.), Problems in measuring change. Madison, WI: University of Wisconsin Press, 1963.

Campbell, D. T. & Stanley, J. Experimental and quasi-experimental design for research on teaching. In N. Gage (Ed.), Handbook of research on teaching. Chicago: Rand McNally, 1963.

Chalmers, T. C., Smith, H., Jr., Blackburn, B., Silverman, B., Schroeder, B., Reitman, D., & Ambroz, A. A method for assessing the quality of a randomized control trial. Controlled Clinical Trials, 1981, 2,(1), 31-49.

Cole, S., Cole, J. R., & Simon, G. A. Chance and consensus in peer review. Science, 1981, 214(4523), 881-886.

Cook, T. & Campbell, D. T. Quasi-experimentation: Design and analysis issues for field settings. Boston: Houghton Mifflin, 1979.

Doll, R. & Hill, A. B. Mortality in relation to smoking: Ten years observations of British doctors. British Medical Journal, 1964, 1, 1399-1410; 1460-1467.

Doll, R. & Peto, R. Cigarette smoking and bronchial carcinoma: Dose and time relationships among regular smokers and life long nonsmokers. Journal of Epidemiology and Community Health, 1978, 32, 303-313.

Einhorn, H. J. & Hogarth, R. M. Behavioral decision theory: Processes of judgment and choice. Annual Review of Psychology, 1981, 32, 53-88.

Ember, L. R. Uncertain science pushes Love Canal solution to political, legal areas. Chemical and Engineering News, August 11, 1980, pp. 22-25.

Final environmental impact statement: Rocky Flats Plant site, Golden, Jefferson County, Colorado (DOE/EIS-0064). Washington, DC: U.S. Department of Energy, 1980.

Fisch, H.-U., Hammond, K. R., Joyce, C. R. B., & O'Reilly, M. An experimental study of the clinical judgment of general physicians in evaluating and prescribing for depression. British Journal of Psychiatry, 1981, 138, 100-109.

Fisch, H.-U., Hammond, K. R., & Joyce, C. R. B. On evaluating the severity of depression: An experimental study of psychiatrists. British Journal of Psychiatry, in press.

Gilensky, V. What is nuclear safety? Chemical and Engineering News, January 7, 1980, p. 5.

Hammond, E. C. Smoking in relation to the death rates of one million men and women. National Cancer Institute Monograph, 1966, 19, 127-204.

Hammond, K. R. Toward increasing competence of thought in public policy formation. In K. R. Hammond (Ed.), Judgment and decision in public policy formation. Boulder, CO: Westview Press, 1978.

Hammond, K. R. & Adelman, L. Science, values, and human judgment. Science, 1976, 194(4263), 389-396.

Hammond, K. R. & Brehmer, B. Quasi-rationality and distrust: Implications for international conflict. In L. Rappoport & D. Summers (Eds.), Human judgment and social interaction. New York: Holt, Rinehart, & Winston, 1973.

Hammond, K. R. & Marvin, B. A. Report to the Rocky Flats Monitoring Committee concerning scientists' judgments of cancer risk (Center for Research on Judgment and Policy, Report No. 232). Unpublished manuscript, University of Colorado, Institute of Behavioral Science, 1981.

Hammond, K. R., McClelland, G. H., & Mumpower, J. Human judgment and decision making. New York: Hemisphere/Praeger, 1980.

Hammond, K. R., Mumpower, J., Dennis, R. L., Fitch, S., & Crumpacker, D. W. Fundamental obstacles to the use of scientific information in public policy making. In F. Rossini, A. Porter, & C. Wolf (Eds.), Integrated impact assessment. New York: Elsevier, in press.

Hammond, K. R., Rohrbaugh, J., Mumpower, J., & Adelman, L. Social judgment theory: Applications in policy formation. In M. F. Kaplan & S. Schwartz (Eds.), Human judgment and decision processes in applied settings. New York: Academic Press, 1977.

Hammond, K. R., Stewart, T. R., Brehmer, B., & Steinmann, D. O. Social judgment theory. In M. F. Kaplan & S. Schwartz (Eds.), Human judgment and decision processes. New York: Academic Press, 1975.

Handler, P. Public policy must rest on sound science. Chemical and Engineering News, September 12, 1979, p. 5.

Handler, P. Public doubts about science. Science, 1980, 208(4448), 1093.

Johnson, C. J. Cancer incidence in an area contaminated with radionuclides near a nuclear installation. Ambio, 1981, 10, 176-182.

Kletz, T. A. Benefits and risks, their assessment in relation to human needs. In R. F. Griffiths (Ed.), Dealing with risk. New York: Halsted Press/John Wiley & Sons, Inc., 1981.

Lilienfeld, A. & Lilienfeld, D. Foundations of epidemiology. New York: Oxford University Press, 1980.

Lowrance, W. W. Of acceptable risk: Science and the determination of safety. Los Altos, CA: Kaufmann, 1976.

Lowrance, W. W. William Lowrance: Probing societal risk. Chemical and Engineering News, July 6, 1981, pp. 13-20.

Mumpower, J., Veirs, V., & Hammond, K. R. Scientific information, social values, and policy formation: The application of simulation models and judgment analysis to the Denver regional air pollution problem. IEEE Transactions on Systems, Man and Cybernetics, 1979, SMC-9(9), 464-476.

Nehnevajsa, J. & Menkes, J. Technology assessment and risk analysis. Technology Forecasting and Social Change, 1981, 19(3), 245-255.

Okrent, D. Comment of societal risk. Science, 1980, 208(4444), 372-375.

Polsby, N. W. Questions of bias. Science, 1975, 190(4215), 665-666.

The report of the National Research Council's committee on the biological effects of ionizing radiation. Washington, DC: National Academy Press, 1980.

Riecken, H., Boruch, R., Campbell, D., Coplan, W., Glennan, T., Pratt, J., Rees, A., & Williams, W. Social experimentation: A method for planning and evaluating social innovations. New York: Academic Press, 1974.

A risk evaluation for the Colorado plutonium-in-soil standard. Denver, CO: Colorado Department of Health, 1976.

Smoking and health: Report of the Advisory Committee to the Surgeon General of the Public Health Service. Washington, DC: HEW, Government Printing Office, 1979.

Spilhaus, A. F., Jr. Societies must shun politics: Act only on scientific issues. Chemical and Engineering News, April 21, 1980, pp. 43-50.

Squire, R. A. Ranking animal carcinogens: A proposed regulatory approach. Science, 1981, 214(4523), 877-880.

Stewart, T. R., Joyce, C. R. B., & Lindell, M. K. New analyses: Application of judgment theory to physicians' judgments of drug effects. In K. R. Hammond & C. R. B. Joyce (Eds.), Psychoactive drugs and social judgment: Theory and research. New York: John Wiley & Sons, Inc., 1975.

Sutherland, J. V. Estimation of lung cancer risk from environmental exposure to airborne plutonium from the Rocky Flats Plant. Unpublished manuscript, University of Colorado, School of Medicine, 1982.

Tweny, R., Doherty, M., & Mynatt, C. On scientific thinking. New York: Columbia University Press, 1981.

SECTION III: Decision and System Science Perspectives

Part I: Theoretical and Methodological Contributions

Issues in Evaluating Risks of Fatalities*

Ralph L. Keeney

Systems Sciences Department, University of Southern California, Los Angeles, CA 90007

INTRODUCTION

Significant attention by both governments and individuals is devoted to addressing problems including the possible loss of life among their concerns. Aspects of the attention concern acquiring an understanding of the problems and identifying manners in which these problems might be resolved. Once a specific problem is reasonably well understood, it can be represented as a decision problem. This would involve a choice among alternatives, each of which would have implications for the potential fatalities that might result. In addition, however, there would be potential economic, social, and environmental implications associated with each of the alternatives. Given that the decision problem is structured, an attempt is made to select the most responsible alternative utilizing the information that one has available at the time the decision must be chosen.

With many of the decision problems described above, the key issue in selecting an alternative is sometimes characterized by the phrase "How safe is safe enough?" This question has been asked with respect to different types of power plants, hazardous waste, the licensing of drugs, safety of aircraft and automobiles, and the list could go on and on. Although some absurdly simple "solutions" to the generic problem have been suggested, such as the appropriate level of risk of possible

*This work was supported by the National Science Foundation under Grant No. PRA-8107319. Any opinions, findings, conclusions, or recommendations expressed in this publication are those of the author and do not necessarily reflect the views of the National Science Foundation. Sections of this paper are liberally adapted from Keeney (1983).

NATO ASI Series, Vol. G4
Environmental Impact Assessment, Technology Assessment, and Risk Analysis. Edited by V.T. Covello et al.

fatalities is zero, specific problems in reality are quite complex. In particular, they are complex enough that systematic analysis has a reasonable potential for providing useful insights to help the decisionmaking process. When analysis is pursued, it will be necessary to analyze explicitly the alternatives in terms of their risks and other implications. To include the risks of fatalities in such analyses, one needs to quantify the possible risks of each alternative and to evaluate those possible risks. The focus of this paper is on the evaluation of fatality risks.

This paper is outlined as follows. The remainder of this section makes some general observations about determining an appropriate level of acceptable risk. Section 2 gives a short survey of previous work on evaluating public fatalities, and Section 3 focuses on some recent results relying on explicit approaches to quantify value judgments. Different assumptions are investigated, and the implications of those assumptions are presented. Section 4 briefly discusses implementation issues, and Section 5 is a conclusion.

An Appropriate Level of Risk

An important point to recognize in determining an appropriate level of risk is that determining "How safe is safe enough?" is not a decision problem. It is a phrase that describes a large number of specific decision problems which include, as one of their components, the possibility of the loss of life. It is the case that the appropriate level of risk in different problems may be very different. Hence, any attempt to find a "solution" to the inappropriately-defined decision problem of "How safe is safe enough?" is doomed to failure. That is not to say that attempts to identify an appropriate level of risk with specific decision problems must also suffer that same fate. Indeed, because such problems are so complex, I feel the potential of analysis for a positive contribution in selecting an appropriate level of risk is high. However, before proceeding, it is useful to point out some observations regarding such analysis from Derby and Keeney (1981).

First, an important point to recognize is that selection of an appropriate level of risk is found by selecting an alternative in a decision problem. This decision problem will involve choosing the best combination of advantages and disadvantages from among the alternatives. Furthermore, this choice will necessarily need to recognize the uncertainties and other complexities that are likely a part of the decision environment. In selecting an alternative, the various pros and cons need to be weighed in some responsible manner, which of course requires value judgments. Then, the level of risk associated with the chosen alternative is, by definition, acceptable. This level of risk is the level which might be characterized as "safe enough."

Second, acceptable risk is not necessarily a level of risk with which we are happy. We would all prefer less risk to more risk given that other consequences were held fixed. If we were not concerned about practicality, then zero risk would be the only risk that would be acceptable. However, there are no alternatives with zero risks. Because we are concerned about practicality, it also follows that when there is no alternative, the level of safety associated with the only course of action is, by definition, acceptable, regardless of how disagreeable that risk is. Stated differently, acceptable risk is the risk associated with the best of the available alternatives; it is not the risk associated with the best alternative which we would ideally like to have available.

Third, the level of acceptable risk is not necessarily the smallest level of risk associated with one of the alternatives. This follows since the best alternative may not be that with the lowest level of risk. If the benefits being derived from an alternative with higher levels of risk than those with the lowest risk available are worth the differential risks between those alternatives, this circumstance would occur. Related to this is the following observation: Simply because one level of risk is acceptable does not imply that a lower level of risk is acceptable, since the level of acceptable risk cannot be interpreted in a context void of the other implications relevant to the alternatives.

Fourth, assessing the structure of values and preferences appropriate for evaluating consequences involving risks is very involved. A key question to be resolved is whose values should be used. Should they come from the elected officials representing the public, from the regulatory authorities charged with various responsibilities to control risks, from the public themselves, or from a combination of the above? How should these values be determined? Perhaps more importantly, regardless of how this is done, there will be major conflicts in values among various people. How should these conflicting values be reconciled? To further complicate this assessment is the possibility that values and preferences for difficult and unpleasant consequences are not clear or well-formulated for many people. Assisting a person or group to understand their own values concerning unfamiliar consequences is as much a challenge as identifying whose values should be used. The results indicated in Section 3 are addressed toward assisting individuals and organizations in formulating and understanding their own values and investigating the implications related to those values. This should be a key element in developing appropriate value structures for selection of alternatives.

SECTION 2: PREVIOUS WORK

In the recent past, numerous scholars, including Starr (1969), Mishan (1971), Zeckhauser (1975), and Lowrance (1976), have recognized the need and potential contributions of explicitly and systematically examining the issue of fatality risks. Several approaches have been suggested in these papers and others for evaluating potential fatalities [see, for example, Acton (1973), Jones-Lee (1974), Linnerooth (1975), Thaler and Rosen (1976), Howard (1979), and Keeney (1980a)]. Zeckhauser (1975), Linnerooth (1979), and Jones-Lee (1982) provide surveys of much of this literature. Raiffa, Schwartz, and Weinstein (1978) present an excellent detailed summary of the numerous crucial value issues involved in evaluating public risks.

One thing is clear: There is no consensus on an appropriate approach for evaluating potential fatalities. In most, if not all, important decision problems involving potential fatalities, a value tradeoff must be made between costs and potential fatalities. Many approaches address this tradeoff by evaluating potential fatalities in terms of a dollar amount. Hence, the term "value-of-life" is often given to this literature. Most of the approaches, including those based on the human capital of individuals, life insurance decisions, job choice and wage differentials, court awards, and governmental safety decisions, involve inferences about the "value of life" on the basis of actions taken. Such approaches are sometimes referred to as revealed preference approaches. The willingness-to-pay and utility analysis approaches address the value tradeoff between costs and fatalities explicitly by requiring expressed preferences. Using utility analysis to address directly the value judgments inherent in evaluating the risks of loss of life, Howard (1979) and Keeney (1980a) both provide utility functions for such evaluations that are consistent with the von Neumann-Morgenstern (1947) assumptions. Howard's approach is to examine the utility function that may be appropriate for an individual, whereas the utility functions utilized by Keeney address the problem from an institutional perspective. This approach will be elaborated in the next section.

SECTION 3: CURRENT RESEARCH

The ultimate intent of the research reported here is to assist in identifying appropriate actions given the risks and costs of various alternatives. Thus, the orientation is prescriptive and specifically concerned with analysis of the relative desirability of spending different amounts of funds to reduce (directly or indirectly) fatality risks. This analysis requires an index of costs and an index of risks as arguments. To structure such indices, the process necessarily involves value judgments. I proceed by postulating various reasonable value judgments and then deriving their implications for the resulting indices. Examining the implications of different basic values provides interesting insights about the evaluation of possible fatalities. Rather than using a formal theorem-and-proof format, I will simply state results, and reference details to other works.

NOTATION

We wish to evaluate fatality risks before any fatalities have occurred and where there will be numerous individuals potentially at risk. Also, initially, our sole interest in fatality risks is because of the resultant fatalities that may occur. Hence, the problem can be characterized by a profile p $= (p_1, p_2, \ldots, p_n)$ where p_i is the risk to individual I_i and n is the total number of individuals at risk. Specifically, p_i will be the probability that individual I_i will be a fatality in the time period of concern due to the cause of concern. This probability is independent of all the other probabilities in the risk profile. To describe circumstances involving independent risks to individuals, lotteries over risk profiles may be utilized. For instance, suppose there was a 0.2 chance of an accident occurring which would result in the risk profile (0.4, 0.4) for a two-person society. If no accident occurred, the risk profile for "normal operations" would be (0.01, 0.01). This circumstance would be represented as a lottery yielding a 0.2 chance at (0.4, 0.4), and a 0.8 chance at (0.01, 0.01). An appropriate manner to measure the desirability of lotteries is a utility function assessed in accordance with assumptions postulated alternatively by von Neumann and Morgenstern (1947) or Savage (1954).

A risk profile characterizes a situation before the time period in which fatalities could occur. At the end of the period, each individual will either be or not be a fatality due to the cause of concern. Thus, we will define attribute X_i to take on levels x_i equal to 0 or 1 depending on whether I_i was not or was a fatality. Then a fatality profile can be represented by $x = (x_1, x_2, \ldots, x_n)$. At times, it may be appropriate to summarize the information about how many fatalities occur. We define the attribute F to be the number of fatalities and f to be a specific level of F. Then clearly $f = x_1 + x_2 + \ldots x_n$.

The cost that one might be willing to expend to reduce fatality risks will be denoted as c. Thus, a consequence characterizing a situation would be the vector $(p;c) = (p_1, p_2, \ldots, p_n; c)$. A basic problem for evaluating fatality risks is then to determine an appropriate utility function u which assigns the utility $u(p;c)$ to each possible consequence $(p;c)$. Based on the assumptions of utility theory, consequences with higher utilities should be preferred to those with lower utilities, and when uncertainty is involved, situations corresponding to higher expected utilities should be preferred.

The observations in this section are appropriate for any separable utility function $u(p;c) = g[u_R(p), u_C(c)]$, where g is any function, u_R is the utility function for fatality risks, and u_C is the utility function for costs. However, for simplicity, we will assume that an appropriate utility function for risks and costs is an additive one,

$$u(p;c) = u_R(p) + \lambda u_C(c), \tag{1}$$

where λ is a scaling constant to assure that u_R and u_C are consistently scaled. For convenience, we will set the origin of all the utility functions by

$$u(0;0) = u_R(0) = u_C(0) = 0 \tag{2}$$

and establish a scale for each of the utility functions by

$$u(1,0,\ldots,0;0) = u_R(1,0,\ldots,0) = u_C(1) = -1. \tag{3}$$

where c is evaluated in millions of dollars.

The Utility Function u_R

To determine the utility function u_R, we can consider utility functions over each component p_i and the integration of these utility functions. Concerning the former problem, it may be reasonable to assume that an organization should evaluate an individual's risks as the individual would want to evaluate them. Assuming that the individual wants to minimize his or her risks, such a utility function should be linear in p_i. This linearity condition also follows from a consistency argument which assumes that the relative utility of fatality risk profiles must equal the expected utility of the implied set of fatality profiles [see Keeney (1980c)]. If the individual had a different value structure, one would need to use a nonlinear utility function $u_i(p_i)$. Since we concentrate on the evaluation of simultaneous individual risks, we employ the linear case for u_i for simplicity. Similar developments are possible for nonlinear cases. We also make the assumption that risks to each individual should be evaluated identically. Furthermore, it may be appropriate to evaluate risks to any subset of individuals in the same manner regardless of fixed risks to those individuals not in the subset. In technical terms, these assumptions are equivalent to the following.

Basic Model Assumptions: The attributes measured by the individual risks are utility independent, all subsets of attributes are preferentially independent, and the utility function for each of these single attributes is linear (namely minus p_i for individual I_i).

As shown in Keeney (1980c), the utility function u_R consistent with the basic model assumptions must either be the additive form

$$u_R(p) = -\sum_{i=1}^{n} p_i \tag{4}$$

or the multiplicative form

$$u_R(p) = -1/k \left[\prod_{i=1}^{n} (1+kp_i)-1\right] \tag{5}$$

where k is a scaling constant. We will now examine further implications on the form of u_R given the basic assumptions above for each of three different objectives: that we want to minimize the expected number of fatalities, that we prefer equitable ex-post distributions of risk, and that we wish to avoid catastrophes. As we will see, these desires are mutually inconsistent.

Minimizing Expected Fatalities: Given specific funds to reduce the risks to individuals, a reasonable guideline might be to minimize the resulting total expected fatalities to save as many lives as possible.

Minimum Fatality Risk Assumption: Fewer expected fatalities are always preferred to more expected fatalities.

The following result directly follows from the fact that the sum of the p_i's is equal to the expected number of fatalities with any fatality risk profile.

> Result 1: Given the basic model assumptions and the minimum fatality risk assumption, the utility function for fatality risks must be the additive form (4).

Recall that f is the number of fatalities resulting from any particular risk situation. To be consistent with the additive u_R, a utility function u_F for fatalities is

$$u_F(f) = -f. \qquad (6)$$

An Equitable Distribution of Risk: We will define consequence $(p_1,\ldots,p_i,\ldots,p_j,\ldots,p_n)$, for any i and j, to be a more equitable distribution of risk than $(p_1,\ldots,p_i+\varepsilon, p_j-\varepsilon,\ldots,p_n)$ if the difference between p_i and p_j is less than the difference between $p_i+\varepsilon$ and $p_j-\varepsilon$. Note that all $p_h, h\neq i,j$ are held fixed in this definition. This definition merely says that if all risks but two are fixed, the better balanced these two are, the more equitable the risk.

Risk Equity Assumption: A more equitable distribution of risk is preferred to a less equitable distribution.

> Result 2: Given the basic model assumptions and the risk equity assumption, the utility function for fatality risks must be the multiplicative form (5) where $0>k>-1$.

As shown in Keeney (1980c), the utility function for fatalities consistent with this preference for risk equity is

$$u_F(f) = -1/k\ [(1+k)^f - 1] \qquad (7)$$

which is risk prone for the allowable values of parameter k.

With a risk prone utility function for fatalities, a lottery with f expected fatalities would be preferred to f fatalities for sure. In simple terms, one would gamble in an attempt to save more than the average possible fatalities. Thus, a risk prone utility function would not always result in evaluating alternatives in the manner to save the most expected fatalities. It follows that to have the original risks more equitably spread among the individuals at risk, one must be willing to allow a few more expected fatalities. Whenever there is a preference for an aspect of risk other than the minimization of fatalities, it may be necessary to give up something in terms of expected fatalities to better achieve the other objective.

Catastrophe Avoidance: A number of papers, such as Ferreira and Slesin (1976) and Slovic, Fischhoff, and Lichtenstein (1977), suggest that a small probability of a catastrophic loss of life is worse than a larger probability of a smaller loss of life, given the expected number of fatalities are the same for each case. To be precise, we will define the following.

Catastrophe Avoidance Assumption: A probability π of having f fatalities is preferred to a probability π' of having f' fatalities for any f less than f' given that $\pi f = \pi' f'$.

Given this assumption, the following is proven using results in Keeney (1980b, c).

> Result 3: Given the basic model assumptions and the catastrophe avoidance assumption, the utility function for fatality risks must be the multiplicative form (5) where k>0.

The corresponding utility function for fatalities is (7) where k>0. As can easily be demonstrated, this utility function is risk averse, which is consistent with a preference for $\bar{f}$ fatalities for sure to any lottery with $\bar{f}$ expected fatalities. With this preference to avoid a large loss of life (i.e., a catastrophe), one must be willing to accept more expected fatalities.

Identifiable Versus Expected Fatalities

Now we will examine the value tradeoff between costs and fatalities. Specifically, suppose we find a consequence (0,...,0;c*) indifferent to consequence (1,0,...,0;0). Then because the first consequence involves no risks and a cost of c*, and the second consequence involves a risk of 1 to one individual and no costs, one could define c* to be the "value of life." The life at risk is clearly identifiable; it is individual I_1's life. Such a circumstance is referred to as an identifiable fatality. Hence, it might be more appropriate to refer to c* as the value of an identifiable life.

Now suppose we determine that (1/n,...,1/n;0) and (0,...,0;c') are indifferent, which gives us another possible value of life. To interpret this, recall that with the first consequence there are n individuals at risk. At the time the evaluation must be made, it is clear that each individual has an equal 1/n chance of being a fatality, but it is not possible to identify who will be fatalities (in fact, there could be zero, one, or more than one fatality). The expected number of fatalities resulting from the risks is 1, so we could say that one expected fatality and no costs is indifferent to 0 expected fatalities and the cost of c'. Thus c' could be referred to as the value of an expected life. Many people feel that the risk (1,0,...,0) is worse than (1/n,1/n,...,1/n) because the former consequence is so unevenly distributed and at least the risk in the latter case is "somewhat equitable." If this is the case, the value c' should be less than the value c*.

In essentially all situations involving an evaluation of fatality risks, we would expect to be dealing with only small risks to each individual. These risks should certainly be much less than 10^{-2} for each person and would much more likely be in the range of 10^{-4} and smaller. However, we could certainly have situations where 20 percent of the individuals had a risk of 5/n each and the other 80 percent had a 0 risk. The expected fatalities corresponding to this risk would also be one, although the risk in some sense would not be as equitable as the situation where each individual had a 1/n risk. In any case, it should be obvious that there is no unique definition for the term value of life. Also, use of the term "value of life" can be misinterpreted to imply that the value of two lives is equal to twice the value of one life. This may be the case but it requires additional value judgments.

Combining the cost utility function in (1) with the utility functions for fatality risks in Results 1 through 3 leads to the following interesting result.

> Result 4: Given the basic model assumptions and (1), the value of an identifiable life equals the value of an expected life if and only if there is a preference to minimize expected fatalities; the value of an identifiable life is greater than the value of an expected life if and only if there is a preference for risk equity; and the value of an expected life is greater than the value of an identifiable life if and only if there is a preference for catastrophe avoidance.

Many individuals might feel that more funds should be expended to eliminate an identifiable risk (1,0,...,0) than an expected risk (1/n,1/n,...,1/n). It follows that the preference for risk equity and the risk prone utility function for fatalities are appropriate, given the basic model assumptions.

Equity of the Process[1]

Broome (1982) recently invoked an objection of Diamond (1967) to Result 2 concerning risk equity. Specifically, he pointed out that a utility function incorporating an equity concern for risk profiles (p_1, p_2) would assume indifference between a 50-50 lottery yielding either $(p_1 = 1, p_2 = 0)$ or $(p_1 = 0, p_2 = 1)$ and the sure consequence $(p_1 = 0, p_2 = 1)$ for a two-person society. Indeed, the consequences in either situation result in one fatality. However, Broome mentions that a preference for equity must lead one to prefer the lottery to the sure consequence. Such a preference may be reasonable, but it concerns a different type of equity. The preference for the lottery over the sure consequence indicates a preference concerning the equity of the process by which the consequences are distributed, not just a preference concerning the equity of the process consequences per se.

To illustrate, we will introduce the concept of marginal risk equity using a two-person society with risk profiles $p = (p_1, p_2)$. Consider the following four lotteries:

I: a 0.5 chance at $p = (1, 1)$ and a 0.5 chance at $p = (0, 0)$.

II: $p = (0.5, 0.5)$.

III: a 0.5 chance at $p = (1, 0)$ and a 0.5 chance at $p = (0, 1)$.

IV: $p = (0, 1)$.

Using the risk equity assumption, one should prefer Lotteries I to II and II to III. Broome (1982) suggests that a desire for equity should also lead one to prefer Lottery III to IV. However, because the risk equity assumption implies that consequences can be characterized by risk profiles, the models discussed in Results 1 through 4 imply indifference between III and IV.

This example suggests that risk profiles (or equivalently the number of fatalities) may not always be an adequate representation of fatality risks. Hence, additional attributes will be needed. This can be addressed by defining a marginal risk profile of marginal risk probabilities $q = (q_1, q_2, \ldots, q_n)$, where q_i represents the marginal probability that individual I_i will be a fatality. Note that q differs from

---

1 The material in this section is condensed from Keeney and Winkler (1982). A related development is found in Sarin (1982).

p in that probabilistic independence of the possible fatalities is not assumed.

For Lotteries I, II, and III, the marginal risk profile is $q = (0.5, 0.5)$, whereas for Lottery IV the marginal risk profile is $q = (0, 1)$. A desire for equity in terms of marginal risk profiles q would lead one to prefer Lottery III to Lottery IV. This notion of marginal risk equity is characterized formally in the following definition. The marginal risk profile $(q_1,\ldots,q_i,\ldots,q_j,\ldots,q_n)$, for any i and j, will be called a more equitable distribution of marginal risk than $(q_1,\ldots, q_i+\varepsilon \ldots,q_j-\varepsilon\ldots,q_n)$ if the absolute difference between q_i and q_j is less than the absolute difference between $q_i+\varepsilon$, and $q_j-\varepsilon$.

It is obvious that marginal risk equity captures aspects of equity different from those captured by the (non-marginal) risk equity. For example, Lottery III is perfectly equitable in a marginal sense but highly inequitable in a non-marginal sense. However, the two types of risk equity are by no means incompatible. Note that Lottery I is perfectly equitable with respect to both types of equity, and Lottery IV is totally inequitable in both senses. In general, if a situation is not perfectly equitable marginally, then it cannot be perfectly equitable in a non-marginal sense, although some of the possible risk profiles p may be equitable.

To further clarify the distinction between the two types of equity, note that a risk profile p has a clear relationship to the eventual fatality profiles since it implies a certain lottery over such fatality profiles. As a result, the risk equity in Result 2 can be thought of as ex-post risk equity, or consequence equity. In contrast, a marginal risk profile indicates the ex-ante probabilities of death for the individuals of concern, and as such it represents a characteristic of the process generating the risks. Thus, marginal risk equity can be defined to address ex-ante risk equity or process equity.

> Marginal Risk Equity Assumption. A more equitable distribution of marginal risk is preferred to a less equitable distribution.

Fishburn (1982) recently proved that there is no utility function over marginal risk profiles for evaluating potential fatalities consistent with such an assumption. However, as shown in Sarin (1982), there are value functions $v_Q(q)$ over marginal risk profiles which rank preferences of those profiles. For a given Σq_i, v_Q must be highest for an equal division of Σq_i in order to satisfy the marginal risk equity assumption. Furthermore, purely in an equity sense, the marginal risk profiles $(0,\ldots,0)$, $(1,\ldots,1)$, $(0.5,\ldots,0.5)$, and so on, might be considered to have equivalent marginal risk equity because all individuals have the same marginal risk in each of these profiles. Thus, excluding any impact of expected fatalities $f = \sum q_i$ from v_Q, we might set $v_Q(q) = 0$ for all equal divisions of q regardless of the value of $\sum q_i$.

> Result 5: A value function v_Q consistent with the marginal risk equity assumption is
> $v_Q(q) \equiv -\sum(q_i - q)^2$
> where $q = \sum q_i/n$.

This function yields $v_Q(q) = 0$ for Lotteries I-III and $v_Q(1) = -1/2$ for Lottery IV for the two-person society.

Identifiable Versus Statistical Fatalities

The general issue related to marginal risk equity has been discussed in the public risk literature for some time. It addresses the distinction between an identifiable and a statistical fatality. Recall that an identifiable fatality can be characterized by the risk profile p = (1,0,...,0), so the value of an identifiable life c* can be defined by two indifferent consequences (1,0,...,0;0) and (0,0,...,0;c*). A statistical fatality results from a situation where it is known that exactly one individual will be a fatality, but the identity of that individual is not known. In terms of risk profiles, a statistical fatality can be represented by a lottery with a 1/n chance of each of the following risk profiles: (1,0,...,0), (0,1,0,...,0),..., (0,...,0,1). If the consequence (0,0,..., 0;c†) is indifferent to the lottery over risk profiles and costs with a 1/n chance at each of (1,0,...,0;0), (0,1,0,..., 0;0),..., (0,...,0,1;0), we could define c† to be the value of a statistical life.

Descriptively, many individuals feel that the value of an identifiable life should be greater than the value of a statistical life. If so, more funds should be utilized to avoid an identifiable fatality than a statistical fatality. To relate the value of identifiable and statistical fatalities to marginal risk equity, we have the following.

> Result 6: If the value of an identifiable life equals the value of a statistical life, the marginal risk equity assumption does not hold; if the value of an identifiable fatality is greater than the value of a statistical fatality, the marginal risk equity assumption does hold.

No one has proposed a preference for the reverse of marginal risk equity, meaning for instance q = (1,0,...,0) preferred to q = (1/n,1/n,...,1/n). Such a preference might correspond to a situation where the psychological fear of dying was proportionally greater for smaller marginal risks and the overall objective was to minimize psychological fear. This preference would be consistent with the value of an identifiable life being less than the value of a statistical life.

SECTION 4: IMPLEMENTATION

A general comment sometimes heard about the "value of life" literature is "Does anybody ever expect to use that?" From a prescriptive viewpoint, the answer is most certainly yes. However, one must be careful about the meaning of "use."

Numerous programs are aimed directly at reducing fatality risks. Other programs are concerned with fatality risks as they are associated with otherwise beneficial actions. To responsibly evaluate decisions on such programs, it is essential to evaluate the various fatality risk profiles. The evaluation might not be conducted in a formal manner, but one must have some concept of what these risk profiles are and make value judgments to appraise them. The choice is whether to make the value judgments formally or informally. Given the complexity of the problem, I think it is appropriate to use both approaches. The formal approach can be used as a device to assist one in understanding and examining various value judgments and their appropriateness for institutional policy regarding fatality risks. But, the purpose of the models and the methodology is to provide critical insights and guidance for action. They should not be a substitute for the decisionmakers.

Use of the models requires the careful consideration of several value judgments. First, one must consider the general structure of the problem. There are implicit value judgments built into this structure, as they are built into any problem structure. Second, one needs to appraise the value judgments necessary to structure the utility function $\underline{u}$. For illustration in this paper, we assumed by (1) that the utility function for fatality risks and costs is additive. This assumption could easily be relaxed and all the ideas in the paper would still be valid. The assumptions leading to functional forms for the fatality risk utility function or the ex-ante equity value function need to be appraised for any specific use. Finally, even if a particular functional form is found to be appropriate, parameters for that utility function or value function must be assessed. This requires the decisionmakers to make a few problem-specific value judgments to identify pairs of indifferent consequences. For such difficult judgments, it is best to ask redundant questions from different perspectives. Inconsistencies will no doubt be observed. However, part of the motivation for the assessment process is to identify such inconsistencies and bring about an internal reconciliation after decisionmakers understand where and why these inconsistencies occurred. In Keeney (1980a), the assessment of some parameters for utility functions involving public risks is illustrated.

SECTION 5: CONCLUSION

In this paper, the risk problem was structured using explicit value judgments and the implications of these judgments for the evaluation of fatality risks was examined. It is evident from the results in Section 3 that seemingly reasonable assumptions can lead to incompatible utility functions for fatality risks. It was shown that a preference to minimize the expected number of fatalities leads to a linear utility function for fatalities; a preference for risk equity leads to a risk-prone utility function for fatalities; and a preference for catastrophe avoidance leads to a risk-averse utility function for fatalities.

We should not, however, mistakenly assume that the inconsistencies among these objectives are unfortunate or wrong. Whenever there is more than one objective in a problem, there will necessarily be some inconsistency in achieving those objectives. If there is no inconsistency, they are de facto one and the same objective. It does not seem unreasonable that for some cases, one may have, among others, the following four objectives: to minimize lives lost, to avoid catastrophes, to maximize ex-post equity, and to maximize ex-ante equity. A four-attribute utility function could be constructed to reflect concern for each of these objectives. Among the individual components there would be inconsistencies, but collectively the resultant utility function could represent our overall values and include the appropriate value tradeoffs to address the inconsistencies. This is indeed one of the major motivations for and advantages of multi-attribute utility.

REFERENCES

1. Acton, J. P., 1973. Evaluating public program to save lives: the case of heart attacks. Santa Monica, California: Rand Corporation Report R-950-RC.

2. Broome, J., 1982. Equity in risk bearing. Operations Research, 30, pp. 412-414.

3. Derby, S. L. and R. L. Keeney, 1981. Risk analysis: Understanding "how safe is safe enough." Risk Analysis, 1, pp. 217-224.

4. Diamond, P. A., 1967. Cardinal welfare, individualistic ethics, and interpersonal comparisons of utility: comment. Journal of Political Economy, 75, pp. 765-766.

5. Ferreira, J., Jr. and L. Slesin, 1976. Observations on the social impact of large accidents. Cambridge, Massachusetts: Operations Research Center, Massachusetts Institute of Technology, Technical Report No. 122.

6. Fishburn, P. C., 1982. Equity axioms for public risk. Murray Hill, New Jersey: Bell Laboratories.

7. Howard, R. A., 1979. Life and death decision analysis. In, Lawrence Symposium on Systems and Decision Sciences. North Hollywood, California: Western Periodicals.

8. Jones-Lee, M. W., 1974. The value of changes in the probability of death or injury. Journal of Political Economy, 99, pp. 835-49.

9. Jones-Lee, M. W., 1982. The Value of Life and Safety. Amsterdam: North-Holland Publishing Company.

10. Keeney, R. L., 1980a. Evaluating alternatives involving potential fatalities. Operations Research, 28, pp. 188-205.

11. Keeney, R. L., 1980b. Equity and public risk. Operations Research, 28, pp. 527-534.

12. Keeney, R. L., 1980c. Utility functions for equity and public risk. Management Science, 26, pp. 345-353.

13. Keeney, R. L., 1983. The analysis of risks of fatalities. In J. Menkes, V. Covello, and J. Mumpower, (eds), Contemporary Issues in Risk Analysis. New York: Plenum Press.

14. Keeney, R. L. and R. L. Winkler, 1982. Von Neumann and Morgenstern utility and equity of public risks. San Francisco: Woodward-Clyde Consultants.

15. Linnerooth, J., 1975. A critique of recent modeling efforts to determine the value of human life. Laxenburg, Austria: International Institute for Applied Systems Analysis, Research Memorandum RM-75-67.

16. Linnerooth, J., 1979. The value of human life: a review of the models. Economic Inquiry, 17, pp. 52-74.

17. Lowrance, W. W., 1976. Of Acceptable Risk. Los Altos, California: Kaufmann.

18. Mishan, E. J., 1971. Evaluation of life and limb: a theoretical approach. Journal of Political Economy, 79, pp. 687-705.

19. Raiffa, H., Schwartz, W., and M. Weinstein, 1978. Evaluating health effects of social decisions and programs. In EPA Decision Making. Washington, D. C.: National Academy of Sciences.

20. Sarin, R. K., 1982. Measuring equity in public risk. Los Angeles: Graduate School of Management, University of California.

21. Savage, L. J., 1954. The Foundations of Statistics. New York: Wiley.

22. Slovic, P., Fischhoff, B., and S. Lichtenstein, 1980. Facts and fears: understanding perceived risk. In R. C. Schwing and W. A. Albers, Jr., (eds.), Societal Risk Assessment: How Safe is Safe Enough? New York: Plenum Press.

23. Starr, C., 1969. Social benefits versus technological risk. Science, 165, pp. 1232-1238.

24. Thaler, R. and S. Rosen, 1976. The value of saving a life: Evidence from the labor market. In Household Production and Consumption. N. E. Terleckyj (ed.), New York: National Bureau of Economic Research, Columbia University.

25. von Neumann, J. and O. Morgenstern, 1947. Theory of Games and Economic Behavior. Princeton, New Jersey: Princeton University Press, 2nd ed.

26. Zeckhauser, R., 1975. Procedures for evaluating lives. Public Policy, 23, pp. 419-464.

Natural and Technical Hazards: Similarities and Differences

Howard Kunreuther

University of Pennsylvania, The Wharton School, Decision Sciences Department, Philadelphia, PA 19104

INTRODUCTION

This paper focuses on two classes of problems associated with low probability events: those caused by natural forces (e.g., floods, earthquakes, and hurricanes) and those caused by new technologies (e.g., nuclear power accidents, liquefied natural gas explosions).

Although these two classes of problems have a number of features in common, there are two distinct literatures which have emerged with little cross reference between them (Covello, 1983). The next section develops a conceptual framework contrasting natural and technological hazards. The framework is designed to integrate descriptive analysis with prescriptive approaches for improving the decision process and final outcomes. Section 3 then discusses the natural hazards problem in more detail by focusing on individual decisionmaking with respect to protection against low probability events. Section 4 focuses on the lessons from a case study of siting of LNG facilities in California with emphasis on the political decision process and alternative policy tools for facilitating negotiations. The final section summarizes the similarities and differences between the two types of hazards and suggests directions for future research in their risk management.

CONCEPTUAL FRAMEWORK

Figure 1 depicts a conceptual framework for structuring the analysis of natural and technological hazards. An appropriate starting point is the problem formulation phase where one addresses the key questions for study. The problem formulation phase is critically important as it forces one to undertake a detailed descriptive or behavioral analysis which guides the formulation of alternative strategies.

NATO ASI Series, Vol. G4
Environmental Impact Assessment, Technology Assessment, and Risk Analysis. Edited by V.T. Covello et al.

In the case of natural hazards this paper focuses on developing hazard mitigation strategies for coping with the potential losses from disasters. Most people located in hazard-prone areas may not recognize the potential consequences to them of a low probability catastrophic event. Hence they are not sensitive to alternative pre-disaster mitigation measures.

With respect to technological hazards the problem of interest in this paper is: "Should one site a facility that may yield potential benefits to society, but may cause possible severe losses to a smaller group of individuals?" In contrast to

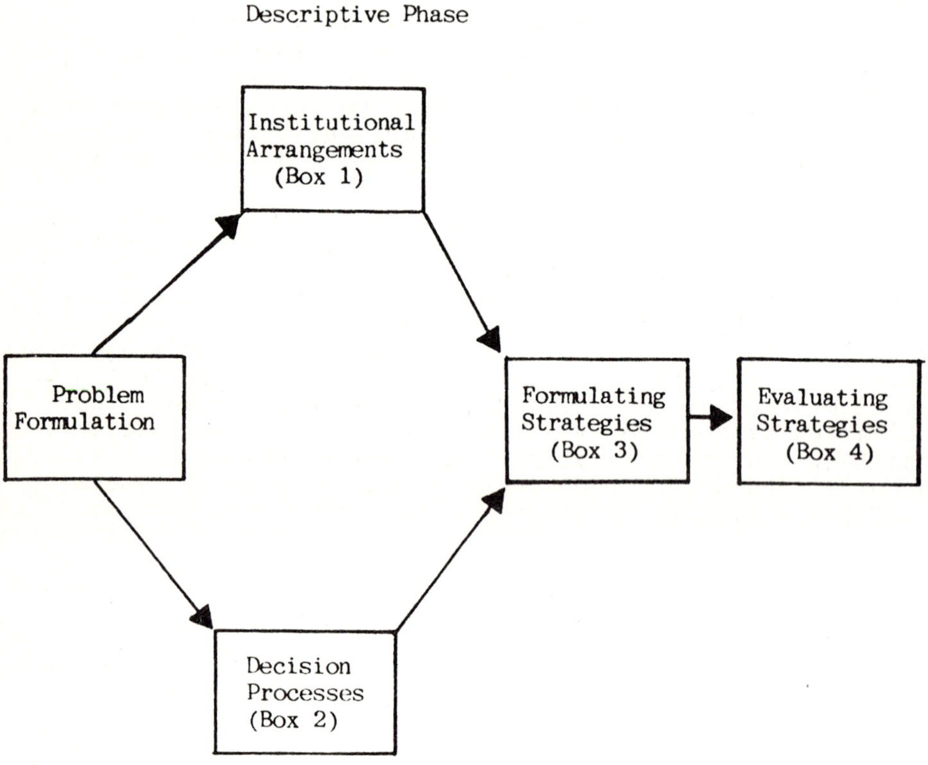

FIGURE 1: Conceptual Framework

natural hazards, individuals in specific areas would only face potential losses if a decision is made to site a facility in their backyard.

Turning to the descriptive phase, there are two components which need to be considered: the institutional arrangements and the decision processes. By institutional arrangements (Box 1 in Fig. 1) we mean the current organizational structure and flow of information between the different actors and stakeholders who are affected by the particular problem. What are the goals and objectives of these interested parties and what kinds of conflicts are likely to emerge?

Residents in hazard prone areas have shown a noticeable lack of interest in protecting themselves against the potential consequences of natural disasters. This paper focuses on insurance as a protective measure so that the relevant interested parties are the insurance industry, the general taxpayer, and government agencies who either subsidize insurance or provide disaster relief. There seems to be a general consensus among all of these parties that it would be useful to reduce the losses from natural disasters. The open questions are: (1) how should this be done? and (2) who should pay for mitigation measures and recovery?

Regarding technological hazards, this paper focuses on the interaction between the many interested parties in the political decision process associated with the siting of a facility. There is normally no consensus on a desirable course of action. Some groups want to site the facility; others do not. The existing institutional arrangements between the parties can be captured with respect to their legal entitlements (e.g., property rights), their formal and informal responsibilities in the decision process, and their right to formally participate in the siting debates.

By decision processes (Box 3) we mean the collection and processing of information by the interested parties concerned with the problem. The natural hazard example will focus on the relevant data used by an individual in determining whether or not to purchase flood or earthquake insurance. The LNG siting example emphasizes the information utilized by each of the interested parties to defend their position and the dynamic interplay among these groups.

With respect to both natural and technological hazards, a key problem centers around the measurement of risk. Specifically, what are the probabilities of different disasters or accidents and the subsequent losses which may result from them? Not only is there a limited historical database on which to make estimates, but there may also be differences between experts on the physical processes causing certain disasters (e.g., earthquakes, LNG explosions). Hence a range of probabilities and losses may emerge for a given event. Each of the interested parties has his own view of an acceptable risk level, so that conflicts are likely to emerge.[1]

The prescriptive phase is concerned with formulating strategies (Box 3) for dealing with the problem at hand. A number of different approaches can be considered. There are market mechanisms such as insurance and compensation; there are incentive systems such as subsidies, or penalties to encourage individuals to undertake action. Finally there are regulations or requirements which restrict or compel certain actions.

In the case of natural hazards, programs have been proposed to prevent losses (e.g., flood control projects, land use regulations or building codes) and mitigate the costs of recovery to victims (e.g., insurance and disaster relief) using a mix of adjustments (White and Haas, 1975, and Burton, Kates, and White, 1978). In the case of technological hazards, the key policy tools have revolved around facilitating the siting and licensing process of different structures. Here the concerns have been with designing programs that ensure that the facility is acceptably safe and that different parties feel that they are sufficiently compensated from the action.

The choice of policies depends on the institutional arrangements which may differ between countries. Some cultures may favor a market system for handling disasters while others may place the government in a central position.

In evaluating strategies, their relative merits are reflected in two considerations. First, there are resource allocation effects which refer to the impact of different programs on aggregate costs and benefits. Second, there are distributional considerations which relate to the relative financial burden of different policies on the different interested parties. Allocation questions are primarily economic in nature while the equity or distributional considerations are principally political ones. The challenge facing the analyst and policymaker is in balancing these two features in designing a strategy. How well different policies perform will be determined by the descriptive aspects of the problem including the relative importance of different stakeholders in the policy process.

---

1 See Fischoff et al. (1982) for a comprehensive discussion of acceptable risk.

ENSURING AGAINST NATURAL HAZARDS[2]

Let us now turn to a more detailed study of homeowner decisions on whether to protect themselves against the consequences of natural hazards and the alternative prescriptive measures for dealing with the lack of interest in voluntary purchase of insurance.

Problem Formulation

The problem of interest is the appropriate role of the public and private sectors in providing insurance protection and disaster relief against the consequences of natural hazards. An historical perspective with respect to this problem is relevant. Expected annual losses from floods, hurricanes, and earthquakes in the United States based on 1970 exposure is estimated at $4.6 billion and is expected to jump to $8.3 billion by the year 2000 (Petak and Atkisson, 1982). Relatively few homeowners have voluntarily purchased insurance against the consequences of floods and earthquakes, even though coverage is easily available and, in the case of floods, highly subsidized by the federal government. In the past, the Small Business Administration (SBA) has responded to the financial plight of the uninsured victims by providing liberal relief in the form of low interest loans and grants to aid the recovery efforts.

Evidence on increased federal disaster relief is provided by comparative data on the SBA disaster loan program. The growth of the program is easily seen in Fig. 2; the increase is particularly significant in the case of home loans where both the total number and total dollar values in the 1966-76 period were more than 25 times what they were in the first 12 years of the program. It is striking that the $1.2 billion approved by the SBA for victims of Tropical Storm Agnes represented almost four times the entire amount allocated by the SBA for all disasters between fiscal years 1954 and 1965. Over $540 million of the amount approved by the SBA for victims of this disaster were in the form of forgiveness grants which did not have to be repaid.

Institutional Arrangements

Insurance against floods is provided by the Federal Insurance Administration with highly subsidized rates on existing property; new property is charged a premium based on estimated

2 This section is based on material in Kunreuther (1981).

SMALL BUSINESS ADMINISTRATION DISASTER LOANS

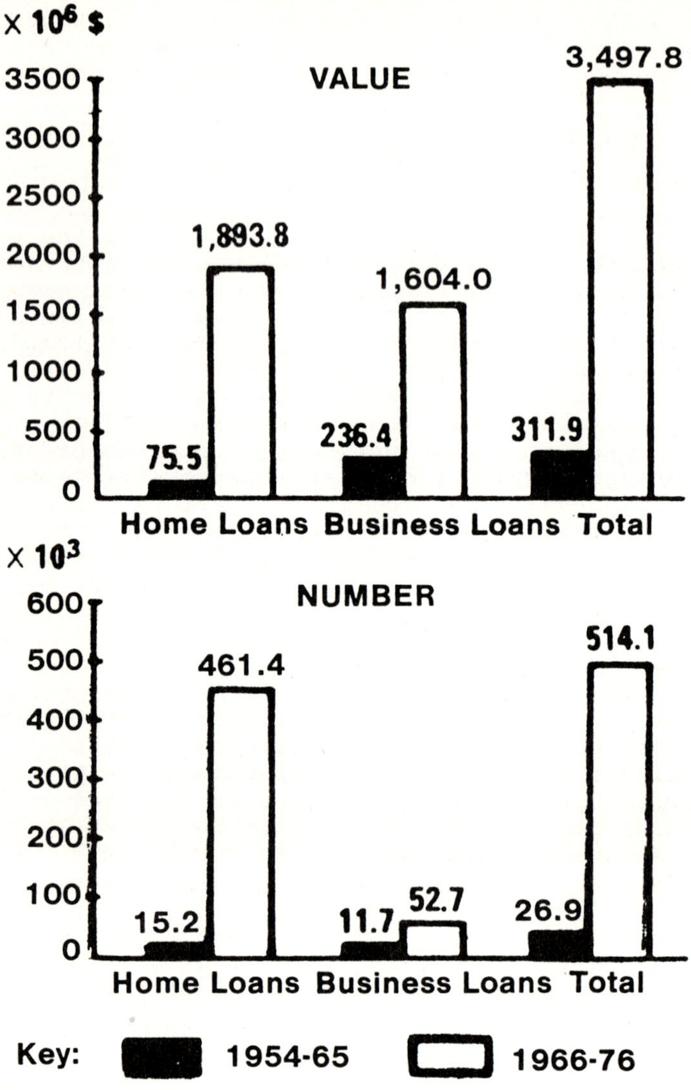

FIGURE 2: Small Business Administration disaster loan from (1954-1965 with (1966-1976)

risk. For insurance to be offered to residents and businesses in a hazard-prone region, the community must agree to adopt land use regulations and building codes to reduce future losses from the hazard.

Earthquake coverage is offered to individuals and commercial firms by private insurance companies. Even though coverage is not expensive ($2 per $1000 coverage on wood-frame homes in California with a 5 percent deductible), less than 3 percent of the homeowners in this earthquake-prone state have bought this insurance. On the other hand, most businesses protect themselves with an earthquake insurance policy in California.

Decision Processes: We now turn to the decision processes of the different interested parties who are affected by this problem. There is substantial knowledge regarding the type of information collected and utilized by some stakeholders. Considerably less is known with respect to others.

Residents of Disaster-Prone Areas: Field surveys of homeowners in hazard-prone areas and controlled laboratory experiments have shed light on the decision process associated with the purchase of disaster insurance[3]. The field survey involved face-to-face interviews with 2,055 homeowners residing in 43 areas throughout the United States subject to coastal and riverine flooding, and 1,006 homeowners living in 18 earthquake-prone areas of California. Half the respondents had purchased flood or earthquake insurance and the other half had not. The controlled laboratory experiments undertaken by Paul Slovic, Baruch Fischhoff, and Sarah Lichtenstein, at Decision Research, investigated the causal relationships between variables entering into the insurance decision. A few of the key findings from this study which relate to individual decision processes are now summarized.

Although most uninsured homeowners interviewed were aware that flood and earthquake coverage existed, the majority were unaware that they were eligible to purchase a policy. Those who were aware had no reliable knowledge of the costs of a policy. The subsidized flood rate is between $2.50 and $3.50 per $1000 coverage depending on the proportion of coverage devoted to structure and contents. The earthquake premium on wood-frame homes in California averages $2 per $1000. Hence any homeowner estimating the respective rates between $2 and $4 for flood coverage and $1 and $3 for earthquake insurance was classified as reasonably accurate. Figure 3 shows that most of the insured homeowners were accurate in their estimate or

---

3 For more details on the survey and laboratory experiments, see Kunreuther et al. (1978), Slovic et al. (1977).

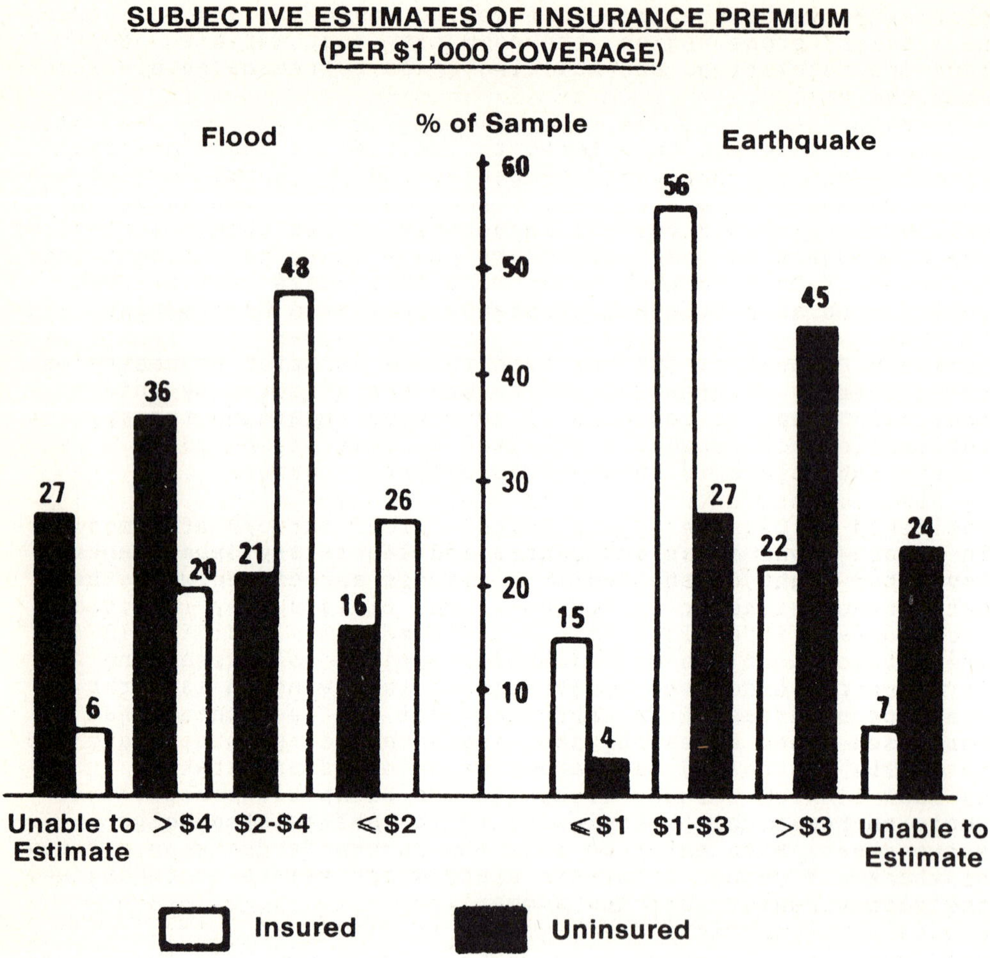

FIGURE 3: Subjective Estimates of Insurance Premium (per $1,000 coverage)

underestimated the premium. On the other hand, a large proportion of the uninsured individuals overestimated the premium. This finding suggests that most non-policyholders expended little energy to obtain information on rates from their insurance agent even if they knew coverage was available.

With respect to the hazard itself, both insured and uninsured individuals had imperfect information on the probability and consequences of a severe flood or earthquake causing damage to their property and contents. When homeowners were asked to estimate the chance of a severe flood or earthquake damaging their property in the next year, 15 percent of the respondents in flood areas and 8 percent of those in earthquake areas, were unable to provide any figure. Others thought the probability of a disaster hitting them was quite high--1 chance in 10--yet they said they had purchased no disaster insurance. Some believed the chances of a disaster affecting them to be almost nil--1 in 100,000--yet they had purchased disaster insurance. It seems clear that many of the individuals participating in the field survey do not understand the concept of probability. The findings are consistent with the heuristics and biases implied by controlled laboratory experiments over the past decade which are summarized in Kahneman, Slovic, and Tversky (1982).

It is tempting to attribute this casual attitude about the risks of natural hazards and protective activities to homeowners' beliefs that the federal government will bail them out in a crisis. Figure 4, however, indicates that the majority of uninsured residents anticipate no aid at all from the government even when they expected to suffer large losses from a disaster. Most of these people were aware that the SBA provides aid to the victims, but they had little knowledge of the loan terms or whether they could receive forgiveness grants. On the basis of these results, one can conclude that most homeowners in hazard-prone areas have not even considered how they would recover should they suffer flood or earthquake damage. Instead they treat such events as being so unlikely that they ignore the consequences altogether.

What variables influence a person's decision to purchase insurance? A key factor is a belief that the hazard is a serious problem. This concern is found primarily among people who have had past experience with the hazard.

> "You ask me why I didn't have insurance before the June 1972 flood," said one homeowner in Norristown, Pennsylvania. "We had the flood in September of '71 and I had two feet of water in my basement. And I felt this I can tolerate and this is probably as high as it will ever get."

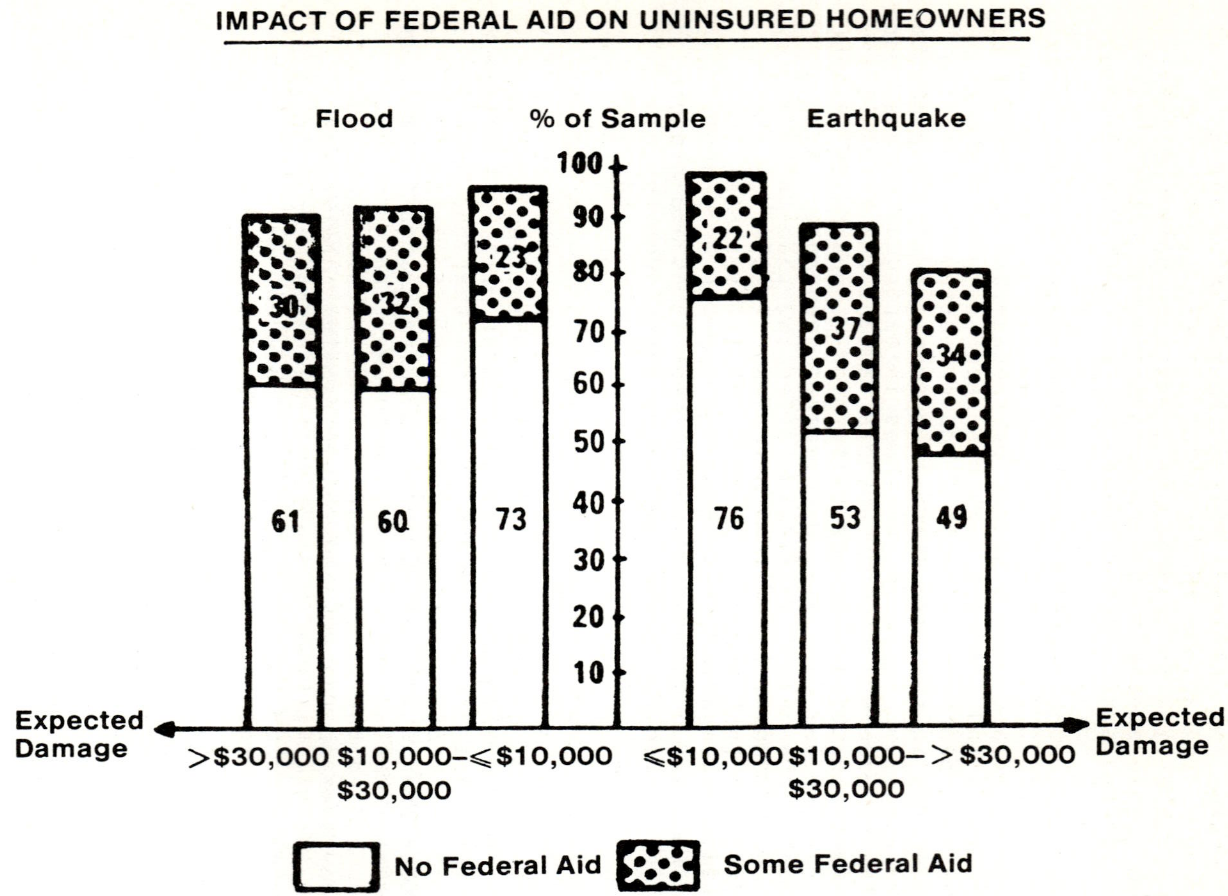

FIGURE 4: Expectation of Federal Aid on Uninsured Homeowners

To his chagrin, this man suffered severe property damage in 1972. Only then did he decide that he needed insurance. Another uninsured flood victim said that his rationale was that:

> The $60 in premiums they could use for something else. But now they don't care if the figure was $600. They're going to take insurance because they've been through it twice and they've learned a lesson from it.

As shown in Fig. 5, another important factor in influencing the purchase of a policy appears to be knowing someone who has purchased coverage or discussing insurance with a friend, neighbor, or relative. The following example graphically illustrates this point. In a pretest of the questionnaire in San Francisco, a homeowner responded to one of the questions by saying that he did not have earthquake insurance. A friend of his who was listening to the interview commented that he had himself purchased such insurance a few years before. The respondent was dumbfounded and asked his friend about the availability of earthquake coverage and how much it cost. "I'm going to have to look into earthquake insurance myself," he added.

The controlled laboratory experiments on insurance undertaken at Decision Research provide further insight into these results. Subjects were exposed to a variety of risks that had different losses and probabilities associated with them. By keeping the premium constant for all risks and varying losses and probabilities in such a way that the expected loss (loss multiplied by probability) was the same, it was possible to test the importance of probability and loss on insurance purchase decisions.

One would expect that individuals would prefer to insure themselves against events having a low probability of occurrence but a high loss rather than against those having a high probability and low loss. The reverse was found to be true for a variety of experimental formats. These results suggest that if the chances of an event are sufficiently low, people do not even reflect on its consequences. In other words, people are primarily interested in buying insurance if they feel the probability of a disaster is high enough for them to stand a good chance of getting a return. Thus, they view insurance as an investment against an event they feel is likely to happen even if the loss is small rather than as protection against unlikely disasters.

One area for fruitful research is to understand more fully why individuals do not purchase insurance. Roberts (1982) contends that the decision on whether or not to purchase insurance depends on other factors besides wealth effects. In this connection Akerlof and Dickens (1982) contend that persons do

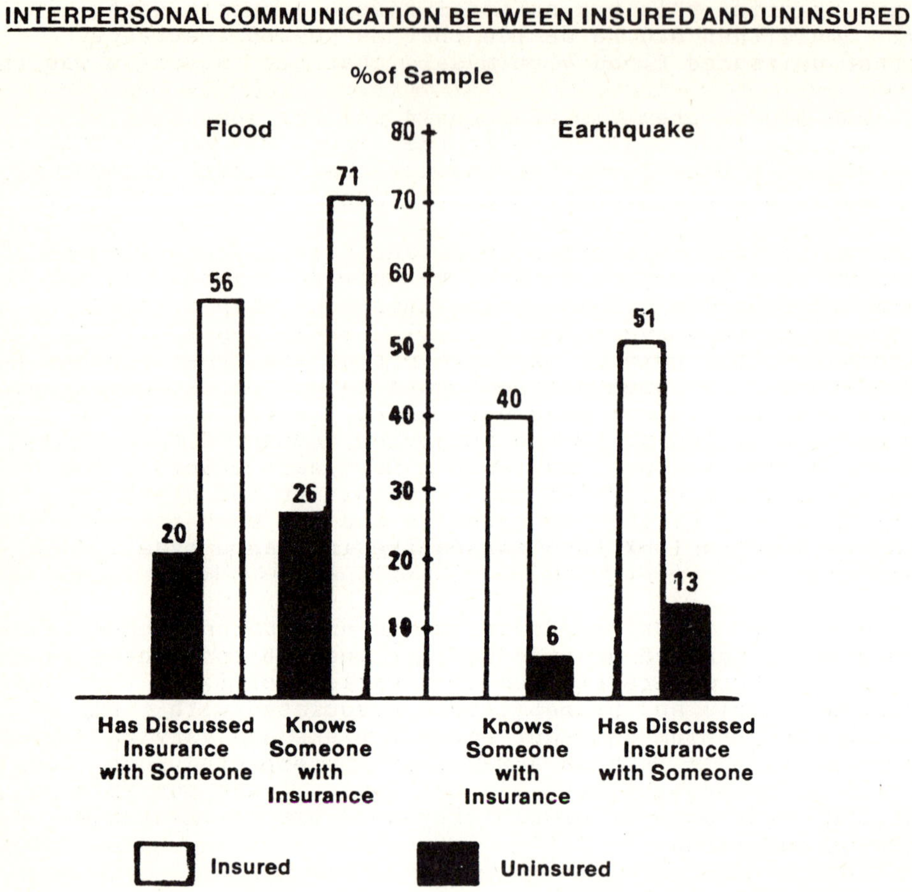

FIGURE 5: Interpersonal Communication Between Insured and Uninsured

not think about buying coverage because by taking such a step they are implicitly acknowledging that they are in a hazard-prone area. They prefer to avoid confronting this issue by believing that they are living in a safe area thus avoiding the unpleasant feelings of constant fear.

Firms and Businesses: We have much less information about the decision processes of the business sector operations in hazard-prone areas. As pointed out earlier, most commercial structures are insured against earthquakes, which suggests that firms have a much greater concern with the hazard than homeowners. Residents appear to focus on the low probability of the event and do not pay attention to the consequences. Risk managers in firms, on the other hand, are primarily concerned with the potential loss and may focus only peripherally on the probability of its occurrence. They see insurance as an attractive buy relative to the costs they would incur after a disaster, should their firm be unprotected (Borkan and Kunreuther, 1979).

Insurance Industry: Until now, there has not been an active interest by insurance firms or their agents in encouraging their clients to purchase earthquake or flood insurance. Coverage is viewed as a special service to customers holding other policies with the company rather than as a profit-making operation. Nevertheless, efforts were made following the San Francisco earthquake of 1971 to make people aware of the availability of this type of protection. For example, the Insurance Company of North America placed newspaper ads in the major dailies, advertised on TV and provided California agents with special brochures and announcements that they could mail to their customers. The month after this campaign began they sold only 61 policies in the entire state and then sales dropped off to an average of seven per month during the next seven months (Syfert, 1972). The Hartford Insurance group and Kemper companies ran similar campaigns to market earthquake insurance. Their efforts also bore little fruit with respect to flood coverage.

A study of the National Flood Insurance Program in New York state found that many agents had little interest in the program because they felt that the volume of business would be low and that they would be unlikely to pick up new clients as a result of the contacts made (Preston, Moore, and Cornick, 1975).

Banks and Financial Institutions: It is not clear from the existing literature how banks and financial institutions in the U.S. feel about requiring flood or earthquake coverage as a condition for a mortgage. If banks were to mandate coverage on new dwellings, this insurance would provide a substitute for liberal disaster relief following a catastrophic disaster. On the other hand, there is no empirical evidence suggesting that people would make different location decisions if they were now forced to take out coverage. The best chance of achieving this sensitivity to the hazard would be for rates to reflect the risk of residing in specific areas. However, given the large monetary transaction in purchasing a house and other factors influencing location, the premium differentials are likely to have a small impact on most buyers' actions.

Residents in hazard-prone areas and the general taxpayer appear to be receptive to some form of insurance requirement as a condition for a mortgage or loan on structures in hazard-prone areas. Over half of the respondents in the field survey felt that it would be fair for banks and financial institutions to impose this condition. Similar findings were reported by Cummins et al. (1974) in a study of consumer attitudes toward insurance. When asked the question, "Would you favor or oppose a law which required all people who live in flood and earthquake zones to have flood and earthquake insurance," almost 60 percent of the respondents replied in favor of such a regulation and only 30 percent were opposed to it. The rest did not have an opinion.

One reason that banks and financial institutions would want to require insurance is that they fear a run on their deposits and liquidity position in the wake of a disaster. Steindl and Weinrobe (in press) have looked at deposit data of seven individual savings and loan associations and eight commercial banks in four disasters,[4] and found no evidence of any run during the month following the disaster. Insurance did appear to make a difference, however. In Xenia, where residents were covered against tornado damage by homeowners coverage, new net sources in the two savings and loan associations studies in the disaster month were between six and seven times higher than normal, presumably because of a rapid settlement of private insurance claims.

Real Estate Agents: Real estate agents can play a critical role in providing potential buyers with key information about the characteristics of their neighborhoods and the type of hazards that they may face. Agents seem to be doing little or nothing to discourage sales in areas susceptible to natural disasters. They are either unaware of the risks associated with portions of the community, or downplay the significance of such warnings as fault zoning (Palm, 1981).

---

4 They are Rapid City flood, Wilkes-Barre flood from Tropical Storm Agnes; the Xenia, Chic tornado and the Johnstown flood.

Results of our field survey confirm this finding. Over 40 percent of the residents interviewed did not believe that there was an earthquake problem in their area at the time they bought their house. Furthermore, 30 percent of the homeowners did not have any idea how far their house was from a fault. In a group focus depth interview, one Bakersfield resident summed up the situation very succinctly:

> When you go out to buy a piece of property, the real estate agent doesn't say, "I want you to understand that there's a fault running right down the middle of this thing." You buy the house, and six years later somebody tells you you're sitting on top of a fault (Kunreuther et al., 1978, p. 107).

Public Sector: Various federal, state, and local agencies have key roles to play by combining insurance with other policy tools in mitigating disaster losses and aiding the recovery process. Currently, the principal roles of state and local agencies are to provide residents and businesses with information on both the hazard itself and available tools for reducing potential losses. The task of educating public officials about the disaster risks in their areas is enormous because of the high turnover in these positions, as well as the low salience of the hazard when contrasted with other issues which demand their attention (Wright et al., 1979).

Formulating Strategies

The performance of different insurance programs will be directly related to the existing institutional arrangements and the decision processes of the different stakeholders. It is thus necessary to understand the type of information that is likely to be disseminated, collected, and processed by the different groups who interact with each other.

It is important to specify in advance the objective of a particular program. For example, if policymakers feel that it is the responsibility of residents in hazard-prone areas to protect themselves against the risks of natural disasters, every effort should be made to provide this group with accurate information on the hazard and alternative ways it can protect itself against the disaster. Furthermore, the government should then not come to the rescue with special relief. If there is concern that the government will succumb to political pressures by providing generous aid to all uninsured victims, then stronger measures may be required. Below we briefly discuss a set of alternatives which could be considered.

Voluntary Insurance: Based on how information is currently presented to consumers on the hazard, there is likely to be a limited interest in coverage until after the next severe disaster. Even then, if the government provides generous relief there will be no incentive for homeowners to buy coverage. Businesses, on the other hand, are likely to be well protected against the financial consequences of a severe disaster because they are primarily concerned with the potential catastrophic loss to them.

Required Insurance: Unless the insurance industry determines that it has the available resources to absorb the consequences of a probable maximum loss, it is likely to be opposed to mandatory insurance. Under current federal income tax provisions, premiums collected by private companies that are put in a reserve fund for catastrophes are treated like excess profits, and hence taxed at a rate of 50 percent. With respect to earthquake coverage, Brown (1982) has concluded that the U.S. government cannot require that existing and future residential homeowners policies include an earthquake clause. The reason that such a proposal would be unconstitutional is that the world-wide financial capacity of the insurance and reinsurance industry is far less than would be needed.

Comprehensive Catastrophic Insurance: Proposals have been made for incorporating this coverage into a catastrophic insurance system covering all natural hazards (Kunreuther, 1968; Anderson, 1975, 1976). The private sector would promote, distribute, and sell the policies, servicing policyholders and adjusting claims; the federal government would establish the essential components of the system and provide reinsurance to the private companies in case of a catastrophic loss. On the surface this concept appears to be attractive due to its simplicity, but Hall (1975) indicates that it raises a set of questions which merit further discussion. Specifically, such coverage would undoubtedly require some form of reinsurance against losses from a catastrophic disaster. This would mean that the government would be injecting itself into areas of property insurance which in the United States have traditionally been in the private domain. In the past, the industry has resisted such encroachment unless it is absolutely essential.

There are lessons from other countries here. Spain has developed compulsory protection against catastrophic disaster losses from natural disasters or riots to property, persons and farming, fishing and forestry. The definition of a "catastrophe" and reimbursement of losses are determined by a central government committee called the Consortium. The catastrophe insurance plan is supported by compulsory surcharges on the premiums of basic policies covering the individuals or their property. Presumably the Consortium may borrow from the Bank of Spain if loss payments exceed the available funds. (Greene 1972).

Creating Concern for the Hazards and Insurance: Results of field survey and controlled laboratory experiments on insurance suggest that persons only will consider protecting themselves against low probability, high consequence events if they are convinced that the chance of occurrence is, in fact, high enough to warrant attention. One way to increase concern with the hazard may be to present information on the probability of a future disaster on a different time interval than the traditional one-year period. Thus, in describing the chances of a flood or quake which has an annual probability of .01, the chance of suffering damage from such a disaster at least once during a 25-year period will be .22. By stretching the time interval in this way the person may then view the probability of loss to be high enough to warrant interest in protection.

The insurance agent can also trigger interest in insurance coverage by stimulating people's awareness of the hazard and providing accurate information on the availability and terms of a policy. Since most individuals appear to treat insurance as an investment rather than a protective mechanism, one of the principal functions of the agent should be to educate his clients that the greatest return on their insurance is not to have any return at all. Such a process of education requires time on the part of the agent, which he may not be willing to give with commissions at their current level.

Insurance as a Coordinating Mechanism: Insurance is a logical mechanism for providing individuals with economic incentives to adopt other measures such as making their structures more earthquake resistant or flood-proof. Pamphlets could be sent to all homeowners indicating the possible measures they could adopt and offering a reduction in annual premiums should they reduce the risk of future damage. In addition, the government might consider providing homes and businesses with low-interest improvement loans to encourage them to undertake such adjustments. If the benefits of the protective measures exceed the costs, the reduction in premiums may more than offset the loan charges.

As pointed out above, there is no guarantee that economic incentives will spur homeowners to adopt protective measures. For this reason, it may be necessary to supplement these incentive systems with land-use controls and building codes such as has been done in the National Flood Insurance Program. In developing these regulations there is a need for accurate information on the hazard and its risk as well as the potential costs and benefits of the rules to the affected individuals and to society.

Even with these combined mitigation and recovery measures, a disaster relief program may be desirable for assisting special groups such as low-income or elderly residents who may be uninsured either because they could not afford coverage or because they did not possess sufficient information on the availability and terms of a policy. Similarly, after a catastrophic disaster, there may be a need to offer special relief to victims to cover the uninsured portion of their loss. If the government deems such measures important, then it would be wise to pass legislation prior to a catastrophe rather than having to react to each situation with a new disaster bill.

THE SITING OF THE CALIFORNIA LNG TERMINAL[5]

This section applies the conceptual framework in Fig. 1 to the siting processes associated with an LNG terminal in California.

Problem Formulation

Liquefied natural gas (LNG) is a potential source of energy which requires a fairly complicated technological process which has the potential, albeit with very low probability, of creating severe losses. For purposes of transportation, natural gas can be converted to liquid form at about 1/600 its gaseous volume. It is shipped in specially constructed tankers and received at a terminal where it undergoes regasification and is then distributed. The entire system (i.e., the liquefication facility, the LNG tankers, the receiving terminal, and regasification facility) can cost more than $1 billion to construct (Office of Technology Assessment, 1977).

Figure 6, which is based on an analysis by Keeney (1980), provides the components of a scenario associated with an LNG accident at some prespecified site. The process begins with some initiating event, such as a vessel accident or a storage tank rupture, which results in an LNG spill. LNG will immediately start to vaporize after a spill, resulting in a vapor cloud which can be carried downwind to a populated area. The mixture of vaporized LNG and air may cause the cloud to

5 The material in this section is based on Kunreuther, Lathrop, and Linnerooth (1982).

ignite if it comes in contact with some ignition source, such as an electric spark. Should the vapor cloud not be ignited within a given period of time then it will mix with a sufficient amount of air so that it is innocuous, and the cloud will disperse. If it does ignite, then there is a chance that it may cause fatalities from fire and explosions in confined spaces and ignitions of materials by radiant heat. There is considerable controversy among the experts on the probabilities of the occurrences of each of these different events.

One way to proceed is to construct different LNG accident scenarios and to assign probabilities to the occurrence of each relevant event. For example, using the components comprising Fig. 6 one can construct a number of scenarios beginning with an initiating event and ending with fatalities from an ignited vapor cloud. One such scenario might be the failure of a storage tank--a vapor cloud forming and traveling 10 km in some specified direction, igniting, killing a certain number of people and causing a specified amount of property damage. It may be difficult to specify the probability of such a sequence of events very precisely.

Institutional Arrangements

The LNG siting process in the U.S. began in September 1974 when Western LNG Terminal Associates (a special utility company set up to represent the siting interests of three gas distribution facilities) applied for approval of three sites on the California Coast--Point Conception, Oxnard and Los Angeles--to receive gas from Indonesia and Alaska. It took until October, 1982, to obtain approval of one of the sites, Point Conception, by which time Western had lost interest in the project.

To structure the siting process we need to have a good understanding of the different concerns of the interested parties. For the LNG problem there are three categories of concern which are relevant: risk aspects, economic aspects, and environmental aspects. Each of these concerns can be described by a set of attributes. Table 1 depicts an interested party/attribute matrix showing the main concerns of each of the relevant groups over this seven-year period in which siting debates took place in California.

The attributes listed have been selected to reflect the attributes as perceived by the parties in the debate. For example, population risk (X_2) involves the risk to life and limb to neighbors of the LNG terminal due to accidents including those induced by earthquakes.

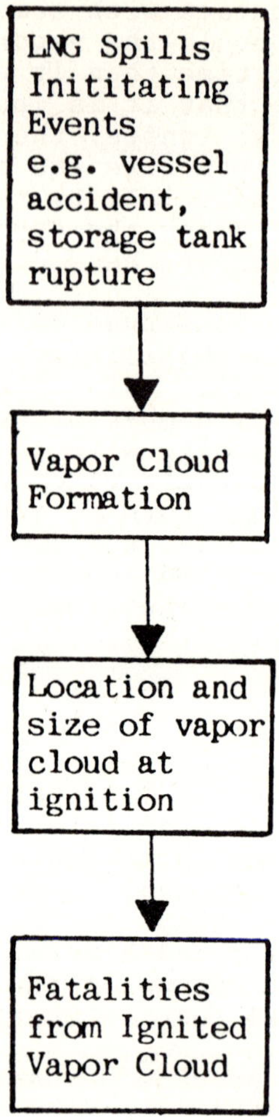

FIGURE 6: Elements of an LNG Accident Scenario

Table 1. Principal Party-by-Attribute Matrix for LNG Siting in California

Source: Kunreuther, Lathrop and Linnerooth (1982)

Principal Party-by-Attribute Matrix for LNG Siting in California

		Principal Parties*							
			Government						
Attributes		Applicant	Federal	State			Local	Interest Groups	
		Utilities P_1	FERC P_2	CCC P_3	CPUC P_4	Legislature P_5	Municipal Government P_6	Sierra Club P_7	Local Citizens P_8
Risk									
supply interruption	X_1	o	o		o	o			
population	X_2		o	o	o	o	o	o	o
earthquake	X_3		o	o	o	o			
Environmental									
air quality	X_4		o		o	o		o	
land use	X_5		o	o	o	o	o	o	o
Economic									
profit consideration	X_6	o							
price of gas	X_7								
local economic benefits	X_8		o		o	o	o		

*Key to party acronyms and abbreviations: FERC: Federal Energy Regulatory Commission, or, in the first two rounds of the process, its preceding agency, the Federal Power Commission; CCC: California Coastal Commission; Legislature: California State Legislature.

arthquake risk (X_3), which involves both population risk and upply interruption risk due to earthquakes, is included as a eparate attribute since it was of particular importance in the alifornia siting process.

he filled cells in Table 1 indicate which parties pay partic- lar attention to which atrributes. Naturally, many of the arties care about all the attributes listed. However, either ecause of the incentives directly felt by the party or because f the role the group plays in society, each party makes its ecisions as a function primarily of a particular subset of the ttributes.

ne applicant, Western LNG Terminal Associates, was a special ompany set up to represent the LNG-siting interests of three as distribution utilities--Southern California Gas Company,

Pacific Gas and Electric, and El Paso Natural Gas Company. As domestic gas supplies seemed to be diminishing in the late 1960s, the gas utilities perceived an increased risk of supply interruption, which could be mitigated by additional supplies such as LNG. Quite naturally the applicant was quite concerned with profitability (X_6) and secure supplies of gas (X_1).

At the various government levels there are five principal parties. The Federal Energy Regulatory Commission (FERC) in the Department of Energy is the principal body at the federal level which determines whether a proposed LNG project is in the public interest and should be allowed. In making its judgment it considers primarily the following attributes: risk factors (X_1, X_2, and X_3), environmental guidelines as reflected in air quality (X_4) and use of land (X_5), and the expected LNG price (X_7).

Let us turn now to state agencies which play a role. The California Coastal Commission (CCC) was created in 1976, and has the responsibility for the protection of the California coastline. Its primary concerns with respect to LNG are with the use of land (X_5) and the associated risks (X_2 and X_3) from building a terminal at a specific site. The California Public Utilities Commission (CPUC) is the principal state body involved in power plant issues and is primarily concerned with the rate-setting process. Hence, the CPUC focused on the provision of energy to California residents and the need for gas (X_1) and the proposed price of the product (X_7).

In addition, it has responsibility for evaluating the impact that a proposed facility would have on the environment and safety. The California State Legislature is ultimately responsible for the outcome of any siting process. It determines which state and local agencies have final authority to rule on the feasibility of a proposed site. In addition, it can set standards to constrain any siting process. Hence, the concerns of the legislators encompass economic, environmental, and safety attributes as shown in Table 1.

At the local level, the city councils evaluate the benefits of a proposed terminal in their jurisdiction in terms of the tax, business revenues, and jobs (X_8) it promises to provide. The councils try to balance this positive feature with the impact that the facility would have on land use (X_5) and risk to the population (X_2). Finally, the public interest groups, represented by Sierra Club and local citizens groups, are primarily concerned with environmental and safety issues. The important message of Table 1 does not lie in the details of exactly which cells are filled, but in the differences between columns of the table. That is, the different parties in the process care about very different subsets of the attributes.

The Decision Process

A multi-attribute multi-party (MAMP) model of choice was developed to determine the siting process for large-scale novel technologies. The approach is a natural extension of the burgeoning literature emphasizing the key roles that limited time, attention, and information processing capabilities play in political decisionmaking when there are uncertain outcomes and likely conflicts among interested parties (Allison 1971, Braybrooke 1974). The siting process in California can be characterized by four rounds of discussions as shown in Table 2, which provides a summary of the entire process. Each round, in turn, contains a summary of how the problem was defined, the initiating event, and how the discussions were concluded. The remainder of this subsection discusses in more detail the decision process within each of the rounds.

Nature of Rounds: Round A began in September 1974, when the applicant filed for approval of three sites on the California Coast--Point Conception, Oxnard, and Los Angeles--to receive gas from Indonesia. The application raised two central questions which defined the problem addressed in round A: Does California need LNG, and if so, which, if any, of the proposed sites is appropriate?

Table 3 specifies the relevant interested parties who were involved in the interaction phase of round A. Those parties which had formal decision power are marked with an asterisk. There were four primary attributes which were utilized in the ensuing debate among the parties. The need for LNG or the risk of an interruption in the supply of natural gas (X) supported the locating of a terminal in at least one of the three proposed sites. While environmental, land-use considerations (X_5) suggested a non-remote site (Los Angeles and Oxnard), the risks to the population (X_2) argued for siting the terminal in a remote area (Point Conception). Finally, concerns about earthquake risk brought about opposition to the Los Angeles site, which was found to be crossed by a significant fault.

The round was concluded with a possible stalemate, at least as perceived by industry (Ahern, 1980). Los Angeles would not receive federal (FERC) approval, Oxnard might not receive state (CCC) approval, and Point Conception would face difficult approval challenges at the county and state (CCC) levels because of its adverse land-use impacts.

TABLE 2

Summary of Rounds In California LNG Siting Case.

Source: Kunreuther, Lathrop and Linnerooth (1982)

Round A		Date
Problem Formulation:	Should the proposed sites be approved? That is: Does California need LNG, and if so, which, if any of the proposed sites is appropriate?	
Initiating Event:	Applicant files for approval of three sites.	September 1974 (34 months)
Conclusion:	Applicant perceives that no site is approvable without long delay.	July 1977
Round B		
Problem Formulation:	How should need for LNG be determined? If need is established, how should an LNG facility be sited?	
Initiating Event:	Applicant and others put pressure on State Legislature to facilitate LNG siting.	July 1977 (2 months)
Conclusion:	A new siting process is set up that essentially assumes a need for LNG, and is designed to accelerate LNG terminal siting.	September 1977
Round C		
Problem Formulation:	Which site should be approved?	
Initiating Event:	Applicant files for approval of Point Conception site.	October 1977 (10 months)
Conclusion:	Site is approved conditional on consideration of additional seismic risk data.,	July 1978
Round D		
Problem Formulation:	Is Point Conception seismically safe?	
Initiating Event:	Regulatory agencies set up procedures to consider additional seismic risk data.	July 1978
Conclusion:	(Round is still in progress).	October 1982

The stalemates of round A formulated the problem for round B. The round was initiated when pressure to change the siting procedure was brought to the State Legislature by the utility companies, the business community, and the labor unions in California. The resulting legislation was a compromise between the environmentalists, who supported consideration of off-shore sites, and those who saw an urgent need for an LNG facility to assure energy and jobs. Indeed, a non-populated area was strictly defined. There could be no more than an average of 10 people per square mile within one mile of the terminal, and no more than 60 people per square mile within four miles of the terminal.

The passage of the Siting Act of 1977 (S.B. 1081) opened up a new procedure for finding an acceptable site and led to round C with the following problem formulation. Which site should be approved? The round was initiated by the CCC which, after considering 82 sites meeting the remote-siting constraint, ranked the top four sites--Camp Pendleton, Rattlesnake Canyon, Point Conception, and Deer Canyon--in that order. The ranking was made on the basis of seismic, soil, wind and wave conditions, rough cost, and coastal resource considerations. (Point Conception was included in the candidate set because S.1081 required that the applied-for site be included.)

The CCC passed these rankings on to the CPUC which chose, by process of elimination, Point Conception; on the grounds that the two higher ranked sites would involve unacceptable delay and would cause unacceptable risk to transients (i.e., campers, swimmers, etc.) at the nearby beaches and public parks. The CPUC, however, could only conditionally approve Point Conception subject to the utility company's ability to show that earthquake faults discovered in the area presented an acceptable risk to the terminal. Round B lasted from July 1978 until October 1982. Only two parties, the FERC and the CPUC, were active in the process, and they considered only one attribute--the seismic risk at Point Conception. By the time the site was approved, Western LNG Associates had lost interest in the project.

Role of Risk Assessments

One reason for the long delay in the siting process was the set of conflicting risk assessments performed by experts, without any reconciliation between them. More specifically, during the course of the LNG debate in California, six studies assessing the safety risks of the proposed terminals were conducted by such interested parties as the utility company, municipal, state, and federal government agencies.

Several risk studies are of particular interest. Western Associates commissioned a consulting firm, Science Application, Inc. (SAI), to do a study and the Federal Energy Regulatory Commission produced its own risk assessment. Both reports showed very low numbers on various probabilistic measures of risk (expected fatalities per year and individual probability of fatality per year). These numbers were interpreted to mean that the risk was acceptable. A risk assessment produced by the Socio-Economic Systems consulting firm (SES) for the Oxnard municipal government, suggested similarly low probabilistic measures of risk (though expected fatalities were 380 times higher than the other assessments). This study pointed out

TABLE 3

Elements of Round A

Problem Formulation:	Should the proposed sites be approved? That is: Does California need LNG, and if so, which, if any, of the proposed sites is appropriate?
Initiating Event:	Applicant files for approval of three sites.
Alternatives:	Site at Point Conception: A^1
	Site at Oxnard: A^2
	Site at Los Angeles: A^3
	Site at any combination of: A^1, A^2, A^3

Interaction:

Involved Parties		Attributes Used as Arguments			
Applicant	P_1	X_1			
*FERC	P_2	X_1		X_3	
*CCC	P_3		X_2		
*City Councils	P_6		X_2		X_5
Sierra Club	P_7		X_2		X_5
Local Citizens	P_8		X_2		X_5

Key Decisions:

1. CCC concerns over population risk implies that A^1 is preferred over the other two sites
2. FERC would not approve A^2 because the seismic risk is greater than a prescribed acceptable level.

Conclusion:

Applicant perceives a stalemate, i.e., that no site is approvable without long delay.

* Interested party with responsibility for decision(s).

Source: Kunreuther, Lathrop and Linnerooth (1982)

that the possible consequences of an accident could be catastrophic.[6]

One explanation lies in the format for presenting the results. The SAI study described maximum credible accidents without accompanying probabilities. Opposition groups interpreted these results as evidence that the terminal was not acceptably safe. The municipal government, originally in favor of the site, began to waver in its support, probably influenced by the apparent uncertainty of the risk and the strength of the opposition groups (Ahern, 1980). In sum, risk assessments did not provide a single, coherent assessment of acceptability of the risk of an LNG terminal; their results were subject to interpretation depending on party position. In fact, risk assessments were used both to promote and to oppose terminal applications.

Formulating Strategies

If each of the individual parties focuses on different attributes in judging the attractiveness of a particular site and laypersons as well as experts differ with respect to their risk assessments, then it is not surprising that proposed projects are rejected. The adversary system which currently exists in the U.S. requires new creative efforts by analysts to resolve these conflicts and reach some possible solutions.

The choice between alternatives can be facilitated by having policy analysts develop programs for sharing gains and losses from a proposed project, thus attempting to negotiate conflicts. By arranging for winners to compensate losers in monetary payments or in payments in kind, such as a recreation park, all parties may feel they are better off with a new technological facility than maintaining the status quo. Alternatively, if the developers are required to internalize the full social costs of their proposed projects by compensating those opposed to large-scale technological developments, they may decide to discontinue specific projects.

A distinction can be made between ex ante compensation which refers to payments in money or in kind at the time a facility is approved or constructed, and ex post compensation which refers to reimbursement to individuals or groups who suffer

6 For more details on the risk assessment studies see Chapter 7 by Mandl and Lathrop in Kunreuther, Linnerooth et al. (1983).

losses from an accident. The most difficult aspect of designing these compensation agreements is the problem of misrepresentation of needs by some of the parties for personal gain. Designing systems which encourage parties to tell the truth independent of what others do (truth dominant procedures) or to tell the truth when others are also telling the truth (incentive compatible procedures) requires some ingenuity (Raiffa 1982). Several types of ex ante and ex post compensation arrangements have been successful in siting facilities in different countries, while others have had their problems.

Ex Ante Compensation Measures: One proposal that has been successful is an agreement by the developer to reduce electricity rates to residents within a certain distance of a new technological facility in order to compensate them for the increased risk or unpleasantness. Such a system has recently been introduced in France with respect to nuclear power plants. People living within approximately 15 km of a facility can apply to the local authority for a reduction of up to 15-20 percent in electricity rates.[7]

Another example of ex ante compensation in kind relates to the construction of a 1500 MA coal-fired power plant in Wyoming. A law suit had enjoined construction of the plant because of its potential damage to the surrounding environment. The suit was settled when the utility companies agreed to set up a $7.5 million trust fund for the express purpose of preserving a 60-mile stretch of the Platt River, the habitat of migratory birds, including the whooping crane. The coal plant was completed in 1981 and is fully operational today.[8]

One case where direct monetary compensation was given to individuals provoked a very strong reaction from others. In the FRG, the utility company STEAG (Steinkohle-Elektrizitatswerke AG) announced plans in 1976 to construct a 1400 MW coal power plant in the city of Bergkamen in the Ruhr area. A citizens' action group protested the project and threatened to delay the licensing procedures. In March 1977 a contract was signed between the utility company and three representatives of the action group: the group would be reimbursed with a DM 1.5 million payment if they agreed to cease protesting the project further. However, a court case

---

7 Personal Communication with Gas de France, 1982.

8 Personal Conversation with Patrick Pateneau, Vice President, Resource Conservation, National Wildlife Federation, September 1982.

was provoked when the city of Bergkamen refused to distribute the money. The federal court decided that the contract was valid because the citizens' group should be compensated for legitimate rights, but the decision was greeted with negative reactions by FRG public opinion. Concerns were voiced in the media that health and safety were citizens' inalienable rights that could not be paid off with money.[9]

Ex Post Compensation: With respect to ex post compensation a key question is: Who is responsible for the damage should an accident occur? To encourage safer designs, and to discourage the construction of unsafe facilities, there is good reason to have the applicant responsible for any damages resulting from an accident. Pfennigsdorf (1979) points out that for ultra-hazardous or abnormally dangerous activities, such as an LNG terminal, public policy supports the doctrine of strict liability whereby the operator of the facility is liable for damages regardless of fault. Whether the developer actually will have to pay for these losses in the event of a catastrophic LNG accident in the future remains to be seen.

It should be noted, however, that the threat of having to compensate individuals for the economic consequences of an accident may be inadequate for dealing with the psychological and social impact of living next to large and potentially hazardous technological facilities. In other words, strict liability requirements, important as they may be, force producers to internalize only one aspect of the opposition to these technologies. As pointed out earlier, this may not be the most important social cost.

If society holds the applicant responsible then some form of public or private insurance appears to be attractive. If premiums are based on risk then there are incentives for firms to make their facilities safer and possibly relieve some of the anxiety of the residents in the community. Yet we have already seen that the experts disagree strongly on the chances and consequences of a catastrophic accident. This places a burden on the insurance industry in setting premiums. In fact, the insurance companies and reinsurance facilities are reluctant to provide protection against these low probability, high consequence events since the premiums are relatively small in relation to the potential losses should an accident occur. We hypothesize that each accident will have a unique set of ex post compensation arrangements among the developers, insurance firms, and government of victims.

---

9 We thank Michael Aschenbrenner for calling this case to our attention. For more details, see Blount (1983).

Rather than making *ex post* compensation payments in direct monetary contributions it may be attractive for all parties to undertake payment in kind. A landmark case in this spirit was the settlement in 1976 by Allied Chemical in Virginia after being found guilty of polluting the James River with the pesticide kepone. Rather than paying a $23.2 million fine, the company proposed paying $5.2 million and establishing an $8 million trust fund to start an environmental fund in Virginia. In essence, the firm provided *ex post* compensation for research to prevent future damage to the environment (Virginia Environmental Endowment, 1977).

These compensation arrangements are important measures for spreading the cost of technological development more equally among those affected. For this reason they may be useful tools in the analyst's repertoire for devising schemes to reach social consensus. An emphasis on negotiation, that is, facing head-on the value conflicts between the different interested parties, bypasses the essentially unsolvable problem of assessing the "acceptability" of risks.

The problems of implementing compensation systems are complex. Can we include all those groups affected by local development? Can we devise bidding schemes for siting hazardous technologies which approximate the strengths of preferences of those groups with different values and interests? What are the costs of enforcing certain property and liability rules? How acceptable are certain proposals likely to be within consensual and adversarial systems? These practical questions need further study. The evidence from the U.S. and European countries suggests that compensation has the potential for aiding negotiations. In fact, it may be the only politically viable alternative to the current strategy of attempting to assess the social acceptability of risks. For this reason these measures deserve careful consideration.[10]

CONCLUSIONS

Based on these two case studies a few tentative conclusions can be drawn on the similarities between natural and technological hazards. In both cases one is faced with potential catastrophic loss with a relatively low probability of occurrence.

---

10 See Kunreuther, Kleindorfer, and Yaksick (1983) for more detail on the design of field and laboratory experiments on the role of compensation in siting hazardous facilities.

The risks are hard to delineate due to limited statistical and historical data as well as conflicting theories about the physical processes causing a disaster.

With respect to the perception of the risk, individuals normally move into an area voluntarily without undue concern for the consequences of natural disasters. Technological facilities are normally imposed on a community so that people view it as an involuntary hazard. They also have an opportunity to block the construction of the facility, thereby preventing future losses.

Natural hazards are viewed as acts of God, while technological hazards are viewed as acts of man. Individuals normally focus on the low probability aspect of a natural disaster, claiming "it won't happen to me." In the case of technological hazards, there is an emphasis on the catastrophic loss with much less consideration given to the probability of the event occurring.

Policy tools are similar for both hazards and revolve around distributional issues. With respect to ex-ante measures, building codes, insurance, and compensation may be useful in reducing the potential loss to victims as well as sharing some gains from potential winners to losers. With respect to ex-post measures, insurance and relief are measures that can deal with the consequences from natural and technological disasters.

There are also differences between the two types of hazards with respect to the use of these strategies. The liability arrangements are much more clearly defined for natural hazards (i.e., earthquake and flood) although disaster relief is still dependent on the nature of the event and the possibility that special Congressional legislation may be passed to deal with a catastrophic event. With respect to technological hazards, it is not clear who pays for the losses if a disaster occurs, with the exception of nuclear power where the Price Anderson Act clearly specifies these arrangements (Wood 1982).
In the case of LNG, we could not determine in any of the four countries we studied who would be responsible for losses should a catastrophic accident happen.

In the final analysis a decision has to be made, even if by default, as to who is responsible for potential losses from disasters. To what extent is this a public or private responsibility? An answer to this question sets the tone for the type of risk management processes that will be undertaken in the future on both technological and natural hazards.

Illusions and Mirages in Public Policy

Richard H. Thaler*

Cornell University, Graduate School of Management, Malott Hall, Ithaca, NY 14853

Like most people, I have always found optical illusions fascinating. Figure 1 is a scale drawing of the world's largest man-made optical illusion--the Gateway Arch in St. Louis. Although it appears to be at least 50 percent taller than it is wide, the height and width are actually equal. This optical illusion is an example of what I will call "judgmental illusions:" Somehow the mind is fooled into making an error of judgment. We all erroneously judge the arch's height to be greater than its width.

Another type of optical illusion is the mirage. We have all experienced the illusion of "seeing" water on a perfectly dry highway on a hot day. Such mirages on the desert are commonplace. Mirages, like judgmental illusions, fool the mind. We are fooled into believing that an object exists when it does not.

The subject of this essay is not optical illusions, but rather the related concept of cognitive illusion. Like optical illusions, cognitive illusions can be of two types. Judgmental cognitive illusions induce people to misestimate magnitudes, and in a specific direction. Other cognitive illusions are like mirages: a situation is structured such that we are fooled into thinking we have many choices when in fact only one really exists. Both kinds of cognitive illusions have powerful influences on how people make choices--in their private lives and in the realm of public policy.

---

* Reprinted by permission from <u>The Public Interest</u>

NATO ASI Series, Vol. G4
Environmental Impact Assessment, Technology Assessment, and Risk Analysis. Edited by V.T. Covello et al.

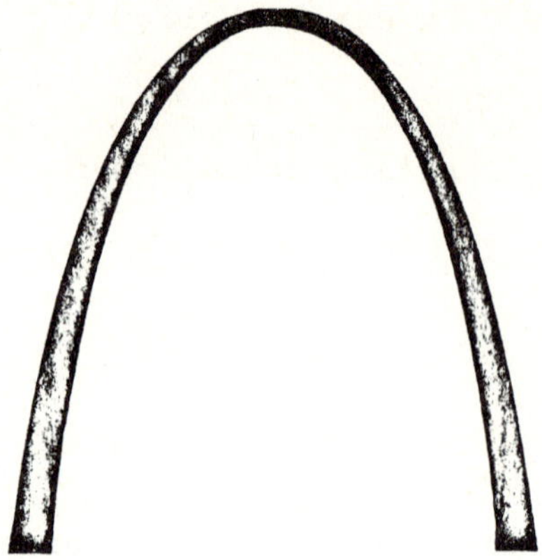

Figure 1. The Gateway Arch in St. Louis (scale drawing).

FOUR ILLUSIONS

The following example of a judgmental cognitive illusion will be familiar to anyone who has taken a course in probability theory. A class has 25 students in it. What is the chance that at least two students will have a common birthday? Most people guess that the chance of a match is pretty small, perhaps one in ten or one in 20. In fact, the chance of at least one pair of students having the same birthday is better than 50-50.

The interesting thing about this problem is that it affects almost everyone the same way. We all judge the chance of a match to be smaller than it really is. This is the defining characteristic of a judgmental cognitive illusion: The problem induces predictable errors in a particular direction.

Can such cognitive illusions that create errors in judgment also create errors in contemporary public policy debates? Yes--and to illustrate this I have selected several examples that relate to our perceptions of risk and uncertainty. The first tests one's ability to estimate magnitudes: Most people think homicides are much more common, when in fact suicides are more common--in one recent year there were 27,300 suicides and 20,400 homicides. (This is true in spite of the fact that the official statistics understate the true level of suicides. Many suicides are classified as accidental deaths.) Why do we guess that homicides are more common? Well, often when we have to estimate the frequency of an event or class of events we do so by judging the ease with which we can recall instances of it. Psychologists Daniel Kahneman and Amos Tversky call this the <u>availability heuristic</u>[1]. The availability heuristic is usually a good way to estimate frequency because ease of recall is usually highly correlated with actual frequency. Sometimes, however, availability and frequency diverge. Since homicides receive more publicity than suicides, they are more available, and thus are erroneously judged to be more common.

The next example, first devised by Daniel Ellsberg, consists of an imaginary lottery:

> You are shown three urns, each containing 100 balls. Urn A has 50 red and 50 black balls, urn B has 80 red and 20 black balls, and urn C has 20 red and 80 black balls. You are given a choice between two lotteries. You can take lottery A, in which a ball will be picked at random from urn A. If the ball picked is red you win $100, otherwise you win nothing. Alternatively, you can take lottery BC. In this case the lottery has two stages. The first step is to flip a coin. If it comes up heads you must choose a ball from urn B. If it comes up tails you must choose from urn C. The second step is to pick a ball from the urn determined by the coin flip. Again, if the ball is red you win $100, otherwise you win nothing. Which lottery do you prefer?

Simple multiplication will confirm that the chance of winning either lottery is 50 percent, yet most people say that they prefer lottery A. Why? Two characteristics of the BC lottery make it unattractive. First, lottery A is simpler. Complexity itself is aversive. Second, the chance of winning the BC lottery is more ambiguous. Depending on the outcome of the coin flip it might be 80 percent or 20 percent. This ambiguity is also aversive. The moral is that most people--if forced to gamble--would prefer a simple, well-defined risk to a complex, ambiguous risk.[2]

The next set of three situations relates to the value people place on their lives. How much would someone pay to avoid a risk to his life? How much would he charge to take an additional risk?

---

1 See Daniel Kahneman, Paul Slovic, and Amos Tversky, <u>Judgment Under Uncertainty: Heuristics and Biases</u> (Cambridge: Cambridge University Press, 1982).

2 Daniel Ellsberg, "Risk, Ambiguity and the Savage Axioms." <u>Quarterly Journal of Economics</u> 75(1961): 643-699.

Risk Situation 1: While attending the movies last week you inadvertently exposed yourself to a rare, fatal disease. If you contract the disease, you will die a quick and painless death in one week. The chance that you will contract the disease is exactly .001--that is, one chance in 1,000. Once you get the disease there is no cure, but you can take an inoculation now which will prevent you from getting the disease. Unfortunately there is only a limited supply of the inoculation, and it will be sold to the highest bidders. What is the most you would be willing to pay for this inoculation? (If you wish, you may borrow the money to pay at a low rate of interest.)

Risk Situation 2: This is basically the same as situation 1 with the following modification. The chance you will get this disease is now .004--that is, four in 1,000. The inoculation is only 25 percent effective--that is, it would reduce the risk to .003. What is the most you would pay for the inoculation in this case? (Again, you may borrow the money to pay.)

Risk Situation 3: Some professors at a medical school are doing research on the disease described above. They are recruiting volunteers who would be required to expose themselves to a .001 (one chance in 1,000) risk of getting the disease. No inoculations would be available, so this would entail a .001 chance of death. The 20 volunteers from this audience who demand the least money will be taken. What is the least amount of money you would require to participate in this experiment?

I have asked these questions of numerous groups. The typical median responses are that people would pay $800 in Situation 1; pay $250 in Situation 2; and charge $100,000 in Situation 3. Obviously, people treat these as three quite different questions. Yet economists would argue that the answers should all be about the same. (They would allow for a small difference between Situation 3 and the other two, but nothing like the magnitude observed.) Essentially each situation presents the subject with a choice between more money or a greater chance of living. If Situations 1 and 2 are compared, we can see that people will pay over three times as much to reduce a risk from .001 to zero than to reduce a risk from .004 to .003. While the increase in the chance of living is the same in each case, the change is more attractive in Situation 1 because the risk is eliminated altogether. Generally, people will pay more to eliminate a risk than to achieve an equivalent reduction of a risk. Daniel Kahneman and Amos Tversky call this the certainty effect.[3]

If we compare the responses to Situation 3 with those given to Situations 1 and 2, we see that the median response to Situation 3 is several times larger. Yet this implies that a typical individual would refuse to pay $5,000 to eliminate a risk, and would refuse to take $5,000 to accept the same risk. How can $5,000 be both better and worse than bearing some risk?

This comparison illustrates another behavioral regularity which I have called the endowment effect.[4] The endowment effect stipulates that an individual will demand much more money to give something up than he would be willing to pay to acquire it. The endowment effect can be observed in cases that do not involve any risk. Suppose you won a ticket to a sold-out concert that you would love to attend, and the ticket is priced at $15. Before the concert you are offered $50 for the ticket. Do you sell? Alternatively, suppose you won $50 in a lottery. Then, a few weeks later, you are offered a chance to buy a ticket to the same concert for $45. Do you buy? Many people say they would not sell for $50 in the first case and would not buy for $45 in the second case. Such responses are logically inconsistent.

POLICY DECEPTIONS

These illusions can affect policy decisions in many ways. Take the endowment effect: Essentially it says that once people have something it is very hard to take it away. Residents of communities with declining school populations know how this has created problems for their school boards. People who would be unwilling to pay for a tax increase to add a school in their neighborhood nevertheless become incensed if an existing school in their neighborhood is closed in order to avoid a tax increase.

---

3 Amos Tversky and Daniel Kahneman. "The Framing of Decisions and the Rationality of Choice." Science 211 (1981) 453-458.

4 Richard Thaler, "Toward a Positive Theory of Consumer Choice." Journal of Economic Behavior and Organization 1 (1980): 39-60.

Similarly, most people know that when social security benefits were indexed, the formula used was inadvertently generous. The result has been that benefits have grown much faster than wages in recent years--clearly an unintended outcome. Nevertheless, any politician who dared to suggest a reduction in the rate at which benefits increase was considered an enemy of senior citizens. Just because it was a mistake to give people something does not mean it can be easily taken away. Even the most recent solution to this problem--a one-time postponement of cost of living increases for six months--is a curious sort of deception. Nothing was taken away--it was simply postponed.

An issue that is particularly interesting in the present context is nuclear power, the debate over which involves all of the illusions discussed so far. The specific question I wish to discuss is: Why is nuclear power so unpopular? I do not intend to evaluate the advantages and disadvantages of nuclear and conventional power plants, since I have no particular insights to offer. The question is why nuclear power generates so much vocal, emotional opposition. This is a puzzle, since the experts seem at least evenly divided. Yet how many demonstrations do we see opposing coal-fired plants? Surely it is not because conventional power sources are without risk to humans or to the environment. Coal mining accidents could be (but are not) attributed indirectly to coal-fired power plants. And acid rain is partly attributed to conventional power plants, but nuclear power is rarely suggested as a solution to the problems of acid rain. So why is nuclear power so unpopular? Each of the four cognitive illusions plays a role.

Availability

As with homicide, the risks from nuclear power are widely publicized, while as with suicides, the risks from conventional sources are not as well known. The Three Mile Island incident, and the popular movie "The China Syndrome" have helped keep the nuclear risks very "available," even though, so far as we know, there has not been a single death related to nuclear power.

Complexity and Ambiguity

As with the Ellsberg urn, the risks from nuclear power are complex and ambiguous. People are not confident that they have been estimated correctly. Future risks, such as those related to waste disposal, are especially ambiguous. The Ellsberg effect shows that such risks are particularly aversive. (It is rather ironic that Ellsberg is now so active in the antinuclear power movement. Has he fallen into his own trap?)

The Certainty Effect and the Endowment Effect

The responses to the three risk situations demonstrated that people are least willing to pay to decrease an existing risk (Situation 2) and demand great compensation for the introduction of a new risk (Situation 3). The way the nuclear power issue is generally discussed, replacing a conventional power plant with a nuclear power plant would reduce an existing risk and add a new risk--the least attractive combination possible.

I am not saying that opposition to nuclear power is silly or irrational. Rather, I have just tried to show how four cognitive illusions all happen to be working to make nuclear power seem highly unattractive, which may help explain why opposition to nuclear power is so widespread.

There is a good book on optical illusions by Stanley Coren and Joan Girgus called Seeing is Deceiving. That title might be applied to thinking: Seeming is deceiving. But like optical illusions, cognitive illusions can be overcome. We can measure the height and width of the Gateway Arch; similarly, we can look in the almanac to find out how many homicides and suicides there are. Nevertheless, there is no practical way of preventing cognitive illusions from influencing policy decisions. Cognitive illusions influence representatives, senators, presidents--even so-called experts are not immune. A physicist may fall for a cognitive illusion just as easily as an economist might fall for an optical illusion. But since informed judgment and explicit analysis can, in principle, mitigate the effects of illusions, we must encourage the use of scientific research, cost-benefit and cost-effectiveness analyses, and expert commissions in the making of policy decisions.

MIRAGES

The second class of policy problems involves mirages. Rather than illusory objects, these mirages are illusory choices, choices we perceive that do not really exist.

Economists tell us that we should keep our options open, and that we should prefer having more choices to having fewer. This is good advice, like "buy low and sell high," but like all good advice it has exceptions. One exception would arise when the costs of deciding are exceptionally high, in which case any decision may be better than more costly pondering.

A second, more interesting, exception is illustrated by the following true story. A group of economists was sitting around having cocktails, awaiting the arrival of dinner. A large can of cashews was placed on the cocktail table, and within 90 seconds one half of the cashews were gone. A simple linear extrapolation would have predicted the total demise of the cashews and our appetites in another 90 seconds. Leaping into action, I grabbed the can and (while stealing a few more nuts on the way) hid it in the kitchen. Everyone seemed relieved, yet puzzled. How could removing the can, and thus removing a choice, have made us better off? Let us analyze this case with the help of the simple decision tree shown in Fig. 2.

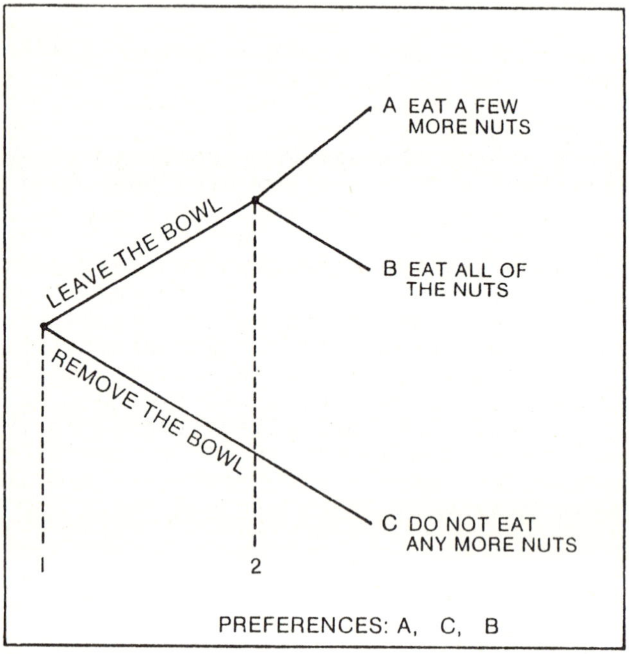

Figure 2. Cashew Nut Decision Tree.

At time 1 we have a decision: to leave the bowl or to remove it. If we remove the bowl, we have no further choices to make, and no more nuts to eat. We obtain Option C. If we leave the bowl, we must then decide how many more nuts to eat. Suppose we would most like to eat a few nuts (Option A) but that we would prefer to stop (Option C) rather than end up eating the whole bowl (Option B). Given these preferences the rational thing would be to leave the bowl and pick Option A. But it is rational only so long as Option A is really feasible. At the dinner party, Option A was a mirage. As much as we might have liked at time 1 to eat only a a few more nuts, at time 2 we would have devoured the whole can. If Option A is a mirage, then Option C becomes the rational choice. I will call taking Option C an act of precommitment: committing oneself to a particular choice in advance.

The first recorded act of precommitment was that taken by Ulysses in his encounter with the Sirens, a popular singing group in Ulysses' time whose songs were highly addictive. Anyone on the seas who heard their songs would feel compelled to draw ever nearer to land, inevitably crashing on the rocks near shore. Ulysses' strategy was to have himself tied to the mast so he could not alter the course of the ship.

Ulysses' method of dealing with the Sirens and my act of removing the cashews were both acts of rational precommitment. They both satisfy the two conditions that are necessary for precommitment to be rational: 1) a change of preferences is anticipated; and 2) the change will be for the worse. Ulysses knew that he would alter the course of his ship if he had the option, and that the change of course would be for the worse.

A common precommitment institution is the Christmas club. Christmas clubs have three distinguishing features: They pay little or no interest; they require weekly deposits at some inconvenience; and they do not allow the customer to withdraw any money before Thanksgiving. This institution clearly seems inefficient. But is it irrational? The decision tree in Fig. 3 shows once again that C is a rational choice only if A is a mirage. But for individuals or families on a tight budget who would otherwise not save enough, a Christmas club could be rational. The fundamental point is that a rule (which is what a precommitment strategy is) must necessarily be crude and inflexible, and may also be inefficient. Nevertheless, these drawbacks, in and of themselves, should not lead us to the conclusion that a particular rule is undesirable. To evaluate a rule we must compare it with the alternatives. Normally a rule specifying that the captain of the ship may alter the ship's course would be judged acceptable. But for Ulysses, that rule was better than any other feasible alternative.

PUBLIC USES OF PRECOMMITMENT

Rational precommitment is highly important in the domain of public policy. Take terrorism or airplane hijacking, for example: As all airline passengers know, we spend millions of dollars and countless hours of time screening airline passengers for bombs, guns, keys, lighters, and calculators. Film is ruined, people miss planes, and we all pay more for plane tickets. What do we gain from this? Since any security system can be beaten, planes are still occasionally hijacked. As an alternative to all the equipment, I propose we adopt the following rule: "Never make concessions to hijackers." Hijackers would be given a choice: Give up or be shot. One might say, "Won't that be a dangerous rule? Won't many innocent people be hurt or killed?" The answer is no. First, if no one ever gets anything from hijacking a plane, then the incidence will fall almost to zero. Second, there is no defense against a crazed terrorist. If someone wants to kill himself and a lot of other people he can do it, and we cannot stop him. Anyone can hijack a bus or blow up an office building at will.

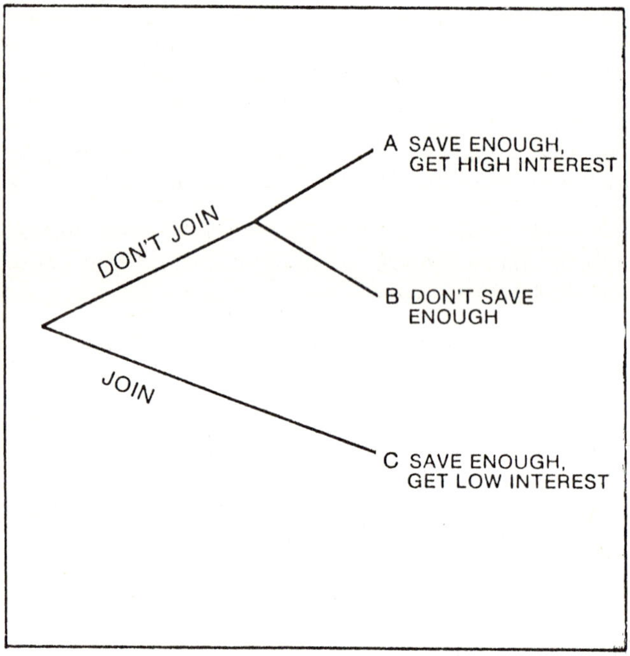

Figure 3. Christmas Club Decision Tree.

A different criticism of this rule is that there might be situations in which society would want to make concessions. Suppose someone hijacks a caravan of school buses--would we not want to make concessions in this case? But such examples simply illustrate the value of the rule: If an exception is made for a school bus once, no school bus would ever be safe again. To enforce this rule, we could decide now that anyone not personally at risk who grants a concession to a hijacker would be guilty of a felony. (A jail sentence would be mandatory, of course.)

Another example of rational precommitment is mandatory retirement. Mandatory retirement rules normally have two separate provisions. First, an organization will announce that it is under no obligation, moral or legal, to keep a worker past the age of 60 (or 65, or 70). One reason organizations adopt such a policy is that for most workers, though not all, productivity reaches its peak some time before retirement. Wages, on the other hand, usually rise throughout a worker's employment. These two factors together suggest that just before retirement most workers are earning more than they produce. Clearly a firm would not want to allow that state of affairs to go on indefinitely.[5]

The second provision, while not universal, frequently accompanies the first. It provides that no worker may remain employed after the normal retirement age; this might be called "mandatory mandatory retirement" or (MMR for short). Since the adoption of MMR eliminates some future options for an organization, it qualifies as precommitment. But is it rational? It is easy to criticize an MMR rule since it might force a university to lose a Nobel laureate who is still active. Nevertheless, I believe such rules make good sense because it is so difficult to tell anyone he is no longer productive. Thus, the option to keep some people can, and will, become a practice of keeping almost anyone who wants to stay on. So for every Nobel laureate who is kept for an extra couple of years, the organization must retain several others who are no longer productive.

I am not advocating any particular age for mandatory retirement; clearly that should differ across jobs. What I do oppose is the recent federal law that declared mandatory retirement illegal for ages under 70. I believe this is an unnecessary and ill-advised intrusion into private organizations' affairs. In many cases these rules were arrived at through an explicit process of collective bargaining. The abolition of mandatory

---

5 See Edward Lazear, "Why Is There Mandatory Retirement?," Journal of Political Economy 87 (1979): 1261-1284.

retirement is sometimes posed, improperly, as a civil rights issue. But old age is a state all hope to attain. Young people benefit when mandatory retirement creates new openings, but those same young people will face those rules later. Mandatory retirement makes sense for some organizations, and they should be given the freedom to enforce their own rules.

THE FEDERAL BUDGET AS DINNER CHECK

I would also encourage the federal government to impose some precommitment rules on itself, in particular fiscal caps such as Proposition 13 in California. Several proposed constitutional amendments would impose such rules on the federal government either by restricting future growth of spending to no more than the growth rate of real GNP, and/or by requiring a balanced federal budget. Such ideas are not new. In 1798 Thomas Jefferson wrote, "I wish it were possible to obtain a single amendment to our Constitution. I would be willing to depend on that alone for the reduction of the administration of our government to the genuine principles of its Constitution; I mean an additional article, taking from the government the power of borrowing."

Since these proposed amendments would greatly restrict the government's flexibility (though exceptions are made for declarations of war), it must be demonstrated that such precommitment would be rational. Though the growth in government's share of GNP in the last 55 years should be evidence enough, let me describe some of the mirages that led to this explosion.

Suppose we consider the behavior of a hypothetical group of 500 people going to an expensive restaurant for drinks and dinner. Compare their expected behavior under two different rules: Each person pays his own bill; or, the total bill is divided evenly. Under which rule will people spend more? The latter of course. If I order a shrimp cocktail for $5.00 and it is split 500 ways, I pay only one cent. I am more likely to order a shrimp cocktail at one cent than at $5.00. This "check-splitting-effect" is what leads to so-called pork barrel legislation.[6] The amount a district will pay toward a particular local project can be very small indeed, so it is in the interest of every Representative to get as many projects into his district--just as it is in my interest to order dessert if everyone else is chipping in.

---

6 See James Buchanan and Gordon Tullock, The Calculus of Consent (Ann Arbor: University of Michigan Press, 1962).

Suppose the larger dinner party comes up with a clever idea: the check will still be divided evenly, but to save time, the ordering will be done by committee. There will be separate committees for drinks, appetizers, entrees, salads, and desserts. Each person can serve on the committee of his choice. How will people allocate themselves to committees? You guessed it--the lushes will be on the drinks committee, the vegetarians on the salad committee, the sweet-tooths on the dessert committee, etc. (After all, we do want the people with expertise on the relevant committees.) Of course such an arrangement will exacerbate the tendency toward over-ordering. The lushes will have a bottle of wine with each course, the sweet-tooths two rounds of dessert, and if by chance a lush finds himself on the dessert committee we can be sure that rum cake will be served. Unfortunately, this metaphor closely resembles the structure of the U.S. Congress, where the committee process reinforces the tendency toward excessive spending.[7]

Suppose further that each of our diners can put his bill on a credit card--a special credit card that does not have to be paid if the diner loses his job or retires. This rounds out the story nicely. To a representative more concerned with getting reelected than with posterity, "buy now, pay later (maybe)" is a very appealing process.

How can we control the budget? I advocate random committee assignments, but that is probably not politically feasible. Failing that, we need both a limit on the rate at which government spending can grow, and an enforced balanced budget.

Such rules have been criticized on numerous grounds: They are rightly described as crude and inflexible, and the definition of a balanced budget is far from obvious. Nonetheless, the spirit of such amendments is worth supporting, and the criticisms are not persuasive. The issue is not whether such rules are perfect, but rather whether they are an improvement over the current situation. As Richard Wagner and Robert Tollison recently wrote: Annual budget balance is a good idea because it places useful and meaningful constraints on political choice. This is not to say that it is a perfect rule for the conduct of government, for there are no perfect rules for the conduct of something as massive as our government. The problem is to search for feasible, workable rules that encourage polit-

---

7 See William Niskanen, Bureaucracy and Representative Government (Chicago: Aldine-Atherton, 1971).

ical decisionmakers to act as if they had good common sense. A rule of annual budget balance and careful up-front monitoring of the variability of long-term government projects seems to be the wisest course of action.[8]

NO THYSELF

By definition, precommitment must be done in advance, before we actually face temptation. We always resolve to go on a diet--next week. "Lord give me strength--but not now." So we should not be surprised to find that we, or our organizations, want to break rules that were established earlier, even if the rules are desirable. Indexation of the federal income tax is a good illustration of this principle.

Because we have a progressive income tax with marginal rates that increase with income, inflation automatically produces real tax increases. In 1981, Congress passed a law that instituted indexing of the personal income tax beginning in 1985. The way the law works, the personal exemption (now $1,000), the zero-tax bracket (now $3,400), and all other bracket levels would automatically rise with inflation. This would be particularly helpful to lower income families, since the brackets are much closer together at lower levels. Not surprisingly, some in Congress now want the law repealed. As Martin Feldstein, chairman of the President's Council of Economic Advisors, argues in a recent Wall Street Journal:

> If tax revenue must be raised, the repeal of indexing isn't a satisfactory substitute for an explicit tax increase. Because the repeal of indexing is a hidden way of increasing taxes, it removes the pressure to choose between spending cuts and more taxes. And unlike voting an explicit tax increase, repealing indexing doesn't provide a fixed amount of additional tax revenues but starts a money machine that will squeeze more and more money from taxpayers in years ahead. The repeal of indexing is politically tempting to many in Congress because it increases revenue without explicitly increasing taxes. But it is the very opposite of responsible budgeting [emphasis added].

---

8 Richard Wagner and Robert Tollison, Alvin Rabuska and John T. Noonan, Jr., Balanced Budgets, Fiscal Responsibility and the Constitution (Washington, D.C.: Cato Institute, 1982).

Feldstein's choice of language is quite apt. First, he describes the tax increases in the absence of indexation as hidden. This is right on target. To keep government spending under control we should want tax increases to be visible, not hidden. Second, he says the repeal of indexation would be tempting to Congress. Congress recognized its addiction to hidden tax increases and resolved to break the addiction--starting in four years. It is important that this resolution be kept. Indexation is particularly important in the absence of spending restrictions or a balanced budget amendment, since it will help keep government from growing.

My proposal can perhaps best be summarized in the dictum Thomas Schelling set down in these pages several years ago: "No Thyself." This is good advice for individuals, organizations, and governments. If we make rules to enforce the "no," we must remember that all rules are imperfect, and that they must not be lightly abandoned. Not even Ulysses could count on self-restraint.

The Multiple Perspective Concept: Bridging the Gap Between Analysts and Decisionmakers

Harold A. Linstone

Portland State University

INTRODUCTION

Modern systems analysis really began as "scientific management" with Frederick Taylor in 1911, gathered steam with World War II's operations research, and was propelled forward by complex weapon system design and strategic analysis in the 1950s. It then burst full force onto the national scene with RAND's systems analysts in the early 1960s. By 1967, Max Ways wrote in FORTUNE's survey "The Road to 1977":

> The further advance of this new style (systems analysis) is the most significant prediction that can be made about the next ten years. By 1977, this new way of dealing with the future will be recognized at home and abroad as a salient American characteristic. (1967: 94)

The optimism exuded by Ways and many others was based not only on the dazzling success of science and technology, but on the ability of systems analysis to deal with complex technological problems--telephone networks, radar, fire control systems, etc. (Fig. 1a). And the development of computers was perfectly timed to provide a powerful modeling instrument for the analyst.

Also in 1967, the well-known Dean of a major engineering school proclaimed, "I doubt if there is any such thing as an urban crisis, but if there were, MIT could lick it in the same way we handled the Second World War." The federal bureaucracy resorted to the planning-programming-budgeting system (PPBS), and industry embarked on cost-benefit studies. Management science and decision analysis flourished. Technology assessment, formalized as a Congressional support function, became another domain of systems analysis. Millions of dollars were spent annually to apply such tools to an ever-widening spectrum of sociotechnical systems to improve decisionmaking (Fig. 1b).

NATO ASI Series, Vol. G4
Environmental Impact Assessment, Technology Assessment, and Risk Analysis. Edited by V.T. Covello et al.

Figure 1—The Technical Perspective

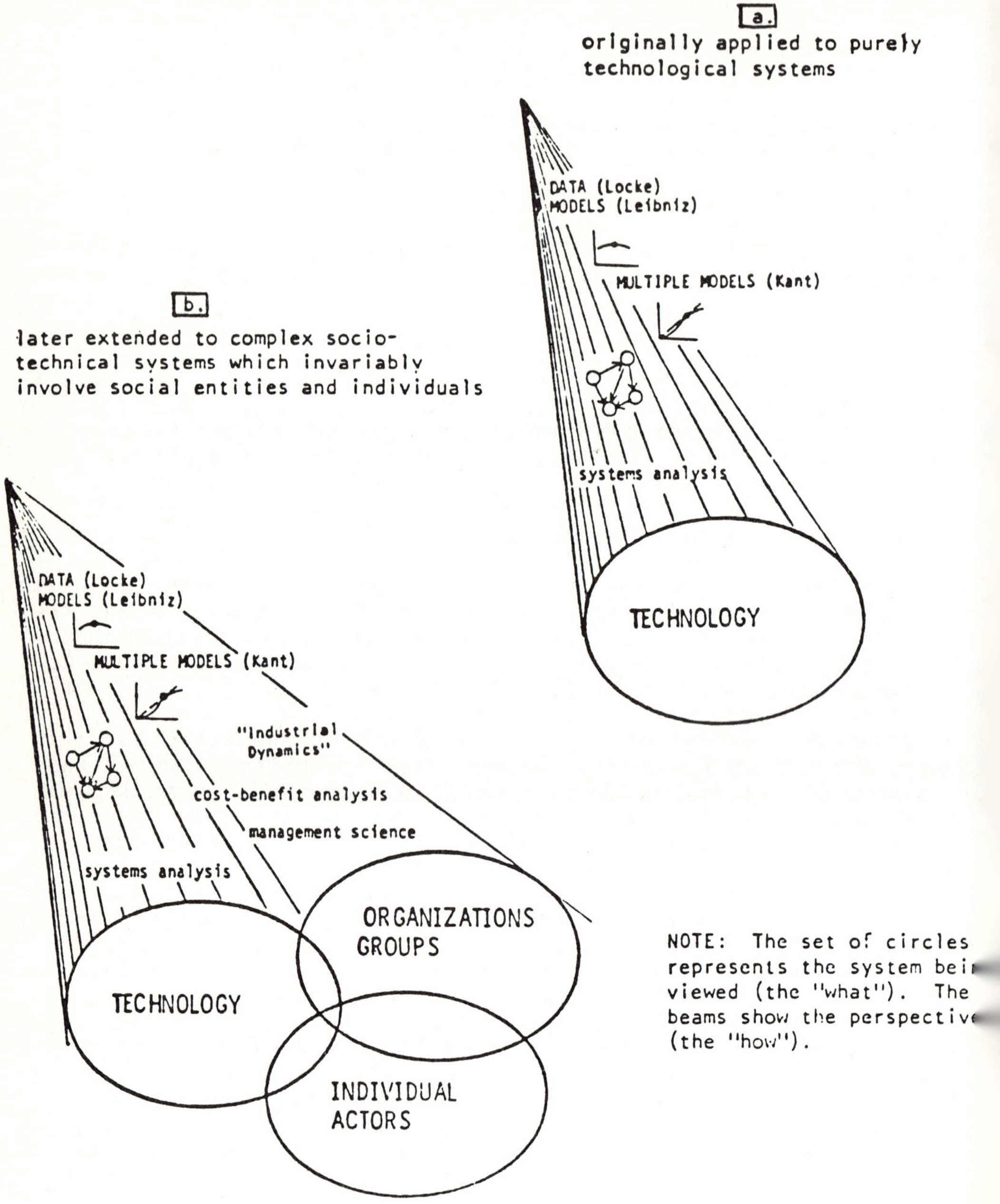

Since the late 1960s, however, there has been a shift from euphoria to skepticism. By 1979, Ida Hoos was indignant:

> In our technical era, the dominant paradigm is so technically oriented that most of our problems are defined as technical in nature and assigned the same treatment--doctoring by systems analysts. The "experts" are methodological Merlins . . . . Technology assessments are conducted by latter-day intellectual condottieri, the brains-for-hire. (1979: 192)

Richard M. Cyert, formerly professor and now President of Carnegie-Mellon University, recently confessed:

> As a professor of organization theory and management, I used to wonder about the practical value of these academic fields. For the last eight years, I've had some first-hand experience finding out . . . and I've concluded that the study of management makes a useful, but only a limited, contribution to the practicing manager. (Wall Street Journal, April 7, 1980)

What are some of the reasons that conventional systems analysis, based on tools and techniques which have yielded stunning successes in science and technology, falls short in the wider realm of application? (1) The "problem-solution" syndrome: Academic training tends to define problems and their solutions neatly. Texts do not point out that beyond the realm of pure technology a solution inevitably generates new problems, that we <u>shift</u> problems rather than solve them. Great achievements in communicable disease control, for example, led to a remarkable decline in the mortality rate--and an explosive global population growth; (2) The "best" solution search: Cost-benefit analysis and linear programming are typical of the search for an optimal solution. It usually comes as a shock to the "technicians" that complex living systems--natural as well as human--have not organized themselves in accordance with this principle. Whereas the technologist seeks a fail-safe design, advanced living systems operate in a safe-fail mode. They trade avoidance of failure for survival of failure, efficiency for resilience; (3) Data and models: The modes of inquiry pushed by the "technicians" focus on data, models, or combinations thereof. Other means of gaining insight are ignored. As with the myth of the Greek sculptor-king Pygmalion, the model becomes the reality; (4) Reductionism: The standard procedure in dealing with complex systems is to divide them into smaller subsystems. Simplification is the price of modeling, but the resulting suboptimization may be misleading in the context of the total system. Problem redefinition in the simplification process may lead to solution of the wrong problem; (5) Quantification: A computer age version of Gresham's Law states that "quantitative analyses tend to drive out qualitative analyses." We have become a measuring and counting society. Whatever is not readily quantifiable is ignored or replaced by something that is: The ubiquitous computer must be fed quantitative input. We love statistics and polls; quality of life is rated by numerical indices, and risks are ranked by expected value of fatalities; (6) Objectivity: In a complex system, virtually "everything interacts with everything," and this includes the analyst. Without the observer, there are no observations; he must therefore be part of the system. Hence, complete objectivity is an illusion; (7) Avoiding the individual: The "technician" dislikes dealing with unique individuals; he prefers types, classes, averages, and aggregates. But, as Churchman tells us, to be able to see the world globally <u>and</u> as a world of unique individuals is crucial in managing complexity; and (8) Time: Neither an organization nor an individual sees time in the steady, unchanging pace of the technician. They discount the future (and the past), albeit to different degrees. They look at the future as if through the wrong end of a telescope: The distant object (i.e., problem or opportunity) looks smaller than it actually is.

Thus, the limitations of the "technical" approach in a world which is not purely technological should hardly be surprising. Paradigms which are perfectly satisfactory in dealing with well-structured or "tame" systems--typically the classical operations research problems--prove inadequate in the realm of ill-structured or "wicked" systems. And the sociotechnical systems which are of most concern to managers and decisionmakers are usually of the latter kind. What can be done?

MULTIPLE PERSPECTIVES: THE CONCEPT

The multiple perspective concept rests primarily on Graham Allison's study of the Cuban missile crisis from three perspectives--the "rational actor," the "organizational process," and the "bureaucratic politics" model (1969, 1971). He found that each perspective contributed important insights not obtainable from the others. For example, it was evident that the transportation of the Soviet missiles to Cuba was accompanied by a high degree of security (e.g., camouflaging lumber ships, emptying a Cuban harbor of all Cubans). Yet, no camouflage was used at the missile construction sites in Cuba which, as the Russians knew, we overflew with high altitude reconnaissance aircraft. From the rational actor perspective, this seemed inexplicable. However, using the organizational process perspective, the reason became clear at once: different Soviet agencies were responsible for transportation and construction, and each had its own standard operating procedures (SOPs). Early extensions of Allison's work were developed in doctoral dissertations by J.D. Steinbruner (1968) and D.F. Andersen (1977) at MIT's Sloan School of Management.

Linstone, et al. (1981a,b) broadened the concept to sociotechnical systems analysis as means for bridging the gap between analyst and decisionmaker created by the limitations previously discussed. Three perspectives are used: the technical (T), the organizational (O), and the personal (P). Each perceives the world through a different filter.

The technical perspective sees the world through quantitative analysis, e.g., the curves (Phillips, Laffer) and computer models (e.g., econometric, system dynamic) so popular with management scientists and modelers. Unquestionably, this perspective is useful, and we do not question its value; our concern is the resulting imbalance. The conscious parallel use of the three perspectives serves to insure a better balance and a recognition that no one perspective should a priori dominate the process leading to management decisionmaking.

The O perspective sees the world through an organizational filter (e.g., a unique manufacturing division or government agency view). There is great concern whether a new policy will threaten the organization's stability, whether it fits the current standard operating procedures and parochial priorities. In this perspective, we deal with power--where is the real leverage? How can conflicts among units be turned to constructive use? There is no intensive search for analytic tools--in fact, often a mistrust of "academic" techniques. They are viewed as unrealistic or incomprehensible and uncontrollable. Machiavelli, often misrepresented, actually presents us with an excellent example of an O perspective. As a modern executive has noted, the principles evolving from this perspective applied to the state in Florence A.D. 1530 are readily transferable to the corporation in New York or London today (Jay, 1968).

The P perspective is the world as seen through the filter of the individual. Intuition, charisma, leadership, and self-interest play vital roles in matters of policy decision.

It is the most subtle and elusive, because it corresponds to the least accessible, the deepest layer of the managerial decision process. Its sensitivity is high, and its replicability low. Does the preferred system "feel" right? What makes Sammy run? Whose instincts are superior? This perspective may be helpful in identifying the hidden movers, facilitating understanding of the O perspective ("knowing the players"), pinpointing opposition and support early, and improving communications. Complex ideas can be most effectively communicated on the level of the personal perspective--as novelists, playwrights, and presidential speech writers have long known.

In color photography, a combination of different color-sensitive layers or filters must be superimposed to produce a natural color image. We face a somewhat analogous challenge. Every individual and organization has a unique world view, its own, often one-sided, blend of T, O, and P. Our theme is that conscious recognition, interplay, and synthesis of these perspective types goes a long way in meeting the needs of the decisionmaker.

The principal characteristics of the three perspectives are listed and compared in Table 1. Figure 2 presents the story schematically and may be contrasted with Fig. 1. We emphasize three points: (1) We must clearly distinguish between <u>what</u> we are looking at and <u>how</u> we are looking at it. The "how" is a question of perspective. Any perspective may illuminate any part of the system under consideration. For example, we may view an organization from a technical perspective and see it in terms of industrial dynamics or cybernetic theory. Alternatively, we may see it in terms of power conflicts and loyalties; (2) Each of the three generic types--T, O, and P--uses very different paradigms and modes of inquiry; and (3) There are many O and P perspectives: Each group or organizational entity and each individual has a unique perspective, and it may change over time.[1]

Not surprisingly, the technicians, who are adept at the T perspective, are often weak in applying an O or a P perspective. They tend to live in a world of theories, data, and models, thriving on S curves, U curves, and matrices. If one computer model proves unsatisfactory, their inevitable answer is another, usually more sophisticated one. Possibly the foremost value of the multiple perspective concept is to

1 This point is schematically suggested in Fig. 2 by the multiple vertices for the O and P perspectives.

Table 1--Multiple Perspectives

	TECHNICAL (T)	ORGANIZATIONAL (O)	PERSONAL (P)
WORLD VIEW	SCIENCE-TECHNOLOGY	SOCIAL INFRASTRUCTURE hierarchical (caste)...egalitarian (sect)	INDIVIDUATION-THE SELF
ETHICAL BASIS	Rationality	Justice/fairness	Morality
GOAL	problem solving product (study, design, explanation)	stability and continuity* process action and implementation	power, influence, prestige status maintenance or improvement
MODES OF INQUIRY	abstraction and modeling data and analysis	dialectic/adversary negotiated reality/consensual	intuition, persona, individual reality experience, learning
TIME CONCEPT	technological time	social time	personal time
PLANNING HORIZON	far often little breadth	intermediate distance intermediate breadth	short distance variable breadth
DISCOUNT RATE	minimal	moderate	high (with rare exceptions)
CONSTRAINTS	problem simplification by limiting variables, relations	fractionating/factoring problems problem delegation to others or avoidance if possible	hierarchy of individual needs (security, acceptance, self-fulfillment)
	cause and effect	agenda ("problem of the moment")* bureaucracy often pervasive*	challenge and response each construes attributes of others
	need for validation & replicability (or "audit trail")	political sensitivity and expediency* loyalties, credentials restricted access by outsiders (caste) or recruits members (sect)	inner world (subjectivity)
	objectivity emphasized	reasonableness, common advantage	
CHARACTERISTICS	prediction	recognition of partial unpredictability long range planning often ritualized	need for certainty, beliefs creativity and vision of the few
	optimization (best solution)	satisficing (first acceptable, rather than best, solution)* incremental change, slow adaptation parochial priorities	cope with few alternatives or variables only filter out images inconsistent with past experience
	feedback loops recognized		game playing ("homo ludens")
	quantification use of averages, probabilities trade-offs	standard operating procedures compromise and bargaining* monitoring and correction	focus on simplistic hypotheses rather than scanning many leaders and followers, mystique
	uncertainties noted: many caveats ("on one hand ...")	uncertainties avoided, fear of error *	fear of change and unknown
COMMUNICATION	technical report, briefing	directive, conference, interview private language with insiders hortatory language with public	narrative (story), discussion, speech importance of personality

* usually applies to hierarchical structures only (see Table 7)

Figure 2--Multiple Perspectives

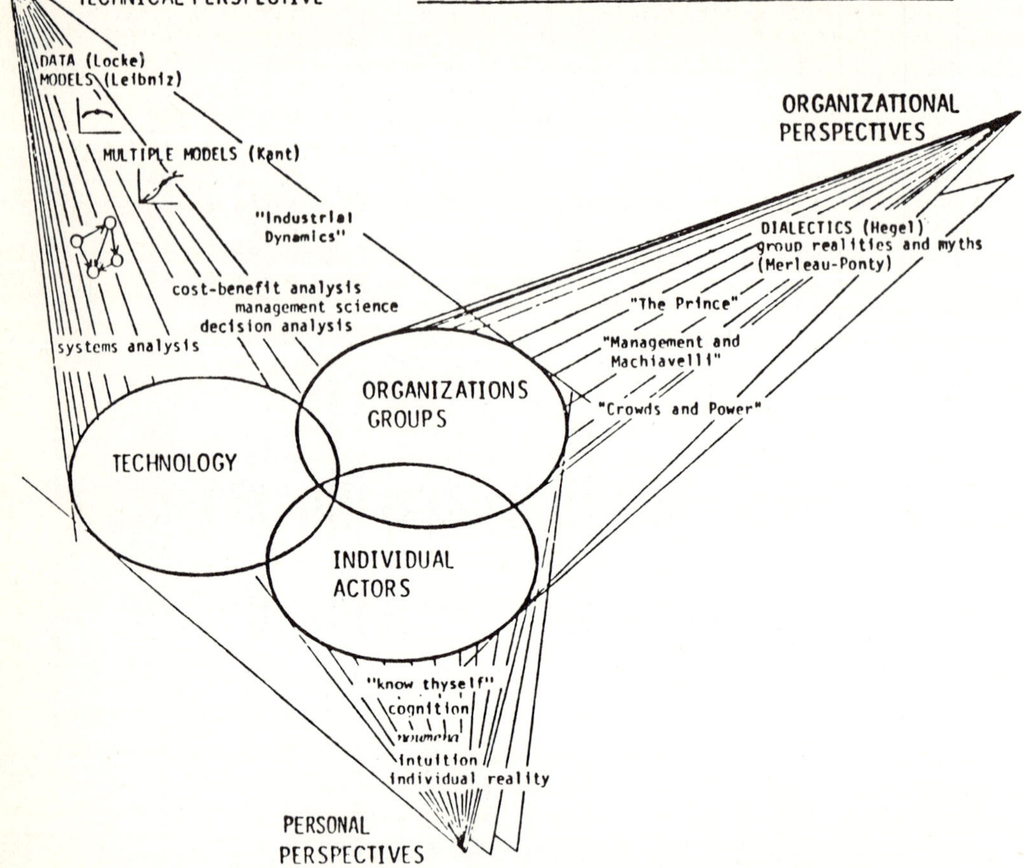

restrain the overselling of the latest "in" model by the experts nurtured in the T perspective. Their technical training may actually prove a disadvantage in dealing with the other perspectives. They search for information in documents and reports, unaware that the most meaningful O and P perspective input may be available only orally. The talents of an investigative reporter may prove more valuable than those of a scientific researcher, a lawyer's courtroom experience more useful than solid engineering know-how. The typical misconceptions as viewed from each perspective are shown in Fig. 3.

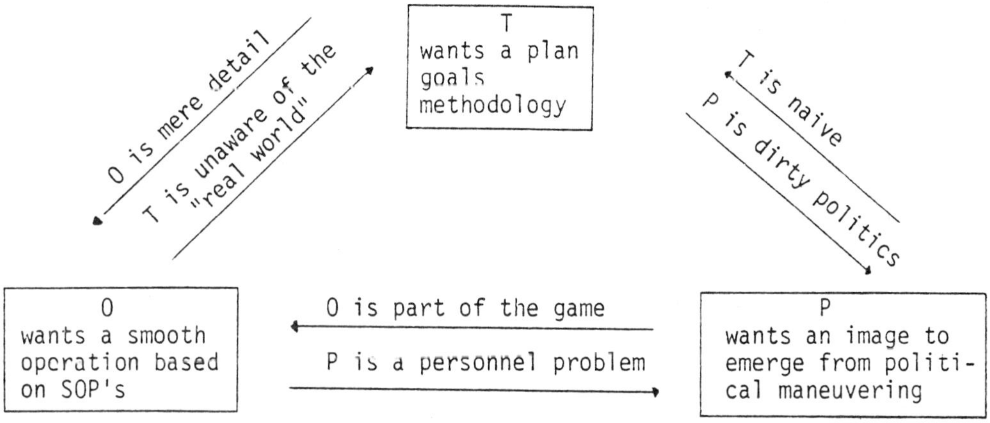

Figure 3: Basis of Miscommunications

Just as three-dimensional space makes us understand the limitations of one-dimensional space, so the O and P perspectives help the manager see the limitations of the T perspective--and deal with them constructively.

Multiple perspectives require something beyond an interdisciplinary team or approach. It will not do to corral, say, an engineer, an economist, a sociologist, and a systems analyst--they are all attuned to the T perspective. One must draw in backgrounds which are more compatible with the other perspectives. The apportioning of work effort among perspectives cannot be fixed a priori. However, as a first approximation, an equal three part split in effort among T, O, and P appears reasonable.

Do the perspectives work at cross-purposes, or do they reinforce each other? They may do either. How are they to be integrated, and by whom? Consider a manager facing a decision on a complex sociotechnical system. He obtains a T-oriented perspective from his engineers and systems analysts, O perspectives from his department heads, and P perspectives from outsiders, workers, and staff whom he queries informally. There will be conflicting input as well as cross-cuing reinforcement. His situation is in some respects similar to that in a court trial: Various witnesses provide different perspectives to the jury. Although the prosecutor and defense attorney offer an integrated version of the testimony to the jury, these decisionmakers have been exposed directly to the diverse perspectives--and must do their own integration. The interaction process is difficult to describe explicitly, although it is recognized by every executive. We have usually found it preferable to present the T, O, and P perspectives and leave the integration to the decisionmaker, rather than to presume to do it for him.

In sum, the essence of the concept is the simultaneous or parallel consideration of perspectives with different paradigms. By going outside the bounds of a perspective, its peculiarities become clear, and this has specifically been the case with systems analysis and related tools (decision analysis, risk analysis, technology assessment, etc.). We find that each perspective has strengths and weaknesses. The T perspective develops replicable models but introduces a plethora of idealizing assumptions; the O perspective reflects bureaucratic realities but shrinks the planning horizon; the P perspective provides insights about the actors, but their partial views encourage distortion. Because the O and P perspectives are not unique or fixed over time, they are sometimes difficult to identify and bound.

How many of these perspectives are enough? By what criteria should their quality be judged? While there are no rigorous answers, it is most encouraging that the use of multiple perspectives to date has consistently provided valuable insights and narrowed the chasm between models and real world decisionmaking.

Finally, it is useful to recall a recent analysis by Nehnevajsa and Menkes (1981). Looking back over a decade of National Science Foundation (NSF) involvement in technology assessment, they see its evolution in terms of several stages. Stage I enlarged the systems analysis, specifically cost-benefit evaluation, to encompass concerns beyond economic feasibility. Stage II was more ambitious, aiming for "comprehensiveness" in identifying and evaluating the full range of social, economic, and environmental impacts. Stage III recognized the futility of this goal and retrenched to the development of "good risk" contingency plans and policy options. Thus, technology assessment was linked much more intimately to the decisionmaking process. Nehnevajsa and Menkes suggest that the use of multiple perspectives may move us into Stage IV of the evolution of technology assessment and that risk analysis may be following similar development stages. In the following sections, we survey some of the considerable number of applications of the multiple perspective concept undertaken to date.

APPLICATIONS TO MILITARY TECHNOLOGY ASSESSMENT

The T perspective has been used extensively in forecasting and assessing military technology since World War II. The study of the impact of a potential new U.S. weapon system on Soviet strategy may not have been labeled a technology assessment, but it essentially is one. In fact, the pioneering work of Allison (1969, 1971) involved an application of multiple perspectives to a military-political problem.

We consider here two innovations in military technology and examine the significance of the O and P perspectives:
(1) introduction of a new ship concept into the U.S. Navy, and
(2) introduction of new rifles into the U.S. Army.

The USS Wampanoag (Based on E.E. Morison, 1966)

The Wampanoag was a 4,200-ton "advanced technology" destroyer built for the U.S. Navy and commissioned in 1868. She had sails and a steam engine, was fast (over 17 knots), and the sea trials proved her to be a magnificent technical achievement--ahead of ships in any navy at that time.

In 1869, all naval steamships were scrutinized by a board of naval officers. The mood of the board is documented. The steam vessel, said the board, was not a school of seamanship for officers or men:

> Lounging through the watches of a steamer, or acting as firemen and coal heavers, will not produce in a seaman that combination of boldness, strength, and skill which characterized the American sailor of an elder day; and the habitual exercise by an

> officer of a command, the execution of which is not under his own eye, is a poor substitute for the school of observation, promptness and command found only on the deck of a sailing vessel. (Morison, 1966: 114)

The board examined the Wampanoag and developed a bill of particulars leading it to the conclusion that the ship was "a sad and signal failure" and could not be made acceptable. They were in a state of peace and opposed building ironclads needed in war to avoid giving unnecessary alarm. There was a large supply of timber in the Navy Yards "which the interests of economy demand should be utilized." They noted the familiarity of the workmen with wooden ship building and their dependence on it for a livelihood.

The ship was laid up for a year and soon sold by the Navy. Morison ponders this strange turn of events:

> Now it must be obvious that the members of this Naval Board were stupid. They had, on its technical merits, a bad case, and they made it worse by the way they tried to argue it . . . . [But after a time] I began to be aware of a growing sense of dis-ease . . . . Could it be that these stupid officers were right? I recalled the sagacious judgment of Sherlock Holmes. The great detective, you will remember, withheld the facts in the incident of the lighthouse and the trained cormorant because, as he said, it was a case for which the world was not yet fully prepared. Was this also the case with the Wampanoag?
>
> What these officers were saying was that the Wampanoag was a destructive energy in their society. Setting the extraordinary force of her engines against the weight of their way of life, they had a sudden insight into the nature of machinery. They perceived that a machine, any machine, if left to itself, tends to establish its own conditions, to create its own environment and draw men into it. Since a machine, any machine, is designed to do only a part of what a whole man can do, it tends to wear down those parts of a man that are not included in the design . . . .

> I don't happen to admire their solution, but I respect their awareness that they had a problem . . . . [It] is not primarily engineering or scientific in character. It's simply human. (116-122)

In these passages, Morison shifts perspectives, from T to O. A decision which appeared "stupid" when viewed from the former suddenly became reasonable when seen from the latter. Thus, a new technological system was held back for reasons which would be exceedingly difficult to uncover with a T perspective.

The M-16 Rifle

The rifle is the most basic weapon system in any army. Hence, decisions to replace a rifle are a very serious matter. The history of the U.S. Army rifle since the Civil War is an interesting one, presenting us with technical, organizational, and personal aspects.

McNaugher (1980) and Fallows (1981) as well as discussions with former key individuals in the Department of Defense (Schwebs and Sprey, 1981) provide an appreciation of the complex decision process which finally led to the M-16 rifle (Linstone et al., 1981b, Fig. 10). The last 100 years can be seen as a struggle between marksman and citizen-soldier weapon proponents, or between the old Army Ordnance/Springfield Arsenal establishment and the newer Infantry Board/Secretary of Defense organizations.

The Civil War inaugurated the era of the bullet. A precursor to the M-16 story is told about the first reliable repeating rifle for military use (Fallows, 1981: 80). When the Civil War began, Union troops carried a cumbersome muzzle-loading rifle. A young inventor named Spencer, who came from outside the Army's Ordnance establishment, had developed the repeating rifle. Because he could not get the Army's attention, he wangled an appointment with President Lincoln. They went behind the White House and the President, who was a good shot, test fired the rifle. The Secretary of War had been invited, but declined to attend. Lincoln then sent him a note ordering the Army to procure the rifle. They bought the minimum amount, assigned them mostly to the cavalry, indicating the rifles were not suitable for infantry use. However, several Union regiments bought the Spencer rifles with their own money rather than use the approved rifle. After Lincoln's assassination, the head of the Ordnance Corps quickly declared the Spencer rifles obsolete and ordered them sold. (Ironically, many of these superior rifles were bought by the Indians and used in the attack on Custer at Little Bighorn.)

After the Civil War, the myth of the Western sharpshooter and the formation of the National Rifle Association (NRA) reflected the development of a uniquely American marksmanship tradition. A marksman firing in long-range rifle competition needed a large, heavy round to maximize steadiness in flight and minimize sensitivity to wind. The Army Ordnance Corps, responsible for small arms, adopted the marksmanship concept as its own creed. Its Springfield arsenal produced the first breech-loading rifle to fill this indicated need in 1867.

Again, in the Spanish-American War, Theodore Roosevelt's Rough Riders with their ordnance-approved rifle found themselves facing troops equipped with a superior rifle--the Mauser. When he became President, Roosevelt ordered the unenthusiastic War Department to buy Mausers.

Every modern conflict has raised doubts about the Ordnance Corps' criteria, i.e., the appropriateness of the marksman's rifle for the citizen-soldier. World War I introduced the machine gun, and the semi-automatic rifle followed; suppression became an important consideration. In Europe the smaller .276 semi-automatic rifle was introduced, and a specially appointed Army Caliber Board approved a similar version in 1932. But the Chief of Staff, General MacArthur, sided with the traditionalists and disapproved. Instead, the .30 caliber M-1 became the standard Army weapon in 1936.

In World War II, assault tactics relying on volume of fire rather than accuracy became important. General Patton wanted "marching fire": Several steps, then a short burst, more steps, another burst. It was recognized that lethality is enhanced by bullet speed more than by caliber size, making a smaller caliber rifle clearly preferable as a weapon. It was also found that few soldiers in combat fired their rifles at targets over 300 yards distant. In fact, nearly 80 percent of the combat soliers never fired their weapons at all during battles. The exception to this rule were groups which carried the Browning automatic rifle--a portable machine gun which could spray continuous fire and "hose down" an area.

After the war the British developed a small, lightweight, .276 caliber assault rifle. The U.S. Infantry Board tested and accepted the British weapon; there was the additional argument of standardization within the North Atlantic Treaty Organization (NATO). Once again, tradition overruled the recommendation, and in 1957 the Army Staff and Ordnance Corps (more precisely the Army Material Command) responded with the M-14. It was an automatic-firing direct descendant of their .30 caliber M-1.

At this time, Eugene Stoner of the Armalite Corporation completed design of the AR-15 which used the smaller, faster, and more lethal .22 caliber bullets as well as a plastic stock; thus, it weighed less. A soldier could carry nearly three times as many rounds as one with the M-14. Extensive Army field tests and analyses led to the conclusion that the AR-15 was superior to the M-14. Nevertheless, full production of the M-14 and its round was ordered.

One individual who did not go along with the "party line" was General Curtis LeMay of the Air Force. In 1962, The U.S. Air Force declared the AR-15 its standard and ordered the rifle.

Another individual now entered the stage: Secretary of Defense NcNamara, a "rational actor" par excellence. Tests in Vietnam conducted by the Advanced Research Projects Agency (not a part of the Army) concluded:

> In overall squad kill potential the AR-15 is up to 5 times as effective as the M-14 rifle. [The AR-15] can be produced with less difficulty, to a higher quality, and at a lower cost than the M-14 rifle. In reliability, durability, ruggedness, performance under adverse circumstances, and ease of maintenance, the AR-15 is a significant improvement over any of the standard weapons including the M-14 rifle. (Fallows, 1981: 84)

The Ordnance establishment countered with its own tests and predictably found the M-14 a "radically" better model (Fallows, 1981: 85). Secretary McNamara overrode the Army and sent more AR-15s to Vietnam, renaming the rifle the M-16. Ordnance shifted ground now and launched development of the Special Purpose Infantry Weapon (SPIW). The Army Staff then proposed a new study, the Small Arms Weapons System Study (SAWS), which "proved" SPIW to be superior to any other rifle. McNamara responded by creating a new Force Planning and Analysis Office (FPAO) which reviewed SAWS and concluded that the study was badly flawed.

Ordnance imposed "modifications" on the M-16s used in Vietnam and replaced Stoner's ammunition. A dramatic rise in the failure rate soon prompted a Congressional inquiry.
Use of multiple perspectives provides us with several important insights: (1) The rational actor's selection of the "best" system hardly reflects the real-life decision process, i.e., the T perspective does not suffice to understand the management of technological change. In this case, the O perspective is central. (2) The Army does not operate as a unitary

decisionmaker. For example, the Infantry Board and Army Staff were on opposite sides of the argument. (3) Individuals play a vital role in the process. For example, civilians in the office of the Secretary of Defense--McNamara, Hitch, and Vance--as well as certain generals--LeMay and Wyman--were instrumental in shifting decisions. (4) The overt and hidden motivations for actions must be clearly distinguished if the decision process is to be understood. Examples: The ordering of the SAWS project by the Army Staff and McNamara's creation of the Force Planning and Analysis Office were indirect means to achieve certain objectives; and (5) The organizational actors may go to remarkable lengths to attain parochial objectives.

A Comment by Borsting

A recent thoughtful discussion of "decisionmaking at the top" by Assistant Secretary of Defense (Comptroller) Jack R. Borsting (1982) also reveals the significance of perspectives other than the technical one. He notes that top-level Department of Defense decisionmaking is

- non-algorithmic, creative and free form (O, P)[2]
- very data-oriented (T)
- basically open-loop, tending to give rudder directions but not to look at the wake (O)
- very short-term oriented (O, P)
- team-work oriented (O)
- very personal: the judgment of one's close advisors is extremely important (P)
- creating a need to budget time for important decisions (O)
- demonstrating that external forces--Congress, the White House--are often more important than a rational systems analysis (O)

[2]Attribution of specific perspectives to Borsting's statements is ours.

- subject to conflicting goals, e.g., between public and private sectors (O)

- dealing with "messes" rather than well-structured problems; they must be treated holistically (T, O, P)

What makes this picture especially interesting is that it is drawn by a past president of the Operations Research Society of America.

APPLICATIONS TO NON-MILITARY TECHNOLOGY ASSESSMENT

In this section we consider in highly condensed fashion applications of the multiple perspective concept to two areas, hydropower development and commercialization of an agricultural product.

Hydropower Development: Private and Public

The first example focuses on a decision in the private sector, the second on a technology assessment in the public sector.

A Utility Decision: Expansion of Hydropower Generating Capacity: In the 1890s, the newly formed Portland General Electric Company erected a power house with three 400 kilowatt generators on the Willamette River in Oregon City. In 1953, it was reconstructed and automated to generate 15 megawatts of power. It has been long known that additional generating capacity is available at the site.

The T Perspective: Since 1958, five major studies have been done (four by an outside consulting engineering firm) to determine the best alternative. The overriding considerations in this technical perspective have been engineering and economics.

The O Perspective: The changed environment of the 1970s as well as internal top management changes led to formation of a Corporate Planning Department and enlargement of Generation Engineering at the expense of the Distribution Department. Their more sophisticated staffs have contributed to the erosion of the prestige and pre-eminent decisionmaking role of the old time units.

The P Perspective: The retirement of the chief executive officer of 30 years and the hiring of his successor from the outside triggered profound changes. New positions have been created, e.g., Vice President for Human Resources, and dynamic younger people brought in from outside to assume these posts.

The P perspective cues the O perspective: New individuals initiate organizational changes. There are new perceptions of the nature of the company and significant changes in the decision process. Specifically, the question of the facility expansion becomes no longer simply an engineering-economics matter. The new Corporate Planning and Human Resources Departments play major roles. A Citizens' Task Force is formed, and eight months of public involvement--most effectively on the part of sports fishermen--results. Community acceptance becomes a decisive criterion in the decision whether or not to proceed.

In sum, the traditional technical perspective could not anticipate the decision--not even the decision process--and vital elements were provided by other perspectives. For more details, see Linstone et al. (1981b, 96-105).

<u>National Hydropower Technology Assessment</u>: In 1981, a technology assessment of three levels of development of U.S. national hydropower was undertaken using multiple perspectives (Linstone et al., 1981c). We confine ourselves here to several examples of quotations from interviews which played a role in clarifying the O and P perspectives. Over 100 interviews were conducted.

> <u>A Cultural Perspective</u>: A Northwest Indian Chief: The Corps has really been largely cooperative. For instance, on 404, that's industrial usage permits, we've been working very closely with them. But to date we have not really been able to get the state to sit down with the tribe. There has been a lot of cooperation for a lot of issues that are still raised because of the threat to industry, and then other local entities get involved. We don't care if there's any kind of development as long as the environments don't lose in exchange. We want to make sure the resources don't lose. Human (industrial) needs come second. First we are interested in the fisheries and the water. Without the water, there just ain't going to be any more humans.
>
> <u>An Entrepreneurial Perspective</u>: (Source: J. McPhee, 1981): You have no lab. You have no inventory. No receivables. No payables. You don't have to have salesmen running around. The utility is your one and only customer. You have a watt-hour meter that tells them what you have generated, and there is no question about their paying their bills. The raw materials are

for free. As the price of oil inflates, the utility will be paying us more, but the creek will still be free. Moreover, there should be quite a bit of satisfaction in producing electricity and producing it clean.

An Institutional Perspective: A Congressional Staffer: The federal government can't do small jobs. It can't work with small interest groups. It can't deal with mayors and city fathers. It is incapable of comprehending what it is they're doing. It is big, aloof, intimidating. Big federal agencies make bad vibes in a small town--they are not institutionally geared to that.

An Agricultural Technology Assessment: Guayule Commercialization

Guayule is a rubber-producing plant suitable to growth in semi-desert areas such as Northeastern Mexico and the Southwestern U.S. The rubber, when processed, is a potential substitute for Southeast Asia's *hevea* rubber, the source of natural rubber since the late 1800s. During World War II, the Emergency Rubber Project planted 32,000 acres of guayule in California. However, from that time to the mid-1970s, only a few hardy researchers, supported by the National Science Foundation, were concerned with guayule.

Interest has recently revived for several reasons: a projected hevea rubber shortfall in the 1990s, dependence on remote and politically unstable areas, projected agricultural water shortages in Southwestern U.S., and Mexican concern with the economy of its Northeastern region (where guayule grows naturally). In 1978, Congress passed a bill authorizing $30 million for guayule research and development (R&D) and creation of a guayule commission. The funds have not been appropriated, however, and the commission has done little.

The University of Arizona's Office of Arid Lands Studies is a recognized center of expertise in the technology of guayule and was chosen by the National Science Foundation to undertake a technology assessment of the commercialization of guayule (Foster et al., 1980). It serves as an excellent example of a T perspective on the subject.

The T Perspective: Guayule uses less water than cotton, corn, and alfalfa, and hence can replace these crops in the water-short Southwest. Extensive research has indicated that a strain with twice the rubber yield of the World War II strain is attainable. Testing has shown the quality of guayule for tires (70 percent of U.S. rubber use) to be entirely adequate. Guayule development can meet 100 percent of the projected natural rubber shortfall by 1991, and no serious adverse side effects from such a policy are envisioned.

The O Perspective: Guayule represents the classic case of a small network of interested and committed partisans with insufficient political and economic resources to fulfill their objectives. It is not primarily a question of deeply vested opposing interests. The proponents had hoped that the Congressional funding would develop a domestic guayule industry up to the point where the private sector would take over production and processing. The old established tire and rubber companies are still waiting and have maintained a conservative posture, relying on defensive research (e.g., experimental planting). This attitude is reinforced by the depressed state of the industry and its principal customer, the auto industry. Inbreeding of industry management may also play a role. The Department of Agriculture is an ancient bureaucracy with little taste for innovation, a greater interest in food than in fiber crops, and an overriding concern with protecting well-established programs from financial erosion. The State of California has pioneered some support for guayule, but this effort is heavily dependent on federal funds.

The key to guayule commercialization is creation of a new coalition at the federal level. As a low priority technology, it must ride piggy-back on some other interests. (In the case of the 1978 Native Latex Commercialization Act, this meant joining forces with Eastern tobacco interests.) The most likely source of such support today comes from the area of national security, just as it did in World War II. At the core of a new coalition would be the Department of Defense (DOD) and the Federal Emergency Management Agency (FEMA). Interest in having control of a vital resource and in building stockpile reserves can provide the motivation for underwriting processing facilities and providing incentives for planters.[3]

Probing of the organizational perspectives also uncovers the possibility of a very different strategy: decentralized development focusing on Indian reservations and small farm organizations. (The Japanese firms once again seem to have taken the lead, contracting with U.S. Indians for the production of jojoba, another commercially useful semi-desert plant.)

---

3 Describing recent developments, Rubber World reported in November 1982: "What has happened with the entrance of the Navy into the guayule field was recommended, in general, by the Linstone (Multiple Perspective) Report."

The P Perspective: There is a large difference between the "advocates" whose fortunes and careers are tied to guayule development and the "judges" who must make implementation decisions about it. The former tend to emphasize the T and P perspectives, the latter the O perspective. The judges see the advocates as self-interested, impatient, unreasonable, and overly optimistic. The advocates see the judges as temporizing, disinterested, devious, and vacillating. There is no mechanism readily available for reconciling opposing perspectives, nor is there much incentive for any third party to intervene to do so--unless a national emergency looms.

Table 2 presents a graphic comparison of insights derived from the three perspectives. It is clear that each perspective sweeps in considerations which are significant for planning and decisionmaking and which differ from those which evolve from the other perspectives.

Table 2--A Comparison of Insights Derived from the Three Perspectives: Guayule

Technical Perspective

1. Tests have shown guayule to be a satisfactory substitute for hevea in automobile and aircraft tires (the primary use of natural rubber).

2. In view of the 300 percent price increase of natural rubber since 1972, guayule is becoming competitive.

3. Guayule development can meet 100 percent of the projected natural rubber shortfall by 1991.

4. Yield per acre is expected to double between 1985 and 2000.

5. Research needs to be done on the use of guayule to retard desertification.

6. The established view on economies of scale is open to question.

Organizational Perspective

7. There is a highly motivated and dedicated nucleus of researchers which, with research grant support, has been keeping guayule alive. There is excellent communication among them (e.g., meetings, newsletters).

8. Research is not the key issue; rather, production start-up raises the question of assumption of financial risk between the tire and rubber companies and the government (federal and state).

9. The Department of Agriculture has not been aggressive. If implementation succeeds it may act as an agent of change within the Department.

10. Inbreeding appears to be a problem in tire and rubber industry management with regard to guayule decisionmaking. Organizational inertia may offer an opening to industry outsiders to jump at guayule commercialization opportunities. The point where control is likely to be exercised is at the processing stage.

11. A number of groups and value systems are not represented in the dialog and decision process (e.g., farmers, Indian tribes, and farm workers).

12. The Guayule Commission could become the lead group but has not led much to date.

13. Mexico has had a long history of interest in guayule (wild natural growth, research institute in Saltillo, pilot processing plant), but motivations are not readily grasped by U.S. interests due to cultural differences, organizational structure differences, and supply/demand differences.

14. National security considerations may act as the decisive catalyst; fear of increased Asian turbulence is spurred by events in Iran and Afghanistan. The Department of Defense and Federal Emergency Management Agency would then become key elements.

Personal Perspective

15. Ed Flynn ("Mr. Guayule Rubber News") is a determined promoter who can be expected to keep the pressure on key actors to move ahead.

16. Representative George Brown (D. California) has been a most effective advocate in the Congress and has been joined more recently by Senator Peter Domenici (R. New Mexico). Texas is lacking a strong Congressional supporter. This suggests potential foci of leverage.

17. Each actor tends to perceive his own prospective behavior and that of his associates quite differently than the others perceive it. Each sees the others' perceptions as distorted. Each therefore interprets a given act differently. This situation discourages formation of intersectoral coalitions and encourages intrasectoral coalitions where perceptions are more similar and mutually comprehensible.

18. None of the present actors with sufficient power to move unilaterally yet sees enough in it for him to have a feasible action strategy. Personnel changes must be watched for indications.

19. There is a crucial difference in perceptions of prospects between those whose careers and fortune are tied to guayule development and those who must make the commercialization decisions (advocates vs. judges).

20. The "top-dead-center" state of guayule commercialization creates an opportunity for entrepreneurial invasion from some adjacent sector. Someone with a unique perspective, technical and political sophistication, and a heretofore excluded constituency (e.g., native American tribes) could precipitate motion along an unexpected trajectory.

APPLICATION TO RISK ANALYSIS

Our initial applications of the multiple perspective concept were in the area of technology assessment. An equally significant field of application appears to be that of risk analysis. Table 3 shows how the three perspectives illuminate different views of risk. The Three Mile Island nuclear accident investigation recognized the inadequacy of the T perspective in risk analysis:

> A truly unexpected result came out of the Kemeny Commission's study of Three Mile Island. A group that set out to investigate a technology ended up talking about people. In the Commission's own words, "It became clear that the fundamental problems were people-related problems." (The Washington Post, October 31, 1979)

While the T perspective focuses on probabilistic calculations and safety design, the O perspective deals with standard operating procedures and avoidance of responsibility, the P perspective with perceptions--personal fears and images of horror. Not surprisingly, there is a dramatic difference between actuarial and personally perceived risk rankings.[4] According to a recent survey, the typical individual views the risk of auto accidents as equal to those of nuclear power, while the actual annual mortality rate of the first is over 500 times that of the second.

From the T perspective, the "distortions" suffered by the P perspective, i.e., reasons for disagreement with the statistical probabilities,[5] can be explained as follows:

(1) Causes of Overestimation of Risk:

- The layman sees little difference between, say 1 percent and 0.001 percent; thus, very low frequency events (e.g., tornado, homicide) are overstressed.
- The media extensively cover dramatic events, even if their likelihood is minute; the image remaining in the mind and in the memory magnify the probability estimate.
- Technological steps taken to reduce a recently publicized risk are ignored.
- Events having some similarity are equated in terms of risk even if the risks are very different (e.g., nuclear weapons and nuclear power stations are assumed to be equally dangerous).

---

4 See particularly the work of Paul Slovic, Baruch Fischhoff, and Sarah Lichtenstein; the new literature review of Covello (1983) is also helpful.

5 Our wording here is meant to reflect the typical T perspective view of the individual's "ignorance" of the "true," i.e., probabilistically derived, risk.

Table 3--Physical Risk as Seen From the T, O, and P Perspectives

T	O	P
Probabilistic	Threat to perpetuation (e.g., of nation, family line)	Type of danger (cancer as the "plague" of this century)
Actuarial	Threat to product line, company image	Time for consequences to materialize (discounting)
Expected value calculations	Ability to avoid publicity	Popular image of horror (nuclear accident)
Fail-safe concept	Political sensitivity (voter anger)	Personal experience (e.g., survival of a hotel fire)
Margin of safety design	Ease of litigation	Age of individual
Applicability of precedent designs	Ease of shifting or spreading blame	Peer esteem (e.g., teenagers and drugs, smoking)
	Anonymity of decision makers	Popularization of risk (e.g., nuclear plants and The China Syndrome)
	Ease of changing decisions	Ethical overlaps (e.g., religion and gene splicing)
	Socioeconomic status as determinant of risk acceptance (life "guaranteed safe" in U.S.)	Archetypal values (e.g., virility and guns, fertility and water)
	Cost of recall	
	Standard Operating Procedure	
	Access to expertise on specific risk	

(2) Causes of Underestimation of Risk:

- High frequency events may not be given much media coverage (e.g., strokes, asthma).[6]

- Low frequency is interpreted as low danger and ignored even when consequences are catastrophic (e.g., reluctance to buy flood or earthquake insurance for residents of floodplain or earthquake regions).

- Long time lag between the event occurrence and the impact (e.g., smoking and lung cancer).

- Misunderstanding of probability, i.e., if a low likelihood event has just occurred, it cannot occur again for a very long time.[7]

- The individual has personal immunity to risks perceived to be under personal control, i.e., voluntary.

Thompson (1982) recognizes the significant contribution of the P perspective in viewing the poor within the framework of risk analysis:

> Oscar Wilde once said: "If only the poor had profiles, there would be no difficulty in solving the problem of poverty." (He meant that) the trouble was that "the poor" could only be perceived by the problem-solvers as an anonymous mass. If only they could discern the features of

---

6 Strokes kill 85 percent more people than do accidents; yet people estimate that accidents take 25 times as many lives as strokes. Asthma kills 20 times as many people as do tornadoes; yet people estimate that tornadoes kill about three times as many people as does asthma (Slovic et al., 1981).

7 Another example of the fallacy: in tossing coins, a strong of five heads in a row means that the chance of a tail on the sixth throw must be greater than 0.5.

> the poor--perceive them in all their individual diversity--the problem would be solved. The irritating thing about Wilde is not that he is so facetious and so witty . . . but that he is usually right as well. (Thompson, 1982: 28)

Thompson not only recognizes the relationship between the social context and risk selection, i.e., the O perspective, but stresses that risk acceptance, risk aversion, and risk absorption are responses to different risk environments which an individual constructs for himself, i.e., a P perspective (Thompson, 1982: 44).

Brim et al. (1962) show that risk perception is at least partially dependent on over 40 personality factors, including sex, economic class, need for achievement, prior experience, obedience, and anxiety. Other studies have indicated highly individualistic responses to semantics and setting, to the information giver, and even to the time of day.

Tversky and Kahneman (1974) cite many of the biases which inexorably dominate individual perceptions of risk and completely ignore the scientist's model of risk = (probability) x (consequence). As Slovic has pointed out, the public understands consequences, but does not understand probabilities the way a scientist does. And this problem in turn affects the O perspective. Meltsner (1977) in his study of California earthquake risk asks: "If the earthquake hazard is not to be believed on an individual basis, why should the hazard become any more credible on a community basis?"

The focus on the distinction between personal, societal, and technical <u>perceptions</u> of risk meshes very well with the multiple perspective approach. For example, we recall that different time horizons are associated with each perspective (see Table 1). It is not surprising to find that risks with distant (long-term) effects, e.g., smoking, are discounted by many individuals in comparison with risks that have immediate impact.

Following are further examples from studies in physical risk acceptance:

> Otway et al. (1975) observe that risk perception is important because the response to risk depends upon how situations are perceived (p 16).
>
> Velimirovic (1975) discusses risk perception in traditional societies and compares risk behavior with modern societies, i.e., an anthropological view.

Fischhoff (1980) concludes that people's risk perceptions are colored by their own prejudices and even the way risks are presented.

Fischhoff, Lichtenstein, Slovic, Derby, and Keeny (1981):

> There are no value-free methods for choosing the most acceptable (risk) option . . . . The expertise needed for (such) decisions is dispersed throughout society . . . . If each new perspective has some unique contribution, we may want to lend an ear to parties not often heard in policy-making circles--the poor, the philosophers, the artists--in hopes that their life experiences have something illuminating to offer. (pp 135-137)[8]

Turner (1982):

> While we cannot throw away our reliance upon systematic, rational thought . . . we need to recognize that many human responses cannot be solely considered in terms of an applied rationality, and that matters of organizational responses to hazard cannot be treated solely as isolated technical problems . . . . Organizational responses . . . are interpenetrated by the inter-organizational political economy of our society, and by our own cultural assumptions and attitudes towards risk and hazard. (p 79)

It is apparent that risk analysis has <u>de facto</u> plunged deeply into multiple perspectives in recent years, and its success in terms of insights gained should give encouragement to analysts elsewhere who are considering the use of multiple perspectives. The revelations drawn from this research may, in fact, prove valuable in their own areas of application.

---

8 Of the seven criteria these authors present for evaluating the suitability of approaches to acceptable risk (53-59), six are T or O focused: T - <u>comprehensive</u>, <u>logically sound</u>, and <u>open to evaluation</u>; O - <u>practical</u>, <u>politically acceptable</u>, and <u>compatible with institutions</u>. The seventh criterion, <u>conducive to learning</u>, may be appropriate for T, O, and P.

ADDITIONAL COMMENTS AND GUIDELINES FOR THE USER

The two preceding sections suggest the kinds of complementary insights obtainable with each perspective. Table 4 lists several other cases included in Linstone et al. (1981b). We conclude with some comments on the role of multiple perspectives in forecasting and planning and a summary of user guidelines.

Comments on Forecasting and Planning

It is a common practice to divide technological forecasting into two complementary types: (1) exploratory or capability forecasts, and (2) normative or needs forecasts (Martino, 1972: 287). The former focus on what can be done in the future, based on the past and present technological capability; the latter on what ought to be done in the future. Exploratory forecasts (e.g., extrapolations of technology trends and analytic models) have a dominant T orientation, whereas needs analyses usually reflect the O perspective of the client organization or the P perspective of the forecaster more strongly (Linstone, 1969).

From 1959 to 1970, Linstone directed four corporate planning "needs analyses" in the area of national defense and space programs. By the time of the fourth of these privately funded projects, our T perspective analysts had made an important discovery: "The gap between what is needed and what is marketable means that a 'needs analysis' is, in fact, a mirage" (Project MIRAGE 85, 1970: 116). Our "need" is determined as much by what we like and are comfortable with as by the purely technical requirements. The case of the USS Wampanoag already made this point apparent.

The familiar adage, "The military is always planning to fight the last war over again with better weapons," underscores the centrality of the O perspective. The construction of the Maginot Line in France in the early 1930s assumed that a future war with Germany would be a war of firepower like World War I, rather than a war of mobility. Yet the technology of tanks and aircraft was clearly advancing at the time. Nor were the slowly changing concepts and standard operating procedures confined to the French military planners. The German High Command also had little faith in highly mobile warfare and tended to see the future through a rear-view mirror. The Blitzkrieg of 1940 was the product of the planning of Hitler and three generals who were outsiders to the vaunted General Staff: von Manstein, von Rundstedt, and Guderian (Rowe, 1959:

Table 4--Further Examples of Insights Derived from T, O, and P Perspectives

(For details, see Linstone et al, 1981b)

	T Technical	O Organizational	P Personal
TECHNOLOGY ASSESSMENT			
Electronic Funds Transfer	Impacts on organizations extensively analyzed (ADL TA). Uncertainties indicated (ADL TA). Impacts of other technologies (telecommunications and computers) on EFT underestimated.	Legislative and administrative recommendations (Commission). Report bought time and public discussion (Commission). New turf arrangements likely; cartels may form around EFT.	McCoy and Fisher have been the driving force in making Banc One a leader in financial services industry (credit card rpocessing, automatic tellers, in-home banking. "(TA's) don't have much value to us."
On-site Solid Waste Treatment	Possible mutual reinforcing effect between reruralization and solid waste treatment systems.	Engineering schools and federal policies favor centralized systems; absence of strong vested interest support (even environmentalists have higher priorities); potential impact of federal sewer funding cut-off; dependence of take-off on real estate developers and home builders.	Deeply ingrained "flush and that's the end of it" attitude; homeowners dislike maintenance; options such as Clivus Multrum attract con men.
HEALTH CARE SYSTEMS			
Nursing Education	Goals: recognition of nursing school as academically credible higher education center; program efficiency and effectiveness.	Change of organizational vision to "community of scholars"; promotion depends on research; PhD credentials.	Struggle of old-timers to survive; political game of department chairpersons; elitism; morale problems for losers.
Perinatal Regionalization	Plan to decrease mortality and morbidity based on levels of risk and care; upgrading of level of practice in hospital; evaluation of each stage. Plan also reduces need for costly duplication of services.	Recognition of opportunities inherent in overlapping interests; maintain relationships with hospital chiefs and administrators; assist public hospital in compensating for decreased resources and negative image.	Development of image of regionalization as personally useful; simultaneously noble and pragmatic; sensitive to Black professionals and their unique, controversial role in community

62, 127). Hitler's charisma and personal power meant that a maverick P perspective overcame an entrenched O perspective.

A priority needs list based solely on a T perspective looks very different from one based on, say, T and O perspectives. The T perspective takes account of King Richard III's lament, "For the want of a nail the shoe was lost . . . ," so that the list would include unglamorous items such as changes in training and maintenance procedures or in communications equipment. However, a list based on both T and O features more prestigious items in the firepower and vehicle areas--glamorous aircraft, ships, and weaponry.

In the nonmilitary area, Wildavsky and Tenenbaum (1981) have shown that the deep division between industrial interests and conservationists on oil and gas resources was already apparent in the early 1900s. Each side seized on the 1908 U.S. Geological Survey (USGS) estimates to confirm its policy stand.[9]

Many forecasts have been made since then. Except for the World War I and II periods, each faction has habitually accused the other of manipulating these forecasts for its own selfish purposes. Table 5 suggests the different views on resource forecasts and the dialectic approach often seen in the O perspective. It becomes clear that the forecasts are the servants of policies already determined or preferred rather than being prerequisites for policy formulation. The T perspective quests for more accurate forecasts in this area are thus rather irrelevant.

In cases where a forecast is done by organization A for client B and few constraints are imposed by B, the biases of A may be decisive for the forecast. If A is experienced in a certain forecasting technique, it is likely to prefer its use in responding to B--whether or not it is most suitable. A may also be subconsciously influenced by the possibility of future grants or contracts from B--and avoid forecasts which may alarm B.

The technological background of the forecaster introduces another bias: It nearly always moves him toward a greater technological sophistication in addressing future needs. It is rare to find other options of satisfying needs given equal consideration.

---

9 At that time, USGS forecasted total U.S. oil resources between 10 and 24.5 billion barrels and indicated that we would run out of oil between 1935 and 1943. In 1974, USGS (Vincent McKelvey) estimated a range between 200 and 400 billion barrels.

Table 5--An O Perspective on Oil Reserve Forecasts

	When prices are high	When prices are low
industrialists favor:	high forecasts "Major new supplies can be found if prices are high."	low forecasts "Higher prices are needed to bring on more supplies."
consumers favor:	low forecasts "Oil is no longer the solution."	high forecasts "No need to raise prices."
conservationists favor:	low forecasts "High prices encourage overproduction."	low forecasts "Low prices encourage overconsumption."

(Source: Wildavsky and Tenenbaum, 1981: 300)

In his careful analysis of the accuracy of forecasts in the population, economic, energy, transportation, and technology areas, Ascher (1978) drew the clear conclusion that the core assumptions underlying a forecast constitute the major determinant of forecast accuracy.

> When the core assumptions are valid, the choice of methodology is either secondary or obvious. When the core assumptions fail to capture the reality of the future context, other factors such as methodology generally make little difference; they cannot "save" the forecast. (Ascher, 1978: 199)

It is our belief that the O and P perspectives are vital in drawing forth the core assumptions. The training of the forecaster, the experience of the forecasting group (e.g., its standard operating procedures), and the tradition and morale needs of the client organization may all play a role in determining the core assumptions.

O and P perspectives may insinuate themselves covertly into exploratory as well as normative forecasts. The analyst's tool kit (e.g., data-based trend extrapolation and growth models) does not force his attention on assumptions and professional biases. These may be based on years of personal involvement in a narrow aspect of technology and communication with peers who think of the future in terms of the same descriptors of capability (see Martino, 1972: 566-571). The effect of multiple perspectives in forecasting should thus prove quite beneficial--at the very least they bring to the surface vital core assumptions and minimize self-delusion.

Strategic planning is another important area where the potential of the multiple perspective concept is apparent. In the 1960s and 1970s, many organizations created planning units, variously termed "long-range planning," "strategic planning," "business planning," etc. (Halal, 1980). There are many models of the planning cycle (e.g., 4, 7, 9, or 12 steps), and much has been written about various analytical aspects, such as forecasting techniques, top down and retrogressive planning, cybernetic modeling of the organization, and environmental scanning. These reflect the T perspective of the analyst. The differential planning horizons associated with the three types of perspective clarify one dilemma in the planning process. Thinking of the perspectives as beams of light directed at the problem, we find that each beam has a different illuminating distance (like high and low headlights in an automobile, Fig. 4). Is it a wonder that the analyst's T perspective dominates in long-range planning and that the effort often proves frustrating?

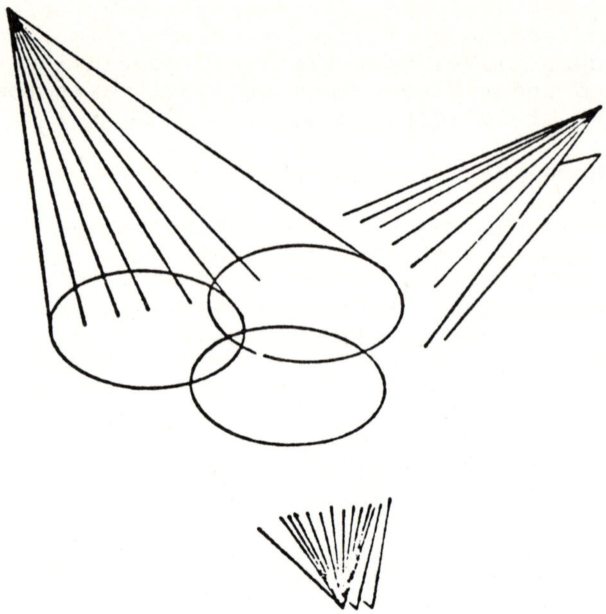

Figure 4: Different Perspectives--Different Planning Horizons

It is increasingly evident that many American and West European corporations have suffered severely from this handicap--one only needs to point to the U.S. shipbuilding and steel sectors (see also Hayes and Abernathy, 1980).

In Japan, the government long ago recognized the limitations of industrial long-range planning. It created the Ministry of International Trade and Industry (MITI) before World War II to formulate and facilitate implementation of industrial policy. Thus, a focal point for long-range planning is provided, and it becomes action-directed rather than ritualistic. The dominant perspectives are two: (1) the technical and (2) the national as organizational perspective. Such planning subordinates any individual corporation's O perspective to that of the country, a situation often characterized as "Japan, Inc."

Guidelines

We complete this discussion with a brief selection of guidelines derived from the experience with multiple perspectives to date. For more details, see Linstone et al. (1981b). (1) Consider using multiple perspectives if a major decision will be required about a problem or system which involves a mixture of technical, social, and human aspects, i.e., one that is not purely in the realm of technology; (2) Recognize that you must draw in talent for O and P that is quite different from that which is satisfactory for T. The management scientist, economist, or engineer only rarely fills this need; (3) Don't confuse <u>what</u> you are looking at with <u>how</u> you are looking at it; (4) "Live" with the issues and get a "feel" for them, i.e., immersion is crucial for O and P; (5) Strive for a good balance among perspectives; (6) Remember that there are nearly always several O and P perspectives--choices will be necessary to limit the number; (7) If you have a team, consider doing the perspectives in parallel; (8) Recognize the inevitable political sensitivity of the O and P perspectives; (9) Fit the mode of communication to the perspective; and (10) If at all possible, leave the integration of the perspectives to the decisionmaker. However, prototypes of the integration process, including cross-cuing of perspectives, may be helpful to the client.

REFERENCES

Allison, G. (1969) "Conceptual Models of the Cuban Missile Crisis," The American Political Science Review, 63 (3), September, 689-718.

Allison, G. (1971) Essence of Decision: Explaining the Cuban Missile Crisis. Boston: Little, Brown, and Company.

Andersen, D.F. (1977) Mathematical Models and Decisionmaking in Bureaucracies: A Case Story Told from Three Points of View. MIT: PhD Dissertation (August).

Ascher, W. (1978) Forecasting: An Appraisal for Policymakers and Planners. Baltimore, MD: The Johns Hopkins University Press.

Borsting, J.R. (1982) "Decisionmaking at the Top," Management Science. 28 (4), 341-351.

Brim, O. Jr., David C. Glass, David E. Lavin, and N. Goodman (1962) Personality and Decision Processes: Studies in the Social Psychology of Thinking. Stanford, CA: Stanford University Press.

Covello, V. (1983) "The Perception of Technological Risks: An Overview," Technological Forecasting and Social Change. 23 (4).

Fallows, J. (1981) National Defense. New York: Random House.

Fischhoff, B. (1980) quoted in "Risk Assessment's Role in Regulation Debated," Chemical and Engineering News. (Oct. 20), 30.

Fischhoff, B., S. Lichtenstein, P. Slovic, S.L. Derby, and R.L. Keeney (1981) Acceptable Risk. Cambridge: Cambridge University Press.

Foster, K., et al. (1980) A Sociotechnical Survey of Guayule Rubber Commercialization. Tucson, AZ: University of Arizona, Office of Arid Lands Studies; Kansas City, MO: Midwest Research Inst. (April).

Halal, W.C. (1980) Strategic Planning in Major U.S. Corporations. Study prepared for the General Motors Corporation at George Washington University, Washington, D.C.

Hayes, R.H. and W.J. Abernathy (1980) "Managing Our Way to Economic Decline," Harvard Business Review. 58 (4) (July-August), 67-77.

Hoos, I. (1979) "Societal Aspects of Technology Assessment," Technological Forecasting and Social Change. 13 (3), 191-202.

Jay, A. (1968) Management and Machiavelli. New York: Holt, Rinehart, and Winston.

Linstone, H.A. (1969) "When is a Need a Need?" Technological Forecasting, 1 (1), 55-71.

Linstone, H.A. et al. (1981a) "The Multiple Perspective Concept," Technological Forecasting and Social Change, 20 (4), 275-325.

Linstone, H.A., et al. (1981b) The Multiple Perspective Concept, Report 81-1, Portland State University, Portland, OR: Futures Research Institute.

Linstone, H.A., et al. (1981c) Technology Assessment of National Hydropower Development. 1. Fort Belvoir: Institute for Water Resources, U.S. Army Corps of Engineers, by Nero and Associates, Inc.

Magaziner, I.C. and R.B. Reich (1982) Minding America's Business: The Decline and Rise of the American Economy. New York: Harcourt Brace Janovich.

Martino, J.P. (1972) Technological Forecasting for Decisionmaking. New York: American Elsevier Publishing Co.

McNaugher, T.L. (1980) "Marksmanship, McNamara, and the M-16 Rifle: Innovation in Military Organizations," Public Policy. 28 (Winter), 1.

McPhee, J. (1981) "Minihydro," The New Yorker. LVII (1),. 44-87.

Meltsner, A.J. (1977) Seismic Safety of Existing Buildings and Incentives for Hazard Mitigation in San Francisco: An Exploratory Study (Report to NSF). UCB/EERC-77/28. Berkeley, CA: Earthquake Engineering Research Center, College of Engineering, University of California (December).

Morison, E.E. (1966) Men, Machines, and Modern Times. Cambridge, MA: The MIT Press.

Nehnevajsa, S. and J. Menkes (1981) "Technology Assessment and Risk Analysis," Technological Forecasting and Social Change. 19 (3), 245-256.

Otway, H.J., P.D. Pahner, and J. Linnerooth (1975) "Societal Values in Risk Acceptance," presented at American Institute of Chemical Engineers, Boston, Sept. 1975 (also IIASA Report RM-75-54).

Project MIRAGE 85 (1970) III, DPR-87. Burbank, CA: Corporate Development Planning Dept., Lockheed Aircraft Corp. April.

Rowe, V. (1959) The Great Wall of France. London: Putnam Press.

Schwebs, D. and P. Sprey (1981) (Formerly OASD/Systems Analysis). Oral communication.

Slovic, P., B. Fischhoff, and S. Lichtenstein (1981) Facts and Fears: Understanding Perceived Risk. Eugene, OR: Decision Research Inc. (manuscript).

Steinbruner, J.D. (1968) The Mind and the Milieu of Policymakers: A Case History of the MLF, PhD Dissertation, MIT.

Thompson, M. (1982) "The Cultural Approach to Risk: The Case of Poverty," in H. Kunreuther (Ed.), Risk: A Seminar Series. Laxenburg, Austria: International Institute for Applied Systems Analysis.

Turner, B.A. (1982) "Organizational Responses to Hazard," in H. Kunreuther (Ed.), Risk: A Seminar Series. Laxenburg, Austria: International Institute for Applied Systems Analysis.

Tverksy, A. and K. Kahneman (1974) "Judgment Under Uncertainty: Heuristics and Biases," Science. 185. (4157) (September 27), 1124-1131.

Velimirovic, H. (1975) An Anthropological View of Risk Phenomena, IIASA Report RM-75-55 (Nov).

Ways, M. (1967) "The Road to 1977," Fortune. 75. (1) (January), 93-95, 194-195.

Wildavsky, A. and Ellen Tenenbaum (1981) The Politics of Mistrust. Beverly Hills, CA: Sage Publications.

The Delphi Technique

Harold A. Linstone

Excerpted from: Handbook of Futures Research, edited by Jib Fowles; Greenwood Press, Westport, CT, and London, England.

The word Delphi refers to the hallowed site of the most revered oracle in ancient Greece. Legend has it that the earth goddess Gaia long ago inhabited this site and was protected by the dragon Pythos. Apollo, the son of Zeus and Leto, slew the dragon and made himself master of Delphi. He was famous throughout Greece not only for his beauty, but also for his ability to foresee the future. The home Apollo chose for himself was located 170 kilometers northwest of Athens on the slopes of Mount Parnassus and in sight of the Gulf of Corinth. The impressive site served not only as an oracular center but also as a kind of art museum. Every important event of Greek history was immortalized there by painting and sculpture. Those who sought to consult the Oracle brought gifts, thereby making Delphi one of the richest and most influential locales in Greece. Forecasts were transmitted from Apollo through intermediaries, women known as Pythia. The focal point for the activity was an inner sanctum of the temple built over a geologic fault through which fumes escaped. These fumes were instrumental in transporting the Pythia into a trance, and their utterances were then interpreted by priests for the petitioners.

In the 1950s, the RAND Corporation adopted the name Delphi for a procedure to "obtain the most reliable consensus of opinion of a group of experts...by a series of intensive questionnaires interspersed with controlled opinion feedback" (Dalkey and Helmer, 1963, p. 458). Of particular interest in the initial endeavor was the application of "expert opinion to the selection, from the point of view of a Soviet strategic planner, of an optimal U.S. industrial target system and to the estimation of the number of A-bombs required to reduce the munitions output by a prescribed amount."

For obvious reasons, this early pioneering work was not publicized. The first widely circulated Delphi study appeared in 1964 as a RAND paper, "Report on a Long-Range Forecasting

NATO ASI Series, Vol. G4
Environmental Impact Assessment, Technology Assessment, and Risk Analysis. Edited by V.T. Covello et al.

Study" (Gordon and Helmer, 1964). Its aim was to assess "the direction of long-range trends, with special emphasis on science and technology, and their probable effects on our society and our world." "Long-range" was defined as a span of 10 to 50 years. Six topics were covered: scientific breakthroughs, population control, automation, space progress, war prevention, and weapons systems. Of 150 persons approached, 82 responded to one or more questionnaires. A panel was selected for each topic, and four sequential questionnaires were distributed to each panel, spaced about two months apart. The average number of completed questionnaires received per panel per round was 14.5. The questionnaires for the first two rounds of the Scientific Breakthrough panel are shown in Fig. 1. The estimates of the year by which there would be a 50 percent chance that a particular development would occur were tallied, and the median year and the interquartile range (25 to 75 percent year span) were calculated. The results of the Scientific Breathrough panel were presented as shown in Fig. 2. The items are ordered according to median year, the first item denoting the earliest median year.

In the mid-1960s, there was a growing interest in technological forecasting, particularly on the part of the advanced technology industry of the United States. The Delphi procedure, as developed by the RAND Corporation, appeared to be a useful device and was soon applied by planners in corporations. TRW's "Probe of the Future" involved 150 of their scientists, engineers, and managers in a Delphi study which was later published in 14 volumes. From America, Delphi spread to Western Europe, Eastern Europe, and the Orient. The largest Delphi ever undertaken was a Japanese project involving several thousand people. Starting in a nonprofit organization, Delphi has found its way into industry, government, and finally, academe. It has simultaneously expanded beyond technological forecasting. The definition used by Linstone and Turoff in their book on the subject (1975) is as follows: "Delphi may be characterized as a method for structuring a group communication process so that the process is effective in allowing a group of individuals, as a whole, to deal with a complex problem."

In the original Delphi process, the key elements were (1) structuring of information flow, (2) feedback to the participants, and (3) anonymity for the participants. Clearly, these characteristics may offer distinct advantages over the conventional face-to-face conference as a communication tool.

LONG-RANGE FORECASTING STUDY

Questionnaire 1

1.1. SCIENTIFIC BREAKTHROUGHS

One of the major problems of conducting a predictive study which poses its questions on the basis of extrapolations of current technology is the almost unavoidable exclusion of discontinuous state-of-the-art advances.

In this current study a period of 50 years is being considered. It is possible that inventions and discoveries not yet visualized could have a major impact on our society during this interval. It is easy to observe that the pace of scientific and technological innovation has been steadily increasing and that the time between origination and application has been decreasing. Therefore we believe that many generations of inventions can find application during the period under study.

Some insight even into discontinuous state-of-the-art advances might perhaps be gained by examining the world's need for such advances, in view of the old truism that necessity is the mother of invention. Therefore, you are asked to list below major inventions and scientific breakthroughs in areas of special concern to you which you regard as both urgently needed and feasible within the next 50 years:

Do you know of the existence of any information, in the form of tabulations or analyses, that might be particularly valuable in reaching projections of the kind requested?

Questionnaire 2

1.2 SCIENTIFIC BREAKTHROUGHS

a. Listed below in Table 1.2a are most of the scientific breakthroughs suggested by the respondents as potentially possible during the next 50 years. Please indicate your judgment of the probability of implementation during each period. Note that the numbers inserted by you in each row should add up to 100. (In the case of items involving gradual development such as synthetic food production or automated education, "implementation" should be interpreted as referring to the time from which the effect on our society will no longer be negligible.)

b. Considering the breakthroughs suggested in Table 1.2a are there other potential breakthroughs which you would care to add? When do you believe they will occur? Please make your additions in Table 1.2b.

TABLE 1.2a

	1963-65	65-68	68-72	72-78	78-86	86-97	97-2013	Later	Not at any time
Sociological									
1. Communication with animals									
2. Breeding of intelligent animals (apes, cetaceans, etc.) for low-grade labor									
3. Education by automation									
4. Education by other means, such as direct information-recording on the brain									
5. Education or conditioning in social behavior to reduce the likelihood of war									
6. Automatic language translators									
7. Efficient idea-coding to convey precise information independent of language									
8. Popular use of personality control drugs									
9. Long-duration coma to permit a form of time travel									
10. Solution to the problem of distribution of goods--computer identification of points of need									
11. Computing machines becoming the most significant source of intelligence on earth									
12. Discovery of life on Mars									
13. Communication with extra-terrestrials									
Physical									
1. Reformation of physical theory, eliminating confusion in quantum-relativity and simplifying particle theory									
2. Experimentation with anti-matter									
3. Control of gravity									
4. Controlled thermo-nuclear power									
5. Commercially efficient transmutation of elements									
6. Focused electromagnetic radiation for power transmission									

Table 1.2a

SUMMARY OF OPINIONS ON SCIENTIFIC BREAKTHROUGHS

	Probability of Implementation During Period								
	1963-65	65-68	68-72	72-78	78-86	86-97	97-2013	Later	Not at any time
Biological									
1. Chemical control over heredity - molecular biology									
2. Biochemical general immunization									
3. Biochemicals to stimulate growth of new organs and limbs									
4. Synthetic generation of protein for food									
5. Oral contraceptive									
6. Other means of fertility control									
7. New organs through transplanting or prosthesis									
8. Use of telepathy and ESP in communications									
9. Understanding of the physiology of mind-brain behavior									
10. Chemical control of the aging process, permitting extension of life span by 50 years									
11. Cancer cure									
12. Man-machine symbiosis, permitting man to extend his intelligence directly through the use of computing machines									
13. Creation of artificial life									

7. Relay of solar energy via satellite									
8. Efficient electric storage device									
9. Limited weather control									
10. Reliable weather forecasts									
11. Miniaturization of electronics carried to the molecular level									
12. Automated highways									
13. Measurement of curvature of the universe									
14. Ballistic transport - 2 hours to anywhere on earth									
15. Theory of the earth's crust permitting accurate earthquake prediction									
16. Development of new synthetic materials for ultra-light construction									
17. Operation of nuclear power systems providing electricity @ 3-4 mills/kw-hr 5-10 year refueling period									
18. Collection and concentration of solar energy, used for power or in man-made organic chemistry manufacturing processes									
19. Operation of a central data storage facility with wide access for general or specialized information retrieval									
Food and Raw Materials									
1. Rise in world agricultural gross yields by a factor of ten									
2. Economically useful desalination of sea water									
3. Economical working of low-grade metal ores									
4. Exploitation of the ocean bottom through farming and mining									

Figure 1. RAND Delphi Questionnaires, Rounds 1 and 2

SOURCE: RAND Paper P-2982. Reprinted by permission of the RAND Corporation.

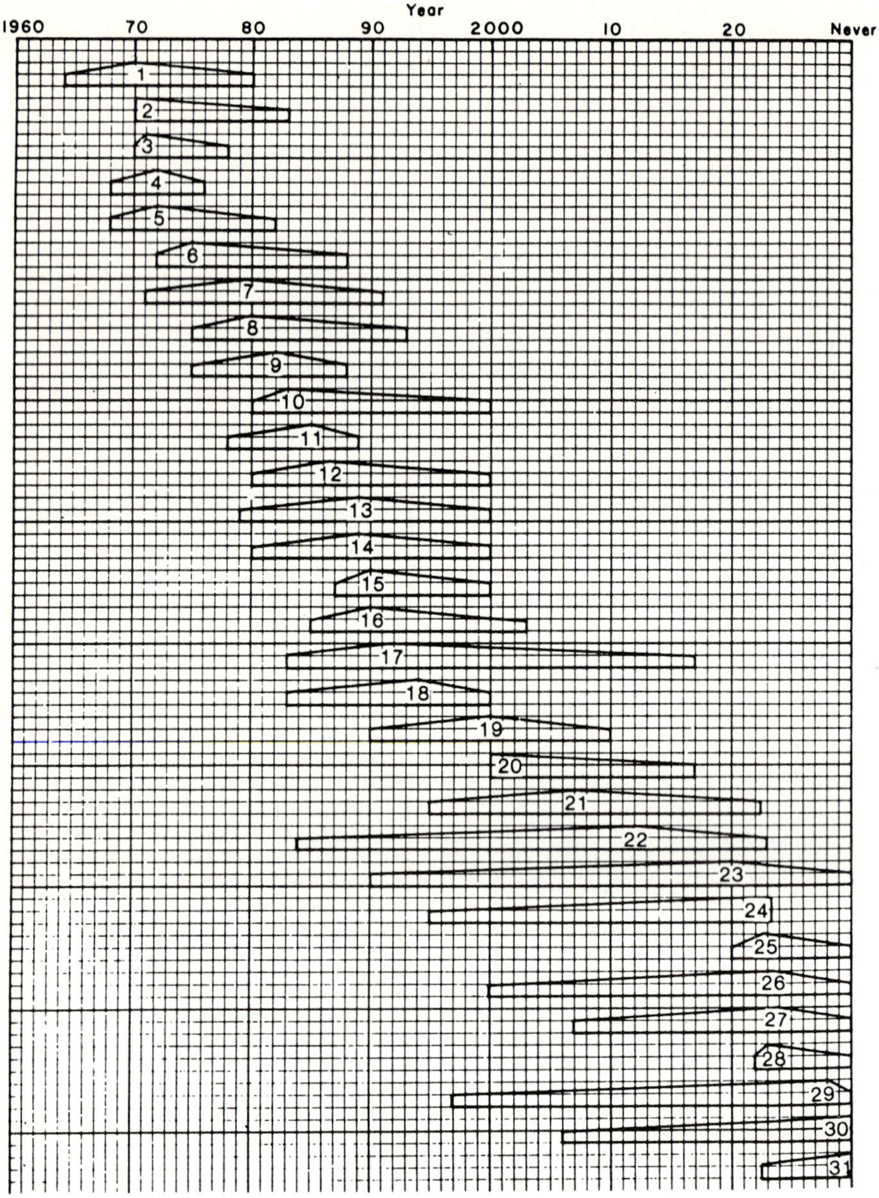

Figure 2. Consensus of Panel 1 on Scientific Breakthroughs (medians and quartiles)

SOURCE: RAND Paper P-2982. Reprinted by permission of the RAND Corporation.

The Delphi Technique

1. Economically useful desalination of sea water

2. Effective fertility control by oral contraceptive or other simple and inexpensive means

3. Development of new synthetic materials for ultra-light construction

4. Automated language translators

5. New organs through transplanting or prosthesis

6. Reliable weather forecasts

7. Operation of a central data storage facility with wide access for general or specialized information retrieval

8. Reformation of physical theory, eliminating confusion in quantum-relativity and simplifying particle theory

9. Implanted artificial organs made of plastic and electronic components

10. Widespread and socially widely accepted use of nonnarcotic drugs (other than alcohol) for the purpose of producing specific changes in personality characteristics

11. Stimulated emission ("lasers") in X and Gamma ray region of the spectrum

12. Controlled thermo-nuclear power

13. Creation of a primitive form of artificial life (at least in the form of self-replicating molecules)

14. Economically useful exploitation of the ocean bottom through mining (other than off-shore oil drilling)

15. Feasibility of limited weather control, in the sense of substantially affecting regional weather at acceptable cost

16. Economic feasibility of commercial generation of synthetic protein for food

17. Increase by an order of magnitude in the relative number of psychotic cases amenable to physical or chemical therapy

18. Biochemical general immunization against bacterial and viral diseases

19. Feasibility (not necessarily acceptance) of chemical control over some hereditary defects by modification of genes through molecular engineering

20. Economically useful exploitation of the ocean through farming, with the effect of producing at least 20% of the world's food

21. Biochemicals to stimulate growth of new organs and limbs

22. Feasibility of using drugs to raise the level of intelligence (other than as dietary supplements and not in the sense of just temporarily raising the level of apperception)

23. Man-machine symbiosis, enabling man to extend his intelligence by direct electromechanical interaction between his brain and a computing machine

24. Chemical control of the aging process, permitting extension of life span by 50 years

25. Breeding of intelligent animals (apes, cetaceans, etc) for low-grade labor

26. Two-way communication with extra-terrestrials

27. Economic feasibility of commercial manufacture of many chemical elements from subatomic building blocks

28. Control of gravity through some form of modification of the gravitational field

29. Feasibility of education by direct information recording on the brain

30. Long-duration coma to permit a form of time travel

31. Use of telepathy and ESP in communications

The method involved the following 10 steps:

1. Formation of a team to undertake and monitor a Delphi on a given subject

2. Selection of one or more panels to participate in the exercise. (Customarily, the panelists are experts in the area to be investigated.)

3. Development of the first round Delphi questionnaire

4. Testing of the questionnaire for proper wording (e.g., ambiguities, vagueness)

5. Transmission of the first questionnaires to the panelists

6. Analysis of the first round responses

7. Preparation of the second round questionnaires (and possible testing)

8. Transmission of the second round questionnaires to the panelists

9. Analysis of second round responses (Steps 7 to 9 are reiterated as long as desired or as necessary to achieve stability in the results.)

10. Preparation of a report by the analysis team to present the conclusions of the exercise

It is not surprising that, as the use of Delphi has spread, many variations of the process have surfaced. One example is the use of the computer to run Delphi as a real-time remote conference. Another is the use of different panels in sequential form, e.g., planning Delphi in two phases: First, several rounds are carried out with research specialists, and then the output of this exercise is presented to development specialists as the starting point for a second sequence.

When we view Delphi as a communication process, the wide spectrum of uses quickly becomes apparent. The process has been applied to exposing priorities of personal values and social goals, explicating the pros and cons associated with potential policy options, evaluating budget allocations, examining the significance of historical events, and distinguishing or clarifying perceived and real human motivations. As with any readily applied methodology, there have been misuses as well as efforts to oversell Delphi. In some respects, Delphi is a method of last resort; it proves particularly useful in the following circumstances:

- The problem does not lend itself to precise analytical techniques but can benefit from subjective judgments on a collective basis.

- The individuals who need to interact cannot be brought together in a face-to-face exchange because of time or cost constraints. Further, a conventional conference tends to be dominated by particularly strong personalities or to give rise to an undesirable bandwagon effect.

A structured communication process has distinct advantages but also limitations, as we shall see. Perhaps the most important advice to be given to the potential user is to suit the method to the problem, not the problem to the method. The second advice is that ease of understanding is no excuse for sloppiness of execution. In the following section, we will discuss some applications and present some comments on the evaluation and analysis of the process itself, and finally, cautions and critiques.

APPLICATIONS

Business

<u>TRW Inc</u>: PROBE II, a version of Delphi, was begun in 1967 with the support of TRW's senior executives and the full cooperation of managers throughout the organization (TRW Probe II, 1969). The categories included in this Delphi were electronics, materials (coatings, fuels, and lubricants), mechanics and hydraulics, information processing, transportation, aerospace, oceans, personal and medical systems, urban and international systems, environmental control systems, and others. From the 7,000 graduate scientists and engineers in TRW's automotive, electronics, equipment, and systems organizations, 140 panelists were selected. A typical panel had 16 members in addition to a chairman. Those appointed were asked to attend one or two briefings at which the instructions were issued, background material describing assumptions was discussed, and questions were answered. These sessions were the first and

last occasion in which panelists met as a group. Subsequently, the contacts were all carried out on an individual basis to preserve the anonymity of the panelists.

In a manner similar to that of the RAND study (see Fig. 1), panelists were first asked to provide a list of technical events which, they felt, were likely to occur under the given assumptions. The participants were encouraged to include items of primary value to other panels, and these were then transferred to the appropriate panel lists. In this manner, 2,500 events were identified.

In connection with each item, the following indications were provided by the panelists in the first rounds:

(a) customer desirability: needed desperately
or desirable
or undesirable but
possible

(b) producer feasibility: highly feasible
or likely
or unlikely but
possible

(c) probable timing: year by which probability is x that the event will have occurred
x = 0.10____
x = 0.50____
x = 0.90____

SAMPLE ITEM: "TV will be available to 90 percent of the world's population"

(a) desirable

(b) highly feasible

(c) x = 0.10 1977
x = 0.50 1980
x = 0.90 1983

Upon receipt of the first round responses, the various events were screened to eliminate duplicate items, ambiguous wording, trivia, and irrelevant events. In this manner, the number of events was reduced to 1,438. One additional item requested of the panelists should be noted: "familiarity" with the item (fair, good, or excellent). This self-evaluation identified those who believed themselves to be experts on a particular question. The third round was limited to a survey of these "experts" in the anticipation that they might have information not available to all panelists which could have a significant bearing on the responses. For this last round, panel chairmen also prepared a series of challenging questions arising from the responses to Round 2. Examples: "Item xxxx: Is there evidence that there are data to support a 10 percent probability in the near future?" Item yyyy: "Why so pessimistic when panel member No. 000 believes 90 percent probability will occur in 1972?" The final report provided integrated indices of desirability, feasibility, and probability, together with the 10, 50, and 90 percent likelihood dates for all items as well as the questions and answers derived in Round 3 (see Fig. 3 and Fig. 4). The entire study took nearly two years to complete.

Bell Canada: Bell Canada is a telecommunications company that provides services to Ontario and Quebec and owns large manufacturing and research and development subsidiaries (Day, 1975). The business planning group at Bell has the responsibility to identify corporate opportunities. At the time of the Delphi, these opportunities focused on merging computer and communication technology, regulatory changes introducing new competitive elements, emerging visual telecommunications markets, perceived and projected social changes, and increasing cost of investment options. The starting point for the Bell Canada Delphi was a definition of market segments that would have the most important impact on future applications of visual and data communications. The aim was to examine applications from the potential user point of view rather than from the direction of technological imperatives. An extensive literature review was undertaken before the initial questionnaires were prepared.

Subjects to be covered in the questionnaires included education, medicine, information systems, and residential markets. Participants on each panel first focused on value trends for the next 30 years. Traditionalism, hard work as a virtue, authoritarianism, involvement in society, acceptance of change, and the like were evaluated as trends, i.e., a significant increase, slight increase, no change, slight decrease, or significant decrease. The education Delphi then examined potential adoption of computerized library systems, computer-aided instruction systems, and visual display systems. The medical Delphi explored the acceptance of multiphasic screening, computer-assisted diagnosis, remote physiological monitoring, computerized medical library systems, and terminal usage. The business information processing technology study considered developments in management information systems, mini and small computers, terminals, and data processing. In addition, a Delphi was undertaken to study the future of communication services in the residential

OCEANS—REGULATIONS AND POLLUTION CONTROL

EVENT DESCRIPTION	SEEAPP	DESIRABILITY D	FEASIBILITY F	PROBABILITY P
414220 FEDERAL LEGISLATION FOR THE STRICT CONTROL OF WATER POLLUTION WILL HAVE BEEN PASSED.		.96	.54	.81
424290 'VACUUM CLEANERS' WILL BE AVAILABLE FOR CLEARING OILS AND OTHER UNDESIRABLE WASTE FROM THE OCEAN'S SURFACE.	•	.44	−.11	.53
424330 SALT WATER PURIFICATION TREATMENT SYSTEMS WILL BE OPERATIONAL FOR THE PROTECTION OF BEACH RECREATIONAL AREAS.	•	.32	−.41	.46
501040 THE OCEAN'S RESOURCES FOR FOOD AND MATERIALS, ITS UTILIZATION, AND THE CONTROL OF ITS POLLUTION (A MAJOR SYSTEM TASK, WITH SOPHISTICATED INSTRUMENTATION AND DATA REDUCTION) WILL BE MANAGED BY AN INTERNATIONAL AGENCY.	•	.19	−.22	.41

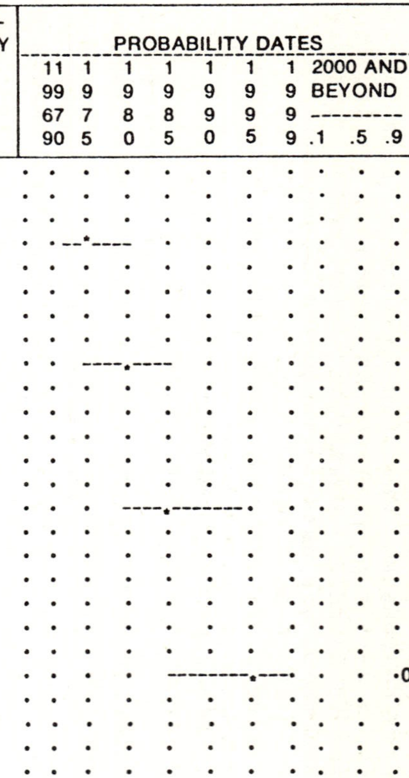

Figure 3. TRW Probe II

Reprinted by permission of The TRW Corporation.

501040 S IN RESPONDING TO ROUND 2, ONE PANELIST WARNED THAT 'SOME NATIONS WILL NOT COOPERATE' AND ANOTHER SUGGESTED 'FUNDING IS THE UNKNOWN WHICH WILL DETERMINE THIS'.
Q PLEASE STATE THE CASE FOR PESSIMISM.
A THE FOOD AND MATERIALS PART WILL MOSTLY BE COASTAL WATERS WHERE EACH COUNTY WILL CONTROL ITS OWN AND INTERNATIONAL COOPERATION ON THE HIGH SEAS WILL TAKE FOREVER TO NEGOTIATE.
Q WHY UNDESIRABLE AND WHY SO FAR OFF
A . . . . FROM EXPOSURE AND ASSOCIATION I BELIEVE THAT IT WILL BE EXTREMELY DIFFICULT TO OBTAIN THE STRONG MANAGEMENT DIRECTION AND CONTROL NEEDED TO PRODUCE TIMELY OBJECTIVE, FOCUSED STUDY RESULTS, FROM AN INTERNATIONAL AGENCY. I RECOMMEND THAT ONE NATION UNDERTAKE THE STUDY AND SUBMIT THE RESULTS FOR INTERNATIONAL CONSIDERATION THROUGH AN INTERNATIONAL AGENCY, E.G., NATO, FAO, ETC.

501050 Q WHY SO OPTIMISTIC
A SOUND ECONOMIC BENEFITS CAN BE DEMONSTRATED. THIS WILL SPEED DEVELOPMENT.
Q PLEASE STATE THE CASE FOR PESSIMISM.
A COMMERCIAL OPERATIONS DO NOT JUSTIFY THE COST OF THE SYSTEM, AND MILITARY REQUIREMENTS FOR THIS ARE MARGINAL TOO.
S IN RESPONDING TO ROUND 2, ONE PANELIST NOTED THE NEED FOR A SHARPER DEFINITION OF THE EVENT, I.E., 'OPEN SEA OR HARBORS—INDIVIDUAL SHIPS OR ALL SHIPS'.

504020 Q PLEASE STATE THE CASE FOR PESSIMISM.
A CARGO HANDLING BY CONTAINER SHIP HAS ALREADY BEEN DEVELOPED AND IS ESPECIALLY PREVALENT IN THE CARIBBEAN. I SEE NO REASON TO COMPLICATE THIS PROCESS WITH BARGES.

504090 Q EXAMPLES (OF SYSTEMS IN OPERATION)
A THE ALPINE GEOPHYSICAL COMPANY WAS SELLING A TRANSIT-COMPATIBLE EQUIPMENT RACK FOR APPROXIMATELY $75,000 TWO YEARS AGO. THE COMPUTER CONTROL NAVIGATION SYSTEMS MAY DELAY THE AVAILABILITY OF THIS SYSTEM TO THE MERCHANT MARINE.

504130 Q PLEASE STATE THE CASE FOR PESSIMISM.
A THE HORSEPOWER REQUIREMENTS OF SHIPBOARD POWER PLANTS INCREASE APPROXIMATELY AS THE SQUARE OF THE SPEED INCREASES. THE NUCLEAR POWER PLANTS MAY SOLVE THE PHYSICAL REQUIREMENTS OF HORSEPOWER VS THE TOTAL SIZE OF THE POWER PLANT. (HOWEVER), AT THE PRESENT TIME THE UNDERWATER VEHICLES CANNOT COMPETE WITH THE SURFACE VEHICLES, SUCH AS THE AIR CUSHION SHIPS IN DEVELOPMENT FOR THE HIGH-SPEED TRANSPORTATION FUTURE IN WATER-BORNE VESSELS.

504190 Q ANY UNSURMOUNTABLE BARRIERS (TO SUPPORT YOUR PESSIMISM)
A AT THE PRESENT TIME, ELECTRO-MAGNETIC ENERGY SEEMS TO BE THE ONLY ALTERNATIVE TO THE

Figure 4. TRW Probe II

market. A question that arises in this context is the definition of an expert. Are housewives or researchers and planners the experts on the adoption of communication services in the home? The issue was resolved by establishing two competing panels, one consisting of housewives, the other of experts in "wired city" services. The steps in the study design of this Delphi were as follows:

1. Literature search

2. Assembly of panels of experts and housewives

3. Design of draft questionnaire

4. Pretesting of this questionnaire

5. Distribution of the revised questionnaire (identical to both groups)

6. Statistical analysis of first run answers

7. Preparation of supporting comments from each group.

8. Design, pretest, and distribution of second round questionnaires, including not only the statistical results from each group but also the supporting comments. There was a specific request for responses on the differences between the panels and possible resolution of these differences.

9. Preparation of final analysis.

Sample results from the study are shown in Fig. 5.

Smith Kline & French Laboratories: This drug company undertook a study of the future of medicine in 1968 (Bender, Strack, Ebright, and VonHaunalter, 1969). This work is divided into five main sections: biomedical research, diagnosis, medical therapy, health care, and medical education. Of the 111 outside individuals contacted by mail at the beginning of the study, 42 responded and 35 supplied usable questionnaires. The selected panel included 21 medical doctors and 14 other experts. The panelists' areas of specialty ranged from internal medicine (10) to biochemistry (5) to nutrition (1) and psychology (1). They had spent an average of 17.5 years in their present occupations. Incidentally, when outside experts are used, the response rate is commonly only a fraction of the number contacted, typically about one-third. Use of inside

SHOP–FROM–HOME SERVICE

In predicting which types of products will be purchased through a Shop–from–Home system, the housewives and experts disagreed on a number of items. The summarized answers are presented below with the answers for the expected costs of such a service. Some typical comments are presented on the facing page. There appear to be significant differences in such items as produce, small and large appliances.

TYPE OF PRODUCT:

E / H	% YES	% NO	over 20% more	5–20% more	0–5% more	same	0–5% less	5–20% less	over 20% less
meat (E)	48	52			E ■	■			
meat (H)	58	42			□	□ H			
produce (E)	**52**	**48**			E ■	■			
produce (H)	**75**	**25**				□ H □			
other perishables (dairy, bread, etc.) (E)	100	0			■	■ E			
other perishables (dairy, bread, etc.) (H)	83	17				□ H □			
groc. dry goods (E)	100	0			■	■ E			
groc. dry goods (H)	96	4				H □	□		
clothing (E)	58	42			■	■ E ■	■		
clothing (H)	48	52				□ H □			
small appliances (E)	**74**	**26**			■	■ E ■	■		
small appliances (H)	**96**	**4**				□ H □			
drugs and cosmetics (E)	90	10			E ■	■			
drugs and cosmetics (H)	75	25				□ H □			
large appliances (E)	**47**	**53**			■	■ E			
large appliances (H)	**79**	**21**				□ H □			

NOTE: Shaded area represents significant differences between panels
E = Expert Panel: median or percentage response.
H = Housewife Panel: median or percentage response.

Do you wish to change any response? Could you comment on the differences between the two groups? Do you have any concluding comments

ROUND II COMMENTS

(Cont'd.)

Figure 5. Results of a Bell Canada Delphi Study

SHOP-FROM-HOME (CONTINUED)

EXPERTS / **HOUSEWIVES**

The panelists had the following generally favorable comments on the service and its expected costs.

Experts

1. "Will succeeds only if it offers an econon ic advantage (savings on sales people, expensive store space, electricity and display, etc.)"
2. "Assumption – meat and produce supplied from a familiar supplier and good previous experience."
3. "Better price and product comparisons can allow the purchaser to save money as well as time."

Housewives

1. "A useful service for those living in suburbs or remote rural areas whose access to cities and stores is difficult, also for elderly people."
2. "I don't think it should cost any more because it would not need a huge store or many salesclerks, and certainly shoplifting would be a thing of the past for anyone participating in the service."
3. "Small or large appliances (if brand name is known) present no problems for Shop-from-Home."

These comments reflect some of the typical reservations about the service:

Experts

1. "Meat, produce – these items are bought by feeling, smelling, and seeing the specific items in question and comparing them to the others available in the display. This cannot be done remotely, hence the continued existance of meat markets, fruit stands, etc."
2. "Insuring quality will raise the price of perishables above attractive cost."
3. "People will expect large cost reductions for appliances in return for not being able to see the merchandise. . .Several visits to warehouses will be typical even with home shopping."

Housewives

1. "For most housewives, shopping is a diversion and a break in a routine."
2. "Meat and produce – Individual specific quantities required; bult packaging not always desirable."
3. "The cost of these items in our daily living budget is expensive enough as it is; if it would cost more to buy these things at home, not many people would take advantage of it unless they were unable to get out of the house. . . Clothing and large appliances would be difficult with this service; with clothing you like to examine the fabric to see how well it is made; with large appliances you would want to discuss with the salesman the pro's-con's of the appliance."

Source: Bedford, Questionnaire, pp. 2-3

experts leads to a much higher fraction since these individuals consider the task as part of their job. Predictably, payment of outside panelists raises the fraction of active participation appreciably. The questionnaires in this Delphi were pretested within the company using 196 members of the R&D staff. In the first round of the extramural Delphi, 867 statements were received. Of these, 209 were used in the second round, and of the 209, consensus was achieved on 87 forecast statements.

The results were presented in a form similar to that of the 1964 RAND Corporation Delphi (see Fig. 6). In an interesting variation, the consensus statements for the 1978-1983 period were combined into a scenario that attempted to provide a more vivid image of the education, training, practice, and research of a student entering medical school in 1978 and beginning practice in the early 1980s. Another useful addition was the comparison of the dates of achievement (median dates) obtained in this study with those of the RAND study (Gordon and Helmer, 1964) and the intramural company study. It was found, for example, that the study with outside experts was consistently more optimistic than the other two, providing earlier dates for the median time estimates of the events described. Agreement between the estimates of the extramural Delphi and the RAND forecast was generally good. On the other hand, the intramural study was considerably more pessimistic.

A possible explanation is that the company personnel have a more realistic appreciation of the difficulties in implementing new concepts and therefore tend to provide more conservative dates.

Government and Institutes

National Materials Advisory Board: An interesting Delphi was conducted in 1970 for the U.S. ferroalloy industry by the National Materials Advisory Board of the National Academy of Science and Engineering (Goldstein, 1975). The board was concerned about a possible shortage of certain strategic and critical materials for the next decade and used Delphi as a means of assessing the implication of technological change on trends of ferroalloys. There were three rounds and three sections in each round: steel, alloys, and key developments.

Trend lines were presented to the respondents in the first two sections. The task of the panelists was to extend the past trends into the future, from 1969 to 1985, and to indicate reasons for the trend behavior (see Fig. 7).

BIOMEDICAL
MEDIAN AND

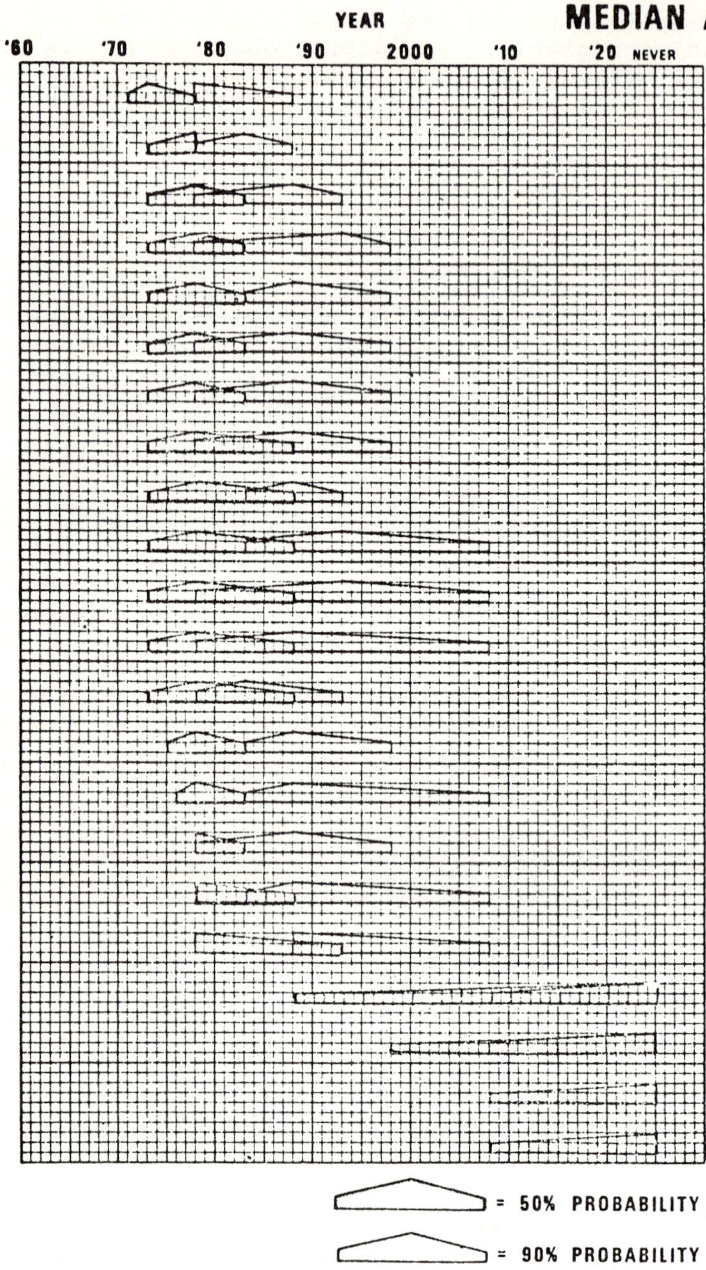

Figure 6. Smith Kline & French Delphi

Reprinted by permission of Smith Kline & French Laboratories.

RESEARCH QUARTILES

5. IDENTIFICATION OF NEW ENZYMES AND ENZYME SYSTEMS INVOLVED IN HEALTH AND DISEASE.*

3. IDENTIFICATION OF A NUMBER OF DISEASES AS AUTOIMMUNE STATES.

4. PRACTICAL KNOWLEDGE OF THE EFFECT OF AGING AND/OR ENVIRONMENT ON IMMUNOLOGIC CAPACITY.

13. SYNTHESIS OF NEW VIRUS FORMS.

23. USEFUL UNDERSTANDING OF THE EFFECT OF CONTACT WITH EXOGENOUS TOXIC BY-PRODUCTS OF CIVILIZATION ON THE LIFE PROCESS.

47. SATISFACTORY METHOD FOR PRESERVATION OF ORGANS, IN VITRO.

66. MECHANISM OF GALLSTONE FORMATION.

7. INDUCTION OF ENZYME SYNTHESIS IN VIVO BY CHEMICALS AND/OR INFECTIOUS AGENTS.

11. TRANSMISSION OF GENETIC INFORMATION VIA VIRUSES.

27. MUCH BETTER UNDERSTANDING OF CAUSAL RELATIONSHIP BETWEEN WHAT WE CHOOSE TO DO AND WHAT DISEASES WE GET, i.e., THE EFFECTS OF SMOKING, DRINKING, ETC.

41. CREATION/SYNTHESIS OF A LIVING VIRUS.

62. CONTROL OF BIOLOGIC SYSTEMS INVOLVED IN BLOOD PRESSURE MAINTENANCE AND HEART RATE.

64. AN UNDERSTANDING OF THE PRECISE ROLE OF ACID IN GI EROSION AND ULCERATION.

12. DETAILED DATA ON CHROMOSOMAL ABNORMALITIES AND CORRELATION WITH DISEASE.

17. COMPLETE UNDERSTANDING OF ACTIVE TRANSPORT PHENOMENA (i.e., SODIUM PUMP, ETC.)

65. AN UNDERSTANDING OF THE INTESTINAL FLORA AND ELECTROLYTE AND WATER TRANSPORT IN THE INTESTINE.

24. DETERMINATION OF THE EFFECT OF SUCH FACTORS AS STRESS, NOISE, RADIATION, CLIMATE AND SOCIAL STATUS ON SUSCEPTIBILITY AND RESISTANCE TO DISEASE.

51. DEFINITION OF NATURE OF RECEPTOR SITES FOR DRUGS.

35. ELECTRONIC CONTROL OF HUMAN BEHAVIOR.*

39. COMPLETE CONTROL OF MENTAL DEVELOPMENT.*

43. CREATION/SYNTHESIS OF A LIVING ORGANISM.*

34. COMPLETE CHEMICAL CONTROL OF HUMAN BEHAVIOR.*

* CONSENSUS AT BOTH THE 50% AND 90% PROBABILITY LEVELS.

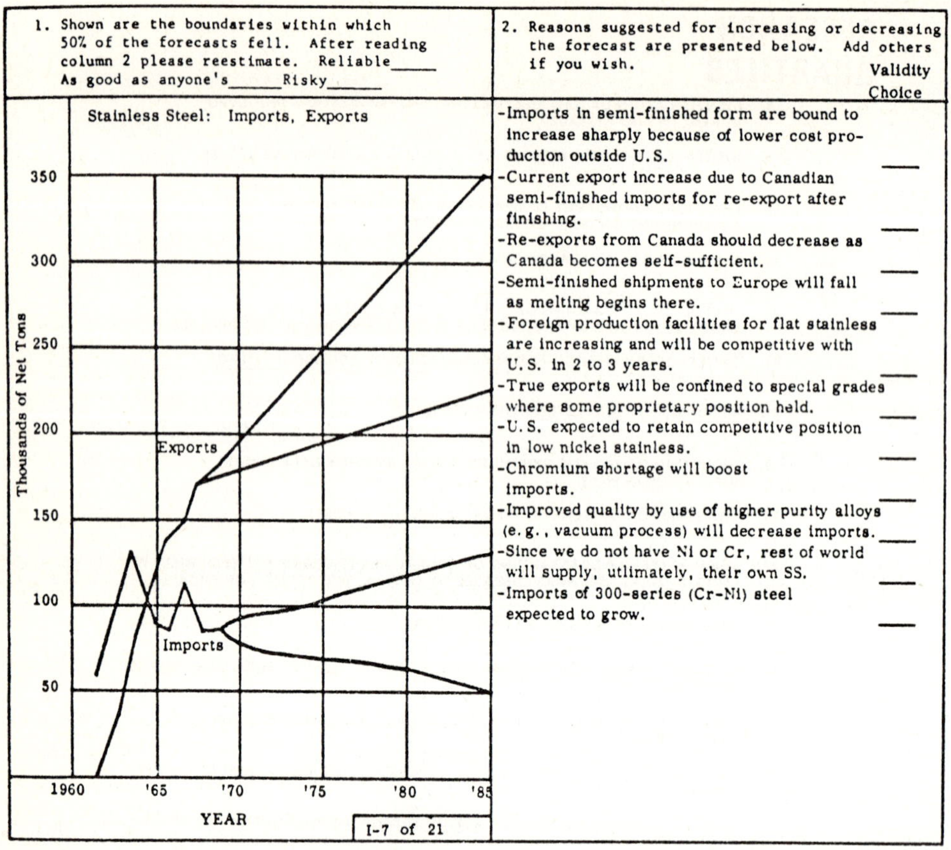

1. Shown are the boundaries within which 50% of the forecasts fell. After reading column 2 please reestimate. Reliable_____ As good as anyone's______Risky______

2. Reasons suggested for increasing or decreasing the forecast are presented below. Add others if you wish.

Validity Choice

- -Imports in semi-finished form are bound to increase sharply because of lower cost production outside U.S. ____
- -Current export increase due to Canadian semi-finished imports for re-export after finishing. ____
- -Re-exports from Canada should decrease as Canada becomes self-sufficient. ____
- -Semi-finished shipments to Europe will fall as melting begins there. ____
- -Foreign production facilities for flat stainless are increasing and will be competitive with U.S. in 2 to 3 years. ____
- -True exports will be confined to special grades where some proprietary position held. ____
- -U.S. expected to retain competitive position in low nickel stainless. ____
- -Chromium shortage will boost imports. ____
- -Improved quality by use of higher purity alloys (e.g., vacuum process) will decrease imports. ____
- -Since we do not have Ni or Cr, rest of world will supply, utlimately, their own SS. ____
- -Imports of 300-series (Cr-Ni) steel expected to grow. ____

Figure 7. Form for Trend Extrapolation

Ref.: Goldstein, 1975.

SOURCE: Reproduced from *The Delphi Method,* edited by Harold A. Linstone and Murray Turoff, with permission of publishers, Addison-Wesley, Advanced Book Program, Reading, Massachusetts, U.S.A.

This format reminds us that it is often convenient to classify forecasts into two types: events and trends. The original Delphi work concerned itself entirely with event forecasts. In this study, we see the application to trend forecasts.

Thirty-four panelists were selected from industry, government, universities, institutes, and trade publications. Specifically, for each graph the first round probed the following questions: (1) How reliable does the respondent consider his graph extension to be? (2) What key developments does the respondent assume in making his extension? and (3) What other developments might result in major changes of the extension? A flowchart of the steel-making process was also presented to the panelists in Section 1, with figures for 1969 and requests to supply corresponding figures for 1980. The elapsed time of the entire three-round effort was 46 weeks. The total senior professional analysis activity for the process took 18 manweeks, professional level work 13, clerical tabulation and curve extrapolations 15, and typing 7. One interesting feature of this effort was the opportunity to compare a Delphi with a more conventional panel study. The same subject was studied by a conventional committtee having no interaction with the Delphi until both exercises were completed. There were several significant differences in the results: (1) the conventional report led to very precise agreement reflected in single valued forecasts, whereas the Delphi presented ranges that suggested the uncertainty, (2) the committee developed consensus recommendations while the Delphi panels did not, (3) the committee limited itself more narrowly to the subject area than the Delphi panel, and (4) the committee presented background information systematically while the Delphi panel provided only random background data.

<u>National Coordinating Council on Drug Education</u>: A Delphi on national drug abuse policy was undertaken in 1974 for the National Coordinating Council on Drug Education (Jillson, 1975). The objectives of the study were (1) to develop the range of possible national drug abuse policy options, (2) to explore application of the Delphi methodology to this and other areas of social policy, and (3) to explore the possibilities of applying the techniques on as-needed and on ongoing bases. The study specifically set out to explore drug abuse policy from three perspectives: (1) the top-down approach, i.e., establishing national drug abuse policy objectives, (2) the bottom-up approach, identifying factors that control the transition between the general population and various degrees of drug use and deciding which of these are important and can be affected, and (3) an issue-oriented approach, deriving policies from issues that are the subject of current controversy. More than 100 potential respondents were listed,

and invitations were sent to 45 persons. Thirty-eight responded positively. The first round questionnaire focused on (1) the development of objectives, (2) the transition matrix, (3) policy statements, and (4) additional items such as self-rating of expertise. Twenty-four of the respondents actually completed the first round. The median time for completion of the round was determined to be 2.25 hours. The second questionnaire included sections on national drug abuse policy objectives and policy issue statements. The third questionnaire focused on national policy objectives and the transition matrix. The general study design is shown in Table 1.

The cost of completing a study of this type ranges from $15,000 to $40,000, assuming a 9- to 12-month effort. This cost is relatively small in terms of the output obtained.

National Commission for the Protection of Human Subjects of Biomedical and Behavioral Research: A very recent effort is a policy Delphi entitled "A Comprehensive Study of the Ethical, Legal, and Social Implications of Advances in Biomedical and Behavioral Research and Technology," conducted jointly by Policy Research Inc. and the Center for Policy Assessment at the New Jersey Institute of Technology. This study is responsive to a section of the National Research Act which calls for "an analysis and evaluation of scientific and technological advances in past, present, and projected biological, behavioral, and medical research and services...and evaluation of implications of public policy of such findings." This Delphi is one of the most sophisticated instruments run in the United States. A single round questionnaire is a looseleaf book, and the respondent is expected to spend six hours in completing a round. He is paid $100 for his efforts per round. In the first round background papers are provided on subjects such as data banks, extension of life, and genetic screening. The evaluation part asks the respondent to consider a number of advances and their implications as well as preferences of policies with regard to research and implementation of research. A set of scenarios, each about a half page in length, is provided, and the respondent is asked to indicate the decision he would make as a policy maker (e.g., should the drugs discussed in the scenario be marketed or prohibited). He is also asked to determine what implications of the scenario influence his decision as well as which groups or mechanisms should participate in the resolution of the decision. We note that a policy Delphi generally cannot be confined to so-called experts. The formulation of national policy must obviously include the public at large. It is therefore important to include in the Delphi representatives of a large or a wide spectrum of vested interests, ranging from bureaucrats to minority groups.

Table 1. The Study Design

Perspective	First Questionnaire	Second Questionnaire	Third Questionnaire	Fourth Questionnaire	Fifth Questionnaire	Final Summary
Objectives ("top down")	List objectives	Rate feasibility, and desirability of objectives	Re-rate feasibility and desirability of selected objectives			
	List key indicators	Expand key indicators, perform initial rating				
Transition Factors ("bottom up")	List factors, indicate direction		Final rating of factors Develop policy areas to affect important transition factors which can be influenced by national policy	Identify any contradiction in objectives and policies formulated by the three approaches	Synthesize a consistent and realistic set of national drug abuse policy options	Write a brief summary of the national drug abuse policy options identified by the study, including the normative forecasts for the key indicators Identify future policy research needs
Policy Issue Statements ("political")	Rate selected policy issues; develop for and against arguments Add other important policy issues	Final rating of selected policy issues; rate for and against arguments Rate and give arguments for and against additional policy issue statements				

Ref.: I Jillson, 1975.

SOURCE: Reproduced from *The Delphi Method*, edited by Harold A. Linstone and Murray Turoff, with permission of publishers, Addison-Wesley, Advanced Book Program, Reading, Massachusetts, U.S.A.

Michigan Sea Grant Program: An example of a regional planning Delphi is the study described by Ludlow focusing on the Grand Traverse Bay watershed region (1975). This area was selected as the focus of a pilot study to "obtain and refine an interdisciplinary group of researchers' judgments about issues and developments that should be considered when planning for intelligent management of the water resources of the Great Lakes." Further, it was considered important to convey the judgment of the researchers to the communities that would benefit from the research. Therefore, influential community leaders were included on the panel, and the Delphi operated as a true communication instrument. Specifically, two of the three panels were made up of researchers, i.e., technicians and behaviorists, and the third group comprised concerned citizens or decisionmakers. Another interesting aspect of this study was the emphasis on interactions between different developments. Let us denote as item A "construction of a spray irrigation system for waste water disposal in the Grand Traverse Bay region" and as item B "requirement by the State, calling for tertiary treatment of municipal sewage for Traverse City." Not content with asking for the probability of occurrence of A and B in the 1971-1980 time period and the 50 percent probability date of each, the questionnaire asked the following question: "If you were certain that B would occur before 1980, your estimates for A would be ------," and "If you were certain that B would not occur in 1971-1980, your estimates for A would be ----." We cannot ignore the fact that individual items in any Delphi are not by any means mutually independent; more likely, there are many interactions between them.

Analyses

The development of Delphi has predictably been accompanied by a growing body of literature devoted to analyses of the method itself. We will briefly summarize some of the significant results.

1. Dispersion as a function of remoteness of estimate: In the original RAND report as well as in subsequent work by Martino (1970), it was found that the remoteness of the forecast date and degree of dispersion are directly related. The further the median date of a forecast lies in the future, the greater is the dispersion around that date. In other words, uncertainty increases consistently with remoteness of the estimate date.

2. Accuracy of forecasts: An inherent bias in forecasting is the tendency to be pessimistic on the long-range estimates and

optimistic on the short-range ones. We do not have to search far for an explanation. In a long-range situation, we assume that, if we do not know an answer to the problem, no answer exists. This is Coates' fallacy of the "argument from incompetence" (1975). On the other hand, for near term items, we do know the answer but make the optimistic assumption that implementation will be smooth and devoid of serious obstacles. Consider supersonic passenger aircraft. In 1930, a forecast would have been pessimistic. At that time, Nevil Shute Norway, a trained aeronautical engineer, forecast a cruising speed of 110 to 130 miles per hour for passenger aircraft in 1980. In 1965, we knew how to build supersonic aircraft, and such aircraft were expected to be operational by 1972--a highly optimistic forecast. Incidentally, this pattern suggests that there should be a time span for which forecasts are most accurate, i.e., at the point from which we shift from being pessimistic to optimistic. Unfortunately, we do not have enough data to give precise guidance on the correction of this bias. This is an appropriate time to remind readers that accuracy is hardly the best measure of the "rightness" of a forecast, Delphi or other; a good forecast is one that provides useful insight and guidance to the planner and decisionmaker. (Some of the most valuable forecasts have been "wrong" by virtue of being self-defeating.)

3. Delphi statements: Inevitably, the statements in a Delphi exercise reflect the cultural attitudes, subjective biases, and knowledge of those who formulate them. In their original study, Gordon and Helmer began the first round with a blank questionnaire (see Fig. 1) to minimize the bias of the exercise monitors. Every high school student knows that multiple choice examinations require insight into the teacher's mode of thought as well as the substance of the question. A student who knows too much is at a disadvantage, just as a student who knows too little is handicapped. In a Delphi exercise, statements may be too concise, leading to excessive variations and interpretations, or they may be too lengthy, requiring the assimilation of too many elements. Salancik, Wenger, and Helfer (1971) have probed this question and have found that there is a "right" length for Delphi statements, i.e., a length that leads to a maximum in the amount of information obtained. The definition of information used here is that of information theory (bits). Low and high numbers of words yield low consensus, with intermediate-length statements producing the highest consensus. In a particular case which they examined, 20- to 25-word statements formed the peak in the distribution. While this result should not be generalized, it does underscore the care that must be exercised in developing Delphi statements for use by the panelists.

4. Basis for respondents' intuitive forecasts: In another study, Salancik (1973) has examined the hypothesis that panelists in a forecasting Delphi assimilate input on feasibility, benefits, and potential costs of an event in an addictive fashion to estimate its probable date of occurrence. He finds that the more feasible, beneficial, or economically viable a concept is judged, the earlier it is forecast to occur.

5. Self-rating of experts: Dalkey, Brown, and Cochran (1969) have tackled another aspect of Delphi: the expertise of the respondents. With a given group of panelists, we might consider two ways of improving its accuracy: iterating responses or selecting a more expert subgroup. The latter process implies an ability to identify such a subgroup, for example, by self-rating. However, the choice of a subgroup itself brings with it a potential degradation in accuracy as a result of the statistically reduced group size. Thus, in resorting to a subgroup we have two counteracting trends. From their experimentation, Dalkey, Brown, and Cochran have found that self-rating is a meaningful basis for identifying expertise, and that selection of expert subgroups improves the accuracy to a somewhat greater degree than does feedback or iteration. Dalkey has found that a suitable minimum panel size is seven; accuracy deteriorates rapidly with smaller sizes and improves more slowly with large numbers.

6. Opinion stability as a method of consensus measurement: Scheibe, Skutsch, and Schofer find that consensus in a Delphi should not be measured by the percentage of votes that fall within a prescribed range, e.g., when the interquartile range is no larger than two units on a 10-unit scale. Rather, the stability of the respondent's vote distribution curve over successive rounds provides a superior measure. The authors determine that marginal changes of less than 15 percent offer a useful definition of stability. If a round provides less than 15 percent change from the preceding round, the exercise should be stopped. Use of this stability measure as a stopping criterion also preserves well-defined disagreements that may exist. Today we are often more concerned with determining the degree of polarization of respondents than consensus. Viewing the technique as a communication concept, we no longer judge success of a Delphi by the degree of consensus obtained (Scheibe, Skutsch, and Schofer, 1975).

7. The role of dogmatism: It might be expected that highly dogmatic individuals would not alter their responses from one round to another. In fact, the situation is more complex. As Mulgrave and Ducanis (1975) have shown, a highly dogmatic individual will remain with his initial response, provided he considers himself an expert on the item. If he does not

consider himself highly knowledgeable, he will very quickly move to the median of the group, i.e., accept the group response as an authoritative one.

CRITIQUE AND FINAL COMMENTS

That Delphi has spawned criticism as well as ardent support is a healthy sign. Ironically, a major critique has emanated from the RAND Corporation--Sackman's report, "Delphi Assessment: Expert Opinion, Forecasting and Group Process." He proclaims that "the future is far too important for the human species to be left to fortunetellers using new versions of old crystal balls. It is time for the oracle to move out and science to move in" (1974).

For Sackman science means psychometrically trained social scientists. His tradition-bound attitude is not uncommon; we also find it in the widely held illusions that science is "objective," that only data-based or model-based inquiring systems are legitimate, and that subjective or Bayesian probability is heretical. Unfortunately, when faced with new paradigms, orthodoxy often responds with sweeping condemnations and unwitting distortions. Sackman cites poorly executed applications of Delphi (and there are many) and ignores significant supportive research such as that briefly outlined in the last section. Coates offers the opinion that Delphi is a method of last resort in dealing with extremely complex problems for which there are no adequate models. As such, one should expect very little of it compared to applicable analytical techniques. One should expect a great deal of it as a technique of last resort in laying bare some crucial issues on a subject for which a last resort technique is required....If one believes that the Delphi technique is of value not in the search for public knowledge but in the search for public wisdom; not in the search for individual data but in the search for deliberative judgment, one can only conclude that Sackman missed the point (1975).

For a recent debate (as well as new work) on the Delphi method, the reader is referred to an issue of the journal <u>Technological Forecasting and Social Change</u> (Volume 7, Number 2, 1975).

Finally, the Delphi designer should keep several major concerns in view.

1. Discounting the future: We all use a discount rate in viewing the future (and the past). A crisis twenty years from

now is not nearly as significant to us as the same crisis next month. We must keep in mind, however, that individuals discount the future at different rates and that this attitude may be reflected in the Delphi responses. A person at the bottom of Maslow's human values pyramid will discount environmental pollution much more heavily than someone near the top. The poor, for whom survival is a daily challenge, are hardly going to lose much sleep over a pollution or population crisis 20 years in the future (Linstone and Turoff, 1975, p. 574). We also find that a Delphi participant or any forecaster is influenced more strongly by recent events than those in the distant past. This problem is discussed in more detail by Tversky and Kahneman (1974).

2. The simplification urge: We prefer certainty to uncertainty and simplicity to complexity. Typically, we forecast by taking one or a few innovations and fitting them in a mental image into an environment set in the familiar structure context of the past and present. We do not visualize a future situation in its own holistic pattern where change has had a pervasive influence. Cross-impact analysis is of some help, but it by no means eliminates the difficulty. Intuitive procedures such as Delphi depend heavily on subjective probability assessments. As Tversky and Kahneman have shown, human beings have a tendency to ignore prior probabilities when worthless or irrelevant information is added. They also confuse desirability and familiarity with probability (i.e., indulge in wishful thinking), so that they find a reasonable sounding scenario with "more of the same" highly likely, even though there is little evidence to support such a forecast.

3. Illusory expertise: We have witnessed many instances in which the alleged expert proves to be a poor forecaster. One reason is obvious. The expert tends to be a specialist and thus views the forecast in a setting that is, in fact, not the most appropriate one. For example, the low aircraft speed projected for 1980 in the 1930s was based on a continuation of the use of reciprocating engines with propellers. The forecasters failed to realize that the task was not to build better reciprocating engines but to move people by air. Technologists have consistently underestimated the complexity and cost of new systems and have ignored nontechnological factors. We find a curious ahistoricity in the outlook of most scientists and technologists, together with a tendency for inbreeding. In Delphi, inbreeding occurs when a panelist is asked to suggest the names of other possible participants and proposes fellow scientists who have the same technological and cultural background.

4. Sloppy execution: The preceding discussion makes it clear that there are many ways to do a poor job. In that respect,

Delphi is no different from other analysis methods. We hope that this brief survey will encourage the potential user to probe the technique carefully before he ventures forward to an application of his own.

5. Format bias: It should be recognized that, particularly for policy Delphis, the format of the questionnaire may be unsuitable to some potential societal participants. The reasoning is similar to that which leads to rejection of the use of IQ tests or other multiple choice tests for certain groups. The Delphi questionnaire, like the typical written test or the conference, assumes a certain cultural background. Those with significantly different backgrounds find it exceedingly difficult to respond as requested. They may give the feedback they believe will please the inquirers, or they may not respond at all. One constructive idea is the use of visual statements, i.e., pictures in the questionnaires (Adelson and Aroni, 1975).

6. Manipulation of Delphi: If Delphi is indeed a communication technique, we should not be surprised that it is possible to manipulate it for deception. The responses of the first round can be altered by the monitors in the hope of moving the second round responses in a desired direction. One possible way to avoid such manipulation is to involve the "subjects," i.e., the panelists themselves, in the staff activities. The difficulty is to do so without compromising the anonymity of the responses.

In conclusion, we suggest that Delphi offers a powerful instrument for group communication. While it is not a panacea, it is a very useful method of last resort for the forecaster.

REFERENCES

Adelson, A. and S. Aroni, 1975. "Differential Images of the Future." In H.A. Linstone and M. Turoff (eds.), The Delphi Method: Techniques and Applications. Reading, Mass: Addison-Wesley Publishing Co., pp. 433-462.

Bender, A.D., A.E. Strack, T.W. Ebright, and G. VonHaunalter, 1969. "A Delphic Study of the Future of Medicine." Philadelphia: Smith Kline & French Laboratories.

Coates, J.F., 1975. "Review of Sackman Report." Technological Forecasting and Social Change, Volume 7, Number 2, pp. 193-194.

Dalkey, N. and O. Helmer, 1963. "An Experimental Application of the Delphi Method to the Use of Experts." Management Science, Volume 9, Number 3, p. 458.

----, B. Brown, and S.W. Cochran, 1969. The Delphi Method, III: Use of Self-Ratings to Improve Group Estimates. Santa Monica, Calif.: RAND Corporation.

Day, L.H., 1975. "Delphi Research in the Corporate Environment." In H.A. Linstone and M. Turoff (eds.), The Delphi Method: Techniques and Applications. Reading, Mass.: Addison-Wesley Publishing Co., pp. 168-194.

Forrester, J., 1971. World Dynamics. Cambridge, Mass.: Wright-Allen Press (particularly Chapters 5 and 6).

Goldstein, N., 1975. "A Delphi on the Future of the Steel and Ferroalloy Industries." In H.A. Linstone and M. Turoff (eds.), The Delphi Method: Techniques and Applications. Reading, Mass.: Addison-Wesley Publishing Co., pp. 210-226.

Gordon, T.J. and H. Hayward, 1968. "Initial Experiments for the Cross-Impact Matrix Method of Forecasting." Futures, Volume 1, Number 2, pp. 100-116.

---- and O. Helmer, 1964. "Report on a Long Range Forecasting Study." RAND Paper P-2982, Santa Monica, Calif.: RAND Corporation.

Helmer, O., 1972. "On the Future State of the Union." Report R-27. Menlo Park, Calif.: Institute for the Future.

Jillson, I.A., 1975. "The National Drug-Abuse Policy Delphi: Progress Report and Findings to Date." In H.A. Linstone and M. Turoff (eds.), The Delphi Method: Techniques and Applications. Reading, Mass.: Addison-Wesley Publishing Co., pp. 124-159.

Kane, J., 1975. "A Primer for a New Cross-Impact Language--KSIM." In H.A. Linstone and M. Turoff (eds.), The Delphi Method: Techniques and Applications. Reading, Mass.: Addison-Wesley Publishing Co., pp. 369-382.

Linstone, H.A. and M. Turoff (eds.), 1975. The Delphi Method: Techniques and Applications. Reading, Mass.: Addison-Wesley Publishing Co.

Ludlow, J., 1975. "Delphi Inquiries and Knowledge Utilization." In H.A. Linstone and M. Turoff (eds.), The Delphi Method: Techniques and Applications. Reading, Mass.: Addison-Wesley Publishing Co., pp. 102-123.

Mulgrave, N.W. and A.J. Ducanis, 1975. "Propensity to Change Responses in a Delphi Round as a Function of Dogmatism." In H.A. Linstone and M. Turoff (eds.), The Delphi Method: Techniques and Applications. Reading, Mass.: Addison-Wesley Publishing Co., pp. 288-290.

Sackman, H., 1974. "Delphi Assessment: Expert Opinion, Forecasting and Group Process." R-1283-PR. Santa Monica, Calif.: RAND Corporation.

Salancik, J.R., 1973. "Assimilation of Aggregated Inputs into Delphi Forecasts: A Regression Analysis." Technological Forecasting and Social Change, Volume 5, Number 3, pp. 243-248.

----, W. Wenger, and E. Helfer, 1971. "The Construction of Delphi Event Statements." Technological Forecasting and Social Change, Volume 3, Number 1, pp. 65-73.

Scheibe, M., M. Skutsch, and J. Schofer, 1975. "Experiments in Delphi Methodology." In H.A. Linstone and M. Turoff (eds.), The Delphi Method: Techniques and Applications. Reading, Mass.: Addison-Wesley Publishing Co., pp. 262-287.

"TRW's Probe of the Future," 1969. 14 Volumes. Redondo Beach, Calif.: TRW Corporation.

Tversky, A. and D. Kahneman, 1974. "Judgment Under Uncertainty: Heuristics and Biases." Science (September 27), pp. 1124-1131.

Expert Judgment for Human Reliability in Risk Analysis[1]

David A. Seaver and William G. Stillwell

The MAXIMA Corporation, 7315 Wisconsin Avenue, Suite 900 North, Bethesda, MD 20814

INTRODUCTION

As more and more attention has been focused on quantitative risk analysis, the role of human reliability in the operation of complex technological systems has become increasingly apparent. The potential for human errors is now widely acknowledged as a significant, if not the major, contributor to risk. Human errors in the accident at Three Mile Island (Kemeny, 1979; Rogovin and Frampton, 1980) have led to considerable attention to human reliability in one of the most visible and controversial technological systems--nuclear power plants.

While we strongly subscribe to the belief that the human cannot be considered in separation from the system, we do believe the focus on the human component is appropriate and necessary. That is, although the term "human error" is a misnomer--it is really a system error (e.g., produced by contributions from

---

1 The work described in this paper has been supported by the U.S. Nuclear Regulatory Commission, Office of Regulatory Research. It is part of a larger effort directed by Sandia National Laboratories on human reliability analysis in nuclear power plant operations. Support has been provided, first under a subcontract to Decision Science Consortium, Inc., and subsequently under a subcontract to General Physics Corporation with The MAXIMA Corporation as a subcontractor to General Physics. Specific acknowledgement is made of Dr. Thomas G. Ryan, the NRC program manager, Dr. Louise Weston, the Sandia National Laboratories technical monitor, and Ms. Kay Comer, the General Physics project manager who have each contributed significantly to this work.

NATO ASI Series, Vol. G4
Environmental Impact Assessment, Technology Assessment, and Risk Analysis. Edited by V.T. Covello et al.

various components--training, human mistakes, person-system design, etc.)--the implication that the human as part of the system must be examined in considering risk is certainly justified. Thus, we too adopt "human error" as a useful term with the understanding that such errors may often be beyond the control of the individuals involved.

Although the need for inclusion of human reliability in risk analysis is clear, the procedure for doing so is not. The human reliability part of a risk analysis can be even more difficult than the rest of the risk analysis. Relatively few data on human errors exist; and human behavior will generally be less consistent than other system components, and, therefore more difficult to include in formal models. The lack of data is, of course, most acute for relatively new systems, where it is most needed for risk analysis.

One relatively easy and inexpensive means to obtain estimates of human error probabilities (HEPs) for use in risk analysis is to use expert judgment. (We note that much of the discussion in this paper on the use of expert judgment need not be limited to human errors. This is simply the context in which we are currently investigating its use.) This may be justified if expert judgment can be demonstrated to produce relatively reliable and valid estimates of error probabilities.

An investigation of this possibility has been undertaken by the U.S. Nuclear Regulatory Commission (NRC). The particular context for this investigation is human errors in nuclear power plant operations. The NRC has recognized the importance of human reliability analysis in its probabilistic risk assessments (PRAs), and has undertaken a broad program of research in this area of which the use of expert judgment is a part.

The use of expert judgment is only one of several procedures being investigated by the NRC for estimating human error probabilities. Other procedures include using: 1) Licensee Events Reports which are filed with the NRC by licensees; 2) Nuclear Power Safety Report System to allow for the anonymous reporting of violations/errors; 3) training simulators to estimate errors during simulated operations; and 4) performance models and their simulation.

In addition, a handbook has been developed (Swain and Guttman, 1983) that describes procedures for human reliability analysis and provides HEP estimates based on currently available data and the (considerable) accumulated experience of the authors. A goal of this broad NRC research program is to identify procedures that both produce valid estimates and are practical to use for estimating the HEPs needed for PRAs.

Expert judgment appears to have the potential for producing HEP estimates, based on its extensive use for estimating probabilities in other contexts (Stillwell, Seaver, and Schwartz, 1982). Such judgments must, however, be obtained systematically; and, because numerous systematic biases exist in judgments (Stillwell, Seaver, and Schwartz, 1982), the procedures used to obtain the judgments must be carefully designed to minimize such biases.

The NRC investigation into the use of expert judgment to estimate HEPs began with a review of its use in other substantive areas in order to make a preliminary determination of feasibility (Stillwell, Seaver, and Schwartz, 1982). When this review suggested there was a potential for use of expert judgment, several particular judgmental procedures were specified in detail (Seaver and Stillwell, 1983). Two of these procedures are now being field tested under a contract with Sandia National Laboratories with General Physics Corporation as the prime contractor and The MAXIMA Corporation as a subcontractor.

Because the results of the field test are not yet complete (completion is expected in early 1984), this paper focuses on the methodologies that are being used to estimate HEPs with expert judgment. It provides a description of these methodologies that we feel based upon our experience and review of other applications, can successfully be used to estimate not only human errors probabilities, but also other probabilities in risk analysis and other contexts.

In the remainder of this paper, we first present a brief overview of five procedures that were developed and reported in Seaver and Stillwell (1983). Then, the two of these procedures that are currently being tested and evaluated are described in more detail. A brief summary of the ongoing implementation and evaluation study is then given, followed by our conclusions regarding these procedures.

OVERVIEW OF PROCEDURES

Five procedures using different types of judgments were described and evaluated by Seaver and Stillwell (1983), where detailed descriptions can be found. We present an overview here of the five procedures: paired comparisons, ranking/rating, direct numerical estimation, indirect numerical estimation, and the success likelihood index method (SLIM), a procedure based on multiattribute utility measurement.

Paired Comparison Procedure

The use of paired comparison judgments and Thurstone's (1927) Low of Comparative Judgment to derive scales has an extensive history in psychological scaling. Torgerson (1958) describes procedures for obtaining scale values from judgments of the type "event i is more likely than event j." The psychological model underlying the use of these procedures is that each stimulus (event) is represented psychologically by a distribution on a subjective scale. When two stimuli are compared, each produces a value selected randomly from its distribution and the stimulus with the higher value is judged to be more likely (larger, more important, etc.).

Several assumptions are then needed to derive quantitative scale values. First, the distributions are assumed to be normal, although other distributions (e.g., logistic; Luce, 1959) can be used to lead to similar results. Thus, the difference in psychological values produced by comparing stimulus i with stimulus j, $d_j - d_i$, is distributed normally with

$$(\mu_{i,j} = d_j - d_i \text{ and } \sigma^2_{i,j} = \sigma^2_i + O^2_j - 2r_{ij}\sigma_i\sigma_j)$$

where r_{ij} is the correlation between momentary values for stimuli i and j. With judges each making all possible paired comparisons, the proportion of times j is judged to be more likely than i, t_{ij} is converted to a standard normal deviate, and a system of equations can be specified to be solved for the unknown parameters.

In this general form, the equations are not solvable, having more unknowns than independent equations. Therefore, additional simplifying assumptions are required. The most frequently used assumption is that ($\sigma^2_{i,j}$) is constant for all i and j (Torgerson, 1958). It can then be shown that the average of all normal deviates for a particular event derived from comparisons with all other events is the scale value for the event.

This scale value must then be transformed into a probability. Based on previous research, Seaver and Stillwell (1983) suggest that the most appropriate transformation is logarithmic:

$$\log p_i = as_i + b$$

where p_i is the estimated probability, s_i is the scale value, and a and b are constants. These constants must be determined from independent probability estimates for at least two of the events under consideration.

In general, the use of paired comparisons is promising because it relies only on qualitative judgments, rather than quantitative judgments that appear more susceptible to biases. People also seem to feel more comfortable with qualitative judgments and accept the results as reasonable. These latter considerations are important in a practical context.

The disadvantages of paired comparisons are the large number of judgments that may be required and the need to transform the scale values into probabilities. If each judge compares each pair of n stimuli, n(n-1)/2 judgments are made. Clearly, n does not need to be very large before the number of judgments required is formidable. Torgenson (1958) and Seaver and Stillwell (1983) suggest some procedures for reducing the number of judgments required.

This transformation of scale values into probabilities is a somewhat arbitrary process based on some empirical support. Additional empirical support is needed and particular attention is being given to this in the current implementation and evaluation study. The two independent probability estimates needed may often be available only from direct judgmental probability estimates. The probabilities resulting from the paired comparison procedure are then really only as good as these two estimates. (This is true for the absolute probabilities themselves, but not necessarily for the relative likelihoods of the various events.)

Ranking/Rating Procedure

Although requiring different types of judgments, ranking and rating are included together because the underlying psychological model is the same and the procedures for deriving probability estimates from the judgments are similar. Ranking requires each expert to rank order the events under consideration according to their likelihood. Rating requires the expert to rate the likelihood of each event on a particular scale, e.g., from one to seven. Ranking and rating are similar in that each rank can be considered a different rating.

The psychological model underlying this procedure is Thurstone's Law of Categorical Judgment (Torgerson, 1958), which is very similar to the Law of Categorical Judgment underlying paired comparison scaling. The primary differences are that the equations solved for the Law of Categorical Judgment deal with cumulative proportions, and each rating category boundary (or boundary between ranks) is also assumed to produce a psychological distribution similar to those for stimuli.

To derive scale values, an event by rating category matrix is specified, with each entry being the proportion of times that the event was judged to be in that category or in a less likely category. These proportions are converted into standard normal deviates. Then, with assumptions similar to those for paired comparisons, the scale value for each event is computed by subtracting the row mean from the grand mean. Column means are the category boundaries.

The scale values derived from the ranking/rating judgments must be transformed into probabilities. The same procedure is used for this transformation as was described above for paired comparisons.

The advantages and disadvantages of ranking/rating are much the same as those for paired comparisons except that many fewer judgments are required and the ranking/rating judgments may be somewhat more susceptible to bias (particularly rating judgments).

Direct Numerical Estimation

This is the simplest of the procedures with experts providing direct probability or odds estimates of the likelihood of events. These individual estimates are then combined to form a single probability (odds) estimate for each event. The combination rule suggested by Seaver and Stillwell (1983) based on previous research (Bordley, 1982; Seaver, 1976, 1978) is

$$p_j = \frac{\left(\prod_{i=1}^{m} p_{ij}\right)^{1/m}}{\left(\prod_{i=1}^{m} (1-p_{ij})\right)^{1/m} + \left(\prod_{i=1}^{m} p_{ij}\right)^{1/m}}$$

where p_j is the probability estimate for event j, p_{ij} is the estimate for event j by expert i, and m is the number of experts.

An important feature of direct numerical estimation appears to be the response mode in which judgments are made. The review by Stillwell, Seaver, and Schwartz (1982) suggested that best results would be obtained by having responses marked on a scale, rather than simply being written or spoken, and by having the scale logarithmically spaced. This latter condition is particularly important in estimating human error probabilities where the estimates can be quite small.

The primary advantage of direct numerical estimation is its simplicity. It requires that each expert make only one judgment for each event and results directly in probability estimates. Its primary disadvantage is that such judgments have been shown to be biased in many circumstances. In addition, it has the practical disadvantage that experts and potential users may not believe that such judgments can be accurate even when they are.

Indirect Numerical Estimation

This procedure requires experts to make ratio judgments regarding the relative likelihood of pairs of events--e.g., "event a is five times as likely as event b." Each event must be compared with at least one other event and all events must be linked, i.e., each event i is compared with event i+1. Thus, n-1 ratio judgments are required.

To convert these ratios into probabilities, an independent probability estimate for one event is needed. Probabilities for other events are then computed by multiplying the ratio judged for the other event and the event with the independently estimated probability by the odds, $p/(1-p)$, for the independent estimate, and then converting these odds into probabilities. Individual estimates can be combined into a single, overall estimate using the same combination rule used for direct numerical estimates.

This procedure is basically a minor variation on the direct numerical estimation procedure with the same advantages and disadvantages. It may be somewhat less susceptible to bias, but the judgments required may be more difficult, requiring quantitative consideration of the likelihood of two events at the same time.

Success Likelihood Index Method (SLIM)

SLIM has been developed by Embrey (1982a,b) specifically for estimating human success probabilities. It is currently being further tested by Embrey with Brookhaven National Laboratory and is described here primarily for completeness, because it cannot be evaluated on the basis of the scant information currently existing regarding its use.

The method is based on multiattribute utility measurement, specifically, the technique advocated by Edwards (1977). The technique first identifies performance shaping factors (PSFs)--e.g., stress level, system design, degree of training, etc.--that have the greatest effect on the likelihood of successfully performing the tasks under consideration. These PSFs are weighted judgmentally with respect to their relative contribution to the success of the tasks.

Each task is then rated, on a 0-to-100 scale, with respect to how much each PSF affects the likelihood of success. For example, a 0 indicates that training increases the likelihood of error for that task, 50 indicates no effect, and 100 indicates that training makes success very likely.

An index of the likelihood of success is then

$$s_i = \sum_{j=1}^{n} w_j \; x_{ij}$$

where s_i is the SLIM index for event i, w_j is the normalized weight for PSF_j, and x_{ij} is the rating for event i on PSF_j. As with paired comparisons and ranking/rating, this index must be transformed, using a similar method, into a probability.

DETAILED DESCRIPTION OF PAIRED COMPARISON AND DIRECT NUMERICAL ESTIMATION PROCEDURES

Rather than describe in detail how to use each of the procedures summarized above, we describe here two procedures that represent the two basic types of procedures previously discussed. These two procedures--paired comparisons and direct numerical estimation--are also the procedures that have been selected for testing in the implementation and evaluation study. Therefore, they are the procedures for which the results of their use in estimating human error probabilities in nuclear power plant operations will soon be available.

The reason for selecting these procedures is that they represent the different advantages and disadvantages associated with using expert judgment to estimate probabilities. Paired comparisons are considered to be the least susceptible to bias, but require more judgments and a transformation from scale values to probabilities. Direct estimation provides probabilities and is relatively simple to use but may be susceptible to biased judgments. Direct estimation has also been used extensively to estimate probabilities in other contexts, whereas paired comparisons have not. Paired comparisons, however, are based on well-established psychological theory, while direct estimates are primarily empirically justified.

In addition to providing probability estimates, these procedures, with some variations, can also lead to uncertainty bounds on the estimates, which can be very important in risk analyses. These procedures, and methods for checking for consistency in judgments for each, are described below.

Paired Comparison Procedure

For the purpose of describing the procedure here, we assume that each expert judges each pair of events and that the judgment "equally likely" is not allowed. Procedures for reducing the number of judgments required are described in Seaver and Stillwell (1983) and Torgerson (1958).

Deriving Probability Estimates: The first step is to obtain an n-by-n matrix of frequencies with entries x_{ij} being the number of times event j was judged more likely than event i. This matrix is then converted into a matrix of proportions, t_{ij}, by dividing x_{ij} by m, the number of experts. A third matrix is then derived with entries z_{ij} equal to the standard normal deviate of t_{ij}.

When t_{ij} is equal to 0.0 or 1.0, the normal deviates would be plus and minus infinity. Although Torgerson (1958) suggests leaving these cells blank, this information is very diagnostic and should not be disregarded. Appendix B of Seaver and Stillwell (1983) offers several suggestions for determining these z values. With a relatively large number of events, the easiest procedure is to use the standard normal deviate of 1/2(m+1) for p=0.0, and the standard normal deviate of (2m+1)/2(m+1) for p=1.0. This results in a reasonable approximation to other techniques.

From the matrix of z_{ij}, scale values are computed as

$$s_j = \left(\sum_{i=1}^{n} z_{ij} \right) / n$$

where s_j is the scale value for event j. These scale values are then converted into probabilities by

$$\log p_j = as_j+b$$

The parameters a and b are determined from independent estimates of the probabilities of two events. These estimates may come from data from other sources or from direct estimates if no such data are available. The known scale values and probabilities for these two events are substituted into the above equation, producing a system of two equations that are then solved for the two unknowns, a and b. If probabilities are known for more than two events, a least squares solution can be used to determine a and b.

Estimating Uncertainty Bounds: Statistical uncertainty bounds can be found for probabilities estimated from paired comparisons by computing standard errors for the estimates. The procedure involves first obtaining a matrix of variances related to the matrix of proportions with entries t'_{ij}. (Primes are used here to indicate quantities estimated from the data.) The entries in this matrix are v_{ij}, the variance of the inverse cumulative normal distribution, F^{-1}, evaluated at t'_{ij}:

$$v_{ij} = \left[\sum_{x=0}^{m} F^{-1}(t'_{ij})\right]^2 P(x|t_{ij}) - \mu^2,$$

$$\text{where } \mu = \sum_{x=0}^{m} F^{-1}(t'_{ij})\, P(x|t_{ij})$$

$$\text{and} \quad P(x|t_{ij}) = \binom{m}{x} (t_{ij})^x (1-t_{ij})^{m-x}, \quad x = 1, \ldots, m.$$

Where t_{ij} is equal to 0.0 or 1.0, the variance is assumed to be the variance associated with the approximate z values discussed above.

These variances can then be used to estimate the standard error, se_j, for each estimated scale value, s_j:

$$se_j = \left(\sum_{i=1}^{n} v_{ij}/n(n-1)\right)^{1/2}.$$

Uncertainty bounds on scale values are then taken to be plus and minus twice the standard error. The uncertainty bounds of

$\log p_j$ are plus and minus two times a times se_j, which can be directly converted into bounds on p_j by taking antilogarithms.

If, in addition, we assume there is also uncertainty in the estimate of p_j contributed by uncertainty in the estimates of a and b (e.g., if a and b are derived through least squares procedures), then the variance of $\log p_j$, $V(\log p_j)$ is computed:

$$V(\log p_j) = V(as_j + b).$$

Assuming a and b are independent (an assumption discussed below),

$$\begin{aligned} V(\log p_j) &= V(as_j) + V(b) \\ &= E(a^2 s_j^2) - E(a)^2\, E(s_j)^2 + V(b) \\ &= E(a^2)\, E(s_j^2) - E(a)^2\, E(s_j)^2 + V(b) \\ &= V(a) + E(a)^2 \quad V(s_j) + E(s_j)^2 \; - E(a)^2\, E(s_j)^2 + V(b) \\ &= V(a)\, V(s_j) + V(a)\, E(s_j)^2 + E(a)^2\, V(s_j) + V(b) \\ &= V(a) \;\; V(s_j) + E(s_j)^2 \; + E(a)^2\, V(s_j) + V(b). \end{aligned}$$

Then, using s_j^2 and a^2 to estimate $E(s_j)^2$ and $E(a)^2$, respectively, and V to indicate other variance estimates,

$$\hat{V}(\log p_j) = \hat{V}(a)\;\; \hat{V}(s_j) + s_j^2 \; + a^2\, \hat{V}(s_j) + \hat{V}(b).$$

$V(s_j)$ is the variance estimate defined above, ie., $V(s_j) = se_j^2$. Variance estimates for a and b may be obtained, for example, from least squares procedures used to estimate them. To make a and b independent in estimating a and b, log probabilities would be regressed on scale values expressed as deviations from their mean. Then

$$\hat{V}(a) = \sigma^2 / \sum_{i=1}^{n} s_i^2, \text{ and } V(b) = \sigma^2/n$$

where σ^2 is the error variance from the regression and n is the number of variables in the regression.

Uncertainty bounds on log p_j can then be taken as plus and minus twice the square root of the variance. Bounds on p_j are found by taking antilogarithms.

<u>Consistency Measures</u>: For paired comparison judgments, two types of consistency may be checked. First, the internal consistency of each expert can be checked based on the number of circular triads in his judgments. Consistency across experts can also be computed to show how well the experts agree on their judgments.

Internal consistency can be measured by the coefficient of consistency (David, 1963). This coefficient varies from a value of 0.0, the expectation for completely random judgments, to 1.0 where there are no circular triads. (A circular triad is one in which a is judged more likely than b, b more likely than c, and c more likely than a.) The number of circular triads, c, is computed as

$$C = n/24\ (n^2-1) - T/2$$

where n is the number of events and

$$T = \sum_{i=1}^{n} (a_j - a)^2, \quad a = (n-1)/2$$

and a_i is the number of times event i was judged more likely than any other event. The coefficient of consistency, k, is then

$$k = 1 - \frac{24c}{n(n^2-1)}, \quad n \text{ odd}$$

OR

$$k = 1 - \frac{24C}{n(n^2-4)}, \quad n \text{ even}$$

The coefficient of concordance (Siegel, 1956) can be used to determine across-expert consistency. The rank orders for each expert needed to compute this coefficient are derived by counting the number of times an event was judged more likely than any other event.

Direct Numerical Estimation

In this procedure, each expert directly estimates the probability of each event in question. As noted above, we recommend that the experts indicate their estimates on a logarithmic probability scale. When dealing with probabilities in risk analysis that may be very small, as is true for HEPs, it is necessary for the logarithmic scale to include several orders of magnitude. The scale used in the implementation and evaluation study for HEPs in nuclear power plant operations, for example, went as low as 0.0000001.

Deriving Probability Estimates: The derivation of probability estimates is a straightforward matter of using the formula

$$p_j = \frac{\left(\prod_{i=1}^{n} p_{ij}\right)^{1/m}}{\left(\prod_{i=1}^{m} (1 - p_{ij})\right)^{1/m} + \left(\prod_{i=1}^{m} p_{ij}\right)^{1/m}}$$

to aggregate individual judgments into a combined estimate. While other aggregation rules may be used, they will generally lead to very similar estimates (Seaver, 1978).

Estimating Uncertainty Bounds: Uncertainty bounds for directly estimated probabilities can be estimated in two ways: 1) They can be estimated judgmentally by the experts; and 2) They can be estimated statistically. Judgmentally estimated uncertainty bounds are obtained by having each expert, in addition to the estimated probability, estimate the upper and lower uncertainty bounds for his judgment. These bounds should be defined for a specific, fixed credible interval, e.g., bounds such that the expert believes the probability is 0.95 that the correct error probability is between the bounds. These estimated uncertainty bounds can be combined across experts using the same aggregation rule used for probability estimates.

Statistical uncertainty bounds can be derived from standard error estimates. To obtain these, the variance across experts of log p_{ij} is calculated for each event j. The standard error for log p_j estimates is then defined as the square root of the quantity, the variance divided by the number of experts. Approximate 95 percent uncertainty bounds for log p_j are plus and minus twice the standard error. Bounds on p_j are found by taking antilogarithms.

Consistency Measures: Because of the type of judgment required, there is no opportunity for observing inconsistency in individual direct estimates. Therefore, the only consistency checks that can be made are across-expert. One such consistency measure is again the coefficient of concordance. This requires the direct estimates to be converted into rank orders. Because the coefficient of concordance is based only on rank orders, it does not use all of the information available in the direct estimates. A similar measure that makes better use of the available information is the intraclass correlation coefficient (Winer, 1971). It represents a measure of the average correlation between each pair of experts.

The intraclass correlation coefficient is estimated using the analysis of variance paradigm (ANOVA). Dependent measures used should be log p_{ij}, rather than p_{ij}. A one factor (events), repeated measures design is used to estimate variations due to differences among events and due to differences among experts. The intraclass correlation coefficient is computed as

$$r = \frac{F-1}{F+(n-1)}$$

where F is the F ratio for the events factor and n is the number of events.

IMPLEMENTATION AND EVALUATION STUDY

As was mentioned in the introduction, an implementation and evaluation study is currently underway to test the procedures described above for estimating HEPs in nuclear power plant operations. This effort will determine whether or not high-quality HEP estimates can be obtained using paired comparison and direct numerical estimation procedures and, thus, if these techniques can be used in nuclear industry PRAs. The data for this effort have been collected and are currently being analyzed. Results will be available in early to mid-1984. Because of the very preliminary stage of analysis, no results are presented here. Rather, we simply describe the procedures and suggest that people interested in the results look for a NUREG report to be published in 1984.

Two types of events are being included in the test. The first type of event is system-level job duties in which an estimate of the HEP for operating a system is obtained. An example of this type of event is:

> One of the main steam relief valves inadvertently opens. The operator, after successfully closing the valve, is monitoring the suppression pool temperature. The indicated temperature of the suppression pool is 95 degrees Fahrenheit. According to procedures, this requires that the residential heat removal (RHR) system be manually placed in the suppression pool cooling mode. What is the likelihood that the operator will fail to activate the suppression pool cooling mode of RHR?

Fifteen events of this type are being considered.

The second type of event is composed of task-level human actions in which equipment components are operated, and task element actions associated with controls, displays, etc. An example of this type of event is:

> An operator chooses the wrong switch from a set of switches that all look similar and are identified only by labels.

Twenty events of this type have been included in the study. All events are for boiling water reactors (BWRs).

The primary criterion for selecting tasks was that they be important to PRA practitioners. In addition, the 20 events of the second type were selected from events for which HEP estimates are available in Swain and Guttman (1983) and in Beare, Dorris, Bovell, Crowe, and Kozinsky (1984). This allows comparison of HEP estimates from expert judgment with those based on available data and the extensive experience of human reliability analysts from the former reference and estimates from training simulator studies from the latter reference.

A total of 19 experts were available to make the required judgments. Each made all paired comparisons for each set of events. (Events of the first type were not compared with those of the second type.) They also made direct numerical estimates including estimates of uncertainty bounds.

The experts were General Physics staff who are currently, or were formerly, licensed BWR control room operators, or are certified as BWR instructors. Each was actively involved in instructing BWR operators at the time of the study. It was decided to use these instructors because they would have a broader perspective on operator error than would actual operators and would be less likely to make biased judgments because of perceived self-control over the events. These instructors have observed operators with a wide variety of backgrounds and training, operating under a diversity of conditions. This experience should help to ensure high-quality judgments.

The process of obtaining the judgments indicated that either procedure is practical. The experts were able to make all required judgments within a period of about three hours, with paired comparisons taking approximately twice as long as direct estimates. Thus, if the results indicate that either (or both) procedure(s) produces reasonable HEP estimates, practical implementation appears to be feasible.

DISCUSSION AND CONCLUSION

We have presented here procedures that can be used to estimate probabilities using expert judgment. The procedures appear to be practically feasible although they require additional study to be sure that the estimates they produce are reasonably valid and reliable. A large step in this direction will be taken as a result of the implementation and evaluation study now underway.

The question remains, however, whether these procedures should be used and, if so, under what conditions? The answer to this question really depends upon one's perspective with respect to quantitative risk analysis. While we believe in the value of quantitative risk analysis, we recognize many of its limitations and accept that many people view quantitative risk analysis, given the current state of its application, as less than desirable. Other papers at this conference clearly address this controversy.

If, however, one accepts the desirability of quantitative risk analysis in a certain context, then there may be a role for these procedures. If quantitative risk analyses are to be performed, probabilities will be needed as inputs. For many such probabilities, there may be no way other than expert judgment to obtain the needed estimates, at least in the short-term.

We do not mean to imply that expert judgment should be used only because no other means of obtaining estimates is available. Clearly, the quality of risk analysis is limited by its inputs and we do not advocate garbage-in garbage-out analyses as better than nothing. Rather, we believe experience with expert judgment in many contexts has demonstrated that such estimates do have a reasonable validity.

Although for quantitative risk analyses in the short-term there may be no alternative to expert judgment, in the longer term efforts put into obtaining data using expert judgment must be compared against other methods for obtaining data. For example, in the HEP context, other methods include field data collection, simulator studies, and performance modeling. Generally, expert judgment will provide a quicker, cheaper method of obtaining data, but validity and acceptability issues must also be considered. There may be trade-offs that have to be made between data collection costs and the validity of data, though we by no means concede that methods other than expert judgment will produce more valid probability estimates.

To the extent that qualitative rather than quantitative risk analyses are desired, these procedures are relatively less useful. They can, however, still be used where qualitative input is needed, particularly the paired comparison (and ranking/rating) procedure.

We acknowledge that additional empirical support is needed to justify use of these procedures. (The current study will contribute to satisfying this need.) These procedures, however, have been designed to take advantage of the information available in the minds of experts, while minimizing the biases that may arise as this information is elicited; and, furthermore, they are practical to use. In conclusion, we suggest that these procedures, used with care (e.g., carefully defined events, carefully selected experts, carefully developed instructions), can provide high-quality probability estimates.

REFERENCES

Beare, A.N., Dorris, R.E., Bovell, C.R., Crowe, D.S., and Kozinsky, E.J. A simulator-based study of human errors in nuclear power plant control tasks (NUREG/CR-3309, SAND 83-7095). Chattanooga, TN: General Physics Corporation, 1984.

Bordley, R.F. A multiplicative formula for aggregating probability assessments. Management Science, 1982, 28, 1137-1148.

David, H.A. The method of paired comparisons, New York: Hafner, 1963.

Edwards, W. How to use multiattribute utility measurement for social decisionmaking, IEEE Transactions on Systems, Man, and Cybernetics, 1977, SMC-7, 326-340.

Embrey, D.E. A new approach to the evaluation and quantification of human reliability in systems assessment. Proceedings of Third National Reliability Conference - Reliability 81, 1981, Birmingham, England, 5B/1/1-5B11/12.a

Embrey, D.C. The use of performance shaping factors and quantified expert judgment in the evaluation of human reliability: An initial appraisal. Skelmersdale, England: Human Reliability Associates, Research Report No. HR-BNL-2, September 1981.b

Kemeny, J.G. Reports of the technical assessment task force (Volume 1). The President's Commission on the Accident at Three Mile Island, Washington, DC: Government Printing Office, 1979.

Luce, R.D. Individual choice behavior. New York: Wiley, 1959.

Rogovin, M. and Frampton, G.I., Jr. Three Mile Island (Volume 1). Washington, DC: Nuclear Regulatory Commission Special Inquiry Group, NUREG/CR-1250, January 1980.

Seave, D. A. Assessment of group preferences and group uncertainty for decision making (SSRI Research Report 76-4). Los Angeles: University of Southern California, Social Science Research Institute, 1976.

Seaver, D.A. Assessing probability with multiple individuals: Group interaction versus mathematical aggregation (SSRI Research Report 78-3). Los Angeles: University of Southern California, Social Science Research Institute, December 1978.

Seaver, D.A. and Stillwell, W.G. Procedures for using expert judgment to estimate human error probabilities in nuclear power plant operations (NUREG/CR-2743, SAND82-7054). Falls Church, VA: Decision Science Consortium, Inc., March 1983.

Siegel, S. Nonparametric statistics for the behavioral sciences. New York: McGraw-Hill, 1956.

Stillwell, W.G., Seaver, D.A., and Schwartz, J.P. Expert estimation of human error probabilities in nuclear power plant operations: A review of probability assessment and scaling (NUREG/CR-2255, SAND81-7140). Falls Church, VA: Decision Science Consortium, Inc., May 1982.

Swain, A.D. and Guttman, H.E. Handbook of human reliability analysis with emphasis on nuclear power plant applications (NUREG/CR-1278). Albuquerque, NM: Sandia National Laboratories, 1983.

Thurstone, L.L. A law of comparative judgment. Psychological Review, 1927, 34, 273-286.

Torgerson, W.S. Theory and methods of scaling. New York: Wiley, 1958.

Winer, B.J. Statistical principles in experimental design. New York: McGraw Hill, 1971.

Scientific and Technological Uncertainty in Quantitative Assessment and Policy Analysis[1]

M. Granger Morgan

Department of Engineering and Public Policy, Carnegie-Mellon University, Pittsburgh, PA 15213

INTRODUCTION

Scientific and technological uncertainty is common to most quantitative assessment and policy analysis that involves problems in technology and public policy. In my three lectures at the NATO Advanced Studies Institute and in this summary paper, my primary concern is with the questions of how, and to what extent, one should characterize and deal with such uncertainty. Of course, these questions are normative in nature and their resolution requires some elaboration of the objectives that are to be served by quantitative policy analysis. In other writings, I have detailed a variety of substance-focused, process-focused, and analyst-focused motivations that can, in varying situations, underlie the commissioning, execution, and use of quantitative assessment and policy analysis (1.2). The interesting feature of these motivations is that despite their diversity it is possible to extract from them a single set of desiderata to characterize "good" risk assessment and other quantitative policy analysis. These are shown in Exhibit 1. Of course, people operating with different motivations are likely to have somewhat different views about the relative importance that should be attached to these different desiderata (2).

---

1 Work discussed in this paper has been supported by a variety of sources including the National Science Foundation, the U.S. Department of Energy, Brookhaven National Laboratory and Carnegie-Mellon University. I am pleased to acknowledge the assistance of M. Henrion, S. Morris, I. Nair, D. Amaral, W. Rish, E. Morgan, P. Steranchak and others.

NATO ASI Series, Vol. G4
Environmental Impact Assessment, Technology Assessment, and Risk Analysis. Edited by V.T. Covello et al.

EXHIBIT 1: Desiderata for risk analysis or other quantitative policy analysis.

A. Clearly define the boundaries of the analysis and provide a careful rationale for their selection.

B. Characterize and deal with all scientific, technological, political, social and economic uncertainties as completely and as explicitly as possible.

C. Explicitly identify all value assumptions and, to the extent possible, treat them in a parametric fashion so that people with different value orientations can use the analysis to reach their own conclusions.

D. Let the policy issues play a central role in driving the development of the models (or analytical tools). Develop these in an iterative fashion so as to assure, through careful sensitivity analysis, that all important variables are identified and properly incorporated and that the models chosen are as robust as possible. Explicitly describe this process and use the findings to justify the choice of models.

E. Keep models (or analytical tools) as simple, clear and understandable as possible. Avoid complexities that are not important to the policy issue(s) of interest. Avoid analytical and other complexity for its own sake.

F. Present results in a clear open manner which makes all assumptions and operations explicit and allows others easily to verify, use, modify and extend the analysis.

G. To the extent possible, describe how the conclusions reached may be affected by the problem boundaries that were selected.

CLASSIFYING AND DEALING WITH UNCERTAINTY

On the basis of the preceding argument, I feel confident in asserting that "good" quantitative assessment and policy analysis must explicitly characterize and deal with scientific and technical uncertainties. The questions of how it should do this, and what constitutes an adequate treatment, are more difficult. Scientific and technological uncertainty can take a number of forms (3), of which the most important for our purposes are situations in which: 1) the functional relation among important variables or coefficients is known but the value of some of these variables or coefficients is not or cannot be known; and 2) the physics, chemistry, biology, and other science and engineering of the problem involved is not fully understood so that the functional relationships among important variables or coefficients are not completely or accurately known.

In the first situation, the functional relationships among the important coefficients and variables in the problem are known so that the policy analyst can build appropriate models. The problem of dealing with uncertainty then becomes one of how to characterize and then propagate the uncertainty in coefficients and variables through these models. Depending upon the extent of the uncertainty, there are a variety of analytical strategies which an analyst can adopt. These include: 1a) Perform single-value-best-estimate analysis and ignore the uncertainty; 1b) Perform single-value-best-estimate analysis, and then acknowledge the uncertainty, perform various sensitivity calculations, and provide a qualitative and/or quantitative discussion of the uncertainty; 2a) Estimate some coefficient of uncertainty, such as the standard deviation, for each important model coefficient and then use analytical procedures for "error propagation" to propagate this uncertainty through the analysis; 2b) Characterize uncertain coefficients as subjective probability distributions and then propagate this uncertainty through the analysis, usually through the use of discretized decision or probability trees, or, if continuous distributions are retained, through the use of stochastic simulation; 3a) Treat some coefficients parametrically, performing the analysis for a variety of plausible values of each of these coefficients; and 3b) Perform order-of-magnitude based bounding analysis which does not produce unique "answers" but rather estimates bounds on the range of possible answers.

For simplicity, we can refer to the first two of these as single-value-best-estimate analysis, to the second two as probabilistic analysis, and to the final two as parametric/bounding analysis. Most quantitative policy analysis, including most risk assessments, performed today use single-value-best-estimate techniques, much of it of Type 1a. A fair number of analyses make some modest use of parametric/ bounding techniques. Only a handful of analyses, including virtually all those called decision analyses, use probabilistic techniques.

Which analytic strategy or mix of strategies is appropriate for a given policy analysis problem? The answer depends largely upon the amount of associated uncertainty. I have listed the strategies roughly in order of increasing problem uncertainty. Thus, Strategy 1a is most appropriate when model coefficients are well known and Strategy 3b is most appropriate when very little is known about the value of these coefficients.

The options that the analyst has available when faced with the second form of technological uncertainty--that is, when there is significant uncertainty about the functional form of the relationships among important coefficients and variables--follows a similar pattern, though there has been considerably less research on this subject and it is considerably harder to point to good examples in the applied literature. The viable strategies include: 1) Perform a single-model-best-estimate analysis, perhaps using uncertainty in the value of model coefficients as a crude way of capturing uncertainty in model form as well; 2) Identify the range of possible alternative models. Perform a separate analysis for each alternative. Then, if one is prepared to make subjective judgments about the relative likelihood of alternative models, combine the results probabilistically, thus at least in operational terms converting the problem to one like 2a or 2b above; 3) Perform order-of-magnitude bounding analysis that is designed to determine how much changes in model form affect the conclusions reached so as to establish bounds on the range of possible answers.

Many real policy problems, of course, involve both uncertainty in model coefficients and uncertainty in model form, and the degree of that uncertainty is often quite varied across different parts of the problem, so that a mixture of analytical strategies is called for.

The obvious next question is: Precisely when is one of these forms of analysis to be preferred to another? To some extent the answer is a matter of professional judgment and aesthetics. Indeed, Baruch Fischhoff has argued that quantitative policy analysis is to a considerable degree a clinical art still evolving into a clinical science. He argues that one may not be able to fully spell out how to do good analysis but that "the finer points of the craft can be learned only in the clinic of a master" (4). Nevertheless, he would agree that much is to be gained by trying to identify, systematize, and verbalize the things a master craftsman does. Indeed, in the same article he argues, "One place to start would be a taxonomy of decision situations indicating which techniques to use where."

While it is in general more difficult and frequently impossible to build causative models that handle the social, political, and economic aspects of policy problems, I believe that uncertainty involving social, political, and economically related variables can be handled in ways that are analogous to those listed above, but we will not be concerned with the problems of dealing with such uncertainties in this paper.

In general, I do not believe that most value issues or decision variables over which the decisionmaker has control (e.g., the rate of investment to prevent accidental death, or the level of control that will be required for a particular pollutant) should be treated in the same way as other uncertain variables. In most cases, I am particularly uncomfortable about characterizing such variables with subjective probability distributions. Rather, it seems to me that the analysis should be set up in such a way as to illustrate the way in which the model results depend upon the values chosen for such variables (e.g., with the use of sensitivity or parametric analysis).

SUBJECTIVE EXPERT JUDGMENTS

For many problems the best way to characterize scientific and technical uncertainties is to ask experts to make careful informed judgments in the form of subjective probability distributions. Over the past 15 years a number of experimental psychologists, including Fischhoff, Kahneman, Lichtenstein, Slovic, and Tversky, have made considerable progress in understanding the way in which people think about and make judgments about uncertain quantities and events, and in identifying the kinds of biases which can develop in such judgments. Because several of these authors are participants in the NATO Advanced Studies Institute, I will not review this literature here but will briefly summarize and refer readers to other papers in this volume for details.

In making judgments about uncertain quantities and events, people appear to use a variety of heuristics, which may produce reasonable results in many day-to-day judgments, but can in some situations give rise to very significant biases (5). While a number of authors have explored the practical and theoretical problems of minimizing these biases when seeking subjective expert judgments for use in quantitative assessment and policy analysis (6), there are not yet any clear solutions. The process of expert elicitation is still more an art than a science, and while they are intuitively appealing the debiasing strategies employed by analysts, these strategies are not yet based upon very much empirical evidence.

It is important to understand that these problems of cognitive biases do not apply uniquely to expert judgments that are being made for use in quantitative policy analyses. The same biases can be expected to emerge in less formal schemes for seeking expert guidance and advice. One advantage of formal elicitation and quantitative analysis is that it can both help one to focus on the problem of minimizing cognitive biases and, through appropriate analysis, help one to systematically understand how conclusions reached are likely to depend upon the possible biases which may exist.

Two basic philosophies have guided all of my own policy analytic work that has made use of expert subjective judgments. First, there is the notion that the science and engineering knowledge and thought patterns of the experts should be allowed to drive the model formulation. Experts should be asked to provide those judgments which they can make most conveniently and naturally and the model should build around these judgments. Analysts should work hard to avoid circumstances in which experts are being asked to fit their answers into an analyst's pre-conceived models. The reason for this is a belief that the more mental gymnastics the expert must perform in making his or her judgments, the more likely these judgments are to be biased.

Second, in most analyses I have done on public sector policy problems, I have attempted, at least in the early stages of analysis, to avoid combining experts' opinions or pushing for consensus. Rather, I have sought to first analytically explore the extent to which the range of serious expert opinions is important in terms of the kinds of conclusions one draws.

In concluding my first lecture, I described in some detail the protocol we adopted in a recent set of interviews of nine of the leading atmospheric scientists and seven of the leading air pollution health effects experts in the United States as part of an analysis we performed of possible health effects of the long-range transport of sulfur air pollutants from large coal-fired power plants. Readers interested in the details of these protocols will find detailed descriptions in references 7 and 8.

A SULFUR AIR POLLUTION EXAMPLE

In my second lecture at the NATO Advanced Studies Institute, I discussed a quantitative policy analysis based on the use of subjective expert judgment. Because this work has been extensively described elsewhere (7,8) I will provide only a brief summary here.

There is considerable uncertainty associated with both the exposure processes (transport, diffusion, oxidation, wet and dry deposition, etc.) and the possible effects processes that may lead to health consequences from human exposure to sulfur air pollution from coal-fired power plants. In this work, we conducted interviews with a set of leading atmospheric scientists and obtained their judgment about how best to model annual average sulfate exposure and the subjective probability distributions necessary to build these models. We used these results to construct separate long-range sulfur mass-balance models for each expert. We used these models to examine the relative impact of the uncertainty expressed by individual experts and the variation between experts on the conclusions one draws about long-range annual average sulfur mass-balance in power plant plumes. Typical results for one expert are shown in Exhibit 2 and a comparison of the results for all experts is shown in Exhibit 3.

We conducted extended interviews with a set of leading health effects experts to obtain their subjective judgments about the possible health impacts of the sulfur air pollutants of coal-fired power plants. These results were used to construct a set of quantitative health damage models and a critical comparison of these models was conducted. The long-range mass-balance models were then linked with a long-range plume trajectory model and these were in turn combined with the various health damage models in order to estimate the health impact of sulfur air pollutants from coal-fired power plants and to explore the relative contributions of the various uncertainties that are involved. Example findings are reproduced in Exhibit 4.

We draw three specific and one general conclusion from this work: 1) While significant uncertainty still exists, there is a considerable consensus developing within the atmospheric science community about many of the basic mechanisms of long-range sulfur transport in power plant plumes. In estimating the annual sulfur mass balance in such plumes, uncertainty expressed by individual experts is always comparable to and sometimes greater than the uncertainty which results from differences between experts; 2) One can set a fairly high upper bound on the possible health impacts of chronic exposure to low-level sulfate air pollution. Between this bound and the lower limit of no effect, there is no agreement across the set of air pollution health effects experts about the likely health impacts of sulfate. Within this group one can get almost any answer, including the answer that with 100 percent probability there are not adverse health

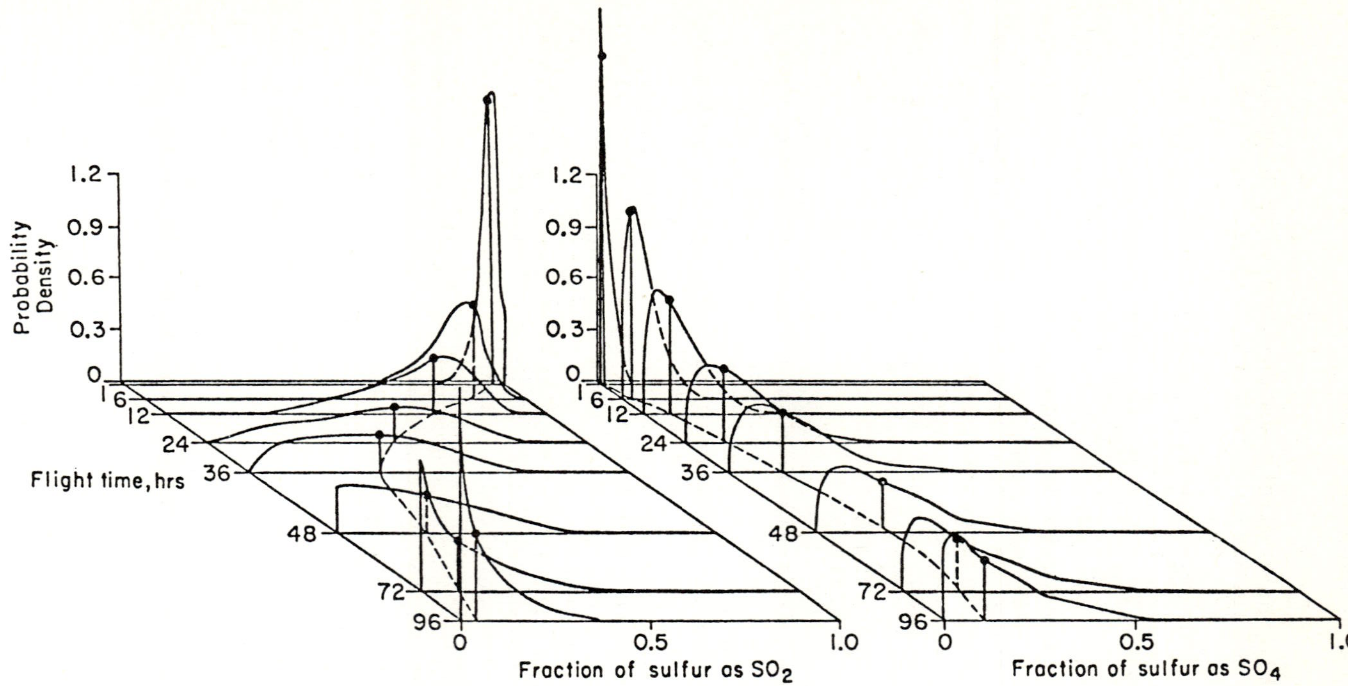

EXHIBIT 2: Probability density functions for the total amount of sulfur emitted from the plant which is present as sulfur dioxide (left) and sulfate aerosol (right) as a function of flight time using the model specifications and subjective probability distributions provided by expert 1.

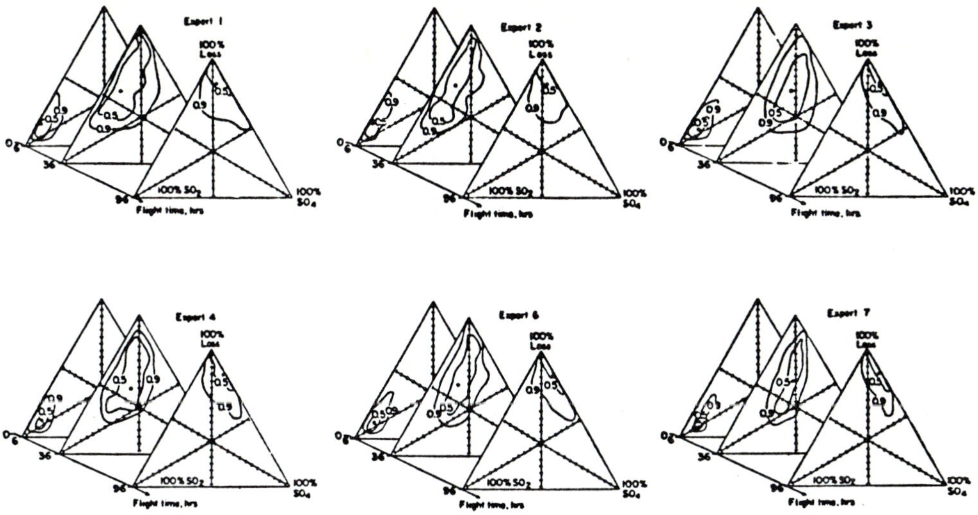

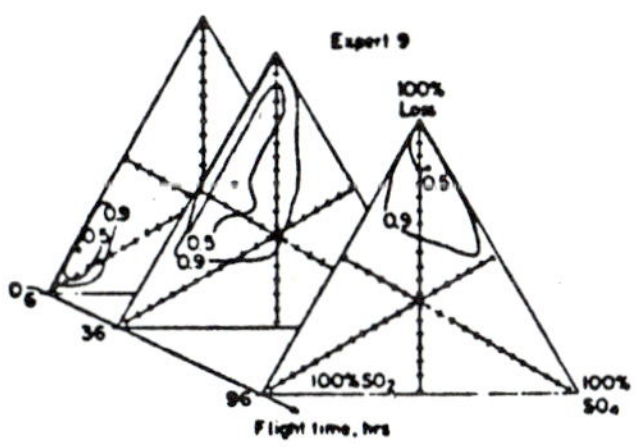

EXHIBIT 3: Probabilistic plots of mass balance as a function of plume flight time based on the models and subject probability distributions provided by each of seven atmospheric science experts. Inner contours report the 50% confidence region. Outer contours report the 90% confidence region.

impacts, depending upon which expert one talks to; 3) Even fairly crude estimates suggest that the potential value of future research in this field is high. Improvement in our understanding of health effects is clearly the most needed, but probably also the most difficult to obtain. Improvement in our knowledge of the values of the coefficients associated with wet and dry loss process would also be desirable in the context of this problem, and probably of much greater value in the companion problem of "acid rain;" 4) Techniques for quantitatively characterizing and dealing with uncertainty in policy analysis of the types that have been demonstrated and refined in this research can offer important insight and illumination in the policy-making process. Such techniques require greater effort and care than traditional single-value-best-estimate analysis but also produce a much

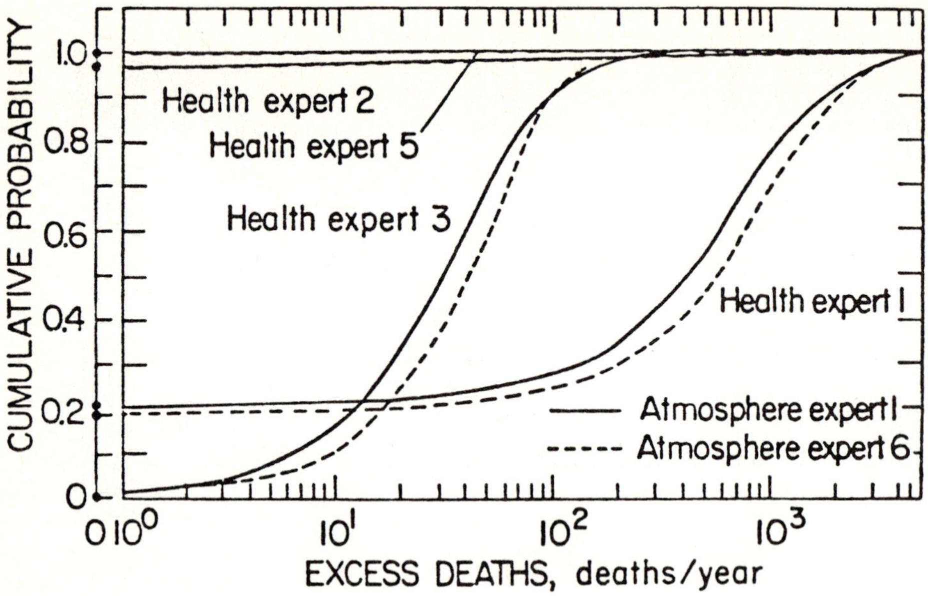

EXHIBIT 4: Estimate, in the form of cumulative distribution functions, of the annual average health impact in excess deaths from exposure to sulfate air pollution from a new 1 Gwe FGD equipped coal-fired power plant located in Pittsburgh, PA for the health effects models of health experts one, two, three and five using the atmospheric models for atmospheric experts one (solid curves) and six (dashed curves) who represent the range of views among atmospheric experts.

clearer statement of our state of understanding. The wider application of such techniques should be vigorously encouraged.

DEMOS...A SOFTWARE TOOL

In Exhibit 1, I listed a set of desiderata for quantitative policy analysis. One way to encourage analysts to do some of the things listed in this exhibit is to provide them with computer software which makes it easier to undertake such analysis. My colleague, Max Henrion, and I have developed such a system, known as Demos, the design goals for which are summarized in Exhibit 5.

EXHIBIT 5: Based on the discussion of the desiderata B to F in Exhibit 1, this table lists the design goals we set out to meet in designing the Demos computer aid.

A computer aid for risk analysis should:

1. provide convenient means for representing and computing values as probability distributions, both discrete and continuous.

2. provide a variety of means for sensitivity analysis and parametric analysis, both deterministic and stochastic, and support management and comparison of alternative model structures.

3. allow both data values and model structure to be defined and modified interactively, supporting both elaboration and simplification of distributions and model structure. The system should automatically check and maintain self-consistency of data values and models structure as they are changed.

4. represent models as a set of modular elements which can be separately specified in non-procedural form using conventional algebraic syntax, so programming in the sense of sequential imperative statements is unnecessary.

5. represent documentary and explanatory text in a form that is integrated with the mathematical model structure to facilitate communication and understanding of the model.

My third lecture at the NATO Advanced Studies Institute provided a summary discussion of design philosophy and operational details of the Demos system. These matters have been discussed at length in previous writings to which interested readers are referred (9,10).

In brief, Demos (for Decision Modeling System) is a non-procedural interactive software environment for structuring, analyzing, and communicating quantitative engineering/economic policy models. It is particularly well suited to problems which involve the treatment of uncertainty. Demos has several characteristics which differentiate it from most other general purpose modeling languages. These are (10):

- Interaction: Model development is usually an interactive process, in which successive versions of a model are constructed, criticized, and modified, as the modellers identify key issues and deepen their understanding of the problem. To support this both the parameter values of the model and the structure of the model itself may be defined and edited interactively. Changes may be permanent or exploratory in a "what if?" mode. Multiple versions may be stored simultaneously.

- Non-Procedural Modelling: The user defines variables and their functional relationships in any convenient order, and does not have to program in the conventional sense of specifying control of execution sequences. Conceptually the elements of the model are declared in parallel with each other. The syntax is checked and compiled incrementally as it is defined bit by bit. The system determines when and how to evaluate the model and how to store intermediate results according to the dictates of efficiency and consistency, relieving the user from having to worry about these issues.
- Representation of Uncertainty: If the value of a quantity is known, then it may be specified as a simple number or an array, as in many conventional languages. If it is uncertain, the value may be a probability distribution, which can be specified by a standard distribution function (uniform, normal, lognormal, etc.) or by an arbitrary cumulative distribution function. Dependent variables are evaluated probabilistically using Monte Carlo simulation. For some quantities, particularly those that represent value judgments (such as the investment rate for life saving), instead of representing uncertainty by a probability distribution, it is often more useful to specify a vector of alternative values. Any variable that depends on multiple-valued quantities will be evaluated simultaneously for each value of each such quantity, thus producing a multi-dimensional parametric analysis.

- o <u>Integrated Documentation</u>: Much of the work in policy modelling and decision analysis consists of understanding models, their assumptions and implications, and communicating them to reviewers and clients. Demos encourages the user to incorporate descriptive and explanatory tests within the computer representation. This can aid understanding of the model by its authors and by reviewers during on-line examination, and it helps in the semi-automatic preparation of comprehensible hardcopy reports. It also helps in maintaining consistency between the model structure and documentation.

In my lecture at the ASI, I provided a rather detailed illustration of a Demos application related to a risk assessment of a hypothetical air pollutant which I called TXC (9). Specifically, we began with a model of the form

$$Xd = Pop*Conc*H$$

where Xd is the annual number of extra deaths, Pop is the number of people exposed to TXC, Conc is the annual average concentration of TXC, and H is the slope of a linear no-threshold health damage function. As an example, if Pop is 30 million people, Conc is 5 $\mu g/m^3$ and H is 10^{-6} excess deaths/person $\mu g/m^3$ year, then Xd=150 extra deaths/year.

An illustration of how such a model is implemented in Demos is provided in Exhibit 6. Exhibit 7 converts several of the deterministic variables in this model to probabilistic form. Exhibit 8 adds a threshold to the effects function. In the lecture, I continued this illustration to deal with issues of control, optimum level of control, expected value of perfect information, and so on. Interested readers can find details elsewhere (9).

CONCLUSIONS

Scientific and technical uncertainty is a common feature of most quantitative assessment and policy analysis that involves problems in technology and public policy. The adequate characterization and treatment of such uncertainty has, until recently, received too little attention. A variety of analytical techniques is available, some of which involve the use of expert subjective judgment. Since the application of many of these techniques can place significant analytical burdens on the analyst, their general use in policy analysis is likely to require the availability of computer software modelling environments such as that provided by Demos (11).

EXHIBIT 6: Transcript of a Demos session in which the model discussed in the text is constructed in the Demos system and evaluated. Underlined text was supplied by the user. Remarks in italics are explanatory comments added after the session.

First we define a model in its simplest form:

```
>Variable Xd
Title: Excess deaths            Demos prompts for the Attributes of the new variable
Units: d/year
Description: Extra deaths per year attributed to exposure to TXC
Definition: Pop*Conc*H

Pop is undeclared, in the Definition of Xd
Do you want to declare it? [Yes]:                          Default "Yes" accepted.

Declaring Variable Pop
Title: Population
Units: people
Description: Population exposed to TXC
Definition: 30M                                             means 30 Million

Conc is undeclared, in the Definition of Xd
Do you want to declare it? [Yes]:

Declaring Variable Conc
Title: Concentration
Units: ug/m^3
Description: Average annual ambient concentration of TXC in the atmosphere
Definition: Lognormal(5, 2)
                   A distribution with median 5 and a factor of uncertainty of 2

H is undeclared, in the Definition of Xd
Do you want to declare it? [Yes]:

Declaring Variable H
Title: Health dam coeff
Units: d/yr/ug/m^3
Description: Slope of health damage function: Increase in annual death rate
      due to an increment in ambient TXC concentration of 1 ug/cubic meter.
Definition: Fractiles[0, 0, 0, .25, .6, 1, 1.5, 2.2, 3, 5, 10]/1M
 These fractiles are the values of H at 0%, 10%, 20%, ... 100% cumulative probabilities.

>XD                          Type the variable name to get its deterministic value.
Xd          : Excess deaths        (d/yr         ) =    150

>Why Xd                        To examine what this value was derived from.
Xd          : Excess deaths        (d/yr         ) =    150
    Description: Extra deaths per year attributed to exposure to TXC
    Xd = Pop*Conc*H

Pop         : Population           (people       ) =    30M
Conc        : Concentration        (ug/m^3       ) =      5
H           : Health dam coeff     (d/yr/ug/m^3) =      1u          means 10^-6
```

EXHIBIT 7: Displays of the cumulative distribution and probability density function for Xd, the excess deaths.

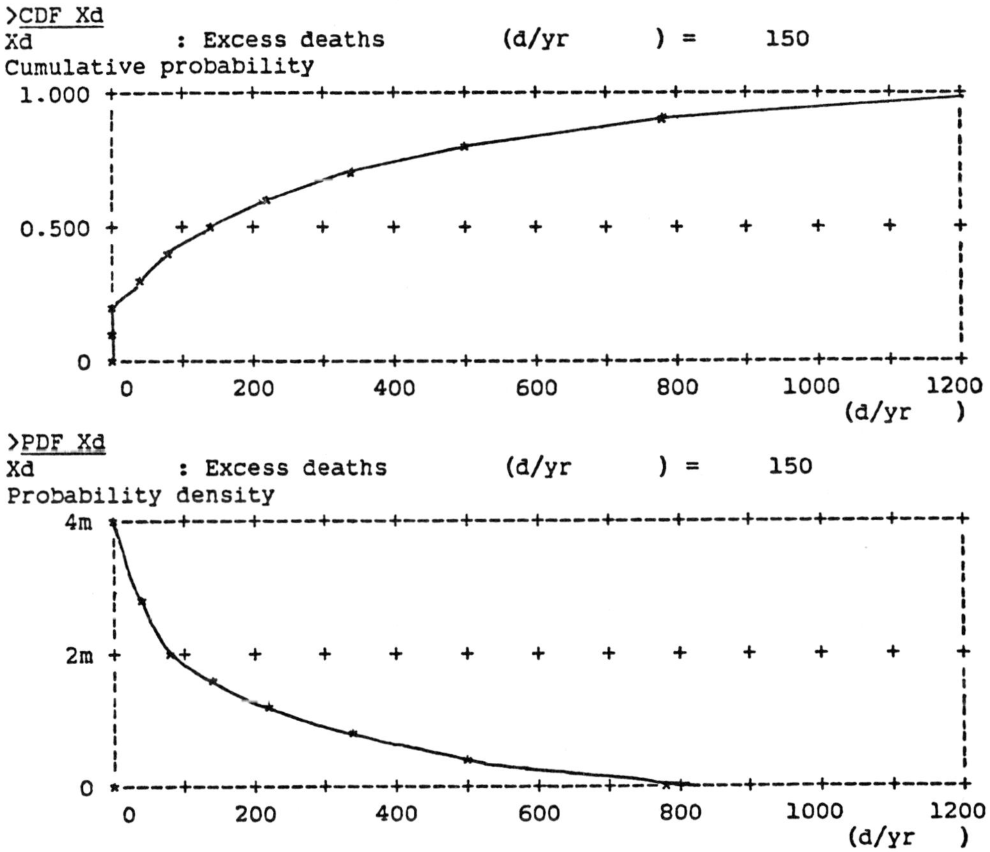

EXHIBIT 8: Interactive Demos session in which the model is extended to explore the effects of a possible threshold below which mortality does not occur.

```
>Xd := Pop*Hdf
```

Assigns a new definition to Xd.

```
Hdf is undeclared, in the Definition of Xd
Do you want to declare it? [Yes]:

Declaring Variable Hdf
Title: Health damage fn
Units: d/yr
Description: Increment in annual death rate per individual from exposure.
Definition: (Conc - Th)*H*IfPos(Conc - Th)
```

If the argument of Ifpos is positive its value is 1 else it is 0

```
Th is undeclared, in the Definition of Hdf
Do you want to declare it? [Yes]:

Declaring Variable Th
Title: Threshold
Units: ug/m^3
Description: Threshold below which TXC has no effect.
Definition: [0, 2.5, 5, 7.5]
```

Try a range of alternative values for a parametric analysis of effect of threshold: inducing a set of corresponding values on Xd.

```
>xd
Xd            : Excess deaths      (d/yr      ) =
   Th              0 2.500      5 7.500
          [    150     75      0      0]
```

Means are derived from a probabilistic computation using the entire distributions. Since the distributions are highly skewed the results are quite diferent from the deterministic values shown above.

```
>Mean(Xd)
   Th              0 2.500      5 7.500
          [ 314.9 182.5 92.78 45.15]
>CDF Xd
Xd            : Excess deaths      (d/yr      ) =
   Th              0 2.500      5 7.500
                   *     O      %     $
```

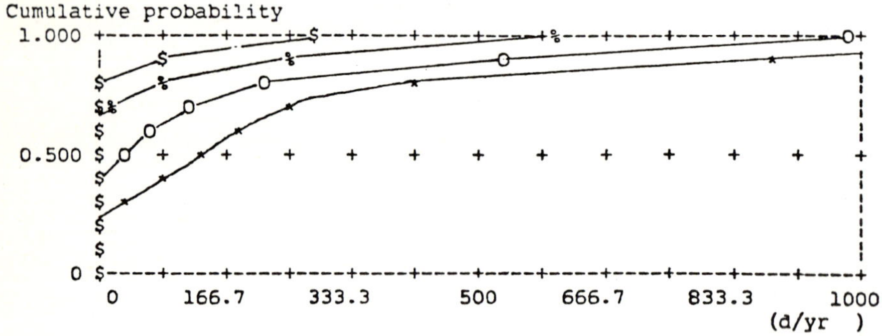

REFERENCES

1. M. Granger Morgan, "Uncertainty and Quantitative Assessment in Risk Management," paper prepared for the American Chemical Society Symposium on Assessment Health Risks from Chemicals, Kansas City, USA, September 15-16, 1983.

2. M. Granger Morgan, "The role of decision analysis and other quantitative tools in environmental policy analysis," Tutorial Paper Prepared for the Chemicals Division of the Environment Directorate, OECD, Paris, January 1983.

3. Most of the discussions of this section quote directly from an earlier discussion provided in reference 2 above.

4. Baruch Fischhoff, "Clinical Decision Analysis," Operations Research, 28, pp. 28-43, January/February 1980.

5. Daniel Kahneman, Paul Slovic and Amos Tversky, (eds.), Judgment under uncertainty: Heuristics and biases, Cambridge University Press, 1982.

6. See, for example, M. Granger Morgan, Max Henrion, and Samuel C. Morris, "Expert judgments for policy analysis," Brookhaven National Laboratory, Upton, NY, 11973, USA, BNL 51353, 1981; Carl-Axel S. Stael von Holstein and James E. Matheson, A Manual for Encoding Probability Distributions, SRI International, Menlo Park, CA, USA, August, 1979, and Dean W. Boyd and Stephen G. Regulinski, Characterizing Uncertainty in Technology Cost and Performance, Decision Focus, Inc. (report 1114), Palo Alto, CA, USA, June 1979.

7. M. Granger Morgan, Samuel C. Morris, Max Henrion, Deborah A.L. Amaral, and William R. Rish, "Technical uncertainty in quantitative policy analysis: A sulfur air pollution example," preprint, Department of Engineering and Public Policy, Carnegie-Mellon University, Pittsburgh, PA, 15213, USA, April 1983.

8. M. Granger Morgan, Deborah Amaral, Max Henrion and Samuel Morris, "Technological uncertainty in policy analysis," final report to the NSF Division of Policy Research and Analysis on grant PRA-7913070, Department of Engineering and Public Policy, Carnegie-Mellon University, Pittsburgh, PA, 15213, August 1983 (available through NTIS).

9. Max Henrion and M. Granger Morgan, "A Computer Aid for Risk and Other Policy Analysis," Risk Analysis, (in press).

10. Max Henrion and Indira Nair, "Demos User's Manual," Department of Engineering and Public Policy, Carnegie-Mellon University, Pittsburgh, PA, 15213, USA, November 1982.

11. A more recent and more detailed discussion of many of the subjects discussed in this paper can be found in M. Granger Morgan, Max Henrion, Indira Nair, Theresa Mullin, and Charlie Wiecha, "A generic 'pre-HEED' on characterizing and dealing with uncertainty in health and environmental risk assessment," prepared for the U.S. Department of Energy, Office of Health and Environmental Research, under contract DE-AC02-83ER60138, NTIS, Washington, D.C., USA, 1983.

Part II: Case Studies

An Approach for Assessing Health Risks Associated With Alternative Ambient Air Quality Standards

Miley W. Merkhofer

SRI International, Menlo Park, CA

INTRODUCTION

This paper proposes a quantitative health risk assessment approach for use by the U.S. Environmental Protection Agency (EPA) in setting national ambient air quality standards.[1] The approach is based on techniques and concepts applied in previous analyses of air pollution control and is compatible with formal risk evaluation and decision analysis methodology.

DESCRIPTION OF EPA'S HEALTH RISK ASSESSMENT PROBLEM

The United States Clean Air Act requires the EPA Administrator to set national primary ambient air quality standards. According to the law, the standards must be set at a level that will "protect the public health" with an "adequate margin of safety" (Clean Air Act, Section 109). A "margin of safety" is desired because the health impacts resulting from a particular standard are highly uncertain due to incomplete and sometimes contradictory scientific evidence, especially in the area of human dose-response relationships. The Administrator is responsible for determining which health effects should be

---

1 An ambient air quality standard is a maximum permissible concentration level for a pollutant or class of pollutants, averaged over some period of time. Both primary standards (set to protect public health) and secondary standards (set to protect public welfare) are proposed by the Administrator. The states are responsible for specifying emission limitations and other programs for individual sources necessary to achieve the standards.

NATO ASI Series, Vol. G4
Environmental Impact Assessment, Technology Assessment, and Risk Analysis. Edited by V. T. Covello et al.

considered adverse and for defining an adequate margin of safety.

"Protecting public health" is interpreted by the EPA to imply standards that prevent the most sensitive group of citizens from experiencing adverse health effects. Limited guidance is available for helping the EPA define the most sensitive population groups. Legislative history makes it clear, however, that standards need not be so stringent as to protect the most sensitive members of the most sensitive group.

Because health is the only basis mentioned in the Clean Air Act, the EPA's interpretation is that economic costs of attainment should not be a factor in standards setting (Gage, 1979).[2] Setting standards at unnecessarily restrictive levels, however, is recognized as producing undesirable economic and social disruption. It might be concluded, therefore, that for each pollutant the Administrator's objective is to identify a least stringent standard that protects public health with a margin of safety. To this end, risk assessment methodology for estimating the health effects resulting from each alternative standard and the uncertainty in the estimates would be useful.

ESSENTIAL CHARACTERISTICS FOR THE DESIGN OF HEALTH RISK ASSESSMENT METHODOLOGY

Recent applications of quantitative methods to air pollution control decisions suggest four considerations important to the design of a health risk assessment approach:

- There is a need to account explicitly for judgmental information in the assessment of health risks. Limited directly applicable hard scientific evidence exists to establish a relationship between ambient concentrations and health effects. However, health experts have a large base of indirectly applicable information. Incorporating this information requires

---

2 This interpretation was upheld by the U.S. Court of Appeals for the District of Columbia in <u>Lead Industries Association v. EPA</u> (1980). Costs and other factors may be considered by the states in developing implementation plans.

assessment of judgmental uncertainties from experts; this is extremely difficult and requires extensive knowledge of, and practical experience with, interviewing and probability encoding techniques.

- Creativity in formulating methods of analysis that effectively use available data is essential. Procedures should be avoided that limit creativity or constrain the assessment by forcing specific input requirements or assumptions, such as specific functional forms for dose-response relationships or the shape of probability distributions.

- Constructing risk assessment models can be complicated and costly. As more information is gathered, these models must be revised. The procedures and models used in the risk assessment should be flexible enough to permit convenient updating to incorporate improved information.

- The approach should reflect a philosophy that the value of formal analysis for public policy decisionmaking derives less from the bottom line produced than from the insights it can create. Therefore, the models and procedures applied should provide the decisionmaker with a clearer understanding of the situation and not merely select an alternative that maximizes some criterion function.

The approach proposed here for the EPA's health risk assessment problem is consistent with the above considerations. The substance of the approach is a framework within which the risk assessment would be conducted rather than a set of specific modeling assumptions and techniques. Where specific models or techniques are described in this paper, they are for illustrative purposes only. The examples have been taken from decision analyses conducted of sulfur oxides and carbon monoxide pollution control problems (North and Merkhofer, 1975; Merkhofer and Korsan, 1978; Merkhofer et al., 1978; Smith, McNamee, and Merkhofer, 1981).

THE PROPOSED HEALTH RISK ASSESSMENT APPROACH

Overview

The risk assessment approach proposed for use by the EPA is based on methods and concepts borrowed from decision analysis, a quantitative approach to decisionmaking that explicitly accounts for risk. With decision analysis, a decision problem is decomposed into simpler parts, each of which falls within a particular field of expertise. Information concerning the parts and their relationships are elicited from experts and captured for the analysis through modeling. The model is then used to estimate the possible outcomes to the decision and measures of the likelihood and seriousness of these outcomes.[3] The proposed risk management approach similarly involves the development of a model. The model quantifies health risk by providing the range of possible health effects resulting from a particular standard, the probabilities of those health effects, and a measure of their seriousness.

Decision analysis is often applied in a procedure called the Decision Analysis Cycle (Howard, 1968). The proposed health risk assessment approach is structured as a similar cycle consisting of the three major phases shown in Fig. 1. The three phases, a deterministic phase, a probabilistic phase, and an informational phase, are briefly described below. More detailed discussions of the steps within each phase of the approach appear in the next section.

In the deterministic phase, a "health effects outcome calculator" is created. The health effects outcome calculator is a set of mathematical equations, implemented in computer code, representing the best currently available information or "best guesses" for the logical relationships between alternative air quality standards and some measure of the overall seriousness of the resulting health effects. No attempt is made to account for judgmental uncertainty in this phase, but statistical uncertainty, such as changes in ambient concentrations due to weather effects, should be included. A sensitivity analysis of the outcome calculator is conducted to

---

3 See R. A. Howard, "An Assessment of Decision Analysis," <u>Operations Research</u> (Vol. 28, No. 1, January-February 1980), pp. 4-27. For a discussion of analysis in environmental protection decisionmaking see R. A. Howard, "Social Decision Analysis," <u>Proceedings of the IEEE</u> (Vol. 63:3, March 1975).

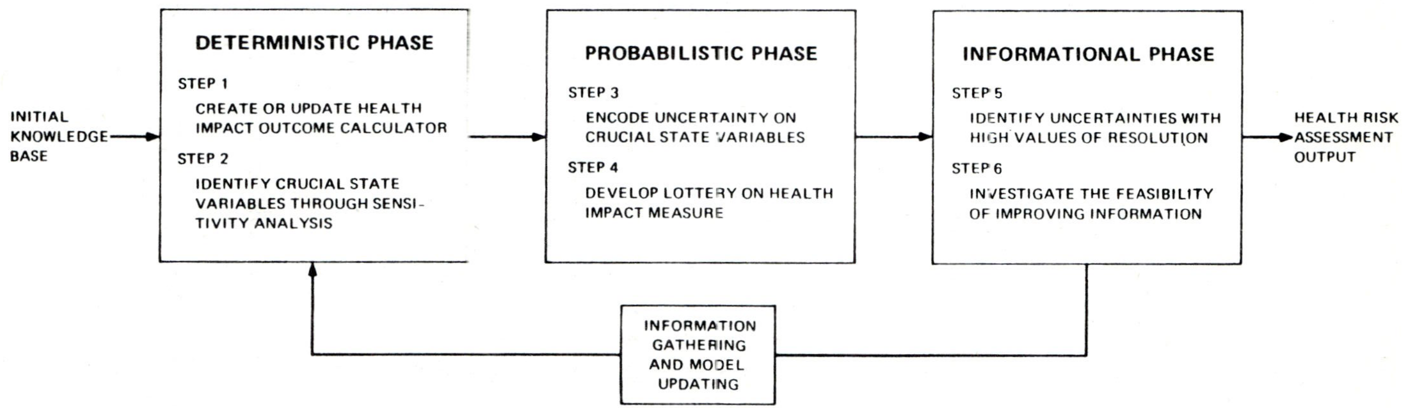

FIGURE 1 HEALTH RISK ASSESSMENT CYCLE

identify those model variables to which the decision is most sensitive. These variables will be referred to as "crucial."

The probabilistic phase begins by encoding judgmental probability on the crucial variables. Probability distributions describing the uncertainty in the overall health impact of alternative standards are then generated using Monte Carlo, decision tree, or similar techniques.

In the informational phase, additional sensitivity analyses are conducted to evaluate the relative importance of reducing uncertainty in each of the crucial variables. This phase also helps determine which elements of the outcome calculator can provide significant improvements in the assessment if they are expanded and refined. Comparing the relative importance of eliminating various uncertainties with the feasibility of improving the corresponding assessments suggests useful sources of additional information.

Several iterations of the risk assessment cycle may be necessary to complete the assessment. As additional information is gathered and the cycle is repeated, elements of the analysis are updated to account for the improved information. In the initial execution of the phases, the outcome calculator is constructed of highly simplified or readily available models. The purpose of the first implementation in the sequence, the pilot model, is to identify the most important problem relationships. Refining the pilot model yields a prototype model--a more detailed representation of the problem--which, although it generally has some aspects that must be improved, demonstrates the overall appearance and performance of the final version. The final model is the most accurate representation of the problem that will be produced as part of the assessment. It should be able to withstand the test of good engineering design--additional modeling resources could be used equally effectively in any part of the model. Conducting the risk assessment according to this process not only provides answers, but also provides considerable insight into the problem.

DETAILED STEPS FOR APPLYING THE HEALTH RISK ASSESSMENT CYCLE

Constructing an appropriate health effects outcome calculator is the most important step in the risk assessment cycle and will likely consume a large portion of the resources available for the assessment. Eliciting judgmental uncertainty is also difficult and time-consuming. The remainder of this paper focuses primarily on these two steps and briefly outlines the others.

Step 1: Develop and Combine Models to Produce an Appropriate Health Effects Outcome Calculator

The models needed to form an outcome calculator for the deterministic phase of the assessment are an emissions and atmospheric dispersion model, an exposure model, health dose-response models, and a significance weighting model (Fig. 2). For efficiency, these models should be developed and implemented as computer code. Although the precise realization for each of the models will depend on the pollutant under consideration, the basic structure and some of the detailed relationships will be similar from pollutant to pollutant. As a result, much of the model development effort for one pollutant will be useful in the analysis of other pollutants as well. Since the EPA and others already have developed models for these individual relationships, much of the modeling necessary for this step may consist of combining relevant elements of existing models.

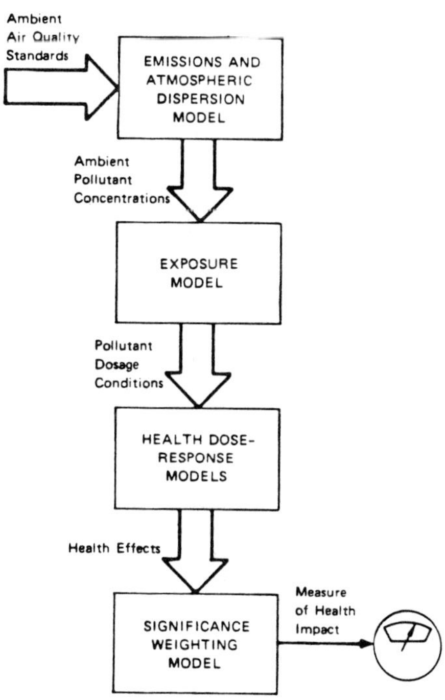

FIGURE 2 MAJOR COMPONENTS OF HEALTH IMPACT OUTCOME CALCULATOR

Emissions and Atmospheric Dispersion Model: Figure 3 shows the proposed structure of the emissions and atmospheric dispersion model. The model must accurately represent the various factors known to affect the relationship between a given standard and the ambient concentrations of the corresponding pollutant.

An ambient air quality standard usually specifies a precise averaging time, level, and expected number of instances in which concentrations are permitted to exceed the level over a given period of time. For example, the standard for ozone is of the form X ppm (parts per million) hourly average concentration, with one or less expected days per calendar year when maximum hourly concentrations rise above X ppm.

Once an ambient standard is promulgated, primary responsibility under the Clean Air Act shifts from the federal government to the states. The states can influence ambient concentration levels either by altering the production and consumption processes that produce the pollutant or its precursors, or by influencing polluters' decisions to contain and control emissions. To predict the effect of a given standard level on concentrations, submodels are therefore needed to represent: (1) the actions taken by state and local regulatory bodies to produce and enforce implementation plans to achieve national standards; (2) the production and consumption processes that produce gaseous residuals to which emissions can be traced; and (3) the emissions containment and control technologies that determine the emissions actually released to the atmosphere.

To reflect the dynamics of bringing an area into compliance with the standard and the fact that concentrations vary dramatically among locations, all submodels must be time-varying and spatially disaggregated. Spatial differences in resulting ambient concentration can be accounted for by dividing regions into geographic cells. Small cell sizes provide greater resolution of differences but produce greater demands on input data and increase computational complexity. In selecting cells, consideration should be given to the geographic basis for available demographic data and the location of ambient monitors. For urban areas where concentrations and population density vary greatly over small distances, smaller cell sizes may be justified, provided that disaggregated estimations of emissions and population growth rates are available. For less populated regions with lower levels of air pollution, larger, non-uniform cells may be more appropriate, such as those defined for an analysis of alternative Florida emissions standards shown in Fig. 4.

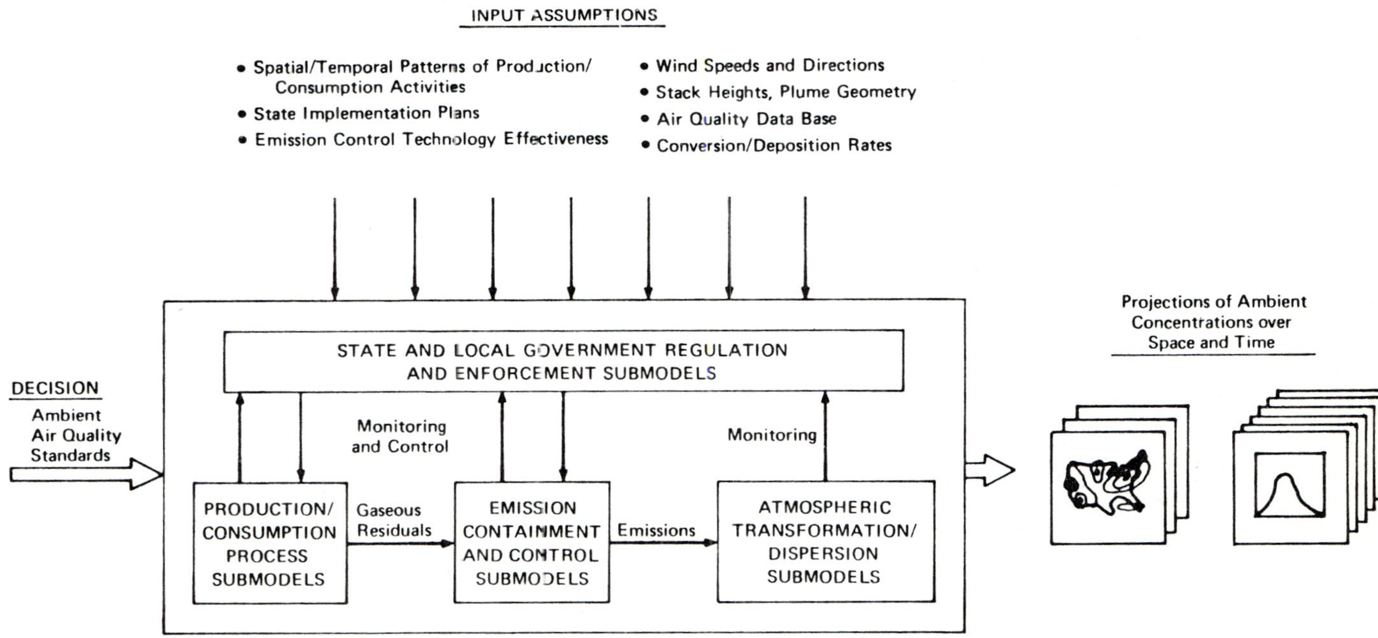

FIGURE 3 EMISSIONS AND ATMOSPHERIC DISPERSION MODEL

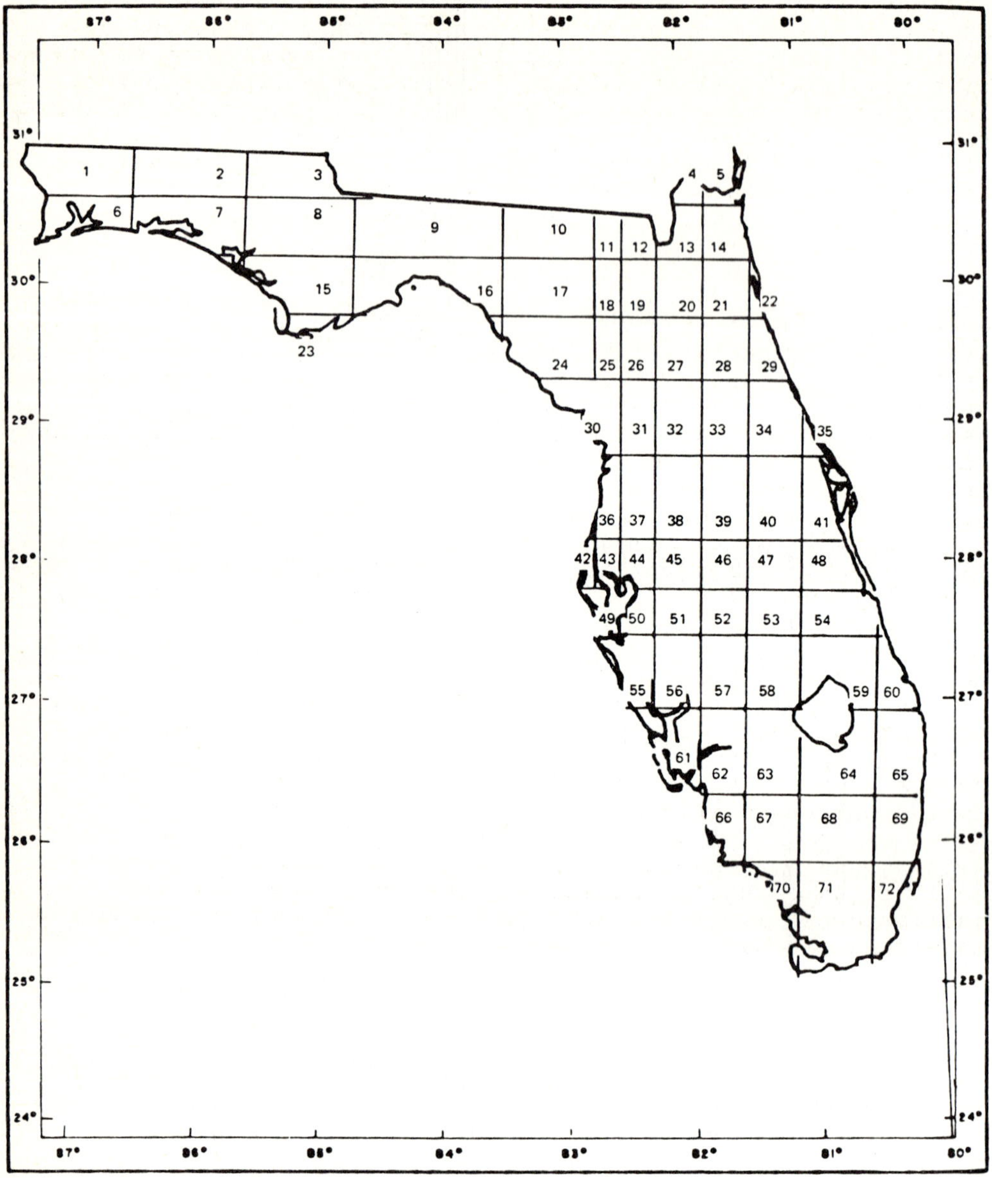

FIGURE 4 CELLS DEFINED FOR ESTIMATION OF FLORIDA SO_2 AND SO_4 LEVELS

Once emissions are estimated, the effect of the emissions on atmospheric concentrations must be modeled. Background or "natural" emissions differ substantially between regions and complicate this relationship. Air quality models of varying levels of sophistication are available for converting emissions patterns to ambient concentrations. One approach is to use empirical data to construct time-series models. Another approach uses causal models to represent processes of atmospheric transport, chemical reactions, and deposition of the pollutant.

Accounting for atmospheric reactions is particularly important whenever it is not the emittant but some product for which the emittant is a precursor that affects health. To represent such processes, appropriate short- or long-term indicators, such as sunlight, humidity, temperature inversions, etc., may be useful. The appropriate modeling technique for estimating the impact on concentrations of alternative emissions assumptions depends on the pollutant, its sources, and atmospheric and geographic conditions of the region being modeled. For example, while causal models have proven very useful in many situations, they have not been as effective in estimating local concentrations of pollutant with complex generation and removal processes, such as ozone.

Regardless of the approach, emissions-to-atmospheric concentration models should be calibrated by geographic area against monitored data before they are used for prediction. Fortunately, many of the areas where data are insufficient for accurate calibration are areas where health risk is small due to low emissions and sparse population densities.

Although causal models can be developed to project short-term daily or even hourly concentrations, simpler models may estimate only long-term average levels. If the dose-response models relate health effects to short-term exposures, it will be necessary to estimate concentration levels for the appropriate short-term averaging intervals. Order of magnitude variations are common in 1-hour and 24-hour averages at a given location due to changes in atmospheric conditions. A simple way to account for these variations is to assign relative frequencies to short-term levels based on estimated long-term averages and frequencies recently observed at selected monitoring sites. Lognormal and Weibull distributions often provide good fits to observed short-term values. However, if local control policies instituted as a result of changing a standard can be expected to alter the shape of short-term frequency distributions, a more sophisticated approach is needed to account for the effect of the change.

The output produced by the emission and atmospheric dispersion model should consist of a series of frequency distributions describing concentrations averaged over one or more time intervals for each geographical cell and for various points throughout the time horizon under consideration. Selecting averaging intervals and the points in time for the forecast requires consideration of several factors. To permit concentrations to be translated into health effects the averaging intervals should be selected to correspond to the exposure data for dose-response relationships and human activity patterns.[4] Whether the estimations should be generated for many points in time (e.g., for each month over the time horizon) or for only a few points (e.g., five year intervals) depends on the rates of change for emissions and on whether seasonal changes in emissions patterns or climatic conditions tend to create seasonal concentration patterns. As Fig. 3 shows, probability distributions and contour plots (Fig. 5) are convenient ways to summarize output data.

Exposure Model: Figure 6 shows a proposed structure for the exposure model used to estimate the conditions of air pollution exposure for sensitive population groups. Population groups typically sensitive to air pollution include bronchial asthmatics, emphysematous people, young children, and the elderly.

Defining a sensitive population group typically entails a trade-off between the size of the group and the severity of the health effects on representative group members. For example, because of underdevelopment of their lungs, prematurely born babies experience more severe reactions to many air pollutants. The total number of reactions experienced by prematurely born infants, however, will obviously be less than that experienced by infants in general. The most sensitive of the sensitive population groups may, in fact, have limited exposures. The weakest premature babies will probably be protected from air pollutants with incubators. To account for such differences, it may be desirable to define complementary population groups. In general, a population group should be defined whenever the health effects experienced by the group are believed to be socially significant, either because of seriousness or because a large number of individuals are affected.

---

4 Air quality concentrations averaged over successive time periods for a given location are highly correlated. Therefore, if concentration frequencies for appropriate averaging intervals must be derived from data on shorter averaging intervals, care must be taken to preserve temporal correlation.

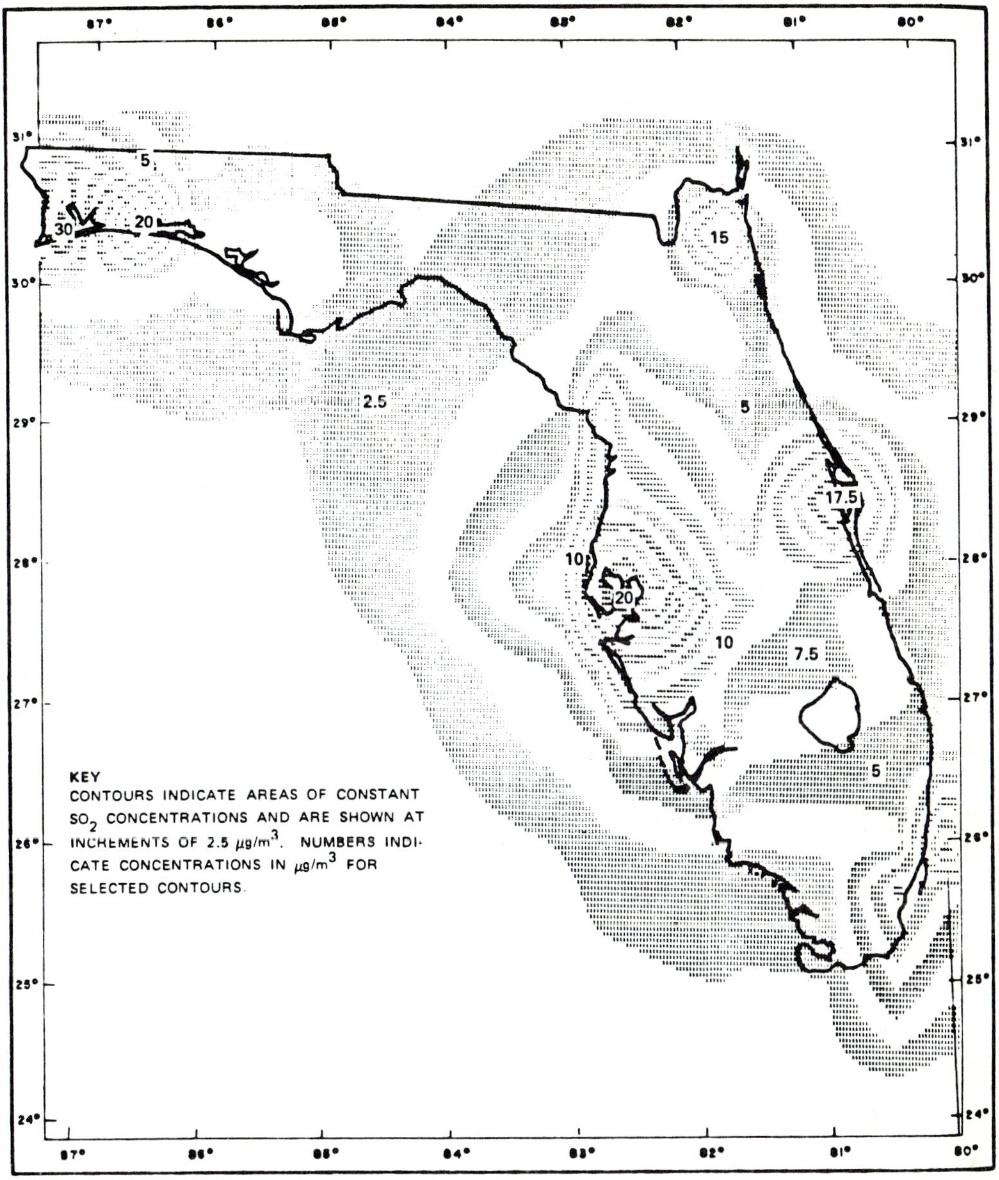

FIGURE 5 ESTIMATED ANNUAL AVERAGE SO_2 LEVELS, 1980

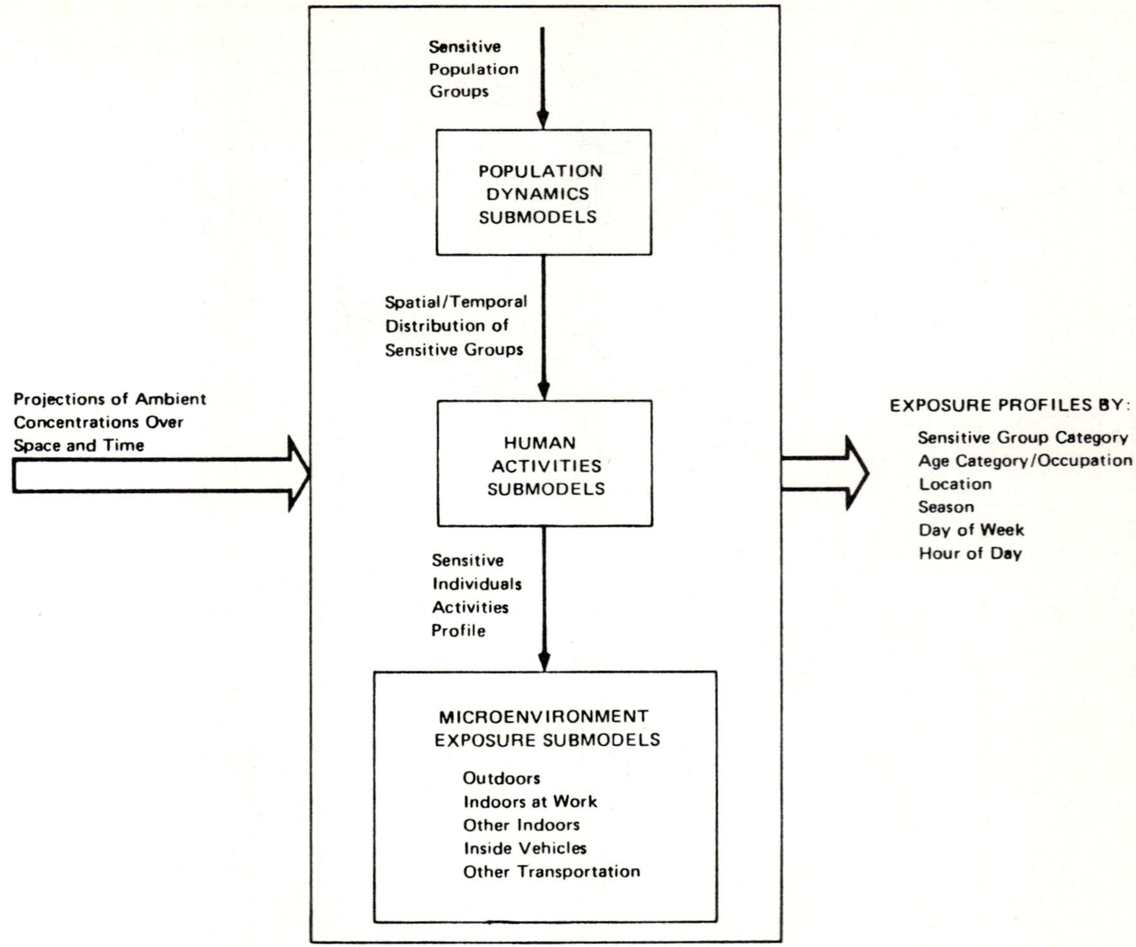

FIGURE 6 EXPOSURE MODEL

Two rules that apply to the development of the emissions and atmospheric dispersion model also apply to the exposure model: Creativity is required in constructing the model so as to make effective use of available data, and simple models should be constructed initially and then tested with sensitivity analysis to determine what level of additional detail is required to accurately represent the essential factors. At a minimum, the exposure model must identify appropriate sensitive groups and estimate the number of individuals in each group (or the base rate for each of the health effects experienced by members of this group) in each geographic cell for which pollution concentrations have been independently estimated. Table 1 illustrates this simple approach with data derived for the cells in Fig. 4.

If sensitivity analyses indicate a need to model the exposure variations that occur throughout a normal day of activities for members of a sensitive group, a more detailed exposure model can be constructed that contains submodels (Fig. 6) to represent the nature, timing, and microenvironments of representative group members' activities. The effects of population dynamics should be considered, since migration of the population and aging affects both the numbers of individuals exposed, the future concentrations, and the characteristics of long-term, cumulative exposures. A human activities submodel can be used to estimate the activities profiles for representative sensitive individuals and to account for preventive behavior taken to avoid harmful exposures during periods of high concentrations. While self-motivated behavior changes serve to reduce adverse health effects, the restrictions produced should themselves be recognized as a social cost of air pollution. The microenvironment exposure submodel would account for local pollutant concentrations (e.g., those occurring indoors or within vehicles) and other environmental factors that determine the health responses of individuals. Microenvironment pollutant concentrations are often modeled more accurately by accounting for pollution sources within the environment, such as type of fuel burned for indoor heating or the presence of a smoker in the household, than by assuming a direct relationship with outdoor concentrations.

The exposure model output must provide all of the inputs required by the dose-response models. In a detailed case, this consists of exposure profiles representing the frequency and duration of exposures to various pollution levels by activity (for example, level of exercise), sensitive group category, and environmental factors that may influence health response.

Table 1

ESTIMATED TOTAL NUMBER OF CASES OF HEALTH EFFECTS, 1990, USED IN FLORIDA SO_x ANALYSIS

			Number of Cases*				
	Cell	Population	Mortality	Person-Days Aggravation of Heart and Lung Disease	Asthma Attacks	Child's Lower Respiratory Disease	Chronic Respiratory Disease
	1	6,420	66	24,185	1,688	138	162
	2	36,934	380	139,137	9,713	794	934
	3	91,086	938	343,138	23,954	1,957	2,305
	4	8,459	87	31,867	2,225	182	214
	5	33,682	347	126,886	8,858	724	852
Pensacola	6	358,056	2,614	621,466	94,162	8,797	10,124
	7	55,139	568	207,719	14,500	1,185	1,395
	8	74,343	766	280,064	19,551	1,598	1,881
Tallahassee	9	225,709	1,535	422,918	59,357	4,945	6,734
	10	47,120	485	177,510	12,392	1,013	1,192
	11	76,550	788	288,378	20,131	1,645	1,937
	12	76,550	788	288,378	20,131	1,645	1,937
	13	7,825	81	29,478	2,058	168	198
Jacksonville	14	771,539	6,867	1,141,309	202,899	18,957	22,320
Small towns	15	157,091	1,618	591,791	41,312	3,376	3,975
	16	35,811	369	134,907	9,418	770	906
	17	17,316	178	65,233	4,554	372	438
	18	48,028	495	180,931	12,630	1,032	1,215
	19	37,689	388	141,982	9,911	810	954
	20	35,408	365	133,389	9,312	761	896
	21	72,428	746	272,850	19,047	1,556	1,832
	22	59,166	609	222,889	15,559	1,271	1,497
	23	8,199	84	30,887	2,156	176	207
	24	23,212	239	87,444	6,104	499	587
	25	1,530	16	5,764	402	33	39
Gainesville	26	168,233	1,565	325,178	44,242	3,509	5,121
	27	33,791	348	127,297	8,886	726	855
	28	62,413	643	235,122	16,413	1,341	1,579
	29	69,332	714	261,187	18,233	1,490	1,754
	30	21,834	225	82,253	5,742	469	552
	31	3,393	35	12,782	892	73	86
	32	6,072	63	22,874	1,597	130	154
	33	119,702	1,233	450,940	31,479	2,572	3,029
	34	38,275	394	144,189	10,066	823	968
Daytona Beach	35	129,331	1,164	545,884	34,011	2,490	3,331
	36	14,811	153	55,796	3,895	318	375
Small towns	37	228,366	2,352	860,298	60,056	4,908	5,778
	38	129,638	1,335	488,371	34,092	2,786	3,280
	39	92,317	951	347,776	24,278	1,984	2,336
Orlando	40	593,213	5,339	1,123,223	156,003	14,658	16,474
Cocoa Beach	41	112,638	721	95,529	29,622	3,170	3,151
Clearwater	42	137,500	2,516	821,730	36,160	2,108	3,313
Tampa	43	493,746	6,172	1,022,532	129,845	11,267	14,159
	44	61,240	631	230,703	16,105	1,316	1,549
Lakeland	45	379,100	4,739	1,353,369	99,696	8,200	9,744
Winter Haven	46	262,989	3,471	944,046	69,161	5,725	6,720
	47	16	0	60	4	0	0
St. Petersburg	48	122,349	1,260	460,912	32,175	2,629	3,096
	49	680,844	8,238	4,001,731	179,048	10,866	16,265
	50	1,531	16	5,768	403	33	39
	51	6,346	65	23,907	1,669	136	161
	52	44,231	456	166,627	11,632	951	1,119
	53	1,017	10	3,831	267	22	26
	54	69,799	719	262,946	18,356	1,500	1,766
Sarasota	55	361,844	4,849	1,527,282	95,158	7,295	9,082
	56	35,031	361	131,968	9,212	753	886
	57	74,177	764	279,439	19,507	1,594	1,877
	58	52,263	538	196,885	13,744	1,123	1,322
	59	19,603	202	73,848	5,155	421	496
Fort Pierce	60	190,115	2,262	682,451	49,996	4,219	4,800
Port Charlotte	61	391,308	4,030	1,474,131	102,906	8,409	9,900
	62	16	0	60	4	0	0
	63	9,309	96	35,069	2,448	200	236
	64	66,568	686	250,774	17,506	1,431	1,684
West Palm Beach	65	836,722	9,288	3,135,586	220,041	16,985	21,929
	66	56,990	587	214,692	14,987	1,225	1,442
	67	38,275	394	144,189	10,066	823	968
	68	16	0	60	4	0	0
Fort Lauderdale	69	1,030,119	11,743	3,657,159	270,901	20,262	27,984
	70	69,358	714	261,285	18,240	1,491	1,755
	71	9,032	93	34,025	2,375	194	229
Miami	72	822,398	8,635	2,416,864	216,274	17,040	23,004

*The figures cited are based on population estimates, health effects prevalence rates, and local mortality prevalence rates. Other prevalence rates used are as follows: (1) prevalence of heart and lung disease in elderly persons, 0.27; average number of aggravated days per day person, 0.20; (2) asthma prevalence rate, 0.036; attacks per day per asthmatic, 0.02; (3) annual incidence rate for lower respiratory disease in children, 0.07; (4) chronic respiratory disease prevalence rate, 0.02 for nonsmokers (62%) and 0.10 for smokers (38%).

Health Dose-Response Models: Dose-response relations are among the most complex and probably the most uncertain of the phenomena represented in health risk estimates. Controlled laboratory exposures typically show a continuum of effects of increasing pollutant concentrations, ranging from subtle physiological responses of doubtful significance to clearly important effects, such as disease and serious discomfort. Exposures to air pollutants at moderate to high concentrations have been shown to aggravate asthma, emphysema, and chronic bronchitis; reduce resistance to bacterial infection; and cause discomforts such as coughing, chest tightness, and irritation of the mucous membranes of the respiratory tract. The effect of a given concentration or dose is known to be a complex function of synergistic, additive, or antagonistic cofactors (e.g., smoking and other pollutants), and distribution of exposure with respect to time (duration, cumulative exposure, and latency periods). Some environmental conditions, such as temperature and humidity, may influence both the ambient concentration (through their effect on atmospheric reactions) and the health response of individuals to that concentration. Such factors should be accounted for in both the emissions and atmospheric conversion model and the health dose-response model.

Unfortunately, the scientific knowledge base is currently insufficient for precise definition of functional relationships among the many factors that determine health response. Much of the applicable research is ambigous, disputed, or contradictory. In some cases, only evidence from animal studies is available on a particular effect. Scientific evidence dealing with human responses is inconclusive. Epidemiological studies are hampered by lack of good exposure data and the difficulty of measuring health response. Toxicology usually involves exposures to much higher dosages than ambient concentrations. Studies on sensitive individuals are lacking (information tends to be on healthy individuals or populations) as are studies on synergistic, additive, and antagonistic effects.

The simplest approach to approximating dose-response relationships is to ignore all factors but concentration and elicit the subjective judgments of experts on alternative functional relationships between exposures and health effects. If there are disagreements on the appropriate functional form and competing forms produce significant differences, the relationship that will ultimately be shown to represent the best approximation can be defined as an uncertain variable. Sensitivity analyses can then be conducted to determine the importance of the uncertainty over alternative functional forms. Health effects may be defined as a coughing spell, chest discomfort for a period exceeding some duration, a given percentage reduction in pulmonary function, and such, so long as the definitions are meaningful to and consistently interpreted by experts and provide a reasonably complete description of the health consequences that may be associated with the pollutant. In this approach, when the expert believes a response occurs for any non-zero dose, it is best not to assume threshold forms because this forces the expert to make a value judgment of how much response should be regarded as a health effect.

Another simple approach is to make direct use of the results of available epidemiological, clinical, or other types of studies. A reasonable functional form for the relationship is assumed, and a mathematical fit is constructed to the available data. Common functional forms for fitting are linear, linear with a zero threshold, and sigmoidal. When data are taken from several studies, reduction of the data to a common set of exposure conditions and measured responses will be difficult. Figure 7 shows that combined data tend not to fall along a smooth curve and thus the functional form assumed can have a large impact on estimated values.

Where sensitivity analysis indicates that a simple approach is inadequate and an exposure model can be developed that stipulates the conditions of exposure (e.g., the concentrations of copollutants with which the given pollutant may be interactive in causing the effect, concentration time patterns the exposure may have, and the state of the subject with regard to possible stresses such as exercise), multi-dimensional dose-response surfaces can be calculated.

The dose-response functions may be applied to convert the exposure profiles generated by the exposure model into health effects. In the simple case, dose-response functions can be integrated over frequency distributions of exposure levels to obtain annual average numbers of cases of each health effect attributable to the air pollutant within each geographic cell. Similar health effects can be summed across cells to obtain the total number of health effects of each type occurring under each alternative regulatory standard. Table 2 shows the results of an analysis in which estimates of annual health effects were generated at 10-year intervals. In this example, three identified health effects (premature mortality, aggravation of heart and lung disease, and asthma attacks) were estimated from dose-response functions integrated over frequency distributions for 24-hour concentrations, while the remaining effects (child's lower respiratory disease and chronic respiratory disease) were assumed to be direct functions of annual average dosages.

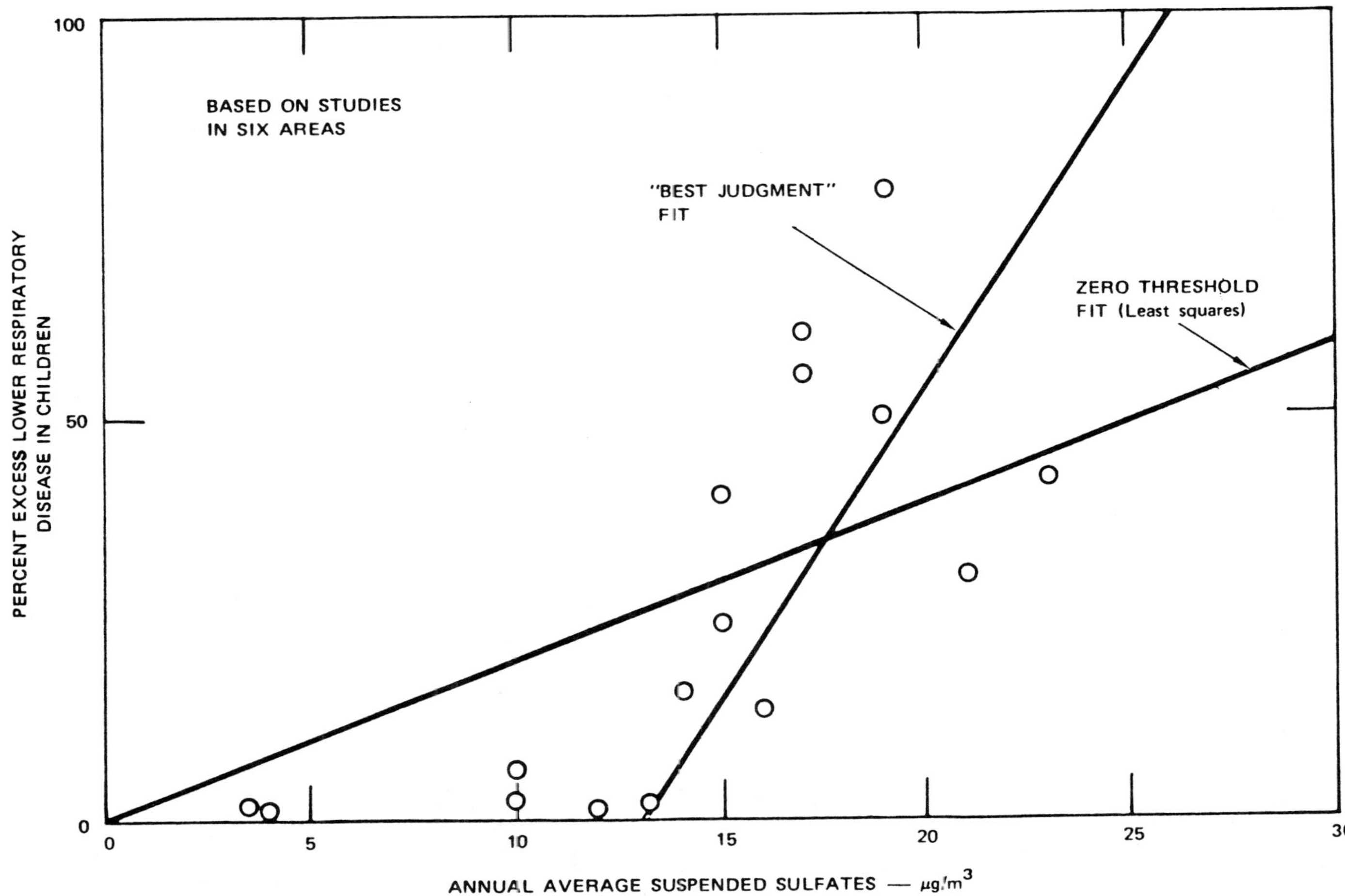

FIGURE 7 ALTERNATIVE DOSE-RESPONSE FUNCTIONS DERIVED FROM EMPIRICAL DATA

Table 2

FLORIDA ANNUAL HEALTH EFFECTS ATTRIBUTABLE TO SO_4 CONCENTRATIONS ASSUMING NOMINAL ELECTRICITY SALES GROWTH

	Air Quality Standard			
	Least Stringent	Current	Moderately Stringent	Most Stringent
Mortality (expected cases)				
1980	4.00	1.33	0.00	0.00
1990	5.66	2.33	0.00	0.00
2000	13.32	8.66	2.33	0.00
Heart and lung (thousands of cases)				
1980	729	447	40	3
1990	987	677	171	11
2000	1553	1273	729	63
Asthma (thousands of cases)				
1980	278	202	39	7
1990	367	287	114	17
2000	516	458	313	60
Lower respiratory (thousands of cases)				
1980	0.00	0.00	0.00	0.00
1990	0.00	0.00	0.00	0.00
2000	0.27	0.00	0.00	0.00
Chronic respiratory (thousands of cases)				
1980	0.20	0.00	0.00	0.00
1990	0.48	0.00	0.00	0.00
2000	5.56	1.60	0.00	0.00

Significance Weighting Model: Spatially and temporally disaggregated estimates of the number of health effects of various types to be expected under alternative standards provide important information for standards setting and should not be obscured by arbitrary methods of aggregation. Nevertheless, a single, overall health impact statistic provides a useful aggregation for a decisionmaker that enables the generation of nationwide estimates of health impact and facilitates sensitivity analyses that may be used to identify crucial uncertainties.

Aggregation requires that health effects be expressed in a common unit of measurement. Obtaining a common scale for measuring health effects is difficult because it requires value judgments on the relative seriousness of different health effects experienced by various individuals. Setting standards requires such value judgments to be made anyway, even if the risk assessment process does not make explicit value comparisons, because the EPA must account for the seriousness of potential health effects when establishing a margin of safety.

The significance weighting model is a means of explicitly incorporating value judgments and enabling the decisionmaker to investigate the sensitivity of the risk assessment to alternative value trade-offs. Estimated health effects must be aggregated along two dimensions. First, health effects occurring in the same time period must be combined into an overall measure of the health impact in that period. Second, the time stream of health impacts must be summarized by combining the health impacts occurring in different periods.

One measure for accomplishing the first aggregation is the "equivalent day of restricted activity (EDRA)" (Jacoby et al., 1973)(8). The Public Health Service defines a day of restricted activity as one on which a person substantially reduces the amount of activity normal for that day because of specific illness or injury. To convert estimated health effects into an overall measure of health impact, each health effect can be expressed using a trade-off ratio (significance weight) as a certain number of EDRAs. For example, for the purpose of risk assessment an asthma attack might be assigned a value of 0.5 EDRAs and an incidence of child's lower respiratory disease may be assigned a value of 3.75 EDRAs. In such weights, the significance of the health effect must be considered; for example, whether it cause irreversible damage or significant pain. Thus, a heart attack might be assigned a larger number of EDRAs than simply the average length of disability suffered.

Assigning an appropriate number of EDRAs for each health effect is a difficult task, and disagreement is likely among medical experts and others whose opinions might be solicited. Obtaining precise trade-off values, however, would not be necessary. The intent in assigning trade-off weights is not to define a health equivalence, but to encode value assignments consistently. A range of possible trade-offs reflecting various points of view could be defined. Sensitivity analysis would then be used to investigate the significance of disagreements over alternative trade-off assumptions. The process of assigning weights would be iterative, and assignments would be altered as understanding grows.

Health effects occurring in different time periods can be added or, if it is desired to reflect the preference expressed by most individuals to weight near-term health impacts higher than effects occurring further in the future, a discount rate can be used. A 1972 Office of Management and Budget directive prescribed a 10 percent rate for discounting future economic impacts. However, this value was not justified, and it is unlikely that any single rate could represent the intertemporal health preferences of society as a whole. In the absence of consensus, the best approach appears to be to treat the discount rate as a variable and to investigate the effect on the decision of alternative assumptions over a range of values. A summary statistic for the total health impact of a given standard (neglecting uncertainty) would thus be discounted EDRAs.

The significance weighting model eliminates the need to draw a precise line between physiological changes that are health effects and those that are not. Because borderline effects will presumably be given relatively little weight, converting every change that may be considered a health effect to a common unit of measurement by an appropriate weight makes the precise definition of health effects less important.

Step 2: Conduct a Sensitivity Analysis of the Health Effects Outcome Calculator to Identify Crucial Uncertainties

In addition to yielding a base-case estimate of the health effects of alternative standards, the outcome calculator can be used to identify which of the many uncertainties make a significant difference in the estimation of health impact and therefore where uncertainty must be explicitly accounted for in the risk assessment.

In the simplest sensitivity analysis, each input variable in the outcome calculator is varied across a range of values selected to approximate its range of uncertainty while holding all other variables constant at their nominal ("best guess") values. Those variables, whose variations produce the greatest change in estimated health impact, are most crucial to the risk assessment and are referred to as "crucial" uncertainties. Those causing only a small change may be ignored. Uncertainty on a variable may be unimportant either because its variation has a small range or because it has little influence on net health impact.

Because functional relationships used in the model will be uncertain, crucial uncertainties may include parameters that specify alternative functional relationships or submodels. If the functional relationship or submodel used is shown to be highly sensitive, it should be refined, if possible, and any simplifying approximations previously introduced should be removed.

Table 3 illustrates the sensitivity of estimates of discounted EDRAs produced by a simple health impact outcome calculator, to uncertainties in health impact outcome calculator, to uncertainties in health dose-response variables. The discount rate assumed was 10 percent. As shown, the impact of current uncertainty varies dramatically. Uncertainties in the dose-response functions for premature mortality, incidence of lower respiratory disease in children, and chronic respiratory disease in adults are apparently of relatively little importance. The most significant uncertainties shown are parameters of the dose-response function for aggravation of heart and lung disease symptoms in the elderly.

Single variable sensitivities may not identify crucial uncertainties whose importance results from probabilistic dependencies. Therefore, joint sensitivities should be measured in which two or more of the input variables are simultaneously varied across their respective ranges of uncertainty. The large number of variables in the outcome calculator makes it impractical to measure all possible combinations of joint sensitivities. Fortunately, knowledge of model structure and experience in conducting sensitivity analyses enable a limited number of necessary joint sensitivities to be identified to avoid missing crucial uncertainties.

<u>Step 3: Encode Uncertainty on Crucial State Variables</u>

The first step in the probabilistic phase is to assign probability distributions to the variables identified as

Table 3

SENSITIVITY OF PREDICTED FLORIDA HEALTH IMPACT TO DOSE-RESPONSE PARAMETERS--
CURRENT AIR QUALITY STANDARD, NOMINAL ELECTRICITY SALES GROWTH

Parameter	Nominal Value	Sensitivity Range	Number of Cases per Year Attributable to SO_4 (1990)	Total Health Impact (discounted EDRA, millions)
Base case				16.5
Thresholds				
Mortality	25 μg/m³	12-50 μg/m³	0-153 deaths	16.4-21.0
Heart and lung	9 μg/m³	5-20 μg/m³	0.019-1.98 million person-days	3.4-42.6
Asthma	6 μg/m³	5-20 μg/m³	3.4-365 thousand attacks	13.8-17.4
Lower respiratory	13 μg/m³	6-25 μg/m³	0-48.5 thousand	16.5-20.2
Chronic respiratory	12 μg/m³	6-25 μg/m³	0-87.8 thousand	16.4-38.6
Slope				
Mortality	0.252*	10-200% of nominal value	0-4.6 deaths	16.4-16.5
Heart and lung	1.41*	10-200% of nominal value	0-1.36 million person-days	3.0-30.4
Asthma	3.35*	10-200% of nominal value	0-574 thousand attacks	13.6-19.4
Lower respiratory	7.69*	10-200% of nominal value	0	16.5
Chronic respiratory	11.1*	10-200% of nominal value	0	16.5
Per-case significance weights†				
Mortality	1,500	150-10,000	2.3 deaths	16.4-16.8
Heart and lung	1	0.25-5	0.678 million person-days	6.4-70.8
Asthma	0.5	0.25-5	287 thousand attacks	15.2-42.4
Lower respiratory	3.75	0.5-15	0	16.5
Chronic respiratory	12.5	1.25-50	0	16.5

*Units are change in percent excess over base rate per μg/m³ change in sulfate level.

†Units are equivalent days of restricted activity.

crucial by the sensitivity analysis. Probability distributions would be elicited from recognized experts using probability encoding techniques. Where differences in scientific opinions exist, experts should be chosen to encompass each credible point of view.

Because most individuals have difficulty understanding uncertainty, the method of extracting the probability distributions is extremely important. Individuals with some experience with probability often attempt to make their distributions look like normal distributions, a characteristic known as "bell-shaped" thinking. Although normal distributions are appropriate in some circumstances, they should not be foregone conclusions.

Experience has shown a number of encoding procedures to be effective. The three basic types of encoding methods are: probability methods, which require the subject to respond by specifying points on the probability scale while the values remain fixed; value methods, which require the subject to respond by specifying points on the value scale while the probabilities remain fixed; and probability/value methods, which ask questions that must be answered on both scales jointly (the subject essentially describes points on the cumulative distribution). Each of these encoding procedures may be presented either in a direct response mode, in which the subject is asked questions that require numbers as answers, or indirect response mode in which the subject is asked to choose between two or more bets. The bets are adjusted until the subject is indifferent in choosing between them. Either external reference events (alternative bets defined on some external event, such as a probability wheel) or internal reference events (events defined on the same value scale as the uncertain quantity) can be used in the indirect mode. Table 4 summarizes the set of available encoding methods.

The proposed method for encoding probabilities uses variations of each of these encoding methods and depends on the variable and the characteristics of the subject being encoded (Stäel von Holstein and Matheson, 1978). The method, conducted in a private interview with the expert, consists of the following five phases:

- Motivating: Rapport with the subject is established and possible motivational biases that must be overcome are explored--for example, the tendency of experts to build safety margins into their judgments and a bias toward what has been scientifically demonstrated rather than what is viewed as most likely.

Table 4

CLASSIFICATION OF PROBABILITY ENCODING TECHNIQUES

Encoding Method	Response Mode: Indirect — External Reference Events	Response Mode: Indirect — Internal Events	Response Mode: Direct
Probability (value fixed)	Probability wheel	Relative likelihoods	Cumulative probability
Value (probability fixed)	Probability wheel Fixed probability events	Interval technique	Fractiles
Probability-Value (neither fixed)	--	--	Drawing graph; Verbal encoding

- o Structuring: The uncertain quantity and the structure underlying it are unambiguously defined.
- o Conditioning: The effect of potential biases is reduced and the subject is conditioned to think fundamentally about his judgment.
- o Encoding: The subject's judgment is quantified in terms of probabilities--responses to questions enable the analyst to draw a cummulative probability distribution for the variable.
- o Verifying: The responses obtained in the encoding are checked for consistency.

When two or more variables are dependent, the risk assessment procedure requires probability distributions that reflect the dependencies. The encoding technique is generally the same, but somewhat more complex, as the subject must condition his answers to interview questions on alternative values of the influencing variables.

Differences in the distributions assessed from various experts can often be resolved by bringing the experts together and allowing them to share information. Nevertheless, in some situations it is impossible to get two or more experts to agree on an appropriate probability distribution. Rather than reconciling divergent probability distributions by assigning weights to each, multiplying, and adding, we suggest grouping divergent probabilities into nominal, optimistic, and pessimistic sets. Whether the expert differences are significant and must be retained can then be determined through sensitivity analysis.

Once initial probability distributions have been obtained, the question arises as to how these probability assignments should be changed in light of new information. Bayes' Theorem can be used to update prior information assignments as new information develops; however, subjective probability assignments are required both in describing the prior information and also in specifying how the new information is related.

Step 4: Develop a Probability Distribution on the Health Impact Measure

Having developed an outcome calculator and obtained probability distributions on the crucial state variables, it is straightforward to design a computer program that will generate the probability distribution over the measure of overall health impact for each alternative standard. This probability distribution summarizes uncertainty in the overall health impact resulting from a given standard.

Step 5: Identify Uncertainties with High Values of Resolution

The informational phase aids the identification of a cost-effective allocation of additional risk assessment resources. Because dollar values have not been assigned to health effects, it is not possible to obtain a dollar value for eliminating uncertainty, as it is in a typical decision analysis. However, analysis can be conducted to identify the relative effectiveness of eliminating various uncertainties to improve the risk assessment.

The first step in the informational phase is a stochastic sensitivity analysis. Each crucial uncertain variable is set equal to a trial value within its range of uncertainty. All other uncertain variables are assigned the appropriate conditional joint probability distributions. Running the computer program will then produce a probability distribution on the health impact measure, and the expected value of this distribution can be calculated. Sweeping the trial value from one end of its range to the other shows how the probability distribution and its expected value change depending on the actual value of the uncertain variable. If the change is small, there is evidence that the analysis may be simplified by eliminating consideration of uncertainty on that variable. As in the case of deterministic sensitivity, stochastic sensitivity can be measured simultaneously on several variables (Howard, 1968).[5]

A number of mathematical calculations using the results of stochastic sensitivities may be used to establish an approximate ranking of the values of reducing various uncertainties. One approach begins by selecting a tentative value "H_A" for the measure of overall health impact that might be considered as acceptable by EPA decisionmakers. A tentative standard S is then selected to yield a probability distribution on overall health impact that exceeds "H_A" with an appropriately small probability. Because the intent of these tentative value judgments is only to rank uncertainties, the choices of "H_A" and "S" are not critical. The expected value of the health impact probability distribution, "H_S," can also be calculated. The difference between "H_A" and "H_S" can be interpreted as a margin of safety necessitated by uncertainty.

Now suppose that the stochastic sensitivity has been calculated for some variable "x" whose uncertainty contributes to overall uncertainty on health impact. For each possible value "$x = x_o$," the standard "$S(x_o)$" may be determined which yields the same probability of exceeding the acceptable health impact "H_A." Integrating the conditional probability distributions produced in the stochastic sensitivity analysis over the probability distribution for "x" yields the probability distribution on health impact given perfect information on "x" (Howard, 1968). The expected value of this

---

5 Where the costs of running complicated models prohibit extensive sensitivity analysis, proximal analysis may be used to obtain approximate results at greatly reduced computational costs. A description appears in R. A. Howard, "Proximal Decision Analysis," Management Sciences (Vol. 17:9, May 1971), pp. 507-541. In proximal analysis, a second order Taylor series expansion forms the basis for approximate expressions for the moments of the outcome lottery. These expressions are used to obtain simplified techniques for approximating stochastic sensitivities.

distribution, "H_I," can also be calculated. Figure 8 shows that with perfect information the same probability of exceeding the acceptable level of health impacts may be obtained, in most cases, with a smaller required margin of safety because the need to protect against the "worst-case" scenario in the setting of the standard will be eliminated. The importance of various uncertainties could therefore be ranked according to the amount by which their resolution reduces the required margin of safety. This approach is admitted ad hoc and probably easily improved upon. The point, however, is that straightforward analysis may be applied to provide a quantitative indication of the approximate importance of resolving the individual uncertainties represented in the assessment.

Step 6: Gather Further Information and Repeat the Cycle as Necessary

Actual information-gathering programs available will seldom completely resolve the uncertainty in a given variable. Selection of specific steps to be taken to improve information must therefore be based not only on calculations such as those described above, but also on the anticipated effectiveness of the gathered data at reducing uncertainty, the costs, and the time required to obtain the information.

When the preferred information-gathering program is performed, it will lead to new probability assignments on the crucial variables and may also indicate necessary changes in the basic structure of the outcome calculator. When all changes motivated by the results of information gathering are incorporated, the deterministic and probabilistic phases are repeated. As the model is improved, the effectiveness of further changes in improving the risk assessment will begin to decline, as measured by deterministic and stochastic sensitivities. The iterative procedure implied by the cycle ensures that resources are effectively allocated by directing the model development and probability encoding to maximize resolution of the probability distribution on health impact.

OUTPUTS OF THE RISK ASSESSMENT

The EPA Administrator must ultimately decide what level of risk is acceptable in setting air quality standards. The output of the risk assessment process should provide the decisionmaker with clear and concise measures of risk that discriminate

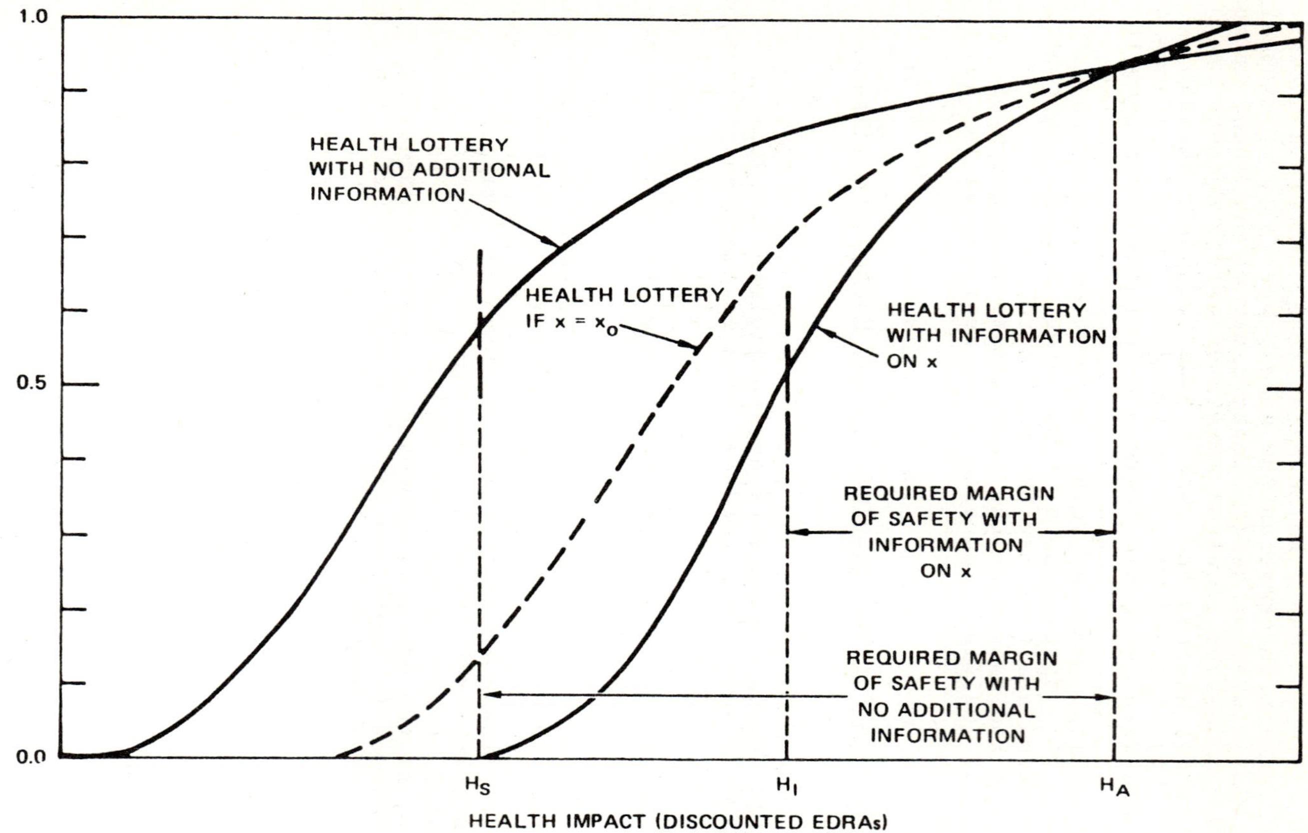

FIGURE 8 ESTIMATING THE EFFECTIVENESS OF RESOLVING UNCERTAINTY

between good and poor standards when judged according to a public health objective, make clear the uncertainties from which the risk measures were derived, and provide supporting documentation that is sufficiently clear and complete to permit peer review.

The proposed method of health risk assessment is explicit, open to public review, and comprehensive enough to provide a broad range of outputs of potential use to the decisionmaker. Its iterative nature is compatible with the five year review cycle of U.S. national ambient air quality standards--each review could consist of a major iteration with the results of information gathered from research suggested by the previous iteration being input at that point. Readily available output from the assessment includes plots of the expected overall measure of health impact versus standard level, the probability that the overall health measure (or the numbers of any specific health effects) for a given standard is less than or equal to any amount, and the probability that the number of persons experiencing each adverse health effect will equal or exceed a given value for a given time period. Specifying a preferred set of outputs and output formats should be an important objective of the risk evaluation methodology. The generality of the risk assessment procedure described herein permits considerable flexibility in the design of the risk evaluation and decisionmaking methodology.

REFERENCES

1. Clean Air Act, Section 109 (42 U.S.C. 7414).

2. Gage, S. J., Statement before the Subcommittee on Health and the Environment, Committee on Interstate and Foreign Commerce, U.S. House of Representatives, November 27, 1979.

3. North, D. W. and Merkhofer, M. W., Analysis of Alternative Emission Control Strategies, Air Quality and Stationary Source Emission Control, U.S. Senate Committee on Public Works, Serials 94-1:540-711 (Washington, D.C., U.S. Government Printing Office, 1975).

4. Merkhofer, M. W. and Korsan, R. J., Florida Utility Pollution Control Options and Economic Analysis--Volume 2: Cost-Benefit Analysis of Alternative Florida Sulfur Oxide Emissions Control Policies. SRI Final Report, Project 5080 (Menlo Park, California, SRI International, 1978).

5. Merkhofer, M. W. et al., Cost-Benefit Analysis of Mobile Source SO Emission Control: An Analysis of the Benefits of Automobile SO Emission Reduction in the Los Angeles Basin (Menlo Park, California, SRI International, 1978).

6. Smith, A., McNamee, P. C., and Merkhofer, M. W., Development of Decision Analysis Methodology for Health Risk Assessment: An Illustrative Application to Alternative Ambient Air Quality Standards for Carbon Monoxide, SRI Final Report, Project 1600 (Menlo Park, California, SRI International, 1982).

7. Howard, R. A., The Foundations of Decision Analysis, IEEE Transactions on Systems Science and Cybernetics (Vol. SSC-4:3, 1968).

8. Jacoby, D. H. et al., Measuring the Value of Emissions Reductions, in William R. Ahern, Jr., ed., Clearing the Air (Cambridge, Ballinger, 1973).

9. Stäel von Holstein, C-A. S. and Matheson, J. E., A Manual for Encoding Probability Distributions, SRI Final Report, Project 7078 (Menlo Park, California, SRI International, 1978).

A Decision-Analytic Technology Assessment of Public-Key Cryptography

Miley W. Merkhofer

SRI International, Menlo Park, CA

INTRODUCTION

Systematic assessment of emerging technologies can be of value to government decisionmakers by providing the information and structure needed for efficient and timely planning. Technology assessments are, however, frequently criticized for not providing information of real value to decisionmakers. This paper describes a unique application of formal decision analysis to produce a "decision-focused" technology assessment. The assessment addresses a new technology for producing secret codes called public-key cryptography (PKC). Unlike the typical technology assessment, this work involved the development and evaluation of a quantitative model of the decision among alternative government policies for dealing with the technology.

OVERVIEW OF PKC TECHNOLOGY AND ISSUES

Conventional cryptographic systems possess a significant disadvantage that restricts their use in commercial communication networks--keys (which specify how messages are encrypted and decrypted) must be distributed to and kept secure by all communicators. In 1975, however, researchers at Stanford University invented PKC, a new type of cryptographic system with the characteristic that encryption keys can be made public without compromising the security of the encrypted data (1). PKC advocates have noted that with such a system a user could place his personal encryption key in a public directory along with his name and address (2). To send a message, a sender would simply look up the appropriate key from the directory and use it to encode the message. Only the intended recipient could decode the message because only he could "undo" the transformation produced by the encryption key. Figure 1 illustrates the PKC communication process.

NATO ASI Series, Vol. G4
Environmental Impact Assessment, Technology Assessment, and Risk Analysis. Edited by V.T. Covello et al.

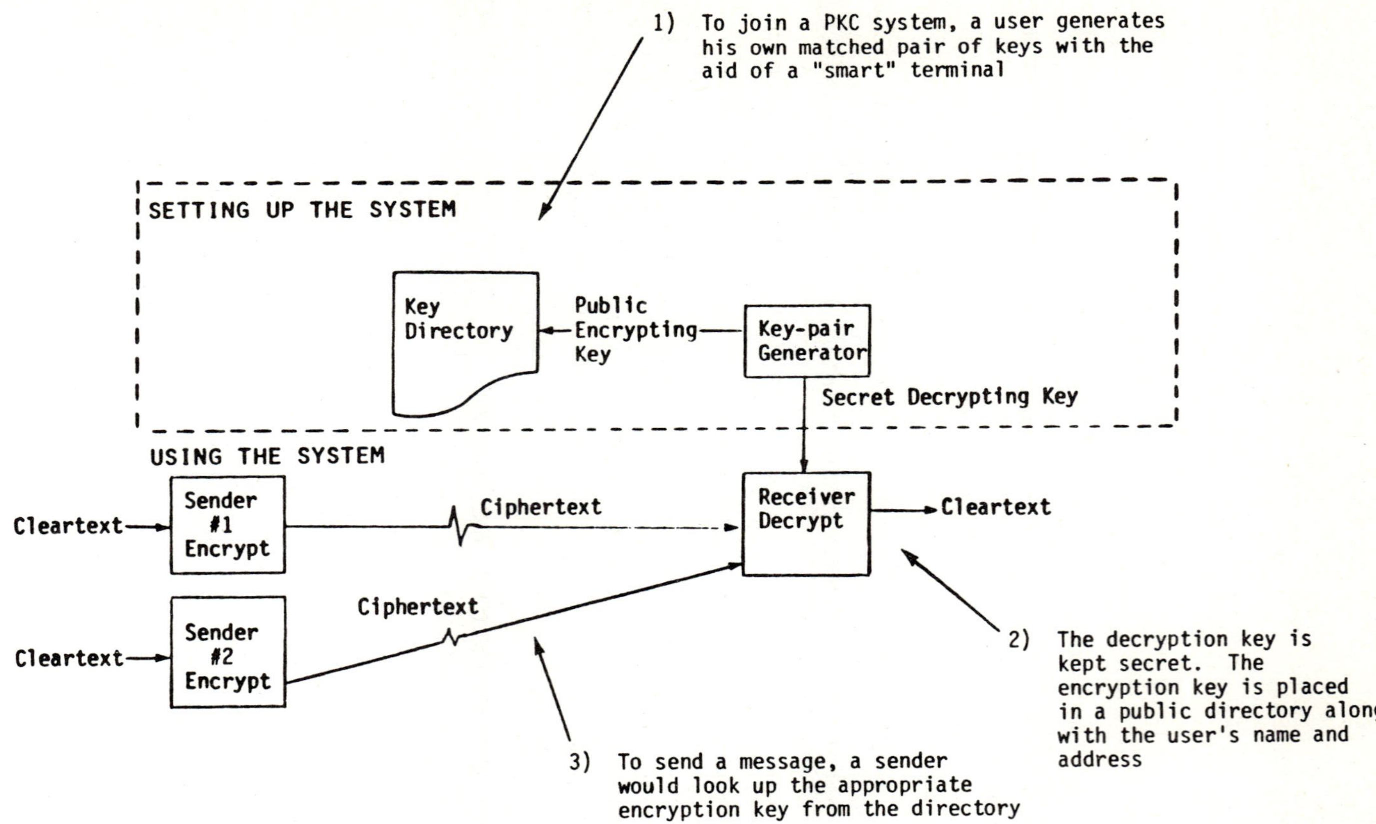

FIGURE 1 A PKC SYSTEM FOR PROTECTING COMMUNICATIONS

The invention of PKC has occurred at a time when a variety of government actions are promoting interest in and demand for civilian-sector cryptography. Information-handling legislation (such as the Privacy Act of 1974, the Freedom of Information Act, and the Foreign Corrupt Practices Act) and regulations specifying circumstances for the use of encryption by federal agencies [(3), for example] produce requirements for the safeguarding of sensitive information. The establishment by the National Bureau of Standards (NBS) of the federal Data Encryption Standard (DES) (4) has further promoted interest in encryption. The DES specifies a conventional cryptographic algorithm for encrypting nonclassified government data. Through standardization, the DES has made available to the public a relatively high level of encryption security at prices well below those for comparable products prevailing prior to the standard.

The special characteristics of PKC and the increasing demand for commercial cryptography suggest that the day is approaching when everyone from the corporate executive to the common criminal will have the ability to communicate and store records using secret codes that, for all practical purposes, are unbreakable. Needless to say, government agencies are interested in PKC technology and the uncertain consequences of increasing civilian-sector development and use of encryption. Some of the difficult social trade-offs that have been raised in the context of PKC include striking a balance between individual privacy versus public access to information, information security versus minimum economic costs, and academic freedom versus national security. The major national security concern relates to government intelligence activities. Extremely secure U.S. commercial cryptography will, presumably, be available to foreign users. Published results and highly secure crypto products may reveal to certain countries that their current codes are insecure and enable them to change their codes to ones that reduce U.S. intelligence-gathering effectiveness.

The remainder of this paper describes the quantitative modeling and analysis that were conducted as part of a technology assessment of PKC. The purpose of the analysis was to provide additional information and understanding through the evaluation of a highly simplified quantitative model of the policy decision. As noted above, the method of analysis was based on principles and techniques from decision analysis (5,6,7).

QUANTITATIVE ANALYSIS

To obtain the information on which the quantitative analysis was based, interviews were conducted with a large number of individuals knowledgeable in the various technical, economic, and social issues surrounding PKC. Quantitative assumptions were based on both empirical and subjective data. In some cases, experts interviewed volunteered a specific quantitative relationship as being appropriate for the analysis. In others, decision analysts developed a set of quantitative assumptions and then reviewed and modified the assumptions as required in subsequent interviews with one or more experts in the relevant area.

Figure 2 illustrates the four major phases of the analysis: alternative and outcome measure specification, deterministic analysis, probabilistic analysis, and informational analysis. Each of the various phases is described below. Conclusions concerning the methodology, as well as conclusions regarding PKC, are presented in the final section of this paper.

Strategic Policy Options and Outcome Measures

From a decision theoretic standpoint, the most fundamental elements of a decision are a set of available alternatives and a criterion for selecting among them. To construct a model of the choice among alternative crypto policies, it is therefore necessary to specify policy alternatives and explicit measures for representing the desirability of any given alternative.

Because the purpose of government intervention is presumably to improve the social consequences of PKC technology, the generation of policy options can be aided by considering the various factors that determine the social impacts of PKC. Figure 3 shows some of the factors that influence the societal impacts of PKC. Beginning at the left of the diagram, research in encryption theory and related areas determines the technical characteristics of commercially available encryption devices and their costs. These technical characteristics, in turn, influence the volume of use of encryption by application area and the level of use of related technologies, such as electronic message systems, that may be affected by encryption. Some societal impacts (labeled Technical/Economic Effects in the diagram) will be a direct result of the use of encryption devices. Other societal impacts (labeled

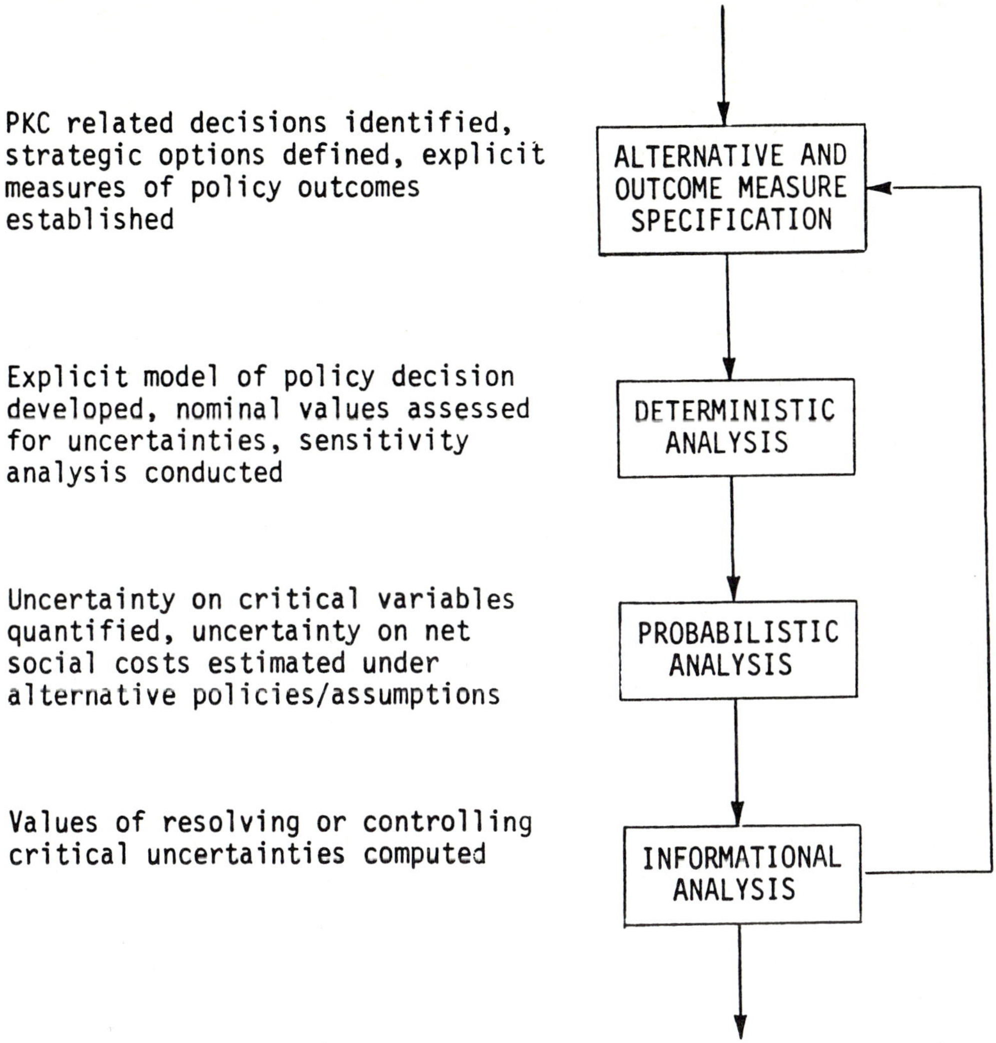

FIGURE 2 PHASES CONDUCTED IN QUANTITATIVE ANALYSIS

Social/Political Effects) will result from the increased confidentiality and security of data that are encrypted. Any policy that alters one or more of the factors shown in the diagram potentially affects social impacts. Thus, policy options may be generated by identifying feasible government actions for altering any of these factors. Table 1 shows potential policy decisions that were so identified, with each listed according to the primary factor that it is designed to influence.

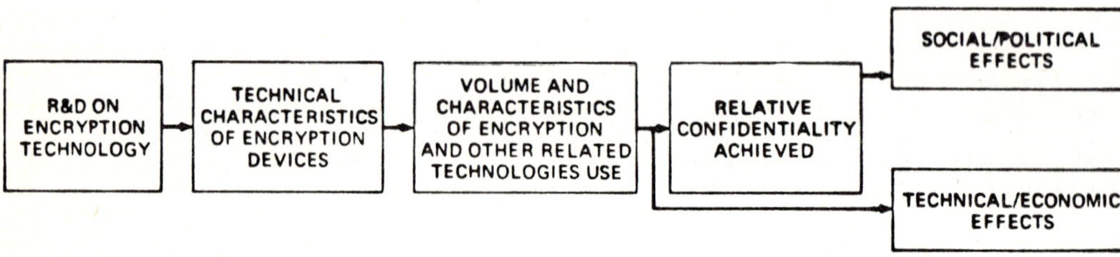

FIGURE 3 FLOW CHART SHOWING RELATIONSHIPS INFLUENCING SOCIAL IMPACTS OF PKC

Table 1

SAMPLE GOVERNMENT POLICY DECISIONS AND THEIR PRIMARY POINTS OF IMPACT

Research and Development	Device Characteristics	Applications and Level-of-Use	Confidentiality Achieved	Impacts
Adjust project funding (amount, source, research area, security classification)	Intensify or relax patents and copyright restrictions	Require, regulate, or prohibit use	Strengthen privacy and other information protection laws	Change legal penalties
Alter skilled labor pool (educational funding, licensing of individuals or facilities, postemployment restrictions, security clearance requirements)	Establish manufacturer licensing and product testing requirements	Subsidize or tax manufacturers or users	Encourage or discourage competing or associated technologies	Promote international treaties regarding transborder data flows
Control information dissemination (workshop and conference sponsorship and participation, publication restrictions)	Promulgate standards (algorithms, protocols, key management)	Provide encryption services		
	Provide government certification service	Encourage or restrict exports		
		Encourage or restrict imports		
		Use or prohibit use in government		

The decisions listed in Table 1 are not completely independent of one another; national objectives might be more effectively achieved if these decisions are coordinated and made consistent with some explicit intent. For the purposes of the quantitative analysis, therefore, the concept of a strategic policy option was defined. A strategic policy option is a combination of specific, consistent actions designed to implement a comprehensive policy for dealing with a technology and its impacts.

Table 2 shows three alternative strategic policy options for dealing with PKC and commercial cryptography. Each option is defined according to specific actions directed at controlling crypto research, device characteristics, and usage. The first option, "encourage PKC," consists of various actions designed to promote the development of PKC and commercial cryptography. The second option, "status quo," assumes the continuation of current policy. The third, "discourage PKC," consists of actions designed to limit and control development and use of PKC and other crypto technologies. These options were defined somewhat arbitrarily, the intent being to make them sufficiently different that the course of the technology and its consequences would likely depend on which policy was instituted, while making them realistic enough to be of interest. Obviously, a great many other options are available for a national crypto policy, and specification of a complete national policy would require a much more detailed specification than merely the set of actions listed in the table. For the purposes of the quantitative analysis, however, it was assumed that policy alternatives were limited to a choice among these three action sets.

Following usual decision analysis practice, evaluating each of the defined alternatives consisted of estimating its effect on a small number of policy outcome variables selected to measure the consequences of the policy in various impact areas. Defining appropriate outcome variables for a decision analysis of alternative national crypto policies is extremely difficult. The variables must be quantifiable, but should address important political and social, as well as more easily quantified, impacts.

Table 3 summarizes the reasoning process applied and the resultant outcome variables chosen for the analysis. The first column in the table lists some of the major areas identified as being potentially impacted by PKC and civilian cryptography. Outcome variables for quantifying the level of impact in three of these areas are shown in the right-hand column. The middle column lists specific issues within each of the impact areas that are of concern.

TABLE 2

STRATEGIC OPTIONS DEFINED FOR QUANTITATIVE ANALYSIS

	REPRESENTATIVE ACTIONS*		
STRATEGIC OPTION	RESEARCH & DEVELOPMENT	DEVICE CHARACTERISTICS	LEVEL OF USE
ENCOURAGE PKC	GOV'T PUBLIC CRYPTO RESEARCH FUNDING DOUBLED, PKC-RELATED RESEARCH EMPHASIZED	PROVIDE COMPENSATION FOR PROFIT LOSS RESULTING FROM PATENT SECRECY ORDERS	LIMIT APPLICATION OF EXPORT CONTROLS, PROVIDE FAVORABLE APPEALS PROCEDURE AND APPROVAL CRITERIA
	NON-DOD AGENCY FUNDING OF CRYPTO RESEARCH ENCOURAGED	ESTABLISH NBS STANDARDS FOR PKC	ESTABLISH STANDARDS AND CONDITIONS FOR APPROVED GOV'T USE OF PKC
	LIMIT NSA REVIEW OF PROPOSALS TO "PEER" CAPACITY ONLY	OFFER NBS CERTIFICATION OF DEVICES AND ENCOURAGE NON-GOV'T CERTIFICATION OF ALGORITHM SECURITY	
STATUS QUO	GOV'T FUNDING OF PUBLIC CRYPTO RESEARCH HELD AT A FEW MILLION $/YR	PATENT SECRECY ORDERS APPLIED OCCASIONALLY TO PROTECT NAT'L SECURITY	EXPORT CONTROLS EMPLOYED OCCASIONALLY TO PROTECT NAT'L SECURITY
	NSF CONTINUES AS MAJOR FUNDING SOURCE FOR PUBLIC CRYPTO RESEARCH		
	PREPUBLICATION REVIEW OF PRIVATE SECTOR CRYPTO RESEARCH BY NSA REMAINS VOLUNTARY		
	GOV'T FUNDED CRYPTO RESEARCH DIRECTLY IMPACTING NAT'L SECURITY FUNDED THROUGH AND CONTROLLED BY THE DOD		
DISCOURAGE PKC	REDUCE GOV'T FUNDING OF PUBLIC CRYPTO BY 1/2	LIMIT LEGAL MANUFACTURE OF LEADING-EDGE CRYPTO DEVICES TO LICENSED FIRMS	PROHIBIT EXPORT OF LEADING-EDGE CRYPTO DEVICES
	ESTABLISH NSA AS MAJOR SOURCE OF GOV'T FUNDED CRYPTO RELATED RESEARCH		REGULATE PRIVATE SECTOR USE OF CRYPTO DEVICES
	REQUIRE PREPUBLICATION REVIEW OF ALL CRYPTO RELATED RESEARCH PREPARED UNDER GOV'T SUPPORT		REITERATE GOV'T FAITH AND COMMITMENT TO DES
	LICENSE INDIVIDUALS AND INSTITUTIONS FOR LEADING-EDGE CRYPTO RESEARCH		REQUIRE USE OF DES FOR PROTECTING PERSONAL DATA HELD BY PRIVATE SECTOR ORGANIZATIONS

* From Section 4.2

Table 3

SOME OJTCOME IMPACT AREAS, ISSUES, AND THE OUTCOME VARIABLES SELECTED FOR QUANTITATIVE ANALYSIS

IMPACT AREA	RELEVANT ISSUES	SELECTED OUTCOME VARIABLE
PERSONAL PRIVACY	SECURITY OF PERSONAL DATA AND COMMUNICATIONS	NUMBER OF UNAUTHORIZED ACCESS TO PERSONAL DATA OR CORRESPONDENCE
BUSINESS SECURITY	SECURITY OF HIGH-VALUE CORPORATE DATA AND COMMUNICATIONS	EQUIVALENT DOLLAR VALUE OF LOSSES SUSTAINED THROUGH WIRETAPS
NATIONAL SECURITY	U.S. INTELLIGENCE GATHERING EFFECTIVENESS	FRACTION OF FOREIGN NATIONS USING "UNBREAKABLE" CODES
	PRIVATE COMMUNICATION NETWORK SECURITY	
SOCIAL FREEDOMS	FREEDOM TO CONDUCT RESEARCH IN, MANUFACTURE, AND USE TECHNOLOGIES THAT PROMOTE PERSONAL PRIVACY	
ECONOMIC	U.S. TELECOMMUNICATION/COMPUTER INDUSTRY INTERNATIONAL COMPETITIVENESS	
POLITICAL	TRANSBORDER DATA FLOWS	
LEGAL	ABILITY TO IDENTIFY AND USE EVIDENCE FOR PROSECUTION	

Because only three outcome variables were defined for quantitative analysis, not all relevant issues are addressed by the quantitative analysis.[1] Those issues that are addressed are the security of personal data and communications, the security of high-value corporate data and communications, and U.S. intelligence-gathering effectiveness. The quantitative variables defined for representing personal privacy and business activity are, respectively, the number of unauthorized accesses to personal data or correspondence and the equivalent dollar value of losses sustained through wiretaps. The quantitative measure selected for U.S. intelligence-gathering effectiveness is the fraction of foreign countries using codes that cannot be broken by U.S. intelligence-gathering agencies. It was felt that the methodological framework would be more usefully demonstrated through the inclusion of even a simplistic measure of national security impact than by ignoring this issue completely.[2]

---

1 The number of outcome variables was limited to three to reduce modeling complexity; it was regarded to be of more value (in terms of likely insights) to devote effort to quantifying considerations in these limited areas than to dilute the effort through an attempt to include a broader spectrum of considerations.

2 Several cautionary comments must be made. First, because not all of the relevant issues are represented by outcome variables in the analysis, quantitative results will not reflect a completely comprehensive consideration of issues. Major issues listed in the table that are not addressed include the national security benefit of improved private-sector network security, freedom of choice, and increased cost-effectiveness of secure communication. Each of these issues tends to argue for policies that encourage, or at least do not discourage, private-sector cryptography. This might seem to provide a systematic bias against the encourage option. On the other hand, there may be other issues arguing for controlled civilian crypto development beyond limiting the fraction of foreign nations whose codes we cannot break. The function of the quantitative analysis is not to identify an optimal policy, but to provide additional information and understanding through the analysis of a highly simplified quantitative approximation of the relations among major factors.

Deterministic Analysis

The deterministic analysis phase was essentially a highly simplified systems analysis. Two main steps were conducted. First, a model was developed to represent the relationships between policy alternatives and outcome variables. Second, a sensitivity analysis was conducted to measure the relative importance of each variable in the model.

Decision Model: Figure 4 shows in flow diagram form the major interactions that were modeled as part of the deterministic phase. There are two major components; a structural model for relating decisions and outcome variables, and a value model for representing social preferences for trade-offs among the outcome variables.

The boxes drawn with heavy lines in the flow diagram represent the major factors in the relationship between crypto research and impacts. (These five boxes correspond to the five elements of the diagram in Fig. 3). As represented by the arrows, each factor impacts, and is impacted by, a number of other factors. Starting at the left of the chart, cryptology research may be regarded as producing PKC and traditional cryptoalgorithms.

The qualities of these algorithms (as measured by "figures of merit") are determined in part by the levels of private and public sector research funding. The figures of merit of available algorithms, in turn, influence the cost-effectiveness of the resulting crypto products, which might also be influenced by government policies such as certifications or limitations on product development. The cost-effectiveness of crypto products, together with government requirements of prohibitions, influence crypto demand. Demand is broken into two market segments: applications that protect business value-sensitive information and those that protect public privacy-sensitive information. Demand is also assumed to be affected by the growth of potential application technologies, such as electronic mail (EM) and electronic funds transfer (EFT), and by the computer crime rate. The level of use of PKC and traditional cryptosystems determines the dollar losses and privacy losses suffered, whereas the fraction of foreign countries using "unbreakable" codes is assumed to be determined by the underlying security of the available crypto products. The value model converts the three outcome variables to a common unit of measurement by assigning equivalent dollar costs per foreign country using "unbreakable" codes and per

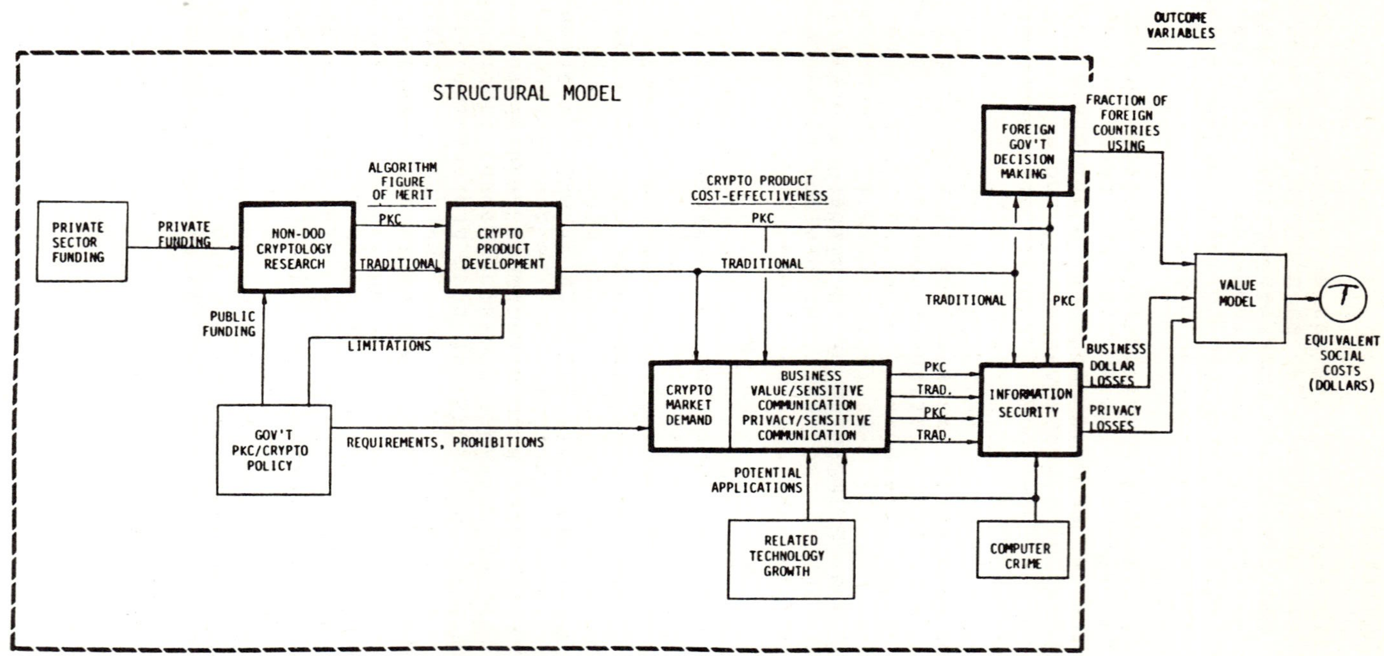

FIGURE 4 DETERMINISTIC MODEL STRUCTURE

unauthorized access to privacy-sensitive information. The purpose of converting outcome variables to a common unit (dollars) was to facilitate sensitivity analysis.

Submodels were constructed to approximate the relationships between inputs and outputs for each of the boxes shown in Fig. 4. Assumptions represented in each of the submodels are briefly described in the paragraphs below.

(1) Government policy and private-sector funding submodels are very simple submodels that have three outputs: research funding level, product development limitations, and requirements for or limitations on use. The values of the outputs are determined by which of the three strategic policy options is selected. If the option is encourage, public crypto research funding is assumed to be doubled, and the assumed rate of growth of the encryption market is doubled. If the discourage option is chosen, funding is halved, the nominal growth in encryption sales is cut to zero, and product development is restricted in such a way that only crypto products that can be cryptanalyzed by U.S. government agencies are produced for commercial use. The private sector funding submodel is simply a variable that may be set to represent any arbitrary funding level. Later, when the deterministic model is used to represent the dynamics of the encryption market, private sector funding is assumed to be proportional to crypto demand.

(2) The non-DoD cryptology research submodel produces as output figures of merit meant to summarize the overall quality of PKC and traditional private-key cryptoalgorithms. In assessing a code's figure of merit, the best available example from each class of algorithms is used. The number assigned to represent the figure of merit is assumed to account for all of the various factors influencing the viability of the algorithm as a basis for commercial products. Thus, the figure of merit is assumed to take into account the difficulty of breaking the algorithm under the best current method of cryptanalysis; the required key length and storage requirements; the speed with which encryption, decryption, and key generation could be carried out in hardware or software; whether data expansion of ciphertext occurs; etc. The figure of merit does not include the economic costs of implementing the algorithm. Costs are accounted for in the product development submodel (described below).

To permit numerical values to be attributed to figures of merit, benchmarks were established as shown in Fig. 5. The DES was assumed currently to represent a figure of merit of three.

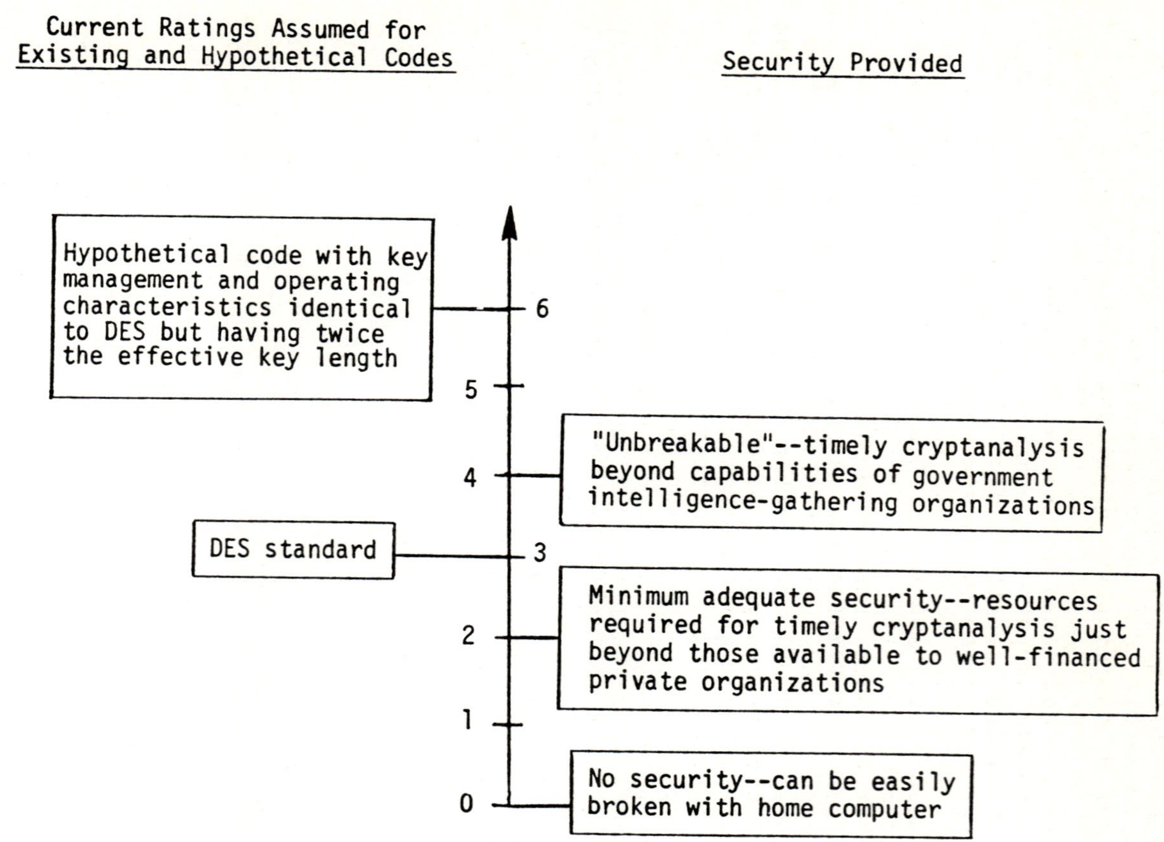

FIGURE 5 A NUMERICAL SCALE FOR A CODE'S FIGURE OF MERIT

If it were possible to double-encrypt with DES (so that breaking the code would be as difficult as it would be if the effective key length were twice as long) without any penalty in speed or key management, such an algorithm was assumed to have a figure of merit of six. Codes with figures of merit of four or above were assumed to be "unbreakable" in the sense that their use by foreign countries would diminish U.S. intelligence-gathering capability. Codes with figures of merit of two or below were assumed to provide inadequate security to protect high-value corporate information from sophisticated attacks by individuals or organizations.[3]

Note that with this formulation, a code's figure of merit depends on the resources and techniques available to the cryptanalyst. As these change, for example, through increases in computer processing speeds that shorten the time required for exhaustive key search, the figure of merit for a given implementation must fall. Thus, a figure of merit that is now an adequate value of 2.5 may be reduced to an inadequate value of 1.5 as cryptanalysis capabilities improve.

(3) The crypto product development submodel produces as output cost-effectiveness ratings for PKC and traditional crypto products. The cost-effectiveness rating is a numerical quantity defined to represent the value of a crypto product, in relation to its price, to potential users. As will be noted in the description of the demand submodel below, two crypto products of identical cost-effectiveness ratings are assumed to achieve equal market penetration.

---

3 Thus, an assessment of a figure of merit of 2.5 for PKC, for example, would represent the judgment that the best current PKC algorithm is, overall, of somewhat less value in commercial applications than the DES, but capable of providing adequate security at present to protect sensitive corporate data. An assessment of a PKC figure of merit of 4.5, on the other hand, would signify the judgment that PKC is of greater value than DES; that is, commercially practicable design characteristics may be set so that breaking the code is beyond the current resources of the most sophisticated cryptanalysis organizations.

A crypto product's cost-effectiveness rating is assumed to be a function of its cost and the figure of merit of the algorithm on which it is based. Figure 6 illustrates the form of the assumed relationship.[4]

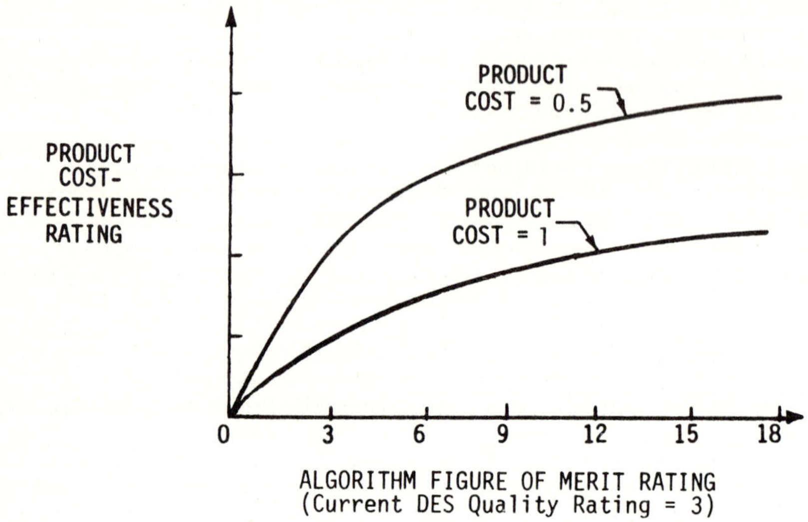

FIGURE 6 RELATIONSHIP ASSUMED BETWEEN ALGORITHM FIGURE OF MERIT AND CRYPTO PRODUCT COST-EFFECTIVENESS

4 In deriving this functional form, the following kinds of questions were used. Suppose as a potential user of cryptography you have a choice between two products. The products have identical operating characteristics, but one product is 25 percent more expensive and has an exhaustive search work factor equivalent to twice the key length of the other one. Is the additional security worth the added cost? Obviously, the answer depends on whether the security provided by the less expensive product is regarded by the user as adequate. As shown in Fig. 6, a doubling of effective key length without operating penalties (which doubles the figure of merit) produces nearly a doubling in the product's cost-effectiveness rating when the code's security is marginal (figure of merit around two), but produces only a small change when the security provided is already likely to be more than adequate for many years (figure of merit around six).

The cost-effectiveness rating increases with the figure of merit; however, doubling the figure of merit (for example, by doubling the time required to perform an exhaustive key search, i.e., the work factor) produces less than a doubling of cost-effectiveness. This represents the assumption that there is declining value to increasing the strength of codes that are already so strong that they cannot be broken. As illustrated in the figure, doubling the cost of a product is assumed to halve its cost-effectiveness. The figure of merit associated with a crypto product is assumed equal to the figure of merit of the underlying algorithm unless the government policy is to discourage PKC. In this case, the figure of merit of commercial products is assumed to be limited so that the products produced can be broken by U.S. intelligence agencies.

(4) Crypto market demand, related technology growth, and computer crime submodels measure the level of use in two market segments: applications by business for protecting value-sensitive communications and applications to protect privacy-sensitive communications. Level of use is represented by an estimate of the fraction of the total number of messages in each market category that are encrypted under each technology. The total number of messages is assumed to be determined by an estimate of the current number of messages sent within each market category and a growth rate. The fraction encrypted by each technology is determined by a nominal forecast and a factor that is applied to the nominal forecast to account for alternative assumptions for the cost-effectiveness of crypto products. The nominal forecast is determined by government policy (the growth rate for the fraction encrypted is doubled if the policy is encourage and cut to zero if the policy is discourage) and a measure of the level of computer crime (if future computer-related crime is higher than expected, encryption growth rates are again doubled).

Figure 7 shows the nature of the factor applied to the nominal fraction of messages encrypted to account for changes in the cost-effectiveness rating of crypto products. The figure illustrates a case in which the cost-effectiveness rating for traditional crypto products is assumed to be three. If the cost-effectiveness of PKC-based products is also three, the figure shows that each technology is assumed to make up exactly one-half the nominal demand forecast. If PKC product cost-effectiveness drops to zero, traditional products are assumed to satisfy all of the nominal demand. If PKC-based product cost-effectiveness increases above three, the fraction of the nominal forecast achieved by PKC increases and the fraction achieved by traditional products decreases. Notice that as the cost-effectiveness rating of PKC increases above three, the sum of the fraction achieved by traditional products plus the fraction achieved by PKC products is greater than one. Thus, should the cost-effectiveness rating of a product be increased, for example through research, the model form is such that total encryption demand will increase above the nominal forecast. This is to represent the assumption that a major improvement in one crypto technology's cost-effectiveness not only enables that product to capture a larger fraction of the encryption market, but also produces additional demand as new customers use encryption in place of or in addition to other means of protecting information.

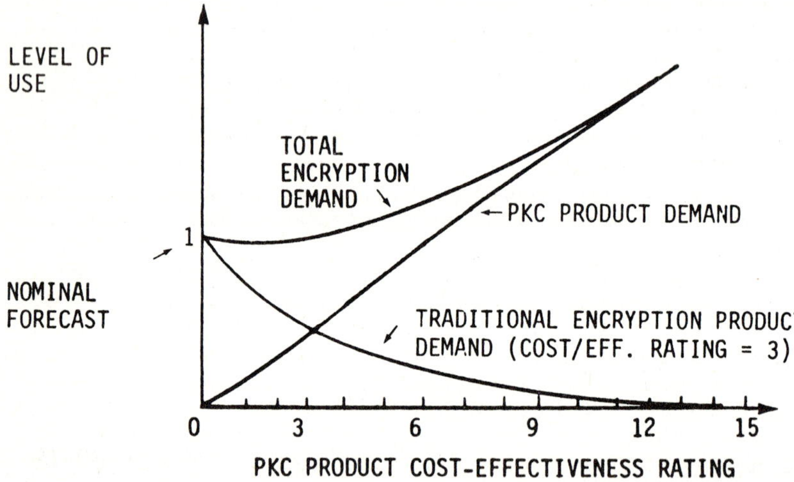

FIGURE 7 ASSUMED EFFECT OF CHANGES IN CRYPTO PRODUCT COST-EFFECTIVENESS ON NOMINAL FORECASTS OF ENCRYPTION DEMAND

(5) The business security and privacy submodel produces as output an estimate of the dollar losses sustained by business as a result of disclosures of value-sensitive communications and an estimate of the number of unauthorized accesses to privacy-sensitive information occurring through successful wiretaps. For value-sensitive messages, encryption with a cryptosystem whose figure of merit is at least two is assumed to prevent all dollar losses that would have occurred without encryption through wiretapping. For privacy-sensitive messages, the assumption is that any encryption (figure of merit above zero) would prevent unauthorized access. The rate at which attempts are made to tamper with or obtain unauthorized access to messages was assumed to be proportional to the computer crime rate.

(6) The foreign government decisionmaking submodel estimates the fraction of foreign countries using codes that cannot be broken by U.S. intelligence agencies at future points in time (five-year intervals). To compute this estimate, the submodel takes an initial fraction assumed to be using "unbreakable" codes and increases this by an amount determined by the higher of the two figures of merit associated with PKC and traditional codes. A figure of merit greater than four is assumed to be computationally unbreakable by the U.S. government (i.e., the time required to break the code is so great as to make the information obtained not worth the computational expenses). If the figure of merit for any publicly available crypto product technology is above four, some foreign countries are likely to use those technologies, and the fraction of foreign countries using "unbreakable" codes is assumed to increase. If the figures of merit of commercial products are below four, there is assumed to be no change in the initial fraction. Figure 8 shows the assumed form of the relationships for three different assumptions for the initial fraction of foreign countries using "unbreakable" codes.

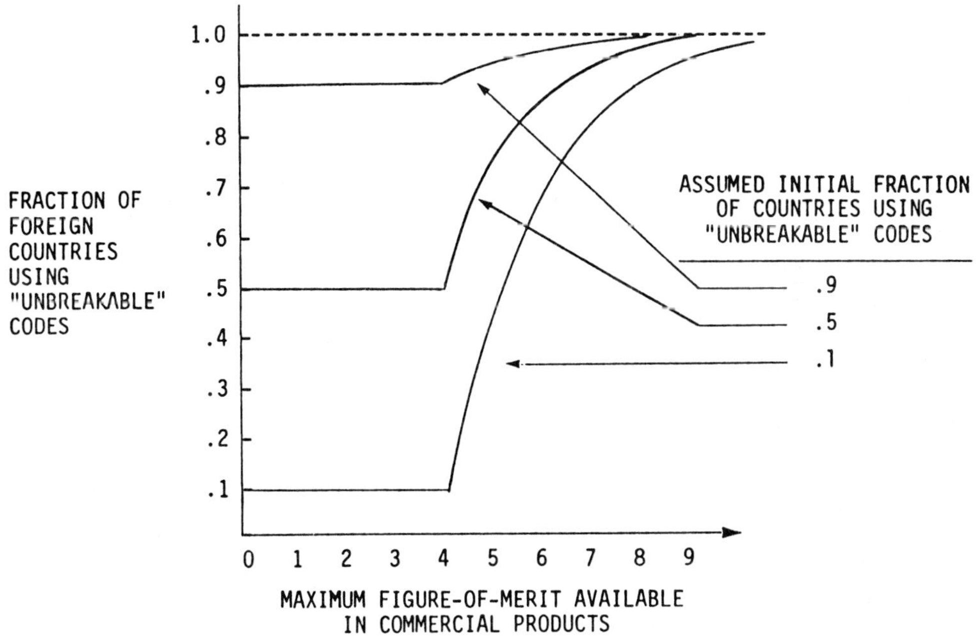

FIGURE 8 ASSUMED RELATION BETWEEN FRACTION OF FOREIGN COUNTRIES USING "UNBREAKABLE" CODES IN 1986 AND THE FIGURE OF MERIT OF COMMERCIAL PRODUCTS IN 1986

Base-Case Model Results and Sensitivity Analysis: The various submodels were implemented in computer code and base-case values assigned to all inputs. Table 3.1 summarizes some of the nominal input value assignments and base-case estimates obtained as a result of running the computer code assuming a five-year horizon.

For the purpose of summarizing outcome estimates in a single measure of equivalent social cost, nominal trade-off weights were assigned as shown in the table. The nominal weights assumed are $50 per unauthorized access to privacy-sensitive information and $100 million per foreign country using "unbreakable" codes. Because the average dollar loss per computer-related crime is assumed to be $500,000, the nominal weights imply the following kinds of difficult trade-off balances: For the purposes of selecting among alternative crypto policies, the loss of signals intelligence from one additional country using "unbreakable" codes is assumed in the base case to produce a social cost comparable to 200 successful computer-related crimes or two million unauthorized accesses to personal privacy-sensitive messages. These nominal trade-off assignments are purely arbitrary. The policy implications of assuming both higher and lower relative values were explored, and the results are described below.

Table 4 shows the most sensitive variables identified through sensitivity analysis and their impacts. The sensitivity analysis consisted of varying each input variable across a range of values selected to approximate its range of uncertainty while holding all other variables constant at their nominal values. For example, in the base case, both PKC and traditional crypto products are assumed to have figures of merit of three. As shown in Table 3.1, running the model with this assumption, base-case values for all other inputs, and nominal trade-off weights for outcome variables produces a total estimated social cost of $102 million for the period 1982-1986. In the sensitivity analysis, the model was run again under various assumptions for PKC and traditional code figures of merit between zero and five, while holding all other variables at their nominal values. Figure 9 shows the resulting variation in total social costs. The difference between the highest and lowest values, $4,550 million, is the value noted in Table 4.

Those variables whose variations produced the greatest change in equivalent social costs were judged to represent critical uncertainties. As indicated by the table, the uncertainties identified as most critical were the figures of merit for the crypto technologies, the incidence of computer-related crime, and market growth rates.

TABLE 3.1

Base-Case Estimates Produced Through Running the Deterministic Model with Nominal Values for Independent Variables

(B = Billion, M = Million)

SAMPLE NOMINAL INPUT VALUES	SAMPLE INTERMEDIATE MODEL VARIABLES	OUTCOME VARIABLES	NOMINAL TRADE-OFF WEIGHTS	TOTAL EQUIVALENT SOCIAL COST (THROUGH 1986)
POLICY = STATUS QUO INITIAL PKC RESEARCH FUNDING PUBLIC = $4M/YR PRIVATE = $1M/YR	1987 POTENTIAL ENCRYPTION MARKET PRIVACY SENSITIVE APPLICATIONS = 3.5B MSG/YR VALUE SENSITIVE APPLICATIONS = 1.75B MSG/YR	FRACTION OF FOREIGN NATIONS USING "UNBREAKABLE" CODES = 0.5 (UNCHANGED)	-$100M/ ADDITIONAL COUNTRY	$102M
ALGORITHM FIGURES OF MERIT PKC = 3 TRADITIONAL = 3 INITIAL PRODUCT COSTS (NORMALIZED) PKC = 10 TRADITIONAL = 1	1987 MARKET FRACTION ENCRYPTED BY CODE PKC-PRIV SENS = .03 PKC-VAL SENS = .05 TRAD-PRIV SENS = .17 TRAD-VAL SENS = .31	NUMBER OF UNAUTHORIZED ACCESSES THROUGH WIRETAPS = 13,980 MSGS (CUMULATIVE THROUGH 1986)	$50/ACCESS	
INITIAL POTENTIAL ENCRYPTION MARKET PRIVACY SENSITIVE APPLICATIONS = 1B MSG/YR VALUE SENSITIVE APPLICATIONS = 2B MSG/YR INITIAL MARKET FRACTION ENCRYPTED BY CODE PKC-PRIV SENS = C PKC-VAL SENS = C TRAD-PRIV SENS = .005 TRAD-VAL SENS = .25 INITIAL FRACTION OF FOREIGN COUNTRIES USING "UNBREAKABLE" CODES = .5 FRACTION OF COMPUTER CRIMES PREVENTABLE THROUGH ENCRYPTION = .01		EQUIVALENT DOLLAR VALUE OF BUSINESS LOSSES = $101M (CUMULATIVE THROUGH 1986)		

Table 4

SAMPLE SENSITIVITY ANALYSIS RESULTS

(Base-Case Equivalent Social Costs: $102 Million)

VARIABLE	NOMINAL VALUE	SENSITIVITY RANGE	RESULTING VARIATION IN EQUIVALENT SOCIAL COSTS (MILLIONS OF DOLLARS)
PKC/TRADITION ALGORITHM FIGURE OF MERIT	3/3	0 TO 5	4550
COMPUTER-RELATED CRIME RATE	30/YR	10 TO 150	938
POTENTIAL MARKET GROWTH RATE	15%/YR	-5% TO 30%	412
AVG COST PER VALUE-SENSITIVE MESSAGE WIRETAP	$500,000	$10K TO $1M	395
NOMINAL FORECAST RATE OF ENCRYPTION MARKET SHARE INCREASE	10%/YR	0% TO 30%	295
RATIO OF PKC PRODUCT COSTS TO TRADITIONAL PRODUCT COSTS	10	1 TO 20	210

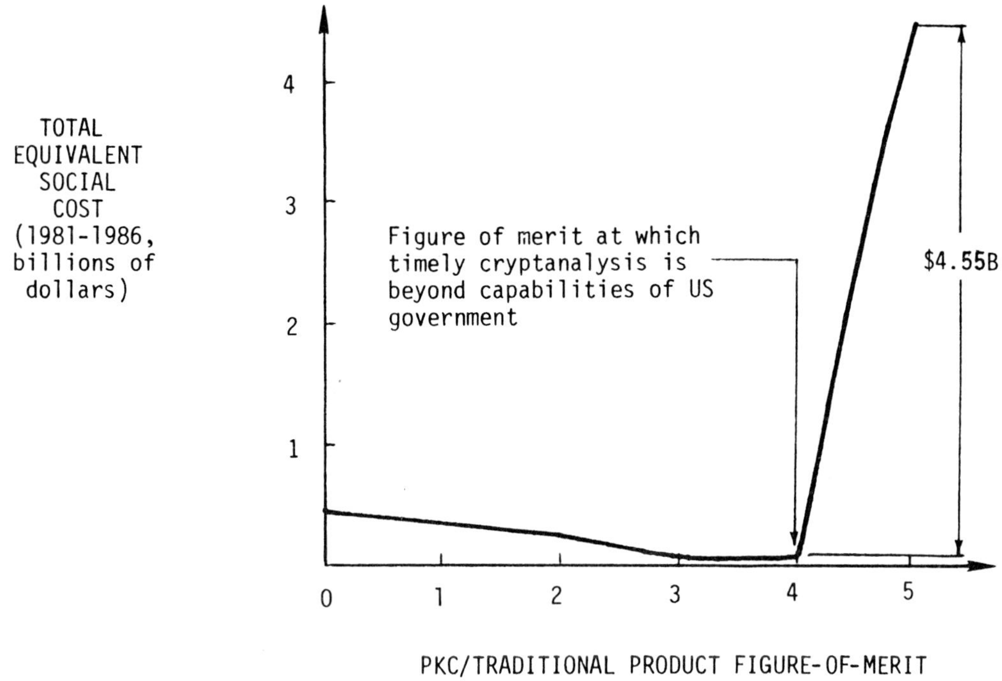

FIGURE 9 ESTIMATED TOTAL SOCIAL COST AS A FUNCTION OF COMMERCIAL CRYPTO TECHNOLOGY FIGURE OF MERIT (WITH BASE-TASK ASSUMPTIONS)

Probabilistic Analysis

Activity in the probabilistic analysis phase consisted of quantifying uncertainty on critical variables and estimating the total uncertainty in net social costs under each policy alternative. To represent dynamics and uncertainty in the decision process, a decision tree was used. Figure 10 shows the decision tree structure using a shorthand notation in which only the variables in the tree and the possible values for each variable are shown. To display the complete tree, each node and its branches in Fig. 10 would be duplicated at the end of each branch emanating from the node to the left of it.

The structure of the tree in Fig. 10 indicates the various assumptions. As shown, time was divided into two discrete five-year intervals: 1982 through 1986 (near term) and 1987

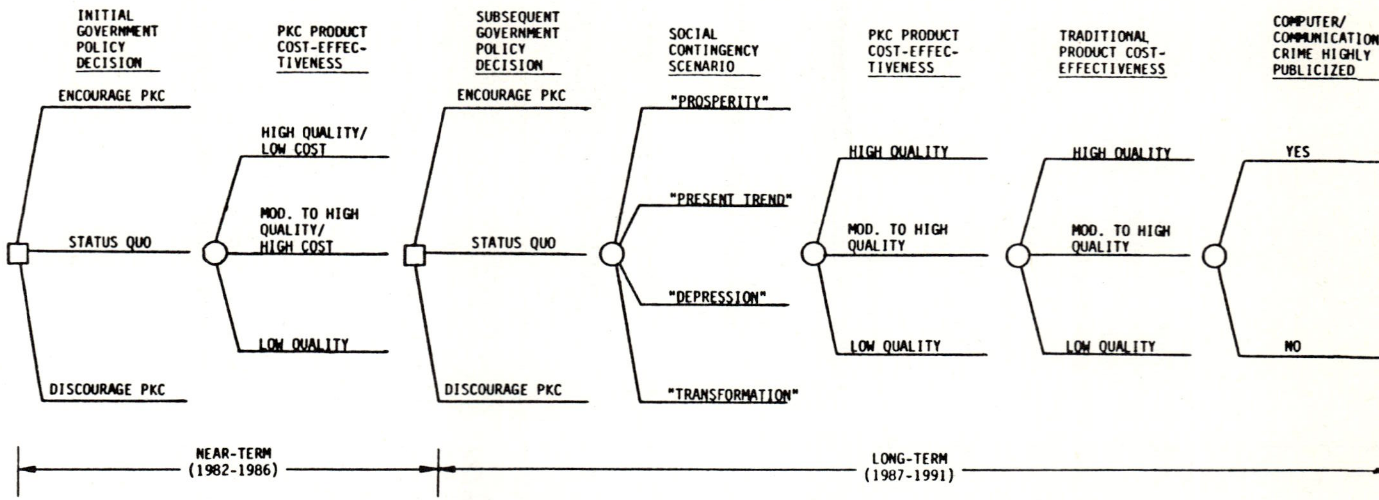

FIGURE 10 DECISION TREE USED TO REPRESENT DYNAMICS AND UNCERTAINTY OF DECISION PROCESS

through 1991 (long term). In 1982 the government is assumed to make an initial policy decision, choosing to encourage PKC and commercial cryptography, to continue current policies (status quo), or to discourage PKC and commercial cryptography. By the end of 1986, the cost-effectiveness of crypto products based on PKC will become known; PKC and traditional codes will each have some portion of the crypto market; and the levels of each of the outcome variables (privacy violations, business losses through wiretaps, and fraction of foreign countries using unbreakable codes) will become approximately known. At this point, the government is assumed to have the opportunity to revise crypto policy.

The impact of the revised government policy depends on several long-term uncertainties. As illustrated in the decision tree, these are a social contingency scenario, the longer term cost-effectiveness of both PKC and traditional products, and whether or not there is a dramatic increase in the number of highly publicized major computer-related crimes. Private research funding for each code in the period 1987-1991 is assumed to be proportional to the level of use achieved for that code during the period 1982-1986. The manner in which uncertainty in the various variables represented in the tree was represented and the process by which the tree was used to estimate overall uncertainty resulting from each policy option is discussed in the subsections below.

Crypto Product Cost-Effectiveness: To represent uncertainty in crypto product cost-effectiveness, a simple stochastic model was hypothesized. The effectiveness of a product was assumed to be determined by the underlying cryptographic algorithm figure of merit. The figure of merit at any future time was assumed to be determined by three factors. First, research was assumed to produce random quantum gains in a technology's figure of merit due to research advances (such as the development of a new algorithm). Second, the figure of merit of any given algorithm was assumed to decay at a fixed rate proportional to the increasing operating speeds of computers. As computers become capable of operating at higher speeds, exhaustive search becomes easier, so figures of merit must decline over time. Third, mathematical breakthroughs occur that produce quantum decreases in the security of an algorithm. For example, publication of a symmetry property in the DES reduced the number of operations that must be performed in the exhaustive key search for the DES algorithm. Mathematical advances can therefore be expected to produce random decreases in figures of merit.

Figures 11(a) and (b) illustrate sample behavior of a cryptographic algorithm's figure of merit over time resulting from the three factors as produced by the stochastic model. As illustrated, the probability of quantum gains in the figure of merit was assumed to be proportional to total research funding (private plus government). The probability of code improvement per unit of research funding, however, was assumed to be higher for PKC than for traditional codes to reflect an assumption of greater opportunity due to the relatively unexplored nature of PKC. The heights (incremental increases) of gains were assumed to be independent of funding. Both the probability and amount of decrease in random drops in figures of merit are assumed independent of funding level.

Crypto product costs were assumed to decline at 15 percent per year. PKC products are assumed in the base case to cost 10 times as much as traditional products. Mass production of PKC was assumed to permit a cost comparable to that of traditional cryptosystems. Therefore, PKC costs were assumed to drop to the same level as traditional product costs when and if the demand for PKC products exceeded a specified minimum value (50 percent of the current demand for traditional products).

The stochastic model for the behavior of algorithm figures of merit and the model for decline in costs were used to obtain high, medium, and low estimates for future product cost-effectiveness and corresponding probabilities for each outcome level. For example, with an initial policy decision of status quo (implying an assumed total annual funding level of $5 million), near-term (1986) PKC product figure of merit was estimated to be 4.5 with probability 0.1, 3.4 with probability 0.8, and 1.1 with probability 0.1.

<u>Computer-Related Crime Rate</u>: Many have suggested that the number of crimes involving computers could grow dramatically in the coming years and, should this occur, the commercial demand for cryptography will greatly increase. Whether such an increase will occur within the next 10 years is still very uncertain.

To represent and quantify subjective uncertainty over future levels of computer-related crime, the following uncertain variable was defined: the total number of computer-related crimes involving losses in excess of $1 million reported in the <u>New York Times</u> during the period 1987 through 1991. Figure 12(a) shows a cumulative probability distribution representing the subjective uncertainty expressed by one individual knowledgeable in issues of computer-related crime for the value of this variable. Decision analysis probability encoding techniques [described in (88)] were used to elicit this

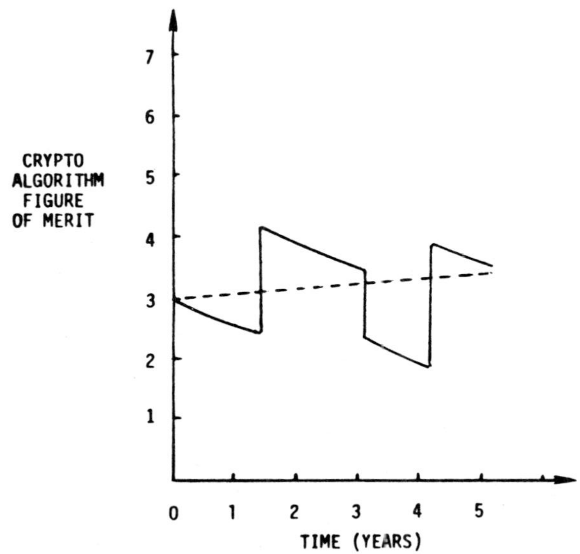

(a) Case 1: Total Research Funding of $5 million/yr

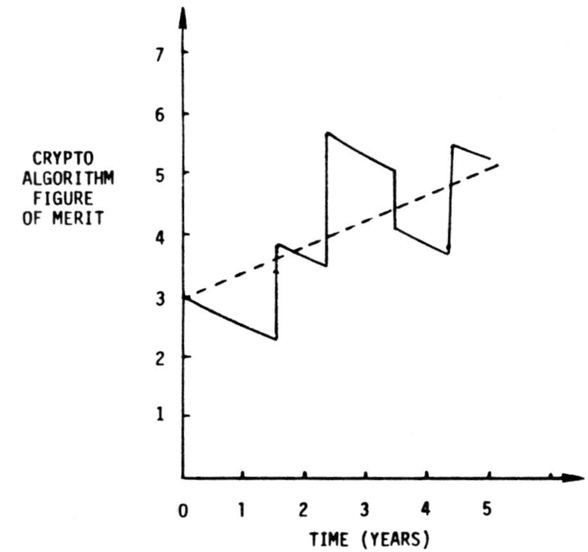

(b) Case 2: Total Research Funding of $10 million/yr

FIGURE 11 SAMPLE BEHAVIOR OF STOCHASTIC MODEL USED TO REPRESENT DYNAMICS OF UNCERTAINTY IN CRYPTO ALGORITHM FIGURE OF MERIT

distribution. The curve shows the probability that the actual number of reported crimes will be less than any value along the x axis. For example, the probability that the number is less than 75 (dashed line) is 0.8. This distribution was used to obtain probabilities for a high and low value for the computer crime rate uncertainty as illustrated in Figure 12(b). These values were subsequently used in the probabilistic analysis.

Social Contingency Scenario: For the purpose of estimating alternative market growth rates, alternative social contingency scenarios were taken from long-range planning forecasts produced by SRI International's Center for the Study of Social Policy [see (89) for an explanation of methodology]. These forecasts recognize four alternative futures for the United States: prosperity, present trend, depression, and transformation. Prosperity and present trend scenarios assume essentially linear extensions of the present, with prosperity being based on a condition in which current fears concerning resource and food shortages, pollution, alienation, and system dysfunctioning are far overstated. A depression scenario involves a failure of the free enterprise system, bringing about a prolonged economic collapse and turn to an increasingly authoritarian form of government. A transformation scenario is characterized by a fundamental shift in values away from conventional industrial views and producing a low economic growth philosophy.

To simplify the analysis, depression and transformation scenarios were assumed to have the same basic effect on commercial cryptography. Prosperity was assumed to imply the high values used for market growth rates in the sensitivity analysis (Table 4), present trend to imply the base-case values, and depression or transformation to imply the low values. A probability of 0.3 was assumed for prosperity, 0.5 for present trend, and 0.2 for depression or transformation.

Tree Analysis: Each path through the decision tree of Fig. 10 (obtained by selecting one of the branches emanating from each of the seven nodes) describes a possible scenario. For example, selecting the second from the topmost branch from each node produces the scenario: initial status quo policy; near-term PKC products turn out to have, as anticipated, a moderate quality level and high cost; selection of a long-term status quo policy; followed by a social contingency characterized by the continuation of present trends; long-term PKC products of moderate quality; long-term traditional crypto products of moderate quality; and no increase in the number of computer-related crimes. A total of 1,458 unique scenarios can be constructed in this way.

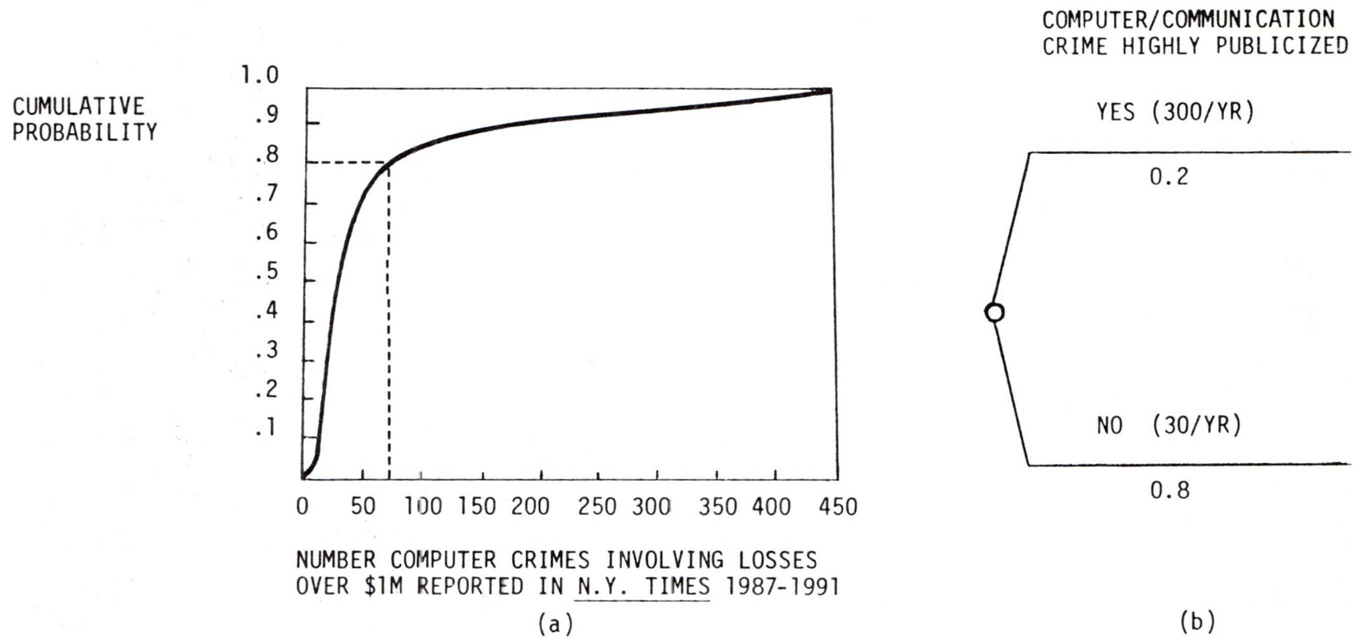

FIGURE 12 PROBABILITY DISTRIBUTION ILLUSTRATING UNCERTAINTY OVER NUMBER OF PUBLICIZED COMPUTER CRIMES (a) AND DISCRETE APPROXIMATION USED IN DECISION TREE (b)

Analysis of the decision tree consisted of running the deterministic model using input values corresponding to each of the possible scenarios through the tree. For each scenario, the deterministic model was run twice, first to simulate the period 1982-1986 and then the period 1987-1991, with the outputs of the first period providing initial conditions for the second period. Table 5 shows the values of the three outcome variables that were generated for a number of selected scenarios.

An examination of Table 5 clarifies the decision model and may provide some insights into possible future behavior of the commercial crypto environment. The first three scenarios each involve high computer crime rates and differ only in the initial government policy. As the policy is made less favorable to PKC, the number of unauthorized accesses and dollar losses increases. This occurs because the less encouraging policies produce a lower level of use both through a direct effect on market growth rates and through an indirect effect of lowered code cost-effectiveness brought about by reduced research funding. It is estimated that by 1991, nearly all foreign countries will be using codes that cannot be broken by U.S. intelligence agencies under either the encourage or status quo policies. This is because simulation under these scenarios produces figures of merit for PKC products that are well above four, the value selected to represent a code that is just incapable of being broken. [Specifically, the 1991 figures of merit calculated are 6.3 and 12.9 for PKC compared with 3.1 and 3.7 for traditional products under status quo (Scenario 2) and encourage (Scenario 1), respectively.] Higher figures of merit are estimated for PKC than for traditional codes because, as noted earlier, the likelihood of improvement through research is assumed higher for PKC. Traditional code figures of merit are strongly tied to DES and other standards that might be promulgated by the NBS. Such standards are assumed to be established in such a way as to encourage the use of private-key codes that do not pose a national security threat. Reflecting these assumptions, figures of merit for traditional codes remain below four.

PKC codes of high figures of merit might be produced to ensure that encrypted information remains secure against advancements in cryptanalysis for longer periods of time; but the higher the figure of merit, the more likely the code is to attract international attention and to be used by other nations. Only the discourage option prevents the commercial sector from producing PKC codes whose quality is so high that they result in significant foreign use. The 1991 figures of merit were 4.1 for PKC and 2.7 for traditional products under the discourage

Table 5

VALUES OF OUTCOME VARIABLES FOR SELECTED PATHS THROUGH DECISION TREE

	Scenario (Path through decision tree)						
No.	Initial Policy	PKC Cost-Effect.	Subsequent Policy	Social Contingency	PKC Cost-Effect.	Trad Cost-Effect.	Crime Rate
1	Enc.	Nom.	Nom.	Nom.	Nom.	Nom.	High
2	S.Q.	Nom.	Nom.	Nom.	Nom.	Nom.	High
3	Disc.	Nom.	Nom.	Nom.	Nom.	Nom.	High
4	S.Q.	V.Good	Nom.	Nom.	Nom.	Nom.	High
5	S.Q.	Poor	Nom.	Nom.	Nom.	Nom.	High
6	Enc.	Nom.	Nom.	Nom.	Nom.	Nom.	Low
7	S.Q.	Nom.	Nom.	Nom.	Nom.	Nom.	Low
8	Disc.	Nom.	Nom.	Nom.	Nom.	Nom.	Low
9	S.Q.	V.Good	S.Q.	Nom.	Nom.	Nom.	Low
10	S.Q.	V.Good	Disc.	Nom.	Nom.	Nom.	Low
11	S.Q.	Poor	S.Q.	Nom.	Nom.	Nom.	Low
12	S.Q.	Poor	Enc.	Nom.	Nom.	Nom.	Low

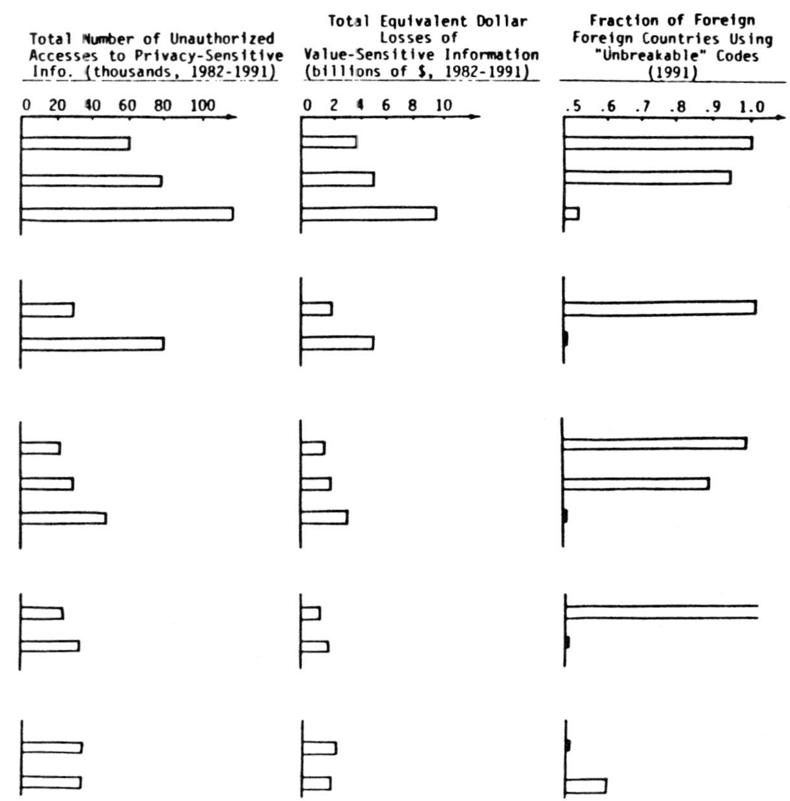

policy (Scenario 3). With the discourage policy, products with figures of merit greater than four are prohibited during the 1982-1986 period.

Scenarios 4 and 5 represent the cases where PKC products are initially significantly more and less cost-effective than currently anticipated under the status quo government policy.[5] As illustrated in the table, uncertainty over PKC cost-effectiveness has a major influence on privacy and dollar losses [because the model predicts a significant market for PKC under the high and moderate initial quality scenarios--about 55 percent (high quality case) and 39 percent (moderate quality case) of value-sensitive messages by 1991]. If near-term PKC product cost-effectiveness is worse than expected (Scenario 5), the ultimate level of PKC use is lower (about 17 percent of value-sensitive messages), and even with the nominal assumption for subsequent long-term PKC cost-effectiveness, no PKC products with high figures of merit are produced. As a result, and because under normal assumptions traditional products will not be unbreakable, the fraction of foreign countries using unbreakable codes remains at the initial value of 0.5. Thus, if PKC turns out not to be cost-effective during the first five years, the model shows PKC products not improving during the second five years to the point that they obtain significant use by foreign nations.

Scenarios 6-8 are identical to Scenarios 1-3 except that crime rate is low (about 1/10 the value assumed in the high crime case and approximately equal to current levels). Notice that the behavior of the estimated outputs is similar, except that the fraction of foreign countries using unbreakable codes is

---

5 The assumptions for these cases were obtained using the crypto product development submodel and the stochastic model for code figure of merit. Both PKC and traditional cryptographic algorithms were assumed to have current (1981) figures of merit of three. As shown in Fig. 11, these figures were assumed to change over the next five years in a random way that depends on research funding level. Resulting 1986 values were grouped into the 80 percent middle values and the 10 percent upper and lower extremes. The average PKC figure of merit values in the upper, middle, and lower groups (4.5, 3.1, and 1.1, respectively) were taken as the high (Scenario 4), nominal (Scenario 2), and low (Scenario 5) cases. As noted previously, PKC product costs were assumed to be 10 times that of traditional code products during the first five-year period.

slightly lower in each case. This is because a lower crime rate reduces demand and research into cryptography, resulting in a slower rate of improvement in the quality of the codes. The number of privacy violations and dollar losses is much more than 1/10 the value computed in the high crime case because there is less use of encryption under the low crime scenarios (about 62 percent of value-sensitive messages in 1991 under Scenario 7 compared with 93 percent under Scenario 2).Scenarios 9-12 illustrate the limited ability of a subsequent policy decision either to retard a very good technology or to improve upon a poor one. In Scenarios 9 and 10, the initial cost-effectiveness of PKC products is assumed to be higher than currently anticipated, the difference in the scenarios being that in Scenario 10 the initial status quo policy is shifted in 1987 to a discourage policy. The effect of the policy shift in this case appears favorable; privacy and value losses increase only slightly, and the problem of foreign countries using the codes is almost completely eliminated. Scenarios 11 and 12 show the opposite case in which PKC products have a cost-effectiveness that is much poorer than anticipated. Here, shifting to an encourage policy in 1987 does not have a significant effect on privacy and value losses and has the detrimental effect of producing a code of high enough quality to result in it being unbreakable by U.S. intelligence and used by a significant number of additional foreign countries. The reason for this is that it is difficult to increase commercial usage by much during the second five-year period if a code is poor during the initial five years. According to the model, it is much less effective to use government policy to improve upon a poor cryptotechnology than to limit production and use of an extremely good technology. In the latter case, adequate security for commercial applications can be obtained without a code that is so secure that it is computationally unbreakable by highly sophisticated cryptanalysis organizations.

Solution of the Decision Tree: A decision tree can be solved to obtain the decision strategy that minimizes expected cost, where expected cost is the average over all possible costs weighted by their probabilities. To permit the decision tree to be solved, the outcome variables must be aggregated into a single measure of social costs expressed in dollars. The nominal trade-off weights given in Table 3 were used for this purpose: $100 million equivalent social cost per foreign country using an unbreakable code and $50 equivalent social cost per unauthorized access to privacy-sensitive information. As noted previously, the selection of these trade-off weights is meant only to serve as an example for analysis. To aggregate outcomes occurring at different points in time, a discount factor of five percent per year was used. A relatively low discount rate was selected because all costs are expressed in constant dollars and therefore do not require an adjustment for inflation.

Figures 13 and 14 show the results of solving the decision tree under two different assumptions. In Fig. 13, the initial fraction of foreign countries using codes that U.S. intelligence cannot break is assumed to be 0.5. In Fig. 14, the fraction is assumed to be 0.9. Arrowheads drawn on a branch emanating from the node associated with each policy decision indicate the alternative that minimizes expected net discounted social costs given the events represented by the branches leading to that node. Thus, for example, if the current fraction is 0.5 (Fig. 13), the best strategy is status quo followed by discourage unless PKC turns out to be less cost-effective than anticipated. The difference between the results in Fig. 13 and 14 point out how crucial is the question of how much intelligence information the U.S. is currently able to obtain through cryptoanalysis. If 90 percent of the foreign countries already use codes that the United States cannot break (or will by 1991, regardless of U.S. commercial cryptography), then, assuming the nominal trade-off weights, the optimal strategy is uniformly to encourage PKC.

Figure 15 shows another result of analysis. The figure shows probability distributions on discounted net social cost computed from the decision tree for the case where the initial fraction of foreign countries using unbreakable codes is assumed to be 0.5. The curves are complementary cumulative distributions--they show the probability that the actual value of social costs will be greater than or equal to any value on the x-axis.[6]

---

6 To obtain these results, the net discounted social cost was computed for each path through the decision tree (assuming contingency decisions as shown in Fig. 13). The probability of each path was then obtained by multiplying the probabilities corresponding to the outcomes associated with the branches along that path. The probabilities for the high, nominal, and low figure of merit branches in each case were assumed to be 0.1, 0.8, and 0.1, respectively, with corresponding numerical values determined by the stochastic model for figure of merit, as described in the previous footnote. The probability of high crime rate was taken as 0.2, as shown in Fig. 12. Knowing the net discounted social cost and the probability of each path, the probabilities of all net social costs greater than or equal to each given amount were identified and summed to obtain the complementary cumulative probability curves.

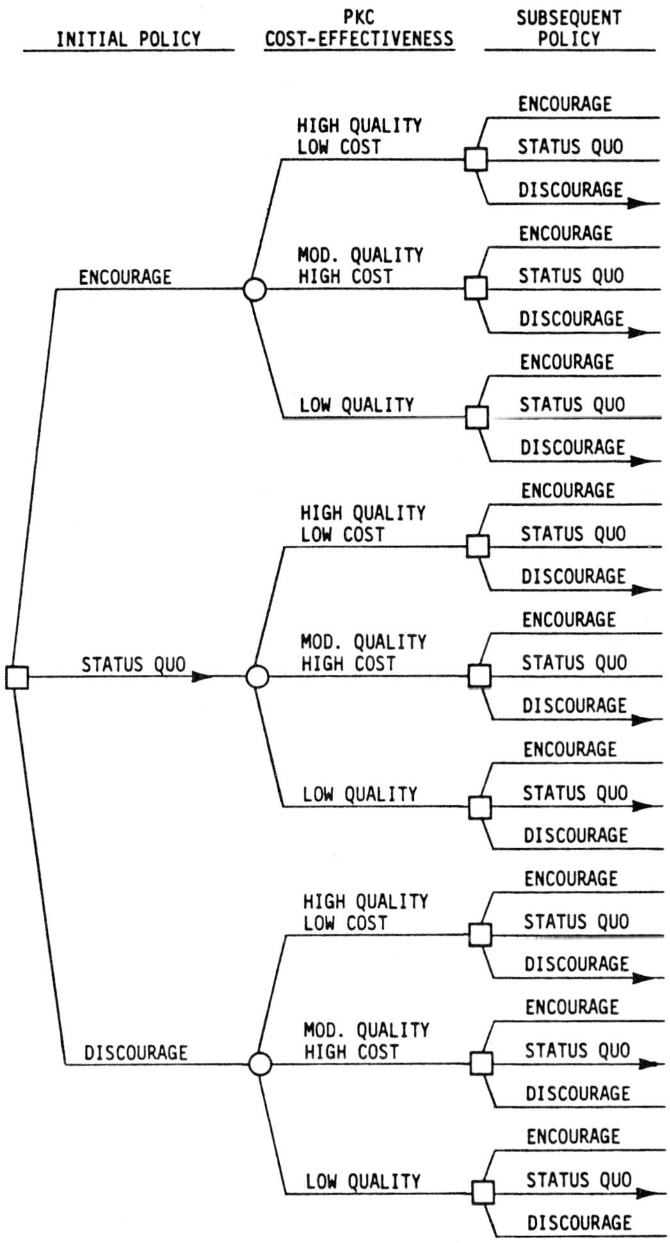

FIGURE 13 DECISION TREE SHOWING WITH ARROWS POLICY STRATEGY MINIMIZING SOCIAL COST ASSUMING 50% OF FOREIGN NATIONS CURRENTLY USE "UNBREAKABLE" CODES

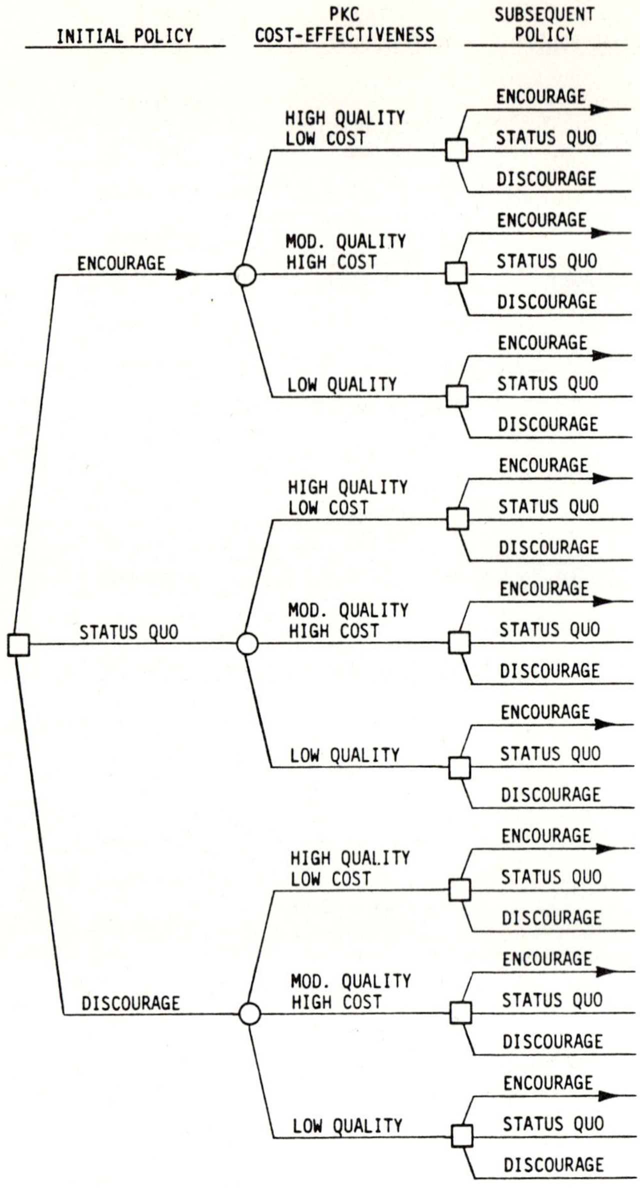

FIGURE 14 DECISION TREE SHOWING WITH ARROWS POLICY STRATEGY MINIMIZING SOCIAL COST ASSUMING 90% OF FOREIGN NATIONS CURRENTLY USE "UNBREAKABLE" CODES

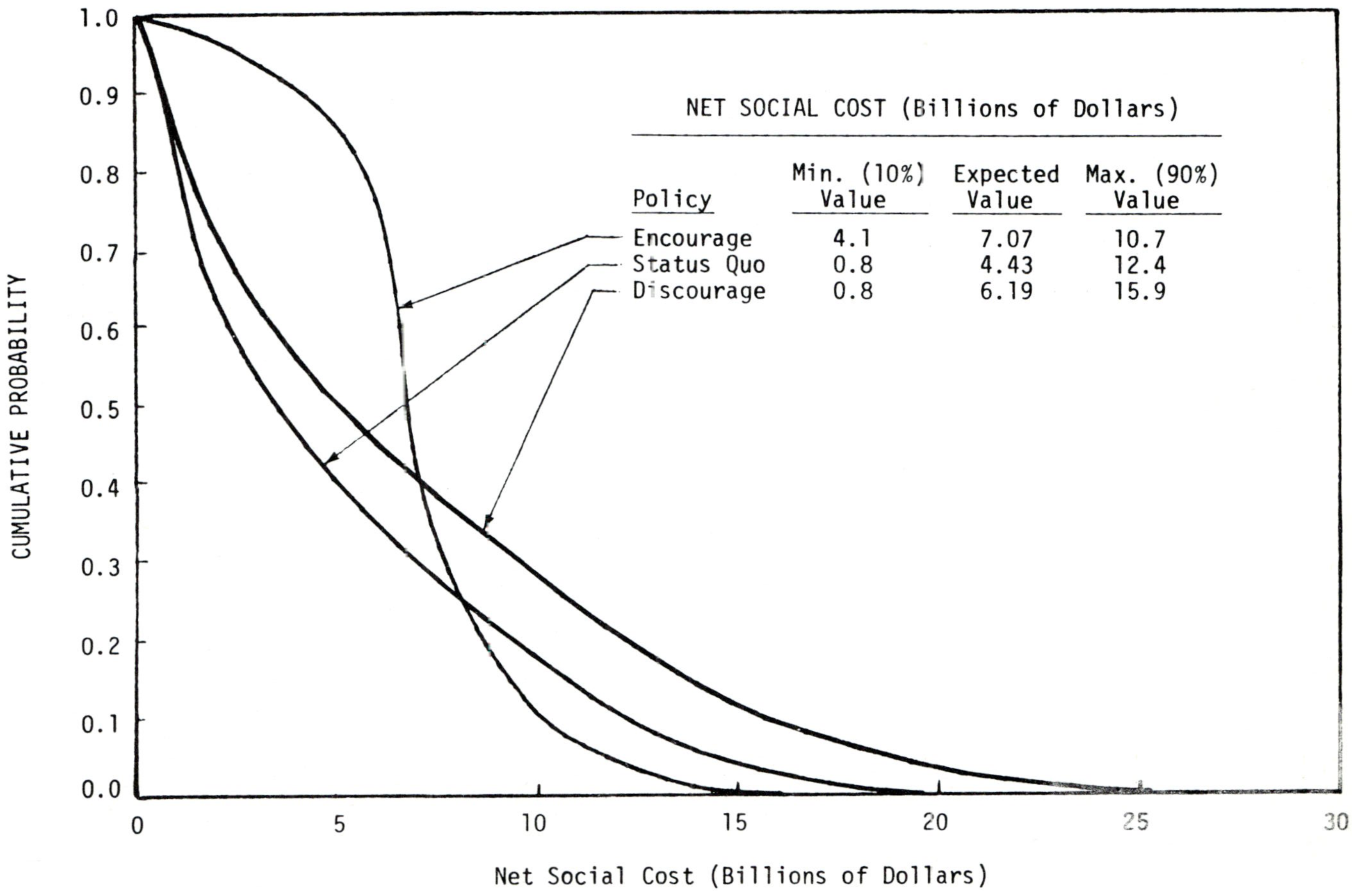

FIGURE 15 ESTIMATED PROBABILITY THAT DISCOUNTED NET SOCIAL COST WILL BE GREATER THAN OR EQUAL TO VARIOUS VALUES UNDER EACH POLICY

The curves shown in Fig. 15 provide a quantitative estimation of the risks associated with each initial policy option. The curve associated with the discourage alternative lies to the right of the curve associated with status quo, indicating that the probability of exceeding any given social cost is higher under the discourage alternative than under the status quo alternative. For example, the probability of producing net social costs greater than $15 billion is shown to be roughly 0.4 under status quo and 0.12 under encourage. Thus, status quo appears uniformly preferable to a discourage policy under the assumptions of the analysis. On the other hand, the encourage and status quo curves cross. Status quo, which has a lower expected cost than the encourage policy ($4.43 billion compared with $7.07 billion), produces a higher probability of very large costs. For example, the probability of social costs greater than $15 billion is nearly zero with the encourage policy. The reason for this is that the highest uncertain social costs are produced for paths through the tree in which the cost-effectiveness ratings of PKC and traditional codes are very low (i.e., the codes are broken) and computer-related crime is very high. With the nominal trade-off weights assigned ($50 per unauthorized access and $100 million per additional foreign country using "unbreakable" codes), these costs are higher than those generated from national security losses. The encourage policy, therefore, reduces risk by making the highest possible losses from computer-related crime less likely. Thus, encouraging PKC produces a higher expected cost than does status quo (due to likely security losses resulting from use by other countries), but less risk of very large social costs (due, for example, to higher than expected computer-related crime rates).

Figure 16 is a policy diagram indicating how the initial crypto policy decision depends on the social cost ascribed to increased security of foreign communications. The curves show how the expected discounted net social cost of each policy option depends on the dollar trade-off weight assigned per additional foreign country using "unbreakable" codes. To obtain these curves, rather than hold the value at the nominal assignment of $100 million, the analysis was repeated assuming a variety of values between $0 and $300 million. The larger the value, the larger the expected net social cost, but, as illustrated in the figure, the policy minimizing expected social costs depends on the value chosen. For value assignments between about $30 million and $200 million, the status quo policy produces minimum expected social costs. Below $30 million, expected social costs are minimized with the encourage policy; above $200 million, expected costs are minimized with the discourage policy. The quantitative

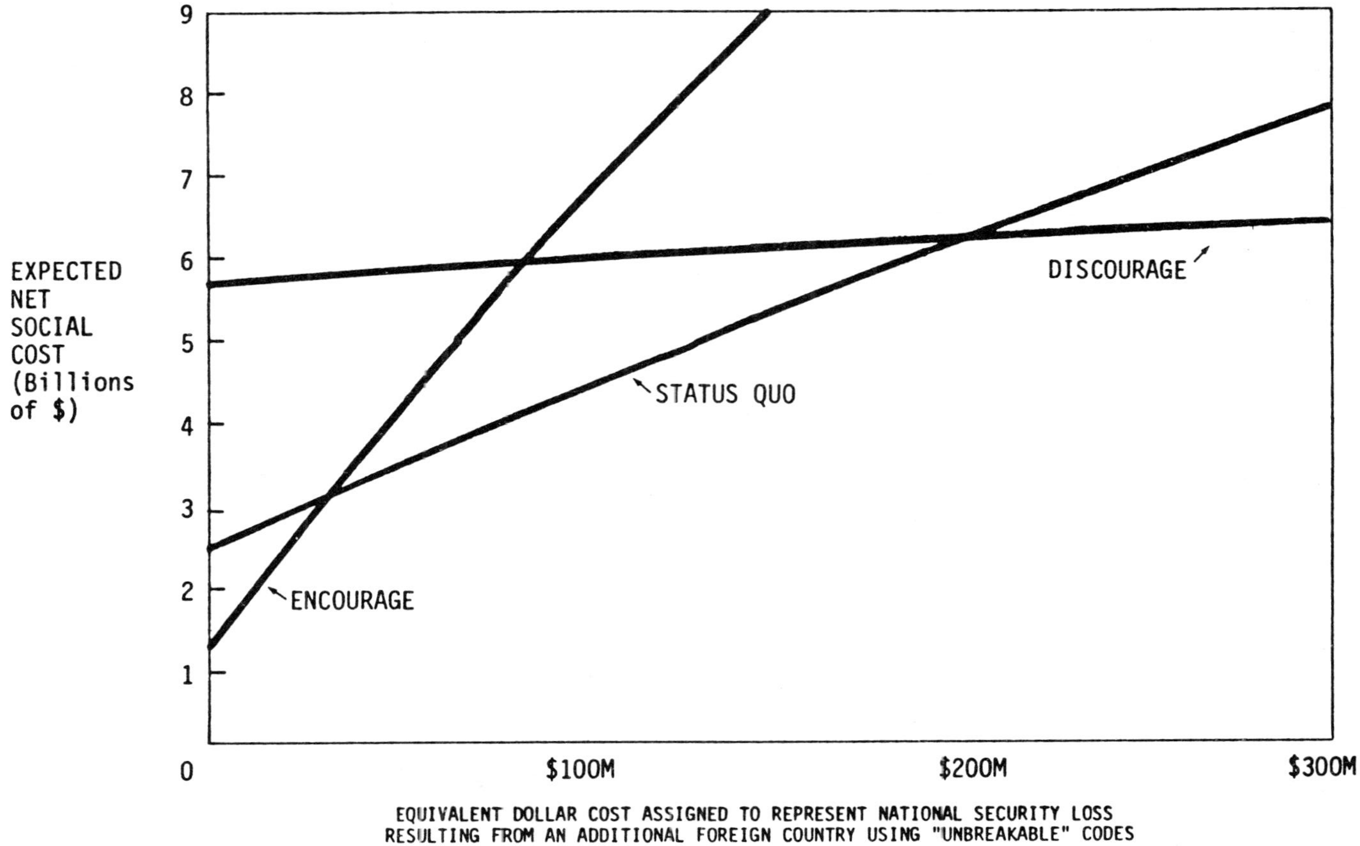

FIGURE 16 POLICY DIAGRAM SHOWING HOW INITIAL CRYPTO POLICY DECISION DEPENDS ON SOCIAL COST ASSIGNED TO INCREASED SECURITY OF FOREIGN COMMUNICATIONS

analysis thus enables the decision among alternative crypto policies to be reduced to a more fundamental social preference question; namely, the relative social cost of being unable to cryptanalyze foreign communications compared with the social cost of privacy and business losses resulting from insecure private-sector communications.

Informational Analysis

The fundamental idea in informational analysis is that of placing a monetary value on gathering additional information. A conceptual aid to computing this value is the concept of a clairvoyant. Suppose someone existed who knows in advance just what value of a particular uncertainty would result in a decision problem. Decision analysis methodology may be applied to estimate how much such information is worth. The approach is to determine by how much expected costs could be reduced if the information were known prior to the decision. This reduction, then, is the maximum amount one would be willing to pay a clairvoyant to obtain the information before having to make the decision. Computing the value of clairvoyance is useful because it places an upper bound on what it might be worth to attempt to gather better information on a given uncertainty before making the decision. Practical information gathering schemes, such as surveys, computer modeling, etc., are costly and yield less than perfect information. The value of perfect information can thus be used to help guide information-gathering decisions.

Another useful concept is that of a wizard. A wizard is defined as someone who can set an uncertain quantity to any specified value. Because in many situations a decisionmaker may have some ability to alter the probabilities of uncertain events, it is of value to know how much this ability is worth. The value of wizardry may be calculated by setting uncertainties to specific values and observing the effect on calculated expected social costs. The difference in the calculated expected social costs yields the value of wizardry.

Table 6 shows the value of clairvoyance and the value of wizardry calculated for the most sensitive uncertainty--computer crime rate. Recall (Fig. 12) that the decision model assumes that there is a 20 percent chance that the rate of major, publicly reported, computer-related crimes will increase by a factor of 10 within the next decade. As indicated, knowing whether this will happen is worth, according to the decision model, $250 million. Being able to guarantee a low crime rate is worth much more--in excess of $2.5 billion.

Table 6

Expected Values of Resolving Uncertainty on and Controlling Computer-Related Crime Rate

	Billions of Dollars
Expected Cost without Information	4.43
Expected Cost with Perfect Information on Crime Rate	4.18
Expected Cost with Crime Rate Low	1.66
Expected Value of Clairvoyance (4.43 - 4.18) (Value of resolving uncertainty on computer-related crime rate)	0.25
Expected Value of Wizardry (4.43 - 1.66) (Value of guaranteeing a low level of computer-related crime)	2.77

The reason for these high values is the potential impact of crime-rate information on the policy decision. With uncertainty over whether the computer-related crime rate will increase, the best strategy is status quo. If it could be determined in advance whether computer crime will increase, the best strategy would depend on the information. If it were known that the crime rate would increase, analysis of the decision tree shows the best strategy to be to encourage commercial cryptography. National security costs would likely be higher, but personal privacy and business losses would be minimized. If the computer-related crime rate remains low, the best strategy is to discourage cryptography, thereby minimizing national security costs. Being able to adjust policy in this way in accordance with knowledge over future crime rate reduces expected social costs calculated from the decision tree by $250 million. On the other hand, ensuring that the crime rate will remain low allows the discourage option to be applied and guarantees that personal privacy, business, and national security losses will be low. The result in this case is a much larger reduction in net social costs of $2.77 billion.

The very large value of controlling computer-related crime derived from the quantitative analysis suggests that it might be of considerable value as part of a national public crypto policy to institute a program of computer crime prevention. It seems likely that, for well under $2.77 billion, steps could be taken to significantly reduce the probability of a major increase in the incidence of computer-related crime. Such actions not only would reduce the commercial need for highly secure cryptosystems (which may threaten national security), but also would produce additional social benefits not represented in the quantitative analysis, such as reducing economic losses from computer-related crimes that cannot be prevented through encryption.

METHODOLOGY CONCLUSIONS

The typical technology assessment describes the technology, its environment, and the issues surrounding it, and then jumps immediately to a statement of policy goals and recommendations. The methodology applied in this study included a formal quantitative analysis of alternative government policies.

Although quantitative analysis unquestionably appears of value to technology assessment, two problems arose in this application of decision analysis techniques. One problem concerned the efficient integration of technical information into the analysis. Decision analysis traditionally relies heavily on the information base which tends to be generated and held by concerned parties to a decision. Efficient integration of this information is accomplished through the involvement and commitment of the concerned parties to support the decision analysis process. In this case, there is little public perception of a need for government action in the area of PKC except within the national security community--a highly knowledgeable and concerned stakeholder in policy decisions. Because the study was unclassified, it was impossible to obtain the sort of intensive involvement of knowledgeable experts that is typical of decision analyses.

The second problem concerned the difficulty of modeling policy outcomes and social preferences over those outcomes. Many of the potential consequences of PKC are extremely difficult to quantify or even make explicit. Personal privacy, social freedoms, and national security, for example, are very broad concepts. Application of decision analysis logic to policy analysis requires that such social concerns be quantified. Perhaps one of the most important conclusions of the study is a demonstration of the importance to policymaking of clarifying social goals and establishing priorities and social preferences for trading off such goals when necessary.

REFERENCES

1. W. Diffie and M. E. Hellman, "New Directions in Cryptography," IEEE Transactions on Information Theory, Vol. IT-22, No. 6, November 1976, pp. 644-654.

2. M. Hellman, "An Overview of Public Key Cryptography," IEEE Communications Society, Vol. 16, November 1978, pp. 24-32.

3. Executive Office of the President, Circular No. A-71, "Responsibilities for the Administration and Management of Automatic Data Processing Activities," Bureau of the Budget, Washington, D.C., March 6, 1965.

4. "The Data Encryption Standard," Federal Information Processing Standards Publication 46, U.S.A., January 1977.

5. R. A. Howard, "The Foundations of Decision Analysis," IEEE Transactions in Systems Science and Cybernetics, Vol. SSC-4, No. 3, September 1968.

6. H. Raiffa, Decision Analysis: Introductory Lectures on Choice Under Uncertainty, Reading, Massachusetts: Addison Wesley, 1968.

7. R. V. Brown, A. S. Kahr, and C. Peterson, Decision Analysis for the Manager, New York: Holt, Rinehart, and Winston, 1974.

8. C.-A. Stäel von Holstein and J. E. Matheson, "A Manual for Encoding Probability Distributions," Final Report, SRI International, Menlo Park, California, August 1979.

9. P. Schwartz and A. Mitchell, "The Art of Exploratory Planning: Identifying the Corporate Long-Range Course," Research Report 582, Business Intelligence Program, SRI International, December 1976.

An Asbestos Hazard Index for Managing Friable Asbestos Insulating Material

James S. Dyer

College of Business Administration, The University of Texas, Austin, Texas

INTRODUCTION

This paper describes the development of an asbestos hazard index to evaluate the potential hazard associated with the presence of asbestos in friable insulating material (FIM) in a building and to recommend the appropriate action to take based on this assessment. This hazard index was developed for use by the U.S. Navy to provide guidance on control measures for any FIM found in Navy shore-based facilities. The index has been incorporated into a comprehensive document for managing friable asbestos insulating material (10).

Asbestos is now recognized as a serious environmental carcinogen [e.g., see (14) and (7)], and its use as an insulating material in construction has been severely curtailed. However, FIM containing asbestos was a popular building material from the mid-1930s until the early 1970s; many buildings containing potentially hazardous FIM are still operated and maintained by the U.S. Navy.

A recent survey of shore facilities of the Navy used bulk sampling to conclusively identify FIM-containing asbestos. On the basis of this survey, it was estimated that 3.2 million square feet of friable asbestos insulating material are in continental United States Naval shore facilities. The costs of immediately removing all of this potentially hazardous material would be prohibitive. Therefore, a hazard index was needed to evaluate the degree of the hazard at each facility containing asbestos FIM and to guide hazard management activities.

The cost of asbestos hazard abatement actions are allocated through the Navy Occupational Safety and Health (NAVOSH) deficiency abatement program. Requests for these funds must compete for a portion of a fixed budget with other asbestos

NATO ASI Series, Vol. G4
Environmental Impact Assessment, Technology Assessment, and Risk Analysis. Edited by V.T. Covello et al.

hazard abatement requests and with all other safety and health deficiency requests, including concerns regarding noise abatement, replacement of rails on ladders and catwalks, and a variety of other concerns. A secondary objective of this hazard index is to aid this major resource allocation decision by NAVOSH.

The organization of the remainder of this paper is as follows. The logic of the asbestos hazard index is explained in the next section. This index is calculated based on consideration of the number of people exposed to the hazard, the level of the exposure, and its time duration. In principle, the level of the exposure to asbestos can be determined by air sampling. We elected to use a subjective scoring system instead, as we will explain later. Next, we present an illustration of the calculation procedure that determines the hazard index, followed by our conclusions.

LOGIC OF THE HAZARD INDEX

Expected Number of Deaths as a Criterion

FIM-containing asbestos can be hazardous if asbestos fibers are released into the air and ingested into the lungs of occupants of the facility. Even a low level of exposure to these fibers can lead to the development of bronchogenic cancer and mesothelioma, a form of cancer that attacks the membrane lining the chest cavity or the abdominal cavity. At low levels of exposure, these generally fatal diseases are the only significant health impacts of asbestos.

Suppose that a decisionmaker is confronted with the information that two hazardous situations exist in the work environment under his control. Further, the potential health effects to those persons exposed to these hazards are dramatic: Either there will be no effect or death will result. That is, there are no non-fatal health effects associated with these hazards, such as physical trauma or hearing loss. Ideally, the decisionmaker would take immediate action to mitigate these hazards, but in practice, the resources available to do so may be limited. Suppose he must choose between these two situations for immediate action. What criterion should be used in determining the priority of each situation?

This is a serious question with philosophical and ethical implications beyond our scope in this paper. The criterion we have elected to use for this application is the expected number of deaths; that is, we propose that the decisionmaker should allocate his resources to minimize the expected number of fatalities caused by the situations under his control. This choice implies a number of assumptions, including: An additive social welfare function is used, the lives of all individuals exposed to the hazard are valued equally, and the problem of "risk equity" can be ignored. Each of these three implicit assumptions will be discussed in turn.

The additive social welfare function for a group of n individuals may be written

$$\sum_{i=1}^{n} u_i$$

where u_i is the utility function of individual i (5). If we also assume that the disutility, or loss in value, from each individual's death is equal, then it is easy to see that minimizing the expected number of deaths is equivalent to minimizing the expected reduction in (1) caused by the hazards under review.

With respect to this specific application, we felt that it was reasonable to assume that the life of each individual exposed to the hazard should be valued equally. Arguments against this assumption are often made on the basis of a perceived public distinction between voluntary and non-voluntary risk [see (16) and (15)]. For example, if a worker voluntarily accepts a hazardous occupation, his death is not considered to be as bad (or perhaps, threatening) to society as the death of an "innocent bystander" who was unaware of the hazard. During the development of the asbestos hazard index, we originally considered different categories for persons exposed voluntarily and non-voluntarily, but decided to avoid this complicating feature because this distinction would be difficult to make for the majority of occupants of Navy facilities.

The issue of risk equity is potentially more serious. For example, the expected number of deaths criterion implies that the decisionmaker is indifferent between a situation where one individual may lose his life with a probability of 0.2, and another situation where two individuals may lose their lives with a probability of 0.1 for each. In the first instance, the single individual is being exposed to "twice the risk" as each individual in the second situation. To some, it will seem unfair to subject the single individual to this higher risk even though the expected number of deaths in each case is the same.

For simplicity, we have ignored the risk equity problem in the construction of the hazard index. This seems justified, in part, because we are dealing with extremely small probabilities of death (on the order of 10^{-5}) for individuals exposed to asbestos containing FIM, and because our estimates of these probabilities are extremely crude. An interesting perspective on the risk equity issue is provided by Keeney (9).

Implementation

The hazard index is determined by the number of persons exposed to asbestos fibers, and the level of their exposure. The level of the exposure is determined by two factors, the intensity of the exposure as measured in asbestos fibers per cubic centimeter (f/cc), and the time duration of the exposure. Shorter time durations of exposure pose less of a threat, and will result in a lower hazard index when all other factors are held constant.

The intensity of exposure can in principle be determined by air sampling techniques. Schirripa (12) has argued that this approach should be used since it does provide an objective method of estimating exposure levels in f/cc. This approach has several drawbacks, however, that argue against its use in this hazard index calculation. Air sampling data is subject to errors if done improperly, and it is relatively expensive and time-consuming. This hazard index is intended as a crude priority guide to help identify potentially hazardous situations. Once these situations have been identified, more elaborate analysis including air sampling can be used advantageously.

Air sampling has an additional limitation as well. Schirripa (12) emphasizes that sampling should be done in a room during normal and routine operations and activities. This ensures that the air sample reading will measure existing exposure levels in the room accurately. But what if these activities should be altered significantly shortly after the air samples are taken? The exposure levels could change dramatically. To avoid the need for continual monitoring of a facility containing asbestos-laden FIM, we preferred an estimate of the potential hazard associated with a situation. This would also provide a margin of safety for the analysis.

Therefore, we chose to develop a "level of exposure factor" that could be related to easily assessed characteristics of the facility and its operations. The objective is to allow individuals to score a situation on several factors that describe the condition of the FIM and the nature of the activities carried on in the facility, and use these scores to determine the appropriate level of the exposure factor. A similar scheme has been suggested by the Environmental Protection Agency (EPA) for health hazard exposure assessment in schools (3).

The first task is to determine the factors that should be considered in estimating the intensity of the exposure to asbestos fibers that actually or potentially exist in the facility being analyzed. Asbestos presents no serious health hazard unless fibers are released into the air and inhaled by individuals. The number of fibers in the air will be determined by the amount of friable asbestos in the area, the condition of the FIM, and the likelihood that it will be disturbed. Each of these characteristics can be evaluated by considering other factors, in turn.

The amount of friable asbestos in the area depends on the concentration of the asbestos in the FIM and its location/accessibility. The concentration of asbestos fibers is measured by the percentage of asbestos in the friable materials. These percentages can range from less than one percent to almost 100 percent. The location/accessibility of the material takes into consideration the percentage of the ceiling and wall surface area within the facility covered by the FIM, and its accessibility to disturbance by the activities of the occupants of the building.

The condition of the FIM is determined by its friability and its state of deterioration. Extremely friable materials break apart and crush into powder very easily and are potentially more hazardous. FIM that has deteriorated so that the adhesive materials are no longer binding them together is more likely to release fibers into the air.

Finally, the activity levels within the facility will determine whether or not the materials will actually be disturbed and release fibers into the air. Loud noises, vibrating machinery, or physical activities by the occupants (in a gymnasium, for example) may disturb friable materials that would otherwise lie dormant. Maintenance activities that contact the materials are another source of disturbances.

The logic of the hazard index may be portrayed graphically, as shown in Fig. 1. We now consider how to actually estimate the required exposure levels.

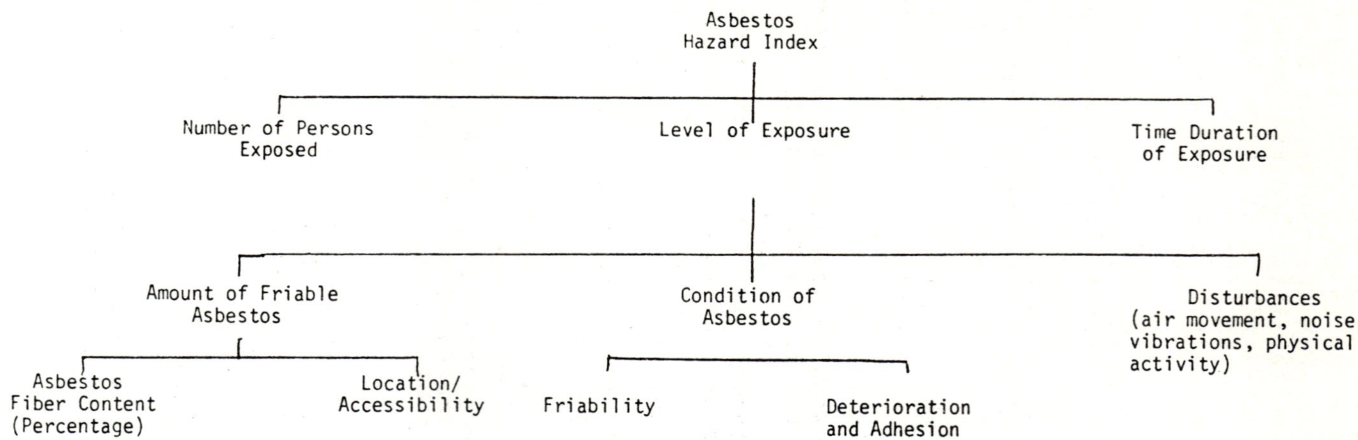

Figure 1. Summary of Factors Included in the Asbestos Hazard Index

ESTIMATION OF EXPOSURE LEVELS

We would like to use ratings on the five factors shown in Fig. 1 to estimate the asbestos fiber exposure level in f/cc. One approach would be to collect both air sample and rating data on a sample of facilities containing FIM, and create a model using multiple regression or some other statistical model. Sufficient data were not available to pursue this option.

As an alternative, we chose to rely on expert judgment. We selected a professor of occupational health in a major medical school with extensive experience in asbestos hazard abatement activities, including the design of a research effort to collect air sample data on asbestos fiber counts associated with various occupational activities requiring contact with asbestos materials. We began this interview by describing what became our "standard environment": a room containing FIM with a 20 percent asbestos fiber content that is in good condition, is moderately friable and moderately accessible, and in which the occupants engage in a moderate level of activity. After defining these terms in more detail, we posed the following question:

> In this environment, what is your best estimate of the level of exposure to asbestos fibers that the normal room occupant will receive?

The response was 0.1 f/cc. This result is consistent with an independent air sampling study by Schirripa (12), who found that 62.5 percent of the facilities in his sample had mean fiber concentrations between 0.01 and 0.1 f/cc.

The next question was more difficult, and less amenable to objective verification. We asked about the likelihood and intensity of exposure for "bursts" of fibers caused by contact with the FIM. These bursts will be sporadic and localized, but will subject those individuals physically close to the disturbance to a much higher concentration of fibers. In the "standard environment," we asked the expert to estimate the full-time equivalent exposure level of occupants that considers a relatively constant exposure level of 0.1 f/cc combined with occasional exposure to bursts of fibers. The expert responded with a caveat that this was a most difficult question, but a range of 1-2 f/cc was his "best guess. . . ." For the purpose of constructing this hazard index, we used 2 f/cc as a conservative estimate.[1]

---

1 This concept of a "conservative estimate" can be given a more precise meaning, as explained in the Appendix.

Next, we explored how changes in the description of the "standard environment" would affect these estimates. He responded, for example, that FIM of low friability would roughly reduce the exposure level by a factor of 0.1, while the exposure with highly friable material would be doubled. Similar estimates were obtained for three descriptions of the accessibility factor, two descriptions of the condition factor, three descriptions of the activity factor, and for three levels of asbestos content in the FIM. This creates 162 possible combinations of factor scores. The expert suggested that these relationships should be similar to those contained in the EPA scheme (3). He also stated that the effects of these factors should be relatively independent, so that the exposure level adjustments could be made by multiplying the f/cc in the standard environment by a series of ratios determined by the description of a particular facility. The set of factors that we eventually used are shown in Table 1.

Table 1. Adjustment Factors for Estimating the Equivalent Exposure Level

Friability Description	Factor	Accessibility Description	Factor	Condition Description	Factor	Activity Description	Factor	Percentage Asbestos Description	Factor
Low	0.1	Low	0.1	Good	1.0	Low	0.5	Low	0.5
Moderate	1.0	Mod.	1.0	Poor	1.5	Mod.	1.0	Mod.	1.0
High	2.0	High	1.5			High	2.0	High	1.5

For example, a room containing FIM with a five percent asbestos fiber content that is highly friable and in poor condition, and that is moderately accessible to occupants engaged in a high level of activity would have an estimated effective exposure level of

$$(2f/cc)\ (2.0)\ (1.5)\ (1.0)\ (2.0)\ (0.5) = 6\ f/cc$$

where the multiplication factors are obtained from Table 1. The exposure level of 6 f/cc is relatively high, as one might intuitively expect given this description of the room, and exceeds the current Occupational Safety and Health

Administration (OSHA) standards for permissible occupational exposure, which is set at 2 f/cc [OSHA (10)].

These exposure levels would have been sufficient to rank facilities on the basis of the potential hazard they create for a single individual exposed to the FIM environment. However, we had selected the expected number of deaths as the decision criterion, so it was necessary to convert these exposure levels into the probability of death for a single individual exposed to the environment for eight hours per day. By multiplying this probability by the number of persons exposed, and adjusting for the time duration of the exposure, an approximation to the desired hazard index could be obtained.

The relationship between the exposure level, or "dose," of asbestos and the latency period for the development of tumors is inverse. Jones and Grindon (8) have proposed a time-to-tumor model which relates the median time for the incidence of a tumor caused by exposure to a carcinogen to the dose of the carcinogen. This model has been applied to the asbestos worker data of Selikoff, Hammond, and Seidman (14) by Enterline and Henderson (4), and the results indicate that the time-to-tumor associated with a dose of asbestos fibers is a log normal distribution with a median time-to-tumor equal to

$$t = 98.65 \left(\frac{1}{D}\right)^{.33}$$

where D is the eight-hour, time-weighted average dose experienced for one year (200 eight-hour days) in f/cc. Hehir, Bayard, and Thompson (6) extended this result by making several assumptions, including: 1) The dose-response relationship between asbestos and bronchogenic cancer or mesothelioma is linear, with approximately 10,145 potential cases/million caused by exposure to each 1 fiber/cc; and 2) Competing risks of death for the first 40 years following exposure are considered to be normal, and the risk of an asbestos-caused death after 40 years following exposure is considered to be zero.

Because of assumption 2, many potential cases never occur because the time-to-tumor exceeds the 40 year cut-off, or because other risks take their toll on a proportion of these potential cases. For example, their model indicates that only 132 actual cases of respiration cancer deaths will be observed from the 10,145 potential cases caused by an exposure to 1 f/cc for one year because of the long median time-to-tumor of 98.65 years associated with this dose.

Bayard (1) reports an excellent fit (r=0.992) between their more complex analysis and the simple equation

$$\log y = 2.081 + 2.213 \log x$$

where x is the average daily dose in f/cc and y is the lifetime respiratory cancer deaths per 1,000,000 persons exposed. We used equation (3) to convert the asbestos exposure levels obtained from 162 combinations of the factors shown in Table 1 into corresponding estimates of the number of deaths/million persons exposed to an FIM-containing facility. These results are shown in Table 2, and can be interpreted as an estimate of the probability of death caused by exposure to the asbestos when multiplied by 10^{-6}. We refer to these estimates as exposure factors.

The choice of a specific time-to-tumor extrapolation model is a potential source of significant error in a risk analysis, and should be explicitly recognized as such. For example, Chand and Hoel (2) and Whittemore and Altschuler (16) have noted that while the log normal and Weibull time-to-tumor models may generate very similar fits for a given data set, they can lead to low-dose risk estimates that differ by several orders of magnitude. While these errors may not alter the rankings of the situations evaluated by our model, they should be considered when comparing risk estimates generated from this model with the results of risk analyses regarding other hazards.

CALCULATIONS AND USE OF THE MODEL

The instructions for the use of the exposure factors in Table 2 are contained in the asbestos hazard handbook by Lory and Cain (9). The first step is to evaluate a facility in terms of the four factors: friability, condition, accessibility, and activities. Instructions and examples are provided. The percentage of asbestos in the FIM must be established by bulk material sampling analyzed at a laboratory. This laboratory analysis is required in any case to establish the existence of asbestos fibers in the FIM.

The exposure factor from Table 2 is multiplied by the number of persons normally occupying the area, and adjusted by an occupancy duration ratio. The occupancy duration ratio is simply an estimate of the mean number of hours the typical occupant is in the area per week divided by 40. Thus, if the occupants only average 20 hours per week in the FIM-containing area, the occupancy duration factor is 20/40 = 0.5.

Table 2. Determining Level of Exposure Factors

Level-of-Exposure Elements				Level-of-Exposure Factors Based on Following Asbestos Contents (Item 5)		
Item 1 Material Friability	Item 2 Occupant Accessibility	Item 3 Material Condition	Item 4 Level of Activity	Low (1%-9%)	Medium (10%-39%)	High (40%-100%)
Low	Low	Good	Low	0.0002	0.0011	0.0031
Low	Low	Poor	Low	0.0056	0.0031	0.0087
Low	Low	Good	Moderate	0.0011	0.0065	0.018
Low	Low	Poor	Moderate	0.0031	0.018	0.050
Low	Low	Good	High	0.0065	0.037	0.102
Low	Low	Poor	High	0.018	0.10	0.28
Low	Moderate	Good	Low	0.065	0.37	1.0
Low	Moderate	Poor	Low	0.18	1.0	2.8
Low	Moderate	Good	Moderate	0.37	2.1	5.8
Low	Moderate	Poor	Moderate	1.0	5.8	16.0
Low	Moderate	Good	High	2.1	12.0	33.0
Low	Moderate	Poor	High	5.8	33.0	92.0
Low	High	Good	Low	0.18	1.0	2.8
Low	High	Poor	Low	0.497	2.8	7.9
Low	High	Good	Moderate	1.0	5.8	16.2
Low	High	Poor	Moderate	2.8	16.3	45.0
Low	High	Good	High	5.8	33.0	92.0
Low	High	Poor	High	16.0	92.0	260.0
Moderate	Low	Good	Low	0.065	0.37	1.0
Moderate	Low	Poor	Low	0.18	1.0	2.8
Moderate	Low	Good	Moderate	0.37	2.1	5.8
Moderate	Low	Poor	Moderate	1.0	5.8	16.0
Moderate	Low	Good	High	2.1	12.0	33.0
Moderate	Low	Poor	High	5.8	33.0	92.0
Moderate	Moderate	Good	Low	21.0	120.0	330.0
Moderate	Moderate	Poor	Low	58.0	330.0	920.0
Moderate	Moderate	Good	Moderate	120.0	690.0	1,900.0

continued

Table 2. Continued

Level-of-Exposure Elements				Level-of-Exposure Factors Based on Following Asbestos Contents (Item 5)		
Item 1 Material Friability	Item 2 Occupant Accessibility	Item 3 Material Condition	Item 4 Level of Activity	Low (1%-9%)	Medium (10%-39%)	High (40%-100%)
Moderate	Moderate	Poor	Moderate	330.0	1,900.0	5,300.0
Moderate	Moderate	Good	High	690.0	3,900.0	11,000.0
Moderate	Moderate	Poor	High	1,900.0	1,100.0	30,000.0
Moderate	High	Good	Low	58.0	330.0	920.0
Moderate	High	Poor	Low	160.0	920.0	2,600.0
Moderate	High	Good	Moderate	330.0	1,900.0	5,300.0
Moderate	High	Poor	Moderate	920.0	5,300.0	15,000.0
Moderate	High	Good	High	1,900.0	11,000.0	30,000.0
Moderate	High	Poor	High	5,300.0	30,000.0	83,000.0
High	Low	Good	Low	0.37	2.1	5.8
High	Low	Poor	Low	1.0	5.8	16.2
High	Low	Good	Moderate	2.1	12.0	33.0
High	Low	Poor	Moderate	5.8	33.0	92.0
High	Low	Good	High	12.0	69.0	190.0
High	Low	Poor	High	33.0	190.0	530.0
High	Moderate	Good	Low	120.0	690.0	1,900.0
High	Moderate	Poor	Low	330.0	1,900.0	5,300.0
High	Moderate	Good	Moderate	690.0	3,900.0	11,000.0
High	Moderate	Poor	Moderate	1,900.0	11,000.0	30,000.0
High	Moderate	Good	High	3,900.0	22,000.0	62,000.0
High	Moderate	Poor	High	11,000.0	62,000.0	170,000.0
High	High	Good	Low	330.0	1,900.0	5,300.0
High	High	Poor	Low	920.0	5,300.0	1,400.0
High	High	Good	Moderate	1,900.0	11,000.0	30,000.0
High	High	Poor	Moderate	5,300.0	30,000.0	83,000.0
High	High	Good	High	11,000.0	62,000.0	170,000.0
High	High	Poor	High	30,000.0	170,000.0	470,000.0

The result is the asbestos hazard index for the facility, which is a crude estimate of the total expected number of deaths that will result from exposing the occupants to this environment for one year. The remaining task is to interpret this hazard index as a guide for action.

Given a health hazard index value, we must decide which risks are "small enough" to be ignored, and which are "big enough" to require immediate action. One approach is to relate these numbers to the risks associated with other activities in the workplace and in our society. These data are commonly expressed in terms of the probability of death per person per year.

A health hazard index value of 100 for a room normally occupied by one person indicates that the probability of his death from asbestos-induced causes is 10^{-4} over his lifespan. The time-to-tumor model (2) can be used to convert this figure into an estimate of the probability of death per year for the individual exposed. First, we note that the model does reflect the latency period associated with tumors resulting from exposure to asbestos fibers, and predicts that only 13.7 of the projected 100 deaths/million will occur in the first 30 years after exposure. In years 31 through 40, the expected deaths/million exposed per year range from 4.1 to 17.4, increasing at an average rate of 1.33 deaths/million each year. The estimated number of deaths/million in a one-year time interval 36 years after exposure to asbestos fibers at this level is 9.9, which translates into a probability of death per year of approximately 10^{-5}. For simplicity, we shall use 10^{-5} as the annual probability of death, even though it grossly overestimates this probability for the first 30 years after the exposure, and underestimates it for 37 or more years after exposure.

Table 3 shows the probability of death per year for a member of the general population from a selection of common hazards (18). From this table, we see that the exposure factor of 100 for a facility occupied by a single individual translated into a probability of death of 10^{-5} is less than the probability of death associated with a machinery accident, 5.1×10^{-5}. Thus, an exposure factor of less than 100 would indicate a risk from the asbestos to occupants of less than the risks accepted by workers in mechanical facilities. The risk from asbestos would also be smaller than the risks of death from falls, fire, drowning, and poisoning, and these latter statistics are averaged over the total population rather than over persons exposed to the hazards. This result argues that the risks associated with other common occupational hazards are relatively more severe than those associated with an environment receiving an exposure factor score of less than 100.

Using (3) we find that the exposure factor of 100 corresponds to an asbestos exposure level of 0.93 f/cc, which is below the permissible OSHA standard of 2 f/cc for occupational environments. There is an alternative way to view the risk associated with an exposure factor score of 100 that makes it even less of a serious concern. Several authors have argued that risks should not be measured simply in terms of lives saved or lost, but that life years saved should be considered as well. Given our examples, a 25-year-old victim of a machinery accident would lose 50.4 expected life years. However, a victim of lung cancer caused by the exposure of 0.93 f/cc of asbestos at age 25 would lose only 19.4 expected life years due to the long latency of the disease [from (2), t = 101 years]. The ratio of expected life years lost from the two causes of death is 0.38.

Table 3. Acute Mortality
in the United States in 1972

Cause of Death	Annual Probability of Death
poisoning (males)	3.7×10^{-5}
falls	8.0×10^{-5}
fire	3.2×10^{-5}
drowning	3.0×10^{-5}
industrial machinery (males)	5.1×10^{-5}
cardiovascular disease	5.1×10^{-3}
cancer	1.7×10^{-3}
chronic respiratory disease	1.7×10^{-4}

Source: World Health Organization (1976).

Following the same analysis, a value of 1,000 in the exposure factors table corresponds to a probability of death per year of roughly 10^{-4}, which is close to the annual probability of death from a fall for a randomly selected individual. In contrast, the probability that an individual will die from cancer during a year from all causes based on 1972 statistics is 1.7×10^{-3}, or almost 17 times the additional risk associated with an exposure factor score of 1,000.

Using (3), however, we find that the exposure factor score of 1,000 corresponds to an asbestos exposure level of 2.321 f/cc, which does exceed the current OSHA standard. However, our estimates were subjectively increased by our expert to allow for localized bursts of asbestos. The OSHA standard is based on an eight-hour time-weighted average. Using (2) the median time-to-tumor associated with the dose of 2.321 f/cc is 75 years, and the expected life years lost for a victim of lung cancer exposed at age 25 is 20.3. On both an absolute and a relative basis, we feel that the risks associated with an exposure factor score of 1,000 or more are cause for legitimate concern.

We do not mean to imply that annual risks less than a probability of 10^{-4} are insignificant and should be totally ignored. Rather, we recognize that other hazards in the typical working environment are likely to create more acute risks for occupants of Navy facilities, so NAVOSH monies would probably save more lives when allocated to other projects than asbestos abatement when the hazard index is in this range.

On the basis of this reasoning, we recommended the action guide summarized in Table 4 for the Navy asbestos hazard management procedures included in Lory and Coin (10). As a rule, facilities with a hazard index of less than 100 will be monitored but no further action will be taken, while those with an index of more than 1,000 will be candidates for asbestos hazard abatement projects, which may require removal or encapsulation of the asbestos-containing FIM. Facilities receiving an index score between 100 and 1,000 are flagged for further review and for actual air sampling studies. The final decision on these structures may include other considerations not measured in the hazard index.

There is one serious caveat to this interpretation. The data are based on the assumption that the typical individual is exposed to the asbestos-containing environment for one year. Cumulative exposure to the same level of asbestos over several years would be potentially more hazardous. In high-turnover situations such as many of those found in Navy facilities, this would not seem to be a serious problem. However, situations involving civilian workers or career service personnel who typically remain in the same FIM environment for several years would require special consideration.

Multiplying their health index value by the mean number of years of exposure would provide a crude, relative adjustment. The implicit assumption of this correction is a linear relationship between risk and length of exposure. If risk is, instead, related to some power of the length of exposure as it appears to be in the use of cigarette smoking, then this procedure would underestimate the true risk. Therefore, it would be prudent to exercise caution and judgment in the applications of this index in long-term exposure situations.

Table 4. Hazard Management Guide

Hazard Index Value	Recommended Action
Less than 100	A long-term corrective measure can usually be deferred; however, the building should be surveyed each year for evidence of change in conditions or occupancy level. Interim control measures should be initiated.*
Between 100 and 1000	Review management special considerations** as necessary to further analyze the situation and conduct air sampling studies; defer unless those considerations justify long-term control measures.
Over 1000	A NAVOSH Asbestos Deficiency Abatement Project for long-term control measures should be initiated.

*Interim control measures include educational programs for the occupants, posting of hazard signs, and special maintenance procedures designed to minimize the release of asbestos fibers.

**Special considerations include the projected remaining life for the structure and its renovation schedule, whether it is utilized by the public, and a general category of miscellaneous concerns.

IMPLICATIONS OF THE HAZARD INDEX AND CONCLUSIONS

Table 2 of exposure factors contains 89 entries less than 100; 24 entries between 100 and 1,000; and 49 entries above 1,000. Recall, however, that the exposure factor is multiplied by the number of occupants of the facility and weighted by an occupancy level factor to determine the hazard index. Further, for exposure factor scores above 1,000, if even one individual is exposed to the environment on a 40-hour per week basis, then the facility would be seriously considered for a NAVOSH Asbestos Deficiency Abatement Project according to Table 4. A question of interest is the following: How many situations are likely to receive hazard indices of 1,000 or more? In order to gain some intuitive insights, let us consider three situations that might be encountered, and calculate their hazard indices.

Examples

Situation 1: A large office area is occupied by 100 workers for 40 hours per week on a regular basis. An inspection determines that the ceiling is covered with FIM containing 20 percent asbestos. It is judged moderately friable, moderately accessible, in good condition, and in a low activity area. From Table 2, the appropriate exposure factor is 120. Multiplying this factor by the occupancy level of 100 persons gives a hazard index of 120 x 100 = 12,000, which clearly exceeds the 1,000 score required as an indicator of a need for action.

This situation is potentially hazardous because the asbestos-containing material is moderately friable and the occupants are not screened from it by any barrier. Suppose we consider the option of installing a suspended ceiling that is only disturbed during infrequent maintenance. This would change the rating of the accessibility factor to "low." Returning to Table 2, the exposure factor would be changed to 0.37, and multiplying by the occupancy level gives a hazard index of 37. This would be sufficient protection according to our evaluation system, since the revised index is less than 100.

Situation 2: A small warehouse is typically occupied by five workers for 40 hours per week. The ceiling is covered with FIM containing 15 percent asbestos as a fire protection. This material is judged to exhibit low friability and is in good condition, but it is highly accessible and in a high activity area. The hazard index is 165, which merits deferral unless special considerations suggest otherwise. In this case, cautioning the workers to avoid contact with the FIM and monitoring their compliance may be considered to be a sufficient action.

Situation 3: The basement of a large office building is occupied by four maintenance personnel on an average of 10 hours per week. The ceiling and walls contain FIM with a five percent asbestos content. This material is highly friable, highly accessible, and in poor condition, but the activity level is low. The hazard index for this situation is 920, just below the 1,000 scale requiring a hazard abatement project.

This situation would require some judgment regarding the relative cost of removing or protecting the FIM, versus cautioning the personnel to minimize their activities and time within the room. The latter strategy would require careful monitoring of the situation.

CONCLUSIONS

As these examples and a study of Table 2 make clear, the key factors in rating the hazard associated with an FIM situation are friability and accessibility. If the material is not friable and not accessible to the occupants of a facility, then it is not hazardous to their health. In fact, the argument can be made that the hazards associated with disturbing this type of FIM when removing or encapsulating it are much more severe than the risks of leaving it undisturbed [e.g., see (12)]. If the majority of the asbestos-containing FIM discovered in Navy facilities is neither friable nor accessible, then the hazard indices will generally be below 1,000, and often below 100. Otherwise, the available evidence combined with professional judgment suggests that there is cause for concern regarding the health risks to occupants, and this will be reflected in the hazard index scores of the facilities.

We emphasize that this asbestos hazard index is based on a series of simplifying assumptions, and it is crude. However, it is presented only as a guide to hazard management decisions, and ample room is left for the use of judgment and additional analysis. Implemented within this context, we believe that this hazard index will play a useful role in improving the difficult NAVOSH resource allocation decisions in the area of occupational safety and health.

APPENDIX

The expert provided a range of 1-2 f/cc as his "best guess" to the full-time equivalent exposure level for occupants of the standard environment. This range could be interpreted as his confidence limits, indicating a high-level of confidence that the "true value" falls within this range.

Thus, the full-time equivalent exposure level may be viewed as a random variable X with an unknown subjective probability density function $f_X(.)$. Suppose we interpret this "best guess" as an 80 percent confidence interval, so that our expert feels there is only one chance in 10 that the true equivalent exposure level is below 1 f/cc, and only one chance in 10 that it is above 2 f/cc. Then his subjective probability that the true equivalent exposure level is less than or equal to 2 f/cc is 0.9.

If the exposure level X is a random variable, then so is the estimate of the lifetime respiratory cancer deaths per 1,000,000 persons exposed, Y, related to X by (3). If $f_X(.)$ is differentiable for all x, then it is a straightforward matter to show that

$$f_Y(y) = x^{1.513} f_X(x)$$

where x = 10 exp (log y - 2.081)/2.513)). Thus, if x = 2 f/cc is the 0.9 fractile for $f_X(.)$, then y = 10 exp (2.081 + 2.513 log 2) = 687.8 is the 0.9 fractile for Y. That is, the expert's subjective estimate that the equivalent exposure level in the standard environment is less than or equal to 2 f/cc with probability 0.9 corresponds to the estimate of less than 687.8 lifetime respiratory cancer deaths per 1,000,000 persons exposed with probability 0.9 if there were no other uncertainty in this risk assessment procedure.

If fact, there is additional uncertainty in this model. First, we obtained the subjective opinion of only one expert regarding the full-time equivalent exposure level. Other "experts" would be likely to disagree, and incorporating their judgments might have widened this range of the confidence limits. Second, the relationship between asbestos exposure and the resulting lifetime respiratory deaths per 1,000,000 persons exposed is based on a set of assumptions that could be debated one-by-one, and changes in each would alter the results. Third, the simple equation we used (3) is the best least squares fit to this data, and introduces additional errors. Fourth, our determination of the exposure factors in Table 2 provides only a crude estimate of actual exposure levels in rooms corresponding to the descriptions.

A thorough risk analysis would explicitly incorporate the uncertainty from these sources if the results were to be used for policy making or inflexible standard setting. Recall, however, that our purpose is to provide a rough guide to action regarding asbestos hazards, and sufficient room is provided for gathering additional information and for judgment. We do feel justified in using the upper limit (2 f/cc) of the expert's confidence limits rather than the midpoint (1.5 f/cc), however. The log transformation in (3) is an order-preserving transformation, but the rankings of rooms would be different after multiplication of the exposure factors by the number of occupants if Table 2 had been generated based on the estimate of 1.5 f/cc for the standard environment rather than 2 f/cc. This sensitivity of the results to the choice of this value encouraged us to opt for the "conservative" estimate of 2 f/cc.

REFERENCES

1. Bayard, Steven P. "A Time-to-Tumor Model for Risk Assessment of Asbestos Induced Death from Respiratory Cancer," Proceedings of the Third FDA Office of Science Symposium, Colorado Springs, Colorado: Air Force Academy, 1978.

2. Chand, Nigand and D. G. Hoel "A Comparison of Models for Determining Safe Levels of Environmental Agents," in Reliability and Biometry, edited by F. Proschan and R.J. Serfling, SIAM, Philadelphia, PA, 1974, pp 382-401.

3. Dorsey, L.C. "Numerical System for Assessing Exposures to Asbestos," U.S. Environmental Protection Agency Office of Toxic Substances Memorandum, Washington, D.C.

4. Enterline, P. and V. Henderson "A Model for Extrapolating to Low Levels of Asbestos Exposure," Pittsburgh, PA: University of Pittsburgh, 1976.

5. Harsanyi, J.C. Rational Behavior and Bargaining Equilibrium in Games and Social Situations, Cambridge, England: Cambridge University Press, 1977.

6. Hehir, Steven Bayard and C. Thompson "CPSC Regulation of Non-Occupational Exposure to Asbestos in Consumer Products," Proceedings of A Workshop on Asbestos: Definitions and Measurement Methods, 1978.

7. Hogan, M.D. and D.G. Hoel "Estimated Cancer Risk Associated with Occupational Asbestos Exposure," Risk Analysis, Vol. 1, No. 1, 1981, pp. 67-76.

8. Jones, H.B. and A. Grendon "Environmental Factors in the Origin of Cancer and Estimate of Possible Hazard to Man," Food, Cosmet., Toxical, Vol. 18, 1975, pp. 251-268.

9. Keeney, Ralph L. "Equity and Public Risk," Operations Research Vol. 28, 1980, pp. 527-534.

10. Lory, Ernest E. and Denise S. Coin "Management Procedure for Assessment of Friable Asbestos Insulating Material," Technical Report R883, Civil Engineering Laboratory, Port Hueneme, California: Naval Construction Battalion Center, 1981.

11. Occupational Safety and Health Administration Title 29, Code of Federal Regulations, Part 1910. Washington, D.C.

12. Schirripa, James T. "Airborne Levels of Asbestos Fibers in General Occupancy Buildings and an Estimate of Carcinogenic Risk," Huntington, New York: Industrial Hygenics, Inc., 1980.

13. Selikoff, I.J. "Cancer Risk of Asbestos Exposure," in Hiatt, Watson, and Winston (eds.), Origins of Human Cancer, Cold Spring Harbor, New York, 1977.

14. Selikoff, I.J., E.C. Hammond and H. Seilman "Cancer Risk of Insulation Workers in the United States," IARC, Biological Effects of Asbestos, Lyon, France, 1973.

15. Slovic, P., B. Fischhoff, and S. Lichtenstein "Perceived Risk," in R.C. Schwing and W.A. Albers, Jr. (eds), Societal Risk Assessment: How Safe is Safe Enough? New York: Plenum Press, 1980.

16. Starr, C. "Social Benefit Versus Technological Risk," Science, Vol. 165, 1979, pp. 1232-1238.

17. Whittemore, A. and B. Altschuler "Lung Cancer Incidence in Cigarette Smokers: Further Analysis of Doll and Hill's Data for British Physicians," Biometrics, 32, 805-816, 1976.

18. World Health Organization World Health Statistics Annual, Vol. 1, Geneva, Switzerland, 1976.

The Social Compatibility of Energy Technologies - Theoretical and Empirical Approaches in West Germany

Helmut Jungermann

Institut für Psychologie, Technische Universität Berlin, Dovestr.1-5, D-1000 Berlin 10

INTRODUCTION

As in many other industrialized countries, in West Germany nuclear energy has been the object of intensive and controversial discussion since the early seventies. This societal conflict has led to a general discussion concerning energy technologies and energy policies: Which path into the future should society follow and, more specifically, which governmental action regarding energy futures should be taken? The scope and the intensity of the conflict, which characterize the public as well as the scientific debate, indicate, on the one hand, the uncertainty about the implications of available policy options and the goals that should be aimed for, and, on the other hand, the willingness of a great number of organizations and individuals to participate in a search for solutions to vital societal problems.

The basic question for policy is: Which energy supply-system is better than alternative systems considering its benefits and costs for society? The problem is how costs arising in different dimensions--such as monetary, environmental, safety--can be legitimately related to one another by political judgment. The German Bundestag (Federal Parliament) charged an Enquête Commission (House Temporary Committee) to study the problem and to make a recommendation concerning the future energy policy. The basic idea was that an agreement on the assessment of costs and benefits of energy supply-systems is not so much a question of different forecasts but of political decisions resulting from public discussion of the advantages and disadvantages of alternative energy futures.

The Commission on Future Nuclear Energy Policy was established on March 29, 1979, exactly one day after the Three Mile Island accident. It consisted of seven members of parliament and

NATO ASI Series, Vol. G4
Environmental Impact Assessment, Technology Assessment, and Risk Analysis. Edited by V.T. Covello et al.

eight experts (four nominated by the ruling coalition parties and four by the opposition). The experts came from engineering, natural and social sciences, and humanities; one of the experts was Wolf Häfele, physicist and Director of the Nuclear Research Center Jülich, and another one was Klaus-Michael Meyer-Abich, physicist, philosopher, and Director of the Working Group Environment, Society and Energy at the University of Essen. The work of the Commission was assisted by a fulltime staff of eight scientists. Two results of the Commission's work, as laid down in a report (1), are relevant in the present context: The development of four scenarios of energy policy and the definition of criteria for evaluating energy policies.

The four possible paths into the future considered by the Commission represent a wide range of expectations and goals with respect to economic and energy developments. They differ in particular in their assumptions regarding economic growth, availability of fossil energy, structural changes in the economy, and energy conservation over the next 50 years. They do not consider cost or economic viability, as their purpose was to facilitate and structure the political debate only. I will not describe the various assumptions but only the energy policies connected with these four paths: Path 1 assumes a strong expansion of nuclear energy, particularly the operation of fast breeders after the year 2000, and gives only little weight to renewable energy sources or energy conservation. Path 4, on the other hand, assumes dispensing with nuclear energy altogether and puts heavy weight on energy conservation, a more efficient use of energy, and the use of renewables. Path 2 and Path 3 lie between these two extremes, differing in their assumptions about the use of nuclear energy conservation.

Table 1: Some major assumptions of four scenarios of energy policy

	Path 1	Path 2	Path 3	Path 4
Economic growth until 2000	3,3%	2,0%	2,0%	2,0%
Economic growth after 2000	1,4%	1,1%	1,1%	1,1%
Structural change of the economic system	medium	medium	strong	strong
Energy conservation	trend	strong	very strong	extreme

The Commission spent much effort on defining criteria for evaluating energy policies. It voted unanimously that the following criteria should be applied in future choices among alternative options: 1) economic efficiency; 2) international compatibility; 3) environmental compatibility; and 4) social compatibility.

The first three criteria are not new. However, the criterion of social compatibility has not always been a criterion of technology or policy assessment, at least not explicitly. Although Weinberg had already called attention to the social commitments that are associated with different energy systems in the early seventies (3), the importance of social acceptability was recognized only toward the end of the seventies when the conflict generalized to technological development as such. The fact that technological progress is increasingly questioned can be interpreted differently: As a phenomenon reflecting saturation in the area of consumption, giving more weight to qualitative values like clean environment or job satisfaction, or as a growing consciousness with respect to the potential impacts of technology on environment and society due to the new quality of these threats, or as a mixture of both. But whatever the roots are, one can safely predict for the foreseeable future that the problem of "social compatibility" of technology policies will remain a major issue.

The question is, however, how do we find out which policy is, or to what degree a policy is, "socially compatible" with the social order and development? The Commission outlined a few aspects that it considered particularly important: Constitutional rights and principles should not be endangered; room should be left for individual styles of living; flexibility for future generations to develop their own values and ways of living should be ensured. But these are still very general and abstract formulations, and certainly do not cover the entire range of relevant aspects of "social compatibility." More important, nothing is said about the <u>procedure</u> for defining, and assessing, "social compatibility." This task was thought to need special investigation.

The Minister for Research and Technology (BMFT) of the Federal Government, then, initiated two research projects on the "Social Compatibility of Energy Supply-Systems." One project is run at the universities of Essen and Frankfurt, the other one at the Nuclear Research Center in Jülich. (I am myself chair of the Scientific Advisory Board set up for the project in Jülich). Next, I will describe and compare the two projects. The focus will be on the methodological approach they have chosen, and only some results will be mentioned for illustrative purposes. I will describe the project in

Essen/Frankfurt first, then the project in Jülich, and finally I will compare both projects and draw conclusions regarding the relevance of the projects for technology and policy assessment in general.

I will draw, of course, heavily on reports and other papers published by the two project groups (mostly in German, unfortunately). Rather than refer each time to some particular publication, I will give a selected number of references to the most important publications of each group at the beginning of each presentation.

THE FIRST PROJECT: WHAT IS ACCEPTABLE?

The first project is directed by Klaus-Michael Meyer-Abich (University of Essen) and Bertram Schefold (University of Frankfurt), and includes eight scientists with differing disciplinary backgrounds. The project began in 1980 and will end in 1983. First reports have been made available (e.g., 2, 4-9).

The Understanding of "Social Compatibility"

Social compatibility, as it is understood in this project, is a _normative_ concept. While empirical research is concerned with describing what is socially _accepted_, the examination of what is socially _acceptable_ must study and evaluate the alternative options in relation to goals, values, and needs which are considered socially more or less desirable.

Due to the plurality of goals and the presently observable conflict about goals, the result of such an examination cannot be a statement that a particular energy supply-system is or is not socially compatible in any absolute sense. The criterion of economic feasibility might be taken as a model to make the point more clearly: The question is not whether certain measures are economically accepted but whether these measures are consistent with specific politically set goals like economic growth, full employment, or monetary stability. In the same sense, social compatibility is a relational concept, definable only with respect to otherwise established or desired goals. One can only state that an energy supply-system is consistent with a particular social order and development, and not with some other order and development. One might find, for example, that neither path 1 nor path 4 is socially acceptable in relation to the goal to protect the individual as much as possible against state control measures: Path 1 might be expected to lead to a "nuclear state" in which security measures would reduce personal freedom considerably, and path 4 might be expected to lead to a "calorie state" in which energy saving constraints would also reduce personal freedom (e.g., by restricting the availability of hot water for showers). In other words, people who favor a particular social, economic, or political development can learn from a social compatibility assessment of energy supply-systems which system is consistent with that development and which is not.

The Methodological Approach

Generally speaking, a methodological approach is required for weighting the potential societal benefits of an energy supply-system against the potential societal costs in relation to certain goals. The project group believes that quantifications will be mostly impossible because the social implications of energy supply-systems can only very imperfectly be described in terms of quantitative social indicators. It is argued that the central question is not answerable primarily by empirical research and simulation studies of the impacts of energy policy decisions, but mainly by reflection of the consistency of energy systems with established, projected, or desired goals. The group believes that, in general, "merely" qualitative answers to the right questions are more useful than quantitative answers to inadequate or irrelevant questions.

The group has formulated three methodological principles on which it based its work (4). The first principle is the analysis of implications. The result of a social compatibility assessment study should be a catalogue of implications in the form of "if - then" statements: If some social, economic, or political order or development x is desired, then energy supply-system E_x can be considered most consistent with x; but if another order or development y is desired, energy supply-system E_y is most consistent with that particular y. For example, if the societal goal is raising the standard of living in terms of consumption, then a nuclear energy system might be most consistent with that goal; but if the goal is to avoid exploiting the natural resources unnecessarily, then a different energy system might be most consistent with this particular goal. But whether goal x or goal y is desired is a matter of political discussion and not a matter of scientific recommendation. Therefore, societal goal conflicts are not affected by an analysis of implications; rather, such analysis might help to identify and articulate conflicts more precisely.

The second principle is called analytical extension. The term is borrowed from the mathematical theory of functions which allows, under certain conditions, a function to be extended beyond the value domain for which it is properly defined. Analytical extension in the present context means to extend the analysis of alternative energy supply-systems beyond the energy domain and examine the consistency of the systems in relation to other social or economic developments. This approach is taken on the grounds that the goals, values, and needs that are directly relevant for assessing the social compatibility of energy supply-systems usually are also indirectly relevant for other political decisions. Therefore, energy policy alternatives are closely interconnected with other alternatives of social, economic, political, and technological development. For example, energy policy decisions in West Germany are directly related to the goal that the country should be militarily defensible or that internal conflict (e.g., with terrorists) should be solvable without destruction. On the other hand, ways of life which are less achievement-oriented or more interested in nature-cures than presently are not directly related to energy supply-systems; yet, one finds considerably more partisans of such "alternative" or "green" orientations among the opponents of nuclear energy than among the proponents. That is to say, there are societal goals which are not directly affected by energy policy decisions and can yet be more or less consistent with a particular energy policy.

The third principle requires the <u>social construction</u> of the alternative energy supply-system, i.e., their description in social terms and categories. The engineering type of definition of an energy supply-system provides no basis for a prognosis of the social characteristics of the system, i.e., its effects on the people living with that system in a particular historical and political situation. The technical components of the entire energy supply-system, i.e., from primary energy to consumption, and the aggregated data of energy statistics have to be decomposed and concretized with respect to the implied social requirements and consequences. Thus, social construction demands the development of social indicators. This task turned out to be much easier for presently available systems (e.g., nuclear energy) than for alternative, not yet available systems (e.g., solar energy) due to the differing degree to which the systems are technologically developed.

The Dimensions for Evaluation of Energy Systems

The project group did not evaluate all four energy supply-system alternatives that had been developed by the Commission. On the one hand, the group considered the two extreme scenarios (1 and 4) as too unrealistic to make a closer analysis worthwhile. On the other hand, they expected a more precise picture if they concentrated on two relatively "pure" cases, i.e., K (for Kernenergie = nuclear energy) and S (for Sonnenenergie = solar energy). These two scenarios were developed on the bases of path 2 and path 3 of the Commission's four paths. The K scenario describes the development of an energy supply-system emphasizing the use of fast breeder reactors, high temperature reactors, and possibly nuclear fusion; light water reactors are neglected because they will or would exploit natural resources too fast. The S scenario emphasizes active use of solar energy (e.g., solar collectors) as well as passive use of solar energy (e.g., by architectural measures). Making various explicit assumptions regarding technical and economic characteristics of both alternatives, and using the figures used and/or assumed by the Commission for path 2 and 3, the K and the S scenario were programmed into a computer model. They have been revised in the meantime due to new figures and have been studied as they are affected by changes in certain assumptions. The K and S scenarios represent the energy policy alternatives that were evaluated with respect to their social compatibility.

The following aspects were defined by the group as being particularly important for comparing both scenarios: law and internal security; defensibility of the Federal Republic; internal peace and societal flexibility; risk; individual concerns and value change; the societal relation to nature; direct economic requirements and consequences; economic style. I will describe three aspects in more detail.

Law and Internal Security: New technical systems require usually new legal rules, and thus alternative energy supply-systems are evaluated with respect to their potentially different legal consequences--e.g., which constitutional norms and corresponding statutes are relevant. In addition, potential changes in federalism, planning law, communal administration, parliamentarism, etc., induced by one or the other system have been considered. Several studies are concerned with these issues.

Another problem thoroughly investigated is which security system might be required by both systems, K and S (7). Because security systems are expected to play a major role in nuclear energy systems, the various comprehensive studies focus on these potential future security systems and their consequences. For example, the group explored in detail what will need security measures in a fully developed K system (nuclear power plants, reprocessing and waste disposal, supply transport), against what threat the security measures are required (e.g., secret production of a nuclear bomb, radiological weapons, release of radioactive material), and against whom (e.g., persons and groups with political, economic, personal motives). The various kinds of potential measures against these threats were outlined and their consequences for the legal system in general and individual rights in particular were analyzed. As one might expect, the K scenario is not considered to be socially compatible with the basic values of our present legal system.

Individual Concerns and Value Change: The alternative energy systems have a multitude of direct and indirect impacts on individuals. This is evident if one looks at the many roles which people have in the long sequence, beginning with producing primary energy over the various steps of transformation and distribution to the consumption of energy. But individuals are also potentially affected, for example, if they live in areas exposed to noise, pollution, or risks in case of an accident in a nuclear power plant.

A second issue is whether, or to what degree, the scenarios K and S are consistent with societal value structures and their expected changes. In one study (8), the following values were distinguished on theoretical grounds and on an analysis of the literature: technical orientation, consumption orientation, achievement orientation, conservation of nature, concreteness and control, freedom and participation, social orientation, and safety. The two energy scenarios were compared on these value dimensions. It was argued, for example, that K and S are equally consistent with a technical orientation, with a consumption orientation, and also with an achievement orientation, although with respect to scenario S this depends to some degree on the yet not quite foreseeable specific forms of solar energy systems. Both scenarios, on the other hand, appear inconsistent with conservation of nature, but scenario S looks here somewhat better than scenario K because it assumes a strong tendency to energy saving. In terms of risk, the conclusion from a comparison depends on the definition and understanding of risk: If risk is defined in the engineering sense, i.e., as expected number of fatalities, both scenarios are equally consistent (or inconsistent) with the goal of ensuring safety. But if risk is given a broader interpretation, including qualitative aspects such as voluntariness, controllability, distribution, etc., scenario K is clearly less consistent with safety than is scenario S. The major result of this study is that scenario S leaves distinctly more room for adjustment to different value priorities than does scenario K. (For example, the adjustment potential of scenario K was shown to be very limited with respect to the job satisfaction dimensions, whereas scenario S could be shaped to increase job satisfaction).

Economic Style

The project group analyzed under one aspect the direct economic requirements and consequences of the two scenarios, primarily by using computer models. Following the principle of analytical extension, however, they also studied the compatibility of the alternative systems with long-range economic tendencies that are not directly induced by any energy system. Such an approach would take into account not only changes in the technical, economic, and social parameters of energy supply-systems, but also changes in other elements of the economic system.

In one study (9) the concept of economic style was revived and further developed to explore the relationship linking conceivable energy systems with the existing socioeconomic system and its future development. The variety of economic styles possible within the present economic system corresponds to the plurality of major ideological positions held in society; ultimately, the economic style of a society expresses the most widely held values articulated within the prevailing political system. The study distinguishes a "hard" and a "soft" economic style to highlight proponents' differing political commitments to rapid or slow economic growth, to token or substantial environmental protection, to low or high levels of unemployment, and to expansionist fiscal policies or redistributive social policy. "The hard economic style is characterized by technocratic division of labor, by monopolistic industrial organization, by centralization of political and economic decisionmaking, by production for exchange in world markets, and by permanent desire for more material well-being in the form of commodities" (2, p. 254). The soft economic style is based on "recognition of the imperative 'small is beautiful,' on acceptance of the assumption that a movement towards regional and individual economic autarchy brings about more social and political autonomy, and on confirmation of the cardinal significance of interpersonal relationships, spiritual values, and cultural traditions as a source of individual satisfaction. More consumption of commodities beyond the current level is categorically rejected as a way to a better life" (2, p. 254). The comparison of the scenarios K and S with these two styles leads to the following result: The nuclear energy system--characterized by "high technical and organizational complexity, need for coordination of decisionmaking at the national level, extensive control apparatus to secure the system against risk of theft, terrorism, and diversion of plutonium for military purposes"--is most consistent with the evolution of a hard economic style. "The nuclear system and the economic style have common ideological roots and are logically compatible with one another" (2, p. 255). "Alternatively, an energy system emphasizing a high degree of energy efficiency and renewable supply sources is a manifestation of the same dominant values that are expressed by a soft economic style. The delivery of energy services is more decentralized thanks to insulation, collectors, on-site conversion, and use of local wastes; administration is at the community level and large additional amounts of energy are no longer needed to support the pattern of consumption" (2, p. 255).

THE SECOND PROJECT: WHAT IS ACCEPTED?

The second project is run under the direction of Wolf Häfele, Erwin Münch, and Ortwin Renn at the Nuclear Research Center in Jülich. The group includes, in addition, four scientists and several consultants. It began its work in 1982 and will finish it in 1984. Several papers are also available from this group (10-14).

The Understanding of "Social Compatibility"

The starting point of the project is, of course, the definition of social compatibility given by the House Committee (1). But the group distinguishes a normative and a descriptive aspect of the concept. The group agreed that the alternative energy supply-systems and their impacts must be checked against normative standards, i.e., goals and values which have been determined as socially and politically desirable. It is the view of the group, however, that it is impossible to make scientific statements about what is actually socially and politically desirable, and also whether an energy supply-system is consistent with certain goal and value structures. This is considered as a genuine task of the political decisionmaking process. The task of scientific research is stated rather to identify the criteria currently considered relevant by society for a comparison of alternative energy supply-systems, to assess the potential impacts of these systems, and to reveal how these impacts are evaluated by society, i.e., whether, or to what degree, the systems are socially accepted. The degree of acceptance of energy systems is considered an important factor for the political evaluation of the acceptability of these systems.

The Methodological Approach

The project takes basically an empirical approach to the problem of social compatibility. I will concentrate on three major steps in this context: the development of a value tree as a basis for evaluating energy supply-systems, the analysis of impacts, and the actual evaluation of alternative energy supply-systems by groups of citizens. Whereas the first project evaluated two scenarios K and S, representing two policy options resembling path 2 and 3 of the set of four paths suggested by the Commission, the second project in Jülich decided to look at the evaluation of all four paths.

The examination of the social compatibility of alternative energy supply-systems is only one, though the major, objective of the research project. The other objective is to study whether the procedure for assessing the social compatibility of (energy) policy options proves itself to be effective and reasonable and could be recommended for use in the political process.

Value Tree Analysis: The first step in the project was to develop a set of criteria by which the alternative systems could be evaluated in the present society. One could have chosen a theoretical approach, e.g., analyzing the political discussion or picking the brains of political and social scientists. But there is a danger that personal preferences could dominate the criteria generation or that important criteria would be omitted because their advocates were outsiders to the political discussion. Therefore, the group chose an empirical approach to the problem and used a technique called value tree analysis by which a logically coherent set of evaluative criteria can be developed (13, 15, 16).

A value tree structures criteria hierarchically such that lower level criteria serve to define what is meant by the upper-level criteria. For example, if a top-level criterion is health and safety of the individual, lower-level criteria might be threats to life, threats to health, injuries, and psychological impacts. A value tree is constructed in an interactive session with an individual or a group. The analyst usually asks two types of questions. One type aims at identifying the criteria that are relevant for comparing the options, in this case energy policy options: What are your main objectives and concerns related to the problem? What criteria, attributes, or measures would be appropriate for differentiating between the given options? Or, why is some option desirable or undesirable? The other type of question aims at logically organizing the criteria: What do you mean by some specific objective? Could any criteria be grouped under some higher label? The final result is a hierarchy of criteria relevant for comparing and evaluating options, e.g., energy systems.

In the specific case of application of this technique--alternative energy supply-systems--one could expect different groups having different criteria. For example, one group might focus on economic criteria, whereas another group might focus on ecological criteria. Therefore, the researchers contacted several important social groups covering a wide range of perspectives on energy problems in the Federal Republic (e.g., Bund der Deutschen Industrie/Federation of the German Industries, Kraftwerk Union/Power Plant United, Deutscher Gewerkschaftsbund/German Labor Federation, Deutsche Katholische Kirche/German Catholic Church, and Bund der Bürgerinitiativen Umwelt/Federation of Citizen Environmental Groups). Separate value trees were developed and the different value trees were then aggregated in a manner that was logically defensible and ensured that the viewpoints of each group were included; the rationale for aggregation and the scientific aggregation procedure are described in (13). The result was a joint value tree which on the top level has the following criteria: financial and material requirements, security of energy supplies, national economic impacts, impacts on the natural environment, health and safety, political impacts, social impacts, and international impacts. These criteria were decomposed to a maximum of four more levels. For example, "social impacts" were decomposed into "quality of life" and "keeping societal options open, ability to change;" quality of life was further decomposed into "cultural and moral quality of life," "social quality of life," and "economic quality of life;" of these criteria, "cultural and moral quality of life" was decomposed into "preservation of cultural goods," "threats to cultural values," and "threats to the human identity."

The value tree developed in this part of the project can serve a number of purposes. For example, it provides a basis for further refining the criteria, e.g., by theoretical analysis or by empirical research with more or other individuals and groups. It might also be used as a basis for communication and discussion about the criteria for evaluating energy policy options. The primary purpose of developing a combined value tree within the project was, however, to provide a set of criteria on which the four energy paths could be compared and evaluated by citizen groups.

<u>Analysis of Impacts</u>: The second major step of the project was to determine the impacts of the four supply systems on the criteria. This implied, first of all, the operationalization of these criteria. To reduce the complexity of the value tree, the hierarchy of criteria was cut off at the second level. For

each of the 31 second-level criteria, 1 to 5 indications were defined. Mostly, the chosen indicators and measures were those that had come up during the development of the tree. In addition, studies on social indicators carried out by the OECD were used as a source for operationalization of criteria. For the remaining criteria, indicators were developed intuitively and then validated through interviews with experts. As a result, a detailed set of partly quantitative, partly qualitative indicators and measurement scales was obtained that represented very closely the criteria of the value tree. For example, the criterion "problems for future generations"--subcriterion of the first-level criterion "health and safety"--was indicated by "long-term consequences" and "long-lasting hazard sources."

The impacts of the four alternative energy options on the set of operationalized criteria were then assessed by an analysis of the literature and by expert inquiries. These assessments were validated with a thorough check by two members of the scientific staff of the Commission (one proponent and one opponent of nuclear energy). The result was an unweighted profile of impacts that served as information material for the citizens who, in the final step, evaluated the policy options.

<u>The Evaluation by Citizens</u>: The third, decisive step was the participation of citizens in what is called "planning cells." The underlying model of public participation was developed and tested in the field of urban planning. A planning cell is "a group of citizens who are selected by a random process and are given paid leave from their workday obligations for a limited period of time in order to work out solutions for given, soluble planning problems with the assistance of advisors on procedure" (14, p. 74).

The group, random selection, time limit, and the advisors on procedure are regarded as the significant structural elements in the planning cell. A group of citizens actually means about 25 people who are selected by a random procedure and thus can be assumed to represent a diversity of knowledge and opinion. The time is limited, of course; in this case, the groups met for four consecutive days. The advisors on procedure should ensure that the necessary and required specialized information is presented and made available and that the group process is as effective as possible. In 1982, 24 such planning cells were formed in various towns and cities of the Federal Republic. They were comprehensively informed about the four energy policy options, about the set of criteria for evaluating these options, and about the impact assessments provided by experts. They discussed these issues and formed their own opinions with respect to the impact of the four energy options on the top-level criteria. They then evaluated the options with a procedure based on multi-attribute utility theory, leading to an overall evaluative measure for each path and thus to a preference order of the paths. One major result is that intuitive preferences were clearly concentrated on path 2 and 3, whereas calculated preferences were concentrated on path 3 and 4; the correlation between intuitive and calculated preferences was only 0.50 [For further details see (12)].

I would like to make some final remarks regarding the potential functions of the planning cell--and their results--in the political decision process. First, it can serve a heuristic function, i.e., goal conflicts and tradeoffs between criteria can be identified and clarified. For example, one might be interested in the tradeoff between personal energy consumption cost and environmental protection measures. Second, there is a survey function due to the random selection of citizens and the large, geographically widespread number of groups. For example, one might be interested in relationships between certain person variables, such as education and income, and characteristics of the evaluation process, such as weighting schemes. Third, we have a problem-solving function because the data should facilitate the political decision process. For example, the knowledge about citizens' weighting schemes and preferences could be used to modify presently discussed options. Fourth, a legitimizing function could be served by the planning cell because policy makers could defend their decisions better if they took the opinions and preferences into account as they were revealed and expressed in such groups which did not represent any particular stake holder or pressure group but relatively unrestrained and informed citizens.

SUMMARY AND CONCLUSIONS

In 1979-1981, an Enquête Commission of the German Parliament on Future Nuclear Energy Policy developed four energy scenarios for the next 50 years, i.e., policy options describing different mixtures of energy supply-systems. The purpose was to facilitate and structure the political debate. The Committee also defined criteria for evaluating these options: economic efficiency, international compatibility, environmental compatibility, and social compatibility. The last criterion is the most unconventional but also the least specified one; the Commission defined it as "compatibility with the social order and development." To obtain a more comprehensive interpretation of "social compatibility" and also an answer to the question regarding the social compatibility of present energy policy options, the Minister for Research and Technology initiated two separate research projects.

The first project may be characterized as asking, "What is acceptable?" This is to say, the emphasis is on studying and evaluating the alternative options in relation to goals, values, and needs which are considered socially more or less desirable. The project thus takes a normative, largely qualitative approach and its results will be statements about the social compatibility of alternative options given certain goals. A number of aspects considered relevant for evaluation have been chosen for the analysis, such as law and internal security, defensibility of the Federal Republic, individual concerns and value change, and economic style. Available reports show unfavorable implications of a nuclear energy scenario, whereas a solar energy scenario seems to have less detrimental implications, in part due to the greater uncertainty about the specific features of such a scenario.

The second project can be characterized as asking, "What is accepted?" The answer to this question is sought by identifying criteria presently considered relevant in the society for evaluating energy options, by assessing the potential impacts of these options, and by revealing how these options are evaluated by the society. The degree and kind of acceptance of energy policy options is considered an important factor for the political evaluation of the compatibility of these options. The project uses a mixture of prescriptive and descriptive methods; it takes a largely quantitative approach to the problem. The major steps are a value tree analysis, the analysis of impacts, and the evaluation of the four energy paths by citizens in planning cells. The results will be data about the assessment of impacts of these options and their evaluation by 24 such cells, each of about 25 persons.

The major differences between the two projects are obvious from the above description. There are some distinguishing features, however, that I would like to stress: First, and most important, the project in Essen/Frankfurt <u>assumes</u> relevant criteria for evaluation by their own expertise, whereas the project in Jülich <u>assesses</u> these criteria from important social groups. This difference manifests itself in several aspects. For example, the nine aspects specified by the Essen/Frankfurt project do not coincide with the major criteria assessed in the value tree analysis. On the one hand, internal peace and social flexibility, the societal relation to nature, and economic style are aspects difficult to identify in the (combined) value tree of the Jülich project. On the other hand, a great number of criteria included in the value tree do not appear to be covered by the nine aspects of the Essen/Frankfurt project, e.g., local, national, and global environmental impacts, health and safety criteria, and international impacts. No reasons are given in the Essen project for their choice of aspects for evaluation, but one can reasonably assume that these aspects were considered particularly important; it appears that aspects have been chosen that have become salient in the controversy about nuclear energy and on which nuclear energy does not score very positively. This is quite legitimate, of course, but it raises the question of how representative the selected criteria are for the set of all relevant criteria. It would be interesting to compare the weights assigned to those criteria in the value tree of the Jülich project with the aspects selected by the other project group. Such a comparison should not imply, however, that aspects weighted heavily by the public are more important. In fact, the public has changed its weights concerning energy issues considerably during the past 10 years in the direction of critical experts' opinions: Who was right 10 years ago? But a comparison, and a finding of differences, would possibly draw the attention to other, unduly neglected aspects and would also have consequences for the significance of the project results in the political decision process.

A second difference between the two projects is that the Essen/Frankfurt project conducts--for the aspects selected--remarkably systematic and comprehensive studies on the assessment and evaluation of energy supply-systems. These are really thorough analyses; for example, the report on future security systems has about 400 pages. The assessment of impacts of the four energy scenarios in the Jülich project, on the other hand, are comparatively global judgments which, of course, can be taken as the quintessential knowledge of the experts who made these judgments. This difference is a consequence of the approaches chosen. In particular, the Jülich project studies the evaluation of energy policies by the public on all apparently relevant criteria. It would have been ridiculous and impossible to confront randomly selected citizens with thousands of pages of information material. To understand the global expert judgments on 31 indicators represented already an extremely demanding task.

A third difference is more generally the methodological approach chosen in the two projects. The Essen project takes a normative approach, applying certain principles which seem to be considered as general methodological principles for examining the social compatibility of technologies. This approach implies that scientific experts study the respective issue and submit their results primarily as input for the politicians' political debate; in this case, the issue is energy policies. The Jülich project, taking an empirical approach, also applies methods that can be generalized to other objects of social compatibility assessment studies. But here, it seems, the focus is on a later step in the political decisionmaking process, namely, when the knowledge provided by experts has been condensed to the relatively global judgments which enter the public political debate. How can one then determine whether, or to what degree, a particular technology is accepted, what features of a technology are perceived as unfavorable, or what measures are desired to compensate unavoidable deficits?

In conclusion, what will be the benefit of the two research projects for the Minister for Research and Technology of the Federal Government? One might ask, first, whether policymakers can learn something for their decision problem concerning which energy policy to choose. They certainly could learn something from both the analytical studies and the empirical data. A more pessimistic answer is, however, that they might use only those findings that support their already made decisions. One might ask, second, whether policymakers can learn something about how to assess social compatibility. Again, they certainly could learn something from the two projects representing quite different but mostly complementary methodological approaches. But here, too, there is a more pessimistic answer, namely, that the appreciation of the approaches will depend to a high degree on the outcome of the assessment, i.e., whether nuclear energy will look socially compatible or not. I do hope that both kinds of pessimism will not be confirmed. The two projects do not deserve it.

REFERENCES

1. Deutscher Bundestag, Bericht der Enquête Kommission Zukünftige Kernenergiepolitik: Drucksache 8/4341 (Bonn, Presse- und Informationszentrum, 1980).

2. Meyer-Abich, K.-M. and Dickler, R.A., Energy issues and policies in the Federal Republic of Germany, Annual Review of Energy, 7, 221-259 (1982).

3. Weinberg, A., Social institutions and nuclear energy, Science, 177, 27-34 (1972).

4. Meyer-Abich, K.-M., Schefold, B. and Weizsäcker, C.F., Zwischenbericht über den Stand der Arbeiten im Forschungsprojekt "Die Sozialverträglichkeit verschiedener Energiesysteme in der industriegesellschaftlichen Entwickling: (Essen, Frankfurt, Starnberg, 1981).

5. Meyer-Abich, K.-M. and Schefold, B., Wie möchten wir in Zukunft leben? Der 'harte' und der 'safte' Weg (München, C.H. Beck, 1981).

6. Bauerschmidt, R., Erste Fortschreibung der Referenzfälle K(ernenergie) und S(onnenenergie) (Essen, 1981).

7. Roßnagel, A., Bedroht die Kernenergie unsere Freiheit. München: C.H. Beck, 1983.

8. Held, M., Die Verträglichkeit der Energieszenarien K(ernenergie) und S(onnenenergie) mit den Wertentwicklungen (Essen, 1982).

9. Schefold, B., in K.-M. Meyer-Abich and B. Schefold, Wie möchten wir in Zukunft leben?-Der 'harte' und der 'sanfte' Weg, 105-194 (München, 1981).

10. Häfele, W., Münch, E. and Renn, O., Analyse von Energieversorgungs-systemen nach Kriterien der Sozialverträglichkeit, Unpublished manuscript (Jülich, 1981).

11. Renn, O., The Jülich social compatibility study, Paper presented at the TIMS/ORSA Joint National Meeting (Chicago, 1983).

12. Renn, O., Albrecht, G. and Kotte, U., The empirical investigation of citizens' preferences with respect to four energy scenarios. Paper presented at the Advanced Study Institute "Technology Assessment, Environmental Impact Assessment, and Risk Analysis: Contributions from the Psychological and Decision Sciences," Les Arcs/France, 1983.

13. Keeney, R.L. and von Winterfeldt, D., Value tree analysis for alternative energy systems, Report (Jülich, 1983).

14. Dienel, P.C., Die Planungszelle (Opladen, 1978).

15. von Winterfeldt, D. and Edwards, W., Decision Analysis and Behavioral Decision Theory, to appear (1983).

16. Keeney, R.L. and Raiffa, H., Decisions with Multiple Objectives (New York, Wiley, 1976).

SECTION IV: Cultural and Philosophical Perspectives; Theoretical and Methodological Contributions

Limits of Rationality[1]

Joshua Menkes

National Science Foundation, Washington, D.C. 20550

INTRODUCTION

It is hard for me to decide whether I should be elated or depressed by the fact that over 2000 years ago, Aristotle had already thought about the problem that I am addressing now, which is more or less the relationship of rationality and ethics. He not only thought about it but also achieved insights to which we have found it difficult to add. In his analysis he separated ethics from politics in explicit recognition of the dialectic tension between the man and the citizen--between the particular and the universal. In Aristotle's own words (or rather in W.D. Ross' translation):

> If the state cannot be entirely composed of good men, and yet each citizen is expected to do his own business well, and must therefore have virtue, still, inasmuch as all citizens cannot be alike, the virtue of the citizen and the good man cannot coincide. All must have the virtue of the good citizen--thus, and thus only, can the state be perfect; but they will not have the virtue of a good man, unless we assume that in a good state all the citizens must be good (1).

In the early nineteenth century Hegel tied the Gordian knot connecting personal morality with social choice by postulating

---

1 Any opinions, findings, conclusions, or recommendations expressed here are those of the author and do not necessarily reflect the views of the National Science Foundation. The help of Professor Stuart F. Spicker is gratefully acknowledged.

NATO ASI Series, Vol. G4
Environmental Impact Assessment, Technology Assessment, and Risk Analysis. Edited by V.T. Covello et al.

that in the rational state, personal morality is preserved (aufgehoben).

Many discussions of rationality leave the impression that there is but one rationality whose meaning encompasses all situations, and that those who would argue otherwise are confused. It is my view that the notion of a free-standing rationality is incoherent. Rationality must be looked upon as embedded in a particular ethical theory that applies to personal morality, public choices, and distributive justice. The qualifier "rational" can be used to describe an act or a preference that is consonant with a system of ethics as long as it is understood that this system has a unique structure. Two key features of this structure are the strong link between rationality and personal morality and the equally strong link between distributive justice and social choice. There is a third but weaker link--consent--which holds personal morality, public morality, and social choices together. This is a peculiarly Hegelian consent that does not emanate from consensus but is embodied in the rationality of the state. Once the principles of personal morality are settled, rationality is determined subject only to compatibility with criteria for social choice; in other words, criteria of rationality cannot be specified independently of personal morality. Similarly, public choices cannot be made in defiance of the accepted principles of distributive justice. If for certain combinations of individual preferences and social choice it proves impossible to reach agreement and establish consent, political adjustment mechanisms have to be brought into play. In a democracy this is ordinarily part of the established political process; in an authoritarian government such adjustments can prove traumatic.

Only when it is assumed that there is but one acceptable system of ethics can one draw the inference that there is only one rationality. Classical utilitarianism does this: It presents a model of a coherent and complete system of ethics which purports to be able to deal consistently with issues raised by questions of rationality. This unified theory, based on a single set of axioms, has a beautiful simplicity; it is achieved, however, by sacrificing much of the ethical content and substance of morality. I intend to show that because the link between personal morality and social choice is weak, it permits a much richer concept of rationality than utilitarianism warrants. Furthermore, it would devalue the attraction and the advantages of a unified theory.

Modifications of classical utilitarianism introduced to cope with the moral neutrality of that theory lead naturally to principles of distributive justice which are based on more than the maximization of personal utility. Rawls comes immediately to mind. His theory of justice is by all accounts a theory of distributive justice which is compatible with personal utilitarian ethics(2). Much of the shadowboxing with respect to Rawls' theory is about whether Rawls proposes a utilitarian theory, an intuitionist view, a strict subjectivism, or none of the above. I believe that his theory of justice is compatible with all of the above: Since the derivation of his principles is almost purely procedural, there is nothing in this theory to prevent a committee of socialist vegetarians from assembling and working out their own principles of justice.

In view of the fact, readily conceded, that not all choices require the explicit sanction of one's moral principles, I shall list a few qualifiers in order to anticipate objections concerning the existence or non-existence of situations which may be conjured in the imagination much like philosophers enjoy conjuring possible worlds. First, I shall assume that there are problems whose moral implications require an agent to make a deliberate and self-conscious decision on the basis of moral choices. Furthermore, I shall postulate that the agent has more than one option to exercise. I shall not, however, insist that an agent must act in the strictest sense; for the difference between acts of omission and commission are here of no consequence.

THE QUEST FOR RATIONALITY

Mankind lives in a world that is for many incomprehensible. The vagaries of the struggle to survive and flourish provide a stimulus to seek explanations that would substitute order for apparent chaos and predictability for uncertainty and risk. Just as the law of gravity allows one to predict how one state of nature will follow necessarily from another, laws of rationality are intended to allow one to predict what acts a rational being is necessarily going to perform in an ever-changing environment. This belief in the predictability of behavior of a rational actor qua rational being in a rational world is deeply embedded in our lives. This belief accounts for the shock bordering on a sense of betrayal that the practitioners of rational decision theory experienced when confronted with the empirical findings of Tversky and Kahneman, published in 1981 in their article "The Framing of Decisions and the Psychology of Choice"(3). Tversky and Kahneman found that people systematically ignore such principles of rational choice as consistency and coherence when the outcomes of their actions are uncertain. The two psychologists asserted, furthermore, that it is possible to predict the manifestations of such "irrational" behavior by manipulating a decisionmaker's conception of the consequences of his choice.

Even Einstein never reconciled himself to a probabilistic description of the physical world, as is attested in his famous letter (of 1944) to Max Born in which he remarks: "You believe in the dice-playing god, and I in the perfect rule of law in a world of something objectively existing...." In another letter, written approximately three years later, Einstein muses on reason's ambiguity: "I cannot substantiate my attitude to physics in such a manner that you would find in any way rational."(4) Centuries before Einstein and Born, philosophers sought to disclose the function of the human mind--its essential rationality--and to understand how it devises general maxims to guide reasonable and moral behavior. Only a rational being, according to Kant, for instance, has the power to act in accordance with principles; these principles, in the form of moral laws, have to be derived from the universal concept of a rational being(5). Again, reliance on rationality reigns supreme.

In the course of our daily lives we make decisions all the time, some so trivial or so thoroughly routinized that we are not even aware of the motivations. I would place in this category decisions about buying groceries or gasoline, decisions concerning what to wear on a given day, and also decisions about whether to feed the wild birds and when to pay the bills lest they (or we) become delinquent. The question of whether these acts are rational or not is hardly ever raised. But when, say, the purchase price of two desired items is so great that it would be impractical to buy them both, then we have to make a more deliberate choice. Decision theorists would advise making the decision on the basis of which purchase will give the greater satisfaction. For most choices, most of the time we will be well served by this advice. This is pretty much an analogue to the flat earth theory: For practically all decisions we can safely assume the earth to be flat, excluding such problems as launching a rocket to the moon, which would require brushing up a little on Newtonian theory of gravity. The moral equivalents of such higher-order problems are found in situations that require us to make self-conscious and deliberate choices that are influenced or determined by what we hold to be moral values.

AUTONOMOUS AGENCY

The age of enlightenment, with its almost religious belief in material progress, released the dormant entrepreneurial spirit of modern capitalism necessary for the success of the industrial revolution and the attendant development of technology. With technology there also came a new version of human misery, attributed to alienation, which found its most dramatic expression in the destruction of machines by Luddites. Simultaneously with the material development arose a new legitimating morality, Aristotle's virtuous "zoon politicon" gave way to the "homo economicus," the maximizer of utility; the embodiment of the spirit of capitalism, to use Max Weber's famous term. This spirit, which continues to stoke the fires of our capitalist economy, is naturally more concerned with matters involving financial risks and benefits than with the societal implications of its ventures. Weber, with an almost clairvoyant perspicacity, observed: "[the modern economic order] is now bound to the technical and economic conditions of machine production which today determine the lives of all the individuals who are born into this mechanism, not only those directly concerned with economic acquisition, with irresistible force. Perhaps it will so determine them until the last ton of fossilized coal is burnt."(6)

There continue to be unwanted, adverse effects to technological growth, effects that individuals feel powerless to control. Recognizing the divergent, often contradictory needs of entrepreneurs and individuals, the government has begun to systematically attend to the projected consequences--foreseen, forecasted, or unexpected of technological development. That concern is reflected in the terms "technology assessment" and "risk analysis."

To the extent that the consequences of technological and other choices affect society, the freedom to select a course of action that will maximize the benefits to the individual and to society becomes a moral concern. One issue is the matter of selection of those decision criteria that would allow an actor to choose from any number of possible actions the one which he deems most rational. These decision criteria are based on moral and normative judgments that are informed by a theory of personal morality. If and when a collection of individual acts eventually affects a particular policy at some level of social aggregation, acceptance, or rejection of the consequences, now or in the future, will thus constitute social choices.

Utilitarianism (whether rule or act utilitarianism) (7) is a theory of personal morality as well as a theory of social choice. Similar criteria are assumed to control and govern both functions. As a theory of personal morality, utilitarianism claims to provide an agent with a paradigm for rational choice. Rationality here is established on a very narrow information base and is presumed to be accessible to the individual and, according to intuitionist interpretations, only to the individual. This presumed parsimony of utilitarian rationality is thought to serve in itself to explain the process of social choice. To the utilitarian, the market is the prototype of processes that link individual preferences to social choices and their legitimation(8). Buyers and sellers in a market are assumed to be motivated only by their self-interest. This self-interest is best served, so goes the argument, by maximizing one's utility in each and every transaction--this surprisingly goes for the buyer as well as the seller.

This may be a good point to recapitulate the structure of the ethical system that I have developed so far, with the aid of the diagram in Fig. 1. I use the term "dyad" to designate the internally linked pairs: Personal Dyad (rationality-personal morality) and Social Dyad (social choice-distributive justice). The pairs are linked through the regulative principles connecting ends and means and also good and right. This clearly constitutes a strong linkage. The external link between the two dyads is contingent on consent, which forms the bridge between the universal and the particular. I maintain that an ethical system possessing this structure obviates the need for a unified theory of the power of classical utilitarianism.

Utilitarianism in its most simple and powerful version postulates that what is right or ethical is what maximizes personal utility. For the collectivity similar principles hold; that is, the state of affairs characterized as the largest net benefit is the desirable state of affairs not in the sense of "being desired" (a mere psychological state) but in the sense of being worthy of achieving (pace J.S. Mill). This theory of morality has had to face the challenge of Kant's deontological, or duty-based, ethical system, which itself was, of course, sharply criticized by Hegel as vacuous(9). This judgment, it appears, is shared by Rawls, as he remarks that "...Kant's doctrine requiring abstraction from all contingency may be unable to distinguish between the lives of the saint and the scoundrel, as long as both are lived by a consistent set of principles freely chosen...." For Rawls, "Moral philosophy must be free to use contingent assumptions and facts as it pleases."(10) This becomes particularly important as we examine utilitarianism as a theory of social choice and not merely a theory of individual moral action.

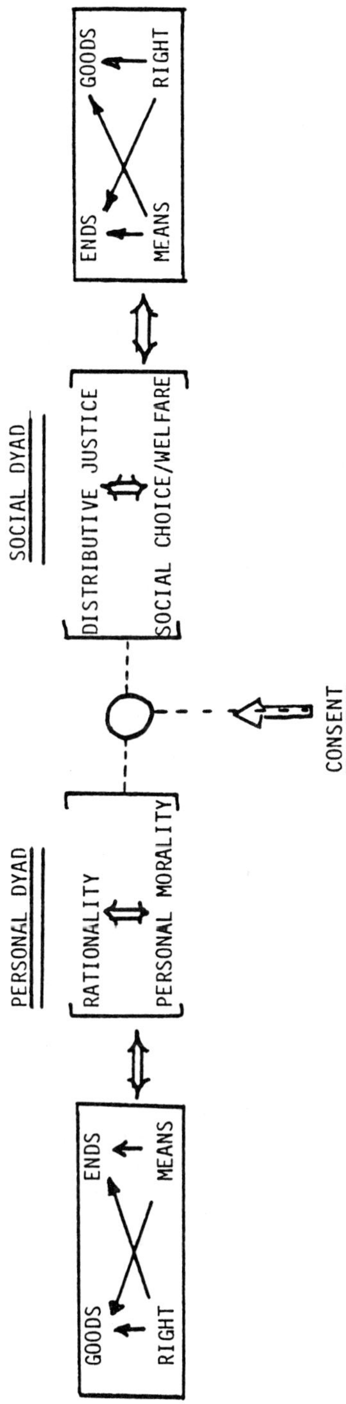
PERSONAL DYAD
GOODS
RIGHT
ENDS
MEANS
RATIONALITY
PERSONAL MORALITY
CONSENT
SOCIAL DYAD
DISTRIBUTIVE JUSTICE
SOCIAL CHOICE/WELFARE
ENDS
MEANS
GOODS
RIGHT

FIGURE 1

When issues of social choice are raised, it is not only the sum total or average of utilities and goods that counts, but also the manner in which the goods are distributed. But given a sum of utilities, there are numerous distributive models that are compatible with the initial aggregate of utilities, e.g., the egalitarian, the economically effective benefit/cost, the Paretian, the Rawlsian. As Nozick has pointed out, for every statement of the form "to each according to his..." there is a distribution of goods which satisfies the postulate(11).

It should, of course, be noted that notwithstanding attacks from many sides, utilitarianism has demonstrated an amazing vitality. In searching to locate its source of strength, Charles Taylor asks rhetorically, "What did utilitarianism have going for it? In the utilitarian perspective, one validated an ethical position by hard evidence. You count the consequences for human happiness of one or another course, and you go with the one with the highest favorable total. What counts as human happiness was thought to be something conceptually unproblematic, a scientifically established domain of facts like others."(12) It turns out that the expectations for unproblematic specifications of rationality, human happiness, and the equitable distribution of goods remain unfulfilled(13).

Kant distinguished two types of rationality--theoretical and practical reason. Simply stated, theoretical reason guides our thinking and as such it is also referred to as cognitive rationality(14). Practical reason, on the other hand, is a guide to conduct and thus carries a moral connotation. Clearly, these two notions are not independent of one another. There must be a sense in which the power of theoretical reason is involved in generating moral maxims germane to human conduct. As H.J. Paton has observed: "In the wide traditional sense reason covers (1) the power of entertaining concepts (or understanding in Kant's terminology); (2) the power of applying concepts to given objects (the power of judgment); and (3) the power of making mediate inferences (the power of reasoning or of reason in a narrower sense)."(15) What we need to consider is the precise part that reason plays in the pursuit of happiness and the pursuit of goodness.

Decision rules based on utilitarianism fail to distinguish between these two functions. Accordingly, cognitive rationality is taken to be rationality "simpliciter," which supplies the necessary and sufficient principles for moral actions. Under this formulation, morality is identified with rationality. John Harsanyi postulates that a revised version of classical utilitarianism is in fact the only ethical theory that is consistent with rational behavior and a commitment to a humanitarian morality(16). This sounds suspiciously like an invisible hand explanation. In fact, Harsanyi in the same paper tellingly uses the subheading "Ethics as a branch of the general theory of rational behavior" for a section in which he defines rationality thus: "The concept of rational behavior arises from the empirical fact that human behavior is to a large extent goal-directed behavior. Basically, rational behavior is simply behavior consistently pursuing some well defined goals, and pursuing them according to some well defined set of preferences and priorities."(17) This is the sort of definition that has made many a moral philosopher's flesh crawl. But help is on the way from unexpected quarters. As it has been said, there is no more melancholic sight than an ugly piece of data on its way to destroy a beautiful theory. The spoilers were the two psychologists Tversky and Kahneman, who introduced what I would call the failure of cognitive rationality(18). In their examples, a subject is offered mathematically equivalent choices that are presented or framed differently. On a purely cognitive basis the subjects should be indifferent to the options, while in fact they exhibited consistently "irrational behavior." The implications for a theory of personal morality based on cognitive rationality are, without exaggeration, devastating.

R.M. Hare rejects such a narrow and morally neutral conception of rational agency as that proposed by Harsanyi. He enlists J.J.C. Smart who, while still basing his view of rationality on a consequentialist interpretation of rational agency, suggests that we use the term "rational action" to characterize actions which on the basis of information available to the agent are likely to produce the best results for society; "right action" on the other hand, connotes action that actually achieves these desirable results(19).

Rationality can be characterized consistently along Smart's lines, leaving the specification of right action to a contingent determination. This definition is even consistent with Harsanyi, as long as preferences and priorities can be generalized to apply to the prevailing ethos (Sittlichkeit). Personal morality, then, need not be based on a utilitarian scheme of maximization but can be derived from principles of morality which are acceptable to most if not all members of a given society and which are compatible with our deeply held moral convictions of what is right action and wrong action.

According to Wolff, the way Rawls conceives the problem is of interest. The bare notion of rational agency as proposed by utilitarian theory is insufficient for a rationally defended morality. This raises the question of what, if anything, can be grafted onto utilitarianism short of the positing of substantive moral claims "(1) that will yield a rational argument for a moral principle that is constructive, (2) that conforms to, or at least does not deviate widely from, our strongest moral convictions, (3) that gives an independent logical status to the notion of the right, and (4) that takes the inviolability and dignity of moral personality as fundamental."(20)

Here it is useful to reiterate that it is neither necessary nor sufficient for a complete moral system (which includes criteria for rationality and determinants of personal morality and of social choice) to be based on a unitary criterion such as the maximization of utility. The price one must pay for this parsimony--namely moral neutrality--is not warranted. I believe that the personal and public dyads can be analyzed according to their constituent parts wth different criteria applying to each.

THE PRISONER'S DILEMMA: AN EXISTENTIAL FAILURE OF RATIONALITY

I referred to the phenomenon described by Tversky and Kahneman as failure of cognitive rationality. It is a failure in a very specific sense: It points to the inability to construct an ethical system on a purely procedural basis. It is interesting to examine why such an ethical system appeared desirable. The enlightenment, having lost faith in the telos of prescriptive ethics, shook the yoke of positive morality. Man was to be the measure of all things; moreover, man had certain inalienable rights which could not be abrogated by other citizens or even the state. Safety was to be found in a system of ethics that depended only on such acts as a rational being would execute. Modern inheritors of this tradition--anarchists like Nozick--simply do not trust the government with anything beyond a "night watchman" function. The mystical view of the state envisioned by Hegel as the abode of the spirit appears to latter-day anarchists as misguided at best but more nearly malevolent. This explains among other things the love affair of anarchists like Nozick with the market. The market left to its own working requires no positive system of ethics; left to its devices it would treat everybody equally and impartially--fulfilling one of the most difficult functions of an ethical system. Yet if the signals sent out and received by the market can be manipulated (by the evil state, no doubt), even this institution is not invulnerable to the meddling interference of governments and their agents.

As if this were not a sufficiently serious problem, there is at least one other failure or rationality which I call existential failure and which is exemplified by the Prisoner's Dilemma(21). In the Prisoner's Dilemma there is no rational choice based on maximization of expected benefits. The market, the model of a non-cooperative conflict resolution, cannot operate here(22).

While the dilemma is somewhat abstract and idealized, it does serve to illuminate some fundamental difficulties posed by searching for rational choice criteria which would allow an agent to choose most prudently in a clearly competitive rather than cooperative context. How should a utilitarian proceed, one might ask? Probably not much differently from a non-utilitarian, that is, prudently. Prudence or practical wisdom is quite likely the typical motivation for selecting one option over another. For most of us, the evaluation of virtually all consequences of our actions is beyond our cognitive capacities.

I have barely mentioned risk up to now in spite of the fact that this paper was motivated by concerns about the ethical nature of managing technological risks. Almost without exception "objective" risk analysis methods have their roots in economic cost/benefit calculations which are closely tied to utilitarian models. For this reason any other doctrine has to locate itself with respect to utilitarian models. In positioning myself I could not avoid sounding at times polemical in my criticism of utilitarianism--this was not my intention.

Just as natural philosophers regarded the laws of mechanics discovered by Isaac Newton to be eternal truths, the appeal of utilitarian rationality was based on the putative permanence of the laws of rational behavior. Newton's revealed truth was eventually supplanted by Einstein's theory of relativity. A similar fate has befallen rationality with a peculiar twist. A rational ordering of choices was assumed to be descriptive of the manner in which people actually make choices. When this proposition failed to reach accord with the facts deduced from observation, rationality became prescriptive. The latter step took utilitarianism down from the pedestal it had occupied, so that it now joined the other contestants in the arena--those who did not lead the charmed existence utilitarianism had led hitherto.

We should at this point note that while a non-utilitarian approach has a very direct and attractive appeal, its theoretical and practical difficulties are serious. In fact, Scanlon conjectures that the appeal of utilitarianism is in no small measure due to the difficulties about the foundation of competing theories(23). One of the difficulties is framed as "the problem of interpersonal comparisons of utility;" another has to do with preferences.

Still another difficulty presents itself when dealing with the intensity of preferences. Problems of an entirely different nature emerge when, as some have proposed, moral choices should not be based on actual desires, but on desires which an agent ought to have, and in fact would have if he were fully informed and critically aware of all the circumstances which bear on the choice to be taken. (This, of course, is practically not possible.) All the factors that I have enumerated have, then, one thing in common: They require a broader information base and understanding of the complex human condition and situation than utilitarianism requires to make moral judgments. The question then is how these "true" desires are to be discovered, and how they are to be legitimated. A strict utilitarian would find answers to this question singularly uninteresting, for in his view the individual agent really does not count here; an agent is merely a vector of utility as measured by his contribution to the happiness of the collective. Individual, autonomous agents play no special role here, for utilitarian moral philosophy is indifferent to the distinction between "true desires" and desires believed to be "true" by individual agents(24).

FROM PRESCRIPTIVE RATIONALITY TO REASONABLENESS

"Governments derive their just powers from the consent of the governed." These are the famous words of Thomas Jefferson telling us something very profound: Only human beings have rights, while institutions, even those created to protect the individual or preserve social justice, have powers that are derivative of human rights. As Stephen Toulmin notes, there are deontological theories of personal morality but not of social choices(25). That implies that in a democracy at least, what is right and good flows from the individual to the collectivity (from the personal to the social dyads according to my Fig.1) but not ordinarily the other way around. This is important when discussing ethical systems which are based on social justice like the one of Rawls.

Rawls recognizes the necessity for a theory of personal morality as a component part of his theory when he asks: "How is a shared understanding of what are appropriate claims possible, in view of citizens' conflicting and incommensurable conceptions of the good?"(26) In his answer he eschews rationality in favor of reasonableness. While one may interpret reasonableness as an approximation of rationality, I think the distinction is fundamental. In support of this assertion I will quote Rawls (or rather paraphrase him, to spare the reader Rawlsian syntax in extenso; the original is given in the reference):

> The notion of justice must be anchored in commonly accepted canons of political reasoning and understanding--a <u>procedural criterion</u>--because only then can justice play its intended ethical role, which is to act as arbiter in the determination of what constitutes sufficient reason for the acceptance of social institutions. For justice to succeed in playing this role, all citizens, irrespective of their social positions and particular interests, must accept the proffered justification on substantive as well as procedural grounds. When the conditions for the existence of a sufficient basis for agreement are met, the legitimization of a particular social order requires that agreement be reached on social arrangements which are most reasonable for the citizen in virtue of how he views himself and the general features of cooperation of like persons(27).

Again, superficially the difference between maximizing utility and maximizing reasonableness may appear to be minor, yet in fact it is fundamental. Reasonableness is grounded in cooperation, whereas utilitarian rationality does not require cooperation; <u>vide</u> the market. Still the Prisoner's Dilemma, briefly discussed earlier, laid bare the difficulties inherent in deriving non-cooperative solutions from a narrow information basis.

CONCLUSIONS

After taking the narrative from Aristotle, Kant, and Hegel, to Nozick and Rawls, the denouement is bound to be anticlimactic--and indeed it is. My conclusion is that a well-informed agent--he may in fact be perfectly well-informed--will not necessarily make his choice on a procedural basis as prescribed by strict utilitarianism, that is, by maximizing his utility; nor would he necessarily under conditions of uncertainty choose the option that maximizes his expected utility. I traced this divergency of observed behavior from the prescribed behavior to the cognitive limitations of the agent as well as to the tendency of an individual to interpret the projected consequences of his actions idiosyncratically.

The failure of self-regulating "invisible hand solutions" which do not require the intervention of a moral person or persons to achieve the desired goal, elevates prudent judgment to the preferred decision paradigm.

Tracing his views to Kant (but the source could equally well be Aristotle's concept of virtue), Rawls states his credo:

> To justify a Kantian conception [of morality] within a democratic society is not merely to reason correctly from given premises, or even from publicly shared and mutually recognized premises. The real task is to discover and formulate the deeper bases of agreement which one hopes are embodied in common sense, or even to originate and fashion starting points for common understanding by expressing in a new form the convictions found in the historical tradition by connecting them with a wide range of people's considered convictions(28).

Indeed, paradise is lost; we have eaten the forbidden fruit and we can tell good from evil. Even the first documented procedural justification for an immoral act, namely Adam's claim that it was Eve who gave him the piece of apple, was found wanting. We cannot claim ignorance, indifference, or lack of interest when confronting moral issues. Nor can we hide behind an invisible hand.

REFERENCES

1. Richard McKeon, ed., The Basic Works of Aristotle (New York: Random House, 1941), p.1180

2. John Rawls, Theory of Justice (Cambridge, Massachusetts: Harvard University Press, 1971). And see, e.g., Norman Daniels, Reading Rawls (New York: Basic Books, 1974) and Robert Paul Wolff, Understanding Rawls (Princeton: Princeton University Press, 1977).

3. A. Tversky and D. Kahneman, "The Framing of Decisions and the Psychology of Choice," Science, 1981, 211: 453-458.

4. Max Born, Natural Philosophy of Cause and Chance (Oxford: Clarendon Press, 1949), pp. 122-123.

5. H.J. Paton, The Categorical Imperative (Philadelphia: University of Pennsylvania Press, 1971), pp. 80-81.

6. Max Weber, The Protestant Ethic and the Spirit of Capitalism (New York: Scribner's, 1958), p. 181.

7. "Act utilitarianism" is the version of utilitarianism in which each individual act must be judged directly in terms of the utilitarian criterion. Thus a morally right act is one that, in the situation the actor is actually in, will maximize social utility. In contrast, "rule utilitarianism" is the view that the utilitarian criterion must be applied, in the first instance, not to individual acts but rather to the basic general rules governing these acts. Thus a morally right act becomes one that conforms to the correct moral rule applicable to this sort of situation, and a correct moral rule is that particular behavioral rule that would maximize social utility if it were followed by everyone in all social situations of this particular type [see John C. Harsanyi, "Morality and the Theory of Rational Behavior," in Utilitarianism and Beyond, Amartya Sen and Bernard Williams, eds. (Cambridge: Cambridge University Press, 1982), p. 41].

8. Philosophers in the 18th century were not reluctant to invoke a little bit of black magic to sustain their sometimes unbelievable explanations. Adam Smith relied on the "invisible hand" to effect the efficient distribution of goods, as Hegel invoked the "cunning of reason" (die List der Vernunft) to make us behave rationally in spite of ourselves. Such processes which lead to social choice and

consequences are referred to by Robert Nozick as "invisible hand processes." Nozick, in fact, attributes "a certain lovely quality to explanations of this sort." He maintains that "they show how some overall pattern or design, which one would have thought had to be produced by an individual's or group's successful attempt to realize the pattern, instead was produced and maintained by a process that in no way had the overall pattern or design 'in mind.'" Robert Nozick, Anarchy, State, and Utopia (New York: Basic Books, 1974), p.18.

9. Charles Taylor, Hegel and Modern Society (Cambridge: Cambridge University Press, 1979), p.77.

10. Rawls, quoted in Michael J. Sandel, Liberalism and the Limits of Justice (Cambridge: Cambridge University Press, 1982), pp. 37-38, 39.

11. Nozick, Anarchy, p.159.

12. Charles Taylor, "The Diversity of Goods," in Utilitarianism and Beyond, p.129.

13. Kenneth Arrow casts some doubts even about the venerable Golden Rule when he cites G.B. Shaw's astute remark: "Do not do unto others as you would have that they do unto you. They may have different tastes." Kenneth J. Arrow, The Limits of Organization (New York: Norton, 1974), p.24.

14. Kant's unequivocal description of those who cannot or will not follow the dictates of theoretical reason is worth quoting: "Deficiency in the faculty of judgment is what we call stupidity, and there is no remedy for that. An obtuse and narrow mind, deficient in nothing but a proper degree of understanding and correct concepts, may be improved by study." Immanuel Kant, Critique of Pure Reason, trans. F. Max Muller (Garden City: Doubleday, Anchor Books, 1966), p. 119n.

Theoretical reason is not to be identified with pure reason. The latter serves to distinguish a priori knowledge from such knowledge as is based on empirical evidence. The transcendental character of pure reason need not concern us here.

15. Paton, Categorical Imperative, p. 79.

16. Harsanyi, "Morality," in Utilitarianism and Beyond, p. 61.

17. Ibid., p. 42.

18. See n. 3 above.

19. J.J.C. Smart, "An Outline of a System of Utilitarian Ethics" in J.J.C. Smart and Bernard Williams, Utilitarianism for and Against (Cambridge: Cambridge University Press, 1973), pp. 46 ff.

20. Wolff, Understanding Rawls, p. 13.

21. I will briefly outline the dilemma: Two prisoners on parole from an earlier conviction were apprehended under suspicion of having committed a crime. The evidence was not as clear cut as one would have liked. The counsel for the defendants asked for the opportunity to plea bargain. The public prosecutor offered the following terms: (a) If both scoundrels admit to having committed the crime, both will get five years in jail; (b) If both deny any guilt, they will each get two years in jail; (c) If one of the two turns state's witness, admits his own guilt, and implicates his partner, then he will get remission of the ten-year jail sentence that he was already serving from an earlier conviction, and the other fellow will get an additional ten-year jail term.

From the perspective of each prisoner, option (b) is not as prudent a strategy as (a), where an admission of guilt brings with it no more than a five-year jail term and possibly no jail term at all, should the other prisoner opt for (c). In other words, option (a) is an enforceable minimax outcome. Here "enforceable" means an outcome that is independent of any cooperation between the two prisoners even if they had made earlier promises to one another. An enforceable minimax outcome (a) is surely worse than the best outcome for both (b), but it is clearly better in this case than (c), the broken promise.

22. Williams discusses the Prisoner's Dilemma from a somewhat different point of view ("A Critique of Utilitarianism" in Smart and Williams, Utilitarianism, p.131). His analysis is concerned more with patterns of social learning described in a very interesting article by Douglas R. Hofstader in reference to the work of Robert Axelrod at the University of Michigan: "Computer Tournaments of the Prisoner's Dilemma Suggest How Cooperation Evolves," Scientific American, May 1983, pp. 16 ff.

23. T.M. Scanlon, "Contractualism and Utilitarianism," in Utilitarianism and Beyond, p.122.

24. In contrast to utilitarianism, the individual agent--conceived as moral agent--plays a central role in the thinking of both Rawls and Jürgen Habermas, the social theorist who is the most articulate contemporary proponent of the so-called Frankfurt School of German philosophers. The question of how "true desires" can be discovered was addressed by the Frankfurt School in their "critical theory," which Habermas has changed somewhat by grafting on a version of transcendentalism. Although Rawls and Habermas approach the discovery problem from quite different ideological standpoints, the similarity of their objectives is striking. With his insistence that an action is morally acceptable or a belief true if and only if arrived at by consensus reached under ideal conditions, Habermas resonates with Rawls.

25. Stephen Edelstone Toulmin, An Examination for the Place of Reason in Ethics (Cambridge: Cambridge University Press, 1958), p. 190.

26. John Rawls, "Social Unity and Primary Goods," in Utilitarianism and Beyond, p.164.

27. John Rawls, "Kantian Constructivism in Moral Theory; The Dewey Lectures 1980," Journal of Philosophy, September 1980, 78: 518:

> ...a conception of justice holds only when a basis is established for political reasoning and understanding within a public culture. The social role of a conception of justice is to enable all members of society make mutually acceptable to one another, their shared institutions and basic arrangements, by citing what are publicly recognized as sufficient reasons, as identified by that conception. To succeed in doing this, a conception must specify admissible social institutions and their possible arrangements into one system so that they can be justified to all citizens, whatever their social position or more particular interests. Thus whenever a sufficient basis for agreement among citizens is the presently known, or recognized, the task of justifying a conception of justice becomes: How can people settle on a conception of justice to serve this social role, that is (most) reasonable for them in virtue of how they

conceive of their persons and construe the general features of social cooperation among persons so regarded?

28. Ibid.

Three Types of Risk Assessment: A Methodological Analysis[1]

Silvio O. Funtowicz[2] and Jerome R. Ravetz

Department of Philosophy, The University, Leeds LS2 9JT, U.K.

INTRODUCTION

The study of risks and the impact of new technological systems in our society and environment is now accepted as a legitimate subject of research.

As a consequence of the dual character of technology, a benefit, but also a threat, policy debates on innovations in technology are increasingly focused on "risk." This is seen most clearly in the case of the nuclear industry, where long and expensive enquiries are now necessary before new installations are permitted (as is currently the case with Sizewell B in the U.K.).

The issues are frequently divisive, involving heavy costs, both economic and political. In such debates, the established methodology of industrial risk assessment (or "Applied Science Methodology") is stretched beyond its limits of applicability and effectiveness. The techniques whereby a particular installation or process can be analyzed and its hazards identified and logically displayed, do not extend to systems which are novel, large, complex, and few in number. The hazards and environmental impact of these systems are incapable of description, measurement, or forecast with the precision experts have, until a few years ago, assumed to be possible and necessary for decisionmaking.

---

1 Research done on the project "Quantitative Reasoning in Risk Analysis," supported by the SERC-SSRC Joint Committee.

2 To whom correspondence should be addressed.

NATO ASI Series, Vol. G4
Environmental Impact Assessment, Technology Assessment, and Risk Analysis. Edited by V.T. Covello et al.

Under these new circumstances, quantitative statements about risks and safety, which by their form seem the product of a rigorous scientific methodology and mathematical techniques, are frequently challenged as dubious or biased, and, sometimes, can be proved totally erroneous.

Now it is not only militant environmentalist groups that challenge the scientific character of risk assessment. Leading experts and theoreticians echo their doubts; the result is a severe loss of confidence in the whole field. The optimism of earlier years has given way to confusion and pessimism among many experts. This new mood may make it more difficult to achieve those genuine results of which the field is capable. This work is intended to remedy this situation by identifying the different types of practice in Risk Assessment, with their characteristic limitations and strengths.

STATE OF THE ART

Because of the lack of an accepted methodology that is both successful and reliable, almost all research work published in the theory of Risk Analysis is about "Foundations." Basic terms such as "probability" and "risk" are argued everywhere (and it is difficult to find the same definitions twice); consequently the estimation techniques differ considerably. Mandl and Lathrop provide an example of this problem in the context of Liquefied Energy Gas Terminals. (1)

> The best way to develop a definition of risk is to start by quoting some of the definitions from the risk assessment literature.
>
> SAI-USA: "Risk is the expected number of fatalities per year resulting from the consequences of an accidental event."
>
> CREM-UK: "Risk is the probability of an injurious or destructive event, generated by a hazard, over a specified period of time."
>
> BATTE2-OTH: "Group risk is defined as 'the frequency at which certain numbers of acute fatalities are expected from a single accident.' The risk to society as a whole is defined as

> 'the expected total numbers of acute fatalities per year resulting from accidental events in the system.'"

Although they emphasize the differences among the measures, we may note that three different terms for "probability" are used, each implying its own underlying theory of the measurement.

The optimistic expectations that sophisticated probabilistic techniques would solve the problem (2) have not been fulfilled; on the contrary, the experience of the last years has been of increasing uncertainty and debate. The concept of risk in terms of probability has proved to be so elusive, and statistical inference so problematic, that many experts in the field have recently either lost hope of finding a scientific solution or lost faith in Risk Analysis as a tool for decisionmaking. A recent authoritative report by an American committee comments (3):

> While, as emphasized in the first part of this report, analysis has a limited role in decisionmaking, it still can be a powerful tool for categorizing risks, assessing them, evaluating a given set of alternative coping strategies, and devising new alternatives. The actual role of risk analysis in particular situations is variable, as are the methods used. There is also the fact that in some decisionmaking on risk, analysis has no role; in other situations analysis is useful only if it is designed so that it provides information that the decisionmaker cannot obtain.

These confusions and disappointments are familiar to many experts in the field; it is enough for us to indicate them briefly. One recent theoretical response to them was an advocacy of "Baysian statistics," where the "subjective probabilities" based on the professional experience of experts, are assigned to those hypothetical events which are incapable of effective empirical study. These are then (in theory) modified by a formal calculus, when experience changes (4). But the variability of such expert judgments (5) and the inadequacy of the formalism in the case of very-low-probability judgments (which are the ones of concern) render this approach less fruitful than was originally hoped. Further, there is the discouraging experience that the serious accidents that do occur have tended to be of the sort where the mathematical theory of risks is irrelevant (Seveso), inadequate (Three Mile Island), or misleading (Salem, New Jersey).

Another response to this crisis in risk assessment has been to re-emphasize the "human" and managerial aspects of the creation of hazards. However, there has also been a tendency to a multiplication of traditional efforts in formal studies of risks. Both are represented in a comment by the New York Times (6):

> The (Nuclear Regulatory) commission has two ways of analyzing safety - field reports of utility performance and probability studies of risk. Recently it has been emphasizing risk assessment, which predicted the automatic scram system would fail once in 33,000 reactor years - hardly a red flag for the problem. But the Salem plant had been cited three times for below-average management. Risk assessment may help compare risks; for real-world accidents, empirical evidence would seem a clearer guide.
>
> Nuclear power is essential to an energy policy that seeks diverse sources. But the public will not respond in measured fashion to another core damage event; it may simply reach for the ax. The safety margin that prevented disaster at Salem is a thread too thin for a whole industry to depend on.

The same point is put even more forcefully in a leader in the New Scientist in London, a journal generally committed to the cause of civil nuclear power. Under the heading "Nuclear Incompetence," it says

> Perhaps the best thing that could happen to the world's nuclear power industry would be the closure of every nuclear power plant in the United States. Then we might see an end to the series of stupid mistakes that cast a shadow over nuclear power in other countries.
>
> Is it any wonder that some level-headed Americans aren't too keen on nuclear power, especially if it is in the hands of tin-pot utilities that cannot even keep track of their mail?... We can only hope that the U.S. does not drag every one down with it in a complete meltdown of nuclear power's already shaky credibility (7).

Such reflections give rise to renewed consideration of the cultural aspects of risk management. While it is only common sense to be cautious about installing nuclear reactors in such places as Zaire or Belfast, it is a sobering thought that the

United States of America may not be an appropriate milieu for what was until recently accepted as a basic, universal energy technology. For an understanding of this phenomenon we need a theory of regulation, which is beyond the scope of this present study (8).

Thus the "art" of the assessment of risks of complex technological systems is at an impasse. The early hopes that it could be reduced to a science are frustrated. And the crisis of confidence among the public in the assessment of such risks inevitably extends to the institutions that regulate the industries and also to the industries themselves. Although some individuals and agencies redouble their efforts in the old ways, others are tending to introduce the "human" and "cultural" factors. The question now becomes, to what extent should these predominate, to the reduction or exclusion of the "scientific" aspects? For, as we shall see, if the perceived phenomena of "risks" are interpreted as lacking all objective content or being merely a small part of some total cultural configuration, then there is no basis for dialogue between opposed positions on such problems. For this reason alone, such "relativist" arguments must be taken seriously. We analyze them in the next sections.

THE SUBJECTIVIST FALLACY

The very first issue of the first journal to be devoted to Risk Analysis contained as its main theoretical contribution a paper on the quantitative definition of risk. Appearing in such a position, the paper inevitably carries the authority of the community of experts who are represented by this journal. As a philosophical contribution, it therefore has an importance for policy debates which does not depend on the scholarly credentials of its authors. We therefore shall discuss it by the criteria of philosophical analysis, although the authors themselves may not have intended it as a contribution to a dialogue on the metaphysics of risks.

By "radical subjectivism" we mean the position which goes beyond the use of subjective probabilities and Bayesian statistics. Its proponents believe that there is a need to justify philosophically the use of subjective techniques, and they think that the justification would consist in showing that risk is a subjective phenomenon. An example is provided by Kaplan and Garrick (9).

> Connected to this thought is the idea that risk is relative to the observer. We had a case in Los Angeles recently that illustrates this idea. Some people put a rattlesnake in a man's mailbox. Now if you had asked that man: "Is it a risk to put your hand in your mailbox?" He

would have said, "Of course not." We however, knowing about the snake, would say it is very risky indeed.

Thus risk is relative to the observer. It is a subjective thing - it depends upon who is looking. Some writers refer to this fact by using the phrase "perceived risk." The problem with the phrase is that it suggests the existence of some other kind of risk - other than perceived. It suggests the existence of an "absolute risk." However, under attempts to pin it down, the notion of absolute risk always ends up being somebody else's perceived risk. This brings us in touch with some fairly deep philosophical matters, which incidentally are reminiscent of those raised in Einstein's theory of the relativity of space and time.

Their argument can be summarized as follows:

1. Example, <u>then</u> Risk is relative to the observer.

2. Risk is relative to the observer, <u>then</u> risk is subjective.

Let us consider first argument 1. It is clear that the letterbox being a hazard depends on the presence of a snake (unless everything is a hazard, making trivial the concept of hazard). Putting a hand in a letterbox will be risky given the presence of a snake; information about the presence of a snake will change the judgment about the situation and will, consequently, produce different behaviors. The letterbox with a snake is "risky" independently of the information possessed by a particular observer. What <u>is</u> dependent on the information available to a particular situation, given that his attention has been drawn to a particular situation, is his judgment and subsequent behavior, <u>not the situaton itself</u>. The snake is in the letterbox, and not in the mind of the victim. Thus this particular example does nothing to establish "relativity" in the notion of risk, at least with the meaning that Kaplan and Garrick assign to "relative."

A simple change in the example provided will show that there is no difference, in principle, between "risk" and "weight" as attributes. Instead of a letterbox, let us assume we have an empty dumbbell. If someone replaces the hollow dumbbell with a similar one, but filled with iron, we have the same situation in the two examples. If you are asked: "Is the dumbbell heavy?" the answer would have been, "Of course not" (to paraphrase our authors). We, however, knowing about the iron, would say it is very heavy indeed.

In the two cases there is a hidden factor, snake and iron. The aprioristic guesses about "riskiness" and "heaviness" will be based on experience and a state of ignorance of the observer about the real situation. An experiment will settle each question; for example, to look inside the letterbox, or to weigh the dumbbell. The aprioristic judgment does not necessarily have any predictive value, whereas the second kind of estimation could be used as a firm basis for future behavior.

We should notice that the experiment need not be a quantitative measurement. Just as we can speak of the "heaviness" of an object without weighing it precisely, so we can speak of the "riskiness" without going through the procedures necessary for a quantitative estimate.

We can now consider the second argument: "If x is relative to the observer" then "x is subjective." The following example will show its falsity.

We have planet 1 (red) with an observer 1 and planet 2 (blue) with an observer 2; there is an astronaut in a rocket, outside the visual range of the observers; both observers are "Ptolomeic," whereas the astronaut is trained in Analytical Philosophy. For observer 1, planet 2 orbits around planet 1, and vice versa. For the astronaut, the problem of which planet is orbiting around which is a pseudo-problem, but as a matter of convenience we could decide that the blue one orbits the red one.

The movement of planet 2 is relative to the position of the observer 1 and vice versa; each observer will judge the situation differently and will behave accordingly, but the movement of the planets is not "subjective."

THE "SOCIAL-REDUCTIONIST" HAZARD

"Social-reductionism" is a methodological position which assumes that every debate over technological risks is really a conflict among contradictory "ways of life" and that the awareness of this would be enough to settle the question. An example is provided by B. Wolfe (10). Someone who believes that the future welfare of society is dependent on new domestic energy supplies will see large advantages to the development of nuclear power, off-shore oil resources, and new sources of coal, even at some risk and inconvenience. Those who believe that society suffers because we already use too much energy will not accept even minimal risk or inconvenience in order to supply more energy. A public discussion of energy development between groups with these opposing views is like a discussion of pork processsing among farmers, meat processors, and orthodox Jews and Muslims. One may talk about humane slaughtering techniques, but the underlying issue is whether or not pork should be eaten. The issue may be couched in technical terms of "spent-fuel disposition," but in fact it is an argument over the morality of eating pork.

The issue could be, of course, "an argument over the morality of eating pork" but the general public and authorities are also concerned with the possibility of "trichinosis" and this concern is shared by the pork industry because cases of disease from eating pork are (at the very least) bad publicity and financial loss for themselves. One may talk about "humane" slaughtering techniques and sometimes "the underlying issue is whether or not pork should be eaten," but it is also about public health, honest or dishonest trade, and controls.

This relativistic position is unlikely to provide a useful solution to the problems faced by risk assessors and decisionmakers. It has not solved the methodological problems of anthropologists, social scientists or philosophers either, and it is in these fields where it is strongly opposed (11). Moreover, it can be used as an excuse to deny the possibility and need of a dialogue and a rational debate on basic problems of technological risks and choices.

The most significant contribution to this approach is the recent book by Douglas and Wildavsky (12). Combining the approaches of political science and anthropology, they produce a powerful case against the pretensions of a "scientific" approach to the management of the many risks to which high-technology civilization is subject. They show that a program of managing all possible risks is impossible; that many significant risks are ill-understood; that the selection of risks for popular concern in America is definitely <u>not</u> done on a scientific basis; and that those who campaign most intensively on particular hazards are motivated by a crusading zeal for a vision of a radically different civilization.

So far their argument expresses, in a particularly erudite and insightful way, the dilemmas of environmental risk management, as now agreed by nearly all who are involved in it. We may say that risks are now revealed to lie outside the province of Science, and are more like those of politics, or of life in general. But Douglas and Wildavsky press their analysis further; they wish to explain <u>why</u> this strident environmentalist movement arose at a particular time in the United States. For this they construct a theory of three sorts of social institutions, "bureaucracy" and "market" (occupying the "center") and "sects" (occupying the "border"). The "center" institutions have failings of their own, familiar to social critics and well detailed here. Less familiar are those of "sects;" their peculiar constructed-world of moral absolutes is derived from the psychodynamics of closed voluntary societies. The history of the classic closed religious communities in North America is here used as the sole evidence of this argument, summed up in the assertion (p. 189), "Border theories emanate from its social predicament of social voluntariness."

The conclusion from this analysis is that "sects" enter into the political process only sporadically in response to a crisis or scandal where the center cannot cope; and there is never any true dialogue between the mutually exclusive cultures. The policy implications of this conclusion are not made explicit by the authors, for their concluding discussions are of philosophical relativism and of "resilience" as a strategy for coping with present and future risks. However, their use of the term "sectarian" for the organized critics gives a hint, as does their comment (p. 11), "The remedies most easily proposed in such organizations are to refuse to compromise with evil and to root it out, accompanied by a tendency towards intolerance and drastic solutions." It is hard to see how they could disagree with the analysis and conclusions of B. Wolfe, mentioned above. Nor would it be easy to escape the practical conclusion that these new "sectarians," like all others in that historic category, are not entitled to respect as participants in a policy debate. For they join it for ulterior motives, and given their absolute and intolerant commitments, will only abuse its procedures for their own ends.

These conclusions (nowhere asserted but nowhere denied by our authors) could have serious consequences for environmental politics, and so deserve careful thought. They also run counter to the recent American experience of pressure-group politics, on the environment and other issues. On this latter ground alone, we are entitled to scrutinize the authors' models as a reflection of social reality. Of course the three sorts of organization are "ideal types," and their authors analyze their real-life manifestations with great insight and subtlety in respect of their internal workings. But their theory of the psycho-dynamics of "sects" is based on closed communities that isolate themselves from society. What resemblance do these have to the politically active national organizations that rely on a mail-order membership, or even local activist groups which are only NIMBY ("Not in My Back Yard") coalitions? Not very much at all, except sometimes in a sort of style of rhetoric, which on occasion is even shared by the United States Congress (p. 163)!

Although the authors discuss the activist groups, their concentration on their own model inhibits any real insight into politics as it is practiced in America. While they analyze the interactions of the two sorts of institutions of the "center," they have very little to say on those between "center" and "border." The reader is left with a picture of total non-communication between a "complacent" center and an "alarmist" (or rather, fanatical) border.

In this way the authors have missed the opportunity to analyze real roots and stylistic tendencies in American political life in relation to their problem. Better than "sects," we can think of such terms as "populist," "evangelical," "crusading," "factional," and "litigious" to describe the traditional style of American reforming politics. Such great historic campaigns as anti-slavery, free-silver, and prohibitionism had their "sectarian" beginnings, and rose to positions of significant strength. More recently, some radical compaigns have failed (as on the hazards of recombinant DNA research), while others have gained surprising strength (as opposition to nuclear weapons strategy). This is the stuff of politics in America; all the skills of political science and cultural anthropology would be relevant to its study. But the authors are quite unconcerned with this level of reality, and their analysis is correspondingly abstract and impoverished.

Seeing the risks agitation nearly as a unique sectarian phenomenon, they tend to become influenced by the cross-cultural study of risks. Fascinated by the bizarre collections of beliefs about risks which are the stock-in-trade of anthropology, they convey the impression that all risks are equivalent, on the classic lines of "pollution is in the nose of the beholder." Thus asbestos hazards are described as follows:

> Why is asbestos poisoning seen to be more fearsome than fire? Asbestos was developed to save people from burning; asbestos poisoning is a form of industrial pollution whose toll of deaths by cancer justified a particular anti-industrial criticism more strongly than does loss of life by fire. (p. 7)

In view of the well-known history of willful neglect of the hazards of asbestos, industrial and environmental, such an analysis is more characteristic of a "border" (though <u>not</u> the one criticized by the authors) than of a "center."

Only an important book merits such detailed criticism; the authors have made a real contribution, through their theoretical insights and case studies. Their solution to the problem of civilized dialogue of risks is, in our view, excessively skeptical. But the problem exists, and they have laid down a challenge to all those who believe in the possibility of a solution. We hope to make a beginning on this in our three-fold division of risks problems in the next section.

A SOLUTION: THREE TYPES OF ENQUIRY

Although no one asserts now that Risk Assessment applied to complex technological systems is a Science, science continues to be the desideratum, as a model of practice, among a majority of experts. But, is the methodology so successfully used in research science suitable to the sort of enquiry found in Risk Assessment? There are two reasons why it is not. First, it is well known that the ideal of a science consisting of exact quantitative "public knowledge" (13) is frequently unattainable in this sort of work. This failure is commonly ascribed to the degree of complexity, uncertainty, and ignorance in crucial problems. But the difference between this field and research science is not merely one of degree. Here, a second dimension, containing the so-called "value aspects," is an essential component of the problems. In this respect this field is qualitatively different from research science.

The "values" dimension does come into scientific practice, as in peer-review of research projects or in general setting of priorities. However, outputs of research as realized in papers, are formally at least uni-dimensional, purely factual. Any methodology that pretends to be successful in Risk Assessment has to take into account dimensions, facts, and values.

We think we can solve the difficulty created by the contradiction between the ideal of "public knowledge" science and the characteristics of the problems encountered in Risk Assessment, without falling into sectarian relativism or social reductionism. The first step is to distinguish among the different types of problems in Risk Assessment and then to apply an appropriate methodology to each kind. The problems would be analyzed as functions of the two dimensions mentioned above: "systems uncertainty" on the technical side, and "decision stakes" on the values side, both varying independently.

"Systems uncertainty" contains the elements of inexactness, uncertainty, and ignorance encountered in the scientific and technical studies; whereas, "decision stakes" involves the costs and benefits to all the interested parties, of all the various available policy options, including delays in decision of some definite or indefinite duration. It also includes possible costs to the consultants themselves of advice which is unwelcome to the client!

The diagram (which should be seen only as a convenient sketch) indicates the realms of application of three different methodologies and the names in it indicate them (Fig. 1). When the uncertainties of the system and of the decision stakes are small, and the data bases are relatively large and reliable, we have the sort of problems where the "Applied Science" methodology has proved to be successful. Even some relatively rare events, involving quite high decision stakes, can be brought under control by "applied science;" aircraft accidents and disasters are a good example of this. One reason for this success is the availability of a good database, in the records of the extremely large number of movements and the large number of recorded incidents of varying degrees of severity. This is supplemented by empirical knowledge of the ratio of non-reported incidents to those reported. In such a case, it is possible to construct a "risks pyramid," giving ratios of empirical frequencies of related sets of incidents, accidents, and disasters. By their means, control-measures may be devised to reduce incidents and their more serious consequences in salient cases (14).

When both the systems uncertainty and the decision stakes are considerable, but where professional expertise can still operate, we will define a different model of practice, the "Technical-Consultancy" methodology; it still involves the use of quantitative tools, but these are explicitly supplemented and interpreted on a qualitative basis by experienced judgment. The result is not intended to function as an element of "public knowledge" science; it is not designed to be fully testable and reproducible, nor to be applied mechanically to other similar problems; and in this sense, this kind of enquiry fails to satisfy the epistemological criteria for research science. Rather, it is an input to a process of decisionmaking in which risks are one among several factors. The values at issue condition the enquiry at every stage, because of its function in a particular decision process.

Outstanding examples of the "technical-consultancy" methodology are the two studies on the petro-chemical installation on Canvey Island, by the Health & Safety Executive (15). There they dealt with a particular site, in terms of its special problems; they claimed no general results, and gave cautions

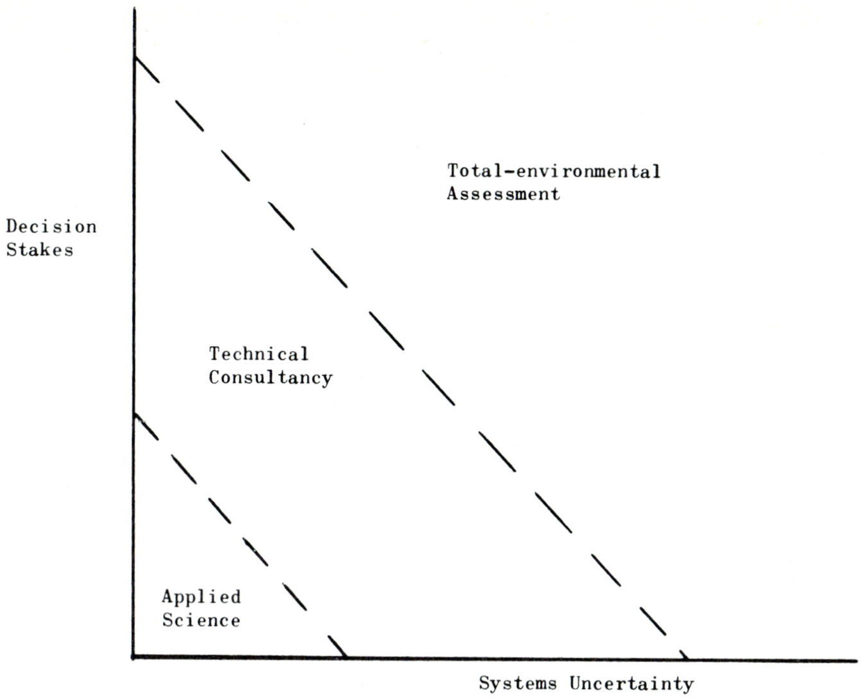

Fig. 1 : Classification of types of enquiry involved in Risk Assessment.

about the inexactness of their probabilistic conclusions, in terms of the varying degrees of reliability of their data. Although they concluded with a policy recommendation, they had sufficient detail on many local hazards for their prescription to be applied to a drastic improvement of safety on the island.

Nowhere would the reports claim that the island could be made "safe," or that their conclusions were "scientific." Indeed, the two reports were commissioned and received in a highly-charged political atmosphere, and criticized in proceedings of an adversarial nature. But in terms of the stated tasks of the enquiries, and also as examples of how such work is to be done, they stand as solid accomplishments.

Finally, when decision stakes and systems uncertainty are very high, we have a "Total environmental assessment," which is permeated by qualitative judgments and value commitments; its result is a contribution to an essentially political debate on larger issues, though no less rational in its own way for that. The enquiry, even into technical questions, takes the form largely of a dialogue, which may be in an advocacy or even in an adversary mode.

Although only a small proportion of risk assessments fall into this category, they are frequently those of greatest practical and political significance. It is to them that the analysis of "social reductionism" is most plausibly applied.

It is very important to realize that a "total-environment" problem is, by definition, <u>not</u> simple or static. Although value-commitments may shape arguments and enquiries on many important aspects, the debates are not all in the realm of metaphysics, nor are the decision-stakes purely in the area of sectional power-struggles. Thus, to take the example cited above, the issue of asbestos hazards is <u>not</u> simply a case of disliking industry more than fire. Even on more complex and speculative issues, as energy supplies, there is a steadily growing area of scientific knowledge and technical expertise, which complements the value-considerations, reducing uncertainties on particular critical decisions, and perhaps securing agreement among experts on all sides concerning particular technical problems. The hearings on the PWR at Sizewell is an example of how, in spite of an overtly adversarial style of debate, common accumulated experience yields an evolution of the issues, if not policy agreement then at least to consensus on salient areas of debate. This last is all we require for a problem in this domain to be excluded from the "social reductionist" model of environmental debates. This evolution, toward rationality and dialogue, may indeed take years to accomplish; in its early stages a "total-environmental assessment" may really seem to be a clash between incommensurable world-views. But such debates tend to stimulate the production of knowledge, of relevant facts and of value-commitments, which eventually enable such problems to be resolved by political debate rather than by civil war. It is important to note that the classification is not static but dynamic, in the sense that any changes in any (or both) of the dimensions could push a particular problem completely from one realm to another, consequently imposing a change in the way it is handled. The most notable recent example of this (welcome) tendency is the problem of possible physical and psychological harm to children resulting from atmospheric pollution by petrol-derived lead.

We should also observe that all these types of problem-solving are different from the practice of "safety management" or "loss control" in an installation that is in operation. There the "scientific" work of survey, analysis, and design is organized around the influencing of human behavior by various means. The problems of this field have been analyzed in terms of a "risk triangle," illustrating the varying valuations and perceptions among those who impose, endure, or regulate the risk (16).

CONCLUSION

This study began with a review of the crisis in confidence in the assessment of technological risks. The earlier optimism in the power of scientific method to assess all risks has given way to confusion and loss of confidence in any risk assessment. This tendency to despair can have serious consequences for the urgent tasks of social management of the technological risks of the present and future. Attempts to identify new philosophical foundations for risk assessment have proved fruitless. We have therefore concluded that the construction of a general theory applicable to all sorts of risks is not the appropriate way forward.

Our path to a solution lies in a distinction among the different types of risk. It's not merely that some are inherently more difficult to assess. Rather, the tasks of assessment for the various cases are qualitatively different. By analyzing the appropriate methodology for each case, we are able to identify the areas that are particularly problematical and challenging, and we can also explain and reinforce successful practice where it exists. Also, by showing what sort of "knowledge" is actually needed from any particular type of Risk Assessment, we avoid the error of excessively severe demands and high expectations; and we thereby prevent the disappointment and confusion that occurs when these are not fulfilled.

We have shown that there is a large and important area where scientific techniques can be successfully applied to complex technological risks even where important policy decisions are at stake; we have called this "technical consultancy." Its success, when properly done, can be understood when we appreciate the nature and function of its outcome. Its conclusions are not intended to be a weaker version of the "public knowledge" resulting from research science. Rather, they are an input to a decision process, constrained by the particular problem set by the client, and validated both by the scientific techniques and by the skill and judgment of those who apply them.

Even in the case of "total-environmental assessment," the problems are not an undifferentiated mass of scientific speculations and cultural values. Such issues always include components capable of study by "technical consultancy" or even "applied science." Also, public debate on such problems stimulates research whereby the "scientific" component is strengthened, perhaps eventually to the point where the value-commitments themselves are changed in the light of new evidence and new experience.

The assessment of technological risks is a very new field. It is only natural that it should go through a cycle of early over-optimism followed by disappointment and pessimism. Our work has been intended to build toward the next phase of this practice, involving a more mature appreciation of its real strengths and limitations. We have done this partly through identifying and reinforcing successful professional practice where it exists, and using this as the basis for further study.

REFERENCES

1. C. Mandl and J. Lathrop: "Comparing Risk Assessments for Liquefied Gas Terminals - Some Results." In The Risk Analysis Controversy. An Institutional Perspective, H. Kunreuther and E.V. Ley (eds.), Springer-Verlag, 1982, pp. 42-43.

2. C. Starr, "Social Benefit versus Technological Risk," Science, Vol. 165, Sept. 19, 1969, pp. 1232-1238. G. Apostolakis: "Probability and Risk Assessment: The Subjectivist Viewpoint and Some Suggestions," Nuclear Safety, Vol. 19, No. 3, May-June 1978, pp. 305-315.

3. Risk and Decisionmaking: Perspectives and Research. Committee on Risk and Decisionmaking Assembly of Behavioral and Social Sciences, National Research Council, National Academic Press, Washington, D.C., 1982, p. 61.

4. G.N. Parry and D.W. Winter, "Characterization and Evaluation of Uncertainty in Probability Risk Analysis," Nuclear Safety, Vol. 22, No. 1, January-February 1981, pp. 28-42.

5. D. Okrent, "A Survey of Expert Opinion on Low-Probability Earthquakes," Annals of Nuclear Energy, 2, 1975, pp. 601-614.

6. "The Axman and Nuclear Power," (Leader) New York Times, March 20, 1983.

7. "Nuclear Incompetence," New Scientist, May 19, 1983, p. 434.

8. J.R. Ravetz, "The Safety of Safeguards," Minerva, Vol. XII, No. 3, July 1974, pp. 323-325.

9. S. Kaplan and F.J. Garrick, "On the Quantitative Definition of Risk," Risk Analysis, Vol. 1, No. 1, 1981, p. 12.

10. B. Wolfe, "Is the Energy Debate Really about Energy?", IAEA Bulletin, Vol. 24, No. 4, 1983, pp. 28-32.

11. M. Hollis and S. Lukes (eds.), "Rationality and Relativism," B. Blackwell, Oxford, 1983.

12. M. Douglas and A. Wildavsky, Risk and Culture, University of California, 1982.

13. J.M. Ziman, Public Knowledge, (Cambridge U.P., 1968).

14. J. Tye, "On the Safety Beat," Safety, (London) February 1982.

15. Health and Safety Executive. "Canvey: An Investigation of Potential Hazards from Operations in the Canvey Island/Thurrock Area," 1978, and "Canvey: A Second Report," 1981 (London: HMSO).

16. J.R. Ravetz, "The Political Economy of Risk," New Scientist, September 8, 1977, pp. 26-27.

From Public Perception of Risk to Cultural Theory of Technology

Brian Wynne

International Institute for Applied Systems Analysis, A-2361
Laxenburg, Austria

INTRODUCTION

Among the various approaches to analyzing people's attitudes to risks, technologies, and their managing institutions, one can see two fundamentally different metaphysics from which nearly all approaches originate. By metaphysics I mean a closed loop--a cosmology--of taken-for-granted views of human nature, social interaction, public life, rationality, values, and ways of observing which confirm our founding assumptions and faiths.

Many modern analysts take it for granted that individuals are discrete beings, morally insulated from their social context and holding integrated values which are complete and more or less stable. Thus people's values are assumed to be scientifically observable, solid, clear (or, at least, capable of being "clarified") and able to be factored into clusters. This usually also entails the view that people interact according to clear-cut self-interests which they are busy optimizing. Social life is just a calculus of rational individual self-interested material optimality. The methods of observation of these aspects of human nature and social behaviour reflect and confirm their basic metaphysics.

An alternative approach--which I confess to find more congenial and plausible, and which I shall use in this paper--is the opposite of all the aspects listed above. Thus people are assumed to be intrinsically social, down to their very identity and "values." Their identity and values are formed by social interaction (which does not mean they are plastic and so present social interactions to the neglect of history, socialization, basic elements of cultural context, etc.). According to this view, human nature is always essentially incomplete, and values are intrinsically vague, unfinished, and available to some significant extent to social negotiation and definition (which is sometimes called "clarification"). On

NATO ASI Series, Vol. G4
Environmental Impact Assessment, Technology Assessment, and Risk Analysis. Edited by V.T. Covello et al.

this view, ambiguity in values can be seen as a rational defense against being inflexible and pinned down when social and other uncertainties would rationally require flexibility. The methodological counterpart is that "scientific" methods, for all their power and precision, capture only one dimension of the way values and attitudes form. Furthermore, they may be no less laden with essentially unproven and metaphysical commitments than other approaches, which may, like the anthropologist's, draw upon a richly varied repertoire of sources of data. The apparently ill-disciplined, often anecdotal character of such observations and data may be worth as much serious attention as the "scientific" methods currently fashionable. Borrowing the words of a famous beer advertisement, this approach may reach the parts that other brews never dreamed existed. In this paper I will attempt to develop the recent argument that rigorous pursuit of the perception of risks leads us toward a political and cultural view of technology as a social institution and indeed as social process. I will then relate this to a sociological, even psychoanalytical, interpretation of attitudes by suggesting that we view technology as a cultural process. I argue that only with this analytical approach to technology can we take seriously the question of public perceptions of risks in technology-dominated societies.

TECHNOLOGY, NATURE, AND CULTURE

Just as there are political implications in the way nature is defined, there are also political implications in the way "technology" is defined. Although there is an apparently irresistible urge to use terms like Nature, Culture, and Technology as if they were unitary entities (and perhaps this is always the fate of potent social symbols), public policy reflection is better served by examining the origins and implications of received definitions and their "interfaces."

There has been a long tradition of research on the social negotiation of nature and its complex relationship with culture (1). What Thompson refers to as an eternal circle of the cultural construction of nature and the natural destruction of culture (2), leads to the apparent conundrum of the cultural destruction (via nature) of culture. The conundrum only appears, however, if we give the floor to the received approach to "Culture," which is to see it as a homogenous, monolithic whole--Western Culture, Islamic Culture, Traditional Culture, etc. By adopting a more modest notion of culture we can attend to the contending differentiations within (and elements of cross-"Cultural" identity between) such abstract monoliths, and link these to real beings, institutions, and issues rather than moral ideals. We can see different social groups--their characteristic customs, belief systems, and social interactions--as more or less discrete local cultures, maintaining their own identity and existence in relation to others within the larger melting pot. Cultures are at the same time destroyed as active social constructs and yet immortalized by being "naturalized" by their proponents.

The distinctive essence of "Culture" as a framework of analysis is the integrated wholeness of cognitive and material social dimensions of existence. Nature is worked on and manipulated through ideas of nature, society, and technology which correspond with basic patterns of social relationships in that "Culture." This is not at all an approach antagonistic to conventional notions of social structure, institutions, power and economic relations; rather it may enlarge our vision of how such material social realities are maintained or changed (3).

In this paper I want to explore the notion of technology as cultural process in this sense, of embodying a differentiated set of cultures, each of which may be essential to the technology, but between which relations of power, communication, and coordination may be problematic. The use of the term "cultural" rather than "social," remains a question, and I have elsewhere tried to suggest the practical importance of seeing technology as social organization (4). However, without wishing to deny the importance of organizational, economic, and physical elements of technology, I am trying here to emphasize the associated attitudes, images, and belief systems which legitimate the social relations of technology. The ultimate goal is to shed a little light on the complex, brittle relationships between alienated "acceptance" and active attempted involvement (often via protest of some form) in the social direction of technology. This relationship is a key node in the dynamics of social and technological change and perhaps in the historic project to re-embed technology in more democratic forms of control. But at the same time it is obscure and highly unstable. It is worth studying by methods less regimented than orthodox attitude surveys can offer (5).

Although the relationships between Technology and Culture have long been a topic of inquiry (6), the notion of Technology as (differentiated) Culture has been of far less concern. Suggestive but neglected work a decade ago by Edge on some cultural implications (e.g. the "dehumanization" question) of technological metaphor acts as a springboard for my explorations (7). I will attempt to relate some of his insights to recent work in cultural anthropology to address some modern policy issues concerning technology (8).

First, however, I will outline a schema for treating culture more realistically, as a differentiated context of competing social-cognitive-metaphysical styles.

TOWARD POLITICAL CULTURES

Several cultural anthropologists associated with Mary Douglas have developed an essentially two-dimensional framework of socio-cultural attributes by which to define basic structural differences and comparisons between cultures. This "grid-group" comparative classification system may be applicable at various levels of aggregation from "national" cultures to individuals within sub-cultures. It has been well articulated elsewhere (9), so that only a brief outline is needed here. My aim is only to use the framework as a way of seeing the relationship between passivism and active protest in context. This may go a little way in adding some needed social dynamics to the framework itself, but that is not my main objective here.

The approach starts from the not unusual premise that ideas of nature held by groups and individuals correspond with basic moral principles crucial to that group's self-maintenance. Egalitarian groups tend to "naturalize" and thus maintain moral equality by seeing biological equality in nature. Hierarchical groups would tend to reflect their social hierarchy in perceptions of hierarchical processes in nature. This is standard fare. Cultural filters shape the perception of nature in systematic ways, blocking inconsistent data and highlighting confirmatory data. These filters are not merely encrusted habits learned by rote and mindlessly enacted from one generation to the next; they are the product of active scheming to maintain a given cultural style or bias in contention with competitors. The theory of the cultural anthropology school associated with Mary Douglas is that from all social contexts there are only a few fundamental types which such cultural biases can take. These can be mapped on orthogonal axes, of "grid" and "group."

High grid social relations involve a high level of external social prescription of the "individual's" role.[1] There is little or no autonomy, and the actor's experience is as a manipulated periphery to someone else's center. Conversely, low (or negative) grid relation involves high autonomy and anomie, and strong prescribing toward others.

---

1 More correctly, the social unit, which may be a group or individual.

High group relations involve strong incorporation into sharply bounded groups. This social demarcation between inside and outside is the key property. "Negative group" would mean active rejection of group boundaries.

It is the orthogonal combinations of these properties which provide empirically recognizable social groups, individuals, and organizations. Thompson has described them as follows:

> The group and grid axes have both positive and negative dimensions. Since group and grid can only be measured on ordinal scales, there are only five distinctions to be made within this social context space--one at the origin and one in each of the four quadrants. In each of these distinct social contexts we find a distinct social type: the hermit, free from coercive involvement in both group-formation and personal network-building; at the bottom left, the entrepreneur, spurning group involvement and central to a large personal network; at the top left, the ineffectual, excluded from social groups and peripheral to the personal networks of others; at top right, the hierarchist, strongly grouped and willingly subject to all the prescriptions that serve to maintain the ranked separation of his group from all the others within the group hierarchy; and at bottom right, the sectist, strongly grouped but rejecting hierarchy and all the prescriptions that are its inevitable accompaniment.
>
> I trace these five stabilizable conjunctions of social context and cultural bias back to three distinctive kinds of organization: the ego-focused network, the hierarchy-nested group, and the bounded egalitarian group. I further argue that this typology of organizations is exhaustive--that these are the only kinds of organization that are socially viable.

This scheme is represented in Fig. 1, where illustrative labels are given for the five basic social types, their typical cultural biases, moral justifications, and ideas of nature.

If we apply this scheme to ideas of technology we can see corresponding patterns. A high grid view would emphasize highly structured forms, and high group would emphasize strongly bounded areas of technical control or consequences, i.e. strong boundaries of responsibility. Thus a combination of high grid and high group would yield a sense of well-ordered technical action with in-principle clearcut boundaries of consequences. If these are not actually clear-cut then better forecasting and assessment can achieve this. Hence there is a sustained concentration bordering on the obsessional, with refined techniques for technological forecasting, risk management, and technological assessment. High grid-low group, on the other hand, would yield an analogous sense of determinism in the direction of technology; but an inaccessible determinism, out of reach of recognizable, organized human perception and management. Technology is well-gridded, but without any recognizable moral community (group) because there is no strong group experience to frame that sense.

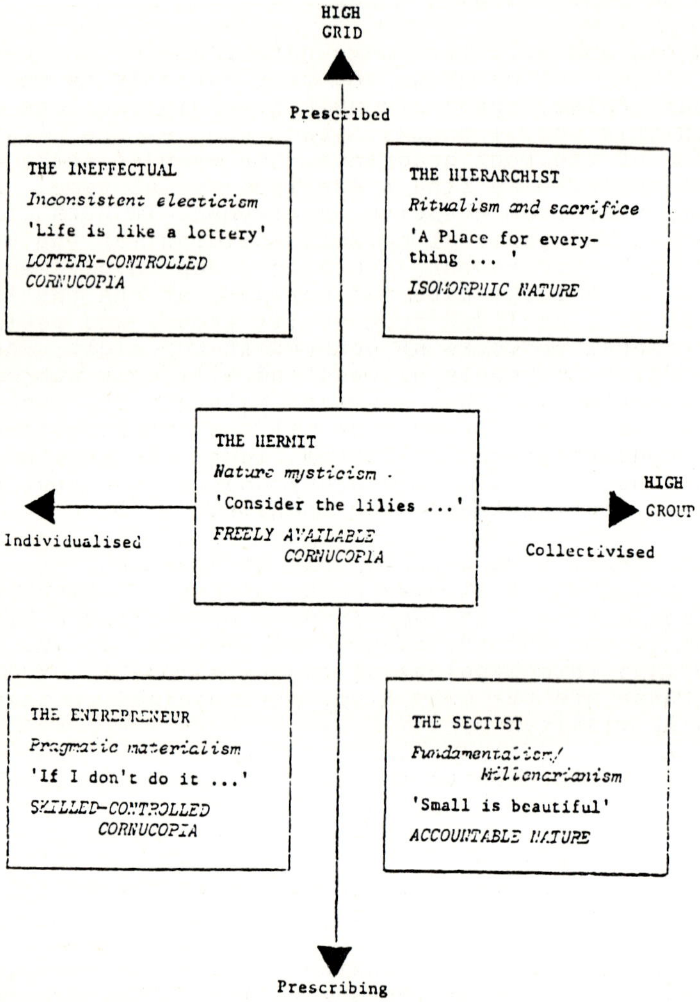

Figure 1. The GRID-Group Typolgy of Cultural Styles (Thompson, 1983): SOCIAL TYPES, *cultural biases*, 'justifications' and *IDEAS OF NATURE*.

Low grid-low group, on the other hand, would entail a similar sense of unpredictability to the low group (top left of Fig. 1,) but this time an unpredictability that was accessible, thus open to exploitation. Sometimes this would pay off, other times it would not. Anticipation would be of limited value; it would be more a case of "ride the tiger" than manage in the conventional sense of cross-impact matrices, n^{th} order consequence probabilities, etc.

Finally, the high group-low grid style would emphasize strong boundaries of responsibility, discontinuous consequence profiles (apocalyptic tendencies), and a low sense of external determination. Hence there would be an emphasis upon normative management, but more via collectivist styles of political organization ("appropriate technology") than by conventional hierarchical forms of management. From this bias, technologies would tend to be evaluated according to perceived intrinsic moral qualities.

One can see how these ideas of technology tend to correspond with ideas of nature. Indeed within each cultural style the ideas of nature and technology interpenetrate and reinforce one another. There is no clear boundary between nature and technology; indeed our publicly certified knowledge of nature, namely science, is nowadays certified only via technology, i.e. as knowledge leading to greater technical control, and nothing else. Truth and manipulation have become culturally confused.

The existence and character of the Hermit type is subject to some conflict--all the basic types are seen as mixed and nested in social reality. The grid-group classification can be applied at various levels of aggregation, from the individual, to specific groups, to whole occupational types, to national political cultures or whole societies. Although this has occasionally been treated as a sign of inconsistency in the schema, it is more relevant to view it as classificatory rather than explanatory. This is a necessary preliminary step toward explanation. The "problem of levels" then becomes less serious, and indeed may be a positive source of development of the approach toward the more complex question of social change via the interactions of such basic types. For example, single organizations may contain a rich blend of entrepreneurs, hierarchs, sectists, and ineffectuals. Within an overall hierarchical formal organization, sectist groups may emerge and operate at a given level, say in response to moves to reorganize or discontinue their work. Entrepreneurs (formal and informal) may also operate at different levels, in constant tension yet in overall unity with the organization as a whole.

Whatever the difficulties of consistent applicability and empirical referents of this rather fundamentalist schema, it does appear to resonate with broader experience and research on organizations and (with more difficulty) political cultures (10). A further criticism, however, is that the basic metaphysics of this theory are a version of "naive pluralism." In other words, although the cultural emphasis usefully reintegrates cognitive dimensions of social behaviour, there is no talk of power, even though the schema pretends to encompass political affairs.

A valuable linkage has been provided, however, by Thompson's suggestion (11) that the diagonal between Hierarchist and Entrepreneur (Bureaucrat and Innovator) can be regarded as a joint axis of power (and complacency) as conventionally treated in social science. In 3-D space, with power as a third dimension, this axis could be regarded as a ridge connecting the Entrepreneurial quadrant with the Hierarchist. Although tensions exist along it, there are many coalitions and elite formations in society which constitute this ridge system. The Ineffectuals can be regarded as a fairly flat landscape, and the Sectists, (for our purposes radical grass-roots labour union sections, environmental, or other activist campaigning groups), a slightly less lowly and, as we shall see, more turbulent landscape. This diagonal axis might be called the axis of instability and powerlessness.

Although this rough schema gives us the opportunity of testing ideas about changing social patterns through the whole system, I am interested here in exploring only one part; namely what makes people and groups move from being passive, alienated, and disoriented "ineffectuals," to becoming active, even zealous, intervenors in the process of technological decision and development. How does this apparently unpredictable sudden process come about? This has great practical importance for a number of groups from government agencies wishing to anticipate and contain such movements within their planning horizons, to activists wondering why they are not being joined by mass uprisings in their cause. Dissatisfied by the simplistic (though no doubt partially true) NIMBY explanation (12), I have tried to dig deeper into the labyrinth of psychic tunnels by which ineffectuals, rather than trying to scale the ridge separating them from activism, instead burrow through like moles to the other side.

THE AXIS OF INSTABILITY

Although the Ineffectuals category is probably the most complex of all those advanced by the grid-group schema, this complexity renders it perhaps the most interesting. By definition, many of the attitudes and beliefs of this category are inarticulate, partial, and latent. This is, after all, the central arena of the perennial "false consciousness" question (13). All the other groups use this passive (if differentiated) majority in their own schemes, involving different versions of "the public interest" and different theories of why the majority is so silent, corresponding with their own cultural bias. This sector could be regarded as a heterogeneous aqueous solution, invisibly super-saturated in parts, where local seeding gives sudden crystallization and an entirely new constellation of phases and interactions. These new phases are our analogy for activist groups with egalitarian sectist properties.

From beaming its ideology one way toward the passive, alienated majority, the axis of power now suddenly has to face the opposite direction too. To maintain power and authority toward this sector with its different rationality may require very different ideological contents, perhaps even ones contradictory to those effective for keeping the ineffectuals quiescent. This is suggested conceptually in Fig. 1, and is borne out in empirical experience.

For example, when controversial policy decisions about complex technological developments are made by institutions like public hearings, legal processes, etc., they are usually described in the public language as expert discovery problems. This description only inflames the often well-informed relevant activist groups, because their disaffection is strongly rooted in antagonism toward expertise and technocracy. They demand more explicit recognition of moral and political choices--a (low grid) language of prescription rather than objective structure. What is good legitimating language for keeping the quiescent majority quiescent, is exactly the opposite for these "sectist" activists. I have elsewhere described in detail this tension in the case of the 1977 Windscale Inquiry and its framing legal rationality (14). Conversely, describing the issue in the language of inevitable expert uncertainty, (thus ["difficult, so you may lose"] political and values choices),--even if the specific decision had gone against them--would have mitigated the impact on many activists because this language caters to their cultural style of moral prescription. By the same token, it would have invited some quiescents to join the fray and take issue.

Seeing this relationship as a fragile balance-in-tension of contradictory ideological tendencies and relationships offers us an analytical framework within which the relatively sudden shifts which are frequently seen in attitudes and levels of conflict (political surprise) can be conceived. Regular symbolic action (15) beamed in one direction and apparently successful at keeping consensus-by-quiescence may conceal the growth of activists popping up threateningly from the view of the power elite. The cultural filters of the elite may allow the activists to develop into significant features of the political landscape with solid connection (e.g. via the skillful use of the media) with the popular culture called "ineffectuals," before they begin to take them seriously.

Once taken seriously, however, some interesting dynamics may emerge. Sectist groups are highly egalitarian and grass-roots in style. They are antagonistic toward leaders, spokespersons, and experts, which is why the leadership rituals of such groups are often more agonizing and bloody than those where at least the notion of leader is accepted (16). Despite this, however, being taken seriously demands that leaders and spokespersons be deputed. In necessary interaction with the institutions of power, such a role demands increasingly expert, technical language of argument. Typically, such leaders move in language, attitudes, and style, toward the axis of power and the elites they begin by rejecting. They find themselves torn between, on the one hand, loyalty to their fundamentalist, uncompromising grass roots with its bosom-like security but politically "outsider" status; and, on the other hand, incorporation into the respectable margins of the policy elite, where status and recognition are traded for the willingness to emasculate original arguments into the narrow technical dicourse controlled by the establishment. They gravitate toward the top-right, Fig. 1, toward cooption.

This kind of metastable state can exist for years, with activist groups in a continual state of crisis and upheaval over their leadership, proper strategies, and styles of argument. If their leaders reduce this tension by becoming too coopted,[2] too drawn toward the hierarchical sector, such groups may simply and quite rapidly dissolve back into anonymity and the majority resume their membership of the ineffectuals. The process of high-group boundary-maintenance, and the fervent articulation of common identity and purpose, fall apart. It may be, of course, that cooption of leaders and the emergence of new leaders from the grass roots is an endless

2 This cooption process, its successes, ebbs, and flows, depends also upon the structure and flexibility of the establishment, which is at least partly relative to the specific issue. Cooption may be more likely with confident establishments (e.g., UK) and less with insecure, thus intransigent, ones.

process, maintaining the active if turbulent existence of vigorous "sectist" groups. If such groups do disappear from view, their members may still make up a latent nucleus--to use the earlier metaphor, a super-saturated area of solution--for later reactivation, perhaps on an adjacent but not identical issue.

This kind of analysis corresponds strongly with the approach to attitudes and behavior which rejects the rational economic individual calculator model of values, goals, and interests as clear, stable, and concrete (17). It supports the view of people and their attitudes as more tentative, experimental, incomplete, and perhaps internally inconsistent; humans as flexible managers of the conflicting, complex grounds of their own being. They may be more ambivalent, "unstable" and open to suggestion of their goals and values by dominant cultural stimuli than more individualistic, rationalistic approaches and methods claim. It is culture, not individuals, which gives these values what consistency and force they may have.

Since technology provides potent experiences and images which shape meanings, perceptions, and behavior, it may be regarded as a key substrate of culture. To the extent that modern technology provides uniform mass experience, it may be a form of common culture cross-cutting, underlying or destroying the differentiations which the grid-group approach posits (18). I shall advance the perspective that such differentiations are underacknowledged and are far stronger than generally assumed; that the experiences, relationships, and their guiding rationalities which people invest in technology are more varied, contradictory, and important than received models of "technology" can accommodate; and that their systematic analysis and recognition is of practical importance to technology policy. It is worth attempting to examine the expression of attitudes and self-images in technological experiences, and then to explore the psychological and sociological undercurrents of these. Technology can be regarded as culture in the sense that it is a potent framework relating dimensions of belief and meaning to social relations and processes; and it is political culture in that the social relations of power embodied by the technology are more or less successfully legitimated by the cognitive structures which are naturalized in the culture, and which thus conceal those underlying structures of power from critical examination and possible change.

In his classic account of the social and psychic devastation caused by the 1972 Buffalo Creek dam failure in the Appalachian mountains (19), Kai Erickson observes that the reaction of the economically and politically-marginal people who were victims of that "point-disaster" was profoundly conditioned by their internalization of the state of "chronic disaster" represented in their long-term neglect and alienation from employers and public authorities. The psychic withdrawal characteristic of extreme traumatic shock was already consolidated on the community scale in the alienation and self-dependence of the community, trusting none of the agencies on whom they nevertheless depended (and thus tolerated) for economic survival. Erickson argues that what was most significant about the social aftermath of the disaster was not the personal trauma--"psychic numbing"--which everyone experienced, but the collective trauma, the inability of the old social networks to reestablish themselves as the framework of personal psychic convalescence and development. The people felt betrayed by the coal company which neglected the dam whose burst caused the disaster, not because they had previously thought it a conscientious company, but because structurally, in their position, they had to trust it, despite realistic appreciation of its selfish motives, past neglects, etc.

In Erickson's perspective (20) the powerless always tend to defend and rationalize, thus consolidate, their own impotence and apathy because to do otherwise is to expose themselves to the greater human damage of explicit neglect and powerlessness. They withdraw, and justify and defend that withdrawal as consistent with cosmic principles. It becomes their culture, integrating their beliefs about cause and effect in the experiences they encounter, with their established social relationships. Erickson saw the classic symptoms of trauma in the ordinary human reactions to "the age we are entering," namely "a sense of cultural disorientation, a feeling of powerlessness, a dulled apathy, and a generalised fear about the state of the universe" (21). These correspond with the features of the "ineffectuals" of the high grid-low group cultural category. They are the symptoms of social experiences and roles which are highly prescribed by others, yet where the structure of such prescriptions--of their own marginality and manipulation--is obscure. The "effective causes" of their powerlessness are socially invisible. What Erickson also saw being enacted in social reality was the tentative, fragile nature of movements as a result of apathy and disorientation. What community developments there had been in that direction were swept away by the flood, which was analogous to the condensation into a single, extreme, dramatic event, of years of non-affirmation (identity-stripping) by the outside world.

I have made this excursion into Erickson's interpretation of a man-made disaster not only to help uncover the complexities of attitudes and some continuities between historical events and historical processes, but also to explore how technology--here a dam central to the community's existence (it was part of the local coal mine system which employed most of the people)--is externalized in images which shape cultural attitudes. These implicitly reflect back people's social relations as alien objects, beyond their control or responsibility to alter. In the Buffalo Creek case, the survivors seemed to have a clear sense of who was responsible for the technology's havoc, but an equal sense of hopelessness that anything might be done about it. The effective cause of their disaster was at least seen to be human agents, even if these were believed to be beyond control. This, however, might be taken as an extreme example of a concretely visible technology with clear lines of control and responsibility. Many other technologies typical of the modern age--nuclear power, genetic engineering, and perhaps most especially computers--lie at the opposite end. Their controlling human agents are invisible, diffuse, and socially remote. It is impossible for ordinary people to identify the effective causes of their confusing and often troubling experience of these technologies, even if they do not produce dramatic interventions in their lives. Yet the importance of these experiences requires that people construct some working explanations so as to rationalize them one way or the other.

One example of the way such effective causes in technology have been mystified, and images cultivated, is given in Fig. 2 (22). The technology--here a nuclear power station of the most "advanced" sort (the Dounreay fast reactor)--is deified to the extent not only of hovering, disembodied above the mere earth, but with a halo to denote its moral purity and magical power. The caption invites people to indulge in awe-struck worship. It dazzles them from all angles, leaving no room for questioning the agencies, interests, uncertainties, and human frailties behind the image. This is symbolic action "in extremis."

An important consequence of this socially-constructed invisibility of effective causes in technology is indicated in a small part of the caption of Fig. 2. Part of the imagery of magic power is the fact that the "fantastic prospect" will emanate mysteriously from a superior force that cannot be seen, heard, or felt. These properties of ionizing radiation, then used to intensify the positive power of the technology, are the very ones which are now regarded as intensifying exaggerated hostility and fear. In other words, legitimation was created by cultivating the idea of awesome, other-worldly power, beyond the bounds of ordinary nature and culture. But this disorienting relationship's corollary is a double-edged instability which can easily and suddenly flip over from benign externality to malign externality.

The point is that with effective causes and structures of responsibility so obscured, the only responses possible are total acceptance (tinged with an ambivalent potential for anxiety in the face of such supernormal power) or total rejection (tinged with fascination at the sheer technical mastery such technology may entail). There is no possibility for measured criticism and conditional, qualified responses--all possible currencies of discrimination have been historically obliterated, leaving behind inflexible absolutes. This is tantamount to primitive thought, where the symbol is collapsed into the word, and no creative tensions exist any longer between the metaphorical skeletons of ideas and literal versions of the metaphor. People behave as if the technology were literally an alien being from space.

Psychoanalysts have examined clinical cases involving similar condensed images of technology which have become central surrogates for explanation of more complex experiences and potential responsibilities which people cannot handle. These images, or specters, are not only psychic simplifiers but also frameworks of social relationships. They are built around technological images, perhaps increasingly so, given the increasingly central role of such experience in daily life.

Daly defines a specter as a kind of potent, artificially created but invisible behavioral force (23):

> A sense of the operation of such forces arises when men find they cannot account for emotionally significant events by ascribing them to the conventional sources of power and efficacy (e.g. human, natural, divine) which are believed to make things happen in the world. When such inexplicable events persist and are experienced by numbers of people, agencies are created to account for these events. These agencies are given names, made into realities, and adapted to as powerful things....
>
> The spectral view of technology arises from a sense of domination by mysterious forces or agencies which are, or were, linked to technological enterprises but which are now apprehended as being beyond the control of any particular man or collection of men....

> (People) behave as if the spirit of meeting specifications in many discreet, limited and finite human ventures had taken flight from the hands of responsible agents and become an independent reality--a reality which has come to overhang the modern world and to enter into the dynamic processes of personality--as a spectral object.

There is, in other words, a ritual defense mechanism--a transference of responsibility for complex and inexplicable experiences which are too emotionally important to be ignored. Daly describes how several patients created such specters of their own biological systems, investing them with powers to decide and cut a clean swath through otherwise overpowering ambiguities. Thus they would obsessively refer to a simple measure such as their pulse rate as a guide to decisionmaking--it was made into a source of "objective decision rules" supposedly reflecting a greater, more powerful but impenetrable biological mechanism. Such agents may become absorbed into part of one's very identity; or more accurately, one's identity may be shaped by, and then absorbed into, the image. One becomes "a cog in a machine," or, with Bettelheim's "Joey: a mechanical boy," an electrical appliance who "plugs himself in" and "switches himself on" before he can speak, and who causes others to behave in parallel fashion in order to relate to him (24).

It is a central point of Daly's analysis of these conditions that they are no longer, if they ever were, restricted to clinically psychotic individuals. They are now in his view mass neuroses, transmitted in normal processes of cultural dissemination. Given the kinds of symbolic action depicted in Fig. 2, this is hardly surprising. Indeed the historical use of images of scientific, technical power as if from outside the realm of human interests and values has ironically cultivated an escalating search for objective decision rules from science, akin to a collective scale version of consulting pulse-rates. An example is the eternal effort to avoid the ambiguity of negotiating acceptability from situation to situation, by instead creating objective scales of "acceptable risk." The artificiality of these entities and the impossibility of their ever providing what they promise may never be apparent to the majority of ineffectuals since they are embedded in a whole labyrinth of dense managerial political language and institutional barriers. But their constant usage in keeping the ineffectuals at bay is just what disaffects and activates the sectists even more (26).

TECHNOLOGICAL ANIMISM AND SOCIAL ACTIVATION

In many cases the creation of such technological specters may ironically be an essentially rational reaction to irrational situations. Most people are fragments of technological systems which entail many connected parts whose coordination is essential, but complex and chronically problematic. However, they never experience the whole system (27). Their experience is fragmentary and bounded by their local organizational and cultural context, within which they have to make out. Finding it impossible to penetrate the boundaries of their local experience and to understand the rationalities, interests, and interactions of those whose activities structure that situation, they create shorthand images to "explain" those external agencies and their frequent unpredictability and apparent malevolence.

A graphic example of this was given by McDermott, who described a specter created by American GIs in Vietnam (28). They were operating in the jungle, constantly being attacked by Vietcong guerillas who could never be identified and pinned down. They were regularly shelled and rocketed, but were never sure it wasn't their own side; and they received orders but never explanations from their superiors. Their experience was frightening, confusing, contradictory, and utterly obscure as to its effective causes. They couldn't find an enemy and they couldn't identify their own side. Yet they received orders and were attacked in equally arbitrary fashion. As part of their rationalization of this (very high-grid) predicament, the GIs had condensed the potent, but diffuse and invisible effective causes onto a single agent--a huge gun which lived in a hollowed-out mountain, and which emerged at whim to unload death and destruction onto them. It was an agent beyond control, imbued with a kind of autonomous malevolent intelligence. In one major sense it was no comfort at all, but in another sense it was, because at least it offered explanation. It was a kind of metaphor representing their social relationships with those elites (and here also enemies) who remotely and invisibly controlled their fate.

Langdon Winner has also discussed this process as technological animism (29). He takes the story of Rudy in Vonnegut's Player Piano. Rudy was a mechanic whose job had been replaced by automation--his skills and experience had been reduced to an algorithm and entirely handed over to a computer. Deeply upset and mystified by this shattering of his very identity, Rudy enacts a scene in a cafe with a doctor friend, where he goes into a frenzy over what he sees as the creepy, superhuman intelligence controlling the keyboard movements on a simple slot machine (player) piano. Perhaps indicating that as a more educated being he sees through this conversion of concrete (if hidden) human goals and interests into extra-human, therefore untouchable intelligences, the doctor friend has to get up and walk out on this pathetic scene. In Winner's words (30), this scene gives:

> . . . a glimpse of the crucial statement and ultimate conclusion of the writings on technological animism. If one asks, Where did this strange life in the apparatus come from? What is its real origin? the answer is clear: it is human life transferred into artifice. Men export their own vital powers--the ability to move, to experience, to work and to think--into the devices of their making. They then experience this life as something alien and removed, something that comes back at them from another direction. In this way the experience of life becomes entirely vicarious....
>
> Man now lives in and through technical creations. The peculiar properties we may notice in these creations are not the result of some spontaneous generation. What we see is human life separated from the directing, controlling positive agency of human minds and souls.

Winner's important insights here must, however, be clarified by one important point. Although men do "export their own vital powers" into the technologies they have created, and reflect them back as aliens beyond control, this falsely implies a lack of any social stratification or cultural differentiation in this process. Elites are also immersed in their myths and fantasies about technological power, and non-elites do make technological creations. But it is also important to see that those ineffectuals are circumscribed by mystifications created through domination by decisionmaking elites, a domination whose arbitrary human structure is increasingly socially complex, remote, and thus "invisible." They therefore transfer responsibility from this frustratingly intangible and impenetrable human complex, into extra-human specters. This is transfer and condensation not so much of their own, small responsibility and power, but of the power of elites in the social structure around them. The myths and fantasies of the axis of complacency (see Fig. 1) actively promote this mystification and concretization of their own power, even if it is not always deliberate.

Not only does this cognitive process artificially consolidate the axis of power by placing it apparently beyond human access, but it encourages a lack of human tolerance for ambiguity, thus a structural brittleness in the system. When responsibility is so condensed into such technological specters whose inner workings are inaccessible, experience has to be interpreted, and life conducted, by either total identification with or total repudiation of such specters. Thus public "debate" and interaction becomes rigid and prone to sudden discontinuities: government itself may become less viable. As Crozier has put it, there is no authority without negotiation (31), and since such fantasies and specters preempt the possibility of negotiation by replacing and "black-boxing" more discriminating perceptions of relationships and causes, they tend to destroy even the possibility of legitimate authority.

A good example of the absolute contradiction in different social perceptions of technology, and the linkages between these and power structures, arose during the 1977 Windscale Inquiry (32). This was a public inquiry into a plan to build a new plant to reprocess spent nuclear fuel from the new generation of reactors, using oxide fuels. This would extract plutonium which could be used in fast breeder reactors or weapons, uranium which could be recycled in further thermal reactors, and radioactive wastes which would ultimately need some safe final disposal. The plant was part and parcel of a longer term historical vision of nuclear development reaching out of colossal past commitments and into future ones. Its go-ahead naturally made all of those envisaged commitments more likely, via institutional momentum and technical-economic logic.

The proponents and the High Court judge in charge defined the issue as the examination of the direct impact only of the reprocessing plant itself, and excluded any question of the implications of future fuel cycle developments which might be entailed by it. These, he argued, would be subject to future separate decisions, and any attempt to cover more comprehensive nuclear futures was "emotive nonsense." Yet many objectors took for granted that the reprocessing plant, being only a part of a historical process, had to be examined as such. Fast reactors, plutonium trading, waste disposal, and reprocessing plants among others, all had to be considered.

This issue was only "resolved" by the fiat of the judge. He found it impossible to negotiate with this alternative definition of the problem, perhaps because it was rooted in objectively different social experience, which he defined as "merely" emotive. To the decisionmaking elite it was logical to say that future plants could be separated as decision issues, because they could identify with the whole process in which those future decisions, as well as the present ones, would be made. They could conceive of decision choice and access to those future steps, which were thus separable from the present issue.

To the powerless however, no such identification could be made, because from their objective social position, consolidated in empirical historical experience, the processes by which the present step might or might not be converted into future elaborations were socially and intellectually impenetrable. From their social position it was therefore entirely logical to reject the equally logical but contradictory definition of the issue by the elite, and to condense all future possibilities into the one present question. It was an undiscriminating, all-or-nothing stance, occasioned by their relationship to the axis of power.

CONCLUSIONS

My own suggested definition of technology is, of course, also political in that it highlights very different questions, and suggests different structures of naturalness and unnaturalness from other definitions, such as technology as "tool," "craft," "package," or "historical dynamo." Although granting that technology does have intrinsic force and that this may well encompass and freeze, in its own way, the whole field of possibility for some societies or groups receiving technology, the cultural process model does not commit the often-ensuing slide into technological determinism as a model of history. Nor does it encourage us to use such terms as "technology" in an undifferentiated way, without attempting to understand people's different perceptions of control and responsibility in relation to it. Just as "nature" acts as a mirror reflecting back our social and moral preoccupations, so too does "technology."

I have tried to sketch a view of technology as a cultural process, attempting to link previous analysis of technology as social organization with ideas about how we structure experience of technology and its imbedded social relations. By exploring the cognitive dimensions of these relations we may approach an understanding of the depth and complexity of the organizational dislocations which frequently beset modern technology. To see these as sociocultural allows us to conceive of them as rooted in cosmological commitments which the language of "management" of "organizational" difficulties may oversimplify. The pattern of possible cosmologies, their associated rationalities, metaphysics, and individual identities and styles of interaction, are suggested by the grid-group cultural hypothesis.

The technological specters, such as those I have discussed, act as a framework of interaction within these cultures, and between them. They also define these cultures by becoming central parts of their very identity. Sherry Turkle has discussed the fact that various technologies invoke strong personal feelings and intense relationships (34). "People develop intense and complex relationships with cars, motorbikes, pinball machines, stereos and ham radios." Computers appear to have particularly strong properties in this direction. Turkle also recognized that such feelings can reflect external social and political concerns. However, what we are discussing here is more than relationships to, but identification with the technology, by fusion of personal or group identity with technological imagery. As we have seen, the purely mechanical technological metaphor can be reanimated by further metaphorical extension into images of intelligent controlling beings; but these are often alien, threatening, and unpredictable--a metaphor for real social relationships.

This cultural process may occur on a microsocial scale in comparison to the overall organizational scale of the technology. Thus nuclear power station laggers are a small if crucial part of the overall system of nuclear power development and use. They install insulation at critical parts of the cooling circuit of reactors, so as to avert catastrophic thermal gradients and stresses which would crack the pipes and release radioactive gases. Their work is arduous and uncomfortable, and they work with protective clothing in a maze of boilers and pipework. Interviews with laggers at the Heysham nuclear station in Lancashire, England (35), revealed that well away from regular supervision as they are, they frequently remove gloves and dust masks to ease working conditions, even though the gloves are supposed to avoid possible corrosion from (acid) perspiration on the stainless steel pipes. When they need to urinate, instead of crawling laboriously back to an exit and then to the site W.C., they find a convenient corner on the job in the pipework system, releasing onto it a potentially corrosive liquid. When they lose a piece of equipment, they are supposed to report it at the end of the shift, and go back down with a supervisor to find it and "sign it off." Instead of subjecting themselves to an open-ended search in their own time, they quietly ignore and cover up the loss, thus leaving the equipment to possibly disrupt the highly-sensitive, precision flow dynamics of the cooling system when the reactor is started up.

Laggers are a culture unto themselves. They see the thing they are building as just a theatre for doing their work and drawing their pay. The identification they have with the technology is as with a white elephant. When asked to justify what looks like their potentially dangerous and irresponsible behavior from the view of nuclear technology as a whole, they do not see it as a piece of nuclear technology. They point in very well-informed fashion to management incompetence on a par at least with their own "irresponsibility," and conclude that the technology will never come into being. Thus, seeing their behavior as irresponsible in the overall technology context is, in their view, irrelevant. Arbitrary forces outside their control completely neutralize the implications of their own behavior.

I would suggest that such cultures as fragments of overall technology systems are commonplace. As the technological division of labor becomes more elaborate and institutionalized, such groups become all the more segmented and isolated. In creating their own cosmologies out of this experience, they create a certain independence from the technology on which they depend. This "independence" is, of course, not total, but gridded by the boundaries of related parts of the overall system. The growth of a quasi-independent cultural identity out of the corresponding social practices may stabilize the boundaries of activism of such groups by "naturalizing" the surrounding social "landscape," to within limits that retain that dependence. However, this deeply ambivalent dependence may be misinterpreted as loyalty from the social distance of the axis of complacency, and the underlying alienation and cultural autonomy of such units never become apparent, except indirectly as technological (and maybe government) systems that do not work.

REFERENCES

1. S.B. Barnes and S.A. Shapin (eds.), Natural Order, London, Sage, 1979. M. Douglas, Implicit Meanings, London, Routledge, 1975.

2. M. Thompson, "The cultural construction of nature and the natural destruction of culture," paper to International Conference on Nature, Culture, Technology, Stockholm, Sweden, September 1983. Mimeo copy, IIASA, Laxenburg, Austria.

3. I have tried to make this connection through a detailed case study, B. Wynne, Rationality and Ritual: The Windscale Inquiry and Nuclear Decisions in Britain, British Society for the History of Science, Chalfont St. Giles, Bucks, 1982.

4. B. Wynne, "Redefining the Issues of Risk and Social Acceptance: The Social Viability of Technology," Futures, 15(1983), 13-32.

5. As offered, for example, by much mainstream risk perception literature. For a critique of this literature, see H.J. Otway and K. Thomas, "Reflections on Risk Perception and Policy," Risk Analysis, Vol. 2, 2(1982), 69-82. Another psychologist has observed that in risk analysis, "we psychologists are a bit trapped by our own proficiency at being good experimentalists. We realize the importance of control and so we are drawn to those tasks in which we can exercise control. Hence our preoccupation with simple, static lotteries" (which are used as if they were real-life risk decisions), L. Lola, Journal of Experimental Psychology: Human Perception and Performance, 9(1983), 137-144. Also, Douglas MacLean, "Is Rationality Extensional?", mimeo, University of Maryland, Dept. of Philosophy and Public Affairs.

6. As indeed in the long-standing, excellent journal of that name, and in classical works in economic history, such as Carlo Cipolla, Clocks and Culture 1500-1700, London, Collins, 1967; Lynn White Jr., Medieval Technology and Social Change, Oxford, Clarendon Press, 1963.

7. David Edge, "Technological Metaphor," in N. Wolfe and D.O. Edge (eds.), Meaning and Control, Essays in the Social Aspects of Science and Technology, London, Tavistock Publications, 1973, pp. 31-64.

8. M. Douglas, Social Factors in the Perception of Risk, Report to the Russell Sage Foundation, New York, 1983; Douglas and A. Wildavsky, Risk and Culture, London, University of California Press, 1982; M. Thompson, "Among the energy tribes," IIASA Working Paper, Laxenburg, Austria, WP-82-59.

9. Ibid. Also: G. Mars, Cheats at Work: An Anthropology of Workplace Crime, George Allen & Unwin, 1981; S. Henry (ed.), Can I have it in Cash: A Study of Informal Institutions and Unorthodox Ways of Doing Things, London, Astragal Books, 1981; M. Douglas (ed.), Essays in the Sociology of Perception, London, Routledge & Kegan Paul, 1981; S. Rayner, The Classification and Dynamics of Sectarian Organisations: Grid/Group Perspective On the Far Left in Britain, Ph.D. Thesis, University College London, Dept. of Social Anthropology, 1979; M. Thompson, Rubbish Theory, Oxford University Press, 1979.

10. See e.g., Douglas, ref. 8, and Thompson's postscript to H. Kunreuther and J. Linnerooth (eds.), Risk Analysis and Decision Processes: The Siting of LEG Facilities in Four Countries, Springer Verlag, Berlin, 1983.

11. Ibid, and ref. 2.

12. NIMBY, the "Not In My Back Yard" theory of social protest, would have it that people only take interest in an issue when their immediate local interests such as property are threatened.

13. And the similar issue of political communication. See, for example, the discussions of "communicative competence," in P. Connerton (ed.), Critical Sociology, Harmondsworth, Penguin Books, 1978.

14. Wynne, op. cit., ref. 3.

15. For the concept of symbolic action, see, e.g., M. Edelman, Politics as Symbolic Action, London, Academic Press, 1976.

16. See, e.g., Douglas and Wildavsky, Risk and Culture, op. cit., ref. 8. P. Lowe and J. Goyder, Environmental Groups in Politics, London, George Allen & Unwin, 1983, describes the strategic tensions for activist group leaders. I can also testify to this process from personal experience of such a (short-lived, but intense) role.

17. This is perhaps the fundamental issue in Western social science; see, e.g., Douglas, op. cit., ref. 8; Thompson, op. cit., ref. 2; Wynne, op. cit., ref. 3, ch. 9; R. Unger, Law in Modern Society, London, Collier MacMillan, 1976.

18. The standard critiques of modern technocracy, e.g. J. Ellul, The Technological Society, New York, Vintage Books, 1964; H. Marcuse, One Dimensional Man, New York, Beacon Press, 1966, tend to encourage such an exclusively monolithic view.

19. K.T. Erickson, Everything in its Path: The Destruction of Community in the Buffulo Creek Flood, New York, Simon & Schuster, 1976.

20. See also R.J. Lifton, The Broken Connection, New York, Simon & Schuster, 1979, for a psychiatric perspective adapting Freud's original concept of instincts and defense or "blocking" mechanisms toward a more central role for images of life and death as explanatory factors for human attitudes and behavior.

21. Erickson, op. cit., ref. 19, p. 258.

22. Taken from The Financial Times, London, Supplement on Atomic Power, October 1956.

23. R. Daly, "The Specters of Technicism," Psychiatry, 33(4), 1970, pp. 417-431 (quote p. 417, 421). For some earlier signs, see e.g. Roger Bastide, Sociologie et Psychoanalyse, Paris, Puf, 1950, describing the dreams of Indian tribes of automobiles breaking down; these are interpreted as a technical metaphor for sexual failure and derangement.

24. B. Bettelheim, "Joey: A Mechanical Boy," Scientific American, March 1959, pp. 2-9. See also Edge's discussion, op. cit., ref. 7, pp. 50-52.

25. H.J. Otway and D. von Winterfeldt, "Beyond Acceptable Risk," Policy Sciences 14(1982), 27-45.

26. B. Wynne, "Institutional mythologies and dual societies in the management of risk", E. Ley and H. Kunreuther (eds.), The Risk Analysis Controversy, Springer Verlag, Berlin, 1982, pp. 127-143.

27. Nor, of course, do the system's "managers;" but they too create elaborate but different myths to fill in the rest in a way consistent with their managerial position. It is just that these myths are different, even if they are more elaborate because they have more money and time spent on their articulation.

28. John McDermott, "Technology: Opiate of the Intellectuals" in A.H. Teich (ed.), Technology and Man's Future, New York, St. Martin's Press, 1974.

29. L. Winner, Autonomous Technology, Cambridge, Mass., MIT Press, 1978, pp. 33-35.

30. Ibid, p. 34.

31. M. Crozier, "Les Developpements Futurs de la Bureaucratie," Courier du Personnel, Commission of the European Communities, 416, 29 July 1980, pp. 13-20.

32. See Wynne, op. cit., ref. 3.

33. See, for example, the work of E. Wenk, Jr., reported in Futures, 15(1), 1983, pp. 87-90, and discussed at a seminar at the International Institute for Applied Systems Analysis, Laxenburg, Austria, September 1983.

34. S. Turkle, "Computers as Rohrschach: Subjectivity and Social Responsibility;" Bo Sundin (ed.), Is the Computer a Tool?, Stockholm, Almquist and Wiksell, 1981, pp. 81-99. Reprinted in Transaction: Social Science and Modern Society, 17(2), 1980.

35. During the research for a doctoral dissertation of Ian Welsh, to whom I am grateful for discussions concerning this point.

SECTION V: Policy Analytic, Administrative, and Legal Perspectives: Theoretical and Methodological Contributions

Public Disputes About Risky Technologies: Stakeholders and Arenas[1]

Ward Edwards and Detlof von Winterfeldt

Social Science Research Institute, University of Southern California

INTRODUCTION

On their way to social acceptance, many technologies experience periods of public concern, controversy, and social opposition. Sometimes these responses to technologies are short episodes, motivated by naive fears or esoteric values of a small segment of the public. As in the case of religious opposition to the implementation of lightning rods (see Mazur, 1981) these episodes may be a little more than amusing footnotes in the history of an ultimately successful technology. In other cases, the public responses may swell to a strong social movement against the technology that leads to its disruption, halting, or eventual abandonment. Water fluoridation, nuclear power, and the SST are recent examples of technologies suffering this fate (see Lawless, 1977; Mazur, 1981; Nelkin, 1979).

The reasons for social opposition to technologies have usually been complex, including worries about employment and

1 Research support for this paper was provided by the Technology Assessment and Risk Analysis Program, Division of Policy Research and Analysis, National Science Foundation, under grant No. PRA-8108638. The opinions and conclusions contained in this document are those of the authors and should not be interpreted as necessarily representing the official policies, either express or implied, of the National Science Foundation or the United States Government. We would like to thank Gerald B. Greenwald, Howard Kunreuther, Jeryl Mumpower, Harry Otway, Ortwin Renn, Chris Whipple, and Aaron Wildavsky for their helpful comments on earlier drafts.

NATO ASI Series, Vol. G4
Environmental Impact Assessment, Technology Assessment, and Risk Analysis. Edited by V.T. Covello et al.

livelihood, moral and religious objections, as well as concerns with industrial growth and pollution (see Otway and v. Winterfeldt, 1982). However, in today's debates, one theme seems to dominate--risks to human health, safety, and the environment. Even in controversies with strong political and ideological undertones, such as the debates about the SST or nuclear power, risk is often a major theme, creating headlines in the newspapers and keeping regulatory agencies busy.

Why has risk become such a dominant factor in technology debates? Four different explanations have been offered. First, as the scale of technologies has grown, some risks may have indeed increased. Some new technologies can produce, or have produced, unprecedented catastrophes and others have produced subtle but long-term health effects.

Second, the modern awareness of risks may simply be a consequence of changed perceptions of risks. Scientific studies have identified technological risks earlier, and media exposure has made them more visible, creating dramatic images of disaster and providing nationwide coverage of scientific estimates of morbidity and mortality rates. Exposure to these publications may have exaggerated public perception of the actual risks. Whether the changed perception is appropriate or not is debatable, but surely depends on the hazard.

The third explanation is that social values and preferences have changed to emphasize risks, not benefits. Perhaps the benefits of technologies are questioned because of satiation effects, diffusion of benefits, and changing of public trade-offs. A radical version of this hypothesis can be found in the "value change" thesis by Inglehart (1977), who claims that modern industrial societies are moving toward adopting "post-materialistic" values.

According to the fourth explanation, institutional changes have caused increased attention to risks. The institutions that plan, promote, regulate, and oppose technologies may be better suited to address risk issues than other concerns, thus shifting the focus of the debate from whatever the "real" issues might be.

These four explanations are neither exclusive nor exhaustive. Beyond the institutional interpretation, one can conceive of cultural explanations for the emergence of the risk debate, as Douglas and Wildavsky (1981) have recently done in a most stimulating book. And most risk researchers would acknowledge that risk debates are probably created by a mixture of reasons, including changes in the production, perception, evaluation, and management of risk, as well as in the cultural context in which the debate takes place.

The four explanations are nevertheless competing, as they generate different interpretations of current risk debates and different expectations about how these debates evolve. Only the "production" and "perception" explanations acknowledge risk as the "real" issue of the debate. Concern about risk leads to the creation of risk-regulating institutions as arenas for technology debates. The rules of these arenas focus attention on risk because they are designed to deal with risk and not with other issues. According to the production and perception explanations, similar risk or similarly perceived risks should create similar controversies, while the value change hypothesis would suggest that technological controversies occur to the extent to which they are appropriate targets and symbols for the expression of changing values and attitudes. The institutional explanation implies that the nature of a risk debate can take radically different turns, depending on which institutions become involved.

In order to assess the four explanations and to better understand the reasons for technology debates, we have studied 10 past, recent, and ongoing technological controversies. By comparing the historic opposition to steamboats, cars, and zeppelins with the recent opposition to the SST, LNG terminal siting, and bullet trains, we attempted to identify historic changes in technology debates. By comparing the controversy about nuclear power with that about genetic engineering we tried to assess the factors that determine the nature of the debate about technically similar risks.

TECHNOLOGICAL CONTROVERSIES IN THE PAST[2]

Early Opposition to the Steamboat

It is difficult today to realize the importance of the steamboat, as it has largely disappeared from our lives. But for a long time steam was the only mechanical power that was capable of moving vessels. The steamboat was, of course, ultimately successful, but its river to success was full of snags and sandbars.

---

2 We gratefully acknowledge Kris Karter's, Rosa Vasquez's, and Mary Stepp's help in collecting the material for this section and drafting some of it.

As usual, the early opponents of steamboats were those whose livelihood was threatened: sailboat owners, sailors, and stagecoach drivers. In France, Denis Papin designed the first steamboat in which an engine pumped water onto a waterwheel that drove two paddlewheels on the same shaft (Dayton, 1925). Other boatmen thought that the boat worked by witchcraft and destroyed it. Jonathan Hulls patented a steamboat in England in 1736 and was severely ridiculed by fellow boatmen.

The first American steamboat designer, Robert Fulton, also experienced opposition from the sailboat business, and reportedly had to guard his boat against sabotage by sailors. Financiers' and friends' attitudes ranged from skepticism to indifference (Hill, 1953). At the launching of his first boat, the Clermont, most people thought it would be a failure; in fact, Fulton had to promise embarrassed friends and investors that their names would not receive publicity in the venture. Passengers were frightened by the unfamiliar noises and the steam.

In spite of all this apprehension, the Clermont was a success. Steamboats soon overcame the opposition by sailors, and their economic prospects convinced investors of their worth. But, simultaneously with the gradual acceptance of the technical and economic feasibility of the steamboat, public fears of its dangers arose. Nicholas Roosevelt constructed the New Orleans, which was to run from Pittsburgh to New Orleans. Despite advertisements, no passengers appeared for its first trip, but Roosevelt and his pregnant wife made the trip together with a crew of 14. The public was outraged that Roosevelt would subject his wife to such danger. Women, fearful for Mrs. Rossevelt's safety, wept and condemned her "headstrong" husband (Drago, 1967).

The New Orleans made its trip safely, but steamboat accidents soon became common. An official government estimate of 1848 listed a total of 233 steamboat explosions between 1816 and 1848, producing 2,563 fatalities and over 2,000 injuries (Burke, 1965).

The technological responses to these disasters were slow and largely inadequate. They included wrought iron construction, lower steam pressure, and safety valves. A curious technological response to the growing public concerns about the safety of steamboats was safety barges, which were towed behind steamboats and advertised widely:

> Passengers on board the safety barges will not be in the least exposed to any accident by reason of the fire or steam on board the

> steamboats. The noise of the machinery, the trembling of the boat, the heat from the furnace, boilers and kitchen, and everything which may be considered upleasant on board a steamboat are entirely eliminated. (Dayton, 1925, p. 38)

The novelty gradually wore off, and the barges were eventually used for cargo only. The need for the use of such devices demonstrates, however, the fears which steamboats had produced.

The frequent disasters, the media attention they received, and the obvious public concerns produced a climate in the U.S. Congress, that favored regulation of steamboat construction and operation. But Congress acted slowly, taking over 30 years to decide on regulations, partly because there were no precedents for regulating a privately developed technology, partly because opposition from the private sector was strong (Burke, 1965). An initial bill that provided for hydraulic testing of all boilers, inspections, and liabilities in case of explosions was finally passed in 1838. This bill was the first regulation of a technological risk in the United States. Because of its lax formulations, it did not, however, change matters much. Steamboat accidents continued and so did the public concern and media outcry against disasters. The debate continued, now focused on the ideological differences concerning the regulation of steamboats. A proponent of a tougher bill argued in Congress:

> I consider that the only question involved in this bill is this: whether we shall permit a legalized, unquestioned, and peculiar class in the community to go on committing murder at will, or whether we shall make such enactments as will compel them to pay some attention to the value of life. (cited after Burke, p. 21)

An opponent replied:

> What will be left of human liberty if we progress on this course much further? What will be, by and by, the difference between citizens of this far-framed Republic and the serfs of Russia? Can a man's property be said to be his own, when you take out his control and put it into the hands of another, though he may be a Federal officer?" (cited after Burke, p. 21)

After much debate in Congress, the proponents of stricter regulation won in 1858. A law was passed that specified tough engineering safety criteria, giving inspectors authority to examine boats and refuse licenses, and creating a regulatory agency, the Board of Directors of Inspectors, supervised by Presidential appointees.

Subsequently, steamboat explosions did, indeed, become less frequent events. Public concerns about steamboats decreased, and steamboats remained the main mode of transportation until the end of the century, when they were replaced by railroads.

Early Opposition to the Automobile

In general, the automobile and the mobility it brought elicited much enthusiasm from most public groups. It was thought to promote health by doing away with unsanitary and noisy horses, end isolation, aid education, and bring families together (Flink, 1970). The scattered opposition occurred for four reasons: threats to the livelihood of special interest groups (farmers, horse breeders, horse carriage associations); safety and risks (publicists and public officials); economic reasons (bankers); moral and cultural reasons (religious groups).

At the start of the automobile age, those who stood to lose their jobs opposed cars most strongly. Horse breeders, livery stable owners, and horse-drawn drivers associations demonstrated opposition and frequently said automobiles were only a passing fad for the wealthy.

Touring was very popular in the early 20th century, but generated substantial rural opposition by damaging crops and endangering livestock. In Rochester, Minnesota, farmers plowed roads, making it impossible for motor vehicles to travel but allowing horses to trot through. But this opposition subsided by 1908 when it became apparent that automobiles would help farmers. With the advent of the tractor and other farm vehicles, the last organized opposition to the motorcar disappeared.

Automotive safety was already recognized as a major problem by the mid-1920s: 5,900 fatalities in 1915, 9,825 in 1919, 10,100 in 1920, 15,000 in 1921 (The Nation, 1922). According to a 1926 article in the New Republic, titled "The Murderous Motor," 22,500 deaths were caused by automobile accidents in 1925. The Nation (1922, p. 79) asked, "How shall we control the modern Juggernaut?" It went on to lament the many accidents caused by motorcars and noted that "in Elmira one woman was killed by a car driven by a man ninety years of age!" But these complaints were more against the drivers than against the technology of the motorcars themselves.

According to Flink (1975) concerns that the car would encourage financial extravagance and economic frivolity started in the 1920s when time payments became common. The Chicago Central Automobile Credit Association wrote on March 20, 1922:

> Widespread conviction on the part of the bankers that the financing of automobile purchases was a more or less hazardous business...a tendency to encourage extravagance was contrary to the thrift ideas which the bankers were so anxious to establish in the minds of the public. (Flink, 1975, p. 148)

Moral attitudes toward the automobile were positive in the beginning, but later turned into suspicions that the vehicle, by permitting individual mobility, threatened the family unit and the moral standards of society. It undercut parental authority and promoted promiscuity. People often found the Sunday drive preferable to going to church (Flink, 1975). In addition, the automobile threatened to tear apart traditional family and friendship units.

> The trend towards decentralization of workers' dwellings means that instead of a family's activities in getting a living, making a home, play, church-going, and so on largely overlapping and bolstering each other, one's neighbors may work at shops at the other end of the city, while those with whom one works may have their homes and other interests anywhere from one to two miles distant. (Flink, 1975, p. 161)

The opposition to the car was neither vehement nor massive, and with the exception of the early opposition by horse-related businesses it was hardly organized. Most opponents knew that the automobile was highly popular, that it was here to stay, and that there was little one could do to stop its expansion.

Opposition to the Zeppelin

The zeppelin had an illustrious and short-lived history. Most of its promoters had military and economic interests. Opponents came from competing industries (mainly the emerging airplane industry), military strategists who did not believe in the zeppelin's ability to attack and defend itself, and journalists and members of the public who stressed its risks.

In 1900 Count von Zeppelin carried out the first successful zeppelin flight. The German government soon recognized the military potential of it, and promoted a zeppelin program prior to World War I. The program almost came to a halt when a zeppelin was completely destroyed during its first flight, killing its crew. Interestingly, the public response in Germany was in favor of continuing the program in spite of this disaster. Count von Zeppelin received large private contributions to continue his work, and the government followed suit (Eckener, 1958).

But the risks and dangers of the zeppelin would soon attract public attention and generate criticism and opposition. In 1922, the French airship Dixmude burned in a thunderstorm off the coast of Sicily, killing its crew. This accident received wide media coverage, and prompted one French journalist to ask: "When will they stop people from flying these criminally dangerous ships?" (Hearnes, 1916) In 1934 the American airship Akron collapsed and sank off the shore of New Jersey, killing 73 people (Newsweek, 1934). The English zeppelin R.101 crashed into a French hillside in 1930, killing 48 of the 54 people aboard. Finally, the tragedy of the Hindenburg in 1937 in New Jersey killed 50 people and highlighted the problems of airships that used hydrogen instead of helium. According to Abbott, this series of accidents was the main reason for the abandonment of zeppelin development: "A tragic series of avoidable accidents caused the entire concept of passenger carrying airships to be abandoned with impatient haste." (Abbott, 1973, p. 152)

There are, of course, other important reasons for the demise of the zeppelin, but they varied from country to country. In Germany, a main reason was the lack of helium, which would have made travel by zeppelin cheaper and safer. The United States had a virtual monopoly on helium, but since World War I helium export was forbidden by law. German lobbyists worked until 1937, when that law was repealed. After a very brief period of helium export, tension grew in Europe, and soon after Hitler's annexation of Austria, helium export to Germany was again halted.

In the United States, the development of the zeppelin was boosted by the economic interests of rubber manufacturers, and supported initially by the Navy and the Federal Aviation Commission. In 1940 the United States had a substantial network of zeppelin transport, but this expansion was shortlived. The risks, combined with improvements in airplanes and the lesser cost effectiveness of the zeppelin, brought about the decline of the zeppelin in the United States (Hearnes, 1916; Eckener, 1958).

The Supersonic Transport (SST)

The SST had a rocky but quiet beginning under the Kennedy administration. In May 1960, the head of the Federal Aviation Agency announced that the government would be willing to pay for half of the costs of developing the SST, but business response was lukewarm at best (Levy, 1973). Nevertheless for nine years Congress voted funds for SST research and development. It was in the late sixties that widespread opposition began. The opposition was based on two concerns: environmental and economic. In 1967, William Schurcliff founded the "Citizens League Against the Sonic Boom," drawing attention to the noise problem of the SST. On April 1, 1970, the "Coalition Against the SST" was formed, backed by 16 national and 15 state and local groups. Its only goal was to stop funding of the SST. The newly founded environmentalists' lobby group, "Friends of the Earth" (FOE), also chose stopping the SST as one of its goals (Levy, 1973). As a major step against the SST, FOE published William Schurcliff's "The SST and Sonic Boom Handbook," which described the possible dangers of the SST to the environment.

The growing concerns about environmental damage were summarized in a recent Los Angeles Times:

> A "greenhouse effect" would raise temperatures around the globe, they said. Radio communications would be disrupted. More people would get skin cancer as jet exhausts depleted the stratosphere's ozone layer and allowed more ultraviolet rays from the sun to penetrate the atmosphere. More species would become extinct as, for example, sonic booms cracked the delicate eggs of the whooping cranes. (Los Angeles Times, 1981)

There was concern that "all animals living out-of-doors during the day would be blinded (by ultraviolet radiation) unless they wore protective goggles." (Chemistry, 1973) And the sonic boom would "break windows, crack walls, and stampede cattle." (U.S. News and World Report, 1970) These arguments were frequently supported by scientific studies commissioned by the environmentalist groups and by regulatory agencies.

The lobbying activities of environmental groups against the SST produced political action. On May 12, 1970, Senator William Proxmire, a long-standing critic of the SST, held hearings. Russell Train, chairman of the Council of Economic Affairs, testified that a fleet of SSTs could warm the atmosphere enough to affect the polar ice caps. Richard Wegman, an assistant for

Proxmire, said "It was the first time any responsible official had suggested that such a far-out scare could be true." (Levy, 1973, p. 34)

The defenders of the SST argued that it was necessary for the United States to develop the plane if it were to remain the leading nation in aircraft technology. The Concorde was being developed in France and the Russians had their supersonic TU-144. Cancellation of the project could result in high unemployment as technicians and workers were laid off. Proponents also argued that a favorable net balance of payments would be reached through sales of the SST to foreign airlines. Finally, the Department of Transportation said that the SST would not cause sonic booms because it would be forbidden to fly faster than sound when it was flying over the nation (U.S. News and World Report, 1970).

However, the anti-SST coalition came up with some interesting responses to these arguments. In 1970 they polled 16 of the nation's top economists, ranging from the very liberal to the very conservative. The economists were asked if the SST would make a profit, and if the government should continue funding the development of the plane. Fifteen economists opposed federal funding of the SST, and did not think it would make a profit. Only one economist supported it, arguing that U.S. technology must keep up with Russia. Milton Friedman expressed the views of many in his reply: "If the SST is worth building, the market will make it in Boeing's interest to build without a subsidy; if a subsidy is needed, the SST should not be built." (Levy, 1973)

Media opposition to the plane was growing as well. Both the New York Times and the Washington Post wrote strong editorials against the SST. One popular argument was that the money for the SST was needed elsewhere. The SST had required $864 million over a nine-year period. In 1970, President Nixon, a supporter of the SST, asked for $290 million for two prototypes of the SST (U.S. News and World Report, 1970). Life Magazine replied: "The Nixon administration's openhandedness with the SST stands in stark contrast to its penny-pinching approach to the neglected parts of the country's transportation system." (Life, March, 1971) And Representative Sidney Yates of Illinois said: "The SST, which looked like a supersonic marvel on paper, looks more and more like a supersonic white elephant as the time approaches to begin prototype construction." (U.S. News and World Report, 1970)

By the end of 1970 the assault against the SST had succeeded. The opposition from economists, environmentalists, politicians, and publicists, the numerous bills against the SST pending in state and local legislatures, and the loss of political support in the Senate and House made the final decision to terminate the project merely a matter of time. On December 3, 1970, the Senate voted to stop government funding of the SST. After a brief compromise, and a last ditch effort by Boeing to revive the project (which included the creation of "American Industry and Labor for the SST"), both the House and Senate voted to cut off future funds. Soon thereafter Boeing announced its cancellation of the SST project.

Opposition to Liquefied Natural Gas (LNG) Terminals[3]

The origin of the following controversy was a proposal by the Pacific Lighting Corporation of California to purchase LNG from Indonesia, and to land and distribute it in California, whose demand for natural gas was predicted (by Pacific Lighting) to outgrow its supply. Western LNG Terminal Corporation, a subsidiary of Pacific Lighting Corporation and of El Paso Natural Gas Company, suggested two possible terminal sites in 1973: Los Angeles and Oxnard. The Oxnard terminal was to be located at Point Conception (Cojo Bay).

A 10-year period of political and legal battles and regulatory investigations over the terminals followed, concentrating on the Point Conception site (Linnerooth, 1980). Advocates of the plan stressed the need for fuel and the benefits of the project in producing economic growth and employment. Opponents were environmentalists, land owners, local residents who feared the risks of explosions and fires, and Indians, who considered the Cojo Bay area sacred.

The debate took place in the media and in licensing and regulatory proceedings. From the beginning, LNG safety was a major issue. The Sierra Club and a group called "Ventura County Concerned Citizens Committee" argued that the entire system of storing and transporting LNG was unsafe (Sierra, 1978). A 1944 LNG disaster in Cleveland, Ohio, which killed 128 people and injured over 300, was widely cited. The environmental and safety concerns were addressed in a series of studies issued by both opponents and proponents of the plan. The studies agreed that the probabilities of disasters were extremely small, but the potential consequences could be

3 This section draws heavily on summaries by Linnerooth (1980) and information provided by L. Philipson.

extremely severe. But the stakeholders drew conflicting conclusions: Opponents stressed the maximum consequences, while proponents pointed to the small probabilities.

The situation was complicated by ill-defined and often diffuse regulatory responsibilities and a lack of clear safety goals. Part of the process therefore consisted of creating a regulatory responsibility and safety guidelines. The 1977 California LNG Terminal Siting Act removed licensing and permit authority from a number of local agencies and gave the California Public Utility Commission (PUC) the main responsibility for regulation (Linnerooth, 1980), with the California Coastal Commission (CCC) in charge of ranking potential sites. Soon after, the Federal Energy Regulatory Commission defined new safety criteria and regulations for LNG storage and transportation. Based on these regulations the PUC chose the Point Conception site from four sites submitted by CCC.

But the safety debate did not stop. During the decision process by the PUC, an earthquake fault, discovered near the proposed site, jeopardized what appeared to be the solution to the siting problem. New studies were needed to examine the safety problem in light of the earthquake fault. In late 1981, the PUC finally accepted the wind/wave seismic studies for Point Conception and concluded that with appropriate engineering fixes the risk to the population was acceptable.

While safety was a major issue in the debate about the LNG site, it was by no means the only one. The Sierra Club and other public interest groups had expressed both economic and safety concerns. They questioned the need for LNG, pointing out that it created another source of energy dependence and that it would be too expensive for the consumer. They also argued that if an LNG terminal turned out to be unavoidable, it should be sited in a remote area. Thus need, economics, and safety were intimately coupled in the environmentalists' arguments.

The opposition by local landowners was based largely on their economic interests in the land on which the LNG terminal was to be built. The Bixby and Hollister Ranch Associations lobbied strongly against the site because the terminal would reduce land value and interfere with development plans (Linnerooth, 1980).

Strong objections also came from American Indians, to whom Point Conception is a sacred place. In May, 1978, 25 Indians invaded Point Conception in a protest against the development (Nation, 1979). One Indian woman said, "That is our sacred ground and if they put that liquefied gas stuff there, something bad is going to happen, you can feel it." (Nation, p. 138, 1979)

Partly because of such strong opposition, partly because of the slackening demand for natural gas, and partly because of the general uncertainty surrounding technological regulation in this area, Pacific Lighting decided in October, 1982, to postpone the development of the LNG plant at Point Conception indefinitely.

Opposition to the Bullet Trains

A number of recent proposals have been made in the United States to develop a high speed train similar to those built by the Japanese and the French to connect large metropolitan areas. Some of the proposals have reached the specific planning stage: The proposed train to link Los Angeles and San Diego is an example. It is therefore of considerable interest to examine the recent opposition to that train in Japan. Opposition in the United States is likely to take similar forms, since the issues are the same, and since the Japanese legal and regulatory mechanisms for coping with risks and environmental hazards are similar to those in the United States. Following is a brief account of the battle Japanese residents fought against the Shinkansen high speed trains. (See also the more extensive treatment in Ikeda and v. Winterfeldt, 1982; Hasimoto, 1978; and Straszak and Tuch, 1981).

In the early days of its development, the Shinkansen was a source of national pride to most Japanese, hailed as a technological breakthrough and as a symbol of Japanese ability to innovate rather than imitate. The Shinkansen built up an immaculate safety and economic record. With speeds of over 200 km/h it was by far the fastest commercial train in the world. And, unlike other railway systems in Japan, it actually produced profits. Thus, when the Tokaido line was opened in 1964 connecting Osaka and Tokyo, the few residents who complained about the noise, vibration, and TV interference were hardly heard.

There is no doubt that the Shinkansens are noisy. At present, trains run at intervals of five minutes along the Osaka-Tokyo line from 6:00 am to 11:35 pm. Noise levels of 85-90 dB are not uncommon and noise can reach over 100 dB on steel girder bridges. Residents at distances of up to 200 m from the tracks can be substantially disturbed; they frequently complain or even stage protests.

In 1975, the Japanese Environment Agency set noise standards well above the residents' requirements (70-75 dB). The noise standards were set on the basis of noise-complaint relationships and consistency requirements with standards for car and airplane noise. Their adequacy is at present a matter of dispute in court, but it seems likely that they will be upheld.

Meanwhile, the battlefronts have widened. Extensions of the Shinkansen system are under heavy attack from environmentalists and local residents. Radical activists have added economic and political objections to the original complaints about noise. The Shinkansen, they say, serves the rich and provides revenue only to the few cities that have stations; the residents along the line and the cities without stations do not reap any benefits, but carry the burden of noise and displacement. With the Narita airport controversy and a recent defeat in a legal case about Osaka airport noise still fresh in their minds, government officials and representatives of JNR are cautious about further expansion plans. A sweeping development plan that was to double the Shinkansen system has already been abandoned and even a small new stretch connecting Tokyo and the new Narita airport is in jeopardy. It appears that, at least for the moment, the Japanese bullet train has lost some of its glamour and prestige.

Comparison of Present with Past Controversies

Having set out to find dramatic differences between the technology debates of the past and the present, we were struck by the surprising similarities in the topics of the debate, in the nature of the arguments, even in some of the rhetoric. The topic of risk, especially to human health and safety, was an important factor in all three historic cases. The worries about steamboat explosions and the fear of zeppelin disasters are reminiscent of today's concerns with LNG and nuclear power safety. And the public outcry against too many automobile-produced deaths that began in the early 20th century has never really stopped.

Just as risk was an important issue in the past, non-risk arguments strongly shaped the more recent debates. Economics ultimately brought down the SST. Political and religious concerns were a large part of the LNG siting controversy. Issues of growth and inequities of risks and benefits became focal concerns in the later stages of the expanding bullet trains conflict. In addition, from the days of the steamboat controversy to today, the arrival of each new technology has stimulated discussions about the necessity and appropriateness of additional governmental regulation.

Changes in topics occurred, of course, reflecting changes in public perceptions and attitudes. The environment was hardly a concern of the past, and little evidence exists of worries about endangered species or about subtle disturbances of the ecological balance, themes that have become so familiar by now. Environmental concerns were usually coupled with people's well being and were not considered as values in themselves. Another novel topic of recent years is concern about long-term impacts on future generations. This is probably motivated by the scale and nature of some new risks, e.g., the risks of genetic research, of nuclear waste disposal, and of fossil fuel burning ("greenhouse effect").

We think that the similarities in topics and arguments far outweigh the differences, and it would be rather difficult to make a case that today's debates are fundamentally different from those of the past. Yet there is little doubt that technologies have fallen on harder times. Our case studies suggest four related reasons:

- Increased scale: Technologies have become larger, more expensive; their complex and time-consuming planning processes offer ample opportunities for public intervention.

- New institutions: Two new types of organizations have emerged that play important parts in technology debates--special-purpose regulatory agencies and powerful, single-issue organizations that can selectively fight technologies.

- Changing arenas: The battles over technologies are fought increasingly in regulatory arenas and in the courts, rather than in political bodies.

- Adversary procedures and expert disagreement: The new arenas have turned technology debates into adversary procedures in which experts take the sides of the technology proponents or opponents.

The scale of past technologies was smaller, they could be developed with private funds, and their success depended crucially on individual entrepreneurship. The early histories of the steamboat, the car, and the zeppelin are examples both of the advantages and the difficulties of small-scale private efforts. Recent controversies that we described were, in contrast, about much larger-scale technologies, involving complex corporate arrangements, and frequently requiring substantial government subsidies. The 10-year planning phases of the SST and of the LNG terminal clearly demonstrate the problems than can result from this process. The necessity of government support and incentives in turn leads to a larger public stake in the technologies and a request for public inputs into planning and decisionmaking. Planning has become a long, arduous process. Many public hearings and discussions make the technology visible and vulnerable in its early stages of development. Debate about the steamboat was mostly after-the-fact; debate about the SST was not.

The case history of the steamboat dramatizes the second factor contributing to the vulnerability of new technologies: the emergence of laws and agencies that monitor, inspect, and license them. If risk regulation was a hotly-debated novelty then, as discomforting to the technology sponsors as it was beneficial in reducing the physical risks, it is commonplace today. Depending on their legislative mandate, regulatory agencies emerged that ranged from critical of new technologies (e.g., Occupational Safety and Health Administration) to relatively supportive (e.g., Nuclear Regulatory Commission). Whether supportively or antagonistically administered, however, the new rules and regulations provided more hurdles and obstacles for emerging technologies. Another institutional novelty is public special-interest groups, usually focusing on a single issue like environmental or individual freedom. Because these issues cut across technologies, representatives of these organizations can pick a fight in regulatory or legal arenas in an attempt to use their resources to their best political advantage.

Regulatory processes provide both the opportunities and the arenas for such fights. Many of the newer regulatory processes and bodies emerged from, and are responsive to, environmental and public interest groups. The direction of political pressure within such groups and processes has significantly changed. Procedural phenomena reflect such changes. Regulatory processes are usually open to intervenors and to opponents of the technology in question--by design. Senator C.P. Anderson argued during a Senate discussion of the Atomic Energy Act:

> I feel so strongly that nuclear energy is possibly the most important thing we are dealing with in our industrial life today, I wish to be sure that the Commission has to do business out of doors, so to speak, where everyone can see them. (Del Sesto, p. 118)

The National Environmental Protection Act and similar acts provided for public hearings concerning licensing and impact assessment. Public financing and strong government involvement in developing large-scale technologies made both regulation and its openness legitimate and necessary.

The legal or quasi-legal nature of planning, licensing, and regulation has turned the stakeholders into adversaries and amplified their differences and disagreements. Experts have begun to play a partisan role in these adversary processes. In the past, experts normally worked for, and sided with, the technology sponsor, but today this picture has changed. Regulators must have the expertise to control and supervise technological developments. Single issue opponents of technologies frequently employ a full-time staff of experts. Experts therefore have a career path that is not directly linked to the success of the technology itself. The fact that experts take sides would, of course, be of little consequence if sufficient data and experience could be accumulated to force expert opinions to converge. Unfortunately, the novelty and scale of many technologies have reduced the role of trial and error and frequently made it impossible to replace expertise with "hard facts." As a consequence many answers to technological questions have become educated opinions rather than scientific truths (Otway and v. Winterfeldt, 1982). Using the same data and different assumptions, experts can arrive at wildly different estimates. The resulting display of expert disagreement has contributed to a growing mistrust of expertise and of the institutions promoting and regulating technologies. As Renn (1982) points out, an expert is more likely to be asked "who paid for your study?" than "what is the scientific basis for your conclusions?"

NUCLEAR ENERGY AND GENE SPLICING: DO SIMILAR RISKS PRODUCE SIMILAR CONTROVERSIES?[4]

Several observers of the nuclear power controversy have pointed out that DNA research and gene splicing technologies may encounter public opposition similar to that of nuclear power, because the risks are similar (Starr and Whipple, 1981; Slovic

4 We gratefully acknowledge Rosa Vasquez's help in collecting the material for this section.

et al., 1982). Gene splicing and nuclear power both may have delayed effects, are relatively unknown, have new risks, have large catastrophe potential, and are severe and dreaded (Slovic et al., 1982). If the debates were purely about risks, therefore, one would expect substantial similarities in their evolution and disposition.

To examine this hypothesis, we compared the debates about nuclear power and gene splicing at comparable stages of development (mid-1950s to mid-1960s for nuclear power, 1970s for gene splicing). Fortunately, the development of nuclear power has been well documented both by proponents and opponents (Del Sesto, 1979; Gyorgy, 1979; Duncan, 1978; Lombardo, 1979). Wade (1977) thoroughly covers the DNA debate up to 1977. We added to this study our own review of all Science articles on the subject between 1977 and 1982. We created a timeline of major events, listing the main stakeholders (proponents, opponents, and regulators) and analyzing their conflicts.

Early Opposition to Nuclear Power

Most reviews credit the beginning of opposition to nuclear power to the environmentalist movements of the late 1960s and the early 1970s, but nuclear power encountered opposition virtually from its very conception as a realistic idea. According to Del Sesto (1979), the AFL-CIO raised concerns about radiation and occupational safety as early as 1949. Only a few years after Eisenhower's famous "Atoms for Peace" speech and the enactment of the Atomic Energy Act, the AFL-CIO fought a legal battle against the AEC over the safety of the Enrico Fermi experimental fast breeder reactor near Detroit. Unions also brought to public attention the early nuclear reactor accidents at the Nuclear Reactor Testing Station in Idaho in 1961 that killed three workers (Del Sesto, 1979). Throughout the 1950s, unions raised accusations that the AEC had suppressed studies and whitewashed reports about incidents (see also Mazur, 1981).

In 1957, the AEC carried out its first reactor safety study (WASH-740), an assessment of the consequences of a major accident. While the report stressed the theoretical nature of the accident, the size of the consequences (thousands of fatalities, up to seven billion dollars in property damage) was staggering. Worse, the liability issues raised by this study became a major obstacle to the economic viability of nuclear power. The Price Anderson Act of 1957 solved that problem by imposing a ceiling on both private and governmental liabilities in the event of a serious accident.

The liability issue was, of course, only one of the many economic problems that early nuclear power development had to overcome. Many utilities were reluctant to replace their cheap and reliable coal- and oil-fired plants with an economically uncertain and still not fully-proven technology. The AEC's nuclear power development program had to provide substantial financial incentives to construction firms and utilities.

As the interest in nuclear power increased in the private sector, public concern began to rise. Various citizens' groups began to fight legal battles against specific plants, usually through the public licensing process. Opposition at that time was, however, still focused on residential and local interests, and did not have a broad political base.

Complaints by unions and local residents broadened through the 1960s to include national and international issues such as plutonium diversion and low-dose radiation. In the late 1960s, the Gofman and Tamplin affair stirred up additional controversy. In 1969 Sternglass published a paper "Infant mortality and nuclear tests" in the Bulletin of the Atomic Scientists, in which he estimated that about 400,000 excess cancer fatalities could be expected from low-dose radiation produced by nuclear fallout. The AEC, attempting to prove that cancer mortality caused by excess radiation from nuclear weapons and nuclear power plants is nonexistent or at least negligibly low, issued studies by Lawrence Livermore scientists Gofman and Tamplin to refute the Sternglass estimates. But, in the process of refuting Sternglass's controversial and staggering numbers, Gofman and Tamplin produced an uncomfortable set of estimates themselves. According to one of their studies, additional cancer fatalities from exposure to radiation could be over 12,000. The AEC and many other scientists quickly contradicted Gofman and Tamplin. In one such study, Bond (1970) estimated nearly .02 (see e.g., Mazur, 1981) additional cancers per year produced by low dose radiation. The public was suddenly confronted with extremely divergent expert assessments of the risks, which, coupled with the often vicious tone of the debate, helped undermine public trust in experts and regulators. Meanwhile, many successful suits by local residents and environmental groups produced costly delays and disruptions in the construction of new nuclear plants.

Thus, even before 1970, nuclear power faced some stiff opposition, starting with union concern about occupational safety. The 1970s, of course, saw an enormous escalation of the conflict, produced by a coalition of environmentalist groups, residents, unions, and political groups. This broad alliance, together with escalating costs, an uncertain regulatory climate, and reduced energy consumption, are the main reasons for the ongoing decline of nuclear power in the United States.

Early Opposition to Gene Splicing Experiments

Criticism of gene splicing experiments began as soon as the technology was invented in 1973. Early warnings came from the scientific community itself. The concerns focused on inadvertent increases of virulence and the potential for epidemics. In June, 1973, a group of scientists gathered at the Gordon conference, and proposed an examination of the safety issues. The group published a letter in Science, suggesting that the National Academy of Sciences (NAS) set up a committee to investigate genetic engineering risks. In April, 1974, the NAS convened an initial meeting that had two outcomes--the first was a proposal for an international conference on the subject; the second was a request for a voluntary partial moratorium on gene-splicing experiments. The proposal was apparently widely acted on not only in the United States but also in foreign countries as scientists all over the world became aware of the potential risks of gene-splicing experiments.

The next important event was an international conference at Asilomar, California, at which risks of genetic engineering were discussed and guidelines for containment procedures were drawn up. The conference apparently favored strict initial regulations, which could later be relaxed. The most critical participants wanted to ban all experiments with E. coli, seen as a particular danger because it colonizes the human colon. Use of E. coli was not banned, but proposals to use "crippled" E. coli strains for experimental purposes (Wade, 1977) were made and approved.

The National Institute of Health (NIH) set up a committee to produce guidelines for gene-splicing experiments based on the Asilomar ideas. The final NIH guidelines required two grades of physical and biological containment, depending on the danger of the experiment. These guidelines were somewhat less stringent than those proposed at Asilomar.

In mid-1976 scientists mounted a new attack against genetic engineering and the NIH guidelines. Two prominent researchers, Chargoff and Sinsheimer, added to the initial list of biological risks the risks of "meddling with evolution" and the deeper ethical problems of genetic engineering. The issue of self-serving regulations was also raised frequently, since most members of the NIH committee had a stake in genetic engineering research (Wade, 1977).

Public involvement was also on the rise. In June, 1976, the Cambridge City Council began investigating the risks of genetic engineering experiments that were to be carried out at Harvard. A public review board examined the issue in 72 hours of hearings, and finally came to the conclusion that the experiments could go on under guidelines somewhat stricter than those of NIH.

Throughout 1977, various environmental and religious groups took positions on genetic engineering, but most of them were surprisingly mild. The Environmental Defense Fund asked for public hearings. The Sierra Club opposed genetic engineering except for small laboratory experiments. Friends of the Earth asked for a moratorium pending risk research. Religious groups were neutral or cautiously supportive. Simultaneously, many local citizens' groups and local regulators entered the debate, since the NIH guidelines applied only to experiments supported by federal grants. Since 1977, Congress has been trying to pass mandatory gene-splicing regulations that would apply to all types of experiments. However, today no such guidelines have yet been passed; the responsibility still belongs to local governments.

In 1978 and 1979 a number of events occurred. An NIH study indicated that the risk of genetic engineering had been exaggerated in the early days. The founding of Genentech and other firms made gene-splicing a business. Environmentalists were still split or uncertain about the issues.

Meanwhile, NIH acted as a strict regulator in two separate incidents. The first occurred after a researcher at the University of California-Los Angeles violated the NIH rules by inserting genetic material into humans. The researcher was asked to resign a number of positions, and was stripped of several federal grants in early 1980. In the second incident a researcher who miscloned a virus at the University of California-San Diego was eventually forced to resign his position.

There still has been no Congressional action on gene-splicing experiments. A September, 1981, Science article noted that NIH may soon relax its guidelines, since it found only four minor violations in five years, and studies indicated that the containment procedures are effective. The media are still quite attentive, but frequently focus on the dramatic benefits of what are now called "designer genes."

Comparison of Gene-Splicing and Nuclear Power

The early opposition to gene-splicing and to nuclear power were similar in content, but different in the processes of their evolution and disposition. Some of the similarities in content are a result of the similarities in the risks. Catastrophe potential, impacts on future generations, and the mysteriousness and unpredictability of the hazards were issues in both cases. The scale and novelty of the risks led to discussion of deeper ethical and moral issues. Scientists were among the most outspoken critics in the early stages, and Sternglass played a role that was not too dissimilar to that of Chargoff and Sinsheimer. Even the rhetoric sounded the same.

The evolution of public concerns about risk was also similar. The nuclear controversy began with union concerns about occupational health and safety. In the DNA case scientists themselves voiced concern. Issues were low dose effect (risks of inadvertent infections--cancer from low dose radiation), disaster (epidemics--meltdowns and massive radiation), war and war-related uses (biological warfare--proliferation of plutonium), and ethical concerns (meddling with evolution--genetic impacts on future generations).

But some pronounced differences also exist in the topics and issues discussed in the two cases. The genetic engineering issues were usually confined to the risks, with an occasional sprinkling of moral and ethical objections, and a recent concern with the commercialization of genetic research. As early as the 1960s, the nuclear debate included economic concerns about the sustainability of nuclear power, decentralization, growth, and conservation. Furthermore, the benefit side was viewed quite differently in the two cases. The benefits of genetic research (for example, the production of insulin and interferon) appear to be tangible and novel contributions to the individual's health and well being. The link between nuclear energy and electricity benefits was, in contrast, much more indirect. Even with the threat of oil embargos and blackouts the benefit of a steady provision of electricity appears merely as a continuation of a status quo that many take for granted.

More important than these differences in the content of the debate are several fundamental differences in institutions and regulatory processes. One difference was the close alliance between nuclear researchers, regulators, and utility companies. The public suspicions of collusion that grew out of this close alliance were amplified by the dual role of the AEC as promoter and regulator of nuclear energy. One commentator on the nuclear debate created an argument against nuclear energy mainly on the basis of this entangled institutional arrangement (Metzger, 1971). And it did not help matters much that most researchers associated with the promotion of peaceful nuclear energy had previously been involved in nuclear weapons development.

The gene-splicing case is quite different. Proponents of genetic research were molecular biologists, chemists, epidemiologists, and physicians who were interested in exploring the potential of DNA research for biological and medical purposes. The commercial interests came at a later stage when it became technically feasible to use gene-splicing for cheap production of drugs and vaccines. The media and some opponents therefore did not identify proponents with big business and big government, but rather with disciplines like medicine and biology.

In addition, the regulator was not an obvious proponent of gene splicing experiments. While some charged the NIH committee with "self-servingness," the conflict of interest was not as clear as in the case of the AEC. Regulatory action supported these different images. The AEC was frequently on the defensive, being accused of "whitewashing" and "cover-ups," and having to tighten standards and regulations after they were re-examined (Mazur, 1981). The scientists and members of the NIH committee, on the other hand, were seen as relatively strict and trustworthy. Their strategy was to set stringent standards initially, which could later be relaxed if the risks were better understood and adequate containment procedures were developed. Trust in the regulation was also aided by the fact that the Cambridge citizens committee essentially endorsed it after an independent review.

We conclude that the risks were similar, that the nature of the debate about risks was similar, but that the opposition to nuclear power, largely through faulty institutional arrangements and inadequate responses to publicly expressed fears, and partly through successful expansion of the issues to environmental and political ones, developed into a much stronger opposition. The conflict about genetic engineering, on the other hand, by creating a more trustworthy regulator and by confining the issues to risk, may today be in a state of de-escalation.

PRELIMINARY CONCLUSIONS

Issues in Disputes About Risky Technologies

In the introduction, we suggested four reasons for the emergence of risk as a major theme of today's technological controversies:

- The risks have increased (risk production)
- The risks are perceived as being larger (risk perception)
- The values have changed (risk evaluation)
- Institutions have changed (risk management)

Having reviewed 10 past, recent, and ongoing technological controversies we come to the not suprising conclusion that changes have taken place in all four areas. But our case studies suggest that institutional changes may be a much more important factor than we initially anticipated.

If we had anticipated this when we started our review, we would have focused more intensively on the detailed institutional structure of the controversies we examined--in particular, on the arenas in which they were resolved, for those that were resolved. We would have done an auxiliary study of the development of regulatory agencies, not only for issues of risk, but also for other forms of government regulation of business. We would have kept an eye out for attempts on the part of participants in controversy to control the arenas in which they fought. And, in consequence, we would have needed to give a much more detailed account of the cut-and-thrust of the processes by which the controversy worked out. Perhaps it is fortunate that we did not; our goal was to write a paper, not a book. Still, what follows is severely limited by our ignorance and by those omissions.

Institutional Changes and Arenas

The major institutional change has been the development of regulatory agencies as arenas for risk controversies. This development can only partially be explained by changes in the risks themselves or in the perception or evaluation of them. Our case studies indicate that risk has always been a major public concern. But in spite of continued public concern with technological risks, Congress created only one major agency whose primary concerns are risk--the Food and Drug Administration--before World War II. The real growth of federal regulation began in the 1950s and accelerated in the

1960s because the other two arenas for risk regulation, the legislative bodies and the courts, were inadequate to the task.

Legislative bodies are the most obvious and important arenas. But on the whole they do not deal with specific risks, but rather (when functioning wisely) define rules and mechanisms by which risk-related issues will be dealt with, in the process specifying objectives that those rules and mechanisms are intended to attain. Recourse to legislatures about specific instances of risk is difficult, expensive, and inconsistent with societal views about how they should perform.

Recourse to the courts is both natural and frequent, but the courts are limited. In principle, they can only enforce laws, not make them. That same principle implies that they can only punish lawbreakers. Any such assertion is a caricature. Injunctions, consent decrees, and similar mechanisms permit courts to demand and forbid particular acts. But such devices are awkward for permanent resolution of a controversy, and have served that purpose rarely, with respect to risks.

Courts and legislatures share two other severe limitations: They are busy and inexpert. The latter problem is at least partly soluble; the former seems not to be.

Our society long ago recognized that a quasi-judicial body capable of reaching binding resolutions of complex technical disputes has many advantages. The three key ones are (1) they can invent, order, and supervise execution of positive solutions, and thus provide specificity and meaning for vaguely worded legislative intents; (2) they can function much more informally than can the courts; and (3) they can combine a quasi-judicial function with in-house expertise continuously on tap, to provide not only advice but also monitoring of problems and supervision of solutions. On the whole, the regulatory agencies can do all this with minimal damage to Constitutional or legal rights, since someone who feels wronged can, at least in principle, appeal to the courts--or even to the relevant legislature to change the law. They are, then, a creative combination of executive-branch and judicial-branch functions.

Regulatory organizations at state and local levels, some called commissions, came into existence during the early 19th century as local and state legislative solutions to public outcries over the misuse of constantly growing business power. The key stimulus to their development was the railroads. The Federal government gave them enormous land grants to encourage westward expansion. Their superiority to other forms of transportation then available enabled them, by controlling fares, freight rates, schedules, and other business practices, to exercise extraordinary power over the economic development of the Midwest and West. The steamboat boiler inspectorate, a minor organization, was created in 1838 and strengthened in 1852, but its functions were technical and it was not politically important. The critical date in the development of federal commissions as a means for controlling U.S. businesses was 1877.

> The Act to Regulate Commerce was...of broadest significance...for it departed from the idea of limited governmental authority that reached private action only to punish wrongful acts after they were committed....(It) substituted...a permanent, continuing, day-to-day surveillance of the private conduct of the railroads. Its central purpose was not to punish...but to regulate...for the future. To practice this type of preventive medicine on the railroads, the acts created a new agency of government, the Interstate Commerce Commission (ICC)....Thus was the form of the regulatory commissions established as independent, expert, nonpartisan agencies wherein the worlds of politics and industrial economics would be reconciled. (Kohlmeier, 1969, p. 12-13)

In fact, the ICC mainly controlled fares and freight rates, with some attention to safety issues.

The next major regulatory agency was the Federal Trade Commission, which came into existence in 1914 to enforce its own act and the Clayton Act, which, together with the earlier Sherman Antitrust Act, define United States antitrust policy.

The decade of the 1920s, dominated by business interests and business-oriented governments, saw only one new regulatory agency come into existence: the Food and Drug Administration (FDA), which started as the Food, Drug and Insecticide Administration in 1928 and became the FDA in 1931. This was the first major federal agency the main agenda of which was health and safety rather than economics.

The combination of an activist Democratic administration eager to innovate and economic conditions ghastly enough to dampen American enthusiasm for laissez-faire economics produced an alphabet-straining rush of regulatory agencies in the 1930s. Examples include the Federal Deposit Insurance Corporation (1933), the Federal Communications Commission (1934), the Securities and Exchange Commission (1934), the National Labor

Relations Board (1935), the Federal Maritime Commission (1936), and the Civil Aeronautics Board (1938). The Federal Communications Commission took over the functions of a 1920s predecessor, the Federal Radio Commission. All of these organizations survive to this day. So by the beginning of World War II, the regulatory commission was well-established as a federal device of choice for regulating the conduct of major business--and also, at that time, for promoting the development of businesses. With the exception of FDA, none of these organizations (not even the Civil Aeronautics Board) had risk as a primary concern.

When the end of World War II brought us into the Nuclear Age, the Atomic Energy Commission (1946) was the natural response to the three interrelated tasks of transferring control of nuclear energy for peaceful purposes into civilian hands, regulating its safety, and facilitating its private development. These functions were separated in 1975 in response to the obvious conflict between promoting an industry and regulating its safety, when the safety issues are so anxiety-provoking as in that case. The Nuclear Regulatory Commission was charged with regulating the safety (and other aspects) of nuclear power, and the Energy Research and Development Agency (ERDA) with promoting it. Later the ERDA was merged into the Department of Energy, when the latter was created during the Carter Administration.

Upswellings of public concerns about specific risk topics have quite naturally produced the invention of new regulations, and regulatory agencies to enforce them, as an adequate legislative response. Examples include the Environmental Protection Agency, the Occupational Safety and Health Administration, the Highway Traffic Safety Administration, and the Federal Aviation Administration. The same phenomenon occurs at state levels; an example is the California Coastal Commission, which regulates development along California's coastline.

Sometimes agencies created for another purpose become the arena for the risk technology debate. For example, the Federal Power Commission (now the Federal Energy Regulatory Commission) was created as an economic regulator rather than as a safety regulator. As an economic regulator, it was granted jurisdiction over natural gas imports and exports. Because of its tradition of public notice and hearings, safety fears about LNG imports had a readily available arena. On the other hand, liquid petroleum gas (LPG) imports and exports are not subject to economic regulation, and as a result, no similar arenas exist for LPG.

Thus we now have three major classes of arenas in which risk-related controversies are fought, the legislative bodies, the regulatory agencies, and the courts. The deficiencies of the others have made the roles of regulatory agencies even more prominent. Still, if the issue is important enough, Congress can and sometimes does preempt the regulatory agencies; it resolved the SST debate, and may well resolve the nuclear-power debate. And if the disputants are combative enough, the fight ends up in the courts--but after the activities of the regulatory agencies have transformed it in major ways.

Values and Strategies

We have identified three classes of stakeholders in technological-risk controversies: proponents, opponents, and regulators. Note that regulators are stakeholders and also manage an arena--a frequent source of complaint.

The values that developers of a technology bring to its introduction and use combine simple desire for profit with genuine concern for wider development and use of the technology about which they are expert. Boeing's development of the SST illustrates the interplay of these motives. Developers frequently have fellow-traveler proponents: The labor unions may support a technology because it will produce employment, and local business and government interests may relish the prospect of local economic growth. These values are best served if the process of introduction is orderly, predictable, and easy to manage. The existence of (surmountable) hurdles matters little, if they do not cost so much to deal with that they change the economics of the technology and if they do not take so long to deal with that they compel modification of its natural schedule. Any incremental costs resulting from regulation can simply be added to the price for which the end-result is sold. Promoters are therefore less concerned with escaping from regulation than with its predictability, cost, and ability to produce delays. Such interests are well served by a regulatory agency. They are less well served by the legislative arena, and are miserably served by the courts. The difference between legislatures and courts arises because technology sponsors typically have much more day-to-day ability to influence legislative action than do technology opponents. This may also be true in the courts, since lawyers' fees are high, but the effect is much less dramatic.

Regulators can tread on the toes of technology promoters in various ways. Worst is unpredictability--but we suspect (in spite of the public statements of technology promoters) that this is a rare event. Technology promoters can afford the time and expertise required to keep detailed track of how regulatory agencies behave in response to various combinations of facts and pressures, and of what facts and pressures in the current conflict are. Delays are more common, since they are easy for technology opponents to produce. Similarly, though the cost of dealing with any one regulatory agency is manageable, the cost of multiple hurdles may be so great as to significantly change the economic advantages of the technological advance.

The values of technology opponents range from tangible self-interests to broad public concerns. Examples of the former are residents' concern with the noise from bullet trains and landowners' worries that an LNG terminal nearby will reduce the value of their land. Examples of more general public concerns are environmental objections to the SST by the Friends of the Earth and Sierra Club's concern with radiation from nuclear power plants. We mentioned earlier that special interest groups with public values of this kind are a relatively recent phenomenon in technological controversies. They usually draw on a (potentially) broad constituency, yet they define their values narrowly, but in a way that cuts across many technological topics.

Such values can be served in either of two ways. One is to set limits on the technology that prevent it from affecting the relevant value (e.g., preservation of the environment) adversely. Since this is what regulatory agencies are set up to do, such technology opponents are well-served if they do it well. But the definition of "well" is specified by statute, and interpreted by the regulatory process. The participants in a controversy may disagree about what the limits should be. In that case, the technology opponents have another option: They can multiply hurdles and seek to increase the cost and time a developer needs to get over each. This means that the developer must fight in every battle, while the technology opponent can choose where to fight and where not. This asymmetry goes a long way toward countering the natural asymmetry in the other direction produced by the developer's longer purse and typically greater technological and legal expertise; we are not wise enough to know on which side the balance comes out, and far too wise to guess.

The values that govern the behavior of a regulating agency are quite a bit more complex than those that govern the behavior of a technology proponent or opponent. Unlike the opponents, and unlike some proponents, the regulator has an explicit, externally-defined mission, specified by statute. The responsibility of a regulating agency is to interpret and implement the intentions, often very vaguely put, of the legislation that created it. This task, of course, must be done with due attention both to changing technological realities and to the current political pressures. Thus regulators are coherent but responsive; not much pressure exists on them to be consistent from one case to another.

Regulatory agencies may be relatively supportive of new technologies, like the Nuclear Regulatory Commission, or less so, like the Environmental Protection Agency. We cannot find a convincing example of a regulatory agency that is highly critical of the technology it regulates. Without evidence, we speculate that a key determiner of the degree of supportiveness of the regulator is the security of the agency. Agencies derive security both from age and history and from being in control of many kinds of business, not just one. If the United States should eliminate nuclear power from its future, the NRC would wither. So long as people eat, we can think of no scenario that would cause FDA to wither.

Far more than opponents or proponents, regulators must tend their images. As adjudicatory agencies, they must be seen as fair by the parties to the disputes and by numerous publics. They must convince experts about the technology that they are expert. In addition, they must be successful. A regulatory agency whose rulings are consistently reversed on appeal to the courts would lose both credibility and all usefulness.

An issue in image-tending for regulatory agencies is the origins of staff members. The NRC is often seen as pro-industry because it is staffed with nuclear engineers, many of whom are AEC veterans. The EPA and OSHA are often seen as anti-industry; they initially attracted many non-industry experts sympathetic to environmental and occupational protection.

Implementing these values under these constraints is tough. Formally, any regulatory agency has an easy out: statutory mandate for procedures and preferences. This out cannot solve the problem of fairness of any agency that is required to promote the technology it regulates; that is why the AEC no longer exists. Insiders in most regulatory agencies say that safety lies in playing it straight. The amount of oversight a regulatory agency receives from Congress, the press, and the courts pretty much requires that if the agency acts at all, it must play by the book. Though the NRC is seen as the most favorably inclined of all regulatory agencies to the technology it regulates, the nuclear industry does not regard the NRC as a friendly party.

But the wide discretion that regulatory agencies typically have makes the formal solution incomplete. It must be supplemented with wisely selected politics initiated by the agency itself. The task of balancing statutory mandate, legislative intent, the need to be perceived as fair, direct business and interest-group pressure, indirect pressure applied by political mechanisms, and the requirements of the problem as understood by the regulating agency's technical people is one of the most interesting wire-walking acts American politics has to exhibit.

The fact that the balancing act is so delicate means that the regulatory agencies have little maneuvering room, and consequently little use for broad strategies of the kind we sketched for technology proponents and opponents. When the political issues are ticklish, the major tool available to the regulatory agency is delay. This serves the opponents well and the developers ill, but it is easy to justify (often in the form of a request for additional information) and it postpones the day when the agency has to take a stand, thereby incurring the wrath of at least some of the contending forces.

The other natural tool available to the regulatory agency is to say no. The legislation leading to the existence of a regulatory agency was invariably the result of concern over hazard. It is very difficult to attack a regulator who identifies a hazard and prohibits it for not having carried out the intent of the legislation, while such attacks are routine when a regulator says yes. But this tool must be used with great discretion; a regulatory agency that consistently says no has no political future.

Issues are substantially less dramatic if the choice confronting the regulator is not yes or no, but rather how much. How much is always a technical question, and arguments can always be made for various values. Whenever possible, therefore, regulators try to avoid phrasing what they do in yes-no terms, and try to find how-much versions of the problem to work on instead. Moreover, how-much questions keep the regulator's technical staff happy and employed.

Rules

Arenas specify rules of procedure and also define audiences; these specifications determine in a relatively detailed way how each class of stakeholder can best set out to implement values in conflicts fought there. We have already mentioned that this fact is more useful to opponents than proponents, since opponents can choose the arena whose rules are most favorable. Strictly worded laws like the Endangered Species Act provide opponents with an opportunity to use a very narrow argument (e.g., that a project endangers the survival of the snail darter fish) to halt a project for which the stakes are, for most, much higher than that.

The legal and quasi-legal rules in regulatory proceedings invite fact finding, examination of procedures, and arguments about what the law says or means. Seldom is that a good context in which to explore or negotiate about issues of fundamental conflicts over values. The greater latitude of regulatory proceedings makes such issues easier to examine there than in courts. The courts differ from the regulatory arenas in being more procedural, more powerful, and less expert. If a regulatory arena exists, the law usually provides that it must be used first.

City councils and legislative procedures at state and national levels follow quite different rules, which shift the emphasis from fact finding to evaluations and from single-issue debates to the task of identifying and packaging options, and then winning acceptance for them from conflicting stakeholders. Negotiating and trading are explicitly required. This of course encourages stakeholders to be persuasive, i.e., to lobby. Technology sponsors can offer tangibles like local economic benefits and jobs. Technology opponents cannot; they must get their clout from voting power, plus as much persuasive skill as they can find. Regulatory agencies enter the legislative arena also, but not ordinarily as advocates. They may simply motivate action and provide expertise. If they have any substantive stake, it is to be given an orderly, straightforward set of rules to enforce---preferably one that focuses on factual issues rather than value conflicts. This assumes that the regulatory agency is already in place; if not, the first issue is whether the risk should be regulated or not. Absent a regulator, some other government body is dealing with it, hampered by a set of rules ill-suited to the task and pressured by the stakeholders to "do something."

NEEDS FOR FUTURE RESEARCH

Our comparative developmental study of old and new risks yielded a conclusion that was new to us: The biggest change from previous to present risk controversies was that present ones fall under the control of regulatory agencies, while most early ones did not. Much of our discussion has focused on the development and characteristics of these agencies. Our ideas about future research also focus on them.

We would like to understand better than we now do the dramatic increase in the number of such agencies, and the development of conflict-reduction tools within each.

We suppose that the enhanced awareness of risks provided by the media leads to both a more rapid and a better-informed development of public concern about them. This offers an obvious explanation of the increase in regulatory agencies, but it also raises some interesting developmental questions. Initial concern about steamboats was economic, but was rapidly supplemented by public fears of their danger. This led to two simultaneous events: development of a rudimentary set of regulatory procedures, and technological development that enhanced safety. The enhanced safety led to decreased public concern, and eventually the regulatory agencies just about vanished.

Is that the normal developmental pattern for the relations among risk production, risk perception, and risk regulation? It seems possible but unlikely. The disappearance of the steamboat regulatory agencies has few parallels; it seems more likely that such organizations will remain in vestigial form long after they outgrow their usefulness. More important, however, is the cycle of risk perception leading to risk regulation, and the combination of risk regulation and natural technical development leading to risk reduction, and risk reduction leading to reduced public concern about risks. That pattern clearly has occurred in the aviation industry; commercial flight was regarded as dangerous in the 1930s, but is considered very safe today, in part because of technical development and in part because of regulation. There does not seem to be any reduction, within the United States, of regulation. But United States technical progress has clearly served to make unnecessary the development of similarly complex regulation in many other parts of the world (Europe is the exception). A better understanding of this kind of developmental link among production, perception, and regulation would be very helpful.

Finally, we are especially interested in developing procedures for eliciting, measuring, and using risk-related values. Such values produce laws, and thus define arenas (courts, hearings) and regulatory procedures. But the arenas and the regulatory procedures alike tend to be formal and confrontational. We believe that more explicit knowledge of values, and better ways for stakeholders to communicate about their value conflicts, would permit the stakeholders themselves to find more constructive compromises, and might permit the development of new kinds of arenas (e.g., mediation) in which more adaptive and flexible compromises could be worked out. It should be easier than it is at present to find solutions to public conflicts about technologies that are acceptable to many stakeholders; after all, so far this has been a game in which just about everyone has been a winner, and we know of no reason to believe that the facts about human inventiveness and flexibility that make that true have changed. We feel that improved ability to elicit and work with stakeholder values should inevitably lead to better risk management.

SUMMARY

Why is risk such an important topic in contemporary debates about new technologies? Four answers have been proposed: Risk may have increased, risk perceptions may have been enhanced; social values and preferences may emphasize risks, not benefits, and institutional changes may have caused increased attention to risks. The purpose of the paper is to explore these explanations by reviewing past and present technological controversies.

An early case was the steamboat. Original opposition, as usual, came from those whose livelihoods were threatened, but accidents soon led to public concern. This led, in 1838, to passage by Congress of the first bill regulating a risk. The regulation, after being toughened in 1858, was effective, and the public concern died down.

Opposition to the automobile because it was risky started in the mid-1920s, and continues to this day. But only the unfortunates whom it made jobless ever wanted to prevent its use.

Opposition to the zeppelin in the 1920s and 1930s, triggered by several disastrous accidents, rose to a peak after the 1937 Hindenburg disaster, and caused abandonment of the concept.

Opposition to the planned SST on environmental grounds grew steadily from 1967 to 1970, and finally its development in the United States was terminated, in part for economic and in part for environmental reasons.

A major public debate over a LNG terminal in California began in 1973; the issues were economic advantage vs. environmental risks. The initial issue was where the import site should be. As demand for natural gas slackened, Pacific Lighting decided to postpone indefinitely development of a terminal at the chosen Point Conception site.

The Japanese bullet train is fast but noisy. Protests about the noise led the Japanese Environment Agency in 1975 to set noise standards. That and other opposition has caused abandonment of sweeping development plans.

The comparison between past and present technology debates showed that they were surprisingly similar; the involvement of risk and other issues in such debates has changed little, although specific concerns about the environment, ecology, and endangered species are new. Technologies have fallen on harder times because of increased scale, the emergence of regulatory agencies and single-issue organizations that can fight them, changes in arenas to emphasize regulatory organizations and courts rather than political bodies, and the transformation of technology debates into adversary procedures in which experts take sides. The scale of technologies involved in controversies is now larger; decisions take longer. New laws and regulatory agencies that enforce them delay or prevent introduction of new technologies, and create new arenas to fight in. Partisan experts work for developers, opponents, and regulators, and disagree in public.

A comparison of the controversy over DNA research with that over nuclear energy shows the opposition to these technologies was similar in content, but different in the processes of resolution. Public concerns about risks evolved similarly in the two cases. But the debates were about different issues, and the institutions and regulatory processes differed. Partly as a result of public mistrust in the regulation of nuclear power, the opposition to nuclear energy was stronger than that to genetic engineering.

Of the four reasons for the importance of risk, all four have changed from past to present, but changes in institutions seem especially important. The important change is the development of regulatory agencies as arenas. Regulatory organizations came into existence during the early 19th century, stimulated by the expansion of the railroads. Economic regulatory agencies like the Federal Trade Commission came next. Many new regulatory agencies came into existence during the 1930s. The Atomic Energy Commission was created in 1946 to regulate the development and safety of nuclear energy. Other risk-regulating agencies emerged later. Regulatory agencies, legislative bodies, and the courts provide the arenas for risk controversies.

Technology developers want to promote their technologies and to make a profit. Regulators can tread on their toes by being unpredictable, by delaying, and by increasing cost.

Technology opponents want to set limits on technologies to prevent them from adversely affecting relevant values, such as preservation of the environment. Their values range from self-interest to broad public concerns.

Regulatory agencies have complex values. They must carry out their missions, tend their images, appear fair, be convincingly expert, and be successful. Implementing these values is tough. Safety lies in playing it straight, because of various forms of oversight. Regulators have two main tools: delay and denial. If a regulatory agency does not say no, it can say how much--a technical question.

Arenas specify rules and audiences, and thus help stakeholders to decide how to implement values in conflicts. Legal rules in regulatory proceedings invite fact finding, examination of procedures, and arguments. That does not provide a good context in which to negotiate about conflicts. Future research should examine the dramatic increase in the number of regulators, should help them develop conflict-reduction tools, should examine the developmental pattern that links risk production, risk perception, and risk regulation. It is important to develop procedures for eliciting, measuring, and using risk-related values.

REFERENCES

1. Abbott, P. Airship, New York: Charles Scribner's Sons, 1973.

2. Automobile Manufacturing Association. Automobile Club of America, Detroit, Michigan: The Wayne State University Press, 1968.

3. Aviation Week and Space Technology. "SST water vapor discounted." August 10, 1973.

4. Bond, V. Radiation standards, particularly as related to nuclear power plants. Raleigh, N.C.: Council for the Advancement of Science Writing, 1970.

5. Burke, T.G. Bursting boilers and the Federal power. Technology and Culture, 1965, 1-23.

6. Chemistry, "Ozone and the SST." April, 1973.

7. Dayton, F.E. Steamboat days. New York: Frederick Stokes Company, 1925.

8. Del Sesto, S.L. Science, politics and controversy: Civilian nuclear power in the United States, 1946-1967. Boulder, Colorado: Westview Press, 1979.

9. Douglas, M., and Wildavsky, A. Risk and culture. Berkeley: University of California, 1982.

10. Drago, H.S. The steamboaters. New York: Dodd, Mead and Co., 1967.

11. Duncan, O.D. Sociologists should reconsider nuclear energy. Social Forces, 1978, 51, 1-22.

12. Eckener, H. My zeppelin. London: Putnam, 1958.

13. Flink, J.J. The car culture. Cambridge, Massachusetts: The MIT Press, 1975.

14. Flink, J.J. America adopts the automobile, 1895-1910. Cambridge, Massachusetts: The MIT Press, 1975.

15. Gyorgy, A. No nukes. Boston: South End Press, 1979.

16. Hashimoto, M. Present status of noise regulation and control in Japan. Inter-Noise, 1975, 719-724.

17. Hearnes, R.P. Zeppelins and super-zeppelins. New York: Lane, 1916.

18. Hill, F.E. The automobile. New York: Cornwall Press, 1967.

19. Hill, R.N. Sidewheeler sage. New York: Rinehart and Co., 1953.

20. Ikeda, S. and v. Winterfeldt, D. Standards against noise pollution: The case of the Shinkansen trains in Japan. Journal of Environmental Management, 1982, 14, 3-16.

21. Inglehart, R. Values, objective needs, and subjective satisfaction among western publics. Comparative Political Studies, 1977, 4, 428-458.

22. Kohlmeier, L.M. The regulators. New York: Harper and Row, 1960.

23. Lawless, E.W. Technology and social shock. New Brunswick, N.J.: Rutgers University Press, 1977.

24. Levy, E. The people's lobby. New York: Delacorte Press, 1973.

25. Linnerooth, J. A short history of the California LNG terminal. Laxenburg, Austria: International Institute for Applied Systems Analysis, October, 1980.

26. Life, "Let's ground the SST," March 19, 1971.

27. Lombardo, T.G. The decision makers: A cacophony of voices. IEEE Spectrum, 1979, 18, 81-91.

28. The Los Angeles Times, January 26, 1982, p. A1.

29. Mazur, A. The dynamics of technical controversy. Washington, D.C.: Communications Press, Inc., 1981.

30. Metzger, H.P. The atomic establishment. New York: Simon and Schuster, 1982.

31. The Nation, "The automobile's death toll," March 8, 1922, pp. 279-281.

32. The Nation, "Last stand at the western gate," August 25, 1979, pp. 136-138.

33. Nelkin, D. (Ed.) Controversy: Politics of technical decisions. Beverly Hills: Sage, 1979.

34. The New Republic, July 7, 1926, pp. 186-280.

35. Otway, H.J. and v. Winterfeldt, D. Beyond acceptable risk: On the social acceptability of technologies. Policy Sciences, 1982, 14, 247-256.

36. Raimundo, J. LNG: Where, when, and why? Sierra, June 1, 1978, pp. 35-38.

37. Renn, O. Analyse der Sozialvertraglichkeit von Energiesystemen. Technical Report, KFA Julich, Julich 1982.

38. Slovic, P., Fischhoff, B., and Lichtenstein, S. Characterizing perceived risk. In R.W. Kates and C. Hohenemser (eds.) Technological hazard management. Cambridge, Mass.: Oelgeschlager, Gunn, and Hain, 1982.

39. Starr, C. and Whipple, C. Risks of risk decisions. Science, 1980, 208, 1114-1119.

40. Sternglass, Infant mortality and nuclear tests. Bulletin of the Atomic Scientists, 1969.

41. Straszak, A. and Tuch, R. The Shinkansen high speed rail network of Japan, New York: Pergamon Press, 1977.

Heuristic Opinion and Preference Evaluation Research for Assessing Technological Options--A User's View

Miller B. Spangler

Special Assistant for Policy Development, Division of Systems Integration, Office of Nuclear Reactor Regulation, U.S. Nuclear Regulatory Commission, Washington, D.C. 20555

ELITISM VERSUS DEMOCRACY IN TECHNOLOGICAL DECISIONS

"Each honest calling, each walk of life, has its own aristocracy based on excellence of performance." - James Bryant Conant

The decade of the 1970s--at least on the American scene as well as that of certain other nations--will no doubt be remembered by historians as the period when the environmentalist movement came to flower along with important citizenry initiatives on behalf of "participatory democracy," one of Naisbitt's 10 megatrends (1). The backlash against representative democracy was in no small measure a revolt against elitist groups in industry and government who lost the trust of important segments of the citizenry for a number of reasons. Included was their failure to consider and, hence, actively plan to mitigate the unwanted risks and other adverse by-product consequences to society and the environment of technological developments as well as inadequate attention by regulatory decisionmakers to equity issues where public interests come into conflict. Communications failures to properly receive public inputs to the decision process and to openly and correctly inform the public of the decision process also contributed to growing distrust. Numerous mid-course corrections in these failures are being made which, when carried to a more advanced level, will constitute the normal democratic response to alleviating the grass-roots pressures of a legitimate reform movement.

It seems clear to this writer that the coming decade or so will, for many advanced nations, usher in a new public concern: namely, the risk of economic stagnation from a variety of potential causes such as shifts of technological

NATO ASI Series, Vol. G4
Environmental Impact Assessment, Technology Assessment, and Risk Analysis. Edited by V.T. Covello et al.

leadership and an inability for certain domestic goods to compete in international markets, the suspect quality of educational programs, the moral decay and the breakdown of traditional institutions such as family and religious organizations, and a general failure of the pursuit of excellence. In a 1981 Science editorial, "In Defense of Elitism," Nancie Gonzalez observes that Americans, to avoid stagnation, need two kinds of educated bodies--a public that appreciates and urges continued support for research and education, and an intellectual elite to provide both scientific and humanistic understanding to help us cope with the harsh realities of the physical world (2).

The above is not intended as a gloomy forecast of the future, but only a dimensioning of a focal issue of risk to which other issues of risk can be related in a hierarchical structure. Thus, the possible outcome of economic stagnation (and the thwarting of much of the social progress that is purchased through economic growth) is hardly a foregone conclusion. Much will depend on the relative success of an advanced society in focusing its intellectual energies on the key issues of the times including the attainment of an appropriate exchange process by which the value inputs of the public in a participatory democracy are blended with the knowledge and judgmental inputs of intellectual leaders or elites, especially those that are collectively conversant in the fuller set of policy and technical issues related to societal options.

Ignorance, of course, is a commodity shared by individual experts and members of the public, for the problems of information overload bear upon us all. Ignorance contributes to uncertainty and uncertainty to risk. Douglas and Wildavsky have explored the question of how society decides which risks to focus on because most people cannot be aware of most dangers at most times (3). Their conclusion is that because no one person can know more than a fraction of the dangers that abound, some sort of priority must be established among dangers. Our ignorance, of course, extends beyond risks associated with technological options, encompassing also the magnitude of societal (and individual) benefits as well as costs or adverse effects not always thought of as risks. If there is sufficient perception of net benefits (e.g., the time and cost savings and comforts of air travel), it is not uncommon to find that otherwise risk averse persons will mount their courage sufficiently to overcome their fears and anxieties.

Thus, without a balanced perception of net benefits (i.e., gross benefits minus non-risk costs), the individual perceptions of risk as the focal point of comparing technological options may distort the consent/dissent attitude of individuals from that which would best serve their true values if only better information of risks, costs, and benefits had been supplied and accepted as valid. The latter condition implies a mutual growth in trust and cooperation between all the parties of interest and responsibility in what has become a highly fragmented (and often poorly coordinated) decision process by which the course of technological progress is charted and accomplished in a modern democratic society.

In achieving an appropriate balance between participatory democracy and elitist inputs to technological decisions that would best serve the public interest, it is important that the public and elitist groups contribute what each can do best. It is the individual in society that is the most expert on his or her own sense of values when confronted with the most accurate assessment of risks, costs, and benefits of technological options of which scientific state-of-the-art and wise resource commitment to information generation and interpretation will permit. Likewise, the collective, interdisciplinary knowledge and insights of policy and technical experts (i.e., the intellectual elites) that can be mustered to assess the risks, costs, and benefits of the options, are what their skills are best able to contribute to the decision process rather than a true, intuitive assessment of values in a pluralistic society (4).

SOURCES OF BIAS IN TECHNOLOGY POLICY DEVELOPMENT AND CURRENT MODES OF PUBLIC RISK PERCEPTION RESEARCH

To design an objective and balanced approach to policy development that would effectively integrate the knowledge and judgmental skills of intellectual elites along with the value inputs of diverse segments of the public regarding choice of technological options, it is essential to understand the potential sources of bias commonly found in current decision procedures and to seek ways to circumvent these shortcomings. There are numerous omissions and flaws in the procedures by which public policy decisions are made affecting technological choice and the regulation of technologies. In a recent article, I have provided an extensive critique of methods in the quantification of risks, costs, and benefits in the societal choice of energy options (5). Among the shortcomings discussed are the following: (1) a failure to include the full scope of risks, costs, and benefits in decision processes of visible concern to the public interest; (2) a pronounced bias toward decision considerations for which numerical data exist that would reflect the magnitude of societal impact, and away from those that are basically intangibles or for which data only reflect indirect indication of their importance to society; (3) a deference toward the Pandora's Box Syndrome in avoiding controversial considerations that encumber or frustrate closure in the decision process (6); (4) a deference toward the Put-It-In-A-Nutshell Syndrome that truncates the information and analytical base supporting the decision process at a level that omits important rationale and key facts of potentially persuasive significance, resulting too often in mere assertions of opinion which fail to stretch the mind; (5) a common tendency, especially on the part of technical experts, to develop or assimilate estimates of the risks, costs, and benefits based only on historical (or rear-view) information rather than forward-looking estimates based on a forecasting of causal factors that collectively affect future consequences to society of technological options; (6) inadequate forecasting methodologies (often a naive extrapolation of past trends) or ones that have the Euphoric Syndrome bias of self-fulfilling prophecies that fail to fulfill; (7) a failure to accompany estimates or forecasts of societal consequences with a quantitative or qualitative description of uncertainties and their causes; (8) a failure to address or adequately treat the inherent conflict between equity considerations and the common good of society in technology policy development; and (9) an assignment of analytical responsibilities into narrowly defined pigeon holes resulting in the weaknesses of multidisciplinary analysis rather than an interactice approach that seeks the larger benefits of interdisciplinary analysis (4).

One of the problems in achieving interdisciplinary analysis is the communication barrier that exists between intellectual elites. There are some who hold the view--not without merit--that science should be kept "value-neutral." Hence, scientific considerations and political considerations involving societal interests should be strictly divorced. These people hold the view that social scientists, management scientists, and behavioral scientists should be the ones to engage in value analysis and the physical scientists and engineers should hold province over technological and related phenomenological considerations. Yet, scientists and engineers employed by industry and governmental regulatory bodies do make nitty-gritty, every-day decisions that are value-laden. Without the benefit of the sort of interdisciplinary analyses in which interrelated uncertainties must be reconciled regarding the risks, costs, and benefits of technological options, we are behaving like the blind men and the elephant in attempting to understand how man-machine systems serve the best interests of society.

The above weaknesses in technology policy development exhibited by technical experts and policy analysts detract significantly from serving the public interests in a fair and balanced manner. Whether this result would be improved upon by the use of voter referenda to decide technological options or through strong reliance by policymakers on survey information on the public perception of the risks, costs, and benefits of technological options is a highly debatable subject. As shown in Table 1, there are numerous gaps in the treatment of technological risks, costs, and benefits by technical experts and the lay public. Although both treatments display weaknesses, whenever a complexity of technological and policy issues is involved in making fair and balanced choices, the availability of information resources that play a crucial role in value formation would appear to be a greater handicap for public perception of the risks, costs, and benefits of technological options than the collective wisdom of experts.

Jim Rubens, who champions the cause of participatory democracy, provides the following commentary on the matter of referenda on complex technical issues (7):

> Referenda, though well suited to value-dependent questions, are entirely unsuited to scientific and technical questions of fact. Experts, representatives of countervailing bodies of thought, and affected parties are the ones who should participate in the decision process on technical questions. How to control acid rain, what methods can control interest rates or inflation, which new energy forms are likely to yield returns on taxpayer research and development subsidies, or what strategic minerals to inventory are examples of primarily, although not wholly, technical questions.

While I have difficulty with Rubens' opinion that these examples are primarily technical questions (because this downplays the importance of inputs of the public's values), there is little question that it is more practical and cost-effective for a body of experts to deal more adequately with the requisite technical information than for each member of the public to gain a similar competence of information. Indeed, the greatest weakness of reliance by policymakers on public opinions regarding technological choices involving complex issues is the inadequate or questionable nature of information premises on which public opinions and preferences are based. Let us be clear that "adequate information" includes the scope of considerations, or kind of public interests, that should be included in policy decisions as well as the level of detail knowledge and analysis that is essential to a sound evaluation of these considerations.

Table 1. Gaps in the treatment of technological risks by technical experts and the lay public[a] (some generalizations having notable exceptions)[b]

Approach	Treatment common to experts	Treatment common to the public
1. *Decision criteria for risk acceptance/rejection*		
(i) Absolute vs relative risk	Risk judged in both absolute and relative terms	Greater tendency to judge risk in absolute terms
(ii) Risk–cost trade-offs	Essential to sound decision-making because of finite societal resources for risk reduction and impractability of achieving zero risk; tends to ignore nondollar costs in such trade-offs	Since human life is priceless, criteria involving risk–cost trade-offs are immoral; ignores risks of no-action alternatives to rejected technology; gives greater weight to non-dollar costs
(iii) Risk–benefit comparisons of technological options	Emphasizes total (net) benefits to society, neglecting benefits that are difficult to quantify; also neglects indirect and certain long-term benefits	Emphasizes personal rather than societal benefits; includes both qualitative and quantitative benefits but tends to neglect indirect and long-term benefits
(iv) Equity considerations	Tends to treat shallowly without explicit decision criteria and structured analyses	Tends to distort equity considerations in favor of personal interests to the neglect of the interests of opposing parties or the common good of society
2. *Risk-assessment methods*		
(i) Expression mode	Quantitative	Qualitative
(ii) Logic mode	Computational ●Risk = consequence × probability ●Fault trees/event trees ●Statistical correlation	Intuitive ●Incomplete rationale ●Emotional input to value judgments
(iii) Learning mode	Experimental ●Laboratory animals ●Clinical data for humans ●Engineering test equipment and simulators	Impressionistic ●Personal experience/memory ●Media accounts ●Cultural exchange
3. *Basis for trusting information*		
(i) Source preference	Established institutions	Non-'establishment' sources
(ii) Source reliability	Qualifications of experts	Limited ability to judge qualifications
(iii) Accuracy of information	Robustness/uncertainty of scientific knowledge	Minimal understanding of strengths and limitations of scientific knowledge
4. *Risk-attribute evaluation*		
(i) Low frequency risk	Objective assessment/conservatisms	Tends to exaggerate or ignore risk
(ii) Newness of risk	Broad range of high and low estimates	Tends to exaggerate or ignore risk
(iii) Catastrophic vs dispersed deaths	Gives equal weight[c]	Gives greater weight to catastrophic deaths
(iv) Immediate vs delayed deaths	Diverse views over treatment of incommensurables and discount rate	Gives greater weight to immediate deaths except for known exposure to cancer-producing agents
(v) Statistical vs known deaths	Gives equal weight	Gives greater weight to known deaths
(vi) Dreadness of risk	Generally ignores	Gives greater weight to dreaded risk
(vii) Voluntary vs involuntary risk	Gives equal weight	Gives greater weight to involuntary risk
5. *Technological considerations*		
(i) Murphy's Law (If anything can go wrong, it will)	Stimulus for redundancy and defense-in-depth in systems design and operating procedures; margins of conservatism in design; quality assurance programs	Stimulus for 'what-if' syndromes and distrust of technologies and technocrats; source of exaggerated views on risk levels using worst case assumptions
(ii) Reports of technological failures/accidents	Valued source of data for technological fixes and prioritizing research; increased attention to consequence mitigation	Confirms validity of Murphy's Law; increased distrust of technocrats

[a] Adapted from Denton[8]

[b] Some of the descriptors in this table unintentionally reflect the image that the "experts are always right". Experts, too, have biases and could benefit from improved information.

[c] For an exception, see the Report of the Advisory Committee on Reactor Safeguards[9]

Often, these evaluations imply the need for forecasts because it is the future consequences of technological options on public interests that are the true determinants of successful decisions. It is, of course, easy to make the case that many elites or experts who contribute to policy development are inexpert in the complex art of forecasting (5). However, this is more a failure of the system to focus on solving this problem than any inherent defect in the proposition that an appropriate body of elites could be selected or trained to cope more effectively with the problems of forecasting and prudent decisionmaking in the face of irreducible uncertainty than could individual members of the public lacking training in these fields and the incentives to accomplish the work to acquire the essential information or prepare the critical analyses and forecasts.

Another weakness in the use for policy development of the opinion survey method (as now commonly practiced) is the failure to come to grips with the twin problems of <u>changing social values</u> and the <u>interactive impact of information premises on the individual expressions of values or preferences</u>. The information resources to which individuals are exposed and responsive will often interact with their intellectual capacities, personality traits, and systems of beliefs and biases to form or reshape personal values through a subjective process that combines reason and emotion (10). The psychodynamic variables in this process often seriously limit the objectivity of opinion survey findings. According to Bowman and Bach (11):

> The problem of ascertaining the ends, or goals, of social policy is a very complex one. How, in a democratic society, can the attitudes of the majority of people be discovered? It is not enough to take a poll of opinions. Many attitudes are only half-conscious, and most people are only partially articulate. And apparently subtle variations in attitudes are sometimes very significant in revealing underlying differences. Moreover, human beings are neither consistent nor fully informed. Each individual has ideas which are mutually incompatible. Conflicts will be even greater between different individuals.

The mutual incompatibility of ideas or values within a given individual is an exceptionally rewarding area for exploration, especially if the analyst-decisionmaker wishes to learn more about how likely it is that individuals will change their beliefs or values, and about how different kinds of information affect individual values and lead to change in social values. Social values and market values both aggregate the individual or personal values of various segments of the society/economy.

The dynamic interaction of social and market values can, in turn, have a lagged effect on personal values through the hopes and/or anxieties that they raise. The reverse is also true. Yankelovich, who analyzes social trends and public attitudes, wrote in his recent book (12):

> We are not accustomed to crediting changing social values with shaping the country's economic and political fortunes. But we had better learn to think in these terms, for many of our troubles--and also the means to alleviate them--are now coming from this unaccustomed source. Generally, we look to economics, or technology or political action for fortuna--the turn of luck or fate that determines our destiny. I want to show, however, that while a prognosis of our future based solely on our political/economic prospects may leave us pessimistic, even desperate, one based on our cultural prospects--our shared social values--may, rather unexpectedly, point the way toward a brighter future.

It should not be concluded that the strategy of using public opinion surveys as an aid to technology policy development should be rejected because of present weaknesses in the way opinion surveys are conducted--namely, an orientation to a static determination of values and preferences based only on present information premises and values of individuals. Rather, it is important to better serve the needs of policy analysts and decisionmakers to develop methods of public opinion and preference evaluation which dynamically explore the interaction of a reasonable range of information premises regarding technological consequences and individual preferences as well as survey methods that provide insights on the likely changes of social values and their implications for policy development.

ALTERNATIVE MODELS OF INFORMATION EXCHANGE BETWEEN THE PUBLIC AND PROFESSIONAL ELITES OR POLICYMAKERS

The Doctor-Patient Model

The doctor-patient model of informed consent is the simplest and most direct form of information exchange that seeks to be responsive to the value system of the individual while making available to the individual access to the experience-based information and expert judgment of the professional/elite. This method of information exchange seeks to rectify the observable deficiencies of information and logic which interact with the more rudimentary values of individuals in determining their option preferences. For example, suppose a physician found that a patient's initial perception of the risks, costs, and benefits of a proposed operation led to a rejection of the operation when it was quite obvious that the patient was poorly informed on certain of these decision attributes. Moreover, on further discussion the physician found that the patient had given little or no thought to the risks, costs, and benefits of having no operation or pursuing other forms of treatment--options which in the best professional judgment of the physician hold a clearly inferior risk benefit prognosis. Would the physician's practice be an ethical one if he or she accepted the initial poorly informed view of the patient as a final determinant of the patient's best interests? I think not. At the very least the physician should provide additional information based on the most reliable scientific sources available including the counsel of other experienced practitioners regarding the most reasonable set of options.

In its idealized structure, the doctor-patient model of information exchange permits a merger of the two most qualified sources of information in arriving at a technical decision: the expert information judgment of the professional in assessing the likely outcomes of optional measures and the values and preferences of the individual who must reap the benefits or pay the penalties of risk-taking decisions that succeed or fail. This system, however, is not without its potential weaknesses: (1) the possible biased interest of a surgeon in performing operations, (2) the failure to get a second or third opinion when the uncertainties are large and the stakes are high, (3) the time constraints on busy doctors that truncate a more desirable level of information exchange or inhibit keeping abreast of the latest medical knowledge, and (4) the failure of patients to keep emotions under sufficient control so that their reasoned values will serve the best long-term interests.

The National Environmental Policy Act (NEPA) Model

Whatever its mix of desirable and undesirable features, the doctor-patient model of information exchange is simply not practicable for making policy decisions that need to reconcile the conflicting interests between different segments of the population or between present and future generations. The ethical principle of "informed consent" and the one-to-one exchange of information between physician and patient is not a real world situation in the social management of most technologies. Regulators will always be confronted with the difficult decisions of exercising, within their legally defined responsibilities and authorities, their best judgment of equitable compromises between different interest groups.

The National Environmental Policy Act (NEPA) of 1969 does provide a mechanism of sorts that facilitates an exchange of information between policymakers and affected segments of the public. The salient features of this mechanism are stated in Section 102(2)(c) of the Act:

> Include in every recommendation or report on proposals for legislation and other major Federal actions significantly affecting the quality of the human environment, a detailed statement by the responsible official on:
>
> (i) The environmental impact of the proposed action,
>
> (ii) Any adverse environmental effects which cannot be avoided should the proposal be implemented,
>
> (iii) Alternatives to the proposed action,
>
> (iv) The relationship between local short-term uses of man's environment and the maintenance and enhancement of long-term productivity, and
>
> (v) Any irreversible and irretrievable commitments of resources which would be involved in the proposed action should it be implemented.

The Council on Environmental Quality has provided specific guidance on the interpretation of the phrase "human environment" which is central to the cost-benefit balancing considerations set forth above (13):

> "Human Environment" shall be interpreted comprehensively to include the natural and physical environment and the relationship of people with that environment. This means that economic or social effects are not intended by themselves to require preparation of an environmental impact statement. When an environmental impact statement is prepared and economic or social and natural or physical environmental effects are interrelated, then the environmental impact statements will discuss all of these effects on the human environment.

Wisely, or unwisely, the NEPA permits a wide measure of flexibility for policymakers in Federal agencies for interpreting or giving specificity to its general provisions and requirements and also regarding certain measures beyond the Environmental Impact Statement (EIS) process that would serve to facilitate information exchange between agency desionmakers and the affected public. Frequent challenges have been made regarding the adequacy of information and scope of considerations presented in Environmental Impact Statements and, in a number of instances, the courts have rendered decisions on these matters (14). The EIS procedure involves the publication of a draft statement followed by a public comment period (including comments from other Federal and State agencies) and then publication of the final statement that includes the verbatim public comments and a staff discussion of them. In a number of cases public hearings are held in which further challenges are made by individuals or other parties of interest in a quasi-legal setting with many thousands of pages of hearing transcripts exploring the minutest of detail in an adversarial confrontation of witnesses and lawyers.

Other measures involving information exchange between affected parties and public decisionmakers include: (1) site visitations with informal discussions with State and local officials and community leaders or informational hearings prior to the analyses and publication of a draft EIS; (2) public workshops with invited panelists representing a wide spectrum of public interests and professional competence to explore policy alternatives; (3) publication in the Federal Register of proposed policy statements inviting public comment prior to final development of policies or policy changes; (4) publication of agency reports or contract studies that provide information and analyses of both technical and policy issues that are pre-decisional; (5) press releases of impending agency actions and general briefings of media personnel to improve their understanding of technical and policy issues; (6) open sessions of agency deliberations under the Sunshine Act and responses to Freedom of Information requests; and (7) response by letter to requests for information from Congressmen and other parties of interest or responsibility regarding proposed technological options.

The total cost of these collective measures of information exchange between experts and the public is often huge with variable cost-effectiveness in serving public interests that, to my knowledge, has not been formally analyzed. My general impression is that various techniques of information exchange required or permissively performed under the NEPA model are in need of significant improvement and/or overhaul.

Heuristic Opinion and Preference Evaluation: A Dynamic Concept for Bridging the Gap Between Public Views and Expert Analyses

An examination of the gaps described in Table 1 of the treatment of technological risks by technical experts and the lay public yields an understanding of why these groups are distrustful of the inputs of each other to public decision processes regarding technological options and the desirability of measures to reduce risk. Their trains of thought simply do not meet because they generally run on different tracks. The doctor-patient model of information exchange on a one-to-one basis is not a practicable solution to bridging these tracks for most technological options. It seems inescapable that experts do experience at least casual one-on-one conversation with friends and other individuals on technical and related policy issues, but scarcely at the in-depth level of analysis and understanding that experts generally indulge. Nor do individuals often care to divulge to such experts their innermost thoughts about personal values and attitudes or candid acknowledgment of their basis for trusting some sources of information rather than others. The public, by and large, has responded reasonably well to the anonymity of opinion survey techniques. However, these are rarely used in the NEPA model of information exchange for arriving at technological or policy related decisions, principally, I believe, because of the static orientation of such surveys to the current attitudes and beliefs of individuals that are overly conditioned by interaction with faulty information premises concerning the scope of valid decision considerations and the estimated levels of societal impacts (good and bad) associated with these considerations.

An expression of individual preference between options implies, by the same token, a form of democratic consent if the preference were accepted by policymakers and implemented. The dilemma this poses for individual views based on faulty information premises is commented on by Benjamin Freedman (15):

> If it is the case that a consent must be informed to be valid, the concept of consent must perforce receive a very limited application. We know that this is so even in the relatively well-defined area of medical care. When we enter the realm of environmental and occupational risks, there are even fewer cases in which the risks which are run are well-enough defined to afford substantial information. Many of the substances in question have not been adequately investigated; needed epidemiological studies have rarely been done, and when done have limited validity. The question of synergistic or antagonistic effects of the variety of substances to which we are exposed in our home and work environments has hardly even been raised. And, as has often been said, the more we know the more we appreciate our ignorance.

As previously noted, ignorance of key technological and policy information is not a monopoly of the lay public but is frequently a serious problem confronting experts. However, we have reason to expect that there will generally be an important difference in the kinds and levels of ignorance of the two groups. Yet, as pointed out by Freedman (15):

> Ignorance does not, by itself and in theory, stand in the way of the consensual enterprise. This is not to say that it is of no significance. One of the characteristics of consent is that it orders relationships over time, it binds to future performance; and when consent has been given without knowledge of some fact which later comes to light, there may be cause for renegotiation.

But therein lies the principle of heuristic opinion and preference evaluation research methods: If individuals would wish to re-examine their preference functions and contributory values in the light of newly developed information, by the same token, the change of personal values to heuristic shifts in information premises can be readily incorporated into opinion survey methods. Superior insight would be provided to technological policy development by public perception research that dynamically explores in the manner of sensitivity analyses the heuristic shifts of values and attitudes of the public as related to variations in information premises of the form: "If attribute 'X,' (or attributes X, Y, etc.) were actually the realized outcome(s) of certain technological/regulatory options, then what would be your valued preferences?" The choice of the X, or Y, outcomes to serve as a basis for these sensitivity analyses should, in my view, be based on the high, low, or medium estimates (or forecasted outcomes) reflecting a reasonable range of uncertainty among experts. This would supply at least some of the beneficial insight to the regulator as is achieved in the information exchanges between physician and patient in which information is supplied by the physician in his or her expert knowledge of the expected range of outcomes of treatment options, and the patient provides information of his or her sense of values as related to these outcomes. This, I believe, would provide a more effective method of assimilating societal values into technological decisionmaking than the present survey methods that basically reflect what societal values and preferences are when based upon a distorted perspective of information premises relative to that which regulators use as a basis for their decisions.

Note that such a use of this heuristic approach to an improved understanding of public perception of the values inherent in more realistic expectations of decision outcomes has nothing to do, in and of itself, with changing peoples' attitudes. However, in my view, perception research along these lines would also potentially benefit educators in various fields, and especially political leaders in their important role of conflict resolution. As stated by Burns (16), the quality of interplay between leadership and followership is highly important:

> Leadership is a process of morality to the degree that leaders engage with followers on the basis of shared motives and values and goals--on the basis, that is, of the followers' "true" needs as well as those of leaders: psychological, economic, safety, spiritual, sexual, aesthetic, or physical. Friends, relatives, teachers, officials, politicians, ministers, and others will supply a variety of initiatives, but only the followers themselves can ultimately define their own true needs. And they can do so only when they have been exposed to the competing diagnoses, claims, and values of would-be leaders, only when the followers can make an informed choice among competing "prescriptions," only when--in the political area at least--followers have had full opportunity to perceive, comprehend, evaluate, and finally experience alternatives offered by those professing to be their "true" representatives. Ultimately the moral legitimacy of transformational leadership, and to a lesser degree transactional leadership, is grounded in conscious choice among real alternatives. Hence, leadership assumes competition and conflict and brute power denies it.

The heuristic opinion and preference evaluation research method, in addition to treating the problem of faulty information premises of the public (as perceived by experts), would also surmount certain intellectual limitations of a certain segment of the public whose basic values are nonetheless of legitimate concern to public decisionmakers. Fischhoff has noted the inability of unselected groups to rationally absorb the available information on complex issues (17):

> People appear to have a great deal of difficulty making proper decisions under conditions of uncertainty, in part because probabilistic processes are counter-intuitive, in part because their values on new and...complex issues are not well formulated, and in part because they lack the cognitive capacity for combining large amounts of information involved in many decisions.

The respectability and acceptance of the heuristic opinion and preference evaluation research method would be enhanced by the careful design of procedures for a determination of the kinds of heuristic information premises and spectrum of magnitudes of forecasted societal effects of technological options (as well as the choice of options themselves) to be used in the public opinion survey. One of the following procedures might be appropriate for this purpose, with each having the common objective of achieving independence from potential biases of the user agency: (1) an agency-sponsored workshop open to the public but conducted by an independent chairman and a panel of outside experts and public interest group representatives; (2) a contract study with a respected scientific agency that would independently select study teams and panelists to accomplish these purposes.

Rubens (7) is of the opinion that an improved value-free decision process could be modeled generally after the "Science Court" concept proposed 15 years ago by Arthur Kantrowitz, an engineer and member of the National Academy of Sciences. Rubens goes on to suggest:

> The Science Court has been used to defuse scientific controversies by having a panel of impartial mediators consider all sides of a question as presented by opposing advocates.
>
> A good value-free decision process must break issues, insofar as possible, into a hierarchy of discrete and relatively independent components. A complex question, such as deciding which, if any, energy forms should be encouraged by tax or other subsidies, would be the product of a succession of underlying value-free and value-laden decisions.

According to Rubens, attempts to use the Science Court concept to resolve technical disputes have met with varying degrees of success. In his view, the Energy Research and Development Administration (ERDA) used the concept successfully in 1977 to select among 11 fusion energy research concepts competing for limited development funds. However, it should be noted that initial uses of the heuristic opinion and preference evaluation research method could (and perhaps should) be explored independently of the Science Court concept, which may appear a rather heady arrangement in the trial phase of heuristic methods of bridging the gap between the treatment of risks, costs, and benefits of technological options by the experts and the lay public.

Moreover, unlike the Science Court concept which would make or recommend decisions among technological options, the heuristic opinion and preference evaluation research method would have its greatest merit as an auxiliary input to aid technology policy development along with more conventional inputs to such decisionmaking. As used here, the term "heuristic" is a logic device relating to an exploratory problem solving or didactic technique that utilizes self-educating techniques to improve performance.[1] A body of experts would be needed to interpret the significance of the results for decisionmaking purposes, and other sources of inputs (including legal constraints or requirements) would have to be reconciled with this source of information on public values and preferences in arriving at technological or policy decisions.

THE ORGANIZATION OF HEURISTIC PREMISES AROUND A HIERARCHICAL STRUCTURE OF TECHNICAL AND POLICY ISSUES

Issues, Consensus, and Conflict Resolution

A well-spring of issues that serve to divide parties of interest regarding the consequences of technological options arises from the limitations of scientific knowledge and of methods for producing hard, non-controvertible facts. But the problem of ignorance or uncertainty is not the only source of division over policy development. Another is the cluster of issues surrounding "What is acceptable risk?" For example, as noted by Peter Hutt, even assuming that an adequate scientific database were available, together with scientific agreement on the meaning and significance of the data, there appears to be no public or scientific consensus today on the risk or uncertainty acceptable to justify the marketing of any substance as a food or drug (20). He goes on to state:

> Attempts to resolve such an issue on the basis of rigorous scientific testing or analytical discourse therefore simply miss the point. A mathematical benefit-risk formula or computer program may eventually be able to quantitate the risk or uncertainty that inheres in a given product, but it is not even relevant to the moral and ethical issues

---

1 This usage of "heuristic" is not to be confused with the concept introduced by Tversky and Kahneman (18) called "availability heuristics" in which the stochastic process of the most recent or vivid experiences of an individual with a type of risk unduly influences perception of its frequency and, hence, riskiness because risk is defined as frequency or probability times the consequence of an action. A more closely related use of heuristic logic in analytical modeling is described by Cooper et al. (19)

involved in deciding whether this degree of risk or uncertainty is acceptable.

This problem arises whenever new doubts or suspicions are cast upon the safety of an already marketed substance. Those who favor a very low public risk demand that the product immediately be removed from the market. Those who advocate a higher risk demand that it remain on the market until it is shown to be unsafe. If, as I suspect will happen, we eventually prove that many of our basic foods and drugs contain at least trace amounts of highly toxic substances--including carcinogens, teratogens, and mutagens--the public simply will have to face these issues in a more forthright way than it has up to now.

One does not need a degree in science to hold and express deeply felt beliefs on the degree of risk or uncertainty society should accept from food and drugs. Nor, indeed, does a scientific background equip anyone with any greater insight into the intricacies of this type of policy issue or any more impressive credentials or greater authority to act as an arbiter in resolving these matters. As long as we remain a free society, these basic philosophical principles will, and properly should remain the subject of intense public scrutiny and debate.

Thus, the point is made that policy issues involving questions of ethics, equity, and personal value systems need to be considered apart from more narrowly defined technical issues if societal consensus or effective conflict resolution is to be achieved in deciding on technological options. Although we separate technical issues from policy issues for purposes of analysis, there are numerous interactions between the two classes of issues. For example, as observed by Pieter Stallen, a paradoxical situation is that although risks are attributed to technology, it is with technology that we try to manage risks (21). If we were to manage technological risks to health and safety by ever increasing expenditure of financial resources to reduce them, we could reach or exceed the point where the marginal cost per life saved by the risk reduction measure is greater than that of alternative opportunities elsewhere to spend these same resources to achieve an even greater saving of lives (22).

This poses an equity issue with important policy implications. For this purpose Margaret Maxey uses the term "equitable management," by which she means policymakers should first be well informed about the broad spectrum of both natural and ordinary hazards that may have health effects for large segments of a population, then make comparisons of actual costs per capita to reduce them, and only then set standards that will get the most public health protection for the many out of a finite amount of money (23). Hazard management is equitable only if it is proportional in relation to the actual harm that can be identified and reduced by expenditures of human effort, time, and money. Maxey feels that if ethical requirements for social justice and equity are to be adequately considered, at least three major problems must be addressed: (1) The first problem is to determine whether current conceptual tools for assessing potential biohazards (that is, basic harms to valued living systems) are ethically fitting and adequate; (2) The second problem is how to set safety standards on the basis of informed consent to scientific evidence presented by experts who disagree in interpreting that evidence; and (3) The third problem is how to deal fairly with expert dissensus, that is to say, how to evaluate the origin of value-conflicts underlying expert-disagreement, namely, conflicting philosophies.

But, of course, the values that need to be addressed in policy development--if conflict resolution is a desirable goal--are much broader than those explicitly examined in safety-cost trade-off considerations. Cotgrove and Duff argue that values play an important role in legitimating social, political, and economic institutions and practices, and therefore changes in values may sometimes have important consequences for the legitimacy of a social system (24). They perceive a neglect of research into the sources of values and value change including the social distribution of values as well as their change over time. Their empirical findings are that senior industrialists are significantly more materialistic than members of the public of the same age group, and that environmentalists are much more post-material than the public even after controlling for age. Furthermore, the shift towards post-materialism in younger age groups is small by comparison with between-group differences.

Moreover, it is my general observation that the procedures by which policy decisions are arrived at encompass values in their own right that are no less consequential than the values implicit in decision outcomes in improving public consensus. That is to say, such methods as consultation with parties of interest at an early stage of the decision process, open and reliable communications of the various stages of analysis and deliberation, the inclusion in the cost-benefit analysis of decision alternatives of the full range of considerations regarded as important by the various parties of interest, plus a fair consideration of equity issues will all contribute to improved acceptance or reduction in the fervor of dissent. If this view is correct, then some such technique as here proposed for heuristic opinion and preference evaluation research could improve the image of trustworthiness of public decisionmakers and acceptance of their decisions on technology options. As it now stands, the assimilation of the public's values into technology policymaking by government agencies is, at best, highly subjective, haphazard, deficient, and largely non-visible to the public.

It is not the author's intent to provide a detailed blueprint in this paper as to how heuristic opinion and preference evaluation research might best be carried out. But certainly an important first step would be a clear definition and hierarchical ordering of both technical and policy issues affecting strategies for an improved public consensus of choice of technological options. For this purpose, "issues" might be thought of as logic tools by which information is filtered and organized for decisionmaking in resolving differences among parties of interest and responsibility. To provide a measure of concreteness to the envisioned methodology, the following section will present, as an illustration of this method, a preliminary hierarchical structure of technical and policy issues for the regulatory treatment of severe nuclear accidents for existing and future plants. In the final section, some limited examples of heuristic premises will be presented to suggest how public values (and insights of value changes) might provide useful inputs to the analysis and resolution of the identified issues.

A Hierarchical Structure of Technical and Policy Issues Pertinent to Regulatory Decisions on Severe Accidents for Existing and Future Nuclear Plants

As noted above in the suggested working definition, issues are regarded as the focal connecting links between a determination of desirable information and the resolution through decisionmaking of differences between parties of interest and responsibility. This concept is illustrated in Fig. 1. A useful starting point in the collection and definition of relevant issues is a canvas of the views of the various parties of interest and responsibility. These are identified for this particular example by the parties shown in Fig. 2. The parties of interest (or stakeholders) in the outcome of severe accident policy decisions for nuclear power plants include: (1) utility companies that are present or would-be licensees of nuclear power plants; (2) the vendor companies that design, manufacture, construct, or deal with the maintenance and safe functioning of nuclear plants and their component equipment; and (3) various individuals or public interest groups. Parties of responsibility (i.e., the elites who exercise roles in safety decisions) include: (1) the U.S. Nuclear Regulatory Commission (NRC), which holds legal responsibility for establishing and administering regulatory policies; (2) the U.S. Congress and other Federal, State, or local agencies (including the Courts) that have related legal responsibilities[2] or otherwise assume a responsible role to protect the interests of their constituents or citizens; (3) licensees and vendors who make value-laden decisions of technological design options and procedures within areas of choice permissible under government laws and regulations; and (4) individuals and public interest groups who assume responsibilities for protecting their own interests or the constituents they represent in active intervention or other contributions to the regulatory decisionmaking process.

An example of the latter is the public comment being received following publication by the NRC on April 13, 1983, of the "Proposed Commission Policy Statement on Severe Accidents and Related Views on Nuclear Reactor Regulation" that invited comment by interested parties (25). Public comment was also invited and received on the NRC's proposed "Policy Statement on Safety Goals for the Operation of Nuclear Power Plants" that would provide an explicit policy on safety philosophy and the role of safety-cost trade-offs in NRC safety decisions, especially those relating to severe (or potentially catastrophic) accidents (26). Other sources of insight as to how elite members of the public perceive severe nuclear accident issues is to be found in their testimonies presented to various Committees of the Congress. Examples include the Oversight Hearing before the House Subcommittee on Energy and the Environment regarding the "Reactor Safety Study (Rasmussen Report)" (27) and the Hearing before the House Subcommittee on Energy Research and Production on "Nuclear Public Information and Rational Public Policy Decisions." (28) The President's Commission on the Accident at Three Mile Island also assembled a large body of public input on severe nuclear accident issues (29).

---

2 In a decision handed down April 20, 1983 (Pacific Gas & Electric Co. v. State Energy Resources Conservation Commission et al, No. 81-1945), the U.S. Supreme Court reaffirmed and declared that "the Federal Government maintains complete control of the safety and 'nuclear' aspects of energy generation, whereas the States exercise their traditional authority over economic questions such as the need for additonal generating capacity, the type of generating facilities to be licensed, land use, and ratemaking."

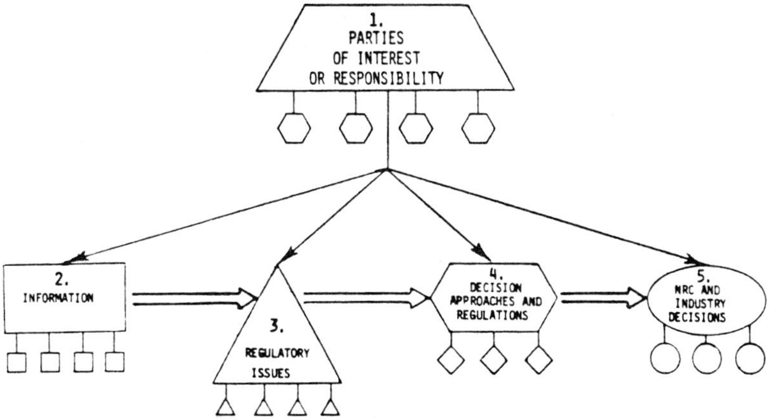

Fig. 1. The focal role of regulatory issues as logic tools by which information is filtered and organized for severe accident decision-making in resolving differences among parties of interest in nuclear power plant regulation.

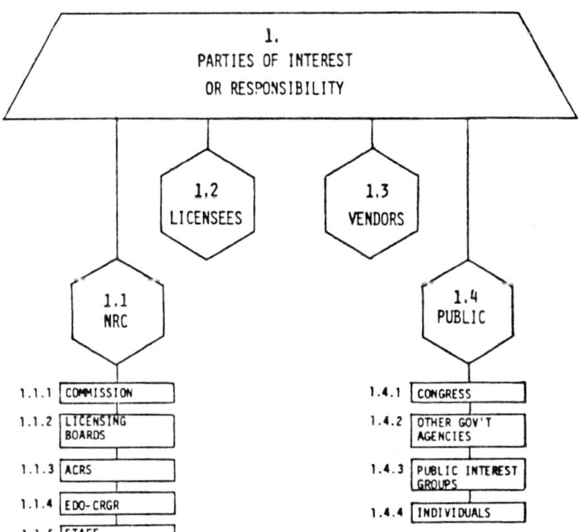

Fig. 2. Parties of interest or responsibility that contribute directly or indirectly to a definition of policy and technical issues related to regulatory decisions on reducing the risks of severe nuclear accidents.

Within the NRC itself there are a number of groups with varying degrees of decisional and analytical responsibilities that are sources of information regarding severe accident technical and policy issues, sometimes of a divisive nature. For example, the aforementioned proposed policy statements on severe accidents and safety goals provide discussions of dissenting Commissioner views and a number of critical views in these matters were provided to the Commission by the Advisory Committee on Reactor Safeguards (ACRS). The latter is a high-level, independent committee of nuclear safety experts appointed by the Commission for terms of four years each to review safety studies and facility license applications referred to it and provide reports that advise the Commission regarding the hazards of proposed or existing reactor facilities and the adequacy of proposed reactor safety standards, among other assigned duties. Other parties of responsibility within the NRC participating in regulatory decisions or policy development regarding severe accident issues include the Atomic Safety and Licensing Board Panel, the Atomic Safety and Licensing Appeal Panel, and the staffs of the Office of General Counsel, Office of Policy Evaluation, Office of the Executive Legal Director, Office of Nuclear Reactor Regulation, Office of Nuclear Regulatory Research, and the immediate staff of the Executive Director for Operations (EDO). The latter includes the recently established Committee to Review Generic Requirements (CRGR), whose basic purpose is to control the number and nature of requirements placed by the NRC on licensees. Their analysis includes an assessment of risk reduction through imposing the proposed safety requirement, the costs to the NRC and the licensees, including occupational dose increase or decrease, added plant and operational complexity, and total financial costs. The CRGR also holds coordinating responsibility for developing the safety goal evaluation plan called for in the Safety Goal Policy Statement (supra).

Table 2 shows a preliminary hierarchical structure of policy and technical issues related to regulatory decisions by the U.S. Nuclear Regulatory Commission on reducing severe accident risk of existing or future nuclear power plants. One important general impression that results from examining this list of policy and technical issues is the inordinate complexity that must be dealt with in regulatory decisionmaking for nuclear power plants to protect the public health and safety while serving the fuller range of societal interests. It serves as additional testimony as to the enormity of the gaps between the treatment of technological risks by experts and the lay public described in Table 1. Moreover, it illuminates the difficulties of participatory democracy in acquiring the requisite information and performing the necessary analyses to resolve these issues in the best interests of society. The heuristic opinion and preference evaluation research method would enhance participatory democracy relative to present regulatory practice in dealing with complex technological and policy issues by keeping the principal burden of information generation and analysis with a team of experts exercising their legally based responsibilities. At the same time, representative members of the public would express their values and preferences relative to a set of heuristic information premises of action consequences supplied by experts.

SOME EXAMPLES OF HEURISTIC INFORMATION PREMISES FOR SURVEYING PUBLIC VALUES AND VALUE CHANGES IN POLICYMAKING FOR TECHNOLOGICAL OPTIONS

Severe Accident Risk and the Public Acceptability of the Nuclear Energy Option

Some limited elements of the heuristic opinion and preference evaluation method relative to the nuclear energy option have already been applied. For example, in several Harris polls, by hypothesizing the cost advantage of nuclear power over other types of power to range from -20 percent to +50 percent, the overall swing of positive response in the general public was +60 percent (i.e., from a low of 24 percent acceptance with a 20 percent negative cost comparison to 84 percent acceptance with a 50 percent cost advantage) (30). These heuristic assumptions of net economic benefit of the nuclear versus other energy options for generating electricity only applied to a single attribute of interest to society and there is no indication that the spectrum of net economic benefits was postulated procedurally by an independent and respected body of experts or panelists to improve legitimation of the technique in the eyes of the public. Nevertheless, this survey technique does provide useful insight that suggests that rising electric generating costs of new nuclear plants to reduce substantially further the risk of severe accidents will likely, by the same token, reduce substantially the public acceptance of the nuclear option.

---

3 Explanation of acronyms for Table 2: LWR - Light Water Reactor; 10 CFR - Code of Federal Regulations, Energy; CRGR - Committee to Review Generic Requirements; EPA - Environmental Protection Agency; NAS - National Academy of Sciences; PRA - Probabilistic Risk Assessment.

Table II. A Preliminary Hierarchical Structure of Policy and Technical Issues Related to Regulatory Decisions by the U.S. Nuclear Regulatory Commission on Severe Accident Risk of Existing or Future Nuclear Power Plants[a]

3.1 POLICY ISSUES

3.1.1 WHAT CRITERIA SHOULD BE USED TO DETERMINE ACCEPTABLE/UNACCEPTABLE SAFETY OF LWRS?

- 3.1.1.1 "NO UNDUE RISK" - 10CFR
- 3.1.1.2 COMPARATIVE RISK CRITERIA
- 3.1.1.3 QUANTITATIVE RISK CRITERIA
- 3.1.1.4 RISK ALLOCATION IN SYSTEM PERFORMANCE
  - 3.1.1.1.1 ACCIDENT PREVENTION
  - 3.1.1.1.2 ACCIDENT MANAGEMENT
  - 3.1.1.1.3 CONSEQUENCE MITIGATION (E.G., CAN A CONTAINMENT PERFORMANCE CRITERION BE SPECIFIED FOR EACH PLANT OR PLANT TYPE?)

3.1.2 WHAT ARE ACCEPTABLE METHODS OF RISK ASSESSMENT?

- 3.1.2.1 DETERMINISTIC/JUDGMENTAL
- 3.1.2.2 DETERMINISTIC PLUS PRA INSIGHTS
- 3.1.2.3 EXCLUSIVE USE OF PRA (PROBABILISTIC RISK ASSESSMENT)
- 3.1.2.4 QUALITATIVE SYSTEMS ANALYSIS
- 3.1.2.5 ASSESSMENT OF UNCERTAINTY IN RISK ESTIMATES (HOW WELL DO WE KNOW THE RISK FROM EXISTING POWER PLANTS?)

3.1.3 HOW SHOULD SPECIAL RISK CONCERNS BE TREATED?

- 3.1.3.1 EXTERNAL EVENTS (EARTHQUAKES, FLOODS, FIRES, HURRICANES, ETC.)
- 3.1.3.2 HUMAN ERRORS (DESIGN, OPERATOR ERRORS OF COMMISSION, ETC.)
- 3.1.3.3 SABOTAGE
- 3.1.3.4 SYSTEMS INTERACTION
- 3.1.3.5 EQUIPMENT QUALIFICATION INCLUDING AGING
- 3.1.3.6 COMMON CAUSE FAILURES

3.1.4 WHAT IS AN ACCEPTABLE/UNACCEPTABLE LEVEL OF UNCERTAINTY IN RISK ASSESSMENT? (HOW WELL DO WE NEED TO KNOW THE RISK FROM EXISTING POWER PLANTS?)

- 3.1.4.1 WHAT ARE THE SOURCES OF IRREDUCIBLE UNCERTAINTY IN RISK ASSESSMENT OF NUCLEAR POWER PLANTS?
  - 3.1.4.1.1 LIMITS OF SCIENTIFIC CAPABILITY
  - 3.1.4.1.2 RESOURCE LIMITATIONS AND SOCIAL CONSTRAINTS ON EXPERIMENTATION OR GENERATING NEW INFORMATION
  - 3.1.4.1.3 GAPS IN KNOWLEDGE (POTENTIAL FOR FUTURE DISCOVERY OF FAILURE PHENOMENA)
- 3.1.4.2 COMPARATIVE UNCERTAINTIES IN RISK ASSESSMENTS (E.G., WHAT ARE THE UNCERTAINTY BANDS OF RISK ESTIMATES FOR VIABLE ENERGY ALTERNATIVES?)
- 3.1.4.3 WHAT NET-BENEFIT/RISK TRADEOFFS ARE THERE FOR ENERGY OPTIONS?
- 3.1.4.4 WHAT TRADEOFFS ARE REASONABLE BETWEEN COST AND UNCERTAINTY REDUCTIONS FOR NUCLEAR POWER PLANTS ?

3.1.5 WHAT ARE ACCEPTABLE STRATEGIES FOR THE USE OF PRA?

- 3.1.5.1 PARTIAL PRAs (EXCLUSION OF SABOTAGE, EXTERNAL EVENTS, ETC.)
- 3.1.5.2 COMPLETE PRAs (BALANCING UNCERTAINTIES)
- 3.1.5.3 EMPHASIS ON DETERMINATION OF DOMINANT-RISK SCENARIOS FOR ACCIDENT INITIATION AND PROGRESSION
- 3.1.5.4 SURROGATE PRAs
- 3.1.5.5 PLANT/SITE-SPECIFIC PRAs
- 3.1.5.6 TREATMENT OF BIAS IN PRAs (ASSESSMENT OF CONSERVATIVE MODELING ASSUMPTIONS)

3.1.6 WHO SHOULD BE RESPONSIBLE FOR THE LEAD ROLES IN RISK ASSESSMENT AFTER THE SEVERE ACCIDENT DECISION?

- 3.1.6.1 LICENSEES
- 3.1.6.2 INDEPENDENT NRC ROLE (E.G., INTEGRATED SAFETY ASSESSMENT PROGRAM -- ISAP)

3.1.7 WHAT IS AN APPROPRIATE TREATMENT OF UNACCEPTABLE RISK AND UNCERTAINTY?

- 3.1.7.1 IMPROVE RISK ESTIMATES (BETTER DATA OR IMPROVED RISK ASSESSMENT MODELS)
- 3.1.7.2 SEQUENTIAL ANALYSIS OF TECHNICAL OPTIONS TO REDUCE EITHER RISK LEVELS OR THE UNCERTAINTY BANDS OF RISK ESTIMATES
  - 3.1.7.2.1 CONSIDER MINOR CHANGES FIRST (DESIGNS, OPERATING AND MAINTENANCE PROCEDURES)
  - 3.1.7.2.2 CONSIDER MAJOR CHANGES FIRST (E.G., THERE MAY BE SUPERIOR OPTIONS TO "BANDAIDING" EXISTING DESIGNS, OR IT MAY BE MORE EFFECTIVE TO PUT EMPHASIS ON MITIGATION BEFORE ALLOCATING THE RESOURCES TO PREVENTION)
- 3.1.7.3 GREATER EMPHASIS ON EMERGENCY PREPAREDNESS PLANS

3.1.8 WHAT IS THE PROPER SCOPE OF RISK-COST-BENEFIT CONSIDERATIONS IN AN EVALUATION OF TECHNICAL OPTIONS FOR RISK OR UNCERTAINTY REDUCTION?

- 3.1.8.1 PUBLIC LAWS AND COURT DECISIONS
- 3.1.8.2 10CFR STATEMENTS (BOTH REQUIREMENTS AND PERMISSIBLE FLEXIBILITY; BACKFIT RULE; ETC.)
- 3.1.8.3 CRGR SCOPE
- 3.1.8.4 REGULATORY ANALYSIS GUIDELINES (NRR OFFICE LETTER NO. 16, REV. 1)
- 3.1.8.5 $1000 PER MAN-REM AVERTED OR OTHER COST-BENEFIT STANDARD
- 3.1.8.6 OTHER BENEFITS OF RISK REDUCTION (E.G., REDUCTION OF GENETIC EFFECTS, ONSITE FINANCIAL LOSS, OFFSITE FINANCIAL AND INCOME LOSSES, ETC.)
- 3.1.8.7 CUMULATIVE EFFECTS ON OVERALL SAFETY LEVELS AND TOTAL GENERATING COSTS OF A LARGE NUMBER OF MEASURES TO TREAT INDIVIDUAL RISK CONTRIBUTORS (I.E., A MACRO SAFETY-COST TRADEOFF PERSPECTIVE)

3.1.9 WHAT ARE PROPER METHODS OF ESTIMATING COSTS AND BENEFITS OF RISK REDUCTION MEASURES?

- 3.1.9.1 RELIANCE ON LICENSEE ESTIMATES OF COSTS AND BENEFITS SUBJECT TO NRC REVIEW
- 3.1.9.2 INDEPENDENT NRC ESTIMATES OF COSTS AND BENEFITS
- 3.1.9.3 USE OF DISCOUNT RATES FOR DOLLAR ESTIMATES AS WELL AS FOR THE EQUIVALENCY VALUES OF LATENT AND PROMPT RADIATION FATALITIES
- 3.1.9.4 USE OF FORECASTING METHODS VERSUS RAW USE OF HISTORICAL DATA FOR EVALUATING COSTS AND BENEFITS
- 3.1.9.5 IMPUTING EXTRA WEIGHT TO CATASTROPHIC RISKS VERSUS RISKS MORE UNIFORMALLY DISTRIBUTED OVER TIME
- 3.1.9.6 USE OF INPUT DATA, MEASUREMENT STANDARDS OR OTHER EVALUATIONS SUPPLIED BY OTHER FEDERAL AGENCIES OR SCIENTIFIC INSTITUTIONS (E.G., EPA, NAS, ETC.)
- 3.1.9.7 USE OF PUBLIC EXPRESSIONS OF VALUE, ESPECIALLY FOR OFFSITE CONSEQUENCES (OBJECTIVE DATA, SUBJECTIVE INTERPRETATIONS, ETC.)

3.2 TECHNICAL ISSUES

3.2.1 WHAT IS THE MOST APPROPRIATE AND BALANCED RESOLUTION OF UNRESOLVED SAFETY ISSUES (USIs) AND GENERIC ISSUES (GIs)?

- 3.2.1.1 FOR ALL PLANTS
- 3.2.1.2 FOR SELECTED CLASSES OF PLANT DESIGNS AND SITE CHARACTERISTICS

3.2.2 WHAT IMPROVEMENTS IN PHENOMENOLOGICAL UNDERSTANDING ARE MOST APPROPRIATE FOR DEVELOPING INTEGRATED STRATEGIES FOR MITIGATING THE CONSEQUENCES OF SEVERE ACCIDENT BEYOND OPERATOP ACTION?

- 3.2.2.1 IMPACT OF TECHNICAL RESOLUTION OF SOURCE TERM ISSUES ON THE INTEGRATED RISK ASSESSMENT OF A VARIETY OF CONTAINMENT PERFORMANCE FUNCTIONS
- 3.2.2.2 QUANTIFICATION OF POTENTIAL CONTAINMENT DISRUPTIVE LOADS
  - 3.2.2.2.1 COOLABILITY OF CORE DEBRIS
  - 3.2.2.2.2 THERMAL AND CHEMICAL INTERACTIONS BETWEEN CORE/DEBRIS AND CONCRETE
  - 3.2.2.2.3 THERMODYNAMIC AND CHEMICAL PHENOMENA IN THE CONTAINMENT ATMOSPHERE INCLUDING FISSION PRODUCT TRANSPORT
  - 3.2.2.2.4 PRODUCTION AND BURNING OR DETONATION OF HYDROGEN AND OTHER COMBUSTIBLE GASES
  - 3.2.2.2.5 LEAKAGE AND INTEGRITY FAILURE CHARACTERISTICS OF DIFFERENT CONTAINMENT DESIGNS AND VESSEL FAILURE MODE ANALYSIS
- 3.2.2.3 OTHER UNDERSTANDING OF SYSTEMS INTERACTIONS AND INTEGRATED STRATEGIES OF CONSEQUENCE MITIGATION FOR INDIVIDUAL PLANTS OR CLASSES OF PLANTS

3.2.3 WHAT IMPROVED UNDERSTANDING CAN BE EXPECTED FROM PLANNED RESEARCH IN DAMAGED FUEL BEHAVIOR AND FISSION PRODUCT TRANSPORT, PLATE-OUT, AND SUPPRESSION POOL SCRUBBING, ETC.? ARE SIMULATED FUEL AND ACCIDENT CONDITIONS OF THESE EXPERIMENTS SUFFICIENTLY CORRELATIVE TO LIKELY REAL WORLD STATES TO YIELD ACCEPTABLE CONFIDENCE LEVELS IN USING THE RESULTS TO GUIDE BACKFITTING DESIGN DECISIONS FOR EXISTING PLANTS? ARE MORE REALISTIC EXPERIMENTS PRACTICABLE? (SEE POLICY ISSUE 3.1.4)

3.2.4 IS THE STRATEGY OF USING DESIGN CONSERVATISMS TO OFFSET WIDE RANGES OF UNCERTAINTY AND LOW CONFIDENCE LEVELS FOR SPECIAL RISK CONSIDERATIONS SUCH AS EXTERNAL EVENTS (SEE ITEM 3.1.3) COST-EFFECTIVE AND SUFFICIENTLY PROTECTIVE OF PUBLIC HEALTH AND SAFETY, OR WOULD FINANCIAL RESOURCES BE MORE PRODUCTIVELY ALLOCATED IN RESEARCH TO REDUCE THE RANGES OF UNCERTAINTY FOR SUCH EVENTS?

3.2.5 ARE TECHNICAL DESIGN OPTIONS AND OPERATING PROCEDURE CHANGES THAT COULD REDUCE THE RISK OF SABOTAGE BEING RESEARCHED WITH SUFFICIENT VIGOR AS WELL AS THE POSSIBLE CONFLICT BETWEEN CERTAIN SABOTAGE RISK REDUCTION OPTIONS AND OTHER SAFETY OBJECTIVES?

3.2.6 WHAT, IF ANYTHING, WILL THE SEVERE ACCIDENT RESEARCH PROGRAM CONTRIBUTE TO AN IMPROVED EVALUATION OF QUALITY ASSURANCE STRATEGIES FOR REDUCTION OF HUMAN ERRORS IN PLANT DESIGN, CONSTRUCTION, OPERATION AND MAINTENANCE?

3.2.7 THE MENU OF TECHNICAL MODIFICATIONS TO BE CONSIDERED AND DECIDED IS:

- 3.2.7.1 ACCIDENT MANAGEMENT/HUMAN FACTORS
  - 3.2.7.1.1 OPERATOR TRAINING INCLUDING USE OF ACCIDENT SIMULATORS
  - 3.2.7.1.2 IMPROVED TRANSIENT INDICATORS AND CONTROL ROOM DATA ACQUISITION AND DISPLAY
  - 3.2.7.1.3 USE OF ARTIFICIAL INTELLIGENCE (COMPUTER AIDED) INCLUDING ATTENTION TO RISK ISSUES IN MAN-MACHINE INTERFACES (MMI)
  - 3.2.7.1.4 USE OF ADVANCED INSTRUMENTATION IMPORTANT TO ACCIDENT MANAGEMENT
  - 3.2.7.1.5 EXTEND EMERGENCY PROCEDURES GUIDELINES TO SEVERE ACCIDENT SEQUENCES
- 3.2.7.2 CANDIDATE HARDWARE IMPROVEMENTS OF SEVERE ACCIDENT SAFETY
  - 3.2.7.2.1 ADDITIONAL CONTAINMENT HEAT REMOVAL
    - 3.2.7.2.1.1 ACTIVE VERSUS PASSIVE
  - 3.2.7.2.2 CONTAINMENT ATMOSPHERE PARTICULATE CAPTURE
  - 3.2.7.2.3 CONTAINMENT ATMOSPHERE MASS REMOVAL
    - 3.2.7.2.3.1 FILTERED VERSUS UNFILTERED
    - 3.2.7.2.3.2 LOW FLOW VERSUS HIGH FLOW
  - 3.2.7.2.4 INCREASED CONTAINMENT MARGINS
    - 3.2.7.2.4.1 INCREASED VOLUME
    - 3.2.7.2.4.2 INCREASED PRESSURE CAPABILITY
    - 3.2.7.2.4.3 PRESSURE SUPPRESSION FEATURES
  - 3.2.7.2.5 COMBUSTIBLE GAS CONTROL
    - 3.2.7.2.5.1 DELIBERATE IGNITION
    - 3.2.7.2.5.2 INERTING (PRIOR/POST ACCIDENT)
    - 3.2.7.2.5.3 FIRE SUPPRESSION (HALON/WATER FOGS)
  - 3.2.7.2.6 CORE RETENTION DEVICES
    - 3.2.7.2.6.1 DRY VERSUS WET VESSEL CAVITY
    - 3.2.7.2.6.2 ACTIVE VERSUS PASSIVE COOLING OR NO COOLING
  - 3.2.7.2.7 MISSILE SHIELDS
    - 3.2.7.2.7.1 IN-VESSEL STEAM EXPLOSIONS
    - 3.2.7.2.7.2 VESSEL THERMAL SHOCK
    - 3.2.7.2.7.3 VESSEL MELT-THROUGH AT HIGH PRESSURE
    - 3.2.7.2.7.4 EX-VESSEL STEAM EXPLOSIONS
    - 3.2.7.2.7.5 COMBUSTIBLE GAS EXPLOSIONS
  - 3.2.7.2.8 BWR CONTAINMENT SPRAY SYSTEM
  - 3.2.7.2.9 PWR PRIMARY SYSTEM DEPRESSURIZATION
    - 3.2.7.2.9.1 AUTOMATIC VERSUS MANUAL
    - 3.2.7.2.9.2 ADDITIONAL RELIEF CAPACITY
    - 3.2.7.2.9.3 PRESSURE SUPPRESSION FEATURES
    - 3.2.7.2.9.4 RADIOACTIVITY REMOVAL SYSTEMS
  - 3.2.7.2.10 ADD-ON DECAY HEAT REMOVAL SYSTEMS
    - 3.2.7.2.10.1 HIGH PRESSURE VERSUS LOW PRESSURE
    - 3.2.7.2.10.2 OPEN LOOP VERSUS CLOSED LOOP
    - 3.2.7.2.10.3 PRIMARY SYSTEM VERSUS SECONDARY SYSTEM
    - 3.2.7.2.10.4 ACTIVE VERSUS PASSIVE
  - 3.2.7.2.11 SPECIFIC PREVENTION CONCEPTS
    - 3.2.7.2.11.1 IMPROVED DRAIN OR VALVE DESIGN
    - 3.2.7.2.11.2 IMPROVED MAINTENANCE PROCEDURES
    - 3.2.7.2.11.3 IMPROVED CONTROL LOGIC
    - 3.2.7.2.11.4 REDUCTION OF COMMON MODE DEPENDENCIES
  - 3.2.7.2.12 ENHANCED EQUIPMENT SURVIVABILITY

[a] For explanation of acronyms; see footnote 3

Other attributes or societal consequences of energy options should also be included in the heuristic survey method such as net comparative effects on health and safety, the ecology, land use, property damage, personal income, national security, energy reliability and security of supply, the interests of future generations and equity considerations in basic differences in the comparative distribution of beneficial and adverse effects. Starr and Whipple have noted that conflicts arise from the balancing of public benefits and involuntary risks to the individual and they proceed to explore the question, "How should decision processes operate to minimize social costs and maximize social benefits?" (31)

> Social costs include intangibles, and the question immediately evident is what costs are included and how are they weighted. It is obvious that if we have a decision process, and if we know how to determine costs to the individual, we still have a problem with the full disclosure of all the social costs. What is full disclosure? Do we include the options for societal risk management as part of full disclosure: that is, the cost of the alternatives for managing the risk? Does it include all present events, future events, the people who get the benefits, and the people who bear the costs? We have geographic distributions, time distributions, demographic distributions--all of these are included by the term full disclosure. Where do we draw the boundaries?

One of the weaknesses common to public opinion survey techniques is that only the briefest of time is permitted for response to questions (and few questions are asked) because this minimizes the cost to the respondee as well as the survey sponsor. However, this detracts from the quality of the value or preference responses of the individual, who is inclined to give top-of-the-mind opinions rather than well-thought-out answers. The full range of the individual's knowledge and cognitive abilities to work out internal conflicts of personal values and what is a fair compromise between these personal values and the overall good of present and future society is seldom exploited in these techniques.

An imaginative step in this direction was pursued by the research team of Fischhoff et al., which asked members of the Eugene, Oregon, League of Women Voters and their spouses (52 women and 24 men) to participate in a study of perceived risk, acceptable risk, and perceived benefit for each of 30 activities and technologies (32). An average of two hours was contributed by the participants in return for a contribution to the organization's treasury. The authors add the following perspective to their approach:

> Although League members and spouses are by no means representative of all American adults, they do constitute an extremely thoughtful, articulate, and influential group of private citizens. If there are systematic relationships between people's judgments of risk and benefit, they should be found in these participants' responses. While the particular relationships found here might differ from those found with other populations, the opinions of League members may be quite similar to those of many of the private citizens most heavily engaged in the public policy-making process.

I find this explanation a rather agreeable one because real world experience suggests that individuals with less-than-average intellectual capabilities and individuals who spend little personal time acquiring knowledge of complex technological and policy issues are generally prone to rely on the trickle-down wisdom or judgment on these matters of persons who hold their respect. There are, of course, many sources of bias in either representative or participatory democracy, not the least of which is the phenomenon of cognitive dissonance, a form of polarization of attitudes in which persons defend their views by a restriction of logical argument and denials against information that contradicts deeply held beliefs (33). Thus, a potential advantage of the heuristic opinion and preference evaluation research method is that it encourages respondents to go along with the spectrum of heuristic information premises supplied by a respected panel of professionals. Moreover, the hierarchical structure of technological and policy issues gives structure to a logical set of survey questions that circumvent an undesirable restriction of logic purview involving equity and other interrelated societal values that may alternatively be contradictory or mutually supportive.

The experiment by Fischhoff et al., whatever its limitations, did have the virtue of requiring respondents to think in broader and more contemplative terms than is generally true of most opinion surveys. For example, the methods used in this study demonstrated the importance of evaluating benefits when rating the social acceptability of risks: It divided respondents into two groups, one of which evaluated risk attributes before evaluating gross benefits, and the second of which used the reverse sequence. Respondents in both groups were asked to provide a numerical "risk adjustment factor" to indicate for any technology (such as nuclear energy) that the respondent did not presently judge to be acceptably safe, that the technology would have to be made X times safer to be acceptable. Trade-offs between dollar costs of improved safety and the increased safety level achievable through modifications were to be included in the process of estimating the risk-adjustment factor. Of course, the value of such assessments by the lay public is seriously questionable, if cost-risk trade-off information has not been provided by qualified technological experts. Nevertheless, it is noteworthy that the group that evaluated risks before benefits provided numerical risk-adjustment factors averaging 23 percent higher than those of the group that evaluated benefits first.

In the case of motor vehicles whose risks and benefits are rather commonplace knowledge, the group that evaluated risks before benefits yielded a risk adjustment factor 49 percent higher than the group that evaluated benefits first. This is striking evidence that the Tversky-Kahneman concept of "availability heuristics" (supra) applies also to the anchoring phenomena of the very sequence by which logical issues are explored. It further demonstrates that a team of responsible experts who spend many hours, days, or months taking a systems analysis view of the full gamut of risks, costs, and benefits of technological options with iterative judgment applied to feedback loops in analysis and debate have a superior opportunity to overcome the inherent distortions of anchoring phenomena demonstrated by the Fischoff et al. experiment.

It also indicates the vital importance of considering benefits in any comparison of the risks associated with technological options. Yet the neglect of technological benefits in such comparisons is widespread and often lacking in imagination. For example, only two of seven institutional reports included a comparison of benefits in their comparative studies of nuclear and alternative energy sources, although all included either quantitative or qualitative comparisons of risks (5).

Although the participants in the Fischhoff experiment felt, on the average, that nuclear power should be made about 29 times safer than at present (i.e., in 1978) to be acceptable, there is no quantitative indication in absolute terms how safe they thought nuclear energy was (e.g., expected early or latent fatalities per billion kilowatt-hours of electricity generated) nor what the cost might be to make nuclear power 29 times safer[4]. Thus, information of this form is about as

---

4 Suppose, for example, that these persons thought nuclear risk at that time was 100 times more risky than a consensus estimate of the experts would yield. This brings us back to the question in the doctor-patient model as to whether the patient's personal interests would be served best by basing the consent decision for an operation on the patient's own initial estimate of the fatality risk of the operation or that of the doctor's informed estimate.

ill-suited to making improvements in criteria of acceptable nuclear risk as the presently vague criterion of "no undue risk" embedded in NRC's Code of Federal Regulations (see item 3.1.1.1 of Table 2). Recognizing the inadequacy of this acceptability criterion, the Commission has proposed a Policy Statement on Safety Goals (supra). In a 1980 paper on "Concepts, Problems, and Issues in Developing Safety Goals and Objectives for Commercial Nuclear Power," Mattson et al. sought to establish criteria by which candidate safety goals might be selected and evaluated, as well as an identification of variations in goal forms worthy of consideration (34). The latter included choices such as single or multiple goals; quantitative or qualitative goals; ends-oriented or process-oriented goals; absolute or relative goals; and individual- or society-oriented goals. In its proposed Policy Statement, the Commission rejected the notion of a single goal for acceptable nuclear power plant safety, but the proposed set of qualitative goals and quantitative design objectives include all of the remaining design forms.

One of the candidate criteria in Table 2 for a determination of acceptable/unacceptable safety of nuclear power plants is that of comparative risk criteria (item 3.1.1.2). One of the Commission's proposed safety goals involves comparative risk:

> Societal risks to life and health from nuclear power plant operation should be comparable to or less than the risks of generating electricity by viable competing technologies and should not be a significant addition to other societal risks.

As noted by the Commission, the comparison to viable competing technologies particularly applies to coal-fired plants. The most recent Environmental Impact Statements relating to the proposed construction of nuclear plants provide mean estimates of excess mortalities (equivalent in concept to life-shortening expectations) to the general public from power generation alone of 0.06 fatalities per year for a nuclear plant of 1000 MWe capacity (80 percent capacity factor) compared to 17.3 fatalities per year for a coal-fired plant of the same capacity and capacity factor (5). If these estimates are correct--and both have a comparably broad range of uncertainty--then over a 30-year assumed plant operating life there would be an expected net savings of about 500 lives from the nuclear choice. If the entire fuel cycle were included for both energy choices (including mining, fuel transportation, fuel processing and enrichment, power generation, etc.), then, according to estimates from the same sources, there would be a net savings of about 1,200 lives from the nuclear over the coal-fired plant option over a 30-year operating period.

Thus, a heuristic information premise might use improved estimates of excess fatalities for these types of nuclear and coal comparisons along with an updated spectrum of estimates for net economic benefits of one fuel choice over the other. The net economic benefits could be in reverse directions for different regions, principally because coal transportation costs vary substantially between regions or also because of high and low projections of relative fuel costs over a 30-year period. The participants in the public survey might be asked, regarding varying combinations of these two types of effects, what their preferred choice would be, taking for the moment no other environmental or societal or equity effects into account. By a method of using diminishing assumptions of _ceteris paribus_ for all remaining effects, a progression of values and preferences could be educed.

One important attribute needing attention in this regard is that of the catastrophic risk of potential nuclear power accidents, which a number of experts as well as many of the public feel deserves an extra measure of weight versus the chronic risks of coal-fired plant pollution that is dispersed more uniformally over space and time (36). In a report on quantitative safety goals, the U.S. Nuclear Regulatory Commission's Advisory Committee on Reactor Safeguards (ACRS) discussed risk aversion, suggesting extra weight be given to potential catastrophic deaths from nuclear accidents versus dispersed deaths from alternative energies (9). Their trial alpha model for assessing equivalent social cost of early deaths from catastrophes computes this cost by raising the actual deaths by an exponential factor of ($\alpha = 1.2$). Thus, 100 catastrophic deaths yield the equivalent social cost of 250 dispersed deaths from noncatastrophic causes, and 1,000 catastrophic deaths equal 4,000 dispersed deaths.

But this formulation poses a serious equity consideration. In the latter scenario it is germane to ask whether it is equitable that 3,000 extra dispersed lives might be lost to spare the 1,000 catastrophic loss of lives in the nuclear option. Do not the deaths represent roughly the same tragic loss to the persons whose lives are cut short and the same grief to their surviving relatives and friends as the catastrophic deaths? Is it not the anxiety costs and societal resiliency problems which the affected community must bear that merit the extra weight to be given to catastrophic risk rather than fatalities per se? These and other questions might be presented to the survey group as explanations and heuristic premises to seek their expressions of values on such matters.

An important set of policy issues for which input of public values would be most appropriate is the determination of the scope and treatment of risk-cost-benefit considerations in an evaluation of technical options for risk or uncertainty reduction (category 3.1.8 in Table 2). One highly controversial item under this category involves the ethical implications of using a quantitative safety-cost trade-off criterion for guiding decisions on the financial resources that are warranted for technological design options or operating procedures in reducing the risk of early or latent fatalities (item 3.1.8.5). Baram notes the wide range of practices by U.S. Federal agencies in safety-cost trade-off criteria: about $100,000 per death averted by the U.S. Department of Transport; $200,000 to $2 million for various analyses by the Consumer Product Safety Commission; and up to roughly $5 million per death averted in the NRC application of $1,000 per manrem averted in the routine release of radioactive material (37). The U.S. Environmental Protection Agency employed a range of $250,000 to $500,000 in the Environmental Impact Statement for establishing 40 CFR 190 (November 1976).

Many would protest that the use of such a decision criterion is immoral because it implies putting a price on human life. It is true that the value of a human life has important intangibles such as the social and personal value of life that goes beyond any income losses due to a life cut short. But this argument misses the point by looking at ethical issues through the wrong end of the telescope. Decision criteria of this kind are not intended to put a price on human life, but rather to encourage an equitable management of finite societal resources among various opportunities for saving lives. An excessive expenditure of resources for risk reduction in one area of technological activity deprives society of the use of these same resources to save lives in other areas of life-saving opportunities (of which there are many) with a possibly greater number of lives saved. The heuristic opinion and preference evaluation research method could be especially valuable to policymakers in dealing with ethical and equity issues of these kinds. The heuristic premises might usefully include a spectrum of safety-cost trade-off values for prompt accidental fatalities as well as for latent cancer fatalities that may take 20 to 40 years following exposure to carcinogens.

There are many other possibilities for using the heuristic method to gain useful public input on values and preferences on complex technological and policy issues such as those illustrated in Table 2. Prominent is the set of issues (item 3.1.4) related to the question, "What is an acceptable/un-acceptable level of uncertainty in risk assessment (i.e., how accurately do we need to know the risks of the options)?" Another (item 3.1.7) is, "What is an appropriate treatment of unacceptable risk and uncertainty?" Still another significant application of the heuristic technique that would cut across a large number of the technical and policy issues involves exploring public views on what methods of assessing risks, costs, and benefits and procedures for making policy and regulatory decisions would improve public trust and confidence in regulating decisionmaking regarding technological options.

Heuristic Methods for Discerning Societal Value Changes

One method of assessing changes in societal preferences and attitudes toward technological options is through comparisons of conventional public opinion surveys conducted at different periods of time. An examination of trends in public evaluations of nuclear power and other energy options was recently made by Melber (38). However, forecasts of changing values from trend analysis alone are probably not a good idea because new events, discoveries, and unexpected cultural developments (see Yankelovich, op. cit.) can reverse or accelerate trends. A range of heuristic information premises or developmental assumptions could provide a most useful input to assessing the potential for value change as well as the forecasting of value change. Postulations of future technological or resource discoveries, the relative trends in fuel prices or other cost components of competing technologies, and certain cataclysmic or momentous events could be used for this purpose in the heuristic opinion and evaluation research method. The latter might include, for example, the advent of another nuclear accident, a scientific determination of the imminence and severity of the greenhouse effect from burning fossil fuels, better cures for cancer, low-cost air pollution control methods, further disruptions of world oil supplies, and many others. One of the key contingency methods of planning in the face of irreducible uncertainties is to anticipate the possibility of both untoward and favorable events and public reaction to them so that a smoother transition to the new circumstances can be made with a minimum of disruption and waste in resource allocations.

SUMMARY

There is a wide gap in the present treatment of the risks, costs, and benefits of technological options by technical experts and their perception by the lay public. With increasing pressures for participatory democracy, it is highly important--if technological progress is to serve the best interests of society--to find improved methods of assimilating public inputs into decisions on technological and related policy issues. Defects in the treatment of these issues by the technical experts or professional elites, as well as the public, have been examined in some detail. The heuristic opinion and preference evaluation research method has been proposed to deal with both kinds of defects. The use of a hierarchical structure of technical and policy issues is a logical first step in applying this method because issues might be thought of as logic tools by which information is filtered and organized for decisionmaking in resolving differences among parties of interest and responsibility.

The greatest weakness of present methods of opinion survey--and the one which most severely limits the application of these inputs to regulatory or other decisionmaking regarding technological options--is the static orientation of such opinion surveys to present value and preferences of the public based upon present information (or misinformation) premises regarding the risks, costs, and benefits of technological choices. It is the author's view that superior insight would be provided to technological policy development by public perception research that dynamically explores, in the manner of sensitivity analyses, the heuristic shifts of values and attitudes of the public as related to variations in information premises of the form: "If attribute 'X,' (or attributes X, Y, etc.) were actually the realized outcome(s) of certain technological/regulatory options, then what would be your valued preferences?" The choice of the X, or Y, outcomes to serve as a basis for these sensitivity analyses should, in my view, be based on the high, low, or medium estimates (or forecasted outcomes) reflecting a reasonable range of uncertainty among experts. This would supply at least some of the beneficial insight to the regulator as is achieved in the information exchanges between physician and patient in which the physicians supply expert information on the expected range of outcomes of treatment options, and the patients provide information on their sense of values as related to these outcomes.

REFERENCES

1. J. Naisbitt, Megatrends: Ten New Directions Transforming Our Lives. New York: Warner Books, Inc., 1982.

2. N. Gonzalez, "In Defense of Elitism," Science 213(4511) (1981).

3. M. Douglas and A. Wildavsky, "How Can We Know the Risks We Face? Why Risk Selection Is a Social Process," Risk Analysis 2(2), 49-51 (1982).

4. M. Spangler, "The Role of Interdisciplinary Analysis in Bridging the Gap Between the Technical and Human Sides of Risk Assessment," Risk Analysis 2(2), 101-114 (1982).

5. M. Spangler, "A Critique of Methods in the Quantification of Risks, Costs and Benefits in the Societal Choice of Energy Options," Annals of Nuclear Energy 10(3/4), 119-151 (1983).

6. M. Spangler, "Syndromes of Risk and Environmental Protection: The Conflict Between Individual and Societal Values," The Environmental Professional 2(3/4), 274-91 (September 1980).

7. J. Rubens, "Retooling American Democracy," The Futurist, 17, 58-64 (February 1983).

8. H. Denton, "Nuclear Power: Epilogue or Prologue?" The Energy Journal, 4(1), 125-141 (1983).

9. "An Approach to Quantitative Safety Goals for Nuclear Power Plants," NUREG-0739, report of the Advisory Committee on Reactor Safeguards, U.S. Nuclear Regulatory Commission, p.p. 65-67 (October 1980).

10. M. Spangler, "Risks and Psychic Costs of Alternative Energy Sources for Generating Electricity," The Energy Journal 2(1), 37-59 (January 1981).

11. M. J. Bowman and G. L. Bach, Economic Analysis and Public Policy. New York: Prentice-Hall, Inc., 1943, p. 140.

12. D. Yankelovich, New Rules: Searching for Self-Fulfillment in a World Turned Upside Down. New York: Bantam Books, 1982, p. xviii.

13. Council on Environmental Quality, "National Environmental Policy Act: Implementation of Procedural Provisions: Final Regulations," Federal Register, Vol. 43, No. 230, November 29, 1978, p. 56004.

14. P. Ricci and L. Molton, "Risk and Benefit in Environmental Law," Science 214(4) 1096-1100 (1981).

15. B. Freedman, Being at Risk and Placing at Risk: A Consensual Guide, a draft study paper for the Protection of Life, Health and the Environment Project of the Law Reform Commission of Canada (October 1982), pp. 273 and 279.

16. J. Burns, Leadership. New York: Harper & Row, 1978, p. 36.

17. B. Fischhoff, "Informed Consent in Societal Risk-Benefit Decisions," Technological Forecasting in Social Change, 1979, p. 349.

18. A. Tversky and D. Kahneman, "Availability: A Heuristic for Judging Frequency and Probability," Cognitive Psychology 4, 207-232 (1973).

19. R. Cooper, L. Coles, W. Rosseau, and D. Meyers, "The Application of Heuristic Modeling to Econometrics," UCRL-51882, a report prepared under contract by the Lawrence Livermore Laboratory, University of California at Livermore, for the U.S. Energy Research & Development Administration, August 11, 1975.

20. P. Hutt, "A Regulator's Viewpoint," How Safe Is Safe? The Design of Policy on Drugs and Food Additives. Washington, D.C.: National Academy of Sciences, 1974, p. 124.

21. P. Stallen, "Risk of Science or Science of Risk?" Society, Technology and Risk Assessment, J. Conrad, ed., New York: Academic Press, 1980, pp. 131-148.

22. E. Siddall, "Risk, Fear and Public Safety," a report issued by Atomic Energy of Canada Limited (April 1981).

23. M. Maxey, "Radiation Health Protection and Risk Assessment: Bioethical Considerations," Perceptions of Risk, Proceedings of the 15th Annual Meeting of the National Council on Radiation Protection and Measurements, National Academy of Sciences, Washington, D.C., March 14-15, 1979, pp. 18-33.

24. S. Cotgrove and A. Duff, "Environmentalism, Values, and Social Change," British Journal of Sociology 32(1), 92-110 (1981).

25. U.S. Nuclear Regulatory Commission, "Proposed Commission Policy Statement on Severe Accidents and Related Views on Nuclear Reactor Regulation," Federal Register 48(72), 16014-16023 (April 13, 1983).

26. U.S. Nuclear Regulatory Commission, "Policy Statement on Safety Goals for the Operation of Nuclear Power Plants," Federal Register 48(50), 10773-10781 (March 14, 1983).

27. "Reactor Safety Study (Rasmussen Report)," Committee Print 94-61, Oversight Hearing before the Subcommittee on Energy and the Environment of the Committee on Insular and Interior Affairs, U.S. House of Representatives, 94th Cong., 2d Sess., (1976).

28. "Nuclear Public Information and Rational Public Policy Decisions," Committee Print 72, Hearing before the Subcommittee on Energy Research and Production of the Committee on Science and Technology, U.S. House of Representatives, 97th Cong., 1st. Sess., (1982).

29. J. Kemeny et al., Report of the President's Commission on the Accident at Three Mile Island: The Need for Change. Washington, D.C.: Government Printing Office, October 1979.

30. M. Firebaugh, "Public Attitudes and Information on the Nuclear Option," Nuclear Safety 2(2), 147-156 (1981).

31. C. Starr and C. Whipple, "Risks of Risk Decisions," Science 208, 1114-1119 (June 6, 1980).

32. B. Fischhoff, P. Slovic, et al., "How Safe Is Safe Enough? A Psychometric Study of Attitudes Towards Technological Risks and Benefits," Policy Science 9, 127-152 (1978).

33. D. Bem, "Self-Perception: An Alternative Interpretation of Cognitive Dissonance Phenomena," Psychological Review 74, 183-200 (1967).

34. R. Mattson, et al., "Concepts, Problems, and Issues in Developing Safety Goals and Objectives for Commercial Nuclear Power," Nuclear Safety 21(6), 703-716 (1980).

35. U.S. Nuclear Regulatory Commission, Safety Goals For Nuclear Power Plant Operations, NUREG-0880, Revision 1, For Comment, May 1983.

36. M. B. Spangler, "Societal Aversion to Catastrophic and Involuntary Risks of Nuclear Power," Transactions, 1982 Annual Meeting of the American Nuclear Society, Los Angeles, CA, June 6-10, 1982, pp. 449-450.

37. M. Baram, "Cost-Benefit Analysis: An Inadequate Basis for Health, Safety, and Environmental Regulatory Decisionmaking," Ecology Law Quarterly 8, 473-531 (1980).

38. B. D. Melber, "Public Evaluations of Nuclear Power and Other Energy Options," Transactions, 1982 Annual Meeting of the American Nuclear Society, Los Angeles, CA, June 6-10, 1982, p. 445.

Canvey Island: The Dynamics of a Major Technological Risk Assessment

P. Lagadec

INTRODUCTION

The industrialized nations, rushing through the last quarter of the twentieth century on the crest of an immense technological wave, find in the wake of that wave risks of unprecedented magnitude. The accident in a chemical plant in Seveso in 1976, for instance, did not only threaten the lives of the Italians living in the adjacent valley, but it also suggested the possibility of evacuating the country's industrial and commercial capital, Milan.

As it becomes increasingly difficult and economically impractical to physically isolate such hazards as industrial plants and nuclear reactors, the potential risks multiply. Public and private commercial interests, the so-called national interest, and the interests of individuals, all become involved in a debate whose outcome must ultimately depend upon societal issues as much as technical data. The social and political dynamics driving such a debate are amply illustrated in the history of the risk assessment of Canvey Island.

CANVEY ISLAND: DESCRIPTION AND EARLY HISTORY

This English island is located in the mouth of the Thames near Southend. Its south side is on the estuary; on its other three sides it is narrowly separated from the mainland by creeks and small rivers, crossed by a single bridge. Canvey and the area immediately adjacent to the west (the Thurrock area, including Stanford le Hope) are the sites of several large oil company facilities: refineries, storage tanks, a methane terminal, and wharves used by the tankers coming from the North Sea. It is a

NATO ASI Series, Vol. G4
Environmental Impact Assessment, Technology Assessment, and Risk Analysis. Edited by V.T. Covello et al.

highly industrialized area, and it has a dense population that is augmented each summer by vacationing Britons.

Residential development has proceeded as follows:

1900: 300 residents living on the island

1920: 1,800 residents

1950: 11,000 residents

1975: 32,000 permanent + 500 occasional residents

1980: 35,000 permanent + 9,000 occasional residents

(within a radius of 10 km: 250,000 persons)

Industrial development has been thus:

1870-1930: petroleum industry develops upstream (to the west of the island)

1936: first petroleum storage facility on Canvey

1950: Shell and Mobil develop upstream

1954: new storage facilities on Canvey

1964: methane terminal built on Canvey (22,000-ton capacity)

1966-1967: methane terminal expanded (100,000-ton capacity)

It can be seen from the above figures that Canvey in the early years of the century was but sparsely populated, having only 1,800 residents by 1920. These residents were solidly opposed to industry on the island, but by 1936 with the backing of the House of Lords, light industry was allowed on the river front. At the same time the population of Canvey Island grew, reflecting the policy of the government to develop new urban areas outside London. As an environmental policy took shape in the nation, Canvey was also developed as a seaside resort, and a protected greenbelt was established on the north of the island.

In 1964 an Industrial Development Certificate was issued to British Gas Corporation by the Government for a new refinery project--a methane terminal. In Parliament the local M.P., Sir Bernard Braine, angrily addressed the Minister responsible, speaking of a "fait accompli." The following years saw planning permissions given to United Refineries in 1965 and Occidental Refinery in 1970. Decisions on these projects were reached following a complex process of public inquiries, which included strong local opposition and even negative recommendations by the inspectors in charge of the inquiries.

BACKGROUND EVENTS: 1970-1974

In 1970 a committee was set up by Lord Robens to study safety in the workplace and "to consider whether any further steps are required to safeguard members of the public from hazards, other than environmental pollution, arising in connection with activities in industrial and commercial premises...." (1, p. 89). The Robens Report, published in 1972, was unequivocal in its call for "radical re-thinking" in dealing with the threats posed to health and safety (1, p. 3). The report recommended new legislation and the establishment of a National Authority for Safety and Health at Work to safeguard not only workers but also the general public (1, p. 91). The report observed:

> Today, materials which are intrinsically dangerous because of their explosive or flammable properties are now being stored and used on a scale undreamed of even a decade ago. Despite the great advances made in control technology, it would be foolish to ignore the potentialities of this relatively new situation, and irresponsible merely to assume that the improbable will not happen. It is a melancholy fact that much of our preventive legislation was introduced following events that were both unforeseen and disastrous.... (1, p.4)

Indeed, a new law was not yet in effect when, on June 1, 1974, the Flixborough Works of Nypro (U.K.) Ltd. were demolished by an explosion of "warlike dimensions"--as the inquiry put it--killing 28 and injuring 36 on site and 53 off site (2). The extent of the disaster was curtailed by the time of the explosion (Saturday) and by the fact that this petrochemical factory was in a nonurban area. However, the inquiry concluded that "no one concerned in the design or construction of the plant envisaged the possibility of a major disaster." (2, par. 217)

As the result of several reports by the Factory Inspectorate, the Robens Report, and the reality of the Flixborough disaster, the Health and Safety at Work Act was legislated in 1974. This law, however, did not aim at suppressing sources of major risks or at forcing industry to apply the best possible means to reduce risks; the key words were "as far as reasonably practicable."

SIR BERNARD BRAINE: 1964-1976

Meanwhile, the M.P. for the district including Canvey, Sir Bernard Braine, was continuing the lone fight he had first launched in 1964 when the Government approved the British Gas methane terminal on the island. This great Conservative figure, eloquent and outspoken, missed no opportunity to harangue the Government, Prime Minister as well as Minister, Tory and Labour alike, about the cause he so ardently championed. His major fear was the methane terminal, but he was also concerned about the totality of risks his constituency faced. In a speech in 1973 he recognized that responsibility for various hazards, including the transport of dangerous cargoes on the Thames, was divided among several Government departments, but there remained an urgent need to establish overall responsibility for the "totality of effect of these cumulative planning decisions on our environment, and on the health and safety of our residential population..." (3, pp. 899-900).

This was a theme he constantly reiterated. In April 1974 he stressed:

> All I am asking is that a total look should be taken of all the planning decisions that are being made one after the other. We had the ridiculous assertion last week by the promoter of the Burmah-Total Refineries Trust Bill that because there were already other refineries and installations near Cliffe it would not matter if another were added. But the whole point is that it would matter very much indeed. (4, pp. 105-106)

Two months later, in a debate in Parliament following the Flixborough disaster, the M.P. from that district commented that while "...the people I represent can sleep soundly," for the plant had already been destroyed, "...we must ensure by any means possible that we prevent such a disaster occurring anywhere else," and he pointedly referred to Canvey, with its much greater "geographical concentration." (5, p. 255) Sir Bernard had earlier in the debate complained, "Why is it that Governments so often fail to see the problem in the round?" Relating his attempts since 1970 to warn of "the dangers--indeed, the folly--of introducing oil refineries into an area where there are already far too many major fire hazards," he concluded: "All that has happened since then is that the population has increased by some 5,000 and we have had two oil refineries forced upon us." (5, p. 252) A month later in the House of Commons Sir Bernard recounted the feeling of "bitterness, despair and frustration" which he and a local deputation of councillors and resistance groups felt when they had fruitlessly pleaded with the Secretary of State that previous year to consider setting up an independent inquiry "into the whole question of industrial development in the Thames Estuary with special reference to the proliferation of oil refineries." He summarized the situation as he saw it: "Despite an existing concentration of high fire risks, and despite intense local opposition, deliberate steps had been taken one after the other by both Labour and Conservative Governments to add to that concentration of risk." (6, p. 1466)

Following the publication of the report of the Flixborough inquiry in 1975, Sir Bernard said:

> For my people at Canvey the point of social tolerance has long been passed. Whatever the excuse for building up hazardous industry there in the past alongside a much smaller population, the situation has now totally changed. Existing hazards must be reduced. No new hazards must be added. (7, p. 838)

1976: THE DECISION TO MAKE A COMPREHENSIVE ANALYSIS OF RISKS

In March 1976 the Labour Government made the decision to launch an in-depth study of the risks facing Canvey Island. In addition to Sir Bernard's vehement attacks, there were several other factors prompting the decision:

1) The Secretary of State, overruling the inspector, had granted permission to United Refineries in 1973, which brought an enormous local protest and a referendum in 1976 registering 10,000-900 against this action.

2) A reopened inquiry to consider whether to revoke the United Refineries permission--undertaken by the new Minister of the Environment and promised before the Labour victory in February 1974--showed the depths of the original opposition and was further strengthened by conclusions of a technical expert, Dr. Napier of London University, who underlined the necessity of an overall study of the risks in Canvey.

3) Oil companies were increasingly hesitant, especially for financial reasons, about developing their projects on Canvey given the ongoing debate.

4) A national technical journal had written about cracks in the ice-wall of the in-ground storage of the methane terminal.

5) A new Government body had been created at the beginning of the year, the Health and Safety Executive (HSE), whose duties included the identification and control of major hazards. Canvey provided the ideal subject for their first big project.

All of these factors interacted to promote the study. There was undoubtedly also the underlying motivation to generate a comprehensive technical study that would break the deadlock between the interests of industry on one side (in the form of British Gas) and the residents of the island on the other (as represented by Sir Bernard Braine). Mediation would supposedly be effected by the "neutral" Government, interested solely in the national interest.

1976-1978: THE HSE REPORT

To assist them, the Health and Safety Executive turned to the Atomic Energy Authority to provide specialist advice about major industrial hazards. In the end, 30 experts worked for two years at a cost of 400,000 pounds. This team was instructed to be "as realistic as possible in its appraisals, and in cases of doubt not to err on the side of optimism." (8, p. 2) To carry out their assessment, the group of experts proceeded:

- To identify any potentially hazardous materials, their location and the quantities stored and in process
- To obtain and review the relevant properties such as flammability and toxicity
- To identify the possible ways in which failure of plants might present a hazard to the community
- To identify possible routes leading to selected failures. Typically, the factors examined included operator errors, fatigue or aging of plant, corrosion, loss of process control, over-filling, impurities, fire, explosion, missiles, and flooding
- To quantify the probability of the selected failures occurring and their consequences (8, p. 9)

Concluding that they could not "assess the possible interaction between a hazardous event occurring at a particular site and the surrounding sites without first appraising the potential hazard from individual events on particular sites" (8, p. 16), the team assessed the various major risks one at a time: ammonia, hydrogen fluoride, liquefied natural gas, liquefied petroleum gas, flammable liquids, and ammonium nitrate. Then the team proceeded to assess the "likelihood and possible effects of various kinds of accidents, including a number resulting from interactions between separate activities and installations in the area." (8, p. 23)

The HSE agreed with the investigating team that "the quantitative approach was the most meaningful way of comparing different risks." (8, p. 11) To this end two different sets of figures were calculated: the level of individual risk and the level of societal risk. The terms used were defined in the report: the word risk was used "to mean strictly the probability of an event occurring, either the probability of the accident itself or the probability of the defined consequence. Thus, the risk to an individual is defined as the probability of his being killed as a result of the types of accidents considered." Societal risk, "a measure of a combination of the likelihood and severity of a single event," expressed the "social consequences of serious accidents in terms of the number of people who could be killed in a single accident, and the probability that this will occur." (8, p. 12)

Before providing the results of the analysis, the team clarified the general level of risk in the nation, which was based mainly on fatalities in road accidents, home accidents, and workplace accidents. As far as societal risk was concerned, they acknowledged that "a single accident causing many deaths arouses greater concern among the community than the same number of deaths distributed widely over the population," yet most of the 18,000 accidental deaths occurring annually arose from a "multitude of events in each of which only a few people are involved." (8, pp. 14-15)

With the probabilities of general individual and societal risks established as reference points, the team then presented their calculations for the present risks of existing activities on Canvey Island and adjacent territory, for the additional risks of proposed activities, and for the level of risks that could be reached after improvements had been made. Table 1 gives the probabilities in units of chances in 10,000 per year. Table 2 indicates some of the facilities and activities at the site mentioned in Table 1.

The investigators were extremely cautious and frank in admitting problems with the reliability of the data collected,

Table 1. Levels of Societal and Individual Risk from Existing Activities in the Canvey Island/Thurrock Area; from Proposed Activities; from Activities after Improvements Achieved (8, p.32)

Annex 1
Societal risks

Existing installations

Probabilities in units of chances in 10 000 a year
Estimated upper limit of probabilities for numbers of casualties exceeding:

	10		1 500		3 000		4 500		6 000		12 000		18 000	
Location	(a)	(b)	(a)	(b)	(a)	(b)	(a)	(b)	(a)	(b)	(a)	(b)	(a)	(b)
Shell UK Oil	7·8	3·3	5	1·9	3·6	1·3	2·7	0·9	1·8	0·5	1·2	0·3	0·8	0·2
Mobil Oil Co Ltd	1·4	0·3	1·1	0·2	0·7	0·1	0·3	0·1	—	—	—	—	—	—
British Gas Corporation	14·7	4	5·4	1·5	3	1	1·6	0·7	0·4	0·4	0·2	0·2	0·1	0·1
Fisons Ltd	3·6	1	2·4	0·6	2	0·5	1·1	0·3	0·8	0·1	0·3	0·1	0·1	—
Texaco Ltd and London & Coastal Oil Wharves Ltd	3·9	—	3·1	—	1·5	—	0·4	—	—	—	—	—	—	—
Totals	31·4	8·6	17	4·2	10·8	2·9	6·1	2	3	1	1·7	0·6	1	0·3
Proposed developments														
Mobil extension	2·8	0·4	1·7	0·2	1·1	0·1	0·7	0·1	0·3	—	0·2	—	0·1	—
Occidental Refinery	2·9	0·9	2·2	0·7	1·7	0·5	1·4	0·4	1·1	0·3	0·8	0·2	0·7	0·2
United Refinery	1·4	0·9	0·9	0·6	0·5	0·3	0·2	0·1	—	—	—	—	—	—
Texaco and London & Coastal Oil Wharves*	9	—	7·3	—	3·6	—	0·9	—	—	—	—	—	—	—
Totals	16·1	2·2	12·1	1·5	6·9	0·9	3·2	0·6	1·4	0·3	1	0·2	0·8	0·2

* **Relates to risks from proposed activities at the jetty of Occidental Refineries Ltd.**

Notes 1 Columns (a) show the assessed risks before improvements are made or, in the case of new developments, on the basis of the proposals as they were understood to be prior to 1 January 1978.

2 Columns (b) show the assessed risks if all the suggested action is taken and improvements are made at the existing installations and proposed developments.

3 Where a number is not shown the risk has been assessed as negligible.

Annex 2
Summary of individual risks

The individual risks were assessed for seven regions (A–G) indicated on the map.

The figures are expressed as chances in 10 000 a year.

The figures are the upper limit of the estimated individual risks.

	Existing installations		*Mobil extension*		*Occidental refinery*		*URL refinery*		*Existing installations + all developments*	
as shown Region	(a)	(b)	(a)	(b)	(a)	(b)	(a)	(b)	(a)	(b)
A	13	6·1	1·6	0·2	5	0·6	6·7	0·8	26·3	7·7
B	4·7	4·1	0·2	—	0·1	—	—	—	5	4·1
C	3·9	0·8	1·9	0·3	1·6	0·7	1·3	0·9	8·7	2·6
D	3·1	1·7	0·3	—	0·2	—	—	—	3·6	1·8
E	1·9	0·6	0·3	—	0·2	—	—	—	2·4	0·7
F	5	1·3	0·2	—	0·1	—	—	—	5·2	1·3
G	1	0·3	0·3	—	0·4	0·1	—	—	1·6	0·4

Notes 1 Columns (a) show the assessed risks before improvements are made or, in the case of new developments on the basis of the proposals as they were understood to be prior to 1 January 1978.

2 Columns (b) show the assessed risks if all the suggested action is taken and improvements are made at the existing installations and proposed developments.

3 Except for the caravan site at region B the numbers are estimated on the assumption that a person is in the region permanently i.e. 24 hours a day and every day of the year.

4 Where a number is not shown the risk has been assessed as negligible.

5 It will be noted that the totals in the final column (b) (i.e. Existing installations and all developments) are not, in all cases, the summation of the previous columns (b). This is because each of the proposed developments has been assessed independently of each other. The figures in the final column (b) reflect the fact that if the two proposed oil refineries were constructed on Canvey Island any release of LPG from the Mobil refinery would almost certainly be ignited as it passed over them.

Table 2. The Canvey/Thurrock area: one of the largest concentrations of petrochemical industries in the United Kingdom

Companies	Facilities; storage capacity; activities
Shell UK Oil	Refinery Petroleum storage: 3,500,000 t Liquefied petroleum gas: 5,000 t Ammonia: 14,000 t Hydrogen fluoride: 20 t
Mobil Oil Co. Ltd.	Refinery Petroleum storage: 1,500,000 t Liquefied petroleum gas: 4,000 t Tankers at the jetty
British Gas methane terminal	Liquefied natural gas: 80,000 t Liquefied petroleum gas: 20,000 t 1 methane carrier a week: 12,000 t
Fisons Ltd.	Ammonia: 1,900 t (pressurized) Ammonium nitrate: 9,300 t (solution)
Texaco Ltd.	Petroleum storage: 80,000 t Tankers at the jetty
London and Coastal Oil Wharves Ltd.	Petroleum storage: 300,000 t Tankers at the jetty
Vehicular traffic	Numerous trucks on the only main street going through Canvey urban center
Boat traffic	Heavy traffic in the Thames estuary, with numerous accidents (e.g., 1970, the liner <u>Monte Ulia</u> smashes through the Mobil jetty; 1972, a German ship collides with a 12,000-t tanker; 1974, a 1,000-t tanker rams a jetty on Canvey; a 12,000-t LNG carrier hits a 1,000-t coaster; an empty ammunition ship crashes into Canvey sea wall; 1976, a 2,500-t oil tanker crashes into the methane terminal jetty; 1978, a 12,000-t LNG carrier breaks adrift from the methane terminal, etc.)

the fallibility of the assessment method, and the difficulty in indicating confidence limits. For example, in the case of the behavior of ammonia, little information existed and the team had to make their own theoretical assessment of limited experimental evidence from the U.S. (8, p. 10). They acknowledged that historical data, on which their method of assessment relied, might be suspect for two reasons: "There is the possibility that new operations may have created new hazards which have not yet produced accidents and secondly, it is possible that the lessons of past accidents have been learned and they will not occur again (8, p. 11).

Finally, the HSE admitted that "it is not possible to establish confidence limits for the conclusions in any clearly defined way." Their judgment was that the team's "figures are realistic, may well overestimate the numbers of deaths by a factor of two or three, but are unlikely to be in error by a factor of as much as ten." (8, pp. 13-14)

In their report, the team made specific recommendations for improvements to plant, equipment, and methods of operation--for example, they suggested that a water spray system be installed for the Shell operation on the jetty where liquefied ammonia is unloaded (8, p. 26). They also stressed that they expected the recommendations, or equally effective alternatives, to be carried out. "The Executive will, if necessary, use the enforcement powers conferred by the 1974 Act to require these improvements." (8, p. 28) They further identified potential hazards which seemed to require future detailed assessments by the firms involved; for instance, it was considered important to do a study of the "chance of a railway accident in the vicinity of the ammonia sphere at Fisons." (8, p. 28)

Once the technical investigation had been done, the HSE could have transmitted the report to the policymaking bodies concerned. But it did more. The HSE chose to make a specific recommendation on the risk policy concerning industrial development in Canvey:

> Provided the improvements asked for or those conferring comparable benefits are carried out, we would not consider the situation such that any of the existing installations should now be required to shut down. We consider that the proposed developments specifically referred to in this report could take place without adding significantly to the risks. We shall continue to keep the whole situation under review and, using all our powers, take steps to secure further improvements where these appear to be necessary and to prevent any future developments which might significantly increase the risk to persons living in the area. (8, p. 26)

At this point the HSE report became more than a technical study; it became a political document.

1978: REACTIONS TO THE HSE REPORT

The publication in 1978 of Canvey: An Investigation of Potential Hazards from Operations in the Canvey Island/Thurrock Area by the Health and Safety Executive brought reactions which were understandably varied. The technical data on risk probabilities, which the Canvey residents were happy to see publicized, were regarded as too pessimistic by industry. The New Scientist wrote that the report demonstrated that "it is safer to work in a coal mine than to sit at home watching television on Canvey Island." (9) Sir Bernard Braine and his constituency, however, deplored the report's conclusion. He wrote to the Prime Minister on June 27, 1978:

> The fact that the Report recommends measures which could reduce the risk by 50% or more is fully appreciated, but there is not a single one of my constituents who believes that in the face of what the Report says about the risks, the Government has any right to permit oil refinery development to take place even after the suggested improvements have been made. (10)

On August 3, Sir Bernard launched an attack in Parliament on the "lunatic conclusion" of a study which "not only confirms all of our worst fears about the existing hazards at Canvey, it actually uncovers others hitherto unknown to us." After describing the worst of the "frightening array of hazards" unveiled, he went on to find "serious flaws" in the report itself, which led him to conclude that "its authors were under considerable pressure from vested interests." In particular he pointed to a missing appendix 10, a paper prepared by British Gas on "The Possibility and Consequences of an Unconfined Explosion Involving LNG," which the Minister had dropped because he thought its inclusion "inappropriate." Sir Bernard found this view understandable, since the paper (which he had subsequently located) displayed how "far more lightly" the corporation takes the dangers of liquefied natural gas "than is currently the case, for example, in the United States." (11, pp. 348-349)

He went on at length to discuss the United States' extensive studies of the safety of liquefied gases. "All I can tell the House is that the criteria of risk to people in the United States, laid down by the Federal Power Commision in turning down an LNG site there, is such that if applied in this country the Canvey methane terminal would be shut down tomorrow." However, Sir Bernard found British Gas Corporation's intentions clear--"to go on importing the same quantities of Algerian LNG for years to come." He then asked a critical question about policy decisions: "Who is to judge the necessity of storing liquefied gases on Canvey? Is British Gas a law unto itself? Or is the question to be determined by the Health and Safety Executive, which is answerable to Ministers?" (11, pp. 349-350)

Sir Bernard found other flaws in the report, the most serious being the "astonishing neglect" of the "human reaction in the event of disaster or even in a limited incident." He found a hint that an additional road would help evacuation from the island but no firm recommendation to have it built. This was one of the seven main points on which he sought assurance from the Minister in summarizing his arguments (11, pp. 350-351).

1979: EXCHANGES WITH LABOUR AND CONSERVATIVE GOVERNMENTS

In a letter to the Prime Minister, James Callaghan, on February 1, 1979, Sir Bernard Braine once again raised the question of removing the main hazard to the islanders, the methane gas plant. "We had the narrowest of escapes a fortnight ago when a terrorist bomb was placed against a near-empty tank at one of our hazardous installations. How many more warnings does the Government need before it takes effective action?" (12)

The Prime Minister replied in Parliament on February 15, saying that he had asked for a "full investigation and report," although "We shall have to consider whether we can totally safeguard the lives of the hon. Member's constituents in all circumstances." (13, p. 1311) His reply was enough to prompt these headlines from the Financial Times the following day: "Callaghan orders review of Canvey Island safety" (14). On April 4, Sir Bernard received word that the Government, upon review, did not think it necessary to set up a further inquiry. The HSE report was "the most thorough and detailed type of assessment relating to industrial operations in a particular location ever carried out in the U.K., and probably in the world. Any further enquiry could only duplicate the work already done." (15, p. 522)

When the new Conservative Government came to power, with Margaret Thatcher as Prime Minister, the Castle Point Refineries Resistance Group wrote her on May 11 and received a reply dated July 12, 1979: "The Government accepts the findings and recommendations of the HSE report, and has decided that none of the existing installations need be required to close down...." "On the two proposed refineries: of course reconsideration of United Refineries' outline planning permission is before the Secretary of State for the Environment, who intends to re-open the adjourned public exploratory enquiry into its possible revocation...and Occidental have decided to withdraw their planning appeal for commercial reasons." (16)

1980-1981: THE WARD REPORT

A year later, on June 5, 1980, the inquiry into the United Refineries' proposed facilities was reopened. The public debate that ensued provided a fresh occasion to examine both the existing risks on Canvey and some of the technical recommendations for improvements that had been made in the HSE report in 1978. The main target was once again the methane terminal, and in the report that resulted from the inquiry (known as the Ward Report), the inspector dealt at some length with problems at the terminal and its "history of concealment, evasion, bad management and bad supervision." He recounted how, on the basis of information provided by the electrical engineer at the terminal,

> H.S.E. Inspectors swooped without warning on the terminal on 6 February 1980 and confirmed that the high level alarms in the tanks were not working and that the high level trips, while said to be operable, were unreliable. Intense pressure was put upon British Gas and three enforcement notices were served. The Terminal had had more enforcement notices in one year than the whole of the rest of the British Gas Corporation's installations. (17, p. 13)

The Ward Report continued, pointing to the main source of danger to Canvey Island:

> In my view a much more drastic step is necessary if the risk to Canvey and others is to be reduced to more reasonable dimensions. As I have already intimated, the greatest potential menace stems from the Methane Terminal. A major spill could cause very high casualties. Such a high risk potential should not be sited so close to highly populated areas and it can only be nullified by a conscious decision of the Gas Board to install a foolproof source of ignition on the terminal's perimeter. If that is

> considered to be too dangerous, the only alternative is to close it down and move it elsewhere. Any other alternative which continues to allow the peoples of Canvey and others to be exposed to what could be an accident of major proportions would be to my mind both foolhardy and irresponsible. (17, p. 76)

To this "catastrophic" report the Conservative Government reacted immediately. On March 24, 1981, the Secretary of State for the Environment stated in Parliament that since the Ward inquiry had not been mandated to deal with the methane terminal, another in-depth inquiry would be launched to study that particular problem, and in the meantime the URL decision would be deferred (18, p. 807). Sir Bernard welcomed this but remained puzzled about the necessity for "an inquiry into an inquiry." (18, p. 808) The Government was obviously acting to dampen the initial provocation of the Ward Report.

1981: THE SECOND HSE REPORT AND THE METHANE TERMINAL STUDY

In the meantime the Health and Safety Executive noted that in the three years that had gone by since its first assessment of Canvey, numerous improvements had been implemented, and it was therefore deemed right to publish a second report describing the effect of these improvements on the numerical risk estimates (19, p. iii). The second study was written on the same lines as the first, examining the risks and activities firm by firm. The results showed that whereas the average individual risk of death or serious injury was 5.3 chances in 10,000 a year in 1978, the situation had improved by 1981 to 0.4 chances in 10,000, a 14-fold reduction. In the area with the highest risk, the improvement was 38-fold (19, p. 13).

It is interesting to note a change in the underlying hypothesis of the 1978 Canvey Report and of the 1981 reassessment. In assigning levels of risk, the first report stressed: "The fact that major industrial accidents resulting in substantial numbers of casualties have not occurred in the past does not mean that they may not occur in the future." (8, pp. 25-26) The second report was based on a contrary assumption: "Experience seems to show that the full lethal potential of a major industrial accident is most unlikely to be realised; the reassessment has reflected this by reducing considerably the predicted risks of accidents with more than 1500 casualties." (19, p. 13)

In 1981 the local task force of the HSE conducted a study of the technical and management aspects of the methane terminal. In both areas the situation was found unsatisfactory, but the central theme of this thorough analysis was that improvements could be made; it appeared that there was no case for shutting down the British Gas terminal (20).

1982-1983: THE PUBLIC INQUIRY ON THE METHANE TERMINAL

The major public inquiry that the Government had promised into the safety of the methane terminal now materialized. The dramatic personae were basically the same in this continuing morality play: The imperiled residents of Canvey Island were represented mainly by Sir Bernard Braine (21); commercial interests were personified by British Gas Corporation (22); and the fair-minded Government appeared in the form of the Health and Safety Executive (23).

The 1982 script, however, had a radical change. British Gas decided that it was time to pull out all the stops, to display the full power of a successful nationalized industry, and to at last have a decisive victory in the eternal Canvey warfare. They marshalled eight specialists and three lawyers, including the task force leader, a lawyer known for his fighting spirit and ruthless cross-examination technique. The opponents, ordered to answer simply "yes" or "no" to questions fired at them by the lawyers and specialists, were no match; this included Sir Bernard, who was more at home on a parliamentary stage. The HSE, who had no intention of "winning" or to humilating a community or its M.P., tried to defuse the confrontation. They chose an open-minded, diplomatic lawyer and prudently maintained their neutrality, refuting Sir Bernard or British Gas when it was auspicious to do so.

On September 1, 1982, the inspector submitted his report on the results of the debate. He found the terminal "acceptably safe," the plant in "good condition," and the management of British Gas displaying the "knowledge, experience, and competence to run the MT efficiently and safely." He pointed out the national importance of this supply of gas, noting that "If closed or curtailed, equivalent facilities would be needed elsewhere...probably on the estuary of the river Thames. The cost of such a move would be large and wholly unjustified." He did suggest that "for historical and psychological reasons the activities at the MT ought not to be expanded and the amount of LNG stored ought not to be enlarged." (24, p. 99)

The inspector's report put forth a set of seven final recommendations, which included having the HSE monitor the operation of the terminal, having comprehensive emergency planning initiated by local authorities "as a matter of top priority," having British Gas repeat publicly the assurances given during the inquiry, and finally, having steps taken "by Government and local government...to inform and reassure the local population that Canvey Island is not a dangerous place in which to live." (24, p. 100) As the final decision, the Secretaries of State accepted the findings of the inspector: "Accordingly, they conclude that at its present level of operation the terminal does not present any justifiable cause for concern on grounds of public safety. They endorse the Inspector's hope that fear and worry will cease to plague any of the local population." (25, p. 3)

CONCLUSIONS

Much can be said in the light of the Canvey Island experience. On technical grounds alone, the work done during the course of the continuing inquiry was a real breakthrough: Many poorly known risks were identified, assessed, and quantified; the risk level in the Canvey area was substantially reduced; and the methodology was developed by which industry could--and did in many cases--carry out its own hazard surveys and assessments of proposed developments elsewhere in the country. ("It is worth observing that all but one of the companies concerned in the Canvey investigation have installations elsewhere in the UK.")(19, p. iv)

As far as the dynamics of the Canvey case--the aspect that we have been concentrating on--the experience proved that the serious issue of risk and safety could be resolved in an ongoing three-sided debate. The choice of risk management strategies, far from being simple "yes" and "no" decisions on the part of the Government, can derive effectively from a dynamic process involving all the participants. The debate revealed risks other than those the Canvey residents were living with. Industry, faced with financial risks and the possibility that Canvey would have repercussions elsewhere, also incurred great expenses in improving their facilities and proving their case during the exercise. In engaging in such an open process, the third member of the debate--government in the form of the Health and Safety Executive--faced the risk of impeding the nation's industrial activities and paradoxically increasing public fears and generating demands for swift and radical decisions.

The HSE not only took the chance, it also provided the mechanism that allowed the debate to evolve constructively--this was the report published in 1978 after a two-year study. The HSE later wrote:

> We regard the report as a watershed--a unique, pioneering exercise which will prove to be a major turning point in risk assessment work. The report aroused extensive interest in the U.K. and around the world, far beyond the sphere of those having a direct local concern. In its range and depth the report provided, and still provides, a potent stimulus to the debate about risk assessment techniques and the practical decisions which have to be made about the relation between potentially hazardous industry and people who live and work nearby. (19, p. iv)

Not only the Canvey report but the lengthy dynamic process surrounding it can be considered a turning point in risk assessment policy.

REFERENCES

1. Safety and Health at Work, Report of the Committee (London: HMSO, 1972).

2. The Flixborough Disaster, Department of Employment, Report of the Court of Inquiry (London: HMSO, 1975).

3. House of Commons, Parliamentary Debates, Official Report (Hansard), Vol. 864, No. 14, Nov. 16, 1973 (London: HMSO).

4. House of Commons, Parliamentary Debates, Official Report (Hansard), Vol. 872, No. 24, Apr. 8, 1974 (London: HMSO).

5. House of Commons, Parliamentary Debates, Official Report (Hansard), Vol. 875, No. 56, June 18, 1974 (London: HMSO).

6. House of Commons, Parliamentary Debates, Official Report (Hansard), Vol. 877, No. 82, July 24, 1974 (London: HMSO).

7. House of Commons, Parliamentary Debates, Official Report (Hansard), Vol. 892, No. 132, May 23, 1975 (London: HMSO).

8. Canvey: An Investigation of Potential Hazards from Operations in the Canvey Island/Thurrock Area, Health and Safety Executive (London: HMSO, 1978).

9. New Scientist, June 22, 1978, p. 802.

10. Sir Bernard Braine, letter to the Prime Minister, James Callaghan, June 27, 1978.

11. House of Commons, Parliamentary Debates, Official Report (Hansard), Vol. 955?, No. 163, Aug. 3, 1978 (London: HMSO).

12. Sir Bernard Braine, letter to the Prime Minister, James Callaghan, Feb. 1, 1979.

13. House of Commons, Parliamentary Debates, Official Report (Hansard), Feb. 15, 1979 (London: HMSO).

14. Financial Times, Feb. 16, 1979.

15. Secretary of State for Employment, written reply, Apr. 4, 1979.

16. Prime Minister Margaret Thatcher, letter to Mr. Haslam, July 12, 1979.

17. Town and Country Planning Act 1971. Re-opened exploratory inquiry into the desirability of revoking the planning permission granted to United Refineries Ltd. on 28 March 1973 for the construction of an oil refinery on Canvey Island.

18. House of Commons, Parliamentary Debates, Official Report (Hansard), Canvey Island (Planning Inquiry), Mar. 24, 1981.

19. Canvey: A Second Report. A Review of Potential Hazards from Operations in the Canvey Island/Thurrock Area Three Years after Publication of the Canvey Report, Health and Safety Executive (London: HMSO, 1981).

20. Report on the Safety of Operations and Management Arrangements, the British Gas Corporation Methane Terminal, Canvey Island, Sept. 1981.

21. Public Inquiry--British Gas Corporation Methane Terminal, Canvey Island. Proof of evidence by Sir Bernard Braine; proof of evidence by D. J. Rasbach, December 1981. Comments on Canvey: A Second Report, by J. A. Fay, Oct. 1981. Comments on Report on the Safety of Operations and Management Arrangements, by J. A. Fay, Nov. 18, 1981. Public Inquiry--Castle Point Refineries Resistance Group, proof of evidence, Feb. 1982.

22. Exploratory Inquiry into the British Gas Methane Terminal, Canvey Island, British Gas Corporation Reports, Jan. 5, 1982: 1. Development of the terminal and description of its facilities; 2. Siting of LNG installations; 3. Site layout and proximity to other development; 4. Assessment of risk to the public; 5. Management of safety; 6. Future proposals.

23. Public Inquiry--British Gas Corporation Methane Terminal, Canvey Island. Proof of evidence by the Health and Safety Executive: P. R. Yeomans, Sept. 1981; A. C. Barrel, Oct. 1981.

24. A. de Piro, Q.C., British Gas Methane Terminal, Canvey Island Report, Sept. 1, 1982.

25. Department of the Environment, Canvey Island: British Gas Terminal, May 10, 1983.

Bibliography. For further reading, see also:

P. Lagadec, Le probleme de la sûreté d'un grand complexe industriel. Le cas de Canvey Island (Grande-Bretagne), Laboratoire d'Econométrie de l'Ecole Polytechnique, Ministere de l'Environnement (S.E.I.), Dec. 1979.

P. Lagadec, Le risque technologique majeur. Politique, risque et processus de développement (Paris: Pergamon Press, Collection Futuribles, 1981).

La civilisation du risque. Catastrophes technologies et responsibilité sociale (Paris: Editions du Seuil, 1981).

P. Lagadec, Le probleme de la sûreté du'n grand complexe industriel. Le cas de Canvey Island (Grande-Bretagne), 1978-1982. Laboratoire d'Econométrie, Ministere de l'Environnement (S.E.I.), May 1982.

Major Technological Hazard (Oxford: Pergamon Press, 1982).

Political Decisionmaking and Scientific Indeterminacy

Timothy O'Riordan

School of Environmental Sciences, University of East Anglia, England

INTRODUCTION

Risk analysis and environmental impact assessment are topics of interest because they have become infused with a modern tendency to politicize what formally used to be the relatively unchallenged domain of experts. Scientists and other experts like to provide "right" answers, but politicans want "clear" answers and tend to become confused when scientific debate becomes suffused with political purpose. Scientists do not function well if 1) they are opposed by other scientists who produce different facts and analyses; 2) the points they wish to argue cannot be proven absolutely; and 3) the analyses they wish to undertake are infused with controversy as to meaning and political purpose. Consequently, politicians tend to use scientific information in different ways to justify taking or avoiding a decision, interest groups align themselves to scientific advice for political purposes, and the bemused general public is usually bypassed as with other matters of great public interest. Case studies to illustrate the more general points raised in this paper will be used to illustrate the actual presentation.

WHY ENVIRONMENTAL RISK ISSUES BECOME POLITICAL ISSUES

From time to time, students of environmental problems wonder why some environmental issues become politically visible and controversial while others remain dormant. For example, many of the consequences of "acid rain" have been known to scientists for over 20 years but widespread political concern is relatively recent. Asbestos was recognized as a killer over 60 years ago (1). In the United Kingdom (UK), it took a highly emotive and controversial television documentary, in which the

NATO ASI Series, Vol. G4
Environmental Impact Assessment, Technology Assessment, and Risk Analysis. Edited by V.T. Covello et al.

dying widow of an asbestos worker was interviewed, to stimulate a halving of the permitted concentration of asbestos fibers in the workplace. Yet even that response is challenged as being medically insufficient.

The removal of lead from petrol (gasoline) became a highly charged political issue only very recently in Britain, largely as a result of prolonged scientific controversy between two warring factions of the medical and chemical community, the intervention of a Royal Commission on Environmental Pollution (2), and the impressive lobbying of a small pressure group, Campaign Against Lead in Petrol (CLEAR). The major achievement of CLEAR was to implant the scientific arguments in the minds of Members of Parliament on all sides of the House of Commons, thereby ensuring that the issue became openly a political matter and not something that could be shunted aside to yet another scientific committee. The British Government, having dragged its feet for over a decade, finally agreed to remove all lead from petrol by 1990 and to take the forefront in encouraging its partner member states of the European Community to do likewise. It remains to be seen whether this commitment will be honored, though the government knows it is being watched.

Another feature of the politicization of environmental risk is the highly publicized cause celebre, the major disaster which becomes a symbol of incompetence and negligence, and brings latent suspicions and fears to the surface. Torrey Canyon, Amoco Cadiz, Seveso, Three Mile Island, and Love Canal are all examples of this genre, cases which become household names but whose scientific details tend to become lost in myth, fantasy, and prejudice.

Environmental pressure groups love to capitalize on such events because they believe that their political cause is advanced by exploiting the symbol rather than just chipping away at principles. In any event, such disasters help to reshape regulatory procedures, heighten the public sense of safety, and can lead to important new legislation. Certainly they create an atmosphere in which environmental risk in all its manifestations [for a good coveage of definitions, see Burton and Whyte (3)], becomes a mixture of science and politics.

By way of a summary, it can be argued that there are five factors which convert certain environmental risk issues into open political controversy; these factors are discussed below.

Establishment and Counterestablishment Science

Establishment science is defined here as officially recognized science as accepted by research establishments, corporate laboratories, research institutes, and prominent individuals. Counterestablishment science is that practiced by non-recognized scientists, sometimes working outside official establishments but usually not members of official scientific committees, most of whom work for non-corporate and non-governmental pressure groups. Those who associated themselves with CLEAR or with the Union of Concerned Scientists fall into this category.

The rise of counterestablishment science is in part associated with the growth of pressure group politics. This in turn is spurred on, particularly in the U.K., by the liberalization of the charitable tax laws, the promotion of specialized scientific branches of environmental law, a more comprehensive scientific training in the universities and colleges, and greater scope for intervention through the medium of the public hearing or other rulemaking device. In the U.K. it is, in part, an outcome of a period of more participatory politics which began in the late 1960s and which has been encouraged by the use of more open review procedures through special commissions of inquiry, parliamentary select committees, and major public inquiries. In part, too, the rise of counterestablishment science is a reflection of a changing mood in Western European democracies, a mood which is suspicious of corporatism in government and business, which is critical of the failure of modern capitalism to provide for what they see as a more caring society enabled to shape its own communities and destiny, and which strongly believes in deploying technology to suit human needs and to improve resource use efficiences rather than to create more unnecessary materialism and environmental damage.

Western European society is becoming more politically restless, more politically active, and more questioning of the conventional, dominant ideologies favoring economic growth and environmental manipulation in which "ordinary people" appear to have no say or understanding. This line of analysis was opened up by Inglehart (4) and again more recently in his reappraisal of post-materialism (5). It has been taken up by the British sociologist Stephen Cotgrove (6) and is the subject of a major forthcoming analysis by Milbrath et al. (7).

What we may be witnessing here is a linkage between societal mobilization and a critique of expertise. The challenge of counterestablishment science is the challenge to authority of the mystique that scientific competence has its own contained logic and purpose (often referred to as rationality) which is supposedly independent of political ideology. Should the scope for this challenge be further enlarged so as to make scientific judgment even more amorphous, the future of risk analysis and environmental impact assessment would be fascinating to witness. But such an enlargement may not be permitted, at least not in an uncontrolled way. This point will be examined in more detail in the companion paper.

Environmental Martyrdom

Environmental martyrdom is a phrase embracing the death, serious injury, or lucky escape of innocent parties, the victims of misjudged environmental risk. The public identifies with such people precisely because anybody could be a martyr, depending on circumstances and ill fortune. Martyrdom can produce hysteria (e.g., the aftermath of Three Mile Island and the almost panicky review of evacuation procedures around any nuclear plant), and certainly can lead to dramatic over-reaction. Children playing on a waste tip near Nuneaton in the English industrial midlands came across 36 drums of used sodium cyanide containers with crystals still attached to the outside. None was affected, but within five weeks Parliament had passed a bill to control the deposit of poisonous waste. This Act subsequently had to be repealed in favor of a more wide-ranging legislation covering all toxic waste disposal, but even today it is possible for warehouses sited adjacent to residential areas to contain unlicensed (and largely unknown) inventories of chemicals which could explode or catch fire. Eventually that problem will be dealt with, probably by an extension of cumbersome paperwork. This may or may not improve practice, and may only be enacted as a result of environmental martyrdom.

Publicity

The media love novelty, human stories, and the tragedy of innocence. This is particularly the target of television and newspapers, which together provide about 90 percent of mass public information. Martyrdom in its various forms is thus meat and drink for the media, which undoubtedly do much to stir up public feeling, in turn translatable into political pressure. Unfortunately, the media also has a habit of distorting "factual" information and hence exaggerating misunderstanding. A classic case recently in the U.K. was a scare story regarding the threat of methaemoglobinaemia (colloquially known as blue baby disease) among infants in rural East Anglia, allegedly as a result of high nitrate levels in the local public water supply. While scientific studies have failed to prove a connection between nitrate concentrations below 100 mg per liter and methaemoglobinaemia, both the local press and a number of prominent local politicians began to demand a massive program of denitrification of the public water supply likely to cost not less than £650 Million ($1,000 million). The revealing aspect of this particular story is that when local water authority officials attempted to calm down these public and political fears, suspicions were strengthened. The next nitrate "scare," which is almost inevitable, will likely inflame these suspicions.

Ubiquitousness

An interesting feature of certain environmental risks is their ubiquitousness, their unselective imposition on the wealthy and the humble. Page (8), in his excellent analysis of the modern characteristics of environmental risk, makes much of this point. Even-handedness of effect means that the rich and powerful cannot buy their way out of an actual or potential nuisance, as they still can with most conventional pollution problems. Therefore, the political pressure for action becomes more widespread partly because there appears to be no escape. The acid rain phenomenon is taking on this characteristic. Originally believed to be a problem afflicting the unbuffered soils of northern Scandinavia and hence local fishing and recreational interests, the dangers of acid rain are now regarded by some scientists almost as universal through Europe, affecting not only lakes but crops and trees. It is now estimated that as much as 80 percent of the German spruce forest could be threatened. This could have almost incalculable implications for the West German quality of life. Gradually, a seemingly scientifically peripheral issue becomes politically central.

Culpability

In an intriguing essay on the culture of environmental risk, Douglas and Wildawsky (1982) suggest that all societies acknowledge that life is full of danger, anxieties, and uncertainties. Normally, they argue, these are incorporated in ways of coping with the external world and are hence culturally containable. But if an agency or an event is for one reason or another to be "blamed," then risk is a convenient device through which to vent feelings and frustrations. As Douglas and Wildawsky (1983, p. 50) argue, "dangers are selected for public concern according to the strength and direction of social criticism." Risk, therefore, has a moral element: It is a manifestation of how society, through its diverse groupings, passes judgment upon itself and its institutions. This is the very essence of politics.

Take, for example, the continuing disquiet over civil nuclear power. The nuclear industry is at last beginning to recognize that almost irrespective of how it tries to persuade the public that nuclear power is safer and cheaper than any other form of energy supply and conservation, the very fact that the message is coming from what is seen as a biased source distorts its content and message. It is, therefore, doubtful that the nuclear industry or its associated research institutions can ever be a successful educator because nuclear power is increasingly seen as a cultural anomaly. The nuclear industry is regarded by many as almost a fifth arm of government, inextricably part of the military, industrial, and financial lobbies which buttress western capitalist ideologies.

This is yet another reason for asserting that facts and values interconnect. Because all facts are couched in ideologies, it follows that if value positions are polarized (or at least unbridgeable) there can be no agreement even over facts. Environmental risk not only provides a convenient vocabulary of such conflicts but in its analysis, may even exacerbate them--as, for example, witnessed in the frenzied arguments in the U.K. over the lead in petrol issue.

POLITICAL TREATMENT OF SCIENTIFIC INFORMATION

Let me summarize the points made so far: 1) The risk debate extends outward to a more general societal critique concerning the future direction of investment, control over social and technological change, and priorities for public policy. Disputes over environmental risk provide avenues for counter-establishment thinking and research to gain public credibility and support; 2) Uneasiness over the future direction of investment is encouraged by changes in the public mood and economic "realities" brought on by prolonged recession and the scope for the deployment of conservation technology and appropriate public behavior. This line of reasoning, for example, permits groups not noted for their antinuclear views to argue against new nuclear power stations on the grounds that public investment in electricity conservation technology and behavior would be far more cost-effective and hence, in the public interest, than net additions to base load power; and 3) Within these nebulae of doubt, expertise itself is questioned, partly because it cannot prove itself to be infallible by the very uncertainties associated with many current environmental risk issues, and partly because the culture of counterexpertise does not permit it.

What then are the circumstances in which politicians, who supposedly must openly mix fact with value, treat scientific indeterminacy? Three factors help to determine the nature of their reactions: 1) Urgency of action: The most desirable escape route for any politician is delay. Disputed evidence, both as to cause and effect of risk, coupled with the general lack of inquiry for new development (the so-called "need" issue) prompt a preference for further study, special committees of investigation, or more research. Scientific indeterminacy is especially conducive to this response. See, for example, the official U.K. response to the acid rain issue, namely, that the matter is receiving further scientific examination and that on the basis of the evidence to date no major change of policy is required (H.M. Government, 1983); 2) Policy ambivalence: The conditions outlined at the outset of this section are leading to reappraisals of policy, notably in key risk areas such as energy waste disposal and toxic chemicals. Where policy is fluid and shifting, scientific indeterminacy can provide a convenient peg on which to hang different hats. Lack of urgency, of course, can encourage this; and 3) Political restlessness: Governments, even strong governments with large party support in the legislature, do not like to act decisively when there is backbench misgiving about a particular direction of policy. This is even more likely in coalition governments or where there are small ruling party majorities.

The nuclear issue is rarely pressed in contemporary western democracies, precisely for these reasons. Either there is no majority in the legislature favoring major new investments and developments, or there is backbench unease over the need for new expenditures when scientific views are still divided. This is certainly the case in the U.K., where, even with a Commons majority of 144, the Conservative Government may find it difficult to persuade some of its monetarist backbenchers of the merits of immediate deployment of at least £1.2 billion ($2.0 billion) in the proposed Sizewell B pressurized water reactor (PWR). This will be especially likely as on the admission of the nation's regulatory body, the Nuclear Installations Inspectorate (1983), all the necessary safety clearances for the PWR will not have been obtained before July 1984 and possibly even later.

Of course, matters become interesting when one or all of these conditions do not apply. How is indeterminate scientific information treated when there is a perceived need for action, when there is political support, and when a clear policy has to be determined? Nowadays these conditions tend to apply more in the military area than in the economic industrial fields.

Politicians are likely to treat scientific information and advice in a number of ways: 1) Given scope for prevarication, politicians tend to segment controversial evidence to apply it favorably to suit their own ends, and to ignore and suppress the remainder. In these circumstances, scientific information can become client-based and the scientist is somewhat irrelevant. This becomes most noticeable when different interest groups hire different scientists to pursue a particular prejudice.

A classic example in the U.K. occurred recently when the Nature Conservancy Council (NCC) was about to declare a nationally important wetland site in Somerset as a Site of Special Scientific Interest (SSSI). Under the Wildlife and Countryside Act of 1981, SSSI areas have an important legal status which requires the NCC to be notified in advance and have effective veto power over potentially damaging agricultural operations likely to disturb the scientific value of the site. Local farmers were naturally anxious to reduce the size of the proposed site to as small an area as possible, and their representative lobbies, plus the Ministry of Agriculture, both brought in ecologists to challenge the NCC's ecologists. The two client scientists produced arguments favoring a halving of the proposed site, but the NCC prevailed and got its whole area protected. While the NCC was acting perfectly legally, in political terms, its action was heavily criticized with representatives of the agriculture lobby threatening dire reprisals.

Two points are important about this particular dispute. First, the scientists were genuinely in dispute and second, public confidence in client-based science was diminished. This further weakens public credibility in expertise and allowed politicians to make what they wanted out of the dispute. Secondly, where an uneasy political decision has to be made, but where the scientific information is controversial, the politician is more likely to become client to the scientist, granting the latter undue power. In the Windscale decision, there were widespread allegations that the British Government of the time had committed itself to a uranium dioxide reprocessing plant for both domestic and foreign policy reasons. The plant was to be a major reprocessing facility for friendly nations who did not necessarily want to deal with France (where there is also such a facility) and who would not get such a plant at home because of domestic political opposition (the U.S. and Japan). Thus, so the story goes, the Government needed Windscale.

Wynne (1983) analyzes the way in which the controversial scientific argument was handled at the public inquiry and concludes that the inquiry enjoyed its own rationality in which scientific fact was distorted to buttress a predicated outcome. Indeed he goes further: He suggests that organizations commited to a particular course of action may suffer from diminished responsibility toward their obligations for public safety due to their own self-delusion. He also asserts that this can be achieved by engaging argument and decision in a ritual to defend credibility and authority. "A rational technology assessment would examine all hidden interests and agendas buried in specific claims and counter claims: but responsible institutions are busy hiding their own agendas." (Wynne, 1983, p.ix)

Compulsive political decisions may be in danger of encouraging organizational aloofness and delusion and of enabling politically acceptable science to have undue influence. An interesting outcome is that political opposition (irrespective of the party or power) may be attracted to counterestablishment science simply to save its own political aims. For example, during the second Windscale debate in the U.K. Parliament in May 1978, the Conservative opposition energy spokesman spoke in favor of conservation as a realistic energy option, citing in part evidence given at the Windscale Inquiry in support of this line. Yet today the roles are reversed. The Conservative Government is playing down the merits of conservation while the two major opposition parties are busy promoting its merits. Doubtless if the party roles were to be reversed, the general thrust of argument might also switch.

Political ambivalence, in part a function of scientific indeterminacy but also a feature of power, is also associated with lobbying by key pressure groups seeking to exploit that ambivalence. This is precisely how the lead petrol issue was "won." As the medical reports flooded in, indicating growing doubt about the lack of any clear connection between relatively low levels of lead concentration and depression of intelligence, so the credibility of the gasoline, automobile, and lead manufacturing companies began to wear thin, their increasingly strident publicity beginning to look more and more socially irresponsible. The Conservative Government, for a long time seemingly immune to pressure on total abolition of lead, quite suddenly changed its stance, admittedly only in expressing a willingness to pursue a zero lead option. The intriguing issue here was how the adroit pressure groups maneuvered, not just in the energy/environment/health ministries but also in transport, treasury, and the foreign office (as EEC connections and export markets were involved). Indeed, it was the concerted attack on no less than seven ministries that eventually won the day. But, partly because seven ministries have always been involved, political inertia has always been enormous.

Scientific ambivalence can also thwart politicians, pressed to act by public opinion. Let us return to the acid rain issue. Not all scientists are convinced by the cause and effect of acid rain, and certainly there is no consensus as to where action would prove most efficacious and cost-effective. Ashby (13) notes that Britian could reduce sulphur emissions by one quarter by 2,000 at a cost of £200 - 300 million ($300 - 400 million) per year. This would hardly benefit the British and would only slightly alleviate the SO_2 loading in Scandinavia. Ashby suggests that more SO_2 deposition comes from sources unknown--truly global sources. Scientists want to wait so as to get the "facts right": The Treasury wants to wait to be sure the SO_2 removal investment is economically justified. Politicians cannot wait. They resort to action based on political hunch rather than scientific advice--and may well waste public money in their anxiety to be seen to care.

THE ALIENATION OF THE INFORMED PUBLIC

Pressure group activism (sometimes involving the specific creation of particular cause groups) undoubtedly heightens the influence of scientific indeterminacy over risk issues. Part of the strategy of the modern pressure group is to produce credible counterestablishment information so as to influence opinion. This is an expensive and time-consuming matter, and may well be crippling some key British groups who simply cannot command the resources to maintain the necessary scientific counteroffensive (or at least rely so heavily on the philanthropy of experts prepared to work voluntarily or at low cost). The future structure and tactics of environmental public interest groups are of relevance to the points raised in this paper but require examination elsewhere.

A related issue, and one which will be dealt with in the companion paper, is the alienation of informed people from any role in influencing debates and eventual decisions over environmental risk. Scientific indeterminacy is bad enough in creating confusion, but when institutions supposedly designed to sort out the arguments also confuse and contradict, then public misgiving is heightened.

In Britain there are two institutions which are supposed to command public respect as providing fair and authoritative resolutions of controversial public questions. One is the public inquiry and the other is Parliament. It is to these two institutions that informed people look for satisfactory performance. It is the contention of the companion paper that as matters currently stand, neither is executing its role adequately and neither is commanding the confidence of the informed and interested public. Does this matter? If it does, how far can these two institutions be reformed, with respect to dealing with environmental risk issues so as to restore a sufficient measure of public confidence? These are the central themes of the companion paper.

REFERENCES

1. A.D. Woolf, "Asbestos Hazards and Standards in a Study in the Arbitrary Acceptance of Hazards," in The Acceptability of Risks, Council for Science and Society, (Barry Rose, London, 1978), 91-96.

2. Royal Commission on Environmental Pollution, Lead in the Environment, 9th Report, Cmnd 8852. (HMSO, London, 1983).

3. I. Burton and A.V. Whyte, Environmental Risk Assessment, Scope 15 (Wiley, Chichester, 1980).

4. R. Inglehart, The Silent Revolution: Changing Values and Political Styles Among Western Publics, (Princeton University Press, New Jersey, 1977).

5. R. Inglehart, Changing Values and the Rise of Environmentalism in Western Societies, Preprint 82-14 (International Institute for Environment and Society, Berlin, 1982).

6. S. Cotgrove, Catastrophe or Cornucopia, (Wiley, Chichester, 1982).

7. C. Milbrath (Ed.), Environmentalism and Social Change, (Sage Publications, Beverley Hills, CA).

8. T. Page, "A Generic View of Toxic Substances and Other Risks," Environmental Law Quarterly, 7 (2), 207-244.

9. M. Douglas and A. Wildawsky, Risk and Culture: An Essay on the Selection of Technical and Environmental Dangers, (University of California Press, Berkeley and Los Angeles, 1982).

10. M. Douglas and A. Wildawsky, "How can we know the risks we face? Why risk selection is a social process," Risk Analysis 2 (2), 49-51.

11. H.M. Government, Coal and the Environment, Cmnd 8877, (HMSO, London, 1983).

12. H.M. Nuclear Installations Inspectorate, Evidence to the Sizewell B Public Inquiry (NII, London, 1983).

13. E. Ashby, How Clean is Clean Enough? Times Higher Education Supplement, 14 January, 1983, 11-12.

PROCEDURAL REFORM

Institutions for resolving, or at least clarifying, complex politico-scientific issues are shaped by the political culture of the country concerned. Political culture refers to the traditions, rules, norms, and expectations of policy formation and decision taking--in essence, to the manner in which people involved expect others to behave. A major distinction separates U.K. and U.S. approaches to risk analysis and environmental assessment. The U.K. customarily operates in a closed, secretive style based upon professional judgment borne of supposed competence and experience, trust in the independence and authority of regulatory officials, acceptance of the political processes through which scientific dispute is translated into political action (or inaction), with a penchant for flexible standards of regulation and self-policing attuned to particular circumstances. The U.S. normally operates in an open adversarial style dominated by enormous amounts of documentation, legally enshrined rules and rulemaking procedures, widespread use of lawyers as intermediaries, and a skeptical distrust between all parties.

RISK-CLARIFYING INSTITUTIONS AND THE POLITICAL CULTURE

A risk-clarifying institution is any formal (i.e., widely but not necessarily universally accepted) mechanism for outlining, debating, and assessing risks and other environmental implications of a proposed policy or project. The public inquiry in the U.K. and the public hearing in North America are two examples used in this paper, though more attention will be given to the U.K. public inquiry. Political culture refers to the manner in which policies are shaped and determined and particular decisions taken. It is a rather nebulous concept, for it is the outcome of conventions of behavior and ways of doing things which are customarily expected and hence which are so intrinsic to day-to-day-activity as to make people unaware of their conformity.

In the U.K., regulation of risk and environmental impact takes place within a set of conventions which have a long and cherished tradition. The four main features are discussed below.

Confidentiality and Selective Consultation

Under the Official Secrets Act (1911) all offical information is treated as secret unless its disclosure is officially authorized (legally by the responsible minister but in practice by the appropriate civil servant). The bias lies with secrecy, not with publication. This is an unwieldy and unworkable arrangement, so it is common for information to be shared among various networks of interested individuals and representative groups and committees even though public disclosure remains illegal. The media may be privy to much of this material but normally in an "off-the-record" basis; the media are trusted to use this material discreetly and responsibly through what is colloquially known as the "lobby system"--informal briefings with ministers and senior officials.

The "Achilles heel" of this fairly cozy arrangement is the deliberate inside leak, from someone who is disaffected and/or a democratic idealist. Nowadays there are moles in the Cabinet and all the way down through official agencies and advisory bodies who feed opposition and the media with often juicy and revelatory material. This can be damaging to the image of trust and competence beloved of officialdom, though the leak must be accurate, reliable, and incisive. In many regulatory agencies and political bodies, the leakage is growing and is resuming ever greater importance, notably for the success of counterestablishment analysis.

Cooperation and Self-Policing

Cooperation between regulator and regulated is one of the principal hallmarks of British regulatory practice. The professional inspectorates in the U. K. regard themselves less as policemen and more as advisers. Self-regulation based on flexibly determined standards and roles is considered more important than externally imposed controls. Sir Peter Hirsch (1), Chairman of the U. K. Atomic Energy Authority, outlines the position with regard to nuclear safety:

> In the UK, the approach to safety regulation of nuclear power is based on the principle that total responsibility for ensuring the safety of an operation is given to the operator, and on having effective watchdog bodies such as the Nuclear Installations Inspectorate to see that he is fulfilling these responsibilities properly. The dialogue between the operators and the regulators in this country is a most important contribution to ensuring the safety of nuclear plants. Public acceptability of nuclear power must, to a large extent, depend upon the degree of public confidence that exists in the rigour of this process.

Sir Walter Marshall (2), formerly Chairman of the UKAEA and now Chairman of the principal nuclear client, the Central Electricity Generating Board, says of the NII: "Its job is to keep the CEGB up to scratch: it never gives its blessing, it only stops nagging. At best it says 'at present we cannot fault your safety arguments but we will review them again and again and again at each stage in the design.'" This approach works well if both sides respect each other and regard themselves as equals.

In recent years there has been some criticism of U.K. regulatory agencies, mostly by outsiders who believe that their work should be more open. This is particularly the case with the Factory Inspectorate (3), though the Alkali and Clean Air Inspectorate is not free from suspicion (4). But it would be wrong to give the impression that this criticism is wide-ranging or widely held. Recent authoritative reports have praised the skills and achievements of British regulatory inspectorates (5, 6).

A recent House of Commons Select Committee (7), while itself convinced of the integrity of the Nuclear Installations Inspectorate, nevertheless noted that "professional respect is not enough--it is a necessary condition but not a sufficient one. The NII must be seen to act strongly and decisively where and when necessary." The Committee recommended that the Inspectorate be permitted to take on more expertise and qualified staff and that, if necessary, it should be granted independent status, free from the Health and Safety Executive in which it currently resides.

The disquiet seems to stem from investigations following up mistakes or near mistakes that capture public attention, suggesting that possibly the cozy client-regulator relationship is perhaps a little too close to ensure good practice. Admittedly, only a few "mistakes" have been uncovered but as is so often the case when one is disclosed, public attention is altered and confidence has to be restored. Openness of procedure, sufficiency of manpower, quality of expertise, and depth of experience are all being demanded of key U.K. inspectorates (and indeed planning departments, about which there is more below). The concern is outlined in this passage from a report by the Council of Science and Society (8):

> Although it is widely known that power corrupts, we as a society are less well aware that impotence also corrupts, especially when it is linked to responsibility. An inspectorate that cannot enforce its requirements must either confess its impotence or conceal it from view. In the latter, natural course, the Inspectorate denies all

cases of abuse except for the most flagrant: in so doing becomes implicated in their contamination. Thus a weak inspectorate is pushed towards identification with those who create the risks to the detriment of those who experience them.

Professional Integrity

Safety regulation and environmental assessment in the U.K. generally operates through a kind of club, the membership of which is highly trained and believes in professionalism and high standards. Judgment borne of experience is a vital element of the regulatory practice, particularly so when there are few other guidelines or rules of cost-benefit analysis. At times this collective self-confidence can be so overwhelming that officials can become immune to what they regard as "outsider criticism." This means that critics of U.K. regulatory practice have to be well-informed, highly respected, and trustworthy. Such people are not easy to find.

There is always the danger of collective delusion--what the Kemeny Commission referred to as the "mindset"--when challenging argument is not given full weight and when regulatory practices are not subject to open and independent scrutiny. The British attempt to avoid this problem by establishing expert task forces or advisory committees. Their job is to sift the relevant evidence and provide independent judgment. But their work is also confidential, so internal memoranda and disagreement are not necessarily recorded. How far these groups--composed normally of reputable people with well-established credentials--are genuinely independent and free to comment remains a matter for speculation. Generalizations would not be fruitful here--much depends upon the scientific and political content of the subject matter and how Chairpeople handle their work and report their discussions.

Gradualism

Institutional policymaking and decision taking is no different in Britain than elsewhere. Decisions are normally made step by step, bearing in mind existing commitments of agencies involved and how they interpret their terms of reference. The British inspectorate likes to build upon existing practice, adding new methodolgies and procedures only when seen as necessary or to have proved themselves. For example, the British were never happy with the Rasmussen approach to nuclear risk assessment. They preferred to develop their own style of probabilistic risk analysis and to concentrate on smaller, more identifiable accident sequences than try to estimate the likelihood of maximum credible accidents. This approach is now more widely accepted in the U.S.

INNOVATIONS IN REGULATORY PRACTICE

The companion paper (Political Decisionmaking and Scientific Indeterminacy: The Analysis) argued that a post-industrializing society that is politically more fissile and democratically more demanding is operating in an economic and political climate when pressures for more specific forms of accountability and more revelatory styles of decision taking are strong. Institutions designed at least to clarify, if not to resovle environmental risk issues, are expected to perform "better" than they have in the past. By "better," one means being more open, more thorough, and more fair, both to the parties concerned and to the arguments they wish to present. This, in turn, means that more and more information has to be provided (because information breeds documentation), counterestablishment groups demand public or philanthropic funds to assist them in preparing and presenting their case, and risk-clarifying procedures become more lengthy, more costly, more adversarial, and more actively inquisitorial. These trends place strains on risk-clarifying institutions not originally designed to cope with the demands now required of them.

Risk and Environmental Assessment and Planning

This means that regulatory practice in the U.K. is gradually becoming more open and hence more publicly accountable, and also judgments made have to be justified more clearly than heretofore. Slowly but surely the Inspectorates are coming out of the darkness into the twilight. For example, the Chief Alkali and Clean Air Inspector whose colleagues are responsible for all aspects of air pollution control from listed works in the U.K., has recently published the most explicit statement yet of the principles and application of best practicable means, the cornerstone of U.K. regulatory practice (9). He makes statements there that might provide grounds for challenge against him and his inspectors in the future.

For example, he insists that current technical knowledge regarding pollution abatement or dispersion must be judged not just to protect human health, but also to take into account effects on "flora, fauna, materials of construction or property, agriculture and amenity." Astute planners responsible for evaluating environmental assessments could make something of this statement, though how far they might do so will depend upon their expertise, commitment and authorization. The Chief Alkali Inspector also commented that "industry wide bpm should be applied to all industry irrespective of the financial circumstances," though he qualifies this by noting that "local circumstances may require some variation in detail."

It remains to be seen how far regulatory authorities, who are not directly politically accountable, will go in attempting to close down or stop the construction of a business enterprise when the local government wishes to preserve or protect jobs. In one recent case where this dilemma was particularly acute, namely the Bedfordshire Brickworks (10)--a major polluting industry proposing to reorganize its works and concentrate its effluent in three tall chimneys--the local planning authority demanded tougher standards of pollution control than the Alkali Inspectorate. But when the Brickworks announced that it would close due to recession and what it claimed as the economic burden of tough regulatory standards, the local authority, pressured by local trade unions, relented and lowered its demands.

This particular issue remains unresolved, but it does pinpoint the dilemma in the U.K. between risk regulation, environmental assessment, and economic planning. A very uneasy relationship exists between the planning authorities and the risk-regulating agencies. The planning authorities are charged with the task of assessing the local consequences of a proposed development in the context of long-term guidelines as set out in a county structure plan, and, in fact, are responsible for evaluating the environmental assessment (11). That assessment, true to U.K. practice, is the responsibility of the developer to provide, though the developer works closely with the planners. For complex polluting or risk creating proposals, the national inspectorates are called in, usually operating at the regional level. Currently there is some doubt whether planners are sufficiently well trained to cope with the technical complexities involved. The problem is complicated because relations between planner and regional inspectorates are very variable indeed.

Because the British government will not accept any form of statutorily imposed environmental impact assessment, even of the highly modified kind proposed by the European Commission (12), responsibility for judging the merits and hence effectiveness of an environmental assessment rests with the local planning authority. The ultimate decision is that of the relevant minister because planning decisions are subject to appeal and ministerial determination. Nevertheless, EIA guidelines are too non-specific in the U.K. at present, procedures are ad hoc, at times being quixotic and almost whimsical, and final decisions rest more on the local political interest in jobs than in any harmonious give and take between economic development and environmental protection.

The Licensing of Nuclear Power Plants

In his study of nuclear plant safety, Chicken (13) notes that when the first U.K. reactors were built in the late 1950s, there were no quantitative reactor safety criteria and the only guidance for designers came from engineers in the chemical industry and from the International Commission on Radiological Protection. The standards that were finally reached were almost in the form of exhortations rather than regulations. This created problems for plant manufacturers and designers who were unsure of what requirements their plant had to satisfy.

In the late 1960s, the British developed the approach of probabilistic risk assessment (PRA) which looks on accident sequences as synthesized from the design and performance characteristics of individual components and operational procedures. Once a PRA has been completed for the system as a whole, for any fault sequence or for any individual component, it can be compared to all safety principles and to criteria set by governments or clients as what are regarded as acceptable levels of risk. However, until now the PRA has still only been a guide: The final safety standard is based upon wider considerations and upon what is technically and economically practicable. This reasoning was echoed by the Director of Health and Safety of the CEGB (10):

> The assessors (NII) will judge that a reactor has an acceptably low risk on the basis of quantitative assessment, supported by numerical analysis, and taking into account not readily amenable to numerical analysis, such as safety administrative procedures, operator recovery and damage control arrangements. Overall acceptance of a design is not based upon a particular numerical standard.

The British treat nuclear plant licensing as a continuing learning process: As there is no point at which safety levels are regarded as immutable, there is no finality about the licensing process.

The important features of British nuclear safety practice have not been fully appreciated (or accepted) either by American advisers working with the U.K. PWR team preparing the Sizewell B design (about which reference was made in the companion paper) or by the Inspector and his Assessors convening the public inquiry into the proposal. The Westinghouse designers prefer to run a series of computerized safety codes set to precise regulatory standards. The British counterparts in CEGB and the construction consortium, the National Nuclear Corporation (NNC), are not entirely happy with this purely quantitative approach and in any case would prefer to redesign and certainly to reappraise all of the computer codes. In any event, they do not accept that the codes provide the final determination. Yet the Vice President of Westinghouse told the House of Commons Select Committee that the British have a tendency "to overindulge in the practice of creating new rules and 'requiring new kinds of analysis' ...I would caution that the risk of adding substantially to the licensing schedule in this way is a real one." (15)

Currently, there is an interesting experience in mutual education between U.S. and U.K. colleagues in PRA and accident sequence assessments. The outcome of this should be valuable for both parties and may well lead to an innovative convergence of safety assessment principles. It may also force the U.S. designers to work more with Nuclear Regulatory Commission (NRC) officials than they are customarily (or indeed legally) inclined. Such moves would likely reassure the British public, but it would be interesting to know whether American public opinion wuld be so favorably impressed.

Another interesting innovation in U.K. regulating practice has been the merging of the nuclear plant licensing process with the public inquiry. The public inquiry in the U.K. is a statutory procedure established under many Acts designed to assist a minister in reaching a fully informed decision (16). The basic format of the inquiry is for a proponent to argue the merits of a case and for objectors to challenge these before an independent inspector. In complex and controversial cases, the proponents and some objectors will be represented by legal counsel responsible for marshalling evidence in its most presentable form and for rigourous cross-examination of opposing witnesses. Also in such cases the Inspector is advised by one or more expert assessors, though the Inspector alone is responsible for the final recommendations to the authorizing minister. The public inquiry is an issue-clarifying device, not a decisionmaking mechanism: Decisions rest with ministers accountable in Parliament. However, objectors see their role more as political than an administrative tribunal and increasingly seek to use it for political and educational purposes.

Up until the early 1970s it was never thought appropriate or necessary that public inquiries into nuclear power plants should cover technical matters such as nuclear fuel cycle risks including plant safety, plant design changes, or management of site construction. In a public inquiry into an advanced cooled gas reactor in 1971, the Inspector commented (17): "Nuclear risk is a subject so sophisticated that we, like the objectors, have no reliable grounds of forming an independent opinion, and we have been obliged to accept both the assurances given by experts called by the Generating Board and those implicit in the government policy of employing nuclear powered electricity generation of an increasing scale in proximity to populated areas."

Two points are worthy of emphasis. One is that safety considerations were seen as too difficult for an independent lay process to pass comment upon, and the other was that government policy (in this case that of favoring nuclear power) was not a matter for discussion at public inquiries.

Today all of this has changed. The current public inquiry into the Sizewell B design (technically to grant the CEGB a site license to construct a PWR on the Sizewell B site in Suffolk) is treating plant safety exhaustively with both the CEGB and the NII fully exposed to detailed proofs of evidence and to cross examination. The CEGB has already provided over 27 volumes of safety case plus hundreds of supporting documents, yet almost daily it is expected to supply additional material at the request of the Inspector and his assessors. Similarly, the NII are under scrutiny far more than ever before. They too have had to provide detailed documentation and have published their correspondence with the CEGB--correspondence that would normally be confidential.

It is almost impossible to ascertain whether the NII is being harder on the CEGB than it would otherwise have been had there not been so much public and political interest in the safety case of the proposed PWR. Informed commentators believe that the NII is being tougher and generally more aggressive in its dealings with the whole PWR task force (which includes representatives from CEGB, NNC, British Nuclear Fuels Ltd, Westinghouse, and Bechtel). One reason is that the NII must be seen to be assertive, knowledgeable, and independent. The interested public demand nothing less nowadays. Indeed, the House of Commons Energy Committee (18) recommends that no holds be barred with respect to providing the NII with sufficient resources "to permit it to carry out the inspection of nuclear sites according to its own criteria, regardless of whether this results in a level of inspection greater than for non-nuclear installations which demand less stringent procedures."

The PWR is truly a foreign plant as far as many British people are concerned, being American technology and of a similar type to the Three Mile Island reactor (albeit of a Westinghouse design). It is not surprising that all the British opposition parties (Labour, Liberal, Social Democratic Party) stated in their election manifestos in June 1983 that they would not allow the Sizewell B PWR to be built. The ruling Conservative government, on the other hand, has frequently stated that it supports a strong nuclear industry and the PWR, so the NII are working in an unusually open and politically charged atmosphere. It is hardly surprising that they are taking their time, drawing upon outside (including U.S.) expertise, and now admit that they will not be able to complete the pre-construction part of the plant licensing process before the Inquiry is over.

The NII refuse to be harried about the completion of the vital pre-construction licensing process (legally and politically the plant could not be built without NII approval). Yet the Inquiry Inspector has urged that the NII satisfy him that they can resolve all outstanding safety issues before the main objectors cross-examine on safety.

The Inspector appears to have accepted that he can do no more to hurry the NII, and has recently turned down, for the time being, a request by one of the leading objectors, U.K. Friends of the Earth, to adjourn the Inquiry until the licensing process is complete. Yet in evidence to the House of Commons Select Committee in March 1980 (19), Mr. Dunster, Deputy Director General of the Health and Safety Executive (the body controlling the NII) stated:

> ... we wish to go into the Inquiry with a complete picture. We wish to appear before it saying that we are now in possession of the information which would enable us to license the plant. That is not a constitutional requirement for the Inquiry: it is obviously an essential requirement in common-sense terms.

It is precisely this that the NII cannot now deliver and on which the NII remain vulnerable to further demands for adjournment of the Inquiry.

This particular matter reveals some interesting features of the changing approach to risk assessment in the U.K.:

1) Cost and safety intertwine: Plant licensing requirements can add to the final cost of a plant and delay its construction. Witnesses appearing for the objectors suggest that these could add up to 30 percent to the final cost of a plant (20).

2) Regulatory delay and indecision can place enormous strains on plant design management, component ordering, and on-site construction: Design teams are not geared to delay, and uncertainty and component manufacturers are most unhappy if specifications have to be changed. The pressures on getting the plant design right prior to tendering final ordering are enormous, yet the U.K. licensing process is not so clear-cut as to permit this.

3) The complexities of safety arguments make it difficult for lay minds to grasp the essence of the dispute: Yet legal lay minds are predominant in the U.K. public inquiry and they certainly have to work very hard to grasp the scientific and technical essentials. Whether cross-examination by legally trained scientists would be more thorough awaits further examination. Experience from the continent suggests it would be.

4) Effective opposition is expensive: The very best expertise is called for and this usually requires high fee consultants (though some will appear for public interest objectors at reduced rates, up to a point). The NII have already estimated that their safety review of the CEGB's PWR design has cost between 5 and 10 million pounds, with the CEGB reckoning on 5 million pounds for the safety case alone. Opposition groups financed by philanthropic donation have to do the best they can on 50 - 75,000 pounds. The debate continues to rage as to whether the NII really do represent the whole public interest as regards nuclear safety as the present Government clearly believes to be the case--hence, its refusal to provide financial aid for objectors to the Inquiry.

5) On three crucial issues, namely plant licensing, post-license design, and on-site management, the Sizewell B Inquiry cannot pass final judgment: This raises important questions as to the relationship between major inquiries and reactor management which have not yet been resolved. It is possible that opponents to the Sizewell B proposal will argue forcefully that the Inquiry will be impotent if these three procedures are not satisfactorily dealt with by the Inspector in his report to the Energy Secretary. Yet this would push the Inquiry far beyond what was intended of it and would make even more uncertain the atmosphere in which an authoritative ministerial decision would be made. Delay and the exacerbation of uncertainty are key weapons in the opposition armoury.

INNOVATIONS TO THE BIG PUBLIC INQUIRY

Currently in the U.K. there is much debate as to whether the existing form of big public inquiry can be extended sufficiently to cope with all the demands for fullness, fairness, and thoroughness now being made upon it, or whether a new form of inquiry should be created. The key to the argument lies in the future relationship between the inquiry and Parliament. At present the inquiry is a creature of a cabinet minister and his executive department. Its inspector reports to the Minister, though in major cases the final decision will be a Cabinet one. Parliament has no power to intervene and can only debate an inspector's report if the Minister so wishes, but normally after a decision has been announced.

It is too early to judge how far the Sizewell B Inquiry will be successful in meeting all the demands made on it. Nevertheless, there have been many interesting procedural innovations. These illustrate the sensitivity of those responsible for risk clarification and environmental assessment to the need to satisfy the public demand for a credible and legitimate inquiry. These innovations can be summarized:

1) Lengthy warning of the Inquiry with six month lead time of key documentation: The objectors had time to marshall their case against the CEGB's arguments, though the CEGB is being forced to reappraise parts of its analysis through additional documentation as the Inquiry proceeds.

2) Three separate preliminary meetings to settle the agenda, delineate the boundaries of the dispute and determine how the cross-examination was to be handled: It is a moot point as to how far any of these objectives were achieved. The agenda is constantly being revised, there are regular requests for more information, and the cross-examination procedures do not seem consistent.

3) Appointment of a counsel to the Inquiry by the Energy Secretary at the Inspector's request: This individual normally reacts to instructions from the Inspector and his assessors, but at times he also asks questions on behalf of objectors and is also free to follow his own train of thought. This is a crucial innovation as the counsel to the Inquiry has a free hand to cross-examine the witnesses and to investigate material that may not have been fully covered by objectors. Normally the Inquiry is passive and awaits arguments to be presented before it. The Sizewell B Inquiry shows signs of being healthily investigatory--though how far this influences its outcome remains to be seen. Nevertheless, in this modest respect the big inquiry is taking on some of the features of the Canadian investigatory commission, and to a lesser extent, the American rulemaking hearing.

4) Policy examination of Government Departments: The Department of Energy and Environment permitted senior officials to be cross-examined by objectors and counsel to the Inquiry as to the substance and (in the case of the Department of Energy), the merits of Government policy toward energy strategy generally and radioactive waste disposal. Although not particularly revealing, the fact that Departments submitted themselves to some public scrutiny represents an important departure for the inquiry which normally does not question policy.

5) Side meetings, held off the record, to facilitate agreement over documents or points of misunderstanding between objectors and proponents: The side meeting is a genuine innovation to sort out difficulties between contending parties. These were shared by assessors, but in one case by the Inspector himself. To date, these have fairly modest objectives but there is scope for developing them into something akin to the American mediation process or the Scandinavian remiss system, given the right incentives and encouragement.

6) Indirect funding of objectors by the CEGB: The CEGB is worried that objectors may say "foul" and hence undermine the legitimacy of the Inquiry because they are so underfunded compared with them and the NII. They are not allowed by law to finance opponents directly, but they are giving indirect assistance by providing free transcripts, working accommodation, and documents. In addition, they are paying for the whole Inquiry (estimated to cost at least 12 million pounds) and for the Councel to the Inquiry. Whether all this keeps the criterion of fairness to be met remains to be seen, but it is certainly a step in the right direction.

All these innovations are stretching the big inquiry way beyond its original procedural boundaries. Yet arguments persist that even these are insufficient to make the risk-clarifying institution totally legitimate in the mind of the informed and interested lay public. The concept of "legitimacy" is an important one here. It means granting the public inquiry the status of effective influence on policy formation and scope for purposeful participation by interested individuals and groups. A legitimate inquiry would grant greater authority to final political decision, assuming that any decision identified clearly how it took inquiry evidence and recommendations into account.

The direct intervention of Parliament, notably to clarify policy priorities and to determine the so-called generic arguments over need and economics of alternative energy supply and demand possibilities and mixes, is therefore likely to remain a serious quest. As it stands, the big inquiry seems to be over-sufficiently legalistic and administrative and insufficiently political. But radical parliamentary reform will be required to meet the demands for a more direct parliamentary role. This would include more powerful policy scrutinizing select committees, more scope for special commissions of investigation possibly involving parliamentarians, experts, and informed lay people, more skilled and knowledgeable MPs, and a Government willing to extend unusual new powers to the backbenchers. These go far beyond anything that any modern government has seriously comtemplated.

Governments like authority and like being in control. It is highly unlikely that any envisageable British government will cede major extension of existing powers to its backbenchers, even less to an extra-parliamentary forum like the public inquiry. Yet in the big, politically sensitive set pieces, such as those dealing with major risk issues and widespread environmental effects, innovations to the major inquiry and greater parliamentary involvement--albeit within controlled bounds--will doubtless continue to occur. After all, flexibility and ad hoc-ery are very British trademarks.

These innovations all move in the direction of openness, accountability, and fairness, and reflect the resilience of British democracy to adjust to new demands for accountability. Yet a price may be paid for this tactic. Smaller, less politically contentious issues may be managed so as to bypass the public inquiry, especially if the proposal being considered is not novel. There are plenty of statutory ruses available to Departments of State to allow this to happen. Alongside responsiveness may emerge a more authoritarian approach.

The three criteria mentioned in the companion paper, namely policy uncertainty, political controversy, and time to delay, may well determine how these two quite different tactics may be deployed.

The ordinary person is still left out of all of this. The big inquiry leaves no room for the interested but uninformed, provides little comfort for those who would like to participate directly, and is either misunderstood or ignored by the masses. Here is where the politically financed Canadian roving commission offers scope for a completely fresh approach. Combine that with modest parliamentary reform, with scope for mediation through more elaborate side meetings, and with a preparedness to create key advisory bodies reviewing generic safety; and economic matters composed of politicians, experts, and informed laypeople, and the procedural innovations, already in train, may need go no further.

Author's Note:
Part of this paper is based upon continuing research into the current Sizewell B Public Inquiry. This research is being financed by the U.K. Social Science Research Council, whose generosity is greatly acknowledged.

REFERENCES

1. P. Hirsch, "Research and Nuclear Power," Atom, 321, July 1983, 142-143.

2. W. Marshall, "Regulation of nuclear power in the U.K." Unpublished talk to a Conference on the Sizewell B PWR Design, University of Birmingham, March 1982. Available from the Central Electricity Generating Board, London.

3. T. Craig, "Who inspects the Factory Inspectorate?" New Scientist, 30 June 1983, 952-957.

4. Social Audit, The Alkali and Clean Air Inspectorate, (Social Audit, London, 1974).

5. E. Ashby and M. Anderson, The Politics of Clean Air, (Clarendon Press, Oxford, 1982).

6. Royal Commission on Environmental Pollution, Air Pollution: An Integrated Approach, (HMSO, London, 1976).

7. House of Commons Select Committee on Energy, The Government's Statement on the New Nuclear Power Programme. H.C. Paper 114-i. (HMSO, London, 1983), 76-77.

8. The Council of Science and Society, The Acceptability of Risks, (Barry Rose, Publishers, London, 1977), 43-44.

9. Health and Safety Executive, Industrial Air Pollution. (HMSO, London, 1982), 14-18.

10. A. Blowers, personal communication, July 1983.

11. C. Miller and C. Wood, Planning and Pollution, (Clarendon Press, Oxford, 1983).

12. House of Lords Select Committe on the European Communities, Environmental Assessment of Projects, H.L. Paper 64. (HMSO, London, 1981).

13. J.C. Chicken, Nuclear Power Hazard Control Policy, (Pergamon Press, Oxford, 1982).

14. R.F. Matthews, "Using quantitative analyses in Britain," Nuclear Engineering International, 25, 45-47.

15. House of Commons Select Committee on Energy (op.cit.), 75.

16. R.F. Wraith and G.B. Lamb, Public Inquiries as an Instrument of Government, (Allen and Unwin, Hemel Hempstead, 1971).

17. Quoted in B. Wynne, Rationality and Ritual: The Windscale Inquiry and Nuclear Decisions in Britain, (The British Society for the History of Science, Chalfont St. Giles, Bucks, 1982), p. 66.

18. House of Commons Select Committee on Energy, (op.cit.), 76.

19. Department of Energy, Nuclear Power, Cmnd b 317, (HMSO, London, 1981).

20. J. Dunster, Evidence before the Commons Select Committee on Energy, H.C. Paper 397 - vi (HMSO, London, 1980), p. 216.

21. Council for the Protection of Rural England, Evidence to the Sizewell B Inquiry by Komanov and Conroy, (CPRE, London, 1983).

Appendix 1

THE LEAD IN PETROL ISSUE IN THE UK

Government Departments involved

"scientific" aspects: Transport, Energy, Environment, Health

"political" aspects: Treasury, Trade/Industry, Foreign Office

NB lack of coordinated departmental policy and inadequate machinery for coordination

Official summaries distorting or eliminating counter-establishment science

political capitulation

US medical psychological studies

1937-1979

Conservation Society working group on lead in petrol

Birmingham "Spaghetti Junction" study

(raised scientific critique)

1979-1980

Lawther Report[1]

Conservation Society response [2]

1982

Political campaign by CLEAR[3]

FOE[4]

1983

Royal Commission on Environmental Pollution[5]

Government response

lead lobby (Associated Octel, Motor Manufacturers and Traders Association, Petrol Companies) argued

(i) lead comes from other sources
(ii) social deprivation intercorrelated
(iii) costs of lead removal too high
(iv) additional energy demand

capitulation

This schematic diagram used for illustrative purposes indicates how scientific disputes become incorporated into political decisionmaking. For five years the UK government backed by the lead lobby and buttressed by official scientific studies blocked any commitment to lead free gasoline. They capitulated partly because the sceintific evidence became embarrassingly difficult to refute and partly because parliamentarians demanded action. The political lobbying effectivness of the campaign against lead in petrol (CLEAR) cannot be underestimated.

Appendix 2

POLITICAL RESPONSE AND SCIENTIFIC INDETERMINACY

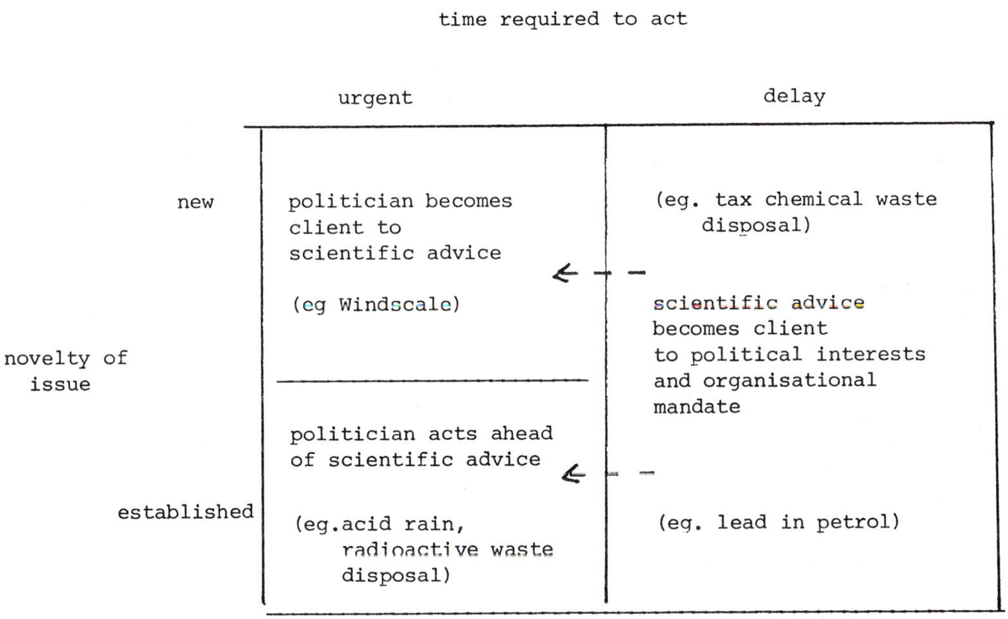

NB This is speculative and does not provide a perfect fit for the circumstances portrayed. Delay can lead to a jump leftward to "inefficient" and possibly "inequitable" political decisions (←----).

Appendix 3

EVOLUTION OF UK PUBLIC INQUIRY INTO RISK/SOCIETAL ISSUES

Stage 1 (up to 1960)

local considerations covered only
inspector's report rarely published
no parliamentary discussion.

Stage 2 (1960-late 70s)

local issues still dominant, very little conservation of wider societal and policy issues.
Inspector's report published with ministerial decision: no parliamentary discussion.

Stage 3 (late 70s early 80s)

rise of national "generic" issues (eg. Windscale Canvey, Belvoir) and covering some policy matters and standards of health and safety
Discussion of policy context and "need" aspects
Inspector's report published prior to Parliamentary debate (Windscale only)

Stage 4 (Sizewell B : early 80s)

limited Parliamentary discussion of policy context	infusion of generic national and local issues (need, safety, siting, fuel cycle local planning)	Inspector's report	Parliamentary decision

Stage 5 (possible future development)

formal parliamentary analysis of policy priorities by select committees plus scientific/informed laypeople commission (infuse scientific & political judgement)	limited inquiry documented evidence precise cross examination coalescence of objections wide public involvement via "travelling" inquiry	Tribunal report	Parliamentary decision

Standard-Setting Standards:
A Systematic Approach to Managing Public Health and Safety Risks[1]

Baruch Fischhoff

Decision Research, A Branch of Perceptronics, 1201 Oak Street, Eugene, OR 97401

INTRODUCTION

Standards are a pervasive feature of modern society. They are set by governments to protect public safety and welfare, by professional organizations to assure consumers of the quality of their work, and by manufacturers to protect themselves against liability claims. Familiar standards include the Code of Federal Regulations, national ambient air quality standards, auto emission limits, the Good Housekeeping Seal of Approval, NCAA helmet specifications, the Geneva Conventions, and Underwriters' Laboratories requirements. The form of these standards is as varied as their source and subject matter. Some refer to allowable effluents; others specify the physical properties of machinery; still others define permissible impacts on the environment. Their one common feature is that each separates activities or technologies into two categories: Those that meet the standards are acceptable; those that do not are unacceptable.[2]

The present analysis proposes a systematic, general approach to setting standards, focused on those standards used to manage risks to public health and safety. It views standard setting as involving three distinct stages. First, there must be a decision to rely on standards. Second, a safety philosophy, balancing risk and other factors, must be determined. Finally, that philosophy must be translated into operational terms. The paper describes problems that arise at each stage and offers guidelines to their resolution. Descriptively, it can be used to predict how successful particular standards will be.

---

1 <u>Management Science</u>, in press.

2 Some pertinent references include: (12), (38), (1), (25).

NATO ASI Series, Vol. G4
Environmental Impact Assessment, Technology Assessment, and Risk Analysis. Edited by V.T. Covello et al.

Prescriptively, it can be used to design sounder standards (and to identify the limits of standards as a managerial tool).

In practice, standards are often applied quite flexibly. Here, they are treated as fixed, explicit rules. Such literal interpretation is essential for determining whether the operational form of a standard is faithful to the intent of the standard setters. The translation process is shown to involve many apparently technical specifications that can make the effective trade-offs realized by the operational standard quite different from the intended ones.

WHEN TO SET STANDARDS

There are two archetypal ways of deliberately managing hazardous activities and technologies--decisionmaking and standard setting. By considering the differences between these two procedures, it is possible to identify the conditions that are most (and least) conducive to standard setting.

A decision is a choice between alternative courses of action. Standards, on the other hand, pass judgment on individual courses of action. Decisionmaking procedures attempt to order options according to relative attractiveness; standards simply categorize them as "pass/no pass." Applied to a set of proposed actions, decisionmaking attempts to identify the one action that is most desirable; reliance on standards may lead to all actions being accepted or to none being judged adequate. If an option has been judged acceptable, it will remain so until either the option or the standard is changed. The judgment rendered by a standard is, thus, irreversible. Decisionmaking is conditional; if a better option comes along, then the selected option may be replaced. With decisionmaking, options must beat the competition; with standard setting, they must beat the standard.

Once set, a standard is meant to be applied repeatedly. The same fixed rule, expressing a particular set of values, is used to judge all proposals falling within its jurisdiction. Decisionmaking is tailored to the particulars of individual choice problems, deriving the implications of the decisionmaker's values for each. The trade-offs between benefits, risks, and non-risk costs are considered only when standards are set. Their application is an administrative act, unlike decisionmaking which is always a political act.

Identification of these differences provides a basis for circumscribing the conditions in which it makes sense to set and apply a standard rather than to make a decision, and when the opposite is true. Table 1 summarizes these differences and derives from them conditions that favor standard setting. In some cases, this derivation is a straightforward, logical deduction; in other cases, it requires consideration of the political realities within which hazardous technologies are managed. When conditions are poorly suited to standard setting, the observer should predict difficulties: with deriving a safety philosophy, with making it operational, with applying it. Hazard managers should make the same predictions. However, they may also be able to change the conditions in order to improve a standard's chances of success.[3]

Standards are appropriate in the following situations:

(1) When no choice between options is possible: Many hazard managers do not have the opportunity to weigh the relative merits of competing options. Rather, they must judge each on its own merits, without explicitly considering what will come instead if it is rejected or what will be foregone if it is accepted. For example, industrial trade associations are seldom empowered to decide which product is best; government regulators typically must act independently upon each option presented to them. A general standard will allow them to treat all options in a consistent manner.

(2) When no choice between options is required: Much of the effort required by decisionmaking schemes is invested in ordering all available options so as to identify the best one. That effort will be wasted if it is possible to choose several options, or none at all. For example, society need not restrict itself to one energy source, nor need it accept any member of a new family of prescription drugs. In such cases, all that is required is an accept/reject rule.

(3) When predictability is important: Standards can simplify life for hazards as well as for their managers. They offer a

3 Although the remainder of this discussion considers actions by policymaking or regulatory bodies, a comparable choice of procedure faces individuals in managing the hazards in their own lives. That is, they can deal with each hazard in isolation or establish a general rule for dealing with many hazards. A parallel analysis generates some interesting hypotheses regarding individual choice behavior.

Table 1

When Are Standards Suitable

Feature	Standards	Decisions	Conditions Favoring Standards
Number of options chosen	None at all	One	(1) No choice possible (2) No selection needed
Task facing options	Satisfy standard	Beat opposition	(3) Predictability important (4) Future options need shaping
Range of application	Category	Single case	(5) Competing technologies in same jurisdiction (6) Category members homogeneous
Expression	Rule	Choice	(7) Explicit policy attractive (8) Value issues sensitive
Application	Technical	Political	(9) Political resources limited (10) Process unimportant
Flexibility	Little	Great	(11) Awkward applications avoidable

clear and fixed target that a technology's designers can strive to meet. Once the standard has been met, approval is irreversible, allowing the designers to concentrate on factors other than safety (e.g., cost containment, market share). The desire for predictability has led the nuclear industry to propose that the U.S. Nuclear Regulatory Commission (50) adopt an overall safety standard. Once a power plant has met this standard, it would no longer be subject to the Commission's case-by-case decisions regarding what designs and procedures to adopt. Although meeting the overall standard might be more expensive than abiding by the local decisions, doing so would eliminate worry about retrofitting and the withdrawal of operating permission.[4]

4 An application of this theory to the Commission effort can be found in (15). As currently proposed, the goals would be an addition to case-by-case decisions, at least for the foreseeable future.

(4) When regulators hope to shape future options: When regulators do not like the choice of options offered to them (or to the public), they may set a standard that is clear, predictable, but out of reach. Such "technology/forcing" standards are intended either to require a particular product or to give a competitive advantage to those who produce it. U.S. automobile fuel economy goals, for example, were intended to do the former, but ended up doing the latter as well. Designed to reduce the environmental and national security risks from excessive gasoline consumption by coercing manufacturers to produce more efficient cars, they inadvertently helped the industry meet foreign competition. On the other hand, when standards mandate a particular technical solution, they can stifle innovation, perhaps even increase monopoly pressures. For example, requiring the catalytic converter as a solution to pollution control greatly benefited General Motors, the leader in its development (51). Standards generally favor large corporations which have greater resources to track, fight, shape, and adjust to them. Anything that decreases competitive forces will also tend to reduce innovation.

(5) When competing technologies fall in the same jurisdiction: Standards are meant to apply to a category of options. If that category includes one method of accomplishing a particular task, it should include its competitors as well (e.g., all ways to generate electricity, all modes of urban transport). If competing options are subject to stronger or weaker standards (or to none at all), then these inconsistencies may be exploited. For example, tough standards on hazards with acute health effects will, over time, encourage the development of competing products and processes with primarily chronic health effects. If a stricter overall safety standard is adopted for nuclear power than for other energy technologies, then it will be at a disadvantage relative to its competitors (which have at the moment no such standard). The range of a standard may change over its working life. Each technology affected by a standard may try to be released from its constraints or to have them extended to its competitors.

(6) When category members are homogeneous: Having one standard for all category members inevitably means that it will be more suited for some than for others. For example, a uniform standard for atmospheric asbestos levels will allow quite different risks in workplaces with high and low proportions of smokers (4). It may exact quite different costs of compliance in different settings (and, hence, implicitly assign different values to human life and health). A standard that allows a level of risk that is commensurate with the benefits derived from the average member will seem unduly restrictive when

applied to a member with particularly high benefits. The extent of such "injustices" should depend upon the heterogeneity of the technologies within its jurisdiction. It is possible, in principle, to develop a standard with correction factors to cope with special cases. However, doing so produces a standard that is more complicated, more discretionary, more situation-specific--and less a standard.

(7) When an explicit policy statement is attractive: When a decision is made, observers may see just the act of choice, which reveals little about its underlying rationale. When that rationale is explicated, it may involve a complex trade-off between diverse considerations. By contrast, many standards offer a tidy statement of principle saying, "This is what we allow." That statement may tell the whole story, or it may just summarize the results of more complex deliberations. For example, after weighing many factors in deciding to allow fishing in Chesapeake Bay but not in the James River, the Environmental Protection Agency announced a standard for Kepone in fish that lay between the levels found in those two bodies of water (13). That statement may be meant literally or figuratively, as an expression of policy makers' concerns. For example, Eastern European countries want strong standards for worker protection on their books, even if economic problems limit their realization (11). The Delaney Amendment made a similarly strong, if impractical, statement of American values toward health and food. Whatever their original motivation, simplifications may develop lives of their own. They create both precedents and expectations (on the part of producers, workers, and consumers).

(8) When value issues are sensitive: Because they always tell some story, but need not tell the whole story, standards can be used to blur value issues (as well as to highlight them). Unlike formal decisionmaking methods, such as cost-benefit analysis or decision analysis, standard setting need not leave an audit trail explaining its rationale. Such vagueness may cover confusion or it may be designed to obscure what the standard was walling in and walling out. It might, then, serve vested interests who want a high-sounding rule to legitimate their actions, or regulators who must make political decisions, but lack the mandate to do so, or warring parties who can negotiate a compromise within the context of a standard that hides what each has conceded.

(9) When political resources are limited: Every decision is a political act, requiring a statement of values regarding the appropriate trade-offs between conflicting objectives. Particularly where the trade-offs are difficult and society is divided, it may be hard to muster the political resources

needed to reach a decision. When many tough decisions must be made, one way to keep the political system from being overwhelmed is to replace some "small" decisions with one big one, by choosing a general standard. That standard can then be applied administratively by technical experts. Three conditions for this strategy to work are: (a) The standard setters tackle and resolve the value issues; (b) The operational standard leaves little room for discretion; and (c) The technicians receive adequate resources to apply the standard faithfully. For example, the Toxic Substances Control Act made a strong value statement, but failed to provide sufficiently precise guidelines or adequate funds for the Environmental Protection Agency to devise a defensible plan ensuring that the tens of thousands of existing chemicals receive equal justice under the standard (42, 49).

(10) When process is unimportant: By concentrating political attention on the standard setting process, standards do away with the repeated discussion of value issues that comes with case-by-case decisionmaking. To realize these potential increases in efficiency, the standard setting process must be able to replace the political and intellectual functions of repeated decisionmaking. Part of the political function can be served by ensuring that varied parties interested in an issue are represented in setting the standard for it. Concentrated participation in a standard is not, however, the same as continued participation in individual decisions, each of which offers an opportunity to educate oneself, monitor the regulators, and generate public discussion. The intellectual functions are most likely to be filled when the issues are well understood in advance, so that standard setting centers on the resolution of known conflicts. The very act of grouping options into categories and searching for a consistent rule may even generate new insights. However, when problems are too new or complex to be understood in advance, they defy the setting of informed, lasting standards. Although it is possible to have a standard evolve through trial-and-error learning, doing so would be contrary to the notion of a general rule. Thus, standards are inappropriate where a political or learning process is essential.

(11) When awkward applications can be avoided: Once any policy has been made (in the form of a decision or standard), those affected adversely may try to overturn it. Standards are inherently vulnerable to attacks that focus on specific cases in which the general rule seems ill suited. For example, standards look bad when they reject otherwise attractive alternatives whose risk levels are just beyond their threshold of acceptability, especially when it would be very expensive to

gain the small decrease in risk needed to achieve compliance (10). Even if they cannot have the standard overturned on the basis of one sticky example, critics may still obtain a precedent-setting relaxation. One way to defend a standard is to avoid awkward applications, perhaps by specifically excluding them or showing how to bend the standard deliberately (to keep it from being broken).[5]

In summary, reliance on standards differs from case-by-case decisionmaking in certain fundamental ways. From these, it is possible to derive a set of conditions that are particularly conducive to standards. When standard setting is mandated, consideration of these conditions can help one to anticipate and perhaps avoid some of the problems that will be encountered. When a choice can be made between a general standard and specific decisions, then this kind of analysis can point to the trade-offs that should underlie that choice. A typical trade-off question would be whether the timeliness and predictability of a standard's application would compensate for its occasional injustices.

JUSTIFYING STANDARDS

Deciding to rely on a standard sets into motion one big decision, choosing a safety philosophy, followed by many small ones, encountered in translating it into operational terms. Often that translation process leads to an operational standard that is expressed in terms of some measure of risk, such as a level of toxic emissions, a failure rate, or expected casualty toll. Because that level expresses the maximum risk that will be tolerated in a technology, it is often termed the "acceptable level of risk." Although perhaps technically correct, that term obscures the fact that benefits and non-risk costs are considered in the setting of most standards (otherwise the tolerated risk level would be near zero). If the standard is not simply a political statement about safety, the focus on risk may reflect the feeling that the other

5 Another part of the defense is being alert to bogus claims, such as that the money required to bring a particular technology in compliance with a safety standard would be better spent on high-sounding and far-flung alternatives (e.g., mobile trauma units), even though there is no chance of funds being transferred from one realm to the other.

consequences are relatively fixed (hence, "here are the risks that those net benefits[6] can justify").

The term "acceptable" can be misleading in other ways as well. It may connote that society is satisfied with that risk level, rather than just resigned to it in the light of current economic and technical realities. It may also connote that everyone exposed to that risk should be content with risk levels that reflect a compromise among the particular interests involved in setting the standard, or are the result of weighing the interests of society as a whole (36).

However it is labeled, the product of the standard setters' deliberations is a safety philosophy, expressing their view of an appropriate trade-off between risks and other consequences. Four generic methods can be identified for justifying such trade-offs. They differ both in the information that they consider and the composition rule that they use to integrate that information. The choice between them is in part an empirical question: Just how well does each fulfill its promise of eliciting and accommodating a particular range of information? The choice is also a political one: Each method embodies a somewhat different perspective on whose views count most and how political processes should be conducted, hence the method chosen can affect society beyond the context of the particular standard.[7]

Formal Analysis

Applying standards rather than making case-by-case decisions means that one cannot apply to those decisions the sophisticated techniques of decision analysis, cost-benefit analysis, and related procedures (e.g., 39, 47). Those techniques might, however, be used to guide the choice among alternative standards. When so applied, they would try to estimate the consequences following from implementation of each possible standard, in order to identify the candidate offering society the best deal. Those consequences would be estimated in terms such as the costs of compliance, the beneficial health effects expected, and the cost-benefits of substandard projects (24).

---

6 Net benefits will be used to represent the difference between benefits and non-risk costs.

7 A fuller general discussion of these arguments, without the focus on their usage for standard setting, may be found in (47). That source also provides extensive references.

Here, as elsewhere, formal analysis is more readily proposed than accomplished. It is hard to anticipate all significant consequences of a standard; it is hard to calculate the magnitude of many consequences that can be anticipated (e.g., reduced incentives to innovation or increased respiratory disease); it is hard to put a consensual (or societal) value on many consequences, such as health effects or loss of life (e.g., 8, 25, 40). These difficulties are, of course, inherent in the problem. By highlighting them, formal analysis may increase the chances of their being solved, or at least of being recognized when the standard is operationalized and applied.

Formal analysis often shapes standards indirectly through various parties' analyses of how different possible standards affect them personally (32). The adopted standard is often the one that looks best on the private balance sheets of the most powerful parties. Modeling these hedonic perspectives is one way to design standards that will be well received.[8]

Professional Judgment

The technical community that creates technologies has always been the primary source of standards for governing them. Builders determine building codes; engineers specify acceptable metal fabrication techniques; medical personnel prescribe the conditions for licensing new drugs. In addition to such explicit expressions, other standards are implicit in the norms of accepted professional practice. The putative advantage of such standards comes from professionals' skill at identifying people's desires and reconciling them with one another and with technical realities. Devising creative solutions lies at the heat of the professionals' craft. Furthermore, their intimate involvement with the technology should ensure that the standard will be translated faithfully into operational standards and applied appropriately.

The trouble with letting professionals set standards is that they are a vested interest. They must balance the public's welfare with their own, that of their immediate employers, that of their industry, and that of the national economy. Depending

---

8 An even more indirect way to rely on formal analysis is possible when the set of future technologies can be anticipated. Analysis could be used to identify the best options. The standard could then be drafted so as to permit those options and no others (i.e., by finding some dimension(s) on which they had the most extreme scores).

upon the situation, there might be temptations to allow too much risk (in order to ensure a technology's survival), to require too much safety (in order to reduce professionals' legal liability), or to design highly technical standards (in order to provide enforcement work for professionals). Professionals' ability to achieve a judicious balance may be greatest when the other vested interests are equally capable of ensuring them: that is, when manufacturers may refuse to employ them, consumers may refuse to buy their creations, legislators may disband (or disregard) their licensing boards, and other professionals can monitor their actions.

Political Processes

Reliance on either formal analysis or professional judgment means entrusting some group of technical experts with setting public policy. The justification for this transfer of power is that they are capable of making decisions in the public's best interest, perhaps even more capable than the public itself, which is often accused of having a flitting involvement with problems, unstable preferences, and limited understanding of risk issues (45, 17). Whether the experts can produce thoughtful, balanced solutions and whether the public really needs their help are empirical questions. If the answers are negative, then the obvious alternative is to leave (or return) primary responsibility for standard setting to political processes such as referenda, legislatures, commissions, and, by some accounts, courts (33, 34).

The laypeople taking part in these processes can solicit expert opinion, but need not be bound by it. For example, they may choose to ignore, override, or adjust the experts' recommendations when the experts have ignored important consequences that are hard to quantify, been insensitive to the public's full risk burden, represented a vested interest, or exaggerated the definitiveness of their analyses (17). Although political processes cannot promise to identify the optimal standard directly, the ruthless criticism of the political arena should reveal the deficiencies of sub-optimal ones. Proposals that survive may have accumulated a considerable constituency.

The credibility of a particular political process depends upon how representative it is, how well it exploits existing technical knowledge, how well its lay participants resist bullying by expert opinion, and how committed participants are to the issue at hand (as opposed to using it as a venue for addressing other problems).

Revealed Preferences

The preceding methods might be called "expressed preferences;" the standard that emerges is what some people say it should be. The revealed preference method substitutes action for words by codifying the trade-offs implicit in past decisions. A weak revealed preference claim is that no new product should be riskier than the riskiest product on the market (unless it had unprecedented benefits).[9] A stronger one is that a new technology can be as risky as existing technologies with comparable benefits. Even more specific would be a safety philosophy that required the adoption of a safeguard only if its cost[10] were less than the market value of its expected reduction in mortality and morbidity.

As the argument goes, the historical record reveals not only real behavior, but also appropriate behavior. Through trial and error, the desires of all citizens involved with the existing technologies have been incorporated in its current cost-risk-benefit trade-offs. If a technology is too risky, then people will stop buying it until it has been made safe. If they have suffered from a technology, then they will turn to the courts for redress, forcing safer products in the future. If a technology is too safe, then either people will use it more dangerously so as to get more benefit out of it (55) or they will turn to riskier, but cheaper alternatives.

In order for a particular trade-off to reveal society's preferences, it must have been achieved in conditions like those needed for an efficient market operation. This means that: 1) Information is shared equally by all actors; 2) All actors optimize their hedonic preferences in each action; 3) A full range of choices, varying in risks and benefits, has been offered; and 4) The risky consequences of technologies can be unambiguously assigned to them.

The preferences so revealed are necessarily those of the past, reflecting its tastes and mores, as well as its distribution

---

9 A non-risk variant of this approach may be found in Dawes' (9) description of an attempt to discourage unpromising candidates to graduate school by suggesting that they apply to themselves a rule specifying the profiles of candidates that had never been chosen in the past.

10 Or, its net cost, being the cost minus any benefits other than risk reduction (e.g., increased productivity due to better working conditions).

of the power to influence social and economic realities. If the past appears to have been unjust, or insensitive to risk issues, then so will be its trade-offs. Changes in taste should either invalidate past experience or require the application of some correction factor to it.

Summary

These four generic approaches to setting standards are summarized in Table 2. They differ in the perspectives that they consider and the method used for integrating them. The choice between them in a particular case is in part an

Table 2

Methods for Setting Standards

How Can Standards Be Justified?

Approach	Locus of Wisdom	Description	Potential Advantages	Potential Disadvantages
Formal analysis	Formalized intellective processes	Choose standard offering highest utility (or best cost-benefit tradeoff)	Systematic explicit sophisticated techniques	Impractical oversold centralizes power
Professional judgment	Intuitive intellectual processes	Let technical experts identity best standard	Realistic implementable creative compromises	Vested interests incomplete perspectives inscrutable
Political processes	Body politic	Have lay groups set standards, informed by technical advice	Broad perspective legitimacy open to criticism	uninformed unrealistic unstable
Revealed preferences	Past social processes	Adopt standard implicitly emerging in actual decisions	Reflects deeds shaped through experience influenced by whole society	Inefficient unfair insensitive to risk

empirical question: To what extent will the potential advantages and disadvantages of each be realized? The choice is also political: How important are these advantages and disadvantages? Hybrids are also possible. For example, the U.S. Nuclear Regulatory Commission (34) has combined the political and professional approaches in its overall safety goals for nuclear power; although prepared by the Commission's technical staff, the proposed goals were presented periodically to a panel representing diverse constituencies for shaping.

MAKING STANDARDS OPERATIONAL

Once a safety philosophy has been determined, it must be translated into operational terms. That act of translation requires answering the questions of formulation discussed below. The answers will determine how readily the standard can be understood and monitored, how likely it is to be overturned by courts or legislatures, and how porous it will prove to the search for loopholes. If the answers are poor, the trade-off emerging from the operational standard may be quite different from that intended by the standard setters. Thus, although their task is ostensibly technical, the translators may deliberately or inadvertently set social policy. The standard formulators must address these issues, or let them be resolved ad hoc.

What Should Be the Range of Application?

The formulator's first task is to define the technologies to which the standard applies. That definition has two aspects: which technologies are in its jurisdiction, and what aspects of each technology are covered by the standard.

Which Technologies Are to Be Included: Whenever compliance with a standard imposes a serious burden, technologies and those responsible for them will be motivated to seek a change of venue. They may hope to be treated by other, more lenient standards, or to avoid regulation altogether. If standards are to work as intended, then their formulators must specify their jurisdiction in such a way that they are not imposed upon technologies for which they impose too great a burden, that they do not become a refuge for technologies escaping more onerous standards, that they do not allow exceptions for all but those technologies for which compliance is trivial, and that they do not inadvertently subject technologies to multiple regulation, whose cumulative burden is much greater than what was intended by the developers of the individual standards.

There are three basic ways to define categories. One is to enumerate all members; that leaves nothing to chance, as long as all potential members are known or can be anticipated. The second way is to specify analytically the properties of category members, in terms of structure (e.g., all polymers), function (e.g., all modes of transportation), or consequences (e.g., all carcinogens). Such a definition can be applied to any present or future technology providing the designers can anticipate jurisdictional conflicts and ambiguities. The third way is to identify the archetypal category member, and then assign technologies to the category whose archetype is most similar to them. The standard itself will be set in terms of that archetype with instructions on how to adjust it in the light of possible deviations. For this strategy to work, competing jurisdictions must also specify archetypes, and a metric for assessing similarity.[11]

It is beyond the scope of the present analysis to provide a more detailed discussion of the problems faced by these three procedures (or, indeed, for many of the other design issues raised below). As but one example of the kind of issue that can arise on the next level of detail, consider McNamara's description of the esoteric, but financially consequential, deliberations regarding whether cosmetic products, such as lip balms, toothpastes, and suntan lotions, are also drugs (hence, subject to the more stringent drug licensing standards).

> <u>Wrinkle removers/smoothers</u>. While it may seem incredible, there have been five published judicial decisions, arising in three different FDA enforcement actions, concerning this class of product. Speaking generally, all of this litigation may be summarized as leading to the following conclusion: If you represent a product as "removing" wrinkles, you are in danger of having it regulated as a drug by the

---

11 The actual standard set here constitutes a decision regarding the disposition of the archetypal member. In some cases, the detailed analysis of that member will improve understanding of it, thereby increasing its safety (by uncovering design problems that are addressed voluntarily); hence, the standard may have to be made more restrictive to ensure that other members are as safe. Alternatively, such close scrutiny will lead to an exaggerated perception of all the things that could go wrong with the archetype. The result would be unduly tight regulations, which should be relaxed for other members.

> FDA; if, on the other hand, you represent the product solely as a wrinkle "smoother," it may be subject to regulation solely as a cosmetic. Note that two such products identical in chemical composition may nevertheless have different regulatory status depending upon their claims (pp. 470-471).

In this light, the seemingly frivolous debate over whether Certs is a candy mint or a breath mint may have serious legal ramifications.[12]

What Aspects Should Be Covered: Technologies typically have both spatial and temporal extent. At any point in time, they may be used in various methods and places. Any given use evolves over time through its life, material, or fuel cycle. In both regards, the formulator must specify what constitutes a technology for the purposes of the standard.

Spatially, the standard could pass judgment on a technology such as asbestos on the basis of each usage, each point of manufacture, each manufacturer, or the entire industry. As the definition broadens, the consequences of failing to meet the standard increase. The probability of that happening may decrease because of the political pressure to keep large economic units alive (even if it means overturning a standard). As the definition narrows, the responsibility for compliance becomes more focused, increasing the chances of individual users being held accountable for excesses. On the other hand, the absolute magnitude of the risk arising from small units may make them too inconsequential to regulate seriously (2).

Temporally, a technology such as asbestos goes from mining to milling to manufacture to usage to intermediate and final disposal. Each state produces some risks; some also produce benefit. An extensively defined standard would consider the trade-offs between cradle-to-grave risks and benefits from a technology. Even though the benefits are most apparent at the stage at which a technology is used, they reflect the cumulative value accruing from the entire production process, hence they should be compared with its cumulative risk. The comprehensive view would give manufacturers the freedom to look through the entire fuel cycle for the most efficient way of achieving the requisite safety (53).

---

12 The Food, Drug, and Cosmetic Law Journal is a rich source of such issues. See, for example, (42), (29), (3).

The comprehensive view is less acceptable when different people receive the risks and benefits accruing from the different stages. Those who bear risks may want risk reduction or compensation and be unmoved by the bargains to be realized by cost-effective expenditure of safety resources elsewhere; those who enjoy benefits may be reluctant to pay or be restricted because of risks occurring far away. If a technology is required to balance risks and benefits at every stage, then it is no more acceptable than its weakest link. Whereas the comprehensive definition allows excess safety at one stage to compensate for inadequate safety at another, the stage-wise definition would require compensation actually to be paid, to sweeten the burden of the riskier stages. This way of defining technologies would likely result in greater expense or less risk.

What Should Be the Locus of Standard?

Just as technologies go through a number of stages from inception to demise, so do hazardous events go through several stages en route from constituting to realizing a threat (22). Figure 1 shows one conceptualization of how automobile accidents go through each stage. Similar figures could trace the path of air pollutants up the smoke stack, though the air, into people's lungs and into health effects, or the path of infectious diseases into and through public water supplies. It is possible, in principle, to block the event by interventions at each stage, designed to keep the incipient threat posed there from developing further. It is also possible, in principle, to measure the magnitude of the threat at each stage. As a result, standards can be set in terms of each stage, measuring there whether technologies exceed the maximum of risk alloted to them by the standard setters (considering their benefits, alternatives, etc.). Although the details vary from case-to-case, each locus has particular problems, prejudices, and loopholes. These are discussed below and summarized in Table 3, which shows the advantages that accrue as the locus of the standard comes closer to the extremes of being set on the technology's hardware or on its averse consequences.

Apply to the Technology: Standards that prescribe the specific hardware to be used in the acceptably safe version of a technology are commonly called design standards. Their strongest advantage is their potential enforceability. Compliance may be checked simply by observing the hardware. When problems are detected, it is relatively clear whom to hold

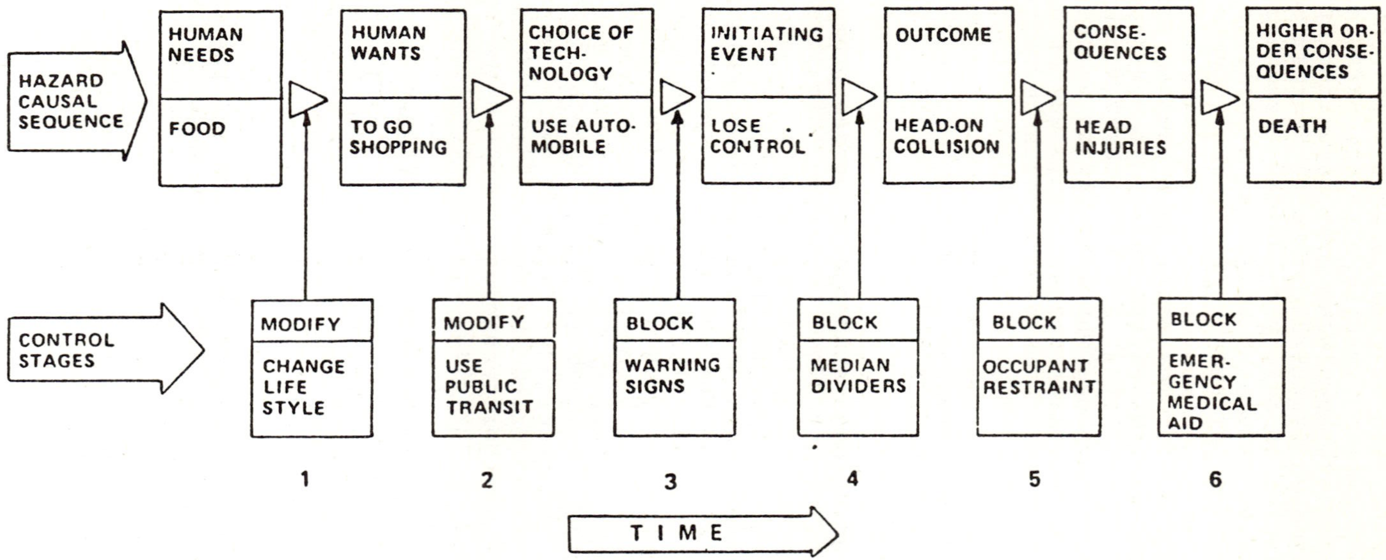

Figure 1. Illustration of the causal chain of hazard evolution. The top line indicates seven stages of hazard development, from the earliest (left) to the final stage (right). These stages are expressed generically in the top of each box and in terms of a sample motor-vehicle accident in the bottom. The stages are linked by causal pathways denoted by triangles. Six control stages are linked to pathways between hazard states by vertical arrows. Each is described generically as well as by specific control actions. Thus, control stage 2 would read: "You can modify technology choice by substituting public transit for automobile use and thus block the further evolution of the motor-vehicle accident sequence arising out of automobile use." The time dimension refers to the ordering of a specific hazard sequence; it does not necessarily indicate the time scale of managerial action. Thus, from a managerial point of view, the occurrence of certain hazard consequences may trigger control actions that affect events earlier in the hazard sequence. (Source: 4)

Table 3

Locus of Standard

Typical Advantages of Placing Standard Close to:	
Hardware	**Aversive Consequence**
Responsibiblity for problems more readily attributed	Greater relevance
Costs of compliance more easily assessed	Allowable risks more readily compared with benefits
Greater predictability	Easier access for measurement
Measurement easier	Creativity in compliance encouraged

responsible. Whenever people must operate and maintain the technology or can frustrate its safety features (e.g., disconnecting seat belt interlocks or emission controls), then specification of their behavior is part of the design. Unlike hardware, which might be evaluated once and for all, behavior requires continual monitoring, greatly increasing the costs of applying a standard (or the opportunities for circumventing it).

Perhaps the weakest point of design standards is the hypotheticality of the risks that they permit. It may be hard for either risk analysts or lay observers to understand the link between the design and actual health effects. As a result, it may be hard to convince the producers of a technology that particular (expensive) design features are needed or to convince those potentially affected by it that their welfare has been protected. Such frustrations have motivated the Nuclear Regulatory Commission's (50) plan to supplement its design standards with standards defining allowable consequences. They are also behind worries about what safety has been achieved by the building codes intended to protect the residents of earthquake country (54).

Design standards effectively ban a technology when they cannot be met with current techniques and within current economic constraints. Such standards are often called "technology forcing" in the hopes that they will prompt the development of better techniques. In practice, however, it is hard to make impossible demands when the link between design and consequences is unclear. The desire for stringency is more likely to express itself in requirements for costly, but feasible safety devices and procedures. When these design requirements are tacked on to an operating technology, they may destroy the integrity of its original design without allowing for (let alone encouraging) the creation of new, more effective designs (20).

Apply to the Initiating Event: Standards set on the remaining loci specify different aspects of a technology's performance. In doing so, they should permit industry technicians to use their creativity in designing cost-effective ways to comply. A logical place to impose performance standards is at the source, constraining the likelihood or extent of initiating events and, thereby, keeping things from getting out of hand. Exemplary initiating-event standards include limits on blood alcohol levels (for automobile accidents), on smokestack emissions (for power plants), or on concentration levels (in pesticide sprays). Standards placed close to the source not only are imposed before any damage is done, but apply to events that are relatively easily measured and traced to their source. Proximity to the source means distance from the (risky) consequence itself, subjecting such standards to attenuated versions of the difficulties facing design standards (e.g., knowing what risk remains).

Two inherent problems of all performance standards emerge with initiating event standards. One is that measuring the performance of a technology is akin to measuring the behavior of an individual; it may fluctuate considerably, requiring either constant monitoring or some sampling scheme (topics discussed below). The second problem is that some of the technicians' creativity may be devoted to detecting and exploiting the standard's loopholes. Harris (19) suggests that emission limits on mercury stimulated the development of new production processes using (and emitting) mercury in smaller quantities. As a result, the yearly increment in mercury emissions has remained constant. The environmental problem may even have become worse, insofar as it is harder to control the more numerous current sources.

Apply to the Outcome: Once the initiating event has occurred, the next step toward realizing a risk is to damage the integrity of the environment, creating an imminent threat to those in it. Outcome standards limit the amount of environmental insult. Examples include ambient air quality standards, motor vehicle rollover requirements, and E. coli limits for drinking water (7). Even more than initiating event standards, outcome standards are set in politically relevant terms and allow freedom in reaching compliance. For example, one option that opens up here is trading in pollution rights, whereby a technology is allowed to pollute an overburdened airshed by reducing the emissions of an existing industry (48).

The debates over pollution rights illustrate one general problem with standards set far from the source: assigning responsiblity for violations. The medium that the monitors must measure is directly accessible, unlike hardware inside an industrial facility where prior warning is needed. However, the diffusion of that medium may make it hard to trace excesses to their source (e.g., 5). When tracing is possible, the question remains of who must cut back (or pay fines or offer compensation) when a threshold is exceeded. Should it be the source that pushed the medium over the limit? The largest source? The most recent source? The source with the lowest ratio of societal benefits to societal costs? The source that is best able to bear the costs? The source with the cheapest control options? All sources in proportion to their contribution to pollution? All sources in proportion to their ability to pay? Without an attributional rule, an outcome standard is meaningless.

Apply to the Consequences: Setting standards in terms of allowable consequences brings to their logical conclusion the advantages and disadvantages that emerge as the locus moves away from the source technology. On the positive side, the standard is set in the most politically relevant terms; it allows maximum latitude in compliance strategies; the relationship between risks and benefits is clearest. On the negative side, it can be most difficult to respond to violations. In some cases (e.g., cancer), it is hard to trace the consequences to a source; in others (e.g., lead poisoning), there are many identifiable sources; in others (e.g., auto accidents), producers of the technology involved can attribute responsibility to intervening events (e.g., poor drivers, bad roads). Where violations cannot be linked to action, consequence standards are not advisable.

The controversy over hazardous waste dumps illustrates some of the issues associated with choosing the locus of standard. Although interest has been stimulated by actual consequences, the situation may be out of hand before health effects can be attributed confidently to a particular dump (26). As a result, a mixture of outcome standards (e.g., polluted water supplies) and consequence standards for less important effects (e.g., chromosomal damage) is used to judge existing dumps, whereas design standards have been devised for new dumps. The superfund for cleaning up abandoned dumps, which punishes the present for the excesses of the past, reflects these attributional problems (48).

How Should Compliance Be Measured?

Once the locus has been selected, a scheme must be created for measuring the technology's design or performance. Although each standard raises special problems in this stage of operationalization, many of those can be grouped under the rubrics of a few basic statistical concepts.

Selecting the Features to Be Measured: Whatever locus is selected, many hazards have several features that might be measured. For example, there are many dimensions to most hardware and operating procedures; each might be specified in a design standard. Industrial processes often emit several pollutants; each could be controlled as the initiator of a hazardous event chain. A chemical may have several sized particles; each could have its own outcome standard: Separate consequence standards might be created for each of the adverse physiological and psychological effects potentially caused by a technology.

All of these features could be measured, or only a subset of representative features, or a single feature. Features might be chosen because they are easily observed, politically sensitive, well-understood, or associated with large health effects. Assuming a reasonable selection criterion, as the number of features increase, so should a standard's comprehensiveness, credibility, explicitness, and expense of application. Depending upon the rule used to aggregate the measurements from the different features, increasing their number may also tend to increase the difficulty and cost of complying with the standard.

Aggregation of Measures Over Features: When several features are measured, the result may be a complex pattern of compliance and non-compliance on different features. An aggregation rule is needed to translate this pattern into an overall assessment of acceptability. Conjunctive rules set a separate cutoff criterion on each feature, all of which must be met for a technology to be acceptable. Strictly interpreted, such rules would fail a generally acceptable technology with a single excess but pass a technology that barely passed on each feature. They might even favor the development of technologies with diverse (but just tolerable) problems. Because each feature provides another opportunity to fail, when a feature is added to the standard, the cutoff criteria on existing features should be relaxed (if the standard's demands are to remain constant). Conjunctive rules are particularly appropriate with design standards in cases where one weak link can threaten the viability of a system and with consequence standards where each affected group's safety must be protected.

Compensatory rules allow more-than-adequate performance on one feature to compensate for less-than-adequate performance on others. These rules are most appropriate with performance standards concerned with the total risk burden and not with upon whom it falls. Here, increasing the number of features may reduce the standard's stringency insofar as additional problems that a technology does not cause are allowed to compensate for those that it does cause.

Handling Statistical Valuation: Any empirical measure will show some variability, due to unreliability in the measurement procedure or to fluctuations in the measured phenomenon. Descriptive statistics summarize distributions of measurements in order to guide actions based upon them. One obvious set of candidate summaries is the statistical moments of a distribution. Figure 2 suggests how each of the first four moments (mean, variance, skew, and kurtosis) might provide the most appropriate way to characterize a risk. The choice among them depends on society's values and the actions it contemplates. Of course, other statistics are possible. One candidate with some uses is an extreme fractile of the distribution, setting a limit on the probability (or relative frequency) of extreme observations.

Determining the Unit of Observation: Each point in the distribution of observations may take in larger or smaller amounts of experience. An elementary result of sampling theory is that as the unit of observation increases, so does the stability of the measure. Conversely, the probability of exceeding any threshold decreases. For example, with fixed limit values, violations are more likely if a design standard is applied to the thickness of each tube, rather than the mean thickness for batches of 100; if an initiating event standard is applied to instantaneous smokestack emissions, rather than daily averages, and so on. Thus, standards that sound the same when expressed in terms of their limit values may require quite different degrees of control. Failure to specify the unit of observation can lead to a standard becoming more rigorous in mid-life simply because finer measurements become possible (35). Smaller units are appropriate when fluctuations are large and the environment is sensitive to local peaks. For example, abundant rain keeps the mean air pollution in the Willamette Valley low; however, making a month or a year the unit of observation would obscure the occasional atmospheric inversion that creates local pollution maxima which pass the threshold for issuing air stagnation advisories or alerts. On the other hand, for pollutants with cumulative effects, long-term averages may be adequate.

Defining a Sampling Scheme: When it is impossible to look at all aspects of the regulated technology (and its byproducts) constantly, a sample must be taken. The definition of the sampling frame must consider the dimensions along which a technology may vary. For example, some hardware may be evaluated when it leaves the plant, with the assurance that it will not change over time; on the other hand, its manufacturing process may be unreliable enough to require intensive sampling of items as they are produced. Other kinds of hardware may be manufactured to precise specification, but be subject to changes as they are used (e.g., disconnecting safety devices); they should be measured over time. Air pollution may vary significantly over altitude, but if ground level concentrations pose the only threat to health, then only spatial variations need be sampled.

Once the frame has been defined, a representative sample must be drawn, through application of statistical theory informed by a substantive knowledge of the phenomenon. The random sampling needed to ensure a representative sample may have the added advantage of being hard to circumvent by those who might try to look especially good at observation time (as might happen with periodic inspections). A key parameter of the sampling scheme is the number of observations to be drawn. Within a fixed budget, the number of observations is inversely related to their intensity. Even when there is reason to get many observations, some resources might be allocated to intensive studies that might uncover biases in the ordinary observations (6). For example, such studies might show that exotic cancers, which are unfamiliar to many physicians, are underestimated in public health statistics that supposedly incorporate the entire population.[13]

Using Surrogate Measures: Selecting a locus was to some extent a choice between monitoring what is important (risk) and monitoring what can be more readily observed (hardware). Design standards are, then, a special case of regulating a surrogate for the actual topic of interest. With surrogates,

---

13 Analogously, anthropological studies of inner-city neighborhoods have identified classes of individuals that are underrepresented in census records of the same neighborhoods (31).

(a) Reliance on mean. Hazards A and B differ only in means. All other things being equal, one would prefer Risk A.

(b) Reliance on variance. Hazards A and B have the same mean, but differ in variance. One might prefer B because it offers some chance of low losses or because its high variance suggests the presence of a variable factor that might be understood and controlled. One might prefer A because its greater predictability allows a standard allocation of resources to compensation adjustment. Indeed, such reasoning might lead one to prefer hazard C, with a higher mean but reduced variability, over B.

(c) Reliance on skew. A and B have similar means and variances, but differ to skew. That is, A has a long, flat tail, reflecting the somewhat remote possibility of a large number of fatalities. This catastrophic potential may induce a preference for B over A, or even a preference for B over C, which has a lower mean and variance but a longer tail.

(d) Reliance on kurtosis. A and B are similar in mean, variance, and skew, but differ in kurtosis (i.e., their flatness or homogeneity of density). One might prefer A because one standard coping behavior is adequate for a larger proportion of problems. Or one might prefer B because its restricted range means that one can more readily prepare for all feasible contingencies. Indeed, the fact that there is hardly a modal consequence with B may force one to be more imaginative and flexible.

Figure 2. Characterizing risks by their statistical moments (of casualties). How one might be concerned primarily with different moments and how choice of moment may affect relative acceptability of hazards.

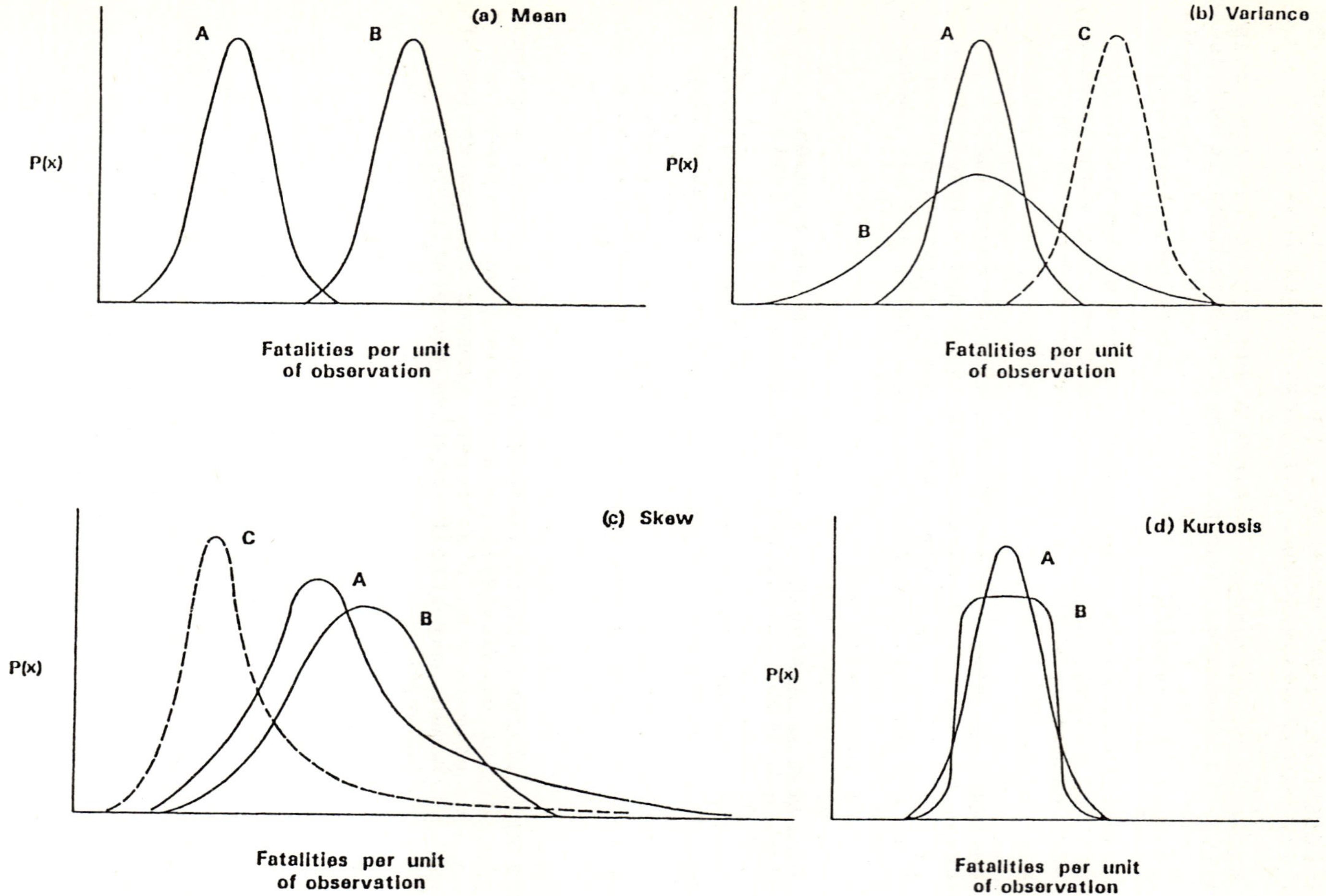
(a) Mean
A
B
P(x)
Fatalities per unit
of observation
(b) Variance
A
C
B
P(x)
Fatalities per unit
of observation
(c) Skew
C
A
B
P(x)
Fatalities per unit
of observation
(d) Kurtosis
A
B
P(x)
Fatalities per unit
of observation

Figure 2

it is harder to prove the standard's faithfulness to the intent of the standard setter. Such standards' defense is practicality and predictability.

The possibility (and problems) of using surrogates may also be found within loci. When a problem affects the whole population in a weak form, it may be easier to monitor the health of a particularly sensitive subgroup (e.g., air quality standards defined in terms of allowable health effects among people with respiratory problems). When a health effect has measurable precursors, a standard defined on them might make control possible, as well as censure (e.g., using interferon levels as a predictor of resilience to medical problems). When accidents are too rare for their rate to be measured, near accidents may allow more stable estimates.[14] When problems have long latency periods, it may be useful to judge current affliction rates even though they reflect the exposure problem of many years ago.

To adopt surrogates, there must be some plausible link between them and the actual effect. If they are adopted, it is important to keep their surrogacy in mind when considering a standard. The expense of air quality standards may seem too high to "save a few asthmatics" but not if their health reflects that of the whole population.[15]

How Should the Standard Be Enforced?

To be more than just an academic exercise, standards must be linked to action. For action to follow in a timely and predictable manner, the standard formulators must anticipate ambiguities in interpreting the standard, provide guidelines for the usage of uncertain evidence, and specify the consequence of noncompliance. Finally, they must review their efforts and estimate how well the resultant standard represents the safety philosophy of the standard setters. If operationalization has introduced some bias, then it may be

14 With a highly politicized technology, like nuclear power, considerable economic stakes may actually ride on the rate of casualtyless near-misses, making a near-miss standard important for non-safety reasons (44, 45).

15 Some of the more facetious criticism of the Delaney Amendment's treatment of saccharin made it seem like a standard protecting laboratory rats from cancer. In principle, a standard might be written to protect the health of street rats if they were particularly sensitive to something that eventually affected humans.

redressed by incorporating a correction factor in the enforcement scheme, making it especially stringent or lax.

Anticipating Problems of Interpretation: The essence of the standard formulators' job is to eliminate any need for interpretation when the standard is applied. Although attention to the details described above should reduce ambiguities, it should be supplemented by a special effort to anticipate surprises. One way is to assume that every ruling of noncompliance will initiate a search for a loophole that will exempt the technology involved. Advance consideration of the validity of such claims can help ensure that the standard is relaxed responsibly.[16] A second strategy is to anticipate novel applications of the standard. That can be done both by trying to stretch the standard to new problems and by identifying possibly related technologies that have less satisfactory (or no) standards. A third strategy is to consider the ramifications of the precedents set by possible applications. For example, when a particular piece of equipment is mandated for a new technology, should it be extrapolated literally to other existing technologies (even if retrofitting is more expensive and less effective than with new designs) or should it be interpreted as embodying a principle (e.g., take any step that will reduce risk by at least amount x for less than $y)? The economic consequences of accidents such as Three Mile Island may vary greatly depending upon how widely the lessons learned are generalized (13).

Finally, because it is impossible to anticipate all contingencies, the standard should include a procedure for resolving interpretive questions, along with a criterion for when recourse to that process is needed. It might specify which parties should participate in resolution and when to go back to the standard setters for guidance.

Anticipating Problems of Incomplete Evidence: The measures chosen to determine compliance with a standard inevitably involve random variables. The variation in these variables creates the risk of Type I and Type II errors: erroneously determining that a technology is in compliance or erroneously determining that it is not. Statistical decision theory shows how to formulate decision rules to achieve different balances between these error rates. Its application requires a value judgment, regarding the relative social cost of the two kinds of errors, being unfair to technologies or to risk bearers

---

16 Just as property owners often hire ex-criminals to diagnose their vulnerability to theft, standard formulators might hire opposing attorneys to expose standards' points of vulnerability.

(37). The decision rule establishes whether the burden of proof is on the technology to prove its innocence or on the standard applier to demonstrate its guilt.

The application of statistical procedures requires assessment of the probability of the measured feature having different values. This presents no problem when a large sample of observations has been drawn with a definitive measurement procedure. When a sample is small or the measurement procedure faulty, some extrapolation procedure is needed. When data have been collected from several sources, perhaps using different research paradigms, an aggregation procedure is needed, such as counting "votes" for or against compliance, pooling the probabilities of compliance (assuming the independence of tests), or computing the mean of best guesses, weighted by the quality of the studies (21). Different extrapolation and aggregation procedures can produce quite different summaries of available evidence. These can affect determinations of compliance and, hence, should be specified in advance.

Specification is also needed regarding two other issues that follow from incomplete evidence. One is how far to exploit the opportunities available to reduce uncertainty at the cost of time and investigatory resources. Value-of-information analysis (39, 52) provides the conceptual framework for addressing these issues systematically. Its application requires (and reveals the role of) assigning explicit costs to information and to misclassification (where inaction implies some classification, depending upon how the burden of proof is defined). With uncertainty, it is inevitable that subsequent evidence will show some classifications to have been in error. The standard must specify whether classifications are irreversible (and whether damages are due). If disapproval prevents a technology from being used, then it may also prevent the collection of information showing that that classification was wrong.

<u>Determining the Rigor of Enforcement</u>: The response to violations has two dimensions, speed and severity. The time allotted to an unacceptable technology to get into compliance (or suffer the consequences) should depend upon the rate at which unacceptable risks (or actual damages) accumulate and the chances that the time will be used successfully to improve the technology (rather than to gain market share or unfair profits by selling a cheaper, riskier product). A wild card with waiting is that political moods may change, generally weakening or strengthening all regulation.

The severity of the response to a non-complying technology (like its speed) is a continuum, along which a number of marker positions may be identified. The most salient of those responses is to ban the technology until it comes into compliance. At times, however, this is impractical; although the technology's benefits are insufficient to justify its risk, they are too great to be forfeited. If these risks are intolerable in part because they are borne by people who do not benefit from the technology, then it may be possible to redress the imbalance by compensating them. That transfer could be viewed as a penalty for violations or as a redesign of the technology to conform to a standard with an equity component. More extreme penalties are also possible, such as accompanying the ban with punitive damages. These could be justified by their signal value, warning potential violators to take the standard seriously; or they could be designed to compensate for known defects in the standard's monitoring system. If only a certain percentage of violations will be detected and proven, then the penalty might be set so that violators' expected penalty (i.e., probability of being penalized x size of penalty) equals what it would be with perfect detection and non-punitive enforcement.

Whatever enforcement policy is written into the standard must be commensurate with the standard appliers' ability to make their determinations stick. If they lack power, then it may be preferable to design an enforcement plan that focuses on encouraging voluntary compliance, eliciting public displeasure, and attacking the clearest excesses.

Including a Correction Factor: The idea of using punitive damages to compensate for imperfect monitoring is one example of the kind of correction factor that standard formulators might use to bring an operational standard into compliance with its underlying safety philosophy. A second is to increase all estimates describing the uncertainty of knowledge about the technology by a fixed factor in order to counter people's tendency to be overconfident in their own knowledge (14, 27). A third is to adjust estimates of risk or benefit upward or downward by a fixed factor to nullify biases introduced to the data at their source. A fourth way would be to disregard mild violations when a standard has been formulated too restrictively. An analogous activity from another domain was the attempt to adjust the results of the 1980 U.S. Census to correct a suspected undercount of certain socio-economic and ethnic groups.

Although a logical possibility, such corrections may be political impossibilities. Especially when they change the fate of a technology, these adjustments may be a lightning rod for criticism. They can be defended only by a technical analysis of the standard safety philosophy match for which political arenas may have little patience. Moreover, the correction itself undermines the credibility of the standard.

The defensibility of a correction can be improved by specifying it in advance, so that it emerges from the standard formulators' deliberations rather than from the standard appliers' disquiet. It may even be requested when the overall bias of a standard is common knowledge to those who provide inputs to it. Rather than participate passively in such an exercise, they may spontaneously adjust their own inputs in order to introduce correction factors from the bottom. For example, scientists may give high-side or low-side summaries of studies if they believe that they will be taken too lightly or too heavily, or inspectors may take their job more or less seriously depending upon the anticipated response to their reports. The cumulative impact of such tinkering may be so hard to assess that an explicit overall adjustment factor may be accepted in return for a promise of candor in local estimates.

Summary

In the end, the acceptability of these or any other injustices associated with a standard may depend upon the alternatives. When the costs of waiting and uncertainty run high, even those treated unfairly may view occasional injustices as a legitimate price to pay for a timely, predictable system. Translating a safety philosophy into an operational standard raises many design issues, whose resolution will determine the practicality and stringency of the resultant standard. If not faced explicitly, these issues will be decided by default as the technicians do their work. Resolving them in the standard setting process can help ensure that they draw the open controversy they deserve. Moderate ignorance about the specific cases to which a standard will be applied may encourage the parties involved to adopt a balanced position and try to design it fairly.

ACKNOWLEDGEMENTS

This research was supported by the Nuclear Regulatory Commission under subcontract 7656 from Union Carbide Corporation, Nuclear Division, Inc., to Perceptronics, Inc.

I would like to thank S.A. Carnes, T.E. Cole, E.D. Copenhaver, H. Inhaber, H. Kunreuther, G. Morgan, D.C. Parzyck, P. Rathbun, P. Slovic, K.A. Solomon, S. Watson, and two anonymous reviewers for their many helpful comments on previous drafts.
Correspondence should be addressed to the author at Decision Research, A Branch of Perceptronics, 1201 Oak Street, Eugene, Oregon 97401.

REFERENCES

1. American Enterprise Institute. Proposals for regulatory reform. Washington, D.C.: Author, 1980.

2. American Public Health Association. Statement on S.2153 Occupational Safety and Health Improvement Act of 1980. Washington, D.C.: Author, 1980.

3. Anastasio, E.L. Restraints on use of the term 'natural' in advertising and labeling of food, drugs, and cosmetics. Food, Drug, and Cosmetics Law Journal, 1981, 36, 424-430.

4. Bick, T., Hohenemser, C. & Kates, R. Target: Highway risks. Environment, 1979, 21(2), 7-15, 29-38.

5. Brombaugh, K.J. & Lee, K.W. Finger-printing organic omissions. Environmental Science & Technology, 1981, 18, 1142-1149.

6. Brooks, A. & Bailar, B.A. An error profile: Employment as measured by the current population survey (Statistical Policy Working Paper 3). Washington, D.C.: U.S. Department of Commerce, 1978.

7. Calabrese, E.J. Methodological approaches to deriving environmental and occupational health standards. New York: Wiley, 1978.

8. Crandall, R.W. & Lave, L.B. The scientific basis of health and safety regulation. Washington, D.C.: Brookings Institution, 1981.

9. Dawes, R.M. A case study of graduate admissions: Applications of three principles of human decision making. American Psychologist, 1971, 26, 180-181.

10. Derby, S.L. & Keeney, R.L. Risk analysis: Understanding "How safe is safe enough?" Risk Analysis, 1981, 1(3), 217-224.

11. Derr, P., Goble, R., Kasperson, R.E. & Kates, R.W. Worker/public protection: The double standards. Environment, 1981, 23(7), 6-15, 31-36.

12. Driver, C. A theory of regulatory enforcement. Public Policy, 1980, 28, 258-259.

13. Evans, N. & Hope, C. Costs of nuclear accidents: Implications for reactor choice. Energy Policy, 1982, 10(4).

14. Fischhoff, B. Debiasing. In D. Kahneman, P. Slovic, and A. Tversky (Eds.), Judgment under uncertainty: Heuristics and biases. New York: Cambridge University Press, 1982.

15. Fischhoff, B. Safety goals for nuclear power. Journal of Policy Analysis and Management, 1983, 3.

16. Fischhoff, B., Lichtenstein, S., Slovic, P., Derby, S.L. & Keeney, R.L. Acceptable risk. New York: Cambridge University Press, 1981.

17. Fischhoff, B., Slovic, P. & Lichtenstein, S. The "public" vs. the "experts": Perceived vs. actual disagreements about the risks of nuclear power. In V. Covello, G. Flamm, J. Rodericks, and R. Tardiff (Eds.), Analysis of actual vs. perceived risks. New York: Plenum, in press.

18. Florman, S.C. Blaming technology. New York: St. Martin's Press, 1981.

19. Harris, R. Mercury: Measuring and managing the hazard. Environment, 1978, 20,(8).

20. Hammer, W. Product safety and management engineering. Englewood Cliffs, N.J.: Prentice-Hall, 1980.

21. Hedges, L.V. & Olkin, I. Vote-counting methods in research synthesis. Psychological Bulletin, 1980, 88, 359-369.

22. Hohenemser, C., Kates, R.W. & Slovic, P. The nature of technological hazard, Science, 1983, 220, 378-384.

23. Johnson, E.L. Risk assessment in an administrative agency. Sixth Symposium on Statistics and the Environment. Washington, D.C.: National Academy of Sciences, 1980.

24. Keeney, R.L. Issues in evaluating alternative standards. San Francisco: Woodward-Clyde Consultants, 1981.

25. Lave, L.B. The strategy of social regulation. Washington, D.C.: Brookings Institution, 1981.

26. Lennett, D.J. Handling hazardous wastes. Environment, 1980, 22(8), 6-15.

27. Lichtenstein, S., Fischhoff, B. & Phillips, L.D. Calibration of probabilities: State of the art to 1980. In D. Kahneman, P. Slovic and A. Tversky (Eds.), Judgment under uncertainty: Heuristics and biases. New York: Cambridge University Press, 1982.

28. McNamara, S.H. When is a cosmetic also a drug--What you need to know and why. Food, Drug, and Cosmetic Law Journal, 1980, 35, 467-478.

29. McNamara, S.H. When does a food become a drug? Food, Drug, and Cosmetic Law Journal, 1982, 37, 222-231.

30. Mahoney, M.T. & Yandle, B. Bubbles and efficiency. Regulation, 1980, May/June, 49-52.

31. Martin, E. Surveys as social indicators: Problems in monitoring trends. Institute for Research in Social Science, University of North Carolina, Chapel Hill, N.C., 1980.

32. May, W.W. Dollars for lives: Ethical considerations in the use of cost/benefit analysis by for-profit firms. Risk Analysis, 1982, 2, 35-46.

33. Mazur, A. The dynamics of technical controversy. Washington, D.C.: Communications Press, 1981.

34. Nelkin, D. & Pollak, M. The atom besieged. Cambridge, Mass.: MIT Press, 1980.

35. Norcross, M.A. Sensitivity of method--A wave of the future. Food, Drug, and Cosmetic Law Journal, 1980, 35, 342-348.

36. Otway, H.J. & von Winterfeldt, D. Beyond acceptable risk: On the social acceptability of technologies. Policy Sciences, 1982, 14, 247-256.

37. Page, T. A generic view of toxic chemicals and similar risks. Ecology Law Review, 1978, 7, 207-243.

38. Portney, P. Efficient use of standards and enforcement: The case of pollution controls. Policy Analysis, 1979, 5, 512-524.

39. Raiffa, H. Decision analysis. Reading, Mass.: Addison-Wesley, 1968.

40. Salem, S.L., Solomon, D.A. & Yesley, M.S. Issues and problems of inferring a level of acceptable risk. R-2561-DOE. Santa Monica, Calif.: RAND Corp., 1980.

41. Selikoff, I.J. & Lee, D.H.K. Asbestos and disease. New York: Academic Press, 1978.

42. Shaw, J.S. & Canape, C. Is the EPA saddled with more than it can handle? Business Week, May 26, 1980, 148.

43. Shindler, J.R. A common sense approach to common or usual names if there be any. Food, Drug, and Cosmetic Law Journal, 1981, 36, 284-292.

44. Slovic, P., Fischhoff, B. & Lichtenstein, S. Perceived risk and quantitative safety goals for nuclear power. Transactions of the American Nuclear Society, 1980, 35, 400-401.

45. Starr, C. Risk criteria for nuclear power plants. Risk Analysis, 1981, 1, 113-120.

46. Starr, C. & Whipple, C. Risks of risk decisions. Science, 1980, 208, 1114-1119.

47. Stokey, E. & Zeckhauser, R. A primer for policy analysis. New York: Norton, 1978.

48. Trauberman, J. Superfund: Legal update. Environment, 1981, 23(2), 25-29.

49. U.S. National Academy of Sciences. Strategies to determine needs and priorities for toxicity testing. Washington, D.C.: Author, Vol. 1, 1981; Vol. 2, 1982.

50. U. S. Nuclear Regulatory Commission. Safety goals for nuclear power plants: A discussion paper. NUREG-0880. Washington, D.C.: Author, 1982.

51. Wall Street Journal. Down with big business. Editorial, 1979.

52. Weinstein, M. C. Decision making for toxic substances control: Cost effective information for the control of environmental carcinogens. Public Policy, 1979, 27. 333-383.

53. Whipple, C. Energy production risks: What perspective should we take? _Risk Analysis_, 1981, _1_, 29-36.

54. Wiggins, J.H. Earthquake safety in the city of Long Beach based on the concept of balanced risk. In _Perspective on benefit-risk decision making_. Washington, D.C.: National Academy of Engineering, 1972.

55. Wilde, G.J.S. The theory of risk homeostasis: Implications for safety and health. _Risk Analysis_, in press.

Law, Risk Management, and the Influence of the Psychological and Decision Sciences

Professor Michael Baram and Robert I. Field

Boston University Center for Law and Health Sciences, 765 Commonwealth Avenue, Boston, MA 02215

THE LEGAL FRAMEWORK FOR RISK MANAGEMENT

This paper discusses certain developments in the psychological and decision sciences which have influenced the legal framework for the management of technological risks in the United States.

We have defined the legal framework as consisting of two parts, a public law system or framework for managing technological risks, and a private law framework.

The public law framework for risk management (1) has developed rapidly and unsystematically over the last two decades. It includes several decision processes:

- o Congressional Legislation--such as the Atomic Energy Act or the Toxic Substances Control Act--which provides agencies with mandates to regulate specific problems, along with diverse requirements and guidance for use in decisionmaking.

- o Agency Regulation--such as the Environmental Protection Agency's regulatory program on toxic substances--which may include ad hoc "adjudicatory" decisions (e.g., a permit decision); generically applicable regulations (e.g., information disclosure requirements for manufacturers of chemicals); numerical standards which apply to a specific risk (e.g., a chemical exposure or emissions standard); and inspection and enforcement decisions.

NATO ASI Series, Vol. G4
Environmental Impact Assessment, Technology Assessment, and Risk Analysis. Edited by V. T. Covello et al.

- Judicial Review of agency regulations, which employs statutory criteria provided by the Administrative Procedure Act and by the statute which authorized the agency regulatory action being reviewed.

The private law framework for risk management (2) provides the principles for governing private activities, many of which involve technological risks:

- Tort Law--which is applied in state courts and involves the judicial application of traditional concepts of rights and duties for resolving conflicts between private parties--such as the various duties imposed by the courts on a chemical company to remedy or prevent harm to the consumers of its products, to its workers, and to the general environment.

- Contract Law--which provides concepts for the judicial resolution of private conflicts growing out of contractual relationships which have a bearing on risk management--such as that branch of contract law which applies to insurance contracts and coverage for liability arising from product, consumer, and environmental risks.

Any private firm whose activities generate technological risks is usually subject to both public and private law requirements; with the former often being of a preventive nature and of considerable specifity, and the latter consisting of a vague set of principles whose applicability and remedial actions can only be conclusively determined by judicial decision after the adverse impacts have occurred.

Although these two frameworks differ in many respects, and considerable diversity in decisionmaking exists within each framework, both frameworks provide for decisionmaking processes on risks which involve several similar functions by each decisionmaking body:

- Determination of which risk problems are to be dealt with or are cognizable (e.g., what types of actual or prospective injury will be dealt with)

- Collection of evidence or factual information, and its analysis, to reach key findings of fact as to the sources of the hazards, causal connections to those at risk, behavior of the parties, and the nature of the actual or prospective injury

- o Assessment of relevant law in order to reach findings of law necessary for resolving the conflict

- o Application of legal findings to factual findings in order to reach an ultimate decision and the provision of available remedies

How have these legal frameworks for risk management responded to developments in the psychological and decision sciences? In this paper, we will deal with this issue by focusing on two important developments:

- o Developments in the psychological sciences which have influenced the private and public law system by bringing about the recognition of new types of cognizable risks in the form of various psychological harms, which had not been previously recognized

- o Developments in the decision sciences which have influenced the public law system by bringing about the use of new techniques for dealing with multiple objectives, and for making findings of fact

The next two sections deal with these major developments only. However, both technical fields, psychology and decision science, have also had numerous other influences--ranging from the selection of jurors, to modification of rules of evidence, to the establishment of presumption, to the setting of agency priorities for research and regulation, to strategic planning for litigation--which will not be dealt with here.

INFLUENCE OF THE PSYCHOLOGICAL SCIENCES[1]

The American legal system has felt the influence of the psychological sciences on risk management most strongly in the

---

1 The authors wish to thank Chendy Kornreich, a second year student at Boston University Law School, for research assistance on this section of the paper.

increasing recognition of psychological harm as a significant consequence of technological and environmental risks. A growing number of both legal and policy decisions on the control of health risks have taken emotional health into account as an important component. In many respects, courts have been in the forefront of this trend, although overall, their rulings have tended to be inconsistent. The frequency with which this issue has arisen in the legal system though, has forced its serious consideration.

Psychological factors have had their greatest impact on the private law system of compensation for harm to individuals. In this context, judges have found themselves faced with claims that psychic injuries resulting from technological and environmental hazards are as significant as physical ones. Varying legal conclusions have resulted. In the public law arena, legal questions have recently been raised about the duty of federal agencies responsible for health risk management to weigh the prospect of emotional harm in their decisions. Again, varying conclusions have resulted. One consequence is an atmosphere of uncertain legal guidance for government officials.

In some ways, the role of the public law system is more intriguing in this area, because of the interplay that it can involve between the three branches of government. Congress has written many laws on the management of health risks, and has the power to specifically include or exclude psychological considerations. As we have seen with other issues, it can also leave crucial questions ambiguous, for courts and agencies to struggle with. Agencies of the executive branch implement the directives contained in Congressional legislation, and are often permitted to use their specialized expertise to determine how new factors, such as findings from the psychological sciences, will be dealt with. The judiciary rules on whether agencies properly interpreted their mandates from Congress, and in instances of statutory vagueness, can interject its own values. This interplay has been crucial in several major legal disputes involving psychological factors in federal health risk management.

Psychological factors have had their greatest impact on two areas of private law compensation: tort law and workers' compensation. Tort law permits recovery in monetary damages for injury resulting from the breach of duty owed toward the plaintiff by the defendant. It is based on findings of fault. Workers' compensation permits employees payment for medical expenses and lost earnings resulting from disease or injury related to their employment. Fault is irrelevant to compensation determinations, which depend on the existence and nature of harm, and its work-relatedness. An administrative process, rather than formal litigation, is used to resolve most disputes.

An example of the difficulties that psychological evidence presents for courts in private law disputes is contained in the following case from the state of Maine. The plaintiff was struck on the head and neck by ice sliding from the roof of a house where he was delivering oil for the defendant, who was his employer. He testified that since the injury, he had experienced a constant dull pain that would cause severe headaches and numbness in his left arm after brief periods of exertion. Three medical experts testified on his condition. The first one stated that the plaintiff's neurological examination and X-rays were normal, and that he had recovered sufficiently to resume part-time work. The second one stated that since muscle strength, range of motion, and reflexes were normal, the plaintiff could resume full-time work. The last one, a clinical psychologist, ran a battery of tests from which he concluded that the plaintiff had developed a chronic neurotic condition with a strong tendency to develop physical symptoms such as headaches, backaches, and weakness after long periods of stress, and that he could not perform his work at all. How chould a judge evaluate these conflicting statements? In this instance, the trial court ruled that the plaintiff was fit for work, but on appeal, the Supreme Judicial Court of Maine found that he was still disabled.

It is only fairly recently that American courts have recognized emotional injury as a compensable form of harm, and it is still not universally accepted (3). The trend is clearly moving, though, toward wider recognition (4). Some states limit recovery to cases in which actual physical harm has caused mental distress, while others now permit compensation for behavior unrelated to physical contact, such as insults and threats. There is also division over whether the conduct must be intentional, or whether emotional harm negligently caused is sufficient for recovery.

The reluctance of many courts to recognize psychological aspects of harm to plaintiffs has been due to a number of stated reasons, including the difficulty of measuring the injury and of establishing its true causes (5). There appears to have been a general feeling among judges that psychological concepts are too difficult to quantify for use in law, and too rudimentary to answer complex questions. There is an often-stated sense that imprecision in the psychological sciences could facilitate large numbers of spurious claims.

On the other hand, consideration of psychological factors is clearly increasing, and is spreading to legal areas beyond traditional tort law. In one recent case, for example, a state court in Oregon ruled that harm resulting from the environmental nuisance of a cement plant can properly include emotional distress (6). A federal appeals court recently ruled that injuries caused by a federal agency's release of confidential information in violation of the Federal Privacy Act (7) can be mental as well as physical(8). A Massachusetts court has held that psychological factors must be considered as a consequence of medical malpractice (8a).

The tort law system of compensation in the United States is one of the central elements of the overall risk management process. It is among the few retrospective mechanisms, which can alleviate the consequences of risks that were not adequately controlled prospectively. It has also proven to be among the most effective means of deterrence for unregulated risk generating behavior. For these reasons, it is crucial that it incorporates important analytical techniques, such as those of the psychological sciences.

Workers' compensation provides a similar retrospective means for dealing with risks of occupational disease and injury. Two basic kinds of psychological harm have been claimed. The first is sudden traumatic reactions, in which an employee experiences an emotional shock on the job. The second is protracted stress, which builds up over time. The first of these varieties has been more readily recognized, although the second has gained substantial acceptance. One major stumbling block has been the issue of whether to define mental harm as a disease or as an injury.

The situation concerning overall recognition of emotional injury has been similar to that in tort law. Recovery has been established slowly, with many misgivings, both by judges who have reviewed compensation awards and by legislatures which have written the statutes authorizing them (9). Many of the same issues have arisen as in tort law, including the need for coincident physical injury, and the need to adequately establish causation. The trend, though, is clearly toward growing acceptance (10).

The direction of the two compensation systems toward greater recognition of emotional injury indicates that contributions from the psychological sciences are clearly being felt. Dramatic increases in work-related stress claims, for instance, have been noted (11). Research has demonstrated strong links between such stress and both physical and emotional ailments (12). At the same time, however, much more of an input is needed. While most courts seem to have been persuaded that psychology is a precise and quantifiable enough tool to measure the harm received from technological and environmental hazards, the legal system is still not fully open to this use. Psychologists face the task of persuading judges and legislatures that input from their techniques is essential to the full resolution of claims involving various kinds of risk-related injury. Until this has been done, many victims of hazards will not be able to fully recover for their injuries.

The public law system has seen the influence of the concept of psychological injury in two major recent controversies. The first involves the parameters of the concept of "health" under the United States Constitution as it applies to state legislation governing abortions. The second involves the same question as it applies to the management of environmental risks under the federal National Environmental Policy Act (NEPA) (13). It is the second of these disputes that has generated the most public debate, although both have important implications for the law governing risk management.

Initial legal attention to psychological factors in a public law context actually substantially predated these disputes. In 1952, the Supreme Court of New Jersey ruled, in a case involving land use planning and zoning, that emotional health was an important consideration in determining minimum living-floor space in housing (14). Although it was subsequently criticized (15), this opinion prompted debate on the role of psychology in public policy decisions.

In terms of the first dispute, in 1973 the United States Supreme Court ruled in the case of Roe v. Wade (16) that states may not restrict a woman's right to an abortion under most circumstances. In a companion case, Doe v. Bolton (17), the court addressed the right to an abortion in situations in which a woman's health is in danger. In doing so, it opened the way for contributions from the psychological sciences in clinical decisionmaking by concluding that the term "health" includes psychological as well as physical well-being. Justice Blackmun, in delivering the opinion declared that in abortion decisions, "...the medical judgment may be expressed in the light of all factors--physical, emotional, psychological, familial, and the woman's age--relevant to the well-being of the patient. All these factors may relate to health." (18)

This opinion means that in this area of constitutional rights, psychological considerations are crucial. It follows a 1971 ruling by Justice Blackmun in a similar dispute that "the general usage and modern understanding of the word health includes psychological as well as physical well-being." (19)

While the ruling applies to clinical rather than to technological or environmental hazards, it is a clear signal as to what the legal system considers a "health risk" to be.

The second controversy, which involves NEPA, has revolved around the licensing of nuclear reactors (20). In particular, possible psychological harm from reopening the ill-fated Three Mile Island reactor has been raised (21). The final Supreme Court ruling on the subject, reached within the past few months held that such considerations are not required, but the dispute highlights the growing importance of this issue.

The appellate court decision by the Circuit Court of Appeals for the District of Columbia in this case had ruled quite forcefully that "health" encompasses psychological health under the meaning of NEPA, which requires agencies to analyze the environmental consequences of major actions (22). The court declared that the lack of quantifiability was irrelevant. It found that the likelihood of emotional stress to those living near the disabled nuclear plant was so large, that such factors could not be ignored in weighing its renewed operation.

The Supreme Court, however, took a different view (23). Justice Rehnquist speaking for the court stated that the existence of a higher level of risk, which might cause great stress for the local residents, was not a sufficient ground for mandatory consideration of possible stress in the relicensing decision. He declared that "a risk of an accident is not an effect on the physical environment." (24) The causal link between the renewed operation of Three Mile Island and psychological injury was not felt to be strong enough to require agency attention.

The opinion expressed other fears which have characterized the legal debate over recognition of psychological injury, namely that claims of emotional harm may often be spurious. Justice Rehnquist observed that "It would be extraordinarily difficult for agencies to differentiate between "genuine" claims of psychological health damage and claims that are grounded solely in disagreement with a democratically adopted policy." (25) He left the issue to Congress to define more explicitly.

The recurring concern over the precision and accuracy of the psychological sciences is one that psychologists should seek to meet. In instances such as nuclear plant licensing proceedings, there may be major contributions to be made to governmental decisionmaking. The costs of emotional stress, for example, may be large enough to outweigh the benefits for many actions. If Congress does not point to such factors in regulatory legislation and courts do not require them in opinions, there will be little incentive for agencies to take them into account. There is a clear need for psychologists to demonstrate in the public policy arena the potential value of their analytical techniques. The legal system can provide a useful forum for presenting such arguments, both in the form of Congressional legislation and judicial rulings. Until this is done, the question of vagueness of results is likely to continue to impede full legal acknowledgment of the contributions that the psychological sciences can make.

Recognition is made more difficult in the public law setting by the lack of a definitive structure for introducing psychological evidence. In private law compensation cases, the testimony of a doctor concerning diagnosis, prognosis, and causation is required. In public law regulation disputes, on the other hand, mental health issues must be raised by citizens' groups on their own initiative. Such evidence is not routinely used by agencies in their probabilistic or cost-benefit assessments.

In many ways, the recognition of psychological factors is on the forefront of the development of risk management. In the area of physical injuries, we generally know what kinds of harm we are looking for, and what kinds of exposure factors are involved, such as chemicals or design defects. The recognition of psychological injury, on the other hand, is new, and we are still searching for the manifestations of illness as well as the corresponding risk factors. The legal system has faced a major challenge in trying to integrate contributions from psychology into decisionmaking concerning health and environmental hazards. The impetus for this integration plainly exists, in the form of citizen lawsuits and expressions of concerns. Techniques to facilitate it are clearly needed for the future.

INFLUENCES OF THE DECISION SCIENCES

Various decision science techniques are now being used for a number of purposes by agencies seeking to manage risks. These techniques include cost-benefit and risk-benefit analyses, various models or structures for dealing with hierarchical ordering and causal linkage between various regulatory options and possible outcomes (e.g., "decision trees"), and various other techniques for making statistical statements of probability regarding future events (26).

Who uses these techniques? For what purposes? Probably all risk managing agencies do, for various purposes: agencies that make decisions about committing public resources to undertake new developments, such as the Corps of Engineers' flood control actions; agencies that essentially make decisions on permit or license applications from private parties, such as the Nuclear Regulatory Commission's licensing of nuclear reactors; and agencies that make decisions of a generic nature to control exposure to hazardous substances and activities, such as the Environmental Protection Agency (EPA) and Occupational Safety and Health Administration (OSHA) programs for controlling human exposure to toxic chemicals in commerce and the workplace, respectively (27).

Most of these uses conveniently fall into two general categories:

1) "Balancing Uses," to Achieve Multiple Objective Decisionmaking

2) "Stochastic Uses," to Reach Credible Findings of Fact, due to the Absence of Complete Data and the Inability to Reach Conclusive Determinations about Causality and Other Factual Matters

Balancing Uses To Achieve Multiple Objective Decisionmaking

These uses of decision science methods, such as cost-benefit analysis, arise from the statutes which mandate agency management of risk. Typically, these statutes, such as the Toxic Substances Control Act, the Atomic Energy Act, and the Occupational Safety and Health Act, provide that each agency decision under their authority must meet multiple objectives. Therefore, the agency decision process must involve the consideration of multiple sets of facts, must reach multiple findings of fact, and must somehow ultimately produce a balanced or integrated decision that meets the multiple objectives (28).

However, Congress has not gone further and told the agency what balancing technique to use, in the usual case; and it has been usually left to agency discretion to choose the particular technique, apply it, and consequently reach defensible decisions. This condition reflects the pluralistic values and interests of American society, and the inability or unwillingness of Congress to ignore or respectively weigh various interests.

For example, the Corps of Engineers is directed by Congressional statute to reach decisions on the basis of safety, cost, and environmental quality; the EPA is told to control hazards from toxic chemicals in commerce by considering the facts relating to health, environment, economic impacts and the availability of substitute products; the OSHA is told to control toxics in the workplace by considering worker health risks and the technical and economic feasibility of various control options.

Thus, Congress has established conditions for each risk management agency which require the use of various "balancing" types of analysis; is cognizant of agency responses through its constant oversight activities; and usually acquiesces to the agency choice of balancing technique, albeit with some occasional quibbles about discount rates used or other features of the balancing method chosen.

However, agency discretion in choosing and using any particular balancing technique is held accountable to some extent through challenges that may be made by affected interests. These interests, if adversely affected, can seek judicial review of agency decisions, and bring several arguments to bear on the balancing technique that was used to reach the final decision. The arguments typically consist of three claims:

1) Agency choice of a particular balancing technique to meet its multiple objective requirements did not conform to its statutory mandate, and is therefore ultra-vires and invalid. This argument has frequently been made, sometimes successfully, when an agency such as EPA or OSHA has sought to use cost-benefit analysis to balance economic findings of fact with health or environmental findings of fact (29).

2) Agency use of the balancing technique was flawed in a significant way. This argument has frequently been made, often successfully, when an agency's conduct of environmental impact analysis under the National Environmental Policy Act was of inadequate scope or failed to adequately consider one of the impact attributes (e.g., impact on historic preservation or rare flora or fauna) (30).

3) Agency choice and use of the balancing technique, although adequate, produced results which do not adequately support its final decision. This argument, often made, is usually based on the provisions of the Federal Administrative Procedure Act (31), which requires that an agency's final action must be

supported by sufficient evidence and evince a "hard look," so that it is not "arbitrary and capricious," or unsupported by substantial evidence (32).

The federal courts have dealt with these arguments about the balancing technique used by an agency in hundreds of cases over the last decade. It is foolhardy to try to summarize the decisions in these cases; nevertheless, some general answers will be attempted.

First, let us deal with the issue of whether or not an agency has properly chosen its balancing technique. This is essentially a question of law for the court--did the agency act within, and consistent with, its statutory authority, or did it act ultra-vires? The answer to this question will be reached by the court's interpretation of the language of the governing statute. (E.g., did the statute call for EPA balancing of economic vs. health facts in setting drinking water standards, or did it require that health be given primacy, with some sort of lesser weight, or no weight, to be accorded economic factors?) If the statutory language is inconclusive, as is the usual case, the court will seek to interpret the legislative history of the law in order to discern what Congress intended for the agency (33).

This is unfortunately a common type of challenge, typically brought after a new regulatory program has made its early decisions. If the federal court decisions on these early actions are consistent on the issue, it usually does not arise in later cases which may involve the same agency program. Obviously, Congress could reduce the incidence of such cases by providing clearer, more specific, guidances to the agencies in their statutory provisions. But Congress would need to be educated as to the available balancing techniques, and in choosing any technique, would implicitly be doing what it has long avoided--providing weights to different societal values and interests, even to the point of possibly ignoring some completely.

Second, the issue of whether an agency's use of a particular balancing technique was flawed in a significant way is really a variation on a traditional theme that the role of the judiciary is "to determine how matters are proven."(34) Most federal courts will deal with this issue qualitatively, by assessing whether or not the agency did a reasonable job in employing the technique, and focus primarily on two attributes of reasonableness--the completeness or the absence of any gaps in the running of the analysis, and the level of effort or rigor exercised by the agency in its analytic conduct (35).

If gaps in the data base or the analysis are found and not satisfactorily explained away by the agency, the matter will usually be remanded back to the agency for remedial action. The courts have been relatively competent in determining the inadequacy of a balancing analysis, when such gaps are present. But if the gaps or inadequacies are defended, or filled with agency assumptions, and these defensive measures are candidly articulated by the agency and the agency evinces some sort of rigor or "hard look" in dealing with the inadequacy and its assumptions, courts have demonstrated in many cases their extreme reluctance to "substitute their judgment for the agency's, on technical matters," "deferred to agency expertise," and accepted the inadequacies and assumptions--even if they are relatively arbitrary and subjective or self-serving (36).

Thus, on matters involving the adequacy of agency conduct of balancing analyses, the courts have essentially deferred to agency judgment and assumptions, and failed to provide the scrutiny and mechanism for agency accountability that many expect the federal courts to provide. To some extent, this condition is based on the volume of data and analytic materials and the technical complexity of the decision science techniques involved, which intimidate the judiciary. Various remedies have been proposed to better equip the courts and improve accountability, but have been deferred largely because of judicial and political reluctance to modify the structure of judicial decisionmaking (37).

Third, whether the results from the running of a particular type of analysis provide adequate support for the agency's ultimate decision is an issue which, in the usual case, has been determined by the outcomes on the prior two issues. Thus, if a court finds an agency's choice of a decisional technique for assessing a multiple objective decision to be appropriate, and further finds no basis for faulting the agency's running of the actual analysis, and the results are used by the agency to support its decision, then the decision will be virtually invulnerable to challenge. This is largely due to the usual rough-hewn and vague criteria employed by the courts on judicial review of a final agency decision: was it arbitrary or capricious? Does it evince a "substantial inquiry" or "hard look" at the problem? Is it supported by substantial evidence? All three types of criteria, based on the provisions of the Administrative Procedure Act, are likely to be satisfied if the prior issues have been resolved in the agency's favor (38).

Thus, agency use of decision science balancing techniques in order to reach final decisions which satisfy the multiple objective attributes of governing statutes has progressed far with only limited judicial scrutiny and virtually no Congressional guidance. Indeed, the Office of the President has superimposed a controversial administrative requirement that each agency subject its proposed regulations to a cost-benefit analysis and review of the results thereof, by the President's Office of Management and Budget (39).

The results for public policy and the affected interests are a mixed bag. The balancing techniques, when responsibly conducted, can lead to decisions which are quite responsive to the multiple-objective mandate and our pluralistic values, and which are perhaps less burdensome on the economy. On the other hand, decision science techniques for balancing, with their methodological limitations and need for assumptions, do not necessarily eliminate the arbitrary, subjective and political features of much agency decisionmaking, and indeed mask such defects in agency decision-processes and reduce the rigor of judicial scrutiny (39a). Further, in practice, they obstruct the achievement of the optimal results sought by important interests (e.g., those who want more effective prevention of public health risks). But from a public policy standpoint, probably the major concerns are the lack of Congressional guidance furnished the agencies, and the erosion and failure of our judicial accountability system (40).

Stochastic Uses to Reach Credible Findings of Fact

Decision-trees and other forms of stochastic analysis have become the techniques of choice and necessity for reaching findings of fact for most regulatory agencies charged with the mandate to reduce and prevent future risks. More traditional approaches to fact-finding, such as the use of direct evidence to establish causation to the extent that particular individuals or other risk receptors can be identified with certainty, are not feasible for several reasons. These reasons include the lack of complete data, an incomplete understanding of multiple causal mechanisms, the scope and scale of the risk scenario involved, and the need for relatively immediate agency action to prevent risk rather than await a "body-count" before acting.

Statutes are usually silent on the issue of whether stochastic techniques should be employed and, if so, which techniques. Virtually all is left to agency discretion. Thus, in trying to reach multiple findings of fact which will then be subjected to some sort of balancing analysis, as previously discussed, agencies employ, out of necessity, various stochastic methods, usually without Congressional guidance.

For example, in setting a "new source air emission standard" or a workplace health standard, a stochastic method may be employed to produce each of several findings required to support a final decision: e.g., the level of toxins emitted, their dispersion or concentration in "hot spots," their transformation in the ambient environment, the populations, and "sensitives" at risk, the health impacts, the technical efficacy of various control options, safety margins, and economic impacts.

If the decision reached by an agency on the basis of findings of fact and balancing has an adverse impact on certain societal interests, the representatives of these interests have a right to judicial review (41). But, once again, the legal response has been shaped by flabby judicial scrutiny and judicial deference.

The essential argument for the opponents of the agency will be the second of the three used to attack a balancing analysis, described above, namely: that agency use of the particular stochastic approach for reaching one or more of its fact-findings was flawed in a significant way, thereby undermining its support for the final decision. For example, it could be argued that the approach taken to determine the dispersion of air pollutants was based on inadequate or incorrect data, such as data taken from monitors improperly or irrationally located, or that the model itself was incorrectly designed.

Just as the argument is similar to what has been discussed earlier, so has the judicial response been similar--namely one of deference to agency expertise, and a willingness to accept assumptions and minor flaws, other than serious gaps in data or analysis, if the agency has evinced rigor in its approach and the model used is one deemed reasonable, though not necessarily the best (42).

The public policy problem is therefore also similar: Without Congressional guidance or meaningful judicial review, agencies are relatively free to adopt and use stochastic techniques for fact-finding without any real mechanism for accountability. To those who believe that technical complexity requires governance by experts, this is acceptable. To those who feel that experts lack objectivity or compassion, or otherwise should not be trusted, this condition is unacceptable, and solutions must be found.

Holding Agencies Accountable in Their Uses of the Decision Sciences

Possible solutions focus largely on the judiciary, whose central function in matters of judicial review is, according to Judge Leventhal, to determine how matters are proven (43). A particularly thoughtful recommendation has recently been offered by Professor Charles Case:

> "The proper role of the court and the appropriate level of judicial review is best expressed in the formulation that the court, given sufficient resources, should sufficiently involve itself in the model data and methodology to ensure a minimal level of meaningful review. This approach must rest on...a significant review of the factual record...[and] may require that the court substitute its judgment for that of the...agency where the inquiry reveals fundamental or uncorrected flaws in the modeling and decisionmaking process...[such as] (1) when the agency uses an incorrect model or a model that...falls short of the modeling art; (2) where the agency uses unexplained or unjustified assumptions or adjustments in the model; (3) where the agency uses data shown to be incorrect or inapplicable...; (4) where the agency fails to properly structure supporting and unsupporting data and methodology in an understandable index...; and (5) where the agency fails...to provide a realistic opportunity for the presentation of...objections." (44)

However, other reforms are also feasible, and could be focused on Congress, the Office of the President, and the agencies themselves, to assure accountability in the use of cost-benefit analysis and other decision science techniques. In his report to the Administrative Conference of the United States, a federal advisory body on regulatory matters, Professor Baram, proposed a three-pronged "process" approach for dealing with the use of cost-benefit analysis, which can be extended to apply to other decision science techniques, as well:

1) Congress should...review...agency implementation of...[their risk-related statutes] to determine the extent to which agency uses of cost-benefit...[and other] techniques are consistent with statutory objectives...also identify conflicts between Executive Order...and statutory requirements...[and] Congress should...conduct a special review of agency implementation of NEPA's (National Environmental Policy Act) requirements for "balancing analysis" and other methodological requirements....Congress should then provide clearer guidance to the agencies...to achieve...clarification of...quantification...

methodology for quantification...discount rate to be used...[and] state expressly when an agency should or should not use a cost-benefit [or other decision science] approach....

2) Office of the President should...provide further guidance...to ensure that...[agencies] implement Executive Order...consistently with legislative requirements...[and] require...Presidential offices to function in open proceedings...in conjunction with the agencies....

3) Each regulatory agency using cost-benefit or other...[decision science technique] should promulgate a generic regulation describing its use of these analytical tools in order to promote clear and consistent regulatory policy.... The regulation should address the sources of agency authority to conduct...analyses, the particular analytical method[s] selected by the agency...methodological limitations...and how they are addressed, the timing and weight afforded the analyses, due process safeguards....In addition, for decisionmaking...each agency should...make publicly available its... findings...and describe fully any...[analyses] conducted.(45)

The Administrative Conference adopted the third recommendation and has conveyed its action to all agencies (46). Several have voluntarily complied, at least in principle.

Finally, there is the issue of professional responsibility for the future uses of decision science, put so succinctly in his novel, The Stochastic Man, by Robert Silverberg:

> Causality is a decent, honorable principle, but it doesn't have all the answers. If we want to make sense of things, we have to move on beyond it. We have to recognize that many important phenomena refuse to be packed into neat causal packages but can be interpreted only by stochastic methods. A system in which events occur according to a law of probability but aren't individually determined in accordance with the principle of causality is a stochastic system....
>
> In the immortal words of Ralph Cudworth (1617-1688), "There is need and use of this stochastical judging and opinion concerning truth and falsehood in human life." Those whose way of life is truly governed by the stochastic philosophy are prudent and judicious, and tend never to generalize from a skimpy sample. As Jacques Bernoulli demonstrated early in the eighteenth century, an isolated event is no

harbinger of anything, but the greater your sampling the more likely you are to guess the true distribution of phenomena within your sample.

So much for probability theory. I pass swiftly and uneasily over Poisson distributions, the Central Limit Theorem, the Kolmogorov axioms, Ehrenhaft games, Markov chains, the Pascal triangle, and all the rest. I mean to spare you such mathematical convolutions. ("Let p be the probability of the happening of an event in a single trial, and let s be the number of times the event is observed to happen in n trials...") My point is only that the pure stochastician teaches himself to observe what we at the Center for Stochastic Processes have come to call the Bernoulli Interval, a pause during which we ask ourselves, Do I really have enough data to draw a valid conclusion?

I'm executive secretary of the Center, which was incorporated four months ago, in August, 2000....Our aim is to find ways of reducing the Bernoulli Interval to zero: that is, to make guesses of ever-increasing accuracy on the basis of an ever-decreasing statistical sample, or, to put it another way, to move from probabilistic to absolute prediction, or, rephrasing it yet again, to replace guesswork with clairvoyance.

So we work toward post-stochastic abilities....In the absolute universe all events can be regarded as absolutely deterministic, and if we can't perceive the greater structures, it's because our vision is faulty. If we had a real grasp of causality down to the molecular level, we wouldn't need to rely on mathematical approximations, on statistics and probabilities, in making predictions. If our perceptions of cause and effect were only good enough, we'd be able to attain absolute knowledge of what is to come. We would make ourselves all-seeing...You tend to be skeptical about such things, don't you? That's all right. You'll change your mind. I know you will. (47)

CONCLUSIONS

The influence of the psychological and decision sciences on the legal framework for managing technological risks can be seen from an overall perspective in terms of the function that courts play in this respect. In terms of the private law system, they interpret case precedents and statutes and make new law in areas that have not yet been legally defined. In terms of the public law system, they review the actions of agencies and give them guidance and interpret the mandates of Congress. Since the psychological and decision sciences have become important aspects of risk management, both in the form of private remedies and of public regulations, courts must find ways of interpreting them in performing these tasks.

Judges must first resolve the level of scientific expertise that they will take unto themselves in making their decisions. To what extent will they try to second-guess technical experts on questions of psychological harm or use of decision analytic techniques? On the other hand, how much deference can they grant to experts, before they have neglected their responsibility to oversee the outcome of disputes? The role of law will revolve in large part on the answers to these questions.

In reviewing compensation claims involving psychological harm, for example, many judges have rejected technical advice and ruled that such claims do not have substance. In overseeing federal agency actions, however, there has been a tendency to defer to decision analysis results based on agency expertise. There is at present, therefore, no overall consistent pattern.

The second task for judges, and perhaps the more difficult one, is to find ways to adapt their functions to a set of disciplines that are still relatively new and rapidly evolving. It is hard in this situation to decide what expertise to accept. This may be one reason why judges have been made uneasy by a perceived lack of quantification in psychology and by an overabundance of quantification in the decision sciences. While the fields are still new, however, courts must eventually recognize their significant influences on risk management.

Continued contributions from these disciplines are clearly essential, if effective legal oversight and implementation of risk management is to take place. The legal system is an integral part of the process of risk control, and it must be receptive to all relevant factors. Practitioners of the psychological and decision sciences, therefore, must see to it that open communication with the world of law is established and maintained. There is much to be gained from such an arrangement for effective public policy decisionmaking.

REFERENCES

1. See discussion in L. Lave, The Strategy of Social Regulation, Washington, D.C. Brookings (1982). Also, M. Baram, "Technology Assessment and Social Control," Science v. 180 (1973), p. 465.

2. See discussion in M. Baram, et al., Alternatives to Regulation: Managing Risks to Health, Safety and the Environment, Lexington, MA.: Lexington Books, D.C. Heath Co. (1982).

3. Smith v. Dexter Oil Co.; 432 A.2d 438 (Me. 1981).

4. W. Prosser, Handbook of the Law of Torts. 4th Edition, St. Paul, Minnesota: West Publishing Co., 1971, p. 49.

5. Ibid., p. 327.

6. Lunda v. Matthews, 613 P.2d 63 (Or. App. 1980).

7. 5 U.S.C. Sec. 552(a).

8. Johnson v. Department of Treasury, ___ F.2d ___ (5th Cir. 1983).

8a. Miles v. Tabor, 387 Mass. 783 (1982).

9. A. Larson, The Law of Workmen's Compensation. New York: Matthew Bender, 1981, Section 42.20.

10. Ibid.

11. B. Lipson, "Legal reform has increased stress-related claims." Boston Globe, Feb. 6, 1983, p.18. See also, B. Rice, "Can Companies Kill?" Psychology Today, June, 1981.

12. "ILO Study of Workplace Stress Sees Links to Physical, Emotional Ailments." Occupational Safety and Health Reporter, Jan. 6, 1983, p. 621.

13. 42 U.S.C. Section 4321 et seq.

14. Lionshead Lake v. Wayne Tp., 89 A.2d 693 (N.J. 1952).

15. See V. Nolan, and F.E. Horack, "How Small a House? - Zoning for Minimum Space Requirements." Harvard Law Review v. 67 (1954), p. 967, and N. Williams, and E. Wacks, "Segregation of Residential Areas Along Economic Lines: Lionshead Lake Revisited." Wisconsin Law Review v. 1969 (1969) p. 827.

16. 410 U.S. 113 (1973).

17. 410 U.S. 179 (1973).

18. 410 U.S. 179 at 192.

19. United States v. Vuitch, 402 U.S. 62 (1971) at 72.

20. NEPA states in part that federal agencies must "utilize a systematic, interdisciplinary approach which will insure the integrated use of the natural and social sciences and the environmental design arts in planning and in decisionmaking which may have an impact on man's environment." NEPA, section 102(2) (A).

21. A previous dispute involving psychological factors under NEPA related to the siting of public housing. Although recognition was not granted, the issue was not addressed head on. See Strycker's Bay Neighborhood Council v. Karlen, 444 U.S. 223 (1980).

22. People Against Nuclear Energy v. U.S. Nuclear Regulatory Commission, 678 F.2d 222 (D.C. Cir. 1982).

23. Metropolitan Edison Co. v. People Against Nuclear Energy, 51 LW 4371 (1983).

24. 51 LW 4371 at 4373

25. 51 LW 4372 at 4374

26. For many examples, see C. Case, "Problems in Judicial Review Arising from the Use of Computer Models and Other Quantitative Methodologies in Environmental Decisionmaking." B.C. Env. Affairs Law Rev., v. 10, n. 2 (1982), p. 251.

27. See Committee on the Institutional Means for Assessment of Risks to Public Health. Risk Assessment in the Federal Government: Managing the Process. Washington, D.C.: National Academy Press, 1983.

28. See discussion in Analysis of the Designation of Chemical Substances for Government Regulation, v. I, ICF, Inc., Washington, D.C., Report to EPA (Dec. 1981); and R. Field, "Patterns in the Laws on Health Risks," Jnl. of Policy Analysis and Management, v. 1 (1982) p. 257.

29. For example, ATMI v. Donovan, 452 U.S. 490 (1981), wherein the Supreme Court held that OSHA is neither required nor permitted to use cost-benefit analysis as a framework for determining standards governing worker exposure to toxic chemicals. Also see Portland Cement Ass'n v. Ruckleshaus, 486 F.2d 375 (D.C. Cir.) 1973; cert. denied, 417 U.S. 921 (1974), rejecting industry's argument that EPA should use quantified cost-benefit analysis in setting new source performance standards under Section 111 of the Clean Air Act of 1970. But the court did permit consideration of economic factors as a "modifier" and Congress affirmed this decision by amending the Clean Air Act in 1977. See Section 317 of the Act as amended.

30. See M. Baram, "Cost-benefit Analysis: An Inadequate Basis for Health, Safety and Environmental Regulatory Decisionmaking," Ecology Law Quarterly, v. 8 (1980) p. 473. For some of the more important decisions on these grounds (e.g., Overton Park, etc.), see Findlay and Farber, Environmental Law: Cases and Materials. St. Paul, Minn.: West Publishing Co. (1981); Sections I-D and II-B.

31. 5 U.S.C. Sections 700-706.

32. H. Leventhal, "Environmental Decisionmaking and the Role of the Courts," U. Pa. L. Rev., v. 122 (1974) p. 509.

33. See Notes 26 and 32, supra.

34. See note 29 supra, and numerous other decisions; e.g. Crowther v. Seabor, 312 F. Supp. 1205 (D. Colo.) 1970; etc. identified in M. Baram, note 30 supra at p. 496.

35. See Leventhal, note 32 supra.

36. See, for example, Baltimore Gas and Electric Co. v. NRDC, U.S.S.Ct., 51 U.S.L.W. 4678 (June 7, 1983), upholding NRC's reliance on a zero-release assumption regarding radioactive waste in its environmental impact analysis. Also see Case article, note 26 supra.

37. The proposed remedies include increasing the scientific competency of members of the bar and bench, alteration of judicial processes; the use of seriated trials wherein expert assessment of causation precedes allocation of responsibility and liability by the judiciary; special science or environmental courts; increased use of court appointed experts and masters; and science clerks for the

judiciary. See note 32, supra. Also see Jasanoff and Nelkin, "Science, Technology and the Limits of Judicial Competence," Science, v. 214 (1981) p. 1211; and Piehler, Twerski, et al., Science, v. 186 (1974) p. 1089.

38. 5 U.S.C. Section 700-706 (the Administrative Procedure Act). Also see Citizens to Preserve Overton Park v. Volpe, 401 U.S. 402 (1971), an authoritative but vague Supreme Court decision on the nature of judicial reviews; and notes 26, 30, and 32, Supra.

39. Reagan Executive Order 12291 (superseding similar Carter Executive Order), (1981). This has stimulated new policy and legal controversies, regarding for example, the intention of Congress in enacting legislation which does not call for cost-benefit analysis, and the Constitutional requirement for the separation of power. See discussion in M. Baram, note 7, supra; and various other materials, such as "Office of Management and Budget Control of OSHA Rule-Marking," Hearings before Subcommittee of the Committee on Government Operations, U.S. H. Rep., 97th Congress, 2d Session, March 11, 18, 19 (1982).

39a. See note 27, supra.

40. See note 30, supra.

41. Note 31, supra.

42. Compare D. Bazelon, "Science and Uncertainty: A Jurist's View," Jurimetrics Journal (1982) p. 372; H. Leventhal, note 32 supra; and the U.S. Supreme Court in Baltimore Gas and Electric, note 35, supra.

43. Note 32, supra.

44. See Case, note 26, supra at p. 359.

45. M. Baram, Regulation of Health, Safety, and Environmental Quality and the Use of Cost-Benefit Analysis (Mar. 1, 1979), final report to the Administrative Conference of the United States.

46. 44 Fed. Reg. 38,817-25 (1979) (to be codified in 1 C.F.R. pt. 305): Section 305.79-4 Public disclosure concerning the use of cost-benefit and similar analyses in regulation (recommendation No. 79-4).

(a) Federal agencies must frequently weigh competing health, safety, resource management, environmental, economic, and other societal interests when seeking to achieve a prescribed statutory objective. Wise decisionmaking presupposes that the potential benefits and costs of the actions under consideration will be identified, will be quantified if feasible, and will be appraised in relation to each other. To give structure to the exercise of this responsibility, agencies sometimes use "cost-benefit" or similar analytic approaches to organize available information to determine the consequences of possible courses of action in terms of their costs, risks, and benefits. Such techniques seek to display the projected net effects of alternative courses of action, and when properly used, can assist the decisionmaker in deciding which of the alternatives is most likely to produce a desired result.

(b) The following recommendation seeks to promote openness in the decisionmaking process, to ensure that agencies' analytic methods are sound and that their assumptions are known, so as to enhance public confidence in the soundness of conclusions finally reached. The recommendation is not intended to promote or to discourage the use of any single kind of analysis as a framework for agency decisionmaking, since this choice is normally a matter of agency discretion. The choice of analytic technique may depend on several factors, including the technical complexity of the problem, the magnitude of the impacts, the time frame for agency action, and the extent to which quantification is possible for the specific costs and benefits to be considered. Any analysis, of course, should be viewed as an aid to rational decisionmaking, and not as an end in itself. The intent of the recommendation will be served by giving the public adequate advance notice of the agency's proposed methodologies, either generically or by means of special notice in a particular proceeding.

Recommendation:

1. Agencies, as general policy though not necessarily by binding rule should adopt the practice of addressing, in their public notices of particular proceedings in which cost-benefit or similar analyses are to be the used, the following points:

a. Any statutory or other legal requirements pertaining to or affecting the agency's conduct of cost-benefit or similar analyses in the proceeding.

b. The particular analytic technique to be followed by the agency (e.g., cost-benefit analysis, cost-effectiveness analysis, qualitative or non-numerative balancing), with a description of the method, including an identification of any analytic models preliminarily determined to be used.

c. Any factual assumptions or preliminary findings of the agency to be utilized in the analyses.

d. The agency's techniques for assessing and revealing uncertainties in its quantitative estimates, and making explicit the range of error associated with particular quantitative estimates.

e. The agency's methods for evaluating intangible costs and benefits, for discounting future costs and benefits, and for taking account of distributional effects arising under the selected methodology, to the extent such issues are involved in the analyses.

f. The stages of the proceeding at which the cost-benefit or similar analyses will be conducted and the results considered.

g. The extent and nature of public participation in the design, conduct, and evaluation of the cost-benefit or similar analyses.

h. The extent and manner in which the public is to be accorded access to assumptions and information used in the analyses.

A statement of the weight given the cost-benefit or similar analyses, and a description of any revisions of assumptions or preliminary findings, should be included in the final agency determination and made available to the public.

2. Where a pattern of recurring decisional problems exists for which a particular analytic technique is appropriate, the agency should consider adopting a generic regulation or policy statement describing the use of that technique with respect to those problems. Agencies that have varied statutory functions may suitably formulate separate regulations or policy statements for different areas of statutory

responsibility. Generic regulations or policy statements so adopted may permit the use of different techniques on an ad hoc basis where the agency determines that to be necessary. Any such regulations or policy statements should address the points listed in paragraph 1.

Id. at 38,826.

47. R. Silverberg, The Stochastic Man, Harper Science Fiction (undated).

INDEX

NATO ASI Series G